CALCULUS

HARCOURT BRACE COLLEGE OUTLINE SERIES

CALCULUS

Scott M. Farrand and Nancy Jim Poxon

Department of Mathematics
California State University at Sacramento

The Dryden Press
Harcourt Brace College Publishers
Fort Worth Philadelphia San Diego
New York Orlando Austin San Antonio
Toronto Montreal London Sydney Tokyo

Requests for permission to make copies of any part of the work should be mailed to:

Permissions Department
Harcourt Brace Jovanovich, Publishers
8th Floor
Orlando, Florida 32887

Printed in the United States of America

Library of Congress Cataloging in Publication Data

Farrand, Scott
College calculus.

(Harcourt Brace Jovanovich college outline series)
(Books for professionals)
Includes index.
1. Calculus. I. Poxon, Nancy, J. II. Title.
III. Series. IV. Series: Books for professionals.
QA303.F3 1984 515 83-22673

ISBN 0-15-601556-0

First edition

Harcourt Brace Jovanovich, Inc.
The Dryden Press
Saunders College Publishing

CONTENTS

PREFACE

This Outline is intended to be used as a supplement to your calculus textbook or as a comprehensive review of calculus techniques. It has three main purposes: (1) to describe the techniques for solving calculus problems, (2) to give you examples of solved problems, and (3) to furnish you with many practice problems. Because these purposes are primarily practical, little attempt has been made to show the derivation of the theorems and formulas used in this Outline. A textbook is the best source for theoretical background on each topic.

In the introductory chapters of this Outline, problems are solved in great detail: Every step in the solution is given. But in later application chapters, we assume that you are able to fill in some of the details for yourself. For example, when integration by parts is introduced, every detail of the technique is shown; but in later chapters, this technique is referred to by name only. You should check the integration for practice and to hone your skills. Similarly, we do not always include algebraic or trigonometric details. This method of presentation should help you discover your mathematical weaknesses. Thus, if algebra is getting in the way of the calculus, you may need to review the algebra. There's no way around it: A solid foundation in algebra and trigonometry is essential for understanding calculus.

We gratefully acknowledge the advice of our colleagues and reviewers. Their suggestions and careful attention to detail were useful in shaping the various drafts of this book.

Our families have shown us tremendous indulgence. We dedicate this book to them.

Sacramento, California

SCOTT FARRAND

NANCY JIM POXON

1 PRELIMINARIES

THIS CHAPTER IS ABOUT

- ☑ Functions and Their Graphs
- ☑ Lines
- ☑ Partial Fractions
- ☑ Trigonometric and Hyperbolic Functions
- ☑ Summation Notation

1-1. Functions and Their Graphs

A. Functions

A relationship between two variables x and y gives y as a **function** of x if to each value of x there corresponds just one value of y.

EXAMPLE 1-1: For which of the following equations is y a function of x: **(a)** $x^2 + y = 3$, **(b)** $y^2 + x = 3$, **(c)** $\sqrt{y+1} = x/(1-x)$, **(d)** $|y| = x$?

Solution:

(a) You can solve this equation for y:

$$y = 3 - x^2$$

Now you can see that a value of x determines a unique value of y.

(b) If you try to solve for y in terms of x,

$$y = \pm\sqrt{3-x}$$

you see that y isn't a function of x. The plus or minus sign gives you two values of y for most values of x. For example, if $x = 2$ then y could be $+1$ or -1.

(c) In spite of the square root, you can solve for y:

$$y + 1 = \left(\frac{x}{1-x}\right)^2$$

$$y = \left(\frac{x}{1-x}\right)^2 - 1$$

and see that y is a function of x.

(d) When $x = 1$, y can be either $+1$ or -1, so y isn't a function of x.

Another notation for a function of x is $f(x)$ (f evaluated at x, or f of x). For example, $f(x) = x^2 - 7$ is a function of x: For any value of x, the value of $f(x)$ is a unique number:

$$f(1) = 1^2 - 7 = -6$$

$$f(-3) = (-3)^2 - 7 = 2$$

$$f(a) = a^2 - 7$$

To **compose** two functions f and g, you apply them successively:

$$(f \circ g)(x) = f(g(x))$$

(f of g of x, or f composed with g).

EXAMPLE 1-2: If $f(x) = x^2 - 3$ and $g(x) = 4 - \sqrt{x}$, find **(a)** $f(g(9))$ and $f(g(x))$, **(b)** $g(f(4))$ and $g(f(x))$.

Solution:

(a) To find $f(g(9))$, find $g(9)$ and then evaluate f at $g(9)$:

$$g(9) = 4 - \sqrt{9} = 1$$

Thus f evaluated at $g(9)$ is just $f(1)$:

$$f(g(9)) = f(1) = 1^2 - 3 = -2$$

For the general case, $f(g(x)) = [g(x)]^2 - 3 = (4 - \sqrt{x})^2 - 3$.

(b) To find $g(f(4))$, evaluate g at $f(4)$:

$$f(4) = 4^2 - 3 = 13$$

$$g(f(4)) = g(13) = 4 - \sqrt{13}$$

You can find a general expression for $g(f(x))$:

$$g(f(x)) = 4 - \sqrt{f(x)} = 4 - \sqrt{x^2 - 3}$$

B. Domain and range

The **domain** of a function $f(x)$ is the set of values of x for which $f(x)$ is defined and real. The **range** of $f(x)$ is the set of all possible values of $f(x)$. Interval notation will help you to describe the domain and range of a function. If a and b are numbers, with $a < b$, then:

"x in $[a, b]$" represents $a \leqslant x \leqslant b$

"x in $[a, b)$" represents $a \leqslant x < b$

"x in $(a, b]$" represents $a < x \leqslant b$

"x in (a, b)" represents $a < x < b$

Thus a square bracket means that the endpoint is included in the interval. This notation is also used for infinite intervals:

"x in $[a, \infty)$" represents $x \geqslant a$

"x in (a, ∞)" represents $x > a$

"x in $(-\infty, a]$" represents $x \leqslant a$

"x in $(-\infty, a)$" represents $x < a$

"x in $(-\infty, \infty)$" represents all x

EXAMPLE 1-3: Find the domain and range of **(a)** $f(x) = \sqrt{x} - 3$, **(b)** $g(x) = (x + 1)^{2/3}$, **(c)** $h(x) = \sqrt{x} + \sqrt{x - 4}$, **(d)** $j(x) = 1/\sqrt{3 - x^2}$.

Solution:

(a) The function $f(x) = \sqrt{x} - 3$ is only defined for $x \geqslant 0$. Also, $\sqrt{x}$ can't be negative, so $\sqrt{x} - 3$ must be at least -3. Thus the domain of f is $[0, \infty)$ and the range is $[-3, \infty)$.

(b) The function $g(x) = (x + 1)^{2/3}$ is defined for all values of x, but any number raised to the 2/3 power is nonnegative. The domain of g is $(-\infty, \infty)$ and the range is $[0, \infty)$.

(c) The function $h(x) = \sqrt{x} + \sqrt{x - 4}$ is only defined for $x \geqslant 4$ because $\sqrt{x - 4}$ requires $x \geqslant 4$. If $x \geqslant 4$, then $\sqrt{x}$ is at least two and $\sqrt{x - 4}$ is at least zero. The domain of $f(x)$ is $[4, \infty)$ and the range is $[2, \infty)$.

(d) In order that $\sqrt{3 - x^2}$ be defined, $3 - x^2$ must be nonnegative:

$$3 - x^2 \geqslant 0$$
$$x^2 \leqslant 3$$
$$-\sqrt{3} \leqslant x \leqslant \sqrt{3}$$

But the denominator mustn't be zero, so $3 - x^2$ can't be zero. Thus the domain of j is $(-\sqrt{3}, \sqrt{3})$. To find the range, notice that $\sqrt{3 - x^2}$ is always between 0 and $\sqrt{3}$. Thus $j(x)$ is at least $1/\sqrt{3} = \sqrt{3}/3$. The range is $[\sqrt{3}/3, \infty)$.

C. The graph of a function

The **graph** of a function $f(x)$ is the set of points (x, y) in the x-y plane for which $y = f(x)$.

EXAMPLE 1-4: Graph the function $f(x) = |x|$.

Solution: You must graph all points (x, y) for which $y = |x|$. The graph will contain points at all distances from the y axis because $f(x) = |x|$ is defined for all values of x (i.e., the domain of f is $(-\infty, \infty)$). If x is positive then $y = x$; if x is negative then $y = -x$. Thus for positive values of x graph $(x, y) = (x, x)$; for negative values of x graph $(x, -x)$, as shown in Figure 1-1.

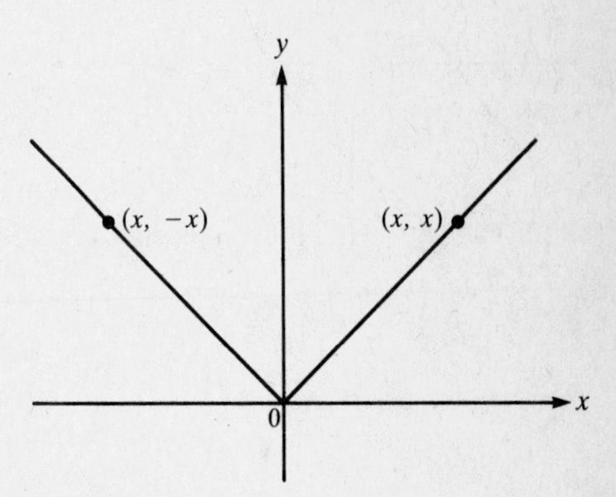

Figure 1-1
$f(x) = |x|$.

D. Inverse functions

If $f(x)$ and $g(x)$ are functions and $f(g(x)) = g(f(x)) = x$, then f and g are **inverse** functions. For example, $f(x) = 2x - 3$ and $g(x) = \frac{1}{2}(x + 3)$ are inverse functions:

$$f(g(x)) = 2g(x) - 3 = 2\left[\frac{1}{2}(x + 3)\right] - 3 = (x + 3) - 3 = x$$

$$g(f(x)) = \frac{1}{2}(f(x) + 3) = \frac{1}{2}[(2x - 3) + 3] = \frac{1}{2}[2x] = x$$

If you are given a function $f(x)$ and asked to find its inverse function, set y equal to $f(x)$ and solve the equation for x in terms of y. If you are successful and if this expression for x in terms of y is a function g of y, then g and f are inverses.

EXAMPLE 1-5: Find the inverse function for $f(x) = 5x^3 + 1$.

Solution: Set $y = f(x)$ and solve for x:

$$y = 5x^3 + 1$$
$$y - 1 = 5x^3$$
$$\frac{1}{5}(y - 1) = x^3$$
$$x = [\tfrac{1}{5}(y - 1)]^{1/3}$$

Thus if $g(y) = [\frac{1}{5}(y - 1)]^{1/3}$, then f and g are inverse functions. To check this, recognize that y is just a variable used in describing the action of g, and you could just as well write $g(z) = [\frac{1}{5}(z - 1)]^{1/3}$ or $g(x) = [\frac{1}{5}(x - 1)]^{1/3}$. Now find $f(g(x))$ and $g(f(x))$:

$$f(g(x)) = f\left(\left[\frac{1}{5}(x - 1)\right]^{1/3}\right) = 5\left\{\left[\frac{1}{5}(x - 1)\right]^{1/3}\right\}^3 + 1$$

$$= 5\left[\frac{1}{5}(x - 1)\right] + 1 = (x - 1) + 1 = x$$

$$g(f(x)) = g(5x^3 + 1) = \left\{\frac{1}{5}\left[(5x^3 + 1) - 1\right]\right\}^{1/3}$$

$$= \left[\frac{1}{5}(5x^3)\right]^{1/3} = (x^3)^{1/3} = x$$

You've verified that f and g are inverse functions.

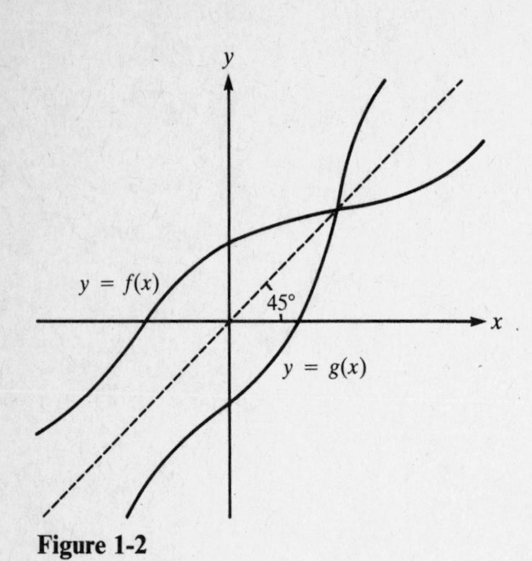

Figure 1-2

Many functions don't have inverse functions. The equation $y = f(x)$ may not give x as a function of y. That is, although each x yields a unique y (because f is a function), it is possible that certain values of y give more than one value of x. For example, $f(x) = x^2$ doesn't have an inverse function because the equation $y = x^2$ describes two values of x for every positive value of y: $x = \pm\sqrt{y}$. In particular, $y = 4$ corresponds to $x = 2$ and $x = -2$.

If you have a graph of $y = f(x)$ and f has an inverse function g, then the graph of g is easy to find: Simply reflect the graph of f about the line $y = x$ (the 45° line through the origin), as in Figure 1-2. This procedure works for a simple reason. The equation $y = f(x)$ relates x and y. The equation $y = g(x)$ is the same equation, with x and y interchanged.

1-2. Lines

A. Slope

Two distinct points in the plane (x_0, y_0) and (x_1, y_1) determine a line. The **slope** m of that line is

SLOPE OF A LINE

$$m = \frac{y_1 - y_0}{x_1 - x_0} \tag{1-1}$$

If this denominator is zero, then the line is vertical and the slope is undefined. You can find the slope of a line using *any* pair of distinct points on the line—you'll always find the same value for m.

The slope measures the tangent of the positive angle that the line makes with the x axis. A positive slope indicates a line that rises from left to right; a negative slope indicates a line that falls from left to right. A line with numerically large slope is steep; a line with slope close to zero is nearly horizontal. If $m = 0$ then the line is horizontal.

EXAMPLE 1-6: Find the slope of the line joining **(a)** (1, 2) and (5, −1), **(b)** (3, −4) and (2, 2), **(c)** (1, 3) and (−4, 3), **(d)** (5, 1) and (5, −2).

Solution:

(a) The line joining (1, 2) and (5, −1) has slope

$$m = \frac{-1 - 2}{5 - 1} = \frac{-3}{4}$$

(b) The line joining (3, −4) and (2, 2) has slope

$$m = \frac{2 - (-4)}{2 - 3} = -6$$

(c) The line joining (1, 3) and (−4, 3) has slope

$$m = \frac{3 - 3}{-4 - 1} = 0$$

and is horizontal.

(d) The line joining (5, 1) and (5, −2) has slope

$$m = \frac{-2 - 1}{5 - 5}$$

which is undefined. This line is vertical.

Two lines are parallel if and only if they have the same slope or are both vertical. Two lines are perpendicular if their slopes m_1 and m_2 are negative reciprocals:

$$m_1 = \frac{-1}{m_2}$$

or if one has slope 0 and the other has undefined slope.

EXAMPLE 1-7: Which of the following lines are parallel and which are perpendicular: L_1 passes through $(1, -4)$ and $(2, -1)$; L_2 passes through $(-3, 2)$ and $(0, 1)$; L_3 passes through $(4, 1)$ and $(-1, 0)$; L_4 passes through $(-3, -3)$ and $(1, 9)$?

Solution: L_1 has slope $m_1 = [-1 - (-4)]/(2 - 1) = 3$. L_2 has slope $m_2 = (1 - 2)/[0 - (-3)] = -1/3$. L_3 has slope $m_3 = (0 - 1)/(-1 - 4) = 1/5$. L_4 has slope $m_4 = [9 - (-3)]/[1 - (-3)] = 3$. So, L_1 and L_4 are parallel and L_2 is perpendicular to L_1 and L_4.

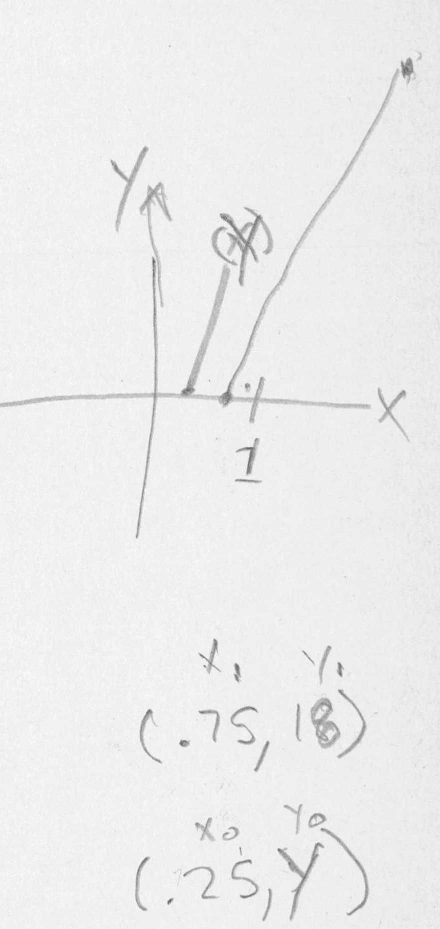

B. Equation of a line

1. Point-slope form If (x_0, y_0) and (x_1, y_1) are two points on a line, then you can find the slope m of that line. For any other point (x, y) on the line, the slope between (x_1, y_1) and (x, y) will also be m:

$$\frac{y - y_1}{x - x_1} = m$$

POINT-SLOPE FORM $$y - y_1 = m(x - x_1) \tag{1-2}$$

This equation in x and y describes the line through the point (x, y) with slope m. This is called the **point-slope** form for the equation of the line.

EXAMPLE 1-8: Find an equation for the line through the point $(7, 2)$ with slope -3.

Solution: Use Equation 1-2

$$y - y_1 = m(x - x_1)$$
$$y - 2 = -3(x - 7)$$

EXAMPLE 1-9: Find an equation for the line that passes through $(1, -2)$ and $(-3, 1)$.

Solution: This line has slope

$$m = \frac{1 - (-2)}{-3 - 1} = \frac{-3}{4}$$

This line has slope $m = -3/4$ and passes through the point $(x_1, y_1) = (-3, 1)$. Use Equation 1-2 and either of the given points to find the equation of the line:

$$y - y_1 = m(x - x_1)$$
$$y - 1 = \frac{-3}{4}[x - (-3)]$$
$$y - 1 = \frac{-3}{4}(x + 3)$$

2. Slope-intercept form If a line intersects the x axis at $(a, 0)$ and the y axis at $(0, b)$, then a is called the x intercept of the line and b is called the y intercept of the line. A line with slope m and y intercept b has equation:

$$y - y_1 = m(x - x_1)$$

$$y - b = m(x - 0)$$

$$y - b = mx$$

SLOPE-INTERCEPT FORM $$y = mx + b \qquad (1\text{-}3)$$

This equation is called the **slope-intercept** form for the equation of the line.

EXAMPLE 1-10: Find the slope-intercept form of the equation for the line with slope 5 that intersects the y axis at $(0, 3)$.

Solution: This line has slope $m = 5$ and y intercept $b = 3$. Use Equation 1-3:

$$y = mx + b$$

$$y = 5x + 3$$

EXAMPLE 1-11: Find the equation in slope-intercept form of the line that passes through $(1, 7)$ and $(-2, 3)$.

Solution: First find the slope of the line:

$$m = \frac{3 - 7}{-2 - 1} = \frac{4}{3}$$

The (point-slope) equation of the line is

$$y - y_1 = m(x - x_1)$$

$$y - 3 = \frac{4}{3}[x - (-2)]$$

$$y - 3 = \frac{4}{3}(x + 2)$$

Now solve for y to obtain the equation in slope-intercept form:

$$y = \frac{4}{3}(x + 2) + 3$$

$$y = \frac{4}{3}x + \frac{8}{3} + 3$$

$$y = \frac{4}{3}x + \frac{17}{3}$$

This equation is in slope-intercept form. You see that the line intersects the y axis at $(0, \frac{17}{3})$. Any equation of the form $Ax + By = C$, where A, B, and C are constants (A and B not both zero), is the equation of a line.

C. Graphs

There are several ways to graph a line when you are given its equation. One way is to find two points on the line—two ordered pairs (x, y) that satisfy the equation—plot them, and draw the line they determine.

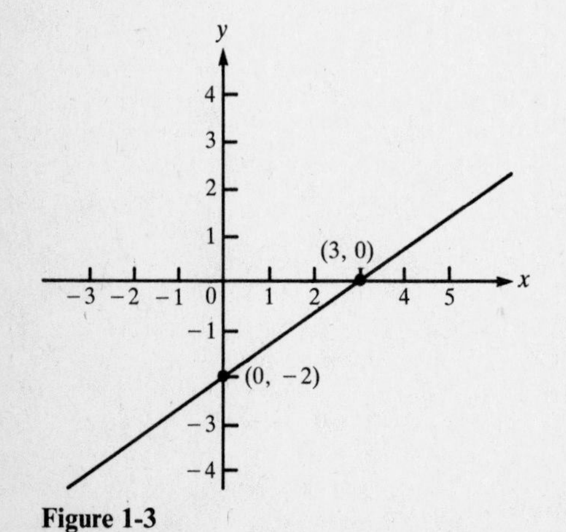

Figure 1-3
$2x - 3y = 6$.

EXAMPLE 1-12: Graph the line $2x - 3y = 6$.

Solution: One way to find two points on the line is to find the two intercepts. When $x = 0$, $-3y = 6$, and $y = -2$, so $(0, -2)$ is on the line. When $y = 0$, $2x = 6$, and $x = 3$, so $(3, 0)$ is on the line. Now plot the points $(0, -2)$ and $(3, 0)$ and the line they determine (see Figure 1-3).

If the equation is in slope-intercept form, you may find it easiest to graph the line by locating the y intercept and drawing the line through that point with the given slope.

EXAMPLE 1-13: Graph the line $y = 2x + 1$.

Solution: This line has y intercept 1 and slope 2, so plot the point (0, 1). Now draw the line through (0, 1) with slope 2—for every change of +1 in the x direction, the change in the y direction is +2—as shown in Figure 1-4.

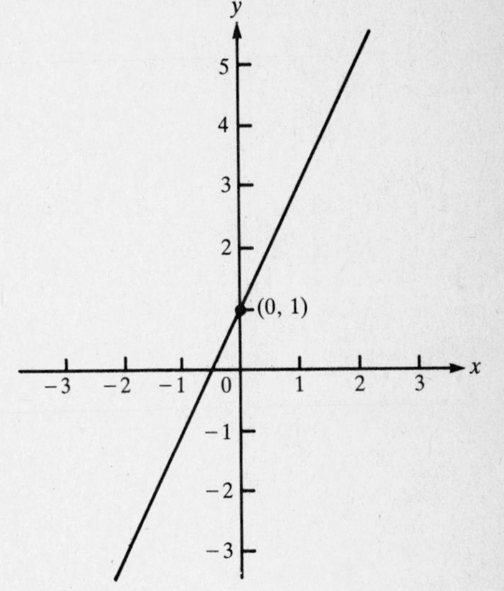

Figure 1-4
$y = 2x + 1$.

D. Linear functions

A **linear** function is any function of the form $f(x) = mx + b$ for constants m and b. It is called linear because the graph of $y = f(x)$ is a line.

EXAMPLE 1-14: Graph the function $f(x) = -2x + 1$.

Solution: The graph of $y = f(x)$ has slope -2 and y intercept 1, as shown in Figure 1-5.

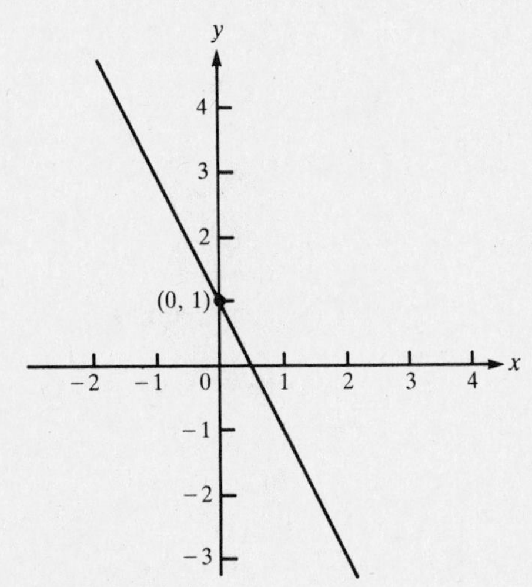

Figure 1-5
$f(x) = -2x + 1$.

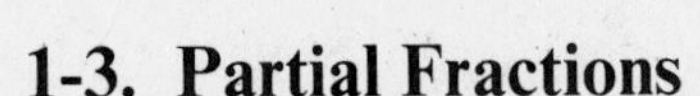

1-3. Partial Fractions

A **polynomial function** is a function of the form $f(x) = a_0 + a_1x + a_2x^2 + \cdots + a_nx^n$, where $a_0, a_1, \ldots, a_n$ are real numbers and the powers of x are nonnegative integers. For example, $f(x) = 2 + \sqrt{3}x - 4x^3$ is a polynomial function while $h(x) = x^2 - x^{1/3} + 2$ isn't ($\frac{1}{3}$ isn't an integer). A **rational function** is a quotient of two polynomial functions, that is, $f(x) = p(x)/q(x)$. For example, $f(x) = (x^3 - 7x + 1)/(x^2 + 3x - 2)$ is a rational function. Also, $g(x) = 2x^2 + [(5x - 3)/(x^2 + 2)]$ is a rational function because it can be expressed as

$$g(x) = \frac{2x^2(x^2 + 2) + 5x - 3}{x^2 + 2} = \frac{2x^4 + 4x^2 + 5x - 3}{x^2 + 2}$$

A rational function is **proper** if the degree of the numerator is less than the degree of the denominator. Otherwise, it is **improper**. By performing division, you can express an improper rational function as the sum of a polynomial and a proper rational function.

EXAMPLE 1-15: Express $f(x) = (x^4 - 2x + 1)/(x^2 + 3)$ as the sum of a polynomial and a proper fractional function.

Solution: Divide the denominator into the numerator:

$$\begin{array}{r} x^2 - 3 \\ x^2 + 3 \overline{) x^4 - 2x + 1} \\ x^4 + 3x^2 \\ \hline -3x^2 - 2x + 1 \\ -3x^2 - 9 \\ \hline -2x + 10 \end{array}$$

Thus $x^4 - 2x + 1 = (x^2 - 3)(x^2 + 3) + (-2x + 10)$, and so

$$\frac{x^4 - 2x + 1}{x^2 + 3} = x^2 - 3 + \frac{-2x + 10}{x^2 + 3}$$

Every proper rational function can be expressed as a sum of other proper rational functions, where each term in the sum has a denominator that is a power of an irreducible polynomial. In order to achieve this *decomposition*, you must first factor the denominator completely, that is, into irreducible factors. Then proceed as follows:

1. If a linear factor $ax + b$ appears in the denominator n times, then the decomposition will include rational terms of the form

$$\frac{A_1}{(ax + b)} + \frac{A_2}{(ax + b)^2} + \cdots + \frac{A_n}{(ax + b)^n}$$

for some constants $A_1, A_2, \ldots, A_n$. In particular, if the linear factor appears only once in the denominator, then the decomposition will simply include a term of the form $A/(ax + b)$, where A is a constant.

2. If the irreducible quadratic factor $ax^2 + bx + c$ $(b^2 - 4ac < 0)$ appears n times in the denominator, then the decomposition will include rational terms of the form

$$\frac{A_1 + B_1x}{ax^2 + bx + c} + \frac{A_2 + B_2x}{(ax^2 + bx + c)^2} + \cdots + \frac{A_n + B_nx}{(ax^2 + bx + c)^n}$$

EXAMPLE 1-16: Find the partial fraction decomposition of

$$f(x) = (8x - 4)/(x^2 + 2x - 3).$$

Solution: First factor the denominator: $x^2 + 2x - 3 = (x + 3)(x - 1)$, so

$$f(x) = \frac{8x - 4}{(x + 3)(x - 1)}$$

Because the denominator consists of linear terms, each appearing only once, you anticipate a decomposition into terms of the form

$$\frac{8x - 4}{(x + 3)(x - 1)} = \frac{A}{(x + 3)} + \frac{B}{(x - 1)}$$

for some constants A and B. In order to find A and B, simply multiply both sides by the denominator $(x + 3)(x - 1)$ in order to clear all denominators:

$$8x - 4 = A(x - 1) + B(x + 3)$$

Combine terms to express the right side as a polynomial in x:

$$8x - 4 = (A + B)x + (3B - A)$$

Because two polynomials are equal only when the coefficients of x^n are equal (for each n), you can equate coefficients:

$$8 = A + B \qquad -4 = 3B - A$$

Now you simply solve these equations simultaneously; you get $A = 7$ and $B = 1$. Thus,

$$f(x) = \frac{8x - 4}{x^2 + 2x - 3} = \frac{7}{(x + 3)} + \frac{1}{(x - 1)}$$

EXAMPLE 1-17: Find the partial fraction decomposition of

$$f(x) = (5x^2 - 3x + 4)/(x^3 - 1).$$

Solution: First factor the denominator. Because 1 is a root of $x^3 - 1$, you find

$$x^3 - 1 = (x - 1)(x^2 + x + 1)$$

a factorization into irreducible factors. So,

$$\frac{5x^2 - 3x + 4}{x^3 - 1} = \frac{5x^2 - 3x + 4}{(x - 1)(x^2 + x + 1)} = \frac{A}{(x - 1)} + \frac{Bx + C}{x^2 + x + 1}$$

for some constants A, B, and C. Multiply both sides by $(x - 1)(x^2 + x + 1)$:

$$\begin{aligned} 5x^2 - 3x + 4 &= A(x^2 + x + 1) + (Bx + C)(x - 1) \\ &= (A + B)x^2 + (A - B + C)x + (A - C) \end{aligned}$$

Equating coefficients you find

$$5 = A + B \qquad -3 = A - B + C \qquad 4 = A - C$$

Solve simultaneously and find $A = 2$, $B = 3$, $C = -2$. So,

$$\frac{5x^2 - 3x + 4}{x^3 - 1} = \frac{2}{(x - 1)} + \frac{3x - 2}{x^2 + x + 1}$$

EXAMPLE 1-18: Find the partial fraction decomposition of

$$f(x) = (6x^2 + x + 10)/(x^3 - 3x - 2).$$

Solution: Factor $x^3 - 3x - 2$ and set up the form for the partial fraction decomposition:

$$f(x) = \frac{6x^2 + x + 10}{(x + 1)^2(x - 2)} = \frac{A}{x + 1} + \frac{B}{(x + 1)^2} + \frac{C}{x - 2}$$

$$\begin{aligned} 6x^2 + x + 10 &= A(x + 1)(x - 2) + B(x - 2) + C(x + 1)^2 \\ &= (A + C)x^2 + (B - A + 2C)x + (C - 2A - 2B) \end{aligned}$$

$$6 = A + C \qquad 1 = B - A + 2C \qquad 10 = C - 2A - 2B$$

So $A = 2$, $B = -5$, and $C = 4$:

$$f(x) = \frac{2}{x + 1} + \frac{-5}{(x + 1)^2} + \frac{4}{x - 2}$$

1-4. Trigonometric and Hyperbolic Functions

A. Trigonometric functions

Unless otherwise noted, angles will be measured in radians in this book, rather than degrees. To convert from degrees to radians and back, use the fact that 2π radians $= 360$ degrees. Thus if an angle θ is $\alpha°$, then $\theta = \alpha(\frac{\pi}{180})$ radians. Similarly, if θ is β radians, then $\theta = \beta(\frac{180}{\pi})$ degrees.

Let θ be a positive number. Rotate the positive x axis about the origin in a counterclockwise direction by the angle θ. If θ is a negative number, rotate in a clockwise direction by the angle $-\theta$. The point $(1, 0)$ will be moved to the

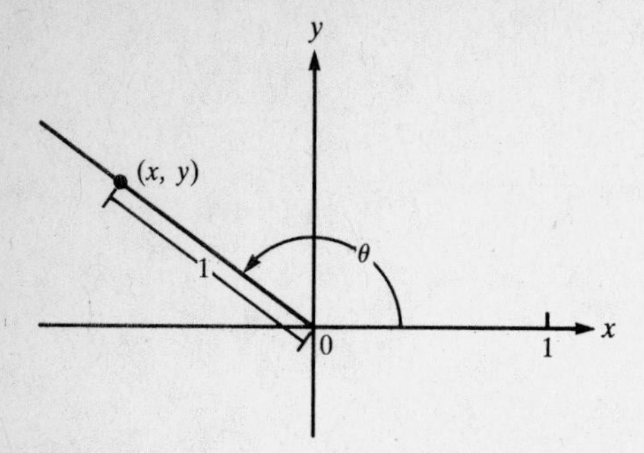

Figure 1-6

position (x, y), where:

$$x = \cos\theta \quad \text{and} \quad y = \sin\theta \tag{1-4}$$

(see Figure 1-6).

Thus $f(\theta) = \sin\theta$ and $g(\theta) = \cos\theta$ have domain $(-\infty, \infty)$ and range $[-1, 1]$. The other trigonometric functions (tangent, secant, cosecant, and cotangent) are defined in terms of sine and cosine:

$$\tan\theta = \frac{\sin\theta}{\cos\theta} \qquad \sec\theta = \frac{1}{\cos\theta}$$

$$\csc\theta = \frac{1}{\sin\theta} \qquad \cot\theta = \frac{\cos\theta}{\sin\theta} \tag{1-5}$$

The graphs of the trigonometric functions are shown in Figures 1-7 through 1-12. You will need to know that $\sin\pi/6 = 1/2$, $\cos\pi/6 = \sqrt{3}/2$, $\sin\pi/4 = \sqrt{2}/2$, and $\cos\pi/4 = \sqrt{2}/2$. If you know these values, then you can readily construct the following chart:

θ	0	$\frac{\pi}{6}$	$\frac{\pi}{4}$	$\frac{\pi}{3}$	$\frac{\pi}{2}$	$\frac{2\pi}{3}$	$\frac{3\pi}{4}$	$\frac{5\pi}{6}$	π
$\sin\theta$	0	$\frac{1}{2}$	$\frac{\sqrt{2}}{2}$	$\frac{\sqrt{3}}{2}$	1	$\frac{\sqrt{3}}{2}$	$\frac{\sqrt{2}}{2}$	$\frac{1}{2}$	0
$\cos\theta$	1	$\frac{\sqrt{3}}{2}$	$\frac{\sqrt{2}}{2}$	$\frac{1}{2}$	0	$-\frac{1}{2}$	$-\frac{\sqrt{2}}{2}$	$-\frac{\sqrt{3}}{2}$	-1
$\tan\theta$	0	$\frac{\sqrt{3}}{3}$	1	$\sqrt{3}$	undefined	$-\sqrt{3}$	-1	$-\frac{\sqrt{3}}{3}$	0

θ	π	$\frac{7\pi}{6}$	$\frac{5\pi}{4}$	$\frac{4\pi}{3}$	$\frac{3\pi}{2}$	$\frac{5\pi}{3}$	$\frac{7\pi}{4}$	$\frac{11\pi}{6}$	2π
$\sin\theta$	0	$-\frac{1}{2}$	$-\frac{\sqrt{2}}{2}$	$-\frac{\sqrt{3}}{2}$	-1	$-\frac{\sqrt{3}}{2}$	$-\frac{\sqrt{2}}{2}$	$-\frac{1}{2}$	0
$\cos\theta$	-1	$-\frac{\sqrt{3}}{2}$	$-\frac{\sqrt{2}}{2}$	$-\frac{1}{2}$	0	$\frac{1}{2}$	$\frac{\sqrt{2}}{2}$	$\frac{\sqrt{3}}{2}$	1
$\tan\theta$	0	$\frac{\sqrt{3}}{3}$	1	$\sqrt{3}$	undefined	$-\sqrt{3}$	-1	$-\frac{\sqrt{3}}{3}$	0

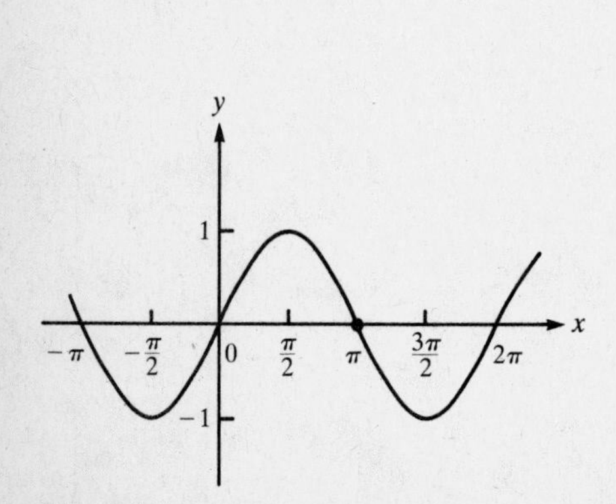

Figure 1-7
$f(x) = \sin x$.

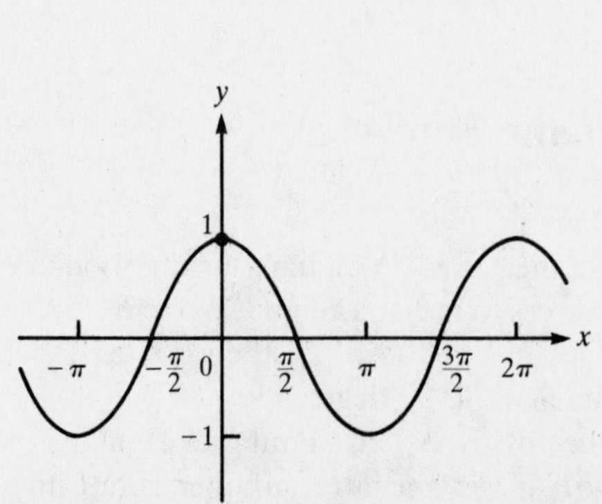

Figure 1-8
$f(x) = \cos x$.

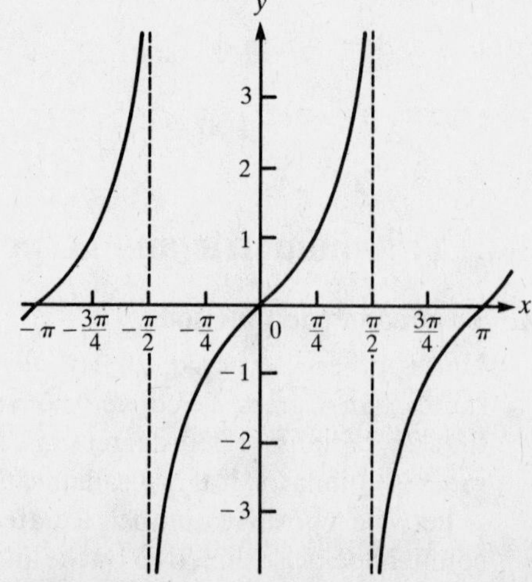

Figure 1-9
$f(x) = \tan x$.

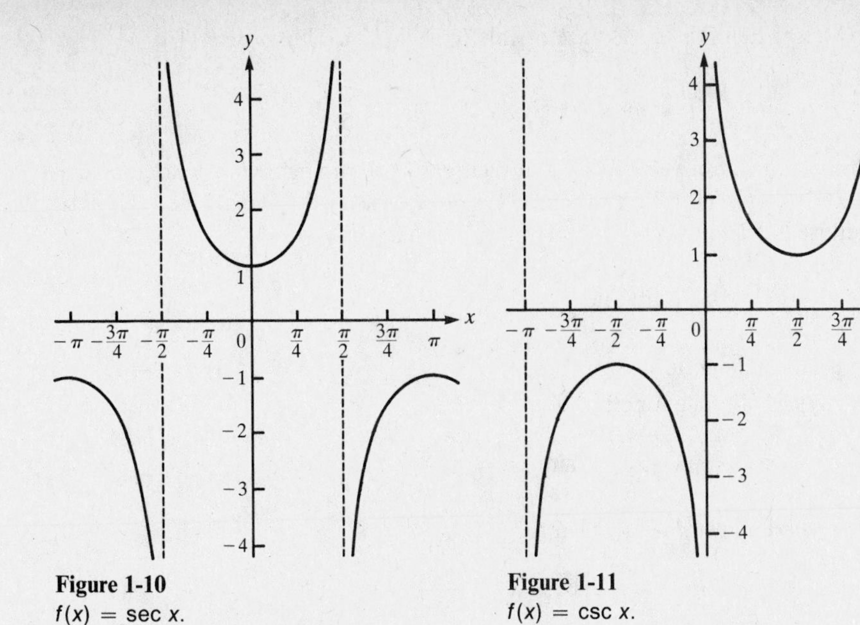

Figure 1-10
$f(x) = \sec x$.

Figure 1-11
$f(x) = \csc x$.

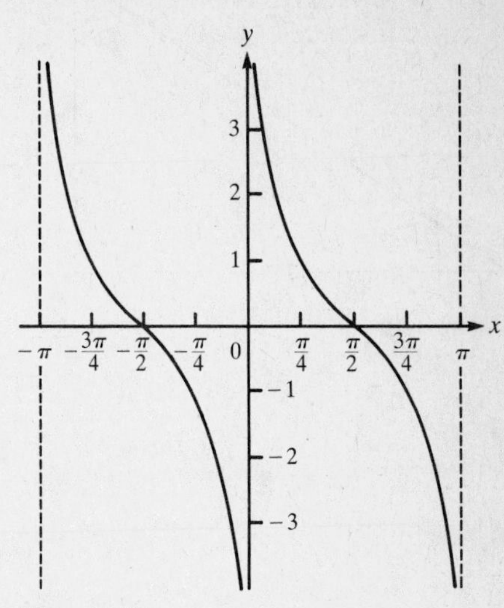

Figure 1-12
$f(x) = \cot x$.

If the angle θ appears in a right triangle, as in Figure 1-13, then the sine and cosine of θ are related to the lengths of the sides of the triangle:

$$\sin\theta = \frac{y}{\sqrt{x^2 + y^2}} = \frac{\text{side opposite } \theta}{\text{hypotenuse}} \quad \textbf{(1-6)}$$

$$\cos\theta = \frac{x}{\sqrt{x^2 + y^2}} = \frac{\text{side adjacent to } \theta}{\text{hypotenuse}} \quad \textbf{(1-7)}$$

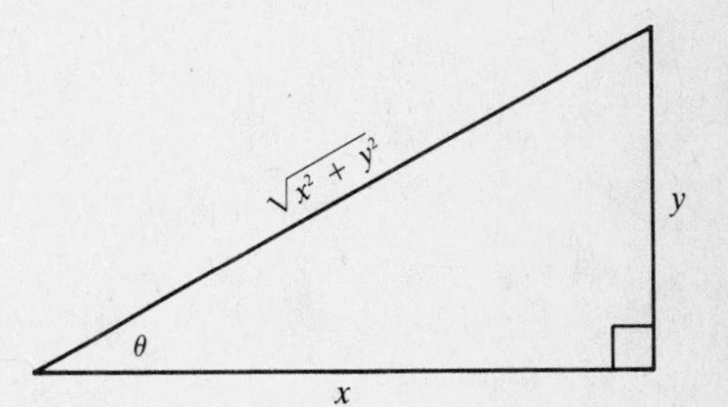

Figure 1-13

The Pythagorean theorem implies that

$$\sin^2\theta + \cos^2\theta = 1 \quad \textbf{(1-8)}$$

for any angle θ. This identity gives rise to several other useful identities:

$$\sec^2\theta = 1 + \tan^2\theta \quad \textbf{(1-9)}$$

$$\csc^2\theta = 1 + \cot^2\theta \quad \textbf{(1-10)}$$

You will also have occasion to use the identities:

$$\cos^2\theta = \tfrac{1}{2}(1 + \cos 2\theta) \quad \textbf{(1-11)}$$

$$\sin^2\theta = \tfrac{1}{2}(1 - \cos 2\theta) \quad \textbf{(1-12)}$$

$$\cos 2\theta = \cos^2\theta - \sin^2\theta \quad \textbf{(1-13)}$$

$$\sin 2\theta = 2\sin\theta\cos\theta \quad \textbf{(1-14)}$$

B. Inverse trigonometric functions

If x is any number between -1 and 1, then arc sin x (also written $\sin^{-1}x$) is the angle between $-\pi/2$ and $\pi/2$ whose sine is x. If x is any number between -1 and 1, then arc cos x ($\cos^{-1}x$) is the angle between 0 and π whose cosine is x. If x is any number, then arc tan x ($\tan^{-1}x$) is the angle between $-\pi/2$ and $\pi/2$ whose tangent is x. If x is a number in $(-\infty, -1]$ or $[1, \infty)$, then arc sec x ($\sec^{-1}x$) is the angle between 0 and π whose secant is x and arc csc x ($\csc^{-1}x$) is the angle between $-\pi/2$ and $\pi/2$ whose cosecant is x. If x is any number, then arc cot x ($\cot^{-1}x$) is the angle between 0 and π whose cotangent is x.

EXAMPLE 1-19: Find $\cos(\arcsin \frac{3}{5})$ and $\sin(\arctan \frac{2}{3})$.

Solution: To find $\cos(\arcsin \frac{3}{5})$, draw a right triangle with an angle whose sine is $\frac{3}{5}$ (see Figure 1-14). You have $\theta = \arcsin \frac{3}{5}$. The Pythagorean theorem allows you

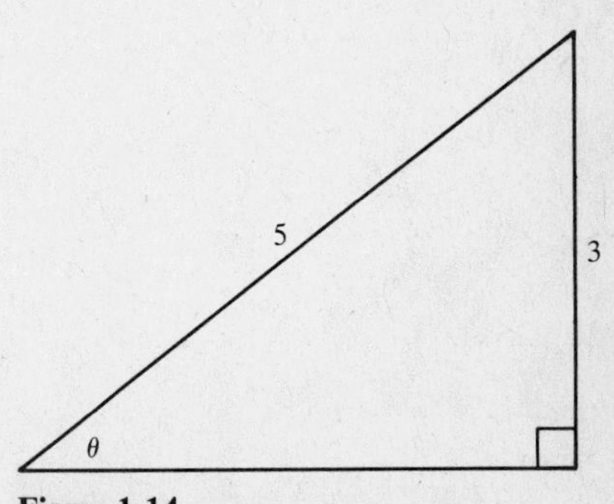

Figure 1-14

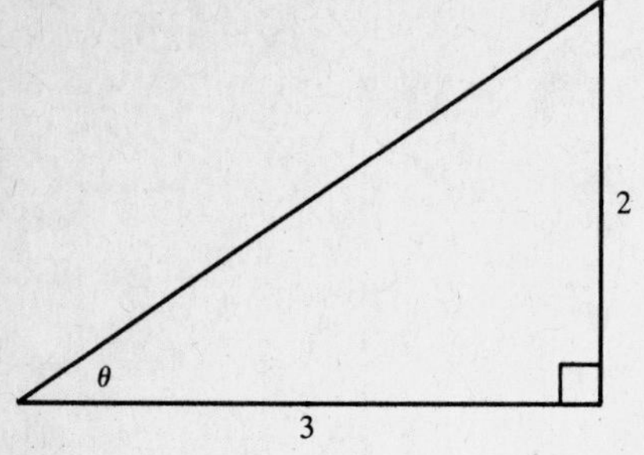

Figure 1-15

to find that the length of the base of this triangle is 4. Thus

$$\cos\left(\text{arc sin}\frac{3}{5}\right) = \cos\theta = \frac{4}{5}$$

To find sin(arc tan $\frac{2}{3}$), draw a right triangle with an angle whose tangent is $\frac{2}{3}$ (see Figure 1-15). You have $\theta = \text{arc tan}\,\frac{2}{3}$. By the Pythagorean theorem, the hypotenuse has length $\sqrt{2^2 + 3^2} = \sqrt{13}$. Thus,

$$\sin\left(\text{arc tan}\,\frac{2}{3}\right) = \sin\theta = \frac{2}{\sqrt{13}} = \frac{2\sqrt{13}}{13}$$

C. Hyperbolic functions

The **hyperbolic sine** function is

$$\sinh x = \frac{e^x - e^{-x}}{2} \tag{1-15}$$

The **hyperbolic cosine** function is

$$\cosh x = \frac{e^x + e^{-x}}{2} \tag{1-16}$$

The other hyperbolic functions are defined in terms of hyperbolic sine and cosine:

$$\tanh x = \frac{\sinh x}{\cosh x} \qquad \text{sech}\, x = \frac{1}{\cosh x}$$

$$\text{csch}\, x = \frac{1}{\sinh x} \qquad \coth x = \frac{\cosh x}{\sinh x} \tag{1-17}$$

1-5. Summation Notation

The symbol $\sum$ is the capital Greek letter sigma. This letter indicates a sum. The symbol $\sum_{k=i}^{j} f(k)$ represents the sum of the numbers $f(k)$ for all integers k running from i through j:

$$\sum_{k=i}^{j} f(k) = f(i) + f(i+1) + \cdots + f(j-1) + f(j) \tag{1-18}$$

For example,

$$\sum_{k=1}^{5} f(k) = f(1) + f(2) + f(3) + f(4) + f(5)$$

$$\sum_{j=3}^{5} f(j) = f(3) + f(4) + f(5)$$

$$\sum_{j=-2}^{2} g(j) = g(-2) + g(-1) + g(0) + g(1) + g(2)$$

$$\sum_{k=1}^{6} k = 1 + 2 + 3 + 4 + 5 + 6 = 21$$

$$\sum_{k=3}^{5} k^2 = 3^2 + 4^2 + 5^2 = 50$$

$$\sum_{k=0}^{2} 1 = 1 + 1 + 1 = 3$$

$$\sum_{k=1}^{3} kf(k) = f(1) + 2f(2) + 3f(3)$$

$$\sum_{k=1}^{4} a_k = a_1 + a_2 + a_3 + a_4$$

$$\sum_{i=2}^{5} a_i^2 = a_2^2 + a_3^2 + a_4^2 + a_5^2$$

EXAMPLE 1-20: Express the sum of the numbers 4 through 28 in summation notation.

Solution: Although there are several ways to express this sum in summation notation, the most natural way is $\sum_{k=4}^{28} k$.

You should know the following summation formulas:

$$\sum_{k=1}^{n} k = \frac{n(n+1)}{2} \tag{1-19}$$

$$\sum_{k=1}^{n} k^2 = \frac{n(n+1)(2n+1)}{6} \tag{1-20}$$

$$\sum_{k=1}^{n} k^3 = \frac{n^2(n+1)^2}{4} \tag{1-21}$$

EXAMPLE 1-21: Find the sum of the integers 1 through 28.

Solution: Write this in summation notation, $\sum_{k=1}^{28} k$, and apply Equation 1-19 with $n = 28$:

$$\sum_{k=1}^{28} k = \frac{28(28+1)}{2} = 406$$

You'll find the following useful:

$$\sum_{k=i}^{j} cf(k) = c \sum_{k=i}^{j} f(k) \tag{1-22}$$

$$\sum_{k=i}^{j} [f(k) + g(k)] = \sum_{k=i}^{j} f(k) + \sum_{k=i}^{j} g(k) \tag{1-23}$$

$$\sum_{k=i}^{j} c = c(j - i + 1) \tag{1-24}$$

EXAMPLE 1-22: Calculate $\sum_{k=1}^{7} (3k^2 - 2k + 4)$.

Solution: Use the appropriate equation:

$$\begin{aligned}
\sum_{k=1}^{7} (3k^2 - 2k + 4) &= \sum_{k=1}^{7} (3k^2 - 2k) + \sum_{k=1}^{7} 4 \\
&= \sum_{k=1}^{7} 3k^2 + \sum_{k=1}^{7} (-2k) + \sum_{k=1}^{7} 4 \\
&= 3 \sum_{k=1}^{7} k^2 + (-2) \sum_{k=1}^{7} k + \sum_{k=1}^{7} 4 \\
&= 3\left[\frac{7(7+1)(2 \cdot 7 + 1)}{6}\right] - 2\left[\frac{7(7+1)}{2}\right] \\
&\quad + 4(7 - 1 + 1) \\
&= 392
\end{aligned}$$

SUMMARY

1. The variable y is a function of x if every value of x gives a unique value of y.
2. The domain of a function f is the set of values of x for which $f(x)$ is defined. The range is the set of possible values of $f(x)$.
3. To find the inverse function to $f(x)$, set y equal to $f(x)$ and solve for x.
4. If you know the slope of a line and a point on the line, you can find the equation of the line.

5. You can express an improper rational function as the sum of a polynomial and a proper rational function.
6. You can express a proper rational function as a sum of proper functions, each with a denominator that is a power of an irreducible polynomial.
7. If you know the value of the trigonometric functions at $\pi/6$ and $\pi/4$ radians, then you can evaluate the trigonometric functions at any integer multiple of $\pi/6$ or $\pi/4$.
8. The right-triangle definition of the trigonometric functions will help you to find the values of *each* of the functions when *one* is known.
9. $\sum$-sum notation is a useful "short-hand" for the representation of sums of a large number (or of a general number, n) of terms, each of which is of the same type.

SOLVED PROBLEMS

PROBLEM 1-1 Is y a function of x: **(a)** $y^3 = x + 2$; **(b)** $y^4 = x + 2$?

Solution: **(a)** Solve for y in terms of x:

$$y = (x + 2)^{1/3}$$

Any value of x gives a unique value of y, so y is a function of x.

(b) If you solve for y in terms of x,

$$y = \pm\sqrt[4]{x + 2}$$

then you see that any value of x greater than -2 gives two values of y, so y isn't a function of x.

[See Section 1-1.]

PROBLEM 1-2 Find the domain and range of $f(x) = 1/\sqrt{x}$.

Solution: This function is defined for any $x > 0$, so the domain is $(0, \infty)$. Because the denominator can be any positive number, the function can take on any value between zero and infinity. The range is $(0, \infty)$.

[See Section 1-1.]

PROBLEM 1-3 Find the domain and range of $f(x) = 1 - \sqrt{2 - x}$.

Solution: The function is defined for $2 - x \geqslant 0$, i.e., for $x \leqslant 2$. The domain is $(-\infty, 2]$. Because $\sqrt{2 - x}$ can be any nonnegative number, $1 - \sqrt{2 - x}$ can be any number less than or equal to one. The range is $(-\infty, 1]$.

[See Section 1-1.]

PROBLEM 1-4 Let $f(x) = 3/(4 - x)$ and find the inverse function for f, if it exists.

Solution: Let $y = f(x)$ and solve for x in terms of y:

$$y = \frac{3}{4 - x}$$

$$4 - x = \frac{3}{y}$$

$$x = 4 - \frac{3}{y}$$

So $x = g(y)$, where $g(y) = 4 - 3/y$. Thus $g(x) = 4 - 3/x$ is the inverse function for f.

[See Section 1-1.]

PROBLEM 1-5 Let $f(x) = 3/x^2$ and find the inverse function for f, if it exists.

Solution: Let $y = f(x)$ and solve for x in terms of y:

$$y = \frac{3}{x^2}$$

$$x^2 = \frac{3}{y}$$

$$x = \pm\sqrt{\frac{3}{y}}$$

So x isn't a function of y. There is no inverse function for f. [See Section 1-1.]

PROBLEM 1-6 Find the slope of the line joining (1, 3) and (−2, 5).

Solution: The slope of the line is

$$m = \frac{y_1 - y_0}{x_1 - x_0} = \frac{5 - 3}{-2 - 1} = \frac{-2}{3}$$ [See Section 1-2.]

PROBLEM 1-7 Which of the following lines are parallel and which are perpendicular: L_1 passes through (3, −4) and (1, 3); L_2 passes through (4, −3) and (4, 2); L_3 passes through (−3, 5) and (2, 5); L_4 passes through (−1, 2) and (6, 4)?

Solution: L_1 has slope $m_1 = [3 - (-4)]/(1 - 3) = -7/2$. L_2 is a vertical line—its slope is undefined. L_3 has slope $m_3 = (5 - 5)/[2 - (-3)] = 0$ (horizontal line). L_4 has slope $m_4 = (4 - 2)/[6 - (-1)] = 2/7$. Thus L_1 is perpendicular to L_4, and L_2 is perpendicular to L_3. None of the lines is parallel to another. [See Section 1-2.]

PROBLEM 1-8 Find the equation of the line through (−3, 2) with slope = $-\frac{1}{3}$.

Solution: Use the point-slope form of the equation:

$$y - y_1 = m(x - x_1)$$

$$y - 2 = \frac{-1}{3}[x - (-3)]$$

$$y - 2 = \frac{-1}{3}(x + 3)$$ [See Section 1-2.]

PROBLEM 1-9 Find the equation of the line that passes through the points (3, 1) and (−1, −2).

Solution: The line has slope $m = (-2 - 1)/(-1 - 3) = \frac{3}{4}$. The line with slope $\frac{3}{4}$ that passes through (3, 1) is

$$y - 1 = \frac{3}{4}(x - 3)$$ [See Section 1-2.]

PROBLEM 1-10 Find the y intercept of the line through the points (−2, 3) and (1, 4).

Solution: The line has slope $m = (3 - 4)/(-2 - 1) = \frac{1}{3}$ and passes through (1, 4):

$$y - 4 = \frac{1}{3}(x - 1)$$

$$y - 4 = \frac{1}{3}x - \frac{1}{3}$$

$$y = \frac{1}{3}x - \frac{1}{3} + 4 = \frac{1}{3}x + \frac{11}{3}$$

This is the slope-intercept form of the equation. The y intercept is 11/3. [See Section 1-2.]

PROBLEM 1-11 Graph the function $f(x) = -3x - 1$.

Solution: The graph of $y = f(x)$ is a line with slope -3 and y intercept -1. The graph is shown in Figure 1-16. [See Section 1-2.]

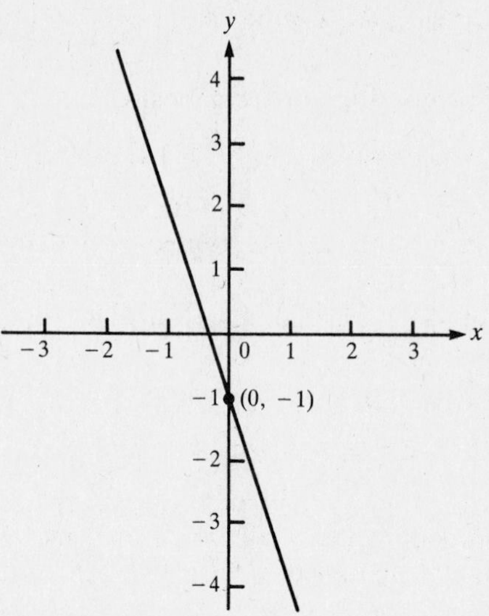

Figure 1-16
$f(x) = -3x - 1$.

PROBLEM 1-12 Express $f(x) = (x^4 - x^3 - 3x^2 - 3x - 1)/(x^2 + x + 1)$ as the sum of a polynomial and a proper rational function.

Solution: Divide $x^2 + x + 1$ into $x^4 - x^3 - 3x^2 - 3x - 1$:

$$\begin{array}{r|l} & x^2 - 2x - 2 \\ x^2 + x + 1 & x^4 - x^3 - 3x^2 - 3x - 1 \\ & \underline{x^4 + x^3 + x^2} \\ & -2x^3 - 4x^2 - 3x - 1 \\ & \underline{-2x^3 - 2x^2 - 2x} \\ & -2x^2 - x - 1 \\ & \underline{-2x^2 - 2x - 2} \\ & x + 1 \end{array}$$

You have

$$x^4 - x^3 - 3x^2 - 3x - 1 = (x^2 + x + 1)(x^2 - 2x - 2) + (x + 1)$$

Divide both sides of this equation by $x^2 + x + 1$:

$$\frac{x^4 - x^3 - 3x^2 - 3x - 1}{x^2 + x + 1} = x^2 - 2x - 2 + \frac{x + 1}{x^2 + x + 1}$$

Thus $f(x)$ is the sum of the polynomial $x^2 - 2x - 2$ and the proper rational function $(x + 1)/(x^2 + x + 1)$. [See Section 1-3.]

PROBLEM 1-13 Find the partial fraction decomposition of $f(x) = (x - 13)/(x^2 - x - 6)$.

Solution: First factor the denominator:

$$x^2 - x - 6 = (x + 2)(x - 3)$$

and then set up the form for the decomposition:

$$\frac{x - 13}{(x + 2)(x - 3)} = \frac{A}{x + 2} + \frac{B}{x - 3}$$

Multiply both sides by $(x + 2)(x - 3)$ to clear the denominator:

$$x - 13 = A(x - 3) + B(x + 2)$$

and combine terms:

$$x - 13 = (A + B)x + (-3A + 2B)$$

Equate coefficients:

$$1 = A + B \qquad -13 = -3A + 2B$$

and solve to find $A = 3$, $B = -2$. Thus $f(x) = [3/(x + 2)] - [2/(x - 3)]$. [See Section 1-3.]

PROBLEM 1-14 Find the partial fraction decomposition of

$$f(x) = (4x^3 + 6x^2 + x + 2)/(x^4 + x^2).$$

Solution: The denominator is $x^2(x^2 + 1)$, the product of a linear term (i.e., x) squared and an irreducible quadratic term, $x^2 + 1$. The decomposition will be of the form:

$$\frac{4x^3 + 6x^2 + x + 2}{x^4 + x^2} = \frac{A}{x} + \frac{B}{x^2} + \frac{Cx + D}{x^2 + 1}$$

Clear denominators and combine terms:

$$\begin{aligned} 4x^3 + 6x^2 + x + 2 &= Ax(x^2 + 1) + B(x^2 + 1) + (Cx + D)x^2 \\ &= (A + C)x^3 + (B + D)x^2 + Ax + B \end{aligned}$$

Equate coefficients:

$$4 = A + C \qquad 6 = B + D \qquad 1 = A \qquad 2 = B$$

and solve to find $C = 3$ and $D = 4$. You have

$$f(x) = \frac{1}{x} + \frac{2}{x^2} + \frac{3x + 4}{x^2 + 1}$$

[See Section 1-3.]

PROBLEM 1-15 Find $\tan(\text{arc cos}\,\frac{1}{5})$.

Solution: Draw a right triangle with an angle whose cosine is $\frac{1}{5}$, as in Figure 1-17. You have $\theta = \text{arc cos}\,\frac{1}{5}$. By the Pythagorean theorem, the remaining side has length $\sqrt{24}$. Thus,

$$\tan(\text{arc cos}\,\tfrac{1}{5}) = \tan\theta = \frac{\sqrt{24}}{1} = 2\sqrt{6}$$

[See Section 1-4.]

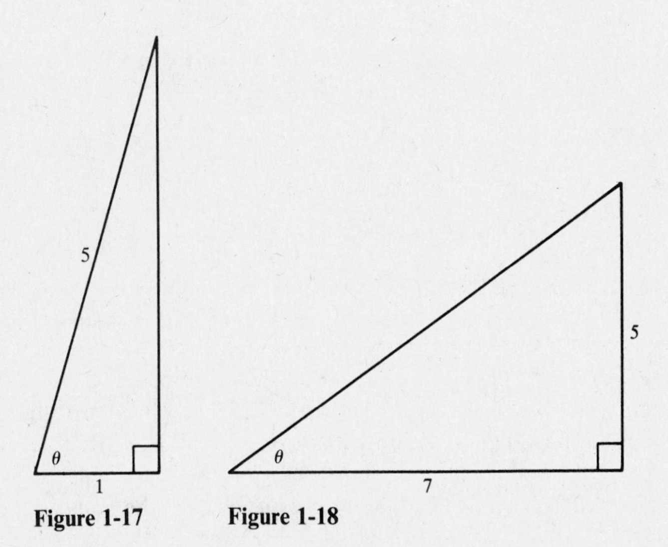

Figure 1-17 **Figure 1-18**

PROBLEM 1-16 Find $\sec(\text{arc cot}\,\frac{7}{5})$.

Solution: Draw a right triangle with an angle whose cotangent is $\frac{7}{5}$, as in Figure 1-18. You have $\theta = \text{arc cot}\,\frac{7}{5}$. By the Pythagorean theorem, the hypotenuse has length $\sqrt{74}$. Thus,

$$\sec(\text{arc cot}\,\tfrac{7}{5}) = \sec\theta = \frac{\sqrt{74}}{7}$$

[See Section 1-4.]

PROBLEM 1-17 Find $\sum_{k=1}^{6} (2k^3 - k)$.

Solution: Separate into two sums and use the appropriate formula:

$$\sum_{k=1}^{6} (2k^3 - k) = \sum_{k=1}^{6} 2k^3 + \sum_{k=1}^{6} (-k)$$

$$= 2 \sum_{k=1}^{6} k^3 - \sum_{k=1}^{6} k$$

$$= 2\left[\frac{6^2(6+1)^2}{4}\right] - \frac{6(6+1)}{2} = 861$$

[See Section 1-5.]

PROBLEM 1-18 Find $\sum_{k=1}^{8} (k^3 + 3k^2 - 6)$.

Solution:

$$\sum_{k=1}^{8} (k^3 + 3k^2 - 6) = \sum_{k=1}^{8} k^3 + 3 \sum_{k=1}^{8} k^2 - \sum_{k=1}^{8} 6$$

$$= \frac{8^2(8+1)^2}{4} + 3\left[\frac{8(8+1)[(2)8+1]}{6}\right] - 6(8 - 1 + 1)$$

$$= 1\,860$$

[See Section 1-5.]

Supplementary Exercises

1-19 Which of the following are functions of x: **(a)** $\sqrt{y+1} = x - 2$; **(b)** $y - 2 = \sqrt{x+1}$; **(c)** $y = 2$; **(d)** $y^2 = \sqrt{x}$; **(e)** $(y+1)^3 = x^3 - 2$?

In Problems 1-20 through 1-24 find the domain and range of f:

1-20 $f(x) = 2 - x^2$

1-21 $f(x) = 1 + \sqrt{x - 3}$

1-22 $f(x) = 1 + \sin x$

1-23 $f(x) = x^{5/2}$

1-24 $f(x) = 1/\sqrt{4 - x^2}$

In Problems 1-25 through 1-28 find the inverse function for f, if it exists:

1-25 $f(x) = 3x + 6$

1-26 $f(x) = x^5 - 2$

1-27 $f(x) = (x - 1)^2 + 1$

1-28 $f(x) = 5/(2x - 1)^3$

1-29 Which of the following lines are parallel and which are perpendicular: L_1 passes through $(-1, 5)$ and $(22, 51)$; L_2 passes through $(1, -1)$ and $(-3, 7)$; L_3 passes through $(6, -1)$ and $(4, 0)$; and L_4 passes through $(4, 1)$ and $(-2, -3)$; L_5 passes through $(3, -2)$ and $(2, -4)$?

In Problems 1-30 through 1-34 find the slope-intercept form of the equation of the line passing through the two points:

1-30 $(2, 1)$ and $(5, 4)$

1-31 $(-1, 3)$ and $(2, -2)$

1-32 $(1, 1)$ and $(3, 1)$

1-33 $(1, 4)$ and $(1, -2)$

1-34 $(2, -6)$ and $(-3, 2)$

In Problems 1-35 through 1-39 express f as the sum of a polynomial and a rational fun

1-35 $f(x) = (x^2 + 2)/(x - 1)$

1-36 $f(x) = x^3/(x^2 + 1)$

1-37 $f(x) = (x^3 + 1)/(x^3 - 1)$

1-38 $f(x) = (x^5 - x)/(x^3 - x - 1)$

1-39 $f(x) = (x^5 - x^3 + 2)/(x^2 + x + 1)$

In Problems 1-40 through 1-47 find the partial fraction decomposition of f:

1-40 $f(x) = (4x + 2)/(x^2 - 1)$

1-41 $f(x) = (3x + 2)/[x(x + 1)^2]$

1-42 $f(x) = (x^2 - 5x - 3)/[(x - 1)(x^2 + 3x + 3)]$

1-43 $f(x) = (2x^2 + 10x + 6)/[(x - 1)(x^2 + 3x + 2)]$

1-44 $f(x) = (2x^3 + 2x^2 - 5x + 3)/(x^4 - x^3)$

1-45 $f(x) = (x^4 + 2x^3 - x^2 + 2x + 1)/[x(x^2 + 1)^2]$

1-46 $f(x) = (x^3 + 3x^2 - 8x - 12)/[x^2(x + 2)^2]$

1-47 $f(x) = (4x^2 + 2x + 6)/(x^4 - 1)$

1-48 Find sin(arc tan (2/5)).

1-49 Find sec(arc sin (4/5)).

1-50 Find sin(arc cot 2).

1-51 Find tan(arc sin (2/5)).

1-52 Find cos(arc csc (3/2)).

1-53 Find cot(arc sin $\frac{1}{2}$).

1-54 Find csc(arc sec 3).

In Problems 1-55 through 1-60 find the sum:

1-55 $\sum_{k=1}^{4} k^2$

1-56 $\sum_{k=1}^{9} (2k^3 - 1)$

1-57 $\sum_{k=1}^{5} (k^2 - 2k + 5)$

1-58 $\sum_{k=1}^{7} (2k^3 - k^2)$

1-59 $\sum_{k=1}^{6} (k^3 + k^2 + k + 1)$

1-60 $\sum_{k=1}^{8} (k - 2k^2)$

Solutions to Supplementary Exercises

(1-19) (a), (b), (c), (e)

(1-20) domain $(-\infty, \infty)$; range $(-\infty, 2]$

(1-21) domain $[3, \infty)$; range $[1, \infty)$

(1-22) domain $(-\infty, \infty)$; range $[0, 2]$

(1-23) domain $[0, \infty)$; range $[0, \infty)$

(1-24) domain $(-2, 2)$; range $[\frac{1}{2}, \infty)$

(1-25) $g(x) = \frac{1}{3}x - 2$

(1-26) $g(x) = (x + 2)^{1/5}$

(1-27) no inverse

(1-28) $g(x) = \frac{1}{2}[(5/x)^{1/3} + 1]$

(1-29) L_1 and L_5 are parallel;
L_3 is perpendicular to L_1 and L_5

(1-30) $y = x - 1$

(1-31) $y = (-5/3)x + 4/3$

(1-32) $y = 1$

(1-33) $x = 1$

(1-34) $y = (-8/5)x - (14/5)$

(1-35) $f(x) = x + 1 + [3/(x - 1)]$

(1-36) $f(x) = x - [x/(x^2 + 1)]$

(1-37) $f(x) = 1 + [2/(x^3 - 1)]$

(1-38) $f(x) = x^2+1+[(x^2+1)/(x^3-x-1)]$

(1-39) $f(x) = x^3-x^2-x+2-[x/(x^2+x+1)]$

(1-40) $f(x) = [1/(x + 1)] + [3/(x - 1)]$

(1-41) $f(x) = (2/x) - [2/(x + 1)] + [1/(x + 1)^2]$

(1-42) $f(x) = [-1/(x - 1)] + [2x/(x^2 + 3x + 3)]$

(1-43) $f(x) = [3/(x-1)]+[1/(x+1)]-[2/(x+2)]$

(1-44) $f(x) = (2/x^2) - (3/x^3) + [2/(x - 1)]$

(1-45) $f(x) = (1/x)+[2/(x^2+1)]-[3x/(x^2+1)^2]$

(1-46) $f(x) = (1/x) - (3/x^2) + [2/(x + 2)^2]$

(1-47) $f(x) = [3/(x - 1)] - [2/(x + 1)] - [(x + 1)/(x^2 + 1)]$

(1-48) $2\sqrt{29}/29$

(1-49) $5/3$

(1-50) $\sqrt{5}/5$

(1-51) $2\sqrt{21}/21$

(1-52) $\sqrt{5}/3$

(1-53) $\sqrt{3}$

(1-54) $3\sqrt{8}/8$

(1-55) 30

(1-56) 4041

(1-57) 50

(1-58) 1428

(1-59) 559

(1-60) -372

2 CONIC SECTIONS

THIS CHAPTER IS ABOUT

- ☑ The General Second-Degree Equation
- ☑ Parabolas
- ☑ Ellipses and Circles
- ☑ Hyperbolas
- ☑ The xy Term and Rotations

2-1. The General Second-Degree Equation

The most general quadratic equation in x and y has the form

$$Ax^2 + Bxy + Cy^2 + Dx + Ey + F = 0 \tag{2-1}$$

where at least one of the first three coefficients is nonzero. The graph of the equation, when it exists, is one of the **conic sections** (a *parabola*, an *ellipse*, or a *hyperbola*), or a **degenerate** conic—a single point (a degenerate ellipse), a pair of intersecting lines (a degenerate hyperbola), or a pair of parallel lines or a single line (a degenerate parabola).

To graph a second-degree equation, you'll find it helpful to know in advance what kind of curve you have. Here is a rule that allows you to recognize the conic section at a glance. For Equation 2-1,

- if $B^2 - 4AC = 0$, then the graph is a parabola.
- if $B^2 - 4AC > 0$, then the graph is a hyperbola.
- if $B^2 - 4AC < 0$, then the graph is an ellipse.

Consider the form of these conditions if $B = 0$ (when the xy term is missing from the equation):

- if $4AC = 0$ (i.e., $A = 0$ or $C = 0$), the graph is a parabola.
- if $4AC < 0$ (i.e., A and C have opposite signs), the graph is a hyperbola.
- if $4AC > 0$ (i.e., A and C have the same sign), the graph is an ellipse.

EXAMPLE 2-1: Identify each of the following as a parabola, an ellipse, or a hyperbola:

(a) $x^2 - 2xy - y^2 + x + 2 = 0$, **(b)** $xy + y^2 - y = 0$,
(c) $3x^2 + 4y^2 + x + 3 = 0$, **(d)** $x^2 - 2xy + y^2 - \sqrt{2}x - \sqrt{2}y = 0$,
(e) $5x^2 + 8x - 3y + 2 = 0$.

Solution:

(a) $A = 1, B = -2, C = -1$	$B^2 - 4AC = 4 + 4 > 0$	hyperbola
(b) $A = 0, B = 1, C = 1$	$B^2 - 4AC = 1 > 0$	hyperbola
(c) $A = 3, B = 0, C = 4$	$B^2 - 4AC = 0 - 48 < 0$	ellipse
(d) $A = 1, B = -2, C = 1$	$B^2 - 4AC = 4 - 4 = 0$	parabola
(e) $A = 5, B = 0, C = 0$	$B^2 - 4AC = 0$	parabola

First consider the special cases in which $B = 0$; in Section 2-5 you'll return to the general situation.

2-2. Parabolas

The rule from Section 2-1 implies that when $B = 0$, the parabola will have equation of the form

$$Ax^2 + Dx + Ey + F = 0 \quad \text{or} \quad Cy^2 + Dx + Ey + F = 0$$

If you complete the square in x (or in y), you get one of the *standard* forms of the equation of a parabola:

PARABOLA: STANDARD FORMS

$$(x - h)^2 = 4a(y - k) \qquad \textbf{(2-2)}$$

$$(y - k)^2 = 4a(x - h) \qquad \textbf{(2-3)}$$

The point (h, k) is called the **vertex** of the parabola.

The **axis** or **line of symmetry** of Equation 2-2 is parallel to the y axis and passes through the vertex. The curve opens upward if $a > 0$ and downward if $a < 0$. The **focus** of the parabola is the point $(h, k + a)$.

In Equation 2-3 the roles of x and y are reversed. The axis of symmetry is parallel to the x axis; the curve opens to the right if $a > 0$, to the left if $a < 0$. The focus is at $(h + a, k)$.

EXAMPLE 2-2: Put each of the following in standard form, identify the vertex and focus of each, and draw the graphs: **(a)** $2x^2 + 4x - 3y + 6 = 0$, **(b)** $y^2 - 4y + 2x = 0$.

Solution:

(a) Complete the square in x (after dividing through by two):

$$x^2 + 2x - \frac{3}{2}y + 3 = 0$$

$$x^2 + 2x + 1 = \frac{3}{2}y - 3 + 1$$

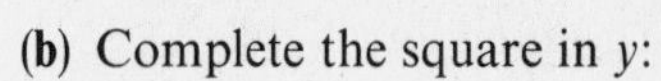

$$(x + 1)^2 = \frac{3}{2}\left(y - \frac{4}{3}\right)$$

Compare with Equation 2-2: You have $h = -1$, $k = \frac{4}{3}$, $4a = \frac{3}{2}$, and $a = \frac{3}{8} > 0$. The curve opens upward from a vertex at $(h, k) = (-1, \frac{4}{3})$. The focus is at $(h, k + a) = (-1, (4/3) + (3/8)) = (-1, 41/24)$. To draw the graph, find one other point and take full advantage of the symmetry of the graph. If $x = 0$, you have $y = 2$. The symmetry tells you that $(-2, 2)$ is also on the graph (see Figure 2-1).

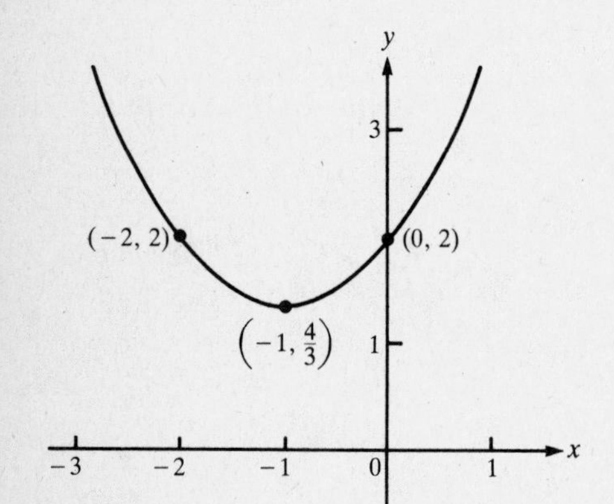

Figure 2-1
The parabola
$(x + 1)^2 = \frac{3}{2}\left(y - \frac{4}{3}\right)$.

(b) Complete the square in y:

$$y^2 - 4y + 4 = -2x + 4$$

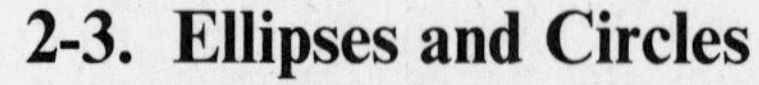

$$(y - 2)^2 = -2(x - 2)$$

Compare with Equation 2-3: You get $h = 2$, $k = 2$, $4a = -2$, and $a = -\frac{1}{2} < 0$. The curve opens to the left, with vertex at $(h, k) = (2, 2)$ and focus at $(h + a, k) = (2 - \frac{1}{2}, 2) = (\frac{3}{2}, 2)$. If $x = 0$, you get $y = 0$. The symmetrically located point is $(0, 4)$ (see Figure 2-2).

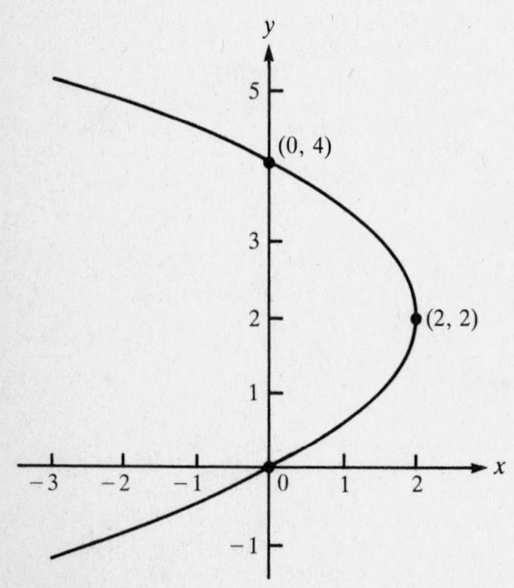

Figure 2-2
The parabola
$(y - 2)^2 = -2(x - 2)$.

2-3. Ellipses and Circles

You know that $Ax^2 + Cy^2 + Dx + Ey + F = 0$ will be an ellipse if A and C are of the same sign. To find a *standard* form of the equation for the ellipse, you complete the square in x and y and then divide both sides of the equation by the constant on the right. You get the form

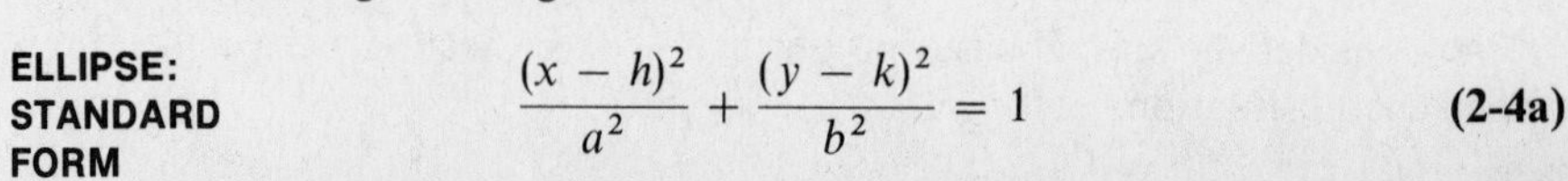

ELLIPSE: STANDARD FORM

$$\frac{(x - h)^2}{a^2} + \frac{(y - k)^2}{b^2} = 1 \qquad \textbf{(2-4a)}$$

In order to help with the graph, adopt the following convention: Of the two numbers in the denominators, agree to let a^2 represent the larger of the two. This means that another standard form is

ELLIPSE: STANDARD FORM

$$\frac{(x-h)^2}{b^2} + \frac{(y-k)^2}{a^2} = 1 \qquad \textbf{(2-4b)}$$

In both cases, the point (h, k) is the **center** of the ellipse. In Equation 2-4a the **major axis** of the ellipse passes through the center, is parallel to the x axis, and has endpoints $(h + a, k)$ and $(h - a, k)$. The **minor axis** is perpendicular to the major axis, passes through the center, and has endpoints $(h, k + b)$ and $(h, k - b)$.

In Equation 2-4b, the roles of x and y are reversed. The major axis is parallel to the y axis and has endpoints $(h, k + a)$ and $(h, k - a)$. The endpoints of the minor axis are $(h - b, k)$ and $(h + b, k)$.

To graph an ellipse, locate the center and determine whether the major axis is parallel to the x axis or y axis. Mark off a units in either direction from the center along the major axis; mark off b units in either direction along the minor axis. Draw a smooth curve joining those four points.

EXAMPLE 2-3: Put each of the following in standard form, locate the center and the endpoints of major and minor axes, and sketch the graph: **(a)** $x^2 + 4y^2 - 2x + 16y + 13 = 0$, **(b)** $2x^2 + y^2 + 12x - 6y + 26 = 0$.

Solution:

(a) Complete the square in x and y:

$$x^2 - 2x + 1 + 4(y^2 + 4y + 4) = -13 + 1 + 16$$

$$(x-1)^2 + 4(y+2)^2 = 4$$

Now divide by four (the constant on the right):

$$\frac{(x-1)^2}{4} + \frac{(y+2)^2}{1} = 1$$

This is the standard form of Equation 2-4a because the larger number is under the x^2 term. The major axis is parallel to the x axis. The center is at $(h, k) = (1, -2)$, with $a^2 = 4$ and $b^2 = 1$. The major axis has endpoints $(h + a, k) = (1 + 2, -2) = (3, -2)$ and $(h - a, k) = (-1, -2)$; The minor axis has endpoints $(h, k + b) = (1, -2 + 1) = (1, -1)$ and $(h, k - b) = (1, -3)$. Draw a smooth curve through these four points, as in Figure 2-3.

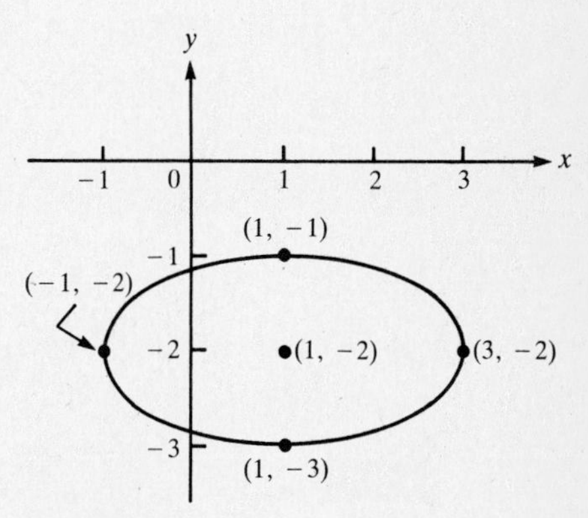

Figure 2-3
The ellipse $\frac{(x-1)^2}{4} + \frac{(y+2)^2}{1} = 1.$

(b) Proceed exactly as in part (a):

$$2(x^2 + 6x + 9) + (y^2 - 6y + 9) = -26 + 18 + 9$$

$$2(x+3)^2 + (y-3)^2 = 1$$

You have a one on the right side, but the equation is not in standard form. Divide the numerator and denominator of the first term by two so that it looks like Equation 2-4b:

$$\frac{(x+3)^2}{\frac{1}{2}} + \frac{(y-3)^2}{1} = 1$$

Under the y^2 term you find $a^2 = 1$, so the major axis is parallel to the y axis. The center is at $(-3, 3)$, $a = 1$, $b = \sqrt{\frac{1}{2}} = \sqrt{2}/2$. The endpoints of the major axis are at $(-3, 4)$ and $(-3, 2)$; the minor axis has endpoints $(-3 + (\sqrt{2}/2), 3)$ and $(-3 - (\sqrt{2}/2), 3)$. Draw the graph, as in Figure 2-4.

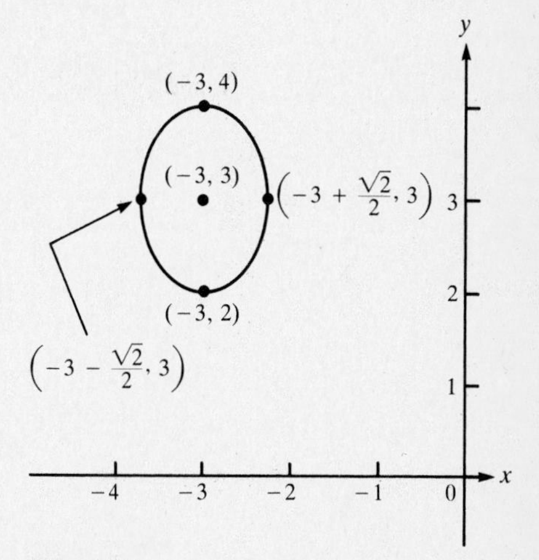

Figure 2-4
The ellipse $\frac{(x+3)^2}{\frac{1}{2}} + \frac{(y-3)^2}{1} = 1.$

EXAMPLE 2-4: Put in standard form and graph

(a) $36x^2 + 36y^2 - 72x + 36y + 29 = 0$,
(b) $x^2 + 4y^2 + 2x - 8y + 9 = 0$,
(c) $2x^2 + y^2 - 12x - 2y + 19 = 0$.

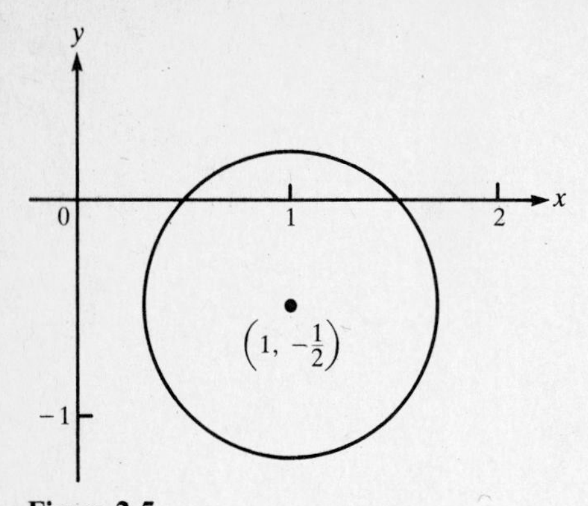

Figure 2-5
The circle
$(x - 1)^2 + (y + \frac{1}{2})^2 = \frac{4}{9}$.

Solution:

(a) Proceed as in Example 2-3:

$$36(x^2 - 2x + 1) + 36(y^2 + y + \tfrac{1}{4}) = -29 + 36 + 9$$
$$36(x - 1)^2 + 36(y + \tfrac{1}{2})^2 = 16$$
$$\frac{(x - 1)^2}{4/9} + \frac{(y + \frac{1}{2})^2}{4/9} = 1$$

This one is different in that you have $a = b = 2/3$. The graph is a circle, center at $(1, -\frac{1}{2})$, radius 2/3 (see Figure 2-5). Remember: An equation with $A = C$ gives a special ellipse known as a circle.

(b) Proceeding as before you get

$$x^2 + 2x + 1 + 4(y^2 - 2y + 1) = -9 + 1 + 4$$
$$(x + 1)^2 + 4(y - 1)^2 = -4$$

Again you have a different situation. The equation should give you an ellipse, but the negative number on the right would require a^2 and b^2 to be negative. Since this isn't possible, there are no points whose coordinates satisfy this equation. This is called an ellipse with *no locus*.

(c) Again complete the square in x and y:

$$2(x^2 - 6x + 9) + (y^2 - 2y + 1) = -19 + 18 + 1$$
$$2(x - 3)^2 + (y - 1)^2 = 0$$

This is yet another exception: It isn't possible to get a positive 1 on the right side. The equation shows the sum of two nonnegative numbers equal to zero. This can be true only when each term is zero: $x = 3$ and $y = 1$. The graph consists of the single point (3, 1).

Note: Example 2-4 points out that not every equation that looks like an ellipse actually gives you an ellipse—it may be a degenerate ellipse. However, you don't have to worry about the exceptions. If an equation looks like an ellipse, you proceed as in each of the previous examples. If the resulting constant on the right is negative, you have no graph; if it is zero, you have a single point.

2-4. Hyperbolas

The equation $Ax^2 + Cy^2 + Dx + Ey + F = 0$ gives you a hyperbola when A and C are of opposite signs. The hyperbola differs from the other conic sections in a fundamental way: Its graph is broken into two parts. The **center** is outside the curves. It's called a center because it's the point of intersection of the two axes of symmetry for the graph. Moreover, each half of the graph has a pair of **asymptotes** that intersect at the center. The graph is easy to obtain if you first find the center and the asymptotes. To do this, you find a standard form for the equation: one for the hyperbola whose two parts open left and right, and one for the hyperbola with parts opening upward and downward.

Complete the square in x and y and get a positive 1 on the right side of the equation, exactly as you did for the ellipse. You'll get

HYPERBOLA: STANDARD FORMS

$$\frac{(x - h)^2}{a^2} - \frac{(y - k)^2}{b^2} = 1 \qquad \textbf{(2-5a)}$$

or

$$\frac{(y - k)^2}{b^2} - \frac{(x - h)^2}{a^2} = 1 \qquad \textbf{(2-5b)}$$

Equation 2-5a is the *standard* form for the hyperbola that opens left and right. It has center (h, k) and asymptotes $y - k = \pm(b/a)(x - h)$. Its vertices lie on a horizontal line through (h, k), at a distance a to the left and right of the center. To graph the equation, first locate the center (h, k) and the vertices $(h + a, k)$ and $(h - a, k)$. Locate the points $(h + a, k \pm b)$. The line containing (h, k) and $(h + a, k + b)$ is one asymptote; the line containing (h, k) and $(h + a, k - b)$ is the other.

Equation 2-5b partly reverses the roles of x and y. The center is still at (h, k), but the curve opens up and down. The graph has the same asymptotes. The vertices lie on the vertical line through the center, with coordinates $(h, k + b)$ and $(h, k - b)$. The steps in graphing are the same.

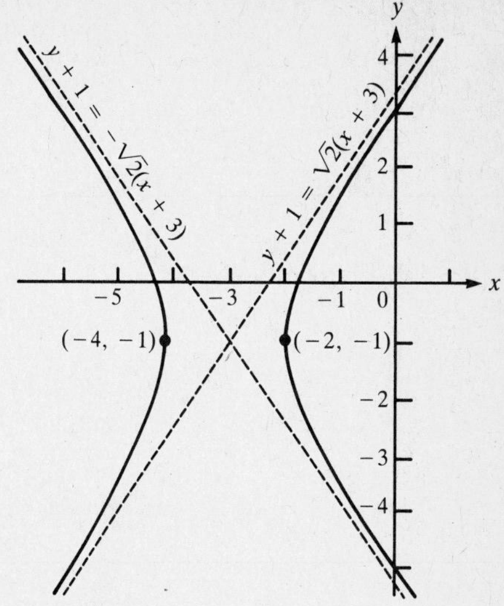

Figure 2-6
The hyperbola
$\dfrac{(x + 3)^2}{1} - \dfrac{(y + 1)^2}{2} = 1.$

EXAMPLE 2-5: Put in standard form, find the center and vertices, find the asymptotes, and graph

(a) $2x^2 - y^2 + 12x - 2y + 15 = 0,$ **(b)** $x^2 - y^2 - 4x + 6y - 1 = 0.$

Solution:

(a) Complete the square and isolate $+1$ on the right:

$$2(x^2 + 6x + 9) - (y^2 + 2y + 1) = -15 + 18 - 1$$

$$2(x + 3)^2 - (y + 1)^2 = 2$$

$$\frac{(x + 3)^2}{1} - \frac{(y + 1)^2}{2} = 1$$

This looks like Equation 2-5a, so the graph opens right and left. The center is at $(-3, -1)$. You have $a = 1, b = \sqrt{2}$. The vertices are one unit to the left and right of $(-3, -1)$. The asymptotes are $y + 1 = \pm(\sqrt{2}/1)(x + 3)$, lines through $(-3, -1)$, with slopes $\pm\sqrt{2}$. Draw two smooth curves: one with vertex $(-2, -1)$, opening right, approaching asymptotes $y + 1 = \pm\sqrt{2}(x + 3)$; the other branch has vertex at $(-4, -1)$, opens left, and has the same two asymptotes (see Figure 2-6).

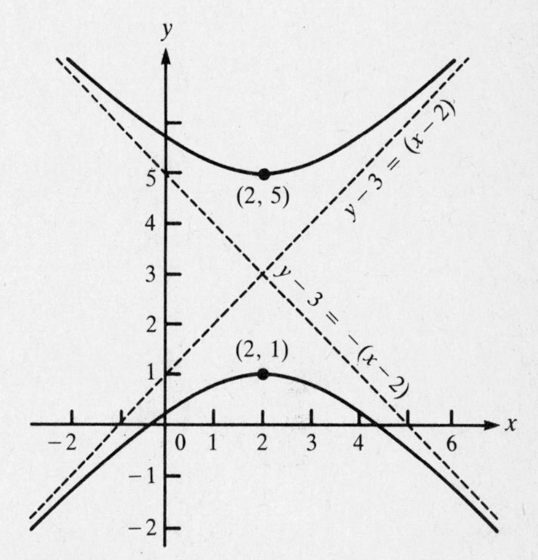

Figure 2-7
The hyperbola
$\dfrac{(y - 3)^2}{4} - \dfrac{(x - 2)^2}{4} = 1.$

(b) The procedure is exactly the same:

$$x^2 - 4x + 4 - (y^2 - 6y + 9) = 1 + 4 - 9$$

$$(x - 2)^2 - (y - 3)^2 = -4$$

The negative sign on the right side is no trouble for the hyperbola—divide by -4:

$$\frac{(y - 3)^2}{4} - \frac{(x - 2)^2}{4} = 1$$

This looks like Equation 2-5b, so the graph opens up and down. The center is at $(2, 3)$ and $a = b = 2$. The vertices are on a vertical line through $(2, 3)$: They are $(2, 5)$ and $(2, 1)$. The asymptotes have equations $y - 3 = \pm(x - 2)$. They are $45°$ and $-45°$ lines through $(2, 3)$. Draw the graph, as in Figure 2-7.

2-5. The *xy* Term and Rotations

Before you return to the general second-degree equation in x and y, consider a simple example: the graph of $y = 1/x$. It is probably familiar to you (see Figure 2-8). It has two branches, with vertical asymptote $x = 0$ and horizontal asymptote $y = 0$. When multiplied by x, the equation is $xy = 1$—a simple second-degree equation, with $A = C = D = E = 0$, $B = 1$, and $F = -1$. Look at the graph again; this time notice the $45°$ and $-45°$ lines through the origin. The curve is a hyperbola with vertices on a line that is neither vertical nor horizontal.

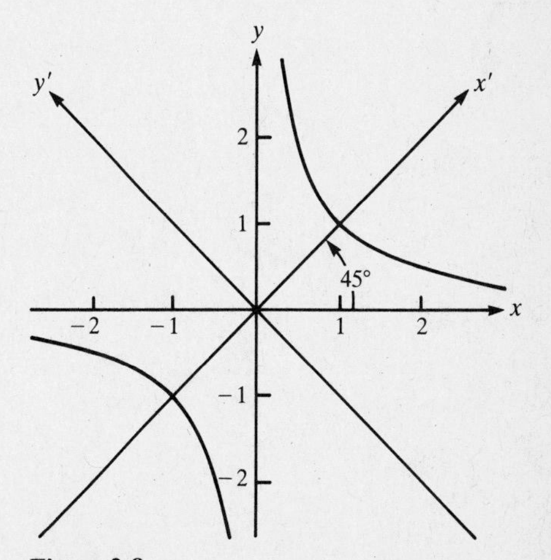

Figure 2-8
$xy = 1.$

In general, it isn't easy to find the graph of a second-degree equation that has an xy term. The problem becomes easier if, after determining which conic section you have, you can eliminate the xy term. You can do this if you find a set of x', y' axes so that the axes of your conic, if it is a parabola or an ellipse, or the line through the vertices, if it is a hyperbola, are parallel to one of the new axes. The problem breaks into two parts.

A. Find the angle between the x axis and the x' axis.

This angle θ is called the **angle of rotation**. You can find this from the formula

ANGLE OF ROTATION

$$\cot 2\theta = \frac{A - C}{B} \qquad \textbf{(2-6)}$$

where A, B, and C are the coefficients of x^2, xy, and y^2, respectively.

EXAMPLE 2-6: Find the angle of rotation that will allow you to eliminate the xy term in $x^2 + xy + y^2 = 1$.

Solution: You have $A = B = C = 1$. Substituting into the formula, you get $\cot 2\theta = 0$. Thus $2\theta = 90°$ and $\theta = 45°$. A rotation of 45° will allow you to put the equation of this ellipse ($B^2 - 4AC = 1 - 4 = -3$) in standard form. Notice that $A = C$ (for any B) tells you to rotate through 45°.

EXAMPLE 2-7: Find the rotation angle θ that will allow the elimination of the xy term in $3x^2 + xy + 2y^2 + x - y + 1 = 0$.

Solution: You have $A = 3$, $B = 1$, $C = 2$, so

$$\cot 2\theta = \frac{3 - 2}{1} = 1$$

$$2\theta = 45°$$

$$\theta = 22.5°$$

Although Examples 2-6 and 2-7 are rather special cases, you should notice that you can always find an angle θ from this formula. It is possible that you'll have to use a calculator to solve

$$2\theta = \text{arc cot}\left(\frac{A - C}{B}\right)$$

B. Find the relationship between (x, y) and (x', y').

Use the rotation formulas:

ROTATION FORMULAS

$$x = x'\cos\theta - y'\sin\theta \qquad \textbf{(2-7a)}$$

$$y = x'\sin\theta + y'\cos\theta \qquad \textbf{(2-7b)}$$

The relationship is shown in Figure 2-9.

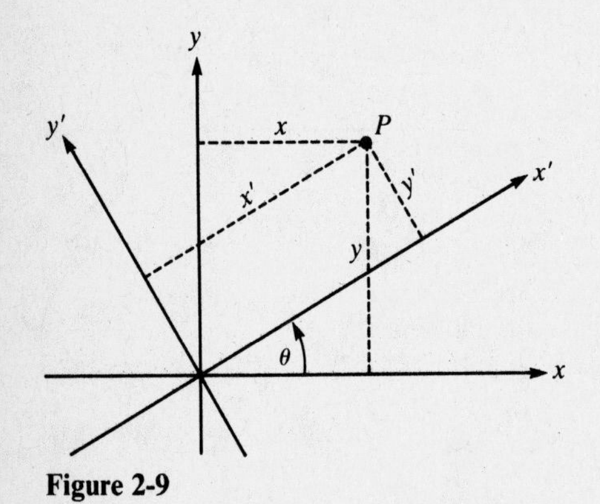

Figure 2-9

EXAMPLE 2-8: A point P has (x', y') coordinates (3, 4). Find $P(x, y)$ if $\theta = 60°$.

Solution: You have

$$x = 3\cos 60° - 4\sin 60° = \frac{3}{2} - \frac{4\sqrt{3}}{2} = \frac{3 - 4\sqrt{3}}{2}$$

$$y = 3\sin 60° + 4\cos 60° = \frac{3\sqrt{3}}{2} + \frac{4}{2} = \frac{3\sqrt{3} + 4}{2}$$

Note that the point P hasn't been moved; the coordinates of P change because you're using a different reference system.

EXAMPLE 2-9: Find the equation of $xy = 1$ if the axes are rotated through an angle of 45°.

Solution: Substitute for x and y in Equations 2-7 and simplify the result:

$$xy = (x' \cos 45° - y' \sin 45°)(x' \sin 45° + y' \cos 45°)$$

$$= \left(\frac{x'}{\sqrt{2}} - \frac{y'}{\sqrt{2}}\right)\left(\frac{x'}{\sqrt{2}} + \frac{y'}{\sqrt{2}}\right) = \frac{(x')^2}{2} - \frac{(y')^2}{2}$$

Thus $xy = 1$ becomes $(x')^2 - (y')^2 = 2$. Note that the graph of the original equation hasn't been altered; its equation changes because the points of the graph are referred to in a different coordinate system.

EXAMPLE 2-10: Given the equation $x^2 + 2xy - y^2 = 1$, (**a**) identify the conic; (**b**) find the angle through which the axes should be rotated; (**c**) find the substitutions for x and y in terms of x' and y'; (**d**) make the substitutions to find the equation of the conic in the x'-y' coordinate system; (**e**) graph the conic.

Solution:

(**a**) You have $A = 1$, $B = 2$, $C = -1$, so $B^2 - 4AC = 4 + 4 > 0$. The conic is a hyperbola.

(**b**) Use Equation 2-6 to find θ:

$$\cot 2\theta = \frac{A - C}{B} = \frac{2}{2} = 1$$

$$2\theta = 45°$$

$$\theta = 22.5°$$

(**c**) The substitutions you need are Equations 2-7:

$$x = x'\cos\theta - y'\sin\theta$$

$$y = x'\sin\theta + y'\cos\theta$$

If $\theta = 22.5°$, you have $\sin\theta \approx 0.38$ and $\cos\theta \approx 0.92$. Your substitutions are

$$x = 0.92x' - 0.38y'$$

$$y = 0.38x' + 0.92y'$$

(**d**) The equation of the conic relative to the rotated axes is

$$(0.92x' - 0.38y')^2 + 2(0.92x' - 0.38y')(0.38x' + 0.92y') - (0.38x' + 0.92y')^2 = 1$$

The angle θ was chosen to eliminate the xy term, so the coefficient of $x'y'$, when all multiplications are done and terms collected, will be zero. Just do the arithmetic for the $(x')^2$ and $(y')^2$ terms:

$$(x')^2[(0.92)^2 + 2(0.92)(0.38) - (0.38)^2] + (y')^2[(0.38)^2 - 2(0.38)(0.92) - (0.92)^2] = 1$$

$$1.4(x')^2 - 1.4(y')^2 = 1$$

$$\frac{(x')^2}{0.7} - \frac{(y')^2}{0.7} = 1$$

You have (approximately) the equation of the hyperbola in standard form!

(**e**) To graph the curve, draw an x-y set of coordinate axes and then the x'-y' set, making an angle of 22.5° with the original set. Now look at your equation. The graph opens right and left, with $a = b = \sqrt{0.7} \approx 0.84$ and center at $(x', y') = (0, 0)$. The vertices are $(x', y') \approx (\pm 0.84, 0)$; the asymptotes are $y' = \pm x'$. Sketch the curve, as in Figure 2-10.

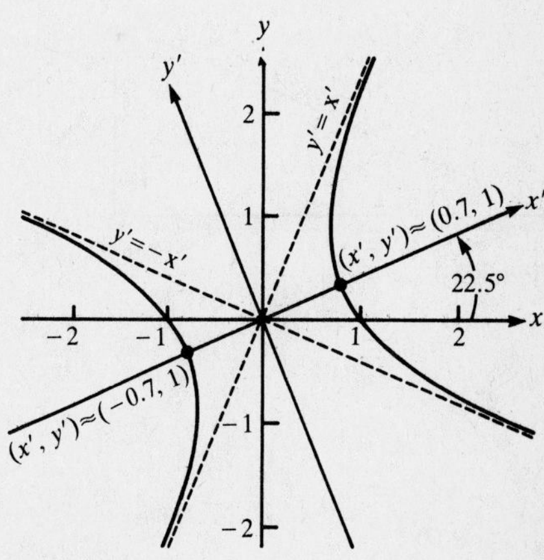

Figure 2-10
The hyperbola $x^2 + 2xy - y^2 = 1$.

Note:

1. In Example 2-10, when you found $\theta = \frac{1}{2}(45°)$, you could have avoided the use of your calculator by using half-angle trigonometric identities and being very careful in the resulting arithmetic. You would have found

$$\sin\left(\frac{45^\circ}{2}\right) = \sqrt{\frac{1 - \cos 45^\circ}{2}} = \sqrt{\frac{1 - \frac{1}{2}\sqrt{2}}{2}} = \frac{\sqrt{2 - \sqrt{2}}}{2}$$

$$\cos\left(\frac{45^\circ}{2}\right) = \sqrt{\frac{1 + \cos 45^\circ}{2}} = \frac{\sqrt{2 + \sqrt{2}}}{2}$$

Similarly, a recognition of identities in a subsequent step could have saved you some calculations. The coefficient of $(x')^2$ is

$$\cos^2\left(\frac{45^\circ}{2}\right) + 2\sin\left(\frac{45^\circ}{2}\right)\cos\left(\frac{45^\circ}{2}\right) - \sin^2\left(\frac{45^\circ}{2}\right)$$

The first and last terms give $\cos[2(45^\circ/2)] = \sqrt{2}/2$. The middle term is $\sin[2(45^\circ/2)] = \sqrt{2}/2$. Thus the coefficient of $(x')^2$ is $\frac{1}{2}\sqrt{2} + \frac{1}{2}\sqrt{2} = \sqrt{2}$. The benefits of these observations are obvious: You can do the work without a calculator, and your results are exact—not calculator approximations!

2. In general, the results of using $\cot 2\theta = (A - C)/B$ will be more accurate if you use a double-angle identity for $\cot 2\theta$:

$$\cot 2\theta = \frac{1}{\tan 2\theta} = \frac{1 - \tan^2\theta}{2\tan\theta} \qquad \textbf{(2-8)}$$

Then solve the equation:

$$\frac{1 - \tan^2\theta}{2\tan\theta} = \frac{A - C}{B}$$

as a quadratic in $\tan\theta$. Once you have $\tan\theta$, you can find $\sin\theta$ and $\cos\theta$ from a right triangle.

In Example 2-10 you had $\cot 2\theta = 1$. Thus,

$$\frac{1 - \tan^2\theta}{2\tan\theta} = 1$$

$$\tan^2\theta + 2\tan\theta - 1 = 0$$

$$\tan\theta = \frac{-2 \pm \sqrt{4 + 4}}{2} = -1 \pm \sqrt{2}$$

Because θ is acute, use the positive value. Draw a right triangle that shows $\tan\theta = \sqrt{2} - 1$ (see Figure 2-11). You get

$$\sin\theta = \frac{\sqrt{2} - 1}{\sqrt{2}\sqrt{2 - \sqrt{2}}} = \frac{(\sqrt{2} - 1)\sqrt{2 - \sqrt{2}}}{\sqrt{2}(2 - \sqrt{2})}$$

$$= \frac{(\sqrt{2} - 1)\sqrt{2 - \sqrt{2}}}{2(\sqrt{2} - 1)} = \frac{\sqrt{2 - \sqrt{2}}}{2}$$

the same expression you found using the half-angle formula.

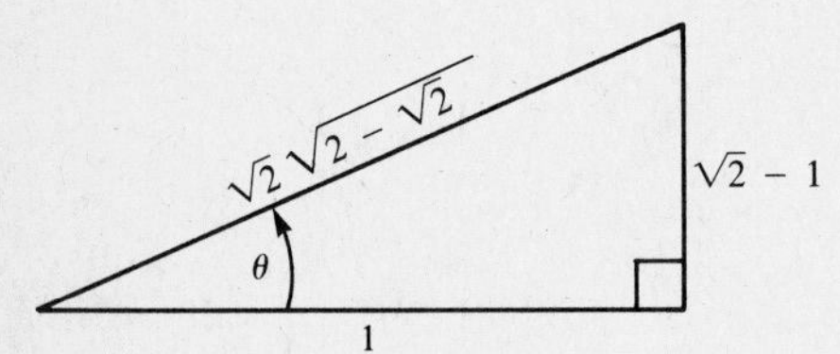

Figure 2-11

SUMMARY

1. The graph of $Ax^2 + Bxy + Cy^2 + Dx + Ey + F = 0$ is always a parabola, an ellipse, a hyperbola, or a degenerate form of one of these: an ellipse without locus, a point, a line, or a pair of lines.
2. An inspection of $B^2 - 4AC$ will help you identify your conic section.
3. The technique of completing the square is used to find the standard form of the equation of each of the conics.
4. The presence of an xy term in the equation indicates that the axis of the conic is rotated with respect to the x and y axes.

SOLVED PROBLEMS

In Problems 2-1 through 2-10 identify the equation as a parabola, ellipse, or hyperbola. Put the equation in standard form and find coordinates of the vertex (or vertices), center, etc. Sketch the graph.

PROBLEM 2-1 $2x^2 - \frac{1}{2}y^2 = 18$.

Solution: You have $A = 2$, $C = -\frac{1}{2}$. A and C have opposite signs, so the curve is a hyperbola. There are no linear terms, so $h = k = 0$. Divide by 18 in order to get a positive 1 on the right:

$$\frac{x^2}{9} - \frac{y^2}{36} = 1$$

The center is (0, 0); $a = 3$ and $b = 6$. The vertices are on the x axis at (3, 0) and (−3, 0), and the asymptotes are $y = \pm(6/3)x$, as shown in Figure 2-12. [See Section 2-4.]

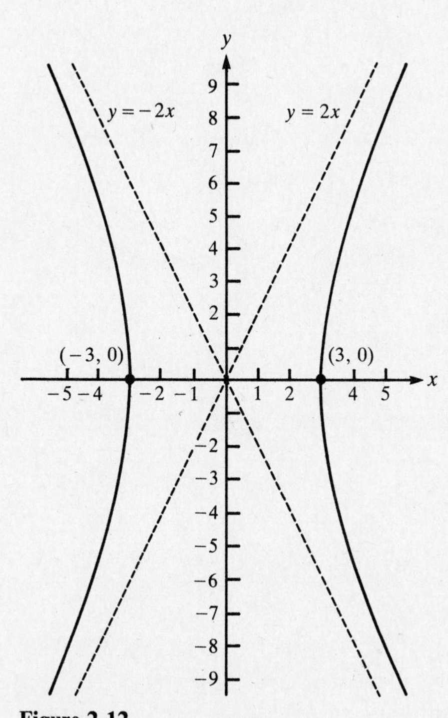

Figure 2-12
The hyperbola $\frac{x^2}{9} - \frac{y^2}{36} = 1$.

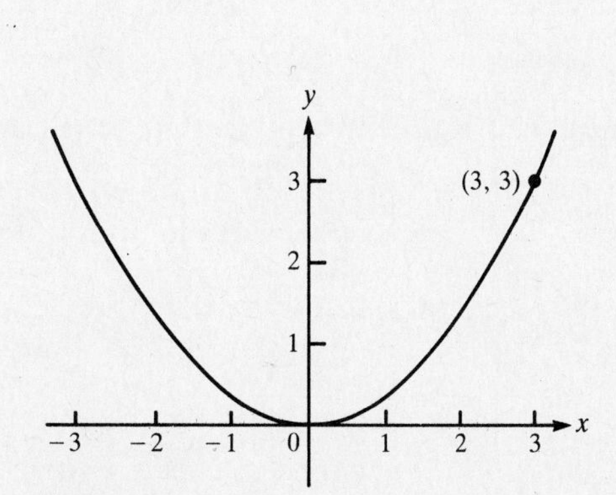

Figure 2-13
The parabola $x^2 = 3y$.

PROBLEM 2-2 $x^2 - 3y = 0$.

Solution: You have $A = 1$, $C = 0$, so this is a parabola. Again, it isn't necessary to complete the square:

$$(x - 0)^2 = 3(y - 0)$$

The vertex is the origin, $4a = 3$, and $a = \frac{3}{4}$. The curve opens upward. You can easily find two additional points on the curve (e.g., if $y = 3$, then $x = \pm 3$) and sketch the graph, as in Figure 2-13.

[See Section 2-2.]

PROBLEM 2-3 $x^2 + 2y^2 + 2x - 8y + 5 = 0$.

Solution: A and C are of the same sign and $B = 0$, so the graph is an ellipse. Complete the square and get $+1$ on the right:

$$x^2 + 2x + 1 + 2(y^2 - 4y + 4) = -5 + 1 + 8$$

$$(x + 1)^2 + 2(y - 2)^2 = 4$$

$$\frac{(x + 1)^2}{4} + \frac{(y - 2)^2}{2} = 1$$

The major axis is parallel to the x axis. The center is $(-1, 2)$; $a = 2$ and $b = \sqrt{2}$. The endpoints of the major axis are $(1, 2)$ and $(-3, 2)$. The endpoints of the minor axis are $(-1, 2 + \sqrt{2})$ and $(-1, 2 - \sqrt{2})$, as shown in Figure 2-14. [See Section 2-3.]

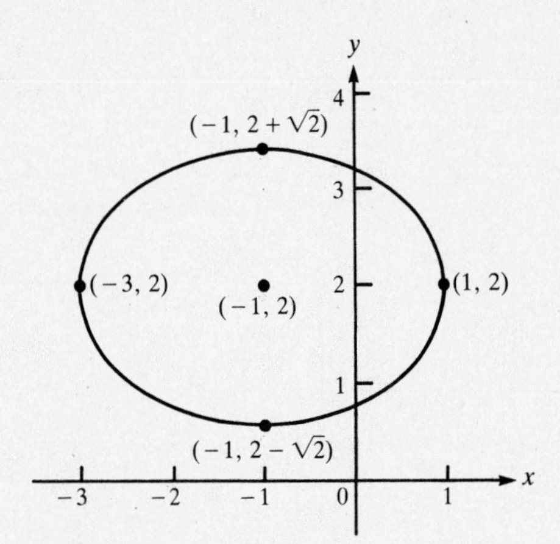

Figure 2-14
The ellipse
$\frac{(x + 1)^2}{4} + \frac{(y - 2)^2}{2} = 1.$

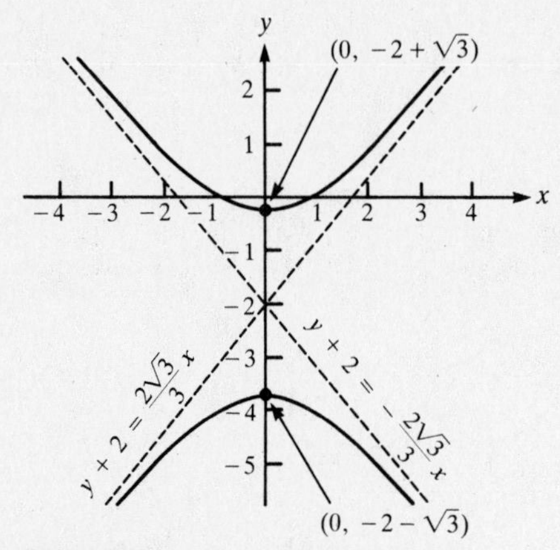

Figure 2-15
The hyperbola
$\frac{(y + 2)^2}{3} - \frac{x^2}{9/4} = 1.$

PROBLEM 2-4 $4x^2 - 3y^2 - 12y - 3 = 0.$

Solution: This is a hyperbola because $B = 0$ and A and C are opposite in sign. Complete the square:

$$4x^2 - 3(y^2 + 4y + 4) = 3 - 12$$

$$4x^2 - 3(y + 2)^2 = -9$$

$$\frac{(y + 2)^2}{3} - \frac{x^2}{9/4} = 1$$

The center is $(0, -2)$, and the curve opens up and down. You have $a = \frac{3}{2}$ and $b = \sqrt{3}$, so the vertices are $(0, -2 + \sqrt{3})$ and $(0, -2 - \sqrt{3})$. The asymptotes are $y + 2 = \pm[\sqrt{3}/(3/2)](x - 0)$, or simply $y = \pm(2\sqrt{3}/3)\,x - 2$. Locate the center on your graph and draw the asymptotes and vertices. Sketch the graph, as in Figure 2-15. [See Section 2-4.]

PROBLEM 2-5 $y^2 - 4x + 14y + 57 = 0.$

Solution: The graph is a parabola, so complete the square and write in standard form:

$$y^2 + 14y + 49 = 4x - 57 + 49$$

$$(y + 7)^2 = 4(x - 2)$$

The vertex is $(2, -7)$, and the graph opens to the right. Choose two values of y near $y = -7$ and find the corresponding values of x:

$$y = -5 \qquad x = 3$$

$$y = -3 \qquad x = 6$$

Using the symmetry of the curve, you find two additional points, $(3, -9)$ and $(6, -11)$. Draw the curve, as in Figure 2-16. [See Section 2-2.]

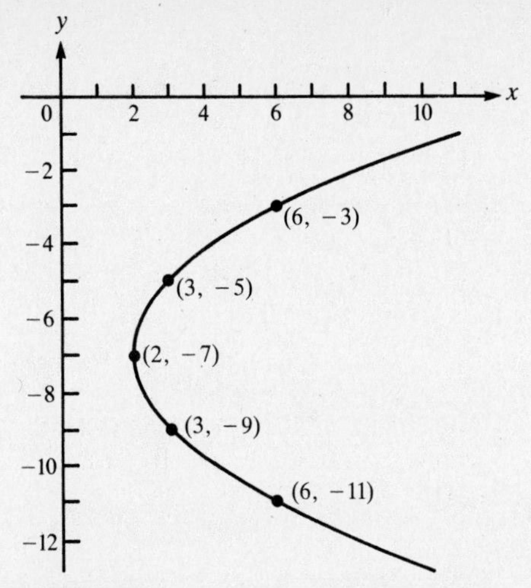

Figure 2-16
The parabola
$(y + 7)^2 = 4(x - 2)$.

PROBLEM 2-6 $2x^2 - 3y^2 + 4x + 12y = 0$.

Solution: A and C have opposite signs; you have a hyperbola:

$$2(x^2 + 2x + 1) - 3(y^2 - 4y + 4) = 2 - 12$$

$$2(x + 1)^2 - 3(y - 2)^2 = -10$$

$$\frac{(y - 2)^2}{10/3} - \frac{(x + 1)^2}{5} = 1$$

The center is $(-1, 2)$, and the curve opens up and down. You have $a = \sqrt{5}$ and $b = \sqrt{10/3}$, so the vertices are $(-1, 2 \pm \sqrt{10/3})$. The asymptotes are $y - 2 = \pm(\sqrt{10/3}/\sqrt{5})(x + 1)$, or more simply, $y - 2 = \pm\sqrt{2/3}(x - 1)$. Locate the center and vertices, draw the asymptotes, and sketch the curve, as in Figure 2-17. [See Section 2-4.]

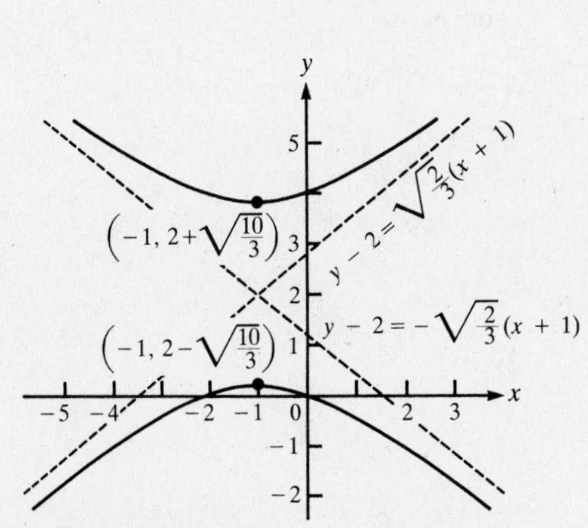

Figure 2-17
The hyperbola
$\frac{(y - 2)^2}{10/3} - \frac{(x + 1)^2}{5} = 1$.

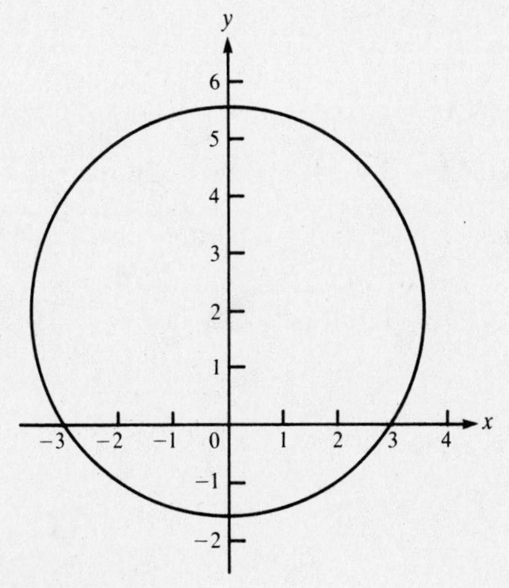

Figure 2-18
The circle $x^2 + (y - 2)^2 = 13$.

PROBLEM 2-7 $x^2 + y^2 - 4y = 9$.

Solution: A and C are equal, so this is a circle:

$$x^2 + (y^2 - 4y + 4) = 9 + 4$$

$$x^2 + (y - 2)^2 = 13$$

The center is $(0, 2)$, and the radius is $\sqrt{13}$, as shown in Figure 2-18. [See Section 2-3.]

PROBLEM 2-8 $3x^2 + 4y^2 - 6x - 8y = 0.$

Solution: A and C have the same sign, so the graph is an ellipse:

$$3(x^2 - 2x + 1) + 4(y^2 - 2y + 1) = 3 + 4$$

$$3(x - 1)^2 + 4(y - 1)^2 = 7$$

$$\frac{(x - 1)^2}{7/3} + \frac{(y - 1)^2}{7/4} = 1$$

The major axis is parallel to the x axis (because $7/3 > 7/4$), and the center is $(1, 1)$. You have $a = \sqrt{\frac{7}{3}}$ and $b = \sqrt{\frac{7}{4}}$, so the endpoints of the major axis are $\left(1 \pm \sqrt{\frac{7}{3}}, 1\right)$; the endpoints of the minor axis are $\left(1, 1 \pm \sqrt{\frac{7}{4}}\right)$. Because a and b are nearly equal, you expect the ellipse to look almost like a circle, as in Figure 2-19. [See Section 2-3.]

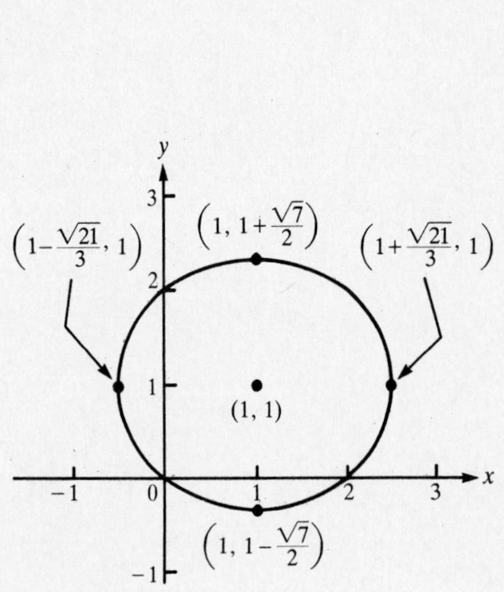

Figure 2-19
The ellipse
$\frac{(x - 1)^2}{7/3} + \frac{(y - 1)^2}{7/4} = 1.$

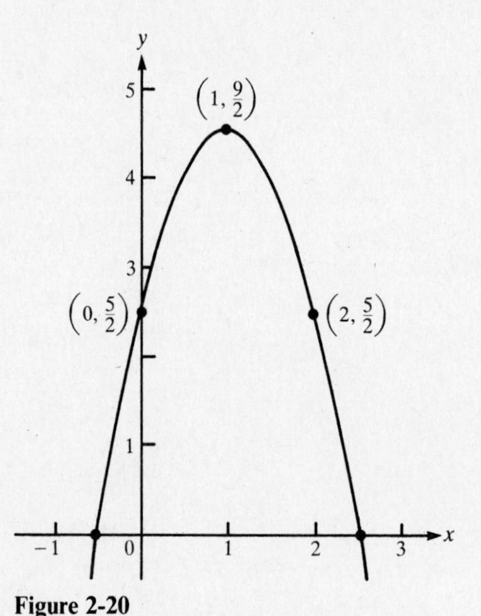

Figure 2-20
The parabola
$(x - 1)^2 = -\frac{1}{2}\left(y - \frac{9}{2}\right).$

PROBLEM 2-9 $4x^2 - 8x + 2y = 5.$

Solution: The curve is a parabola because $C = 0$:

$$4(x^2 - 2x + 1) = -2y + 5 + 4$$

$$4(x - 1)^2 = -2y + 9$$

$$(x - 1)^2 = -\frac{1}{2}y + \frac{9}{4}$$

$$(x - 1)^2 = -\frac{1}{2}\left(y - \frac{9}{2}\right)$$

The vertex is $(1, \frac{9}{2})$, and the curve opens downward. If $x = 0$, $y = \frac{5}{2}$. By symmetry, $(2, \frac{5}{2})$ is another point on the curve. You might also find the x intercepts: If $y = 0$, then $4x^2 - 8x - 5 = 0$. This factors as $(2x + 1)(2x - 5) = 0$, so $x = -\frac{1}{2}$ or $x = 5/2$. Draw these points and the graph, as in Figure 2-20. [See Section 2-2.]

PROBLEM 2-10 $9x^2 - 4y^2 + 18x + 24y - 27 = 0.$

Solution: Because A and C have opposite signs, this is a hyperbola. Complete the squares and put this in standard form:

$$9(x^2 + 2x + 1) - 4(y^2 - 6y + 9) = 27 + 9 - 36$$
$$9(x + 1)^2 - 4(y - 3)^2 = 0$$

The zero on the right tells you that this is a degenerate hyperbola, that is, two lines. Solve to find the equations of the two lines,

$$9(x + 1)^2 = 4(y - 3)^2$$
$$3(x + 1) = \pm 2(y - 3)$$
$$y - 3 = \pm\frac{3}{2}(x + 1)$$

and graph, as in Figure 2-21. [See Section 2-4.]

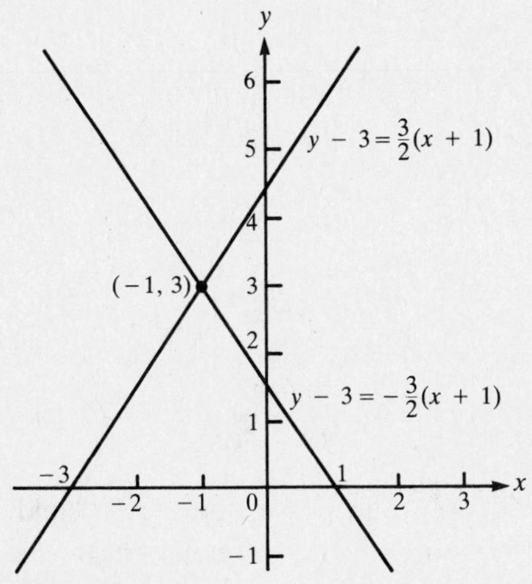

Figure 2-21
The degenerate hyperbola
$9(x + 1)^2 - 4(y - 3)^2 = 0$.

In Problems 2-11 through 2-13 identify the type of conic section, determine the angle of rotation, write the substitutions that will eliminate the xy term, and find the new (x', y') equation. Sketch the graph.

PROBLEM 2-11 $x^2 + xy = 10$.

Solution: $A = 1$, $B = 1$, $C = 0$; $B^2 - 4AC = 1 > 0$, so the graph is a hyperbola. Find the rotation angle:

$$\cot 2\theta = \frac{A - C}{B} = 1$$
$$2\theta = 45^\circ$$

Thus $\cos 2\theta = \sqrt{2}/2$. You need $\sin\theta$ and $\cos\theta$, which you can find from the identities:

$$\sin\theta = \sqrt{\frac{1 - \cos 2\theta}{2}} = \sqrt{\frac{1 - \frac{1}{2}\sqrt{2}}{2}} = \frac{\sqrt{2 - \sqrt{2}}}{2}$$
$$\cos\theta = \sqrt{\frac{1 + \cos 2\theta}{2}} = \frac{\sqrt{2 + \sqrt{2}}}{2}$$

The x' and y' substitutions are

$$x = x'\cos\theta - y'\sin\theta = x'\frac{\sqrt{2 + \sqrt{2}}}{2} - y'\frac{\sqrt{2 - \sqrt{2}}}{2}$$
$$y = x'\sin\theta + y'\cos\theta = x'\frac{\sqrt{2 - \sqrt{2}}}{2} + y'\frac{\sqrt{2 + \sqrt{2}}}{2}$$

Substitute into the original equation $x^2 + xy = 10$:

$$(x')^2\frac{2+\sqrt{2}}{4} - 2x'y'\left(\frac{\sqrt{2+\sqrt{2}}}{2}\right)\left(\frac{\sqrt{2-\sqrt{2}}}{2}\right) + (y')^2\frac{2-\sqrt{2}}{4} + (x')^2\left(\frac{\sqrt{2+\sqrt{2}}}{2}\right)\left(\frac{\sqrt{2-\sqrt{2}}}{2}\right)$$
$$+ x'y'\frac{2+\sqrt{2}}{4} - x'y'\frac{2-\sqrt{2}}{4} - (y')^2\left(\frac{\sqrt{2-\sqrt{2}}}{2}\right)\left(\frac{\sqrt{2+\sqrt{2}}}{2}\right) = 10$$

Grouping like terms, you get

$$(x')^2\left(\frac{2+\sqrt{2}}{4} + \frac{\sqrt{2}}{4}\right) + x'y'\left(\frac{-\sqrt{2}}{2} + \frac{2+\sqrt{2}}{4} - \frac{2-\sqrt{2}}{4}\right) + (y')^2\left(\frac{2-\sqrt{2}}{4} - \frac{\sqrt{2}}{4}\right) = 10$$

The coefficient of $x'y'$ is $(-\sqrt{2}/2) + \frac{1}{2} + (\sqrt{2}/4) - \frac{1}{2} + (\sqrt{2}/4) = 0$. If this hadn't been zero, you would know that you made a mistake: x' and y' were chosen to make the $x'y'$ term disappear! You have

$$(x')^2\left(\frac{1+\sqrt{2}}{2}\right) + (y')^2\left(\frac{1-\sqrt{2}}{2}\right) = 10$$

or, in standard form:

$$\frac{x^2}{\dfrac{10(2)}{1+\sqrt{2}}} - \frac{y^2}{\dfrac{10(2)}{\sqrt{2}-1}} = 1$$

$$\frac{x^2}{20(\sqrt{2}-1)} - \frac{y^2}{20(\sqrt{2}+1)} = 1$$

With respect to the x', y' axes, the curve opens right and left with $a = \sqrt{20(\sqrt{2}-1)}$ and $b = \sqrt{20(\sqrt{2}+1)}$. For a roughly accurate sketch, use $a \approx 3$ and $b \approx 7$. Locate the vertices (on the x' axis), draw the asymptotes, and sketch the curve, as in Figure 2-22. [See Section 2-5.]

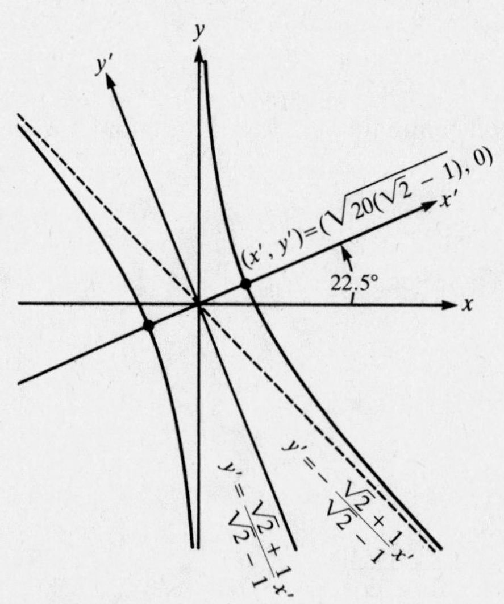

Figure 2-22
The hyperbola $x^2 + xy = 10$.

PROBLEM 2-12 $x^2 - 3xy + y^2 = 5$.

Solution: You have $A = 1, B = -3$, and $C = 1$, so $B^2 - 4AC = 9 - 4 = 5 > 0$. The graph is a hyperbola. Find the angle of rotation:

$$\cot 2\theta = \frac{A - C}{B} = 0$$

$$2\theta = 90°$$

So $\theta = 45°$, and you have

$$x = x'\frac{\sqrt{2}}{2} - y'\frac{\sqrt{2}}{2}$$

$$y = x'\frac{\sqrt{2}}{2} + y'\frac{\sqrt{2}}{2}$$

because $\sin 45° = \cos 45° = \sqrt{2}/2$. Substitute into the original equation:

$$\frac{(x')^2 - 2x'y' + (y')^2}{2} - \frac{3}{2}[(x')^2 - (y')^2] + \frac{(x')^2 + 2x'y' + (y')^2}{2} = 5$$

Again, as anticipated, the $x'y'$ terms cancel, and you have

$$-(x')^2 + 5(y')^2 = 10$$

$$\frac{(y')^2}{2} - \frac{(x')^2}{10} = 1$$

The center is at $(x', y') = (0, 0)$, and the curve opens up and down (with respect to the x', y' axes). You have $a = \sqrt{10}$ and $b = \sqrt{2}$, so the vertices are $(x', y') = (0, \pm\sqrt{2})$. The asymptotes are $y' = \pm(\sqrt{2}/\sqrt{10})x'$, or more simply, $y' = (\pm\sqrt{5}/5)x'$. Draw the graph, as in Figure 2-23.

[See Section 2-5.]

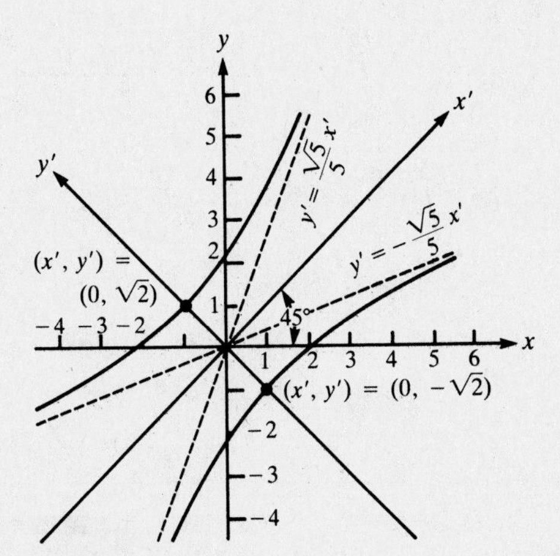

Figure 2-23
The hyperbola $x^2 - 3xy + y^2 = 5$.

PROBLEM 2-13 $2x^2 + \sqrt{3}xy + y^2 - 2x = 0$.

Solution: You have $A = 2$, $B = \sqrt{3}$, $C = 1$, so $B^2 - 4AC = -5 < 0$. The graph is an ellipse. Find the angle of rotation:

$$\cot 2\theta = \frac{A - C}{B} = \frac{1}{\sqrt{3}}$$

Thus $2\theta = 60°$ and $\theta = 30°$. The substitution equations are

$$x = x'\frac{\sqrt{3}}{2} - y'\frac{1}{2} = \frac{\sqrt{3}x' - y'}{2}$$

$$y = x'\frac{1}{2} + y'\frac{\sqrt{3}}{2} = \frac{x' + \sqrt{3}y'}{2}$$

Substitute into the original equation:

$$2\left(\frac{3(x')^2 - 2\sqrt{3}x'y' + (y')^2}{4}\right) + \frac{\sqrt{3}}{4}(\sqrt{3}(x')^2 + 2x'y' - \sqrt{3}(y')^2)$$
$$+ \frac{(x')^2 + 2\sqrt{3}x'y' + 3(y')^2}{4} - (\sqrt{3}x' - y') = 0$$

Collect terms and check to be sure that the $x'y'$ term disappears:

$$(x')^2\left(\frac{3}{2}+\frac{3}{4}+\frac{1}{4}\right)+(y')^2\left(\frac{1}{2}-\frac{3}{4}+\frac{3}{4}\right)-\sqrt{3}x'+y'=0$$

$$\frac{5}{2}(x')^2+\frac{1}{2}(y')^2-\sqrt{3}x'+y'=0$$

$$\frac{5}{2}\left((x')^2-\frac{2\sqrt{3}}{5}x'+\frac{3}{25}\right)+\frac{1}{2}((y')^2+2y'+1)=\frac{3}{10}+\frac{1}{2}$$

$$\frac{5}{2}\left(x'-\frac{\sqrt{3}}{5}\right)^2+\frac{1}{2}(y'+1)^2=\frac{4}{5}$$

$$\frac{\left(x'-\frac{\sqrt{3}}{5}\right)^2}{8/25}+\frac{(y'+1)^2}{8/5}=1$$

The center is at $(x', y') = (\sqrt{3}/5, -1)$, with major axis parallel to the y' axis (because $8/5 > 8/25$); $a = \sqrt{\frac{8}{5}} = 2\sqrt{10}/5$, $b = \sqrt{\frac{8}{25}} = 2\sqrt{2}/5$. Plot the endpoints of the minor axis $(x', y') = (\sqrt{3}/5 \pm (2\sqrt{2}/5), -1)$, and the endpoints of the major axis $(\sqrt{3}/5, -1 \pm (2\sqrt{10}/5))$. Draw the ellipse, as in Figure 2-24.

[See Section 2-5.]

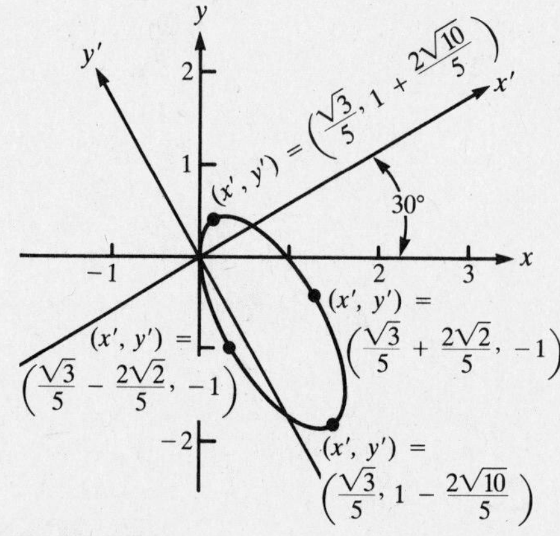

Figure 2-24
The ellipse
$2x^2 + \sqrt{3}xy + y^2 - 2x = 0$.

Supplementary Exercises

In Problems 2-14 through 2-21 find the coordinates of the vertices and the center and sketch the graph:

2-14 $x = 4y^2$

2-15 $5x^2 + 2y^2 = 50$

2-16 $4x^2 + y^2 + 6y + 5 = 0$

2-17 $-x^2 + 2y^2 + 2x + 8y + 1 = 0$

2-18 $3y + 4x^2 + 8x = 5$

2-19 $2x^2 - 3y^2 - 4x - 6y = 13$

2-20 $x^2 + 3y^2 - 4x + 6y = -1$

2-21 $x^2 + 4y^2 + 2x - 24y + 33 = 0$

2-22 $9x^2 + 18x + 4y^2 + 24y + 45 = 0$

In Problems 2-23 through 2-25 identify the conic section, determine the angle of rotation, write the substitution equations, find the equation in terms of x' and y', and draw the graph:

2-23 $xy = -3$

2-24 $2x^2 + \sqrt{3}xy + y^2 = 20$

2-25 $3x^2 - 2\sqrt{3}xy + y^2 - 2x - 2\sqrt{3}y - 4 = 0$

Solutions to Supplementary Exercises

(2-14) parabola; vertex (0, 0); Figure 2-25

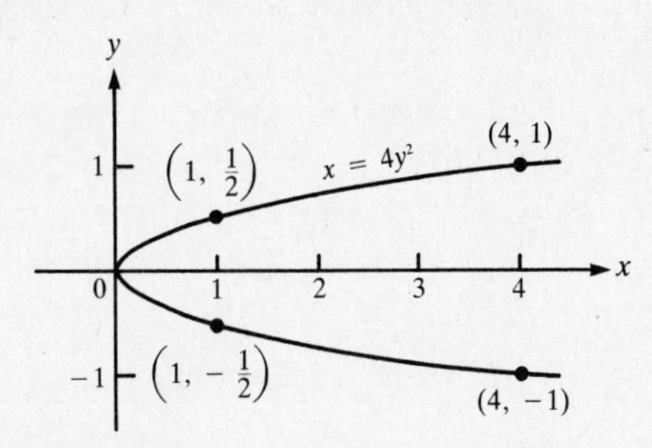

Figure 2-25
Problem 2-14

(2-15) ellipse; center (0, 0); endpoints of major axis (0, −5) and (0, 5); endpoints of minor axis $(\sqrt{10}, 0)$ and $(-\sqrt{10}, 0)$; Figure 2-26

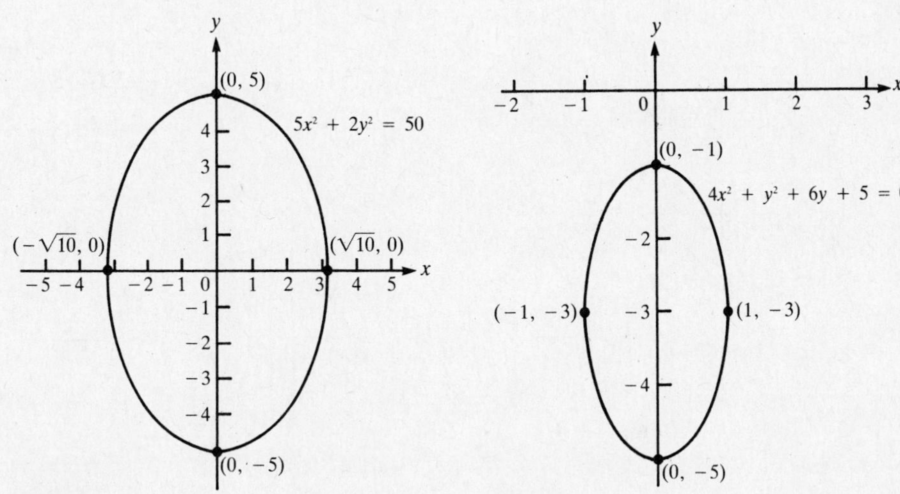

Figure 2-26
Problem 2-15

Figure 2-27
Problem 2-16

(2-16) ellipse; center (0, −3); endpoints of major axis (0, −1) and (0, −5); endpoints of minor axis (1, −3) and (−1, −3); Figure 2-27

(2-17) hyperbola; center (1, −2); vertices $(1, -2 + \sqrt{3})$ and $(1, -2 - \sqrt{3})$; asymptotes $(y + 2) = \pm(\sqrt{2}/2)(x - 1)$; Figure 2-28

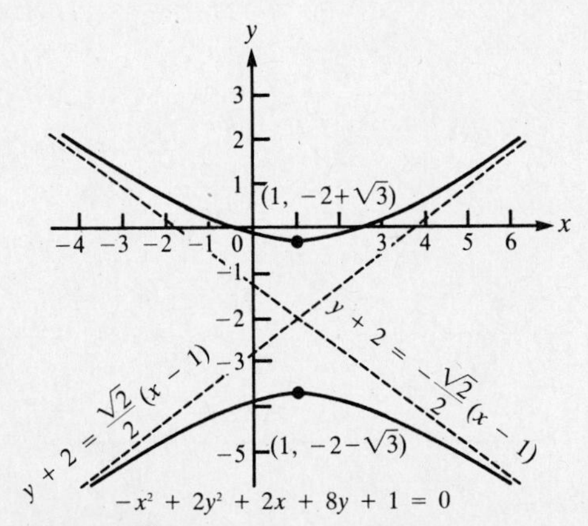

Figure 2-28
Problem 2-17

(2-18) parabola; vertex $(-1, 3)$; Figure 2-29

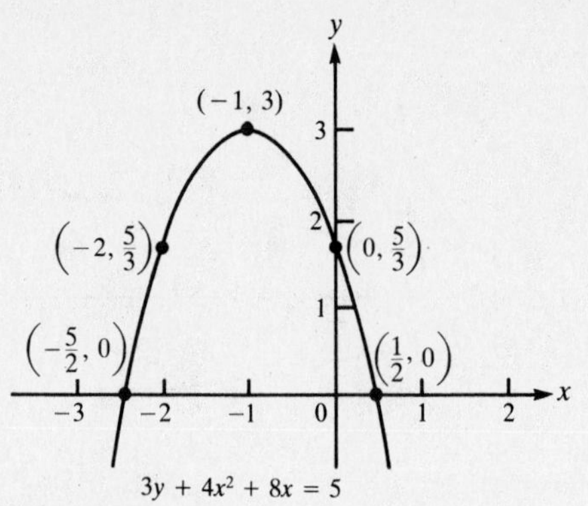

Figure 2-29
Problem 2-18

(2-19) hyperbola; center $(1, -1)$; vertices $(1 + \sqrt{6}, -1)$ and $(1 - \sqrt{6}, -1)$; asymptotes $(y + 1) = \pm(\sqrt{6}/3)(x - 1)$; Figure 2-30

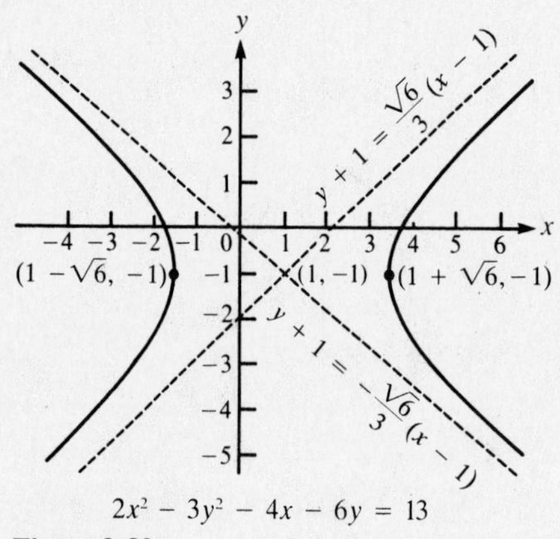

Figure 2-30
Problem 2-19

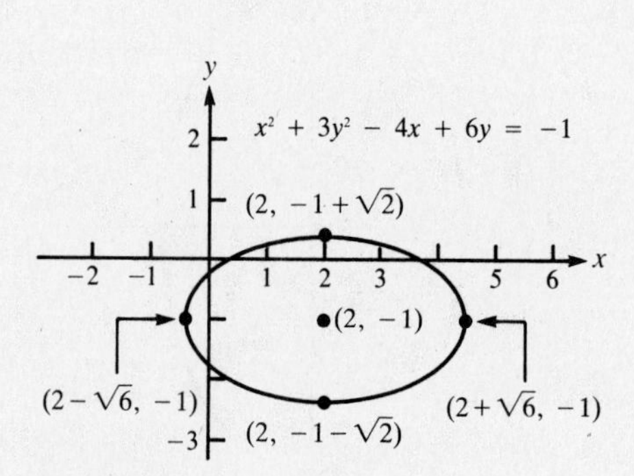

Figure 2-31
Problem 2-20

(2-20) ellipse; center $(2, -1)$; endpoints of major axis $(2 + \sqrt{6}, -1)$ and $(2 - \sqrt{6}, -1)$; endpoints of minor axis $(2, -1 + \sqrt{2})$ and $(2, -1 - \sqrt{2})$; Figure 2-31

(2-21) ellipse; center $(-1, 3)$; endpoints of major axis $(1, 3)$ and $(-3, 3)$; endpoints of minor axis $(-1, 4)$ and $(-1, 2)$; Figure 2-32

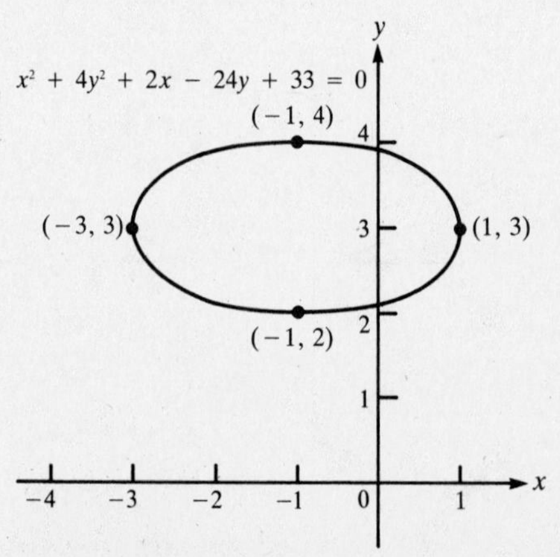

Figure 2-32
Problem 2-21

(2-22) degenerate ellipse; the graph consists of the point $(-1, -3)$

(2-23) hyperbola; angle of rotation $\theta = 45°$; $x = (\sqrt{2}/2)(x' - y')$, $y = (\sqrt{2}/2)(x' + y')$; $[(y')^2/6] - [(x')^2/6] = 1$; vertices $(x', y') = (0, \pm\sqrt{6})$; Figure 2-33

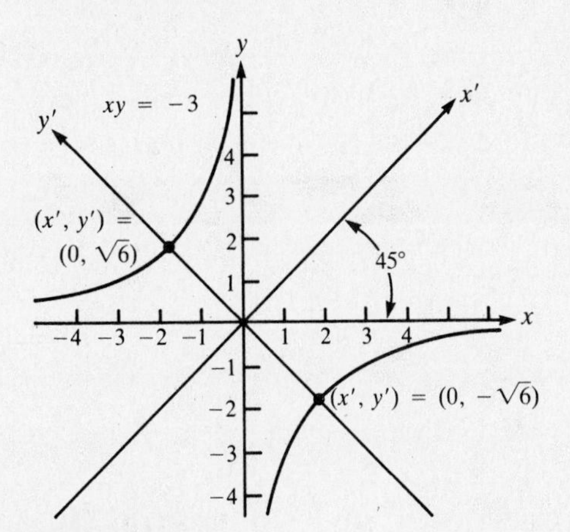

Figure 2-33
Problem 2-23

(2-24) ellipse; angle of rotation $\theta = 30°$; $x = (\sqrt{3}/2)x' - \frac{1}{2}y'$, $y = \frac{1}{2}x' + (\sqrt{3}/2)y'$; $[(x')^2/8] + [(y')^2/40] = 1$; endpoints of major axis $(x', y') = (0, \pm 2\sqrt{10})$; endpoints of minor axis $(x', y') = (\pm 2\sqrt{2}, 0)$; Figure 2-34

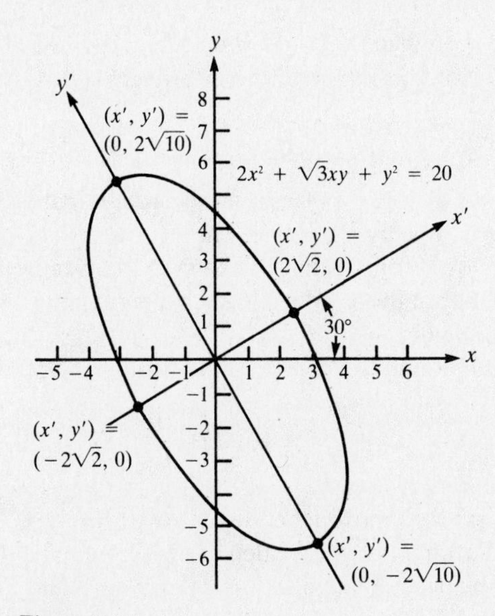

Figure 2-34
Problem 2-24

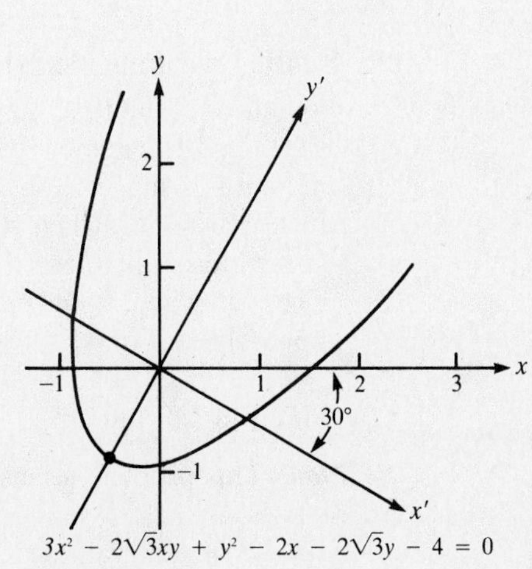

Figure 2-35
Problem 2-25

(2-25) parabola; angle of rotation $\theta = -30°$; $x = (\sqrt{3}/2)x' + \frac{1}{2}y'$, $y = -\frac{1}{2}x' + (\sqrt{3}/2)y'$; $y' + 1 = (x')^2$; vertex $(x', y') = (0, -1)$; Figure 2-35

3 LIMITS AND CONTINUITY

THIS CHAPTER IS ABOUT

3-1. The Intuitive Concept of Limit

The concept of limit is important enough in the study of calculus that you should try to get as good an understanding of it as possible, and intuitively, that's easy to do.

A. Limits reveal the behavior of a function near a point.

You know that a function may be thought of as a set of ordered pairs (x, y), with the y value related to the x value by some rule. The notion of the limit of a function is suggested by the question, "What happens to the y values (the functional values) when you choose the x values in such a way that they are getting closer and closer to some number a?"

That question, stated mathematically, becomes, "Find the limit of $f(x)$ as x approaches a," or, "Find $\lim_{x\to a} f(x)$."

EXAMPLE 3-1: Find $\lim_{x\to 1} (x^2 + 1)$.

Solution: This problem seems easy; if x approaches $x = 1$, then surely $x^2 + 1$ approaches $1^2 + 1$. A few calculations on your calculator will support your guess. Take several values of x near $x = 1$ (some less than $x = 1$ and some greater than $x = 1$) and find the value of $x^2 + 1$:

x	$x^2 + 1$
0.9	1.81
0.95	1.9025
0.99	1.9801
1.1	2.21
1.05	2.1025
1.01	2.0201

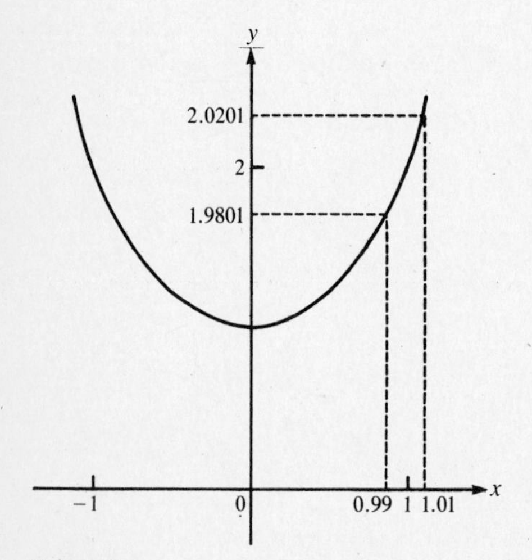

Figure 3-1
$f(x) = x^2 + 1$.

A graph of $y = x^2 + 1$ will also support your conclusion (see Figure 3-1).

Let's get a bit more formal in our description of "limit." Consider a function $f(x)$, with a number $x = a$ that may or may not be in the domain of f; that is, $f(a)$ may not be defined at all. If there exists a number L such that $f(x)$ gets closer and closer to L when x gets closer and closer to a, then you say "the **limit** of $f(x)$, as x approaches a, is L," and you write "$\lim_{x \to a} f(x) = L$."

EXAMPLE 3-2: Let $f(x) = (x^2 - 1)/(x - 1)$ and find $\lim_{x \to 1} f(x)$.

Solution: You notice, first of all, that $f(1) = 0/0$, which is not defined; and having noticed that, you move on to the problem of what happens to $f(x)$ if x is *near* $x = 1$. You could use the calculator again, but there's a neater approach for this problem—simplify it, algebraically:

$$\frac{x^2 - 1}{x - 1} = \frac{(x + 1)(x - 1)}{x - 1}$$
$$= x + 1 \quad \text{for all } x \neq 1$$

Therefore

$$\lim_{x \to 1} \frac{x^2 - 1}{x - 1} = \lim_{x \to 1} (x + 1)$$
$$= 2$$

An extremely important point has just been made. It is repeated for emphasis: The statement "$\lim_{x \to a} f(x) = L$" is not equivalent to the assertion "$f(a) = L$." Look at Example 3-2 again: $f(1)$ is not defined, but $\lim_{x \to 1} f(x) = 2$, a perfectly well-defined number!

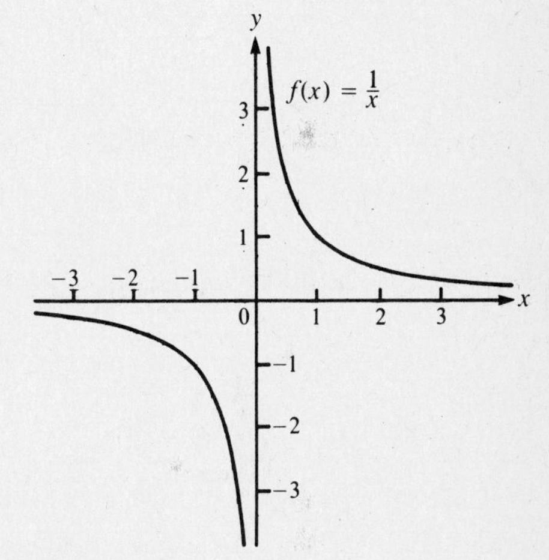

Figure 3-2
$\lim_{x \to 0} \left(\frac{1}{x}\right)$ does not exist.

B. The limit of $f(x)$ as x approaches a may not exist.

For several reasons, $\lim_{x \to a} f(x)$ might not exist. Consider three cases.

1. The values of $f(x)$ may become increasingly large as x approaches a.

EXAMPLE 3-3: Find $\lim_{x \to 0} 1/x$.

Solution: This is, at first glance, similar to Example 3-2: The function $1/x$ is not defined at $x = 0$. There the similarity ends, however; as x gets closer to $x = 0$ through positive values, $1/x$ gets larger and larger. If x approaches $x = 0$ through negative values, $1/x$ is negative, but becomes larger numerically. Look at the graph of $1/x$ in Figure 3-2; it should help you visualize this kind of functional behavior.

2. The limit of $f(x)$ may assume different values when x approaches a from different directions.

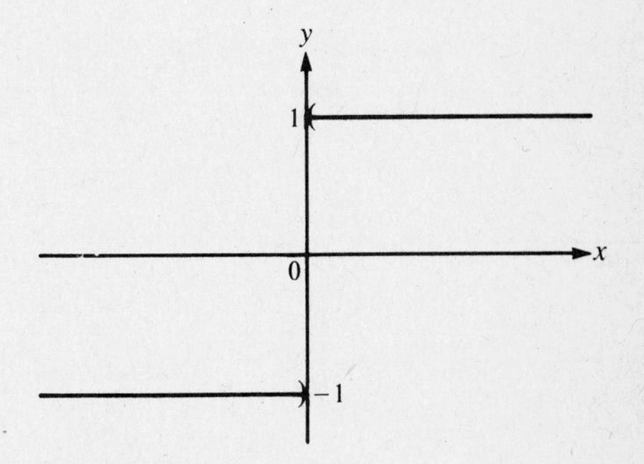

Figure 3-3
$\lim_{x \to 0} \frac{\sqrt{x^2}}{x}$ does not exist.

EXAMPLE 3-4: Find $\lim_{x \to 0} \sqrt{x^2}/x$.

Solution: Recall that by definition $\sqrt{x^2} = |x|$; this allows you to rewrite the given function:

$$\frac{\sqrt{x^2}}{x} = \frac{|x|}{x} = \begin{cases} 1 & \text{when } x > 0 \\ -1 & \text{when } x < 0 \end{cases}$$

Now you can see that the limit as x approaches 0 depends on *how* x approaches 0. Approaching $x = 0$ through positive values, the limit is 1; approaching $x = 0$ through negative values, the limit is -1. Consequently, you say that the limit, as x approaches 0, doesn't exist (see Figure 3-3).

3. The values of $f(x)$ may fluctuate as x approaches a.

EXAMPLE 3-5: Examine the behavior of $\sin 1/x$ as x approaches 0. Build a table that shows x, $1/x$, and $\sin 1/x$, and choose your x values approaching 0 in such a way that $\sin 1/x$ is easy to compute. For example, choose $x = 2/k\pi$, starting with $k = 1$ and continuing through integer values of k:

x	$\frac{1}{x}$	$\sin \frac{1}{x}$
$\frac{2}{\pi}$	$\frac{\pi}{2}$	1
$\frac{1}{\pi}$	π	0
$\frac{2}{3\pi}$	$\frac{3\pi}{2}$	-1
$\frac{1}{2\pi}$	2π	0
$\frac{2}{5\pi}$	$\frac{5\pi}{2}$	1
$\frac{1}{3\pi}$	3π	0
$\frac{2}{7\pi}$	$\frac{7\pi}{2}$	-1
$\frac{1}{4\pi}$	4π	0

As x approaches 0, $\sin 1/x$ just bounces through all its possible values between -1 and $+1$ and starts all over again; it never "settles down:" Therefore, $\lim_{x\to 0} \sin 1/x$ doesn't exist.

3-2. Limits: The Formal Definition

The discussion of the previous section was designed to give you an intuitive understanding of the limit concept. A formal definition makes precise the meaning of "closer and closer."

Definition: $\lim_{x\to a} f(x) = L$ if and only if, for each arbitrarily chosen positive number ε, there exists a positive number δ such that if $0 < |x - a| < \delta$, then $|f(x) - L| < \varepsilon$.

A reasonable paraphrase, which uses more words and fewer symbols, would be as follows: The limit of $f(x)$ as x approaches a is L if the difference between the numbers $f(x)$ and L may be made as small as you like by choosing x sufficiently close to a.

EXAMPLE 3-6: If $f(x) = 3x - 1$, find $\lim_{x\to 1} f(x)$.

Solution: It seems that, as x gets close to 1, $f(x)$ should get close to $3(1) - 1 = 2$. To show that this is in fact true, choose an ε—say, $\varepsilon = 0.01$—and find δ so that $|f(x) - 2| < 0.01$ when $|x - 1| < \delta$. Notice that

$$\begin{aligned} f(x) - 2 &= (3x - 1) - 2 \\ &= 3x - 3 = 3(x - 1) \end{aligned}$$

so

$$|f(x) - 2| = 3|x - 1| \quad \text{and} \quad |f(x) - 2| < 0.01$$

when

$$3|x - 1| < 0.01 \quad \text{or} \quad |x - 1| < \frac{0.01}{3}$$

Choose $\delta = 0.01/3$; if $|x - 1| < 0.01/3$, then

$$\begin{aligned} 3|x - 1| &= |3x - 3| \\ &= |3x - 1 - 2| \\ &= |f(x) - 2| < 0.01 = \varepsilon \end{aligned}$$

There was nothing special about the choice of $\varepsilon = 0.01$. The same line of reasoning would produce $\delta = \varepsilon/3$ for any ε. Thus, $\lim_{x \to 1} f(x) = 2$ because $|f(x) - 2| < \varepsilon$ whenever $|x - 1| < \varepsilon/3$.

EXAMPLE 3-7: Show $\lim_{x \to 1}(x^2 + 1) = 2$.

Solution: You need to find δ so that, for a given ε, $|x^2 + 1 - 2| < \varepsilon$ for $|x - 1| < \delta$. Now

$$\begin{aligned} x^2 + 1 - 2 &= x^2 - 1 \\ &= (x + 1)(x - 1) \end{aligned}$$

This function requires a two-step approach to δ: First choose $|x - 1| < 1$ so that

$$\begin{aligned} -1 < x - 1 < 1 \\ 0 < x < 2 \\ 1 < x + 1 < 3 \end{aligned}$$

and

$$|x + 1| < 3$$

You have

$$\begin{aligned} |x^2 + 1 - 2| &= |x + 1||x - 1| \\ &< 3|x - 1| \quad \text{for } |x - 1| < 1 \end{aligned}$$

Now $|x^2 + 1 - 2| < \varepsilon$ if $3|x - 1| < \varepsilon$ or $|x - 1| < \varepsilon/3$.
You now have two conditions on x:

$$|x - 1| < 1 \quad \text{and} \quad |x - 1| < \frac{\varepsilon}{3}$$

Choose $\delta = \min\{1, \varepsilon/3\}$ (the smaller of the two numbers, 1 and $\varepsilon/3$), and the demonstration is complete: For any given ε, choose $\delta = \min\{1, \varepsilon/3\}$; then when $|x - 1| < \delta$, it will be true that $|x^2 + 1 - 2| < \varepsilon$.

3-3. Rules for Calculating Limits

Not every limit problem can be done by formula; however, there exists a set of rules (theorems) that you use over and over again. In this section you'll learn the rules, and in Section 3-4 you'll learn the methods (which can be traced to the rules) for calculating limits.

1. If $f(x) = c$ (a constant) then

$$\lim_{x \to a} f(x) = c \quad \text{for any number } a$$

Assume that the limits of $f(x)$ and $g(x)$, as x approaches a, exist, with $\lim_{x \to a} f(x) = L$ and $\lim_{x \to a} g(x) = M$. Then:

2. $\lim_{x \to a}(f(x) + g(x)) = \lim_{x \to a} f(x) + \lim_{x \to a} g(x) = L + M$

3. $\lim_{x \to a}(f(x) \cdot g(x)) = \lim_{x \to a} f(x) \cdot \lim_{x \to a} g(x) = L \cdot M$

4. If c is a constant,

$$\lim_{x \to a} cf(x) = c \lim_{x \to a} f(x) = cL$$

(You can see that this follows from rules 1 and 3.)

5. If $M \neq 0$,

$$\lim_{x\to a}\frac{f(x)}{g(x)} = \frac{\lim_{x\to a} f(x)}{\lim_{x\to a} g(x)} = \frac{L}{M}$$

6. If $\lim_{x\to a} f(x) = L$ and $\lim_{x\to L} g(x) = Q$, then

$$\lim_{x\to a} g(f(x)) = \lim_{y\to L} g(y) = Q$$

3-4. Methods for Calculating $\lim_{x\to a} f(x)$

A basic consideration in the calculation of $\lim_{x\to a} f(x)$ is whether or not $x = a$ is in the domain of $f(x)$.

A. If $f(a)$ is defined

If $x = a$ is in the domain of $f(x)$, and a is not an endpoint of the domain, and if $f(x)$ is defined by a single expression, then

$$\lim_{x\to a} f(x) = f(a)$$

(See Section 3-4 C for functions that are defined by two or more expressions.)

EXAMPLE 3-8: Find $\lim_{x\to 1}(x + 3)$.

Solution:

$$\lim_{x\to 1}(x + 3) = 1 + 3 = 4$$

EXAMPLE 3-9: Find $\lim_{x\to 1} 1/(x + 2)$.

Solution:

$$\lim_{x\to 1}\frac{1}{x + 2} = \frac{1}{1 + 2} = \frac{1}{3}$$

EXAMPLE 3-10: Find $\lim_{x\to 8}(x^2 - 7x + 5)$.

Solution:

$$\lim_{x\to 8}(x^2 - 7x + 5) = 8^2 - 7(8) + 5 = 13$$

B. If $f(a)$ is not defined

If $f(x)$ is a ratio of two functions, $\dfrac{g(x)}{h(x)}$, with $\lim_{x\to a} h(x) = 0$, then the limit of $f(x)$ as x approaches a depends on $\lim_{x\to a} g(x)$.

1. If $f(x) = \dfrac{g(x)}{h(x)}$, with $\lim_{x\to a} h(x) = 0$ and $\lim_{x\to a} g(x) \neq 0$, then $\lim_{x\to a} f(x)$ doesn't exist. (Notice that rule 5 for the limit of a quotient doesn't apply here because the limit of the denominator is zero.)

EXAMPLE 3-11: Find $\lim_{x\to 1}(x^{3/2} - 2)/(x^2 - 1)$.

Solution:

$$\lim_{x\to 1}(x^{3/2} - 2) = 1 - 2 = -1$$

$$\lim_{x\to 1}(x^2 - 1) = 1 - 1 = 0$$

Therefore $\lim_{x\to 1}(x^{3/2} - 2)/(x^2 - 1)$ doesn't exist.

2. If $f(x) = \dfrac{g(x)}{h(x)}$, with $\lim_{x \to a} g(x) = \lim_{x \to a} h(x) = 0$, then the limit as x approaches a may or may not exist. There are several techniques that will help you find the limit, when it exists.

- $g(x)$ and $h(x)$ may have a common factor of $x - a$.

EXAMPLE 3-12: Find $\lim_{x \to 2}(x^2 - 4)/(x - 2)$.

Solution: Notice that $x^2 - 4 = (x + 2)(x - 2)$, and so for all $x \neq 2$:

$$\frac{x^2 - 4}{x - 2} = x + 2$$

and

$$\lim_{x \to 2} \frac{x^2 - 4}{x - 2} = \lim_{x \to 2}(x + 2) = 4$$

EXAMPLE 3-13: Find $\lim_{x \to 3}(x^2 - 5x + 6)/[(x - 3)^2]$.

Solution:

$$\frac{x^2 - 5x + 6}{(x - 3)^2} = \frac{(x - 3)(x - 2)}{(x - 3)(x - 3)}$$

$$= \frac{x - 2}{x - 3} \quad \text{for } x \neq 3$$

$$\lim_{x \to 3} \frac{x^2 - 5x + 6}{(x - 3)^2} = \lim_{x \to 3} \frac{x - 2}{x - 3}$$

and this limit doesn't exist: The numerator approaches one; the denominator approaches zero.

EXAMPLE 3-14: Find $\lim_{x \to -2}[(x + 2)^2/(x^2 + 3x + 2)]$.

Solution:

$$\frac{(x + 2)^2}{x^2 + 3x + 2} = \frac{(x + 2)(x + 2)}{(x + 2)(x + 1)}$$

$$= \frac{x + 2}{x + 1} \quad \text{for } x \neq -2$$

$$\lim_{x \to -2} \frac{x + 2}{x + 1} = \frac{0}{-1} = 0$$

EXAMPLE 3-15: Find $\lim_{x \to -1}(x^2 - 1)/\sqrt[3]{x + 1}$.

Solution:

$$\frac{x^2 - 1}{\sqrt[3]{x + 1}} = \frac{(x + 1)(x - 1)}{(x + 1)^{1/3}}$$

$$= (x + 1)^{2/3}(x - 1) \quad \text{for } x \neq -1$$

Therefore

$$\lim_{x \to -1} \frac{x^2 - 1}{\sqrt[3]{x + 1}} = \lim_{x \to -1} (x + 1)^{2/3}(x - 1)$$

$$= 0(-2) = 0$$

- An algebraic simplification (including addition of fractions) may reveal a factor of $x - a$ or otherwise help you find the limit.

EXAMPLE 3-16: Find $\lim_{x\to 1}[1 - (1/x)]/(x - 1)$.

Solution:

$$\frac{1 - (1/x)}{x - 1} = \frac{(x - 1)/x}{(x - 1)}$$

$$= \frac{1}{x} \quad \text{for } x \neq 1$$

Thus

$$\lim_{x\to 1} \frac{1 - (1/x)}{x - 1} = \lim_{x\to 1} \frac{1}{x} = \frac{1}{1} = 1$$

- Rationalization may be helpful.

EXAMPLE 3-17: Find $\lim_{x\to 4}(x - 4)/(\sqrt{x} - 2)$.

Solution:

$$\frac{x - 4}{\sqrt{x} - 2}\left(\frac{\sqrt{x} + 2}{\sqrt{x} + 2}\right) = \frac{(x - 4)(\sqrt{x} + 2)}{x - 4}$$

$$= \sqrt{x} + 2 \quad \text{for } x \neq 4$$

and

$$\lim_{x\to 4} \frac{x - 4}{\sqrt{x} - 2} = \lim_{x\to 4}(\sqrt{x} + 2)$$

$$= \sqrt{4} + 2 = 4$$

EXAMPLE 3-18: Find $\lim_{h\to 0}(\sqrt{x + h} - \sqrt{x})/h$.

Solution:

$$\left(\frac{\sqrt{x + h} - \sqrt{x}}{h}\right)\frac{(\sqrt{x + h} + \sqrt{x})}{(\sqrt{x + h} + \sqrt{x})} = \frac{(x + h) - x}{h(\sqrt{x + h} + \sqrt{x})}$$

$$= \frac{1}{\sqrt{x + h} + \sqrt{x}} \quad \text{for } h \neq 0$$

Therefore

$$\lim_{h\to 0} \frac{\sqrt{x + h} - \sqrt{x}}{h} = \lim_{h\to 0} \frac{1}{\sqrt{x + h} + \sqrt{x}}$$

$$= \frac{1}{\sqrt{x} + \sqrt{x}} = \frac{1}{2\sqrt{x}}$$

3. If $f(x) = g(x) - h(x)$ and $\lim_{x\to a} g(x)$ doesn't exist and $\lim_{x\to a} h(x)$ doesn't exist, than the limit of $f(x)$ as x approaches a may or may not exist. If you can combine $g(x)$ and $h(x)$ into a single expression, you may be able to find the limit.

EXAMPLE 3-19: Find $\lim_{x\to 0}[(1/x) - (1/x^3)]$.

Solution:

$$\frac{1}{x} - \frac{1}{x^3} = \frac{x^2 - 1}{x^3}$$

$$\lim_{x\to 0}\left(\frac{1}{x} - \frac{1}{x^3}\right) = \lim_{x\to 0} \frac{x^2 - 1}{x^3}$$

This limit doesn't exist (the denominator approaches 0, the numerator approaches -1).

EXAMPLE 3-20: Find $\lim_{x \to 1} \left(\dfrac{x}{x-1} - \dfrac{1}{x-1} \right)$.

Solution:

$$\frac{x}{x-1} - \frac{1}{x-1} = \frac{x-1}{x-1}$$

$$= 1 \quad \text{for } x \neq 1$$

Thus

$$\lim_{x \to 1} \left(\frac{x}{x-1} - \frac{1}{x-1} \right) = \lim_{x \to 1} 1 = 1$$

C. Functions defined by more than a single expression

Suppose that $f(x)$ is defined by one expression for $x < a$ and by a different expression for $x > a$. Then the question of the existence (and the value) of the limit as x approaches a requires that you examine the behavior of f on both sides of $x = a$.

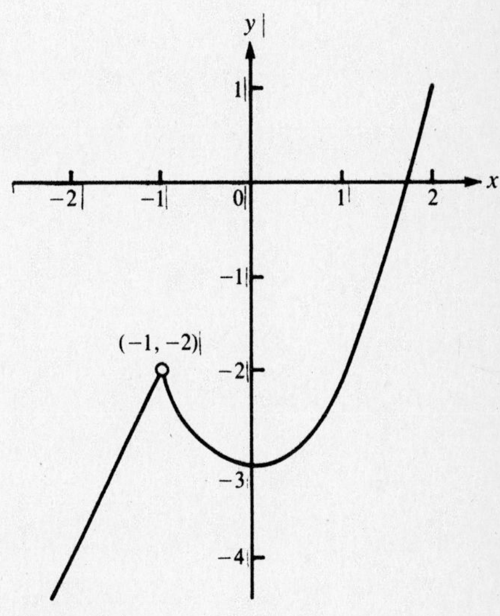

Figure 3-4

EXAMPLE 3-21: Find $\lim_{x \to -1} f(x)$, if

$$f(x) = \begin{cases} 2x & \text{for } x < -1 \\ x^2 - 3 & \text{for } x > -1 \end{cases}$$

Solution: If x approaches -1 from the left (i.e., through values less than -1), the limit of $f(x)$ is found by looking at $\lim_{x \to -1} 2x = -2$. If x approaches -1 through values of x greater than -1, the limit of $f(x)$ is found by looking at $\lim_{x \to -1}(x^2 - 3) = -2$. Since these limits are the same, you have $\lim_{x \to -1} f(x) = -2$ (see Figure 3-4). Section 3-5 examines these **one-sided limits** more closely.

3-5. One-Sided Limits

Some functions, by their nature, demand that the limit problem be broken into two steps (see Examples 3-4 and 3-21). For notation, mathematicians modify their symbolic representation of "the limit as x approaches a:" "The limit as x approaches a from the left" is denoted by "$\lim_{x \to a^-}(\ )$"; "the limit as x approaches a from the right" is denoted by "$\lim_{x \to a^+}(\ )$."

If the values of $f(x)$ become increasingly large (and positive) as x approaches a from the right (or left) then you say that the one-sided limit is infinity and write

$$\lim_{x \to a^+} f(x) = \infty \qquad \text{or} \qquad \lim_{x \to a^-} f(x) = \infty$$

If, as x approaches a from the right (or left), the values of $f(x)$ become increasingly large in absolute value while $f(x)$ is negative, then you say that the one-sided limit is negative infinity and write

$$\lim_{x \to a^+} f(x) = -\infty \qquad \text{or} \qquad \lim_{x \to a^-} f(x) = -\infty$$

If, for some function $f(x)$, $\lim_{x \to a^+} f(x) \neq \lim_{x \to a^-} f(x)$ then you say $\lim_{x \to a} f(x)$ doesn't exist. To put that another way: If $\lim_{x \to a^+} f(x) = \lim_{x \to a^-} f(x) = L$; then $\lim_{x \to a} f(x) = L$ (L can be ∞ or $-\infty$). To find one-sided limits, you use the same techniques that you use to find ordinary limits.

Now that you have one-sided limits, you can examine limits of the form

$$\lim_{x \to a} \frac{g(x)}{h(x)}$$

where

$$\lim_{x \to a} g(x) \neq 0$$

and

$$\lim_{x \to a} h(x) = 0$$

more closely. There are three possibilities:

$$\lim_{x \to a} \frac{g(x)}{h(x)}$$

might not exist (if one of the one-sided limits is $+\infty$ and the other is $-\infty$), or

$$\lim_{x \to a} \frac{g(x)}{h(x)}$$

can be $+\infty$, or it can be $-\infty$.

EXAMPLE 3-22: Find $\lim_{x \to 0^+} \sqrt{x}$.

Solution: Since $\sqrt{x}$ is real only for $x \geq 0$, the one-sided limit from the right is the only limit that makes sense.

$$\lim_{x \to 0^+} \sqrt{x} = \sqrt{0} = 0$$

EXAMPLE 3-23: Find $\lim_{x \to -1^+} \sqrt{1 - x^2}/(1 + x)$.

Solution: The function $\sqrt{1 - x^2}/(1 + x)$ is defined only for $-1 < x \leq 1$; thus x must approach -1 from the right.

Since both numerator and denominator have limits of 0 as x approaches -1^+, you should seek an algebraic manipulation that will allow you to solve the problem:

$$\frac{\sqrt{1 - x^2}}{1 + x} = \frac{\sqrt{1 + x}\sqrt{1 - x}}{1 + x}$$

$$= \frac{\sqrt{1 - x}}{\sqrt{1 + x}}$$

$$\lim_{x \to -1^+} \frac{\sqrt{1 - x^2}}{1 + x} = \lim_{x \to -1^+} \frac{\sqrt{1 - x}}{\sqrt{1 + x}}$$

and this limit is infinite: The numerator approaches $+\sqrt{2}$; the denominator approaches 0 through positive values. Thus you have

$$\lim_{x \to -1^+} \frac{1 - x^2}{1 + x} = +\infty$$

EXAMPLE 3-24: Show that $\lim_{x \to 0} |x|/x$ doesn't exist.

Solution: Notice that

$$\frac{|x|}{x} = \begin{cases} \dfrac{x}{x} = 1 & \text{for } x > 0 \\ \dfrac{-x}{x} = -1 & \text{for } x < 0 \end{cases}$$

That is, you seek a limit at $x = 0$ of a function that is defined differently on either side of $x = 0$; you must use one-sided limits:

$$\lim_{x \to 0^-} \frac{|x|}{x} = \lim_{x \to 0^-} (-1) = -1$$

$$\lim_{x \to 0^+} \frac{|x|}{x} = \lim_{x \to 0^+} (1) = 1$$

Since the two one-sided limits are different, $\lim_{x \to 0} |x|/x$ doesn't exist.

3-6. Some Special Limits Involving Trigonometric or Exponential Functions

The techniques and rules for finding limits discussed in the previous sections apply to trigonometric and exponential functions as well as to algebraic functions. Two additional limits (given without proof) may prove useful to you:

1. $\lim_{x\to 0}(\sin x)/x = \lim_{x\to 0}x/(\sin x) = 1$
2. $\lim_{x\to 0}(1 - \cos x)/x = 0$

EXAMPLE 3-25: Find $\lim_{x\to\pi/4} (\sin x)/\cos x$.

Solution: Since $\pi/4$ is in the domain of $f(x) = (\sin x)/\cos x$, you find

$$\lim_{x\to\pi/4} \frac{\sin x}{\cos x} = \frac{\sin(\pi/4)}{\cos(\pi/4)} = 1$$

EXAMPLE 3-26: Find $\lim_{x\to 0}(1 - \cos x)/\sin x$.

Solution:

$$\lim_{x\to 0}(1 - \cos x) = 1 - 1 = 0$$

and

$$\lim_{x\to 0}(\sin x) = \sin 0 = 0$$

The rule for quotients doesn't apply. To investigate the limit as x approaches 0, you use an identity:

$$\begin{aligned}\sin^2 x &= 1 - \cos^2 x\\ &= (1 + \cos x)(1 - \cos x)\end{aligned}$$

If you multiply numerator and denominator by $1 + \cos x$, you discover a common factor of $\sin x$:

$$\begin{aligned}\frac{1 - \cos x}{\sin x}\left(\frac{1 + \cos x}{1 + \cos x}\right) &= \frac{1 - \cos^2 x}{\sin x(1 + \cos x)}\\ &= \frac{\sin^2 x}{\sin x(1 + \cos x)}\\ &= \frac{\sin x}{1 + \cos x} \quad \text{for } \sin x \neq 0\end{aligned}$$

Therefore

$$\begin{aligned}\lim_{x\to 0}\frac{1 - \cos x}{\sin x} &= \lim_{x\to 0}\frac{\sin x}{1 + \cos x}\\ &= \frac{0}{2} = 0\end{aligned}$$

EXAMPLE 3-27: Find $\lim_{x\to 0}(\sin 3x)/x$.

Solution: Multiply the numerator and denominator by three:

$$\frac{\sin 3x}{x} = 3\left(\frac{\sin 3x}{3x}\right)$$

Notice also that if x approaches 0, then $3x$ approaches $3(0) = 0$. Therefore

$$\begin{aligned}\lim_{x\to 0}\frac{\sin 3x}{x} &= \lim_{x\to 0} 3\left(\frac{\sin 3x}{3x}\right)\\ &= 3\lim_{3x\to 0}\frac{\sin 3x}{3x} = 3(1) = 3\end{aligned}$$

EXAMPLE 3-28: Find $\lim_{x\to 0}(1 - \cos 2x)/\sin 3x$.

Solution: Use the technique of Example 3-27: multiply and divide by $2x$ (because of the $1 - \cos 2x$ term) and by $3x$ (because of the $\sin 3x$ term). You have

$$\frac{1 - \cos 2x}{\sin 3x} = 2x\left(\frac{1 - \cos 2x}{2x}\right)\frac{3x}{\sin 3x}\left(\frac{1}{3x}\right)$$
$$= \frac{2}{3}\left(\frac{1 - \cos 2x}{2x}\right)\frac{3x}{\sin 3x}$$

and

$$\lim_{x\to 0}\frac{1 - \cos 2x}{\sin 3x} = \frac{2}{3}\left(\lim_{x\to 0}\frac{1 - \cos 2x}{2x}\right)\left(\lim_{x\to 0}\frac{3x}{\sin 3x}\right)$$
$$= \frac{2}{3}(0)1 = 0$$

Two limits involving e and the exponential function e^t are also useful in finding limits:

3. $\lim_{x\to 0}(1 + x)^{1/x} = e$
4. $\lim_{x\to 0}(1 + tx)^{1/x} = e^t$

EXAMPLE 3-29: Find $\lim_{x\to 1} x^{1/(x-1)}$.

Solution: Let $y = x - 1$. Then as x approaches $x = 1$, $y = x - 1$ approaches 0:

$$\lim_{x\to 1} x^{1/(x-1)} = \lim_{x\to 1}(1 + x - 1)^{1/(x-1)}$$
$$= \lim_{y\to 0}(1 + y)^{1/y} = e$$

EXAMPLE 3-30: Find $\lim_{x\to 0}(1 + x)^{3/x}$.

Solution: Recall (see Section 3-3, Equation 6) that, as long as both limits exist, $\lim_{x\to a} g(f(x)) = g(\lim_{x\to a} f(x))$.

$$\lim_{x\to 0}(1 + x)^{3/x} = \lim_{x\to 0}[(1 + x)^{1/x}]^3$$
$$= \left[\lim_{x\to 0}(1 + x)^{1/x}\right]^3 = e^3$$

3-7. Limits at Infinity

In many instances you'll be interested in what happens to the values of $f(x)$ as x gets larger and larger. This problem is considered as an extension of the limit problem discussed in previous sections. "The limit of $f(x)$ as x gets larger and larger" is $\lim_{x\to\infty} f(x)$. As in the case in which x approached a finite limit a, the limit of $f(x)$ might be finite ($\lim_{x\to\infty} f(x) = L$) or not ($\lim_{x\to\infty} f(x) = \infty$).

A. $\lim_{x\to\infty} 1/x = 0$; $\lim_{x\to -\infty} 1/x = 0$.

These limits may be verified (with your calculator or without) by assigning numerically larger and larger values to x and computing the resulting reciprocals:

x	$\frac{1}{x}$
10	0.1
100	0.01
1000	0.001

When x is negative, $1/x$ is also negative, but numerically the results are the same: as x approaches $-\infty$, $1/x$ approaches 0 through negative values. Refer again to Figure 3-2; for large $|x|$, the graph of $1/x$ approaches the x axis, that is, y approaches 0.

B. The limit at infinity of a rational function.

A **rational function** is the quotient of two polynomials:

$$f(x) = \frac{p_m(x)}{q_n(x)}$$

where m and n are the *degrees* of the two polynomials.

1. If $m < n$, $\lim_{x\to\infty} f(x) = 0$.

EXAMPLE 3-31: Find $\lim_{x\to\infty}(x + 1)/(x^2 + 4)$.

Solution: The degree of the numerator is one; the degree of the denominator is two. Therefore

$$\lim_{x\to\infty} \frac{x+1}{x^2+4} = 0$$

This can be verified by dividing numerator and denominator by x^2:

$$\frac{x+1}{x^2+4} = \frac{\dfrac{1}{x} + \dfrac{1}{x^2}}{1 + \dfrac{4}{x^2}}$$

$$\lim_{x\to\infty} \frac{x+1}{x^2+4} = \frac{0+0}{1+0} = 0$$

2. If $m > n$, $\lim_{x\to\infty} f(x) = \pm\infty$. (The proper sign is determined by the polynomials $p_m(x)$ and $q_n(x)$; if they are of the same sign as x gets larger, the quotient is positive; if they are of opposite sign, the quotient is negative.)

EXAMPLE 3-32: Find $\lim_{x\to\infty}(x^3 - 2x^2 + 3x + 4)/(3x + 5)$.

Solution: The degree of the numerator (three) is greater than the degree of the denominator (one); therefore the limit is $+\infty$.

3. If $m = n$, $\lim_{x\to\infty} f(x) = a/b$, where a is the coefficient of x^m in the numerator and b is the coefficient of x^n in the denominator.

EXAMPLE 3-33: Find $\lim_{x\to\infty}(x^3 - 4x + 1)/(3x^3 + 2x + 7)$.

Solution: The degree of the numerator (three) is equal to the degree of the denominator (three); therefore

$$\lim_{x\to\infty} \frac{x^3 - 4x + 1}{3x^3 + 2x + 7} = \frac{1}{3}$$

Again, you can verify this rule by dividing numerator and denominator by x^3:

$$\frac{x^3 - 4x + 1}{3x^3 + 2x + 7} = \frac{1 - \dfrac{4}{x^2} + \dfrac{1}{x^3}}{3 + \dfrac{2}{x^2} + \dfrac{7}{x^3}}$$

$$\lim_{x\to\infty} \frac{x^3 - 4x + 1}{3x^3 + 2x + 7} = \frac{1 - 0 + 0}{3 + 0 + 0} = \frac{1}{3}$$

3-8. Continuity

A function $f(x)$ is continuous at $x = a$ if:

- $f(a)$ is defined
- $\lim_{x \to a} f(x)$ exists
- $\lim_{x \to a} f(x) = f(a)$

Notice that, to be continuous at $x = a$, all three conditions must be satisfied. If at least one condition fails, f is said to have a discontinuity at $x = a$. On the other hand, if f is continuous at every point of its domain, you say that f is continuous.

In terms of the graph of $f(x)$, the definition means that the curve is unbroken at $x = a$; you can move along the curve, from points corresponding to x values less than a, through $f(a)$, to points corresponding to x values greater than a, without lifting your pencil.

EXAMPLE 3-34: Determine whether $f(x) = x^2 + 1$ is continuous at $x = 1$.

Solution: From the discussion of limits in Section 3-4, and because $x = 1$ is in the domain of $f(x)$, you know that

$$\lim_{x \to 1} (x^2 + 1) = f(1) = 1^2 + 1 = 2$$

Therefore $f(x) = x^2 + 1$ is continuous at $x = 1$.

EXAMPLE 3-35: Show that $f(x) = a_0 + a_1x + a_2x^2 + \cdots + a_nx^n$ is continuous at $x = x_0$, where x_0 is any real number.

Solution: The domain of a polynomial is the set of all real numbers. As in Example 3-34, $x = x_0$ is in the domain of $f(x)$ and hence $\lim_{x \to x_0} f(x) = f(x_0)$.

EXAMPLE 3-36: Determine whether $f(x) = |x|/x$ is continuous at $x = 0$.

Solution: Since $f(0)$ isn't defined, $f(x)$ is not continuous at $x = 0$.

EXAMPLE 3-37: Determine whether

$$f(x) = \begin{cases} \dfrac{|x|}{x} & \text{for } x \neq 0 \\ 0 & \text{for } x = 0 \end{cases}$$

is continuous at $x = 0$.

Solution: Note that $f(0)$ is now defined. First, you must examine $\lim_{x \to 0} f(x)$. Using the definition of $|x|$, you get

$$f(x) = \begin{cases} 1 & \text{for } x > 0 \\ -1 & \text{for } x < 0 \\ 0 & \text{for } x = 0 \end{cases}$$

Then, $\lim_{x \to 0} f(x)$ must be considered in two steps:

$$\lim_{x \to 0^-} f(x) = \lim_{x \to 0^-} (-1) = -1$$

$$\lim_{x \to 0^+} f(x) = \lim_{x \to 0^+} (1) \quad = 1$$

Since the two one-sided limits aren't the same, $\lim_{x \to 0} f(x)$ does not exist and $f(x)$ is not continuous at $x = 0$.

EXAMPLE 3-38: Is

$$f(x) = \begin{cases} \dfrac{x^2 - 4}{x - 2} & \text{for } x \neq 2 \\ 4 & \text{for } x = 2 \end{cases}$$

continuous at $x = 2$?

Solution: Note that $f(2)$ is defined; you must examine $\lim_{x\to 2} f(x)$:

$$\lim_{x\to 2} \frac{x^2 - 4}{x - 2} = \lim_{x\to 2} \frac{(x + 2)(x - 2)}{(x - 2)} = \lim_{x\to 2} (x + 2) = 4$$

You have $\lim_{x\to 2} f(x) = 4 = f(2)$, so $f(x)$ is continuous at $x = 2$.

EXAMPLE 3-39: Determine whether

$$f(x) = \begin{cases} \dfrac{1}{x} & \text{for } x \leqslant -1 \\ \dfrac{x-1}{2} & \text{for } -1 < x < 1 \\ \sqrt{x} & \text{for } x \geqslant 1 \end{cases}$$

is continuous at $x = -1$ and at $x = 1$.

Solution: Since $f(x)$ is defined differently to the left and to the right of $x = -1$ (and of $x = 1$), you must use one-sided limits to find $\lim_{x\to -1} f(x)$ and $\lim_{x\to 1} f(x)$:

$$\lim_{x\to -1^-} f(x) = \lim_{x\to -1^-} \left(\frac{1}{x}\right) = -1$$

$$\lim_{x\to -1^+} f(x) = \lim_{x\to -1^+} \left(\frac{x-1}{2}\right) = -1$$

$$f(-1) = \frac{1}{-1} = -1$$

Thus, $f(x)$ is continuous at $x = -1$.

$$\lim_{x\to 1^-} f(x) = \lim_{x\to 1^-} \frac{x-1}{2} = 0$$

$$\lim_{x\to 1^+} f(x) = \lim_{x\to 1^+} \sqrt{x} = 1$$

The one-sided limits are not the same; $f(x)$ is not continuous at $x = 1$ (see Figure 3-5).

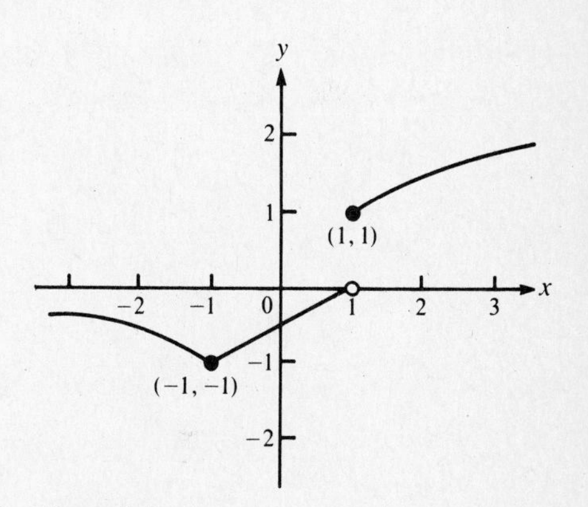

Figure 3-5

To summarize, discontinuities of functions may occur at points where a denominator is zero or at "meeting points" of a function that is defined by more than one expression.

SUMMARY

1. $\lim_{x\to a} f(x)$ reveals the behavior of $f(x)$ in the vicinity of a, but not necessarily at $x = a$.
2. $\lim_{x\to a} f(x)$ may fail to exist for any one of several reasons.
3. To calculate limits, you must often use significant algebraic manipulation, such as factoring, rationalizing the numerator, rationalizing the denominator, and combining fractions.
4. Limits of functions defined by more than a single expression require the evaluation of one-sided limits.
5. To calculate one-sided limits, you use the same techniques you use to calculate two-sided limits.
6. The limit of a rational function as x approaches infinity depends on the degrees of the numerator and denominator.
7. Limits at infinity also often require significant algebraic manipulation.
8. The continuity of a function is defined in terms of limits.
9. Possible points of discontinuity are zeros of denominators and "meeting points" of functions defined by more than a single expression.

SOLVED PROBLEMS

For Problems 3-1 through 3-26 calculate the limits.

PROBLEM 3-1 $\lim_{x\to 2} (x + 2)/(x^2 + 4)$

Solution: Because $x = 2$ is in the domain of $f(x) = (x + 2)/(x^2 + 4)$, you simply have

$$\lim_{x\to 2} \frac{x + 2}{x^2 + 4} = \frac{2 + 2}{2^2 + 4} = \frac{1}{2}$$

[See Section 3-4.]

PROBLEM 3-2 $\lim_{x\to 2} (x + 2)/(x^2 - 4)$

Solution: In this case $f(x) = (x + 2)/(x^2 - 4)$ isn't defined at $x = 2$. Because the numerator is nonzero and the denominator is zero when $x = 2$, you know that $\lim_{x\to 2} f(x)$ does not exist. Note that $\lim_{x\to 2^+} f(x) = \infty$ and $\lim_{x\to 2^-} f(x) = -\infty$. [See Section 3-4.]

PROBLEM 3-3 $\lim_{x\to 2} (x - 2)/(x^2 - 4)$

Solution: $f(x) = (x - 2)/(x^2 - 4)$ isn't defined at $x = 2$. Because both the numerator and denominator are zero when $x = 2$, you can factor $(x - 2)$ out of both:

$$\lim_{x\to 2} \frac{x - 2}{x^2 - 4} = \lim_{x\to 2} \frac{x - 2}{(x - 2)(x + 2)} = \lim_{x\to 2} \frac{1}{x + 2} = \frac{1}{4}$$

[See Section 3-4.]

PROBLEM 3-4 $\lim_{t\to 3}(t^2 - 6t + 9)/(t^2 - 2t - 3)$

Solution: Both numerator and denominator are zero when $t = 3$. Factor:

$$\lim_{t\to 3} \frac{t^2 - 6t + 9}{t^2 - 2t - 3} = \lim_{t\to 3} \frac{(t - 3)^2}{(t - 3)(t + 1)} = \lim_{t\to 3} \frac{(t - 3)}{(t + 1)} = \frac{0}{4} = 0$$

[See Section 3-4.]

PROBLEM 3-5 $\lim_{x\to 16} (x - 16)/(\sqrt{x} - 4)$

Solution: Rationalize the denominator:

$$\lim_{x\to 16} \frac{x - 16}{\sqrt{x} - 4} = \lim_{x\to 16} \frac{(x - 16)(\sqrt{x} + 4)}{(\sqrt{x} - 4)(\sqrt{x} + 4)} = \lim_{x\to 16} \frac{(x - 16)(\sqrt{x} + 4)}{(x - 16)}$$

$$= \lim_{x\to 16} (\sqrt{x} + 4) = 8$$

[See Section 3-4.]

PROBLEM 3-6 $\lim_{x\to 3}(\sqrt{x^2 + 7} - 4)/(x^2 - 3x)$

Solution: Both numerator and denominator are zero when $x = 3$. To factor, you rationalize the numerator:

$$\lim_{x\to 3} \frac{\sqrt{x^2 + 7} - 4}{x^2 - 3x} = \lim_{x\to 3} \frac{(\sqrt{x^2 + 7} - 4)(\sqrt{x^2 + 7} + 4)}{(x^2 - 3x)(\sqrt{x^2 + 7} + 4)}$$

$$= \lim_{x\to 3} \frac{x^2 + 7 - 16}{(x^2 - 3x)(\sqrt{x^2 + 7} + 4)}$$

$$= \lim_{x\to 3} \frac{(x + 3)(x - 3)}{x(x - 3)(\sqrt{x^2 + 7} + 4)}$$

$$= \lim_{x\to 3} \frac{x + 3}{x(\sqrt{x^2 + 7} + 4)} = \frac{3 + 3}{3(\sqrt{3^2 + 7} + 4)} = \frac{1}{4}$$

[See Section 3-4.]

PROBLEM 3-7 $\lim_{t\to 0}\left(\frac{1}{t\sqrt{t+1}} - \frac{1}{t}\right)$

Solution: Combine the fractions to obtain a more manageable expression:

$$\lim_{t\to 0}\left(\frac{1}{t\sqrt{t+1}} - \frac{1}{t}\right) = \lim_{t\to 0}\left(\frac{1-\sqrt{t+1}}{t\sqrt{t+1}}\right)$$

Now rationalize the numerator:

$$= \lim_{t\to 0}\frac{(1-\sqrt{t+1})(1+\sqrt{t+1})}{(t\sqrt{t+1})(1+\sqrt{t+1})} = \lim_{t\to 0}\frac{1-(t+1)}{t\sqrt{t+1}(1+\sqrt{t+1})}$$

$$= \lim_{t\to 0}\frac{-1}{\sqrt{t+1}(1+\sqrt{t+1})} = \frac{-1}{\sqrt{0+1}(1+\sqrt{0+1})} = -\frac{1}{2}$$ [See Section 3-4.]

PROBLEM 3-8 $\lim_{x\to 3} f(x)$, where $f(x) = \begin{cases} x^2 - 4 & \text{for } x \leqslant 3 \\ \dfrac{x+2}{x-2} & \text{for } x > 3 \end{cases}$

Solution: You must examine the limit from both sides:

$$\lim_{x\to 3^+} f(x) = \lim_{x\to 3^+}\frac{x+2}{x-2} = \frac{3+2}{3-2} = 5$$

$$\lim_{x\to 3^-} f(x) = \lim_{x\to 3^-}(x^2 - 4) = 3^2 - 4 = 5$$

Because both one-sided limits agree, the two-sided limit exists: $\lim_{x\to 3} f(x) = 5$.

[See Sections 3-4 and 3-5.]

PROBLEM 3-9 $\lim_{x\to 4}(x - \sqrt{x})/(x + \sqrt{x})$

Solution: $x = 4$ is in the domain, so

$$\lim_{x\to 4}\frac{x-\sqrt{x}}{x+\sqrt{x}} = \frac{4-\sqrt{4}}{4+\sqrt{4}} = \frac{1}{3}$$ [See Section 3-4.]

PROBLEM 3-10 $\lim_{t\to 3}\sqrt{9-t^2}/(3-t)$

Solution: Notice that the domain of this function is $-3 \leqslant t < 3$ because $9 - t^2$ must be positive and $3 - t$ mustn't be zero. Thus, what appears to be a two-sided limit is actually a one-sided limit:

$$\lim_{t\to 3^-}\frac{\sqrt{9-t^2}}{3-t} = \lim_{t\to 3^-}\frac{\sqrt{3-t}\sqrt{3+t}}{3-t} = \lim_{t\to 3^-}\frac{\sqrt{3+t}}{\sqrt{3-t}}$$

which is infinite because the numerator is nonzero and the denominator is zero when $t = 3$. As t approaches 3 from the left, the numerator is positive and the denominator is positive, so the quotient is positive:

$$\lim_{t\to 3^-}\frac{\sqrt{9-t^2}}{3-t} = +\infty$$

PROBLEM 3-11 $\lim_{x\to 1^+}\sqrt{x-1}/\sqrt{x^2-1}$

Solution: Factoring the denominator, you find

$$\lim_{x\to 1^+}\frac{\sqrt{x-1}}{\sqrt{x^2-1}} = \lim_{x\to 1^+}\frac{\sqrt{x-1}}{\sqrt{x-1}\sqrt{x+1}} = \lim_{x\to 1^+}\frac{1}{\sqrt{x+1}} = \frac{1}{\sqrt{2}}$$ [See Section 3-5.]

PROBLEM 3-12 $\lim_{x\to\infty}(x^2 + 3x + 1)/(3x^2 + 2x)$

Solution: Both numerator and denominator have degree two, so the limit is the quotient of the leading coefficients:

$$\lim_{x\to\infty} \frac{x^2 + 3x + 1}{3x^2 + 2x} = \frac{1}{3}$$ [See Section 3-7.]

PROBLEM 3-13 $\lim_{x\to\infty} (8x^3 + 5x - 2)/(3x^2 + x + 1)$

Solution: The degree of the numerator exceeds the degree of the denominator; both numerator and denominator are positive when x is positive, so the limit is $+\infty$. [See Section 3-7.]

PROBLEM 3-14 $\lim_{x\to-\infty} (8x^3 + 5x - 2)/(3x^7 + x + 1)$

Solution: The degree of the denominator exceeds the degree of the numerator, so

$$\lim_{x\to-\infty} \frac{8x^3 + 5x - 2}{3x^7 + x + 1} = 0$$ [See Section 3-7.]

PROBLEM 3-15 $\lim_{x\to\infty} (4x^3 + 8x^2 - 1)/\sqrt{x^6 + 1}$

Solution: Although this fraction isn't a rational function, the technique of dividing by a power of x still works:

$$\lim_{x\to\infty} \frac{4x^3 + 8x^2 - 1}{\sqrt{x^6 + 1}} = \lim_{x\to\infty} \frac{(4x^3 + 8x^2 - 1)(1/x^3)}{\sqrt{x^6 + 1}(1/x^3)} = \lim_{x\to\infty} \frac{4 + (8/x) - (1/x^3)}{\sqrt{1 + (1/x^6)}}$$

$$= \frac{4 + 0 - 0}{\sqrt{1 + 0}} = 4$$ [See Section 3-7.]

PROBLEM 3-16 $\lim_{x\to0} (1 - \cos 2x)/x$

Solution: Multiply and divide by 2

$$\lim_{x\to0} \frac{1 - \cos 2x}{x} = \lim_{x\to0} 2\left(\frac{1 - \cos 2x}{2x}\right) = 2\left(\lim_{x\to0} \frac{1 - \cos 2x}{2x}\right) = 2(0) = 0$$ [See Section 3-6.]

PROBLEM 3-17 $\lim_{x\to0} [\cos x - 1]/\tan x$

Solution: Use the identity $\tan x = (\sin x)/(\cos x)$; then divide numerator and denominator by x:

$$\lim_{x\to0} \frac{\cos x - 1}{\tan x} = \lim_{x\to0} \frac{\cos x[\cos x - 1]}{\sin x} = \lim_{x\to0} \frac{\cos x[(\cos x - 1)/x]}{\frac{\sin x}{x}}$$

$$= \frac{1(0)}{1} = 0$$ [See Section 3-6.]

PROBLEM 3-18 $\lim_{x\to0} (\sin^2 2x)/8x^2$

Solution: You find

$$\lim_{x\to0} \frac{\sin^2 2x}{8x^2} = \lim_{x\to0} \frac{1}{2}\left(\frac{\sin 2x}{2x}\right)^2 = \frac{1}{2}\left[\lim_{x\to0}\left(\frac{\sin 2x}{2x}\right)\right]^2 = \frac{1}{2}(1)^2 = \frac{1}{2}$$ [See Section 3-6.]

PROBLEM 3-19 $\lim_{x\to0} [1 + (x/3)]^{2/x}$

Solution: You find

$$\lim_{x\to0}\left(1 + \left(\frac{x}{3}\right)\right)^{2/x} = \lim_{x\to0}\left[\left(1 + \frac{1}{3}x\right)^{1/x}\right]^2 = \left[\lim_{x\to0}\left(1 + \frac{1}{3}x\right)^{1/x}\right]^2$$

$$= [e^{1/3}]^2 = e^{2/3}$$ [See Section 3-6.]

PROBLEM 3-20 $\lim_{x\to 0}(1-x)^{1/x}$

Solution: You find

$$\lim_{x\to 0}(1-x)^{1/x} = \lim_{x\to 0}\left(1+(-1)x\right)^{1/x} = e^{-1}$$

[See Section 3-6.]

PROBLEM 3-21 $\lim_{x\to -\infty}\sqrt{x^2-1}/(4x+3)$

Solution: Dividing numerator and denominator by x, you see

$$\lim_{x\to -\infty}\frac{\sqrt{x^2-1}}{4x+3} = \lim_{x\to -\infty}\frac{\sqrt{x^2-1}(1/x)}{(4x+3)(1/x)}$$

For $x<0$, notice that $1/x = -1/\sqrt{x^2}$, so the limit becomes

$$\lim_{x\to -\infty}\frac{-\sqrt{1-(1/x^2)}}{4+(3/x)} = -\frac{1}{4}$$

[See Section 3-7.]

PROBLEM 3-22 $\lim_{x\to 7}[(1/\sqrt{x})-(1/\sqrt{7})]/(x-7)$

Solution: Clear fractions in the numerator and rationalize it:

$$\lim_{x\to 7}\frac{(1/\sqrt{x})-(1/\sqrt{7})}{x-7} = \lim_{x\to 7}\frac{\sqrt{7}-\sqrt{x}}{(x-7)\sqrt{7x}} = \lim_{x\to 7}\frac{(\sqrt{7}-\sqrt{x})(\sqrt{7}+\sqrt{x})}{(x-7)\sqrt{7x}(\sqrt{7}+\sqrt{x})}$$

$$= \lim_{x\to 7}\frac{7-x}{(x-7)\sqrt{7x}(\sqrt{7}+\sqrt{x})} = \lim_{x\to 7}\frac{-1}{\sqrt{7x}(\sqrt{7}+\sqrt{x})}$$

$$= \frac{-1}{7(2\sqrt{7})} = -\frac{\sqrt{7}}{98}$$

[See Section 3-4.]

PROBLEM 3-23 $\lim_{x\to\infty}\sqrt{3x+2}-\sqrt{3x}$

Solution: Rationalizing, you see

$$\lim_{x\to\infty}\sqrt{3x+2}-\sqrt{3x} = \lim_{x\to\infty}(\sqrt{3x+2}-\sqrt{3x})\frac{(\sqrt{3x+2}+\sqrt{3x})}{(\sqrt{3x+2}+\sqrt{3x})}$$

$$= \lim_{x\to\infty}\frac{3x+2-3x}{\sqrt{3x+2}+\sqrt{3x}} = \lim_{x\to\infty}\frac{2}{\sqrt{3x+2}+\sqrt{3x}} = 0$$

[See Section 3-7.]

PROBLEM 3-24 $\lim_{x\to\pi/2}\cos x/[x-(\pi/2)]$

Solution: Recall from trigonometry that $\sin((\pi/2)-x) = \cos x$:

$$\lim_{x\to\pi/2}\frac{\cos x}{x-\frac{\pi}{2}} = \lim_{x\to\pi/2}\frac{\sin(\frac{\pi}{2}-x)}{x-\frac{\pi}{2}}$$

Now let $y = x-(\pi/2)$:

$$= \lim_{y\to 0}\frac{-\sin y}{y} = -1$$

[See Section 3-6.]

PROBLEM 3-25 $\lim_{t\to 0}\dfrac{4-(1/t^2)}{3+(2/t)+(1/t^2)}$

Solution: Multiply numerator and denominator by t^2:

$$\lim_{t\to 0}\frac{4-(1/t^2)}{3+(2/t)+(1/t^2)} = \lim_{t\to 0}\frac{(4-(1/t^2))t^2}{[3+(2/t)+(1/t^2)]t^2} = \lim_{t\to 0}\frac{4t^2-1}{3t^2+2t+1} = -1$$

[See Section 3-4.]

PROBLEM 3-26 $\lim_{x\to 0} f(x)$ where $f(x) = \begin{cases} x^3 - 2 & \text{for } x \leqslant 0 \\ \dfrac{\sqrt{x}+1}{3\sqrt{x}-2} & \text{for } x > 0 \end{cases}$

Solution: You find

$$\lim_{x\to 0^-} f(x) = \lim_{x\to 0^-} (x^3 - 2) = -2$$

$$\lim_{x\to 0^+} f(x) = \lim_{x\to 0^+} \frac{\sqrt{x}+1}{3\sqrt{x}-2} = -\frac{1}{2}$$

Because the one-sided limits have different values, $\lim_{x\to 0} f(x)$ does not exist. [See Section 3-4.]

PROBLEM 3-27 Find the discontinuities of $f(x) = \begin{cases} \dfrac{x^2-1}{x^2+x} & \text{for } x > -3 \\ 1 & \text{for } x \leqslant -3 \end{cases}$

Solution: The possible discontinuities are at $x = -1$ and $x = 0$ (the zeroes of $x^2 + x$) and at $x = -3$. Since f isn't even defined at $x = -1$ and $x = 0$, f is discontinuous at $x = -1$ and $x = 0$.

$$\lim_{x\to -3^+} f(x) = \lim_{x\to -3^+} \frac{x^2-1}{x^2+x} = \frac{4}{3}$$

$$\lim_{x\to -3^-} f(x) = \lim_{x\to -3^-} 1 = 1$$

So f is also discontinuous at $x = -3$. [See Section 3-8.]

PROBLEM 3-28 Find the discontinuities of

$$f(x) = \begin{cases} x/(x^2+1) & \text{for } x > -1 \\ (x+1)/(x^2-1) & \text{for } x < -1 \\ \frac{1}{2} & \text{for } x = -1 \end{cases}$$

Solution: Although $x = 1$ is a zero of $x^2 - 1$, f is defined by $x/(x^2 + 1)$ in the vicinity of $x = 1$, so the only possible discontinuity is at $x = -1$:

$$\lim_{x\to -1^+} f(x) = \lim_{x\to -1^+} \frac{x}{x^2+1} = -\frac{1}{2}$$

$$\lim_{x\to -1^-} f(x) = \lim_{x\to -1^-} \frac{x+1}{x^2-1} = \lim_{x\to -1^-} \frac{1}{x-1} = -\frac{1}{2}$$

Although the one-sided limits agree, they don't equal $f(-1)$, so f is discontinuous at $x = -1$. [See Section 3-8.]

Supplementary Exercises

3-29 $\lim_{t\to 2} (t^3 - 8)/(t - 2)$

3-30 $\lim_{t\to 0} (t^3 - 8)/(t - 2)$

3-31 $\lim_{x\to 0} (\sqrt{3+x} - \sqrt{3})/x$

3-32 $\lim_{x\to 0} \sin 3x/(1 - \cos 2x)$

3-33 $\lim_{t\to\infty} (8t^2 + 4)/(3t^2 - t + 2)$

3-34 $\lim_{x\to\infty} 4x(\sqrt{x} - \sqrt{x-1})$

3-35 $\lim_{x\to 0} (1 + (x/2))^{3/x}$

3-36 $\lim_{x\to 1} (1 + x)^{1/x}$

3-37 $\lim_{x\to 0^+} (\sqrt{x} - 2)/(3\sqrt{x} - 4x)$

3-38 $\lim_{x\to 1^-} \sqrt{x^2 - 1}/\sqrt{x^2 + 1}$

3-39 $\lim_{x\to \infty} (\sqrt{x} + 3)/(x - 4)$

3-40 $\lim_{x\to 0} \tan^2 x/(1 + \cos x)$

3-41 $\lim_{x\to -2} (x^2 + 6x + 8)/(x^2 - 5x - 14)$

3-42 $\lim_{x\to -\infty} (x^2 + 6x + 8)/(x^2 - 5x - 14)$

3-43 $\lim_{x\to 2} f(x)$, where $f(x) = \begin{cases} x^2 - 4x + 4 & \text{for } x \geqslant 2 \\ \dfrac{x^2 - 4x + 4}{x^2 - 3x + 2} & \text{for } x < 2 \end{cases}$

3-44 $\lim_{x\to 3^-} \sqrt{3x - x^2}/\sqrt{9 - x^2}$

3-45 $\lim_{x\to 5} (x^2 - 8x + 15)/(x^2 + 10x + 25)$

3-46 $\lim_{x\to \infty} x^2[\sqrt{2 + (1/x^2)} - \sqrt{2}]$

3-47 $\lim_{x\to 0} (1 + x^2)^{1/x^2}$

3-48 $\lim_{x\to 0} (2 - \cos x)/x$

3-49 $\lim_{s\to \infty} (s^3 - 8s + 5)/\sqrt{s^7 - 2s}$

3-50 $\lim_{x\to 5} e$

3-51 $\lim_{x\to 0} (\sqrt{x^2 + 7} - 4)/(x^2 - 3x)$

3-52 $\lim_{x\to 0} x/(\cos x - 1)$

3-53 $\lim_{x\to 5} (\sqrt{5x} - 5)/(\sqrt{x + 4} - 3)$

3-54 $\lim\limits_{x\to 2} \left(\dfrac{1}{x - 2} - \dfrac{x - 1}{x - 2}\right)$

3-55 $\lim\limits_{x\to 0^-} \dfrac{\sqrt[3]{x^3}}{x}$

3-56 $\lim_{x\to 2} (|x| - 2)/(x - 2)$

3-57 $\lim_{x\to 0} (\sin 4x)/\sin 3x$

3-58 $\lim_{x\to \infty} (\sqrt{\pi x^2} - 1)/(x + 3)$

3-59 $\lim_{x\to \infty} (x - \sqrt{x^2 - x})$

3-60 $\lim_{x\to 3^+} (\sqrt{x - 2} - 1)/\sqrt{x - 3}$

3-61 $\lim_{t\to 0} e^{(t-3)/(t+1)}$

3-62 $\lim_{x\to 0} [(3 + x)^2 - 9]/x$

3-63 $\lim_{x\to \infty} x/[x - (2/x)]$

3-64 $\lim_{x\to 0} [\sqrt{x} - (6/x)]/[(3/x) + \sqrt{x}]$

3-65 $\lim_{x\to 0} [1 + x(e)]^{1/x}$

3-66 $\lim_{x\to \infty} [\sqrt{x} - (6/x)]/[(3/x) + \sqrt{x}]$

3-67 $\lim_{x\to 3} (x - \sqrt{x} - 6)/(x - 3)$

3-68 Find the discontinuities of $f(x) = \begin{cases} 1/(x-3) & \text{for } x > 3 \\ 1/(x+3) & \text{for } x \leqslant 3 \end{cases}$

3-69 Find the discontinuities of $f(x) = \begin{cases} (x^2 + x)/\sqrt{x} & \text{for } x > 0 \\ 0 & \text{for } x = 0 \\ x^2/(x^2 + x) & \text{for } x < 0 \end{cases}$

3-70 Find the discontinuities of $f(t) = \begin{cases} \sqrt{t} - \sqrt{t-1} & \text{for } t \geqslant 1 \\ (1-t)/(t-2) & \text{for } t < 1 \end{cases}$

Solutions to Supplementary Exercises

(3-29) 12

(3-30) 4

(3-31) $\sqrt{3}/6$

(3-32) doesn't exist

(3-33) 8/3

(3-34) $+\infty$

(3-35) $e^{3/2}$

(3-36) 2

(3-37) $-\infty$

(3-38) 0

(3-39) 0

(3-40) 0

(3-41) $-2/9$

(3-42) 1

(3-43) 0

(3-44) $\sqrt{2}/2$

(3-45) 0

(3-46) $\sqrt{2}/4$

(3-47) e

(3-48) doesn't exist

(3-49) 0

(3-50) e

(3-51) doesn't exist

(3-52) doesn't exist

(3-53) 3

(3-54) -1

(3-55) 1

(3-56) 1

(3-57) 4/3

(3-58) $\sqrt{\pi}$

(3-59) $\frac{1}{2}$

(3-60) 0

(3-61) e^{-3}

(3-62) 6

(3-63) 1

(3-64) -2

(3-65) e^e

(3-66) 1

(3-67) doesn't exist

(3-68) $x = 3$ and $x = -3$

(3-69) $x = -1$

(3-70) $t = 1$

THE DERIVATIVE

THIS CHAPTER IS ABOUT

- ☑ Definition of the Derivative
- ☑ Using the Definition to Calculate Derivatives
- ☑ Derivatives of the Basic Functions
- ☑ Derivatives of Sums, Products, Quotients, and Composite Functions
- ☑ Derivatives Involving Repeated Applications of the Product, Quotient, and Chain Rules
- ☑ The Differentiation Formulas: Using the Chain Rule
- ☑ Higher Order Derivatives

4-1. Definition of the Derivative

The **derivative** of a function $y = f(x)$ is a function $f'(x)$ defined as a limit:

THE DERIVATIVE

$$f'(x) = \lim_{\Delta x \to 0} \frac{f(x + \Delta x) - f(x)}{\Delta x} \tag{4-1}$$

where Δx (read "delta x") represents a change in x. Your textbook may use the equivalent definition

$$f'(x) = \lim_{h \to 0} \frac{f(x + h) - f(x)}{h}$$

The domain of f' is the set of all numbers for which the limit exists; at those values of x, f is said to be **differentiable**. The process of finding a derivative is known as **differentiation**.

The derivative of $y = f(x)$ is denoted in several ways:

$$f'(x) \qquad Df \qquad D_x f \qquad \frac{dy}{dx} \qquad \frac{df}{dx} \qquad \frac{d}{dx}(f(x)) \qquad y' \qquad \text{or} \qquad \dot{y}$$

The presence of x in $D_x f$, dy/dx, and df/dx is to remind you that x is the independent variable (and you speak of the derivative *with respect to* x). If $y = f(t)$, then you would write dy/dt, df/dt, and so on and speak of the derivative *with respect to t*.

You may want to find the value of the derivative for a particular value of x. The value of $f'(x)$ at $x = a$ is denoted (as you might guess) by $f'(a)$,

$$f'(x)\bigg|_{x=a} \qquad \text{or} \qquad \frac{dy}{dx}\bigg|_{x=a}.$$

4-2. Using the Definition to Calculate Derivatives

You can use the definition of the derivative to find the derivative of simple functions.

EXAMPLE 4-1: Differentiate $f(x) = x^2$.

Solution: First find the *difference quotient*

$$\frac{f(x + \Delta x) - f(x)}{\Delta x}$$

and simplify it algebraically:

$$\begin{aligned}\frac{f(x + \Delta x) - f(x)}{\Delta x} &= \frac{(x + \Delta x)^2 - x^2}{\Delta x} \\ &= \frac{x^2 + 2x\,\Delta x + (\Delta x)^2 - x^2}{\Delta x} \\ &= \frac{\Delta x(2x + \Delta x)}{\Delta x}\end{aligned}$$

Then apply the definition of the derivative:

$$\begin{aligned}f'(x) &= \lim_{\Delta x \to 0} \frac{f(x + \Delta x) - f(x)}{\Delta x} \\ &= \lim_{\Delta x \to 0} \frac{\Delta x(2x + \Delta x)}{\Delta x} \\ &= \lim_{\Delta x \to 0} (2x + \Delta x) = 2x\end{aligned}$$

EXAMPLE 4-2: Differentiate $f(x) = 7x^3$.

Solution: Proceed as before:

$$\begin{aligned}\frac{f(x + \Delta x) - f(x)}{\Delta x} &= \frac{7(x + \Delta x)^3 - 7x^3}{\Delta x} \\ &= \frac{7[x^3 + 3x^2(\Delta x) + 3x(\Delta x)^2 + (\Delta x)^3] - 7x^3}{\Delta x} \\ &= \frac{7[3x^2 + 3x(\Delta x) + (\Delta x)^2]\Delta x}{\Delta x}\end{aligned}$$

and so

$$\begin{aligned}f'(x) &= \lim_{\Delta x \to 0} \frac{f(x + \Delta x) - f(x)}{\Delta x} \\ &= \lim_{\Delta x \to 0} \frac{7[3x^2 + 3x(\Delta x) + (\Delta x)^2]\,\Delta x}{\Delta x} \\ &= \lim_{\Delta x \to 0} 7[3x^2 + 3x(\Delta x) + (\Delta x)^2] \\ &= 21x^2\end{aligned}$$

EXAMPLE 4-3: Differentiate $f(x) = 1/x$.

Solution: Construct the difference quotient:

$$\begin{aligned}\frac{f(x + \Delta x) - f(x)}{\Delta x} &= \frac{\dfrac{1}{x + \Delta x} - \dfrac{1}{x}}{\Delta x} \\ &= \frac{\dfrac{x - (x + \Delta x)}{x(x + \Delta x)}}{\Delta x} \\ &= \frac{-\Delta x}{(\Delta x)x(x + \Delta x)}\end{aligned}$$

Then take the limit:

$$f'(x) = \lim_{\Delta x \to 0} \frac{-\Delta x}{(\Delta x)x(x + \Delta x)}$$

$$= \lim_{\Delta x \to 0} \frac{-1}{x(x + \Delta x)} = \frac{-1}{x^2}$$

You should begin to see that using Definition 4-1 to find derivatives has definite disadvantages: The algebra can be tedious and the calculation of the limits may be nearly impossible. To avoid repetition of these calculations, you apply the definition of the derivative to a *generic* function of a particular type in order to develop a general formula.

EXAMPLE 4-4: Differentiate $f(x) = x^n$, where n is a positive integer.

Solution: The difference quotient is

$$\frac{f(x + \Delta x) - f(x)}{\Delta x} = \frac{(x + \Delta x)^n - x^n}{\Delta x}$$

Use the binomial formula

$$(x + y)^n = x^n + nx^{n-1}y + \frac{n(n-1)}{2}x^{n-2}y^2 + \cdots + y^n \qquad \textbf{(4-2)}$$

to simplify this quotient. The difference quotient becomes

$$\frac{f(x + \Delta x) - f(x)}{\Delta x} = \frac{x^n + nx^{n-1}(\Delta x) + \frac{1}{2}n(n-1)x^{n-2}(\Delta x)^2 + \cdots + (\Delta x)^n - x^n}{\Delta x}$$

$$= \frac{[nx^{n-1} + \frac{1}{2}n(n-1)x^{n-2}(\Delta x) + \cdots + (\Delta x)^{n-1}]\Delta x}{\Delta x}$$

and so

$$f'(x) = \lim_{\Delta x \to 0} \frac{[nx^{n-1} + \frac{1}{2}n(n-1)x^{n-2}(\Delta x) + \cdots + (\Delta x)^{n-1}]\Delta x}{\Delta x}$$

$$= \lim_{\Delta x \to 0} [nx^{n-1} + \tfrac{1}{2}n(n-1)x^{n-2}(\Delta x) + \cdots + (\Delta x)^{n-1}]$$

$$= nx^{n-1}$$

You have a formula!

$$\frac{d}{dx}(x^n) = nx^{n-1} \quad \text{for } n \text{ a positive integer}$$

EXAMPLE 4-5: Differentiate $y = x^7$ with respect to x.

Solution: Use the formula from Example 4-4 with $n = 7$,

$$\frac{dy}{dx} = \frac{d}{dx}(x^7) = 7x^{7-1} = 7x^6$$

4-3. Derivatives of the Basic Functions

If you apply the definition of the derivative to the basic functions, you will find the following formulas. It is very important that you familiarize yourself with this list.

1. $\frac{d}{dx}(c) = 0$ for any constant c
2. $\frac{d}{dx}[cf(x)] = cf'(x)$ for any constant c
3. $\frac{d}{dx}(x^n) = nx^{n-1}$ for any real number n
4. $\frac{d}{dx}(\sin x) = \cos x$

5. $\frac{d}{dx}(\cos x) = -\sin x$

6. $\frac{d}{dx}(\tan x) = \sec^2 x$

7. $\frac{d}{dx}(\sec x) = \sec x \tan x$

8. $\frac{d}{dx}(\csc x) = -\csc x \cot x$

9. $\frac{d}{dx}(\cot x) = -\csc^2 x$

10. $\frac{d}{dx}(a^x) = a^x \ln a$ for any number $a > 0$

11. $\frac{d}{dx}(e^x) = e^x$

12. $\frac{d}{dx}[\log_a(x)] = \frac{1}{x} \cdot \frac{1}{\ln a}$ for $a > 0, a \neq 1$

13. $\frac{d}{dx}(\ln x) = \frac{1}{x}$

14. $\frac{d}{dx}(\text{arc sin } x) = \frac{1}{\sqrt{1 - x^2}}$

15. $\frac{d}{dx}(\text{arc cos } x) = \frac{-1}{\sqrt{1 - x^2}}$

16. $\frac{d}{dx}(\text{arc tan } x) = \frac{1}{1 + x^2}$

17. $\frac{d}{dx}(\text{arc sec } x) = \frac{1}{|x|\sqrt{x^2 - 1}}$

18. $\frac{d}{dx}(\text{arc csc } x) = \frac{-1}{|x|\sqrt{x^2 - 1}}$

19. $\frac{d}{dx}(\text{arc cot } x) = \frac{-1}{1 + x^2}$

20. $\frac{d}{dx}(\sinh x) = \cosh x$

21. $\frac{d}{dx}(\cosh x) = \sinh x$

22. $\frac{d}{dx}(\tanh x) = \text{sech}^2 x$

23. $\frac{d}{dx}(\text{sech } x) = -\text{sech } x \tanh x$

24. $\frac{d}{dx}(\text{csch } x) = -\text{csch } x \coth x$

25. $\frac{d}{dx}(\coth x) = -\text{csch}^2 x$

EXAMPLE 4-6: Differentiate $y = 7x^3$ with respect to x.

Solution: Simply use formula 2 and then formula 3:

$$\frac{d}{dx}(7x^3) = 7 \cdot \frac{d}{dx}(x^3) = 7 \cdot 3x^{3-1} = 21x^2$$

Note: Replace radical signs with fractional exponents when using the formulas.

EXAMPLE 4-7: Differentiate $y = \sqrt{x}$.

Solution:

$$\frac{d}{dx}(\sqrt{x}) = \frac{d}{dx}(x^{1/2}) = \frac{1}{2}x^{(1/2)-1} = \frac{1}{2}x^{-1/2} = \frac{1}{2\sqrt{x}}$$

EXAMPLE 4-8: Differentiate $y = 4\sqrt[3]{x^5}$.

Solution:

$$\frac{d}{dx}(4\sqrt[3]{x^5}) = 4 \cdot \frac{d}{dx}(x^{5/3}) = 4 \cdot \frac{5}{3}x^{(5/3)-1} = \frac{20}{3}x^{2/3}$$

Recall that $1/(x^n) = x^{-n}$. You may use negative exponents when using the formulas.

EXAMPLE 4-9: Differentiate $y = 2/x^3$.

Solution:

$$\frac{d}{dx}\left(\frac{2}{x^3}\right) = 2 \cdot \frac{d}{dx}(x^{-3}) = 2(-3)x^{-3-1} = -6x^{-4}$$

EXAMPLE 4-10: Differentiate $y = 3/\sqrt{x^5}$.

Solution:

$$\frac{d}{dx}(3/\sqrt{x^5}) = 3\cdot\frac{d}{dx}(x^{-5/2}) = 3\left(-\frac{5}{2}\right)x^{(-5/2)-1} = \frac{-15}{2}x^{-7/2}$$

EXAMPLE 4-11: If $f(x) = \sqrt{x}/\sqrt[3]{x}$, find $f'(1)$.

Solution: Express $f(x)$ as a power of x:

$$f(x) = \frac{x^{1/2}}{x^{1/3}} = x^{(1/2)-(1/3)} = x^{1/6}$$

$$f'(x) = \frac{1}{6}x^{-5/6}$$

$$f'(1) = \frac{1}{6}(1)^{-5/6} = \frac{1}{6}$$

4-4. Derivatives of Sums, Products, Quotients, and Composite Functions

Most of the functions you'll have to differentiate won't be as simple as those listed in Section 4-3. Instead, you'll encounter sums, products, quotients, and composite functions.

A. Derivative of a sum

The derivative of the sum (or difference) of two or more functions is the sum (or difference) of the derivatives of the functions.

DERIVATIVE OF A SUM

$$\frac{d}{dx}[f(x) + g(x)] = f'(x) + g'(x) \qquad \textbf{(4-3)}$$

As an immediate result of this formula and the first three formulas of Section 4-3, you can now find the derivative of a polynomial.

EXAMPLE 4-12: Differentiate $y = 2x^3 - 4x^2 + 7x + 6$.

Solution: The derivative of a sum is the sum of the derivatives:

$$\frac{d}{dx}(2x^3 - 4x^2 + 7x + 6) = \frac{d}{dx}(2x^3) - \frac{d}{dx}(4x^2) + \frac{d}{dx}(7x) + \frac{d}{dx}(6)$$

The derivative of a constant times a function is the constant times the derivative of the function, so

$$\frac{d}{dx}(2x^3 - 4x^2 + 7x + 6) = 2\frac{d}{dx}(x^3) - 4\frac{d}{dx}(x^2) + 7\frac{d}{dx}(x) + \frac{d}{dx}(6)$$

$$= 2(3x^2) - 4(2x) + 7(1) + 0$$

$$= 6x^2 - 8x + 7$$

EXAMPLE 4-13: Differentiate $y = 4x^7 + 12x^5 - 14x^2 + 7x - 5$.

Solution:

$$\frac{dy}{dx} = 4(7x^6) + 12(5x^4) - 14(2x) + 7 - 0$$

$$= 28x^6 + 60x^4 - 28x + 7$$

EXAMPLE 4-14: Differentiate $f(x) = 5x^3 + x + (1/x) - (3/\sqrt{x})$.

Solution: First express the function in a form that allows you to apply the formulas:

$$f(x) = 5x^3 + x + x^{-1} - 3x^{-1/2}$$

Now differentiate:

$$\begin{aligned} f'(x) &= 5(3x^2) + 1 + (-1)x^{-1-1} - 3(-\tfrac{1}{2})x^{-1/2-1} \\ &= 15x^2 + 1 - x^{-2} + \tfrac{3}{2}x^{-3/2} \end{aligned}$$

B. The product rule

The derivative of the product of two functions $f(x)$ and $g(x)$ is (in words) $f(x)$ times the derivative of $g(x)$ plus $g(x)$ times the derivative of $f(x)$. Symbolically,

PRODUCT RULE

$$\frac{d}{dx}[f(x)g(x)] = f(x)g'(x) + g(x)f'(x) \tag{4-4}$$

EXAMPLE 4-15: Differentiate $y = (x^3 + 3x - 1)(x^2 - x - 1)$.

Solution:

$$\begin{aligned} &\frac{d}{dx}[(x^3 + 3x - 1)(x^2 - x - 1)] \\ &\qquad = (x^3 + 3x - 1)\frac{d}{dx}(x^2 - x - 1) + (x^2 - x - 1)\frac{d}{dx}(x^3 + 3x - 1) \\ &\qquad = (x^3 + 3x - 1)(2x - 1) + (x^2 - x - 1)(3x^2 + 3) \end{aligned}$$

EXAMPLE 4-16: Differentiate $y = x^2 \sin x$.

Solution:

$$\begin{aligned} \frac{d}{dx}(x^2 \sin x) &= x^2\frac{d}{dx}(\sin x) + \sin x\frac{d}{dx}(x^2) \\ &= x^2 \cos x + \sin x\,(2x) \\ &= x^2 \cos x + 2x \sin x \end{aligned}$$

EXAMPLE 4-17: Differentiate $f(x) = (\cos x)/x^3$.

Solution:

$$\begin{aligned} \frac{d}{dx}\left(\frac{\cos x}{x^3}\right) &= \frac{d}{dx}(x^{-3}\cos x) = x^{-3}\frac{d}{dx}(\cos x) + \cos x\frac{d}{dx}(x^{-3}) \\ &= x^{-3}(-\sin x) + \cos x(-3x^{-4}) = \frac{-\sin x}{x^3} - \frac{3\cos x}{x^4} \end{aligned}$$

C. The quotient rule

The derivative of the quotient of two functions

$$\frac{f(x)}{g(x)}, \qquad \text{where } g(x) \neq 0,$$

is more easily expressed symbolically than it is with words:

QUOTIENT RULE

$$\frac{d}{dx}\left(\frac{f(x)}{g(x)}\right) = \frac{g(x)f'(x) - f(x)g'(x)}{[g(x)]^2} \tag{4-5}$$

However, it will help you to remember the formula if you verbalize it as follows: The derivative of the quotient is the denominator times the derivative of the numerator minus the numerator times the derivative of the denominator all divided by the square of the denominator.

EXAMPLE 4-18: Differentiate $y = (x + 2)/(x - x^2)$.

Solution:

$$\frac{d}{dx}\left(\frac{x+2}{x-x^2}\right) = \frac{(x - x^2)\frac{d}{dx}(x + 2) - (x + 2)\frac{d}{dx}(x - x^2)}{(x - x^2)^2}$$

$$= \frac{(x - x^2)(1 + 0) - (x + 2)(1 - 2x)}{(x - x^2)^2}$$

$$= \frac{(x - x^2) - (x + 2)(1 - 2x)}{(x - x^2)^2}$$

$$= \frac{x^2 + 4x - 2}{(x - x^2)^2}$$

EXAMPLE 4-19: Differentiate $f(x) = x^2/\sin x$.

Solution:

$$\frac{d}{dx}\left(\frac{x^2}{\sin x}\right) = \frac{\sin x\left[\frac{d}{dx}(x^2)\right] - x^2\left[\frac{d}{dx}(\sin x)\right]}{(\sin x)^2}$$

$$= \frac{(\sin x)2x - x^2 \cos x}{\sin^2 x}$$

EXAMPLE 4-20: Differentiate $y = (\ln x)/(x^2 - 2x + 1)$.

Solution:

$$\frac{d}{dx}\left(\frac{\ln x}{x^2 - 2x + 1}\right) = \frac{(x^2 - 2x + 1)\frac{d}{dx}(\ln x) - \ln x\left[\frac{d}{dx}(x^2 - 2x + 1)\right]}{(x^2 - 2x + 1)^2}$$

$$= \frac{(x^2 - 2x + 1)(1/x) - \ln x(2x - 2)}{(x^2 - 2x + 1)^2}$$

D. The chain rule

You'll use the chain rule to find the derivative of a composition of two functions, $f(g(x))$.

Note: Your text may use the notation $(g \circ f)(x)$ to describe composite functions. Simply remember that

$$(g \circ f)(x) = g(f(x))$$

1. The first difficulty you'll encounter is recognizing a composite expression when you see one.

EXAMPLE 4-21: Consider the function $h(x) = (3x - 2)^4$. This function is the composition of two functions $g(x) = 3x - 2$ and $f(x) = x^4$, so

$$h(x) = f(g(x))$$

Think about the steps you'd perform to evaluate $(3x - 2)^4$ at $x = 5$. First you'd find $3 \times 5 - 2 = 13$, and then you'd raise 13 to the fourth power. Thus, the *inner* function is $3x - 2$ and the *outer* function is the fourth power function, x^4.

EXAMPLE 4-22: Consider the function $f(x) = \sin(x^2 + 1)$. This function is a composition of $\sin x$ and $x^2 + 1$. To find $f(2)$ you'd first find $2^2 + 1$ and then apply the sine function to 5. Thus, the inner function is $x^2 + 1$ and the outer function is the sine function.

2. The symbolic statement of the **chain rule** for $f(g(x))$ is

CHAIN RULE
$$\frac{d}{dx} f(g(x)) = f'(g(x))g'(x) \tag{4-6}$$

If you think of $f(g(x))$ as a composition of outer function $f(x)$ and inner function $g(x)$, then the derivative of the composition is the product of the derivative of the outer function, evaluated at the inner function, and the derivative of the inner function.

EXAMPLE 4-23: Differentiate $h(x) = (3x - 2)^4$.

Solution: First recognize that $h(x) = f(g(x))$, where the outer function is $f(x) = x^4$, and the inner function is $g(x) = 3x - 2$. The derivative of the outer function is

$$f'(x) = 4x^3$$

So the derivative of the outer function, evaluated at the inner function, is

$$f'(g(x)) = 4(g(x))^3 = 4(3x - 2)^3$$

The derivative of the inner function is

$$g'(x) = \frac{d}{dx}(3x - 2) = 3$$

By the chain rule, the derivative of $h(x)$ is

$$h'(x) = f'(g(x))g'(x) = 4(3x - 2)^3(3) = 12(3x - 2)^3$$

EXAMPLE 4-24: Differentiate $h(x) = \sin(x^2 + 1)$.

Solution: The outer function is $f(x) = \sin x$; the inner function is $g(x) = x^2 + 1$. The derivative of the outer function is $f'(x) = \cos x$, so

$$f'(g(x)) = \cos(x^2 + 1)$$

The derivative of the inner function is $g'(x) = 2x$, so

$$h'(x) = f'(g(x))g'(x) = \cos(x^2 + 1)2x = 2x\cos(x^2 + 1)$$

EXAMPLE 4-25: Differentiate $y = (x^2 + 5)^3$.

Solution: The outer function is x^3; its derivative is $3x^2$. Thus, the derivative of the outer function, evaluated at the inner function, is $3(x^2 + 5)^2$. The derivative of the inner function is simply $2x$, so

$$\frac{dy}{dx} = 3(x^2 + 5)^2(2x) = 6x(x^2 + 5)^2$$

EXAMPLE 4-26: Differentiate $f(x) = e^{5x-1}$.

Solution: The inner function is $5x - 1$; the outer function is the exponential function, e^x. The derivative of the outer function is e^x, which yields e^{5x-1} when evaluated at the inner function. The derivative of the inner function is 5, so

$$f'(x) = e^{5x-1}(5) = 5e^{5x-1}$$

EXAMPLE 4-27: Differentiate $f(x) = \ln(2 + \sin x)$.

Solution: The outer function is $\ln x$; the inner function is $2 + \sin x$.

$$f'(x) = \begin{pmatrix}\text{derivative of outer function}\\ \text{evaluated at inner function}\end{pmatrix}(\text{derivative of inner function})$$

$$= \left(\frac{1}{2 + \sin x}\right)(\cos x)$$

EXAMPLE 4-28: For $f(x) = (x^{1/2} + x)^3$, find $f'(4)$.

Solution:

$$f'(x) = 3(x^{1/2} + x)^2 \frac{d}{dx}(x^{1/2} + x)$$

$$= 3(x^{1/2} + x)^2\left(\frac{1}{2}x^{-1/2} + 1\right)$$

$$f'(4) = 3(4^{1/2} + 4)^2\left(\tfrac{1}{2}4^{-1/2} + 1\right)$$

$$= 3(2 + 4)^2\left(\frac{1}{2}\cdot\frac{1}{2} + 1\right) = 135$$

EXAMPLE 4-29: Differentiate $f(x) = \sin e^{x^2+2x+5}$.

Solution: Note that this is a composition of more than two functions, but don't panic: It simple requires more than one application of the chain rule. Let the notation guide you. At each stage, simply identify the outer function. The outer function is $\sin x$; the inner function is e^{x^2+2x+5}. Thus,

$$\frac{d}{dx}\sin e^{x^2+2x+5} = \cos e^{x^2+2x+5}\left[\frac{d}{dx}(e^{x^2+2x+5})\right]$$

To find $(d/dx)(e^{x^2+2x+5})$, use the chain rule a second time, with outer function e^x and inner function $x^2 + 2x + 5$:

$$\cos e^{x^2+2x+5}\left[\frac{d}{dx}(e^{x^2+2x+5})\right] = \cos e^{x^2+2x+5}\left[e^{x^2+2x+5}\frac{d}{dx}(x^2 + 2x + 5)\right]$$

$$= \cos e^{x^2+2x+5}[e^{x^2+2x+5}(2x + 2)]$$

EXAMPLE 4-30: Differentiate $y = \cos^5(x^3 + 1)$.

Solution: The outer function is x^5, and the inner function is $\cos(x^3 + 1)$, so

$$\frac{d}{dx}(\cos^5(x^3 + 1)) = 5\cos^4(x^3 + 1)\frac{d}{dx}(\cos(x^3 + 1))$$

Now the outer function is $\cos x$ and the inner function is $x^3 + 1$, so

$$\frac{dy}{dx} = 5\cos^4(x^3 + 1)[-\sin(x^3 + 1)]\frac{d}{dx}(x^3 + 1)$$

$$= -5\cos^4(x^3 + 1)\sin(x^3 + 1)[3x^2] = -15x^2\cos^4(x^3 + 1)\sin(x^3 + 1)$$

4-5. Derivatives Involving Repeated Applications of the Product, Quotient, and Chain Rules

Let's consider a few problems that, at first glance, look more difficult than those we've considered so far. Although you must use more than one rule for each expression, you won't need any new rules or techniques.

EXAMPLE 4-31: Differentiate

$$f(x) = x^2e^x - [(x + 2)/(x - x^2)] + \sin(x^2 + 1).$$

Solution: First recognize $f'(x)$ as a sum of derivatives:

$$f'(x) = \frac{d}{dx}(x^2e^x) - \frac{d}{dx}\left(\frac{x + 2}{x - x^2}\right) + \frac{d}{dx}[\sin(x^2 + 1)]$$

Then use the product rule on the first term, the quotient rule on the second term, and the chain rule on the last term:

$$f'(x) = x^2\left(\frac{d}{dx}e^x\right) + e^x\left(\frac{d}{dx}x^2\right)$$

$$- \frac{(x - x^2)\dfrac{d}{dx}(x + 2) - (x + 2)\dfrac{d}{dx}(x - x^2)}{(x - x^2)^2}$$

$$+ \cos(x^2 + 1)\frac{d}{dx}(x^2 + 1)$$

$$= x^2e^x + 2xe^x - \frac{(x - x^2) - (x + 2)(1 - 2x)}{(x - x^2)^2} + 2x\cos(x^2 + 1)$$

EXAMPLE 4-32: Differentiate $f(x) = \sin\left(\frac{x^2 + 2x}{x - 1}\ln x\right)$.

Solution: In this example, you will find compositions, quotients, and products of functions. It isn't necessary that you envision all of the rules and steps at once to differentiate the function. Rather, you can differentiate the function one step at a time. You must merely recognize the next step, letting the notation guide you.

The function f is a composition of $\sin x$ (the outer function) with $[(x^2 + 2x)/(x - 1)]\ln x$ (the inner function). Apply the chain rule as your first step toward finding $f'(x)$:

$$\frac{d}{dx}\sin x = \cos x$$

so

$$f'(x) = \cos\left(\frac{x^2 + 2x}{x - 1}\ln x\right)\frac{d}{dx}\left(\frac{x^2 + 2x}{x - 1}\ln x\right)$$

Now to differentiate $[(x^2 + 2x)/(x - 1)]\ln x$, you use the product rule, as this is a product of two functions:

$$f'(x) = \cos\left(\frac{x^2 + 2x}{x - 1}\ln x\right)\left[\frac{x^2 + 2x}{x - 1}\frac{d}{dx}(\ln x) + (\ln x)\frac{d}{dx}\left(\frac{x^2 + 2x}{x - 1}\right)\right]$$

To differentiate $(x^2 + 2x)/(x - 1)$ you must employ the quotient rule, as this is a quotient of two functions:

$$f'(x) = \cos\left(\frac{x^2 + 2x}{x - 1}\ln x\right)\left[\frac{x^2 + 2x}{x - 1}\left(\frac{1}{x}\right)\right.$$

$$\left. + (\ln x)\frac{(x - 1)\dfrac{d}{dx}(x^2 + 2x) - (x^2 + 2x)\dfrac{d}{dx}(x - 1)}{(x - 1)^2}\right]$$

$$= \cos\left(\frac{x^2 + 2x}{x - 1}\ln x\right)\left[\frac{x^2 + 2x}{x^2 - x} + (\ln x)\frac{(x - 1)(2x + 2) - (x^2 + 2x)}{(x - 1)^2}\right]$$

EXAMPLE 4-33: Differentiate $f(x) = x^3/\ln(e^x + 1)$.

Solution: The first feature of $f(x)$ that you must recognize is that $f(x)$ is a quotient. Accordingly, the first step toward finding $f'(x)$ is an application of the quotient rule:

$$f'(x) = \frac{\ln(e^x + 1)\dfrac{d}{dx}(x^3) - x^3\dfrac{d}{dx}\ln(e^x + 1)}{[\ln(e^x + 1)]^2}$$

Now to differentiate $\ln(e^x + 1)$ you need the chain rule:

$$f'(x) = \frac{\ln(e^x + 1)3x^2 - x^3\left[\frac{1}{(e^x + 1)}\frac{d}{dx}(e^x + 1)\right]}{\ln^2(e^x + 1)}$$

$$= \frac{3x^2 \ln(e^x + 1) - x^3e^x/(e^x + 1)}{\ln^2(e^x + 1)}$$

EXAMPLE 4-34: Differentiate $f(x) = \sqrt{\sin x \cos x + 2}$.

Solution: You can replace the square root sign with an exponent. Now recognize f as a composite function with outer function $x^{1/2}$ and inner function $\sin x \cos x + 2$:

$$f(x) = (\sin x \cos x + 2)^{1/2}$$

$$f'(x) = \frac{1}{2}(\sin x \cos x + 2)^{1/2}\frac{d}{dx}(\sin x \cos x + 2)$$

Now use the product rule:

$$f(x) = \frac{1}{2}(\sin x \cos x + 2)^{-1/2}\left[\sin x\frac{d}{dx}(\cos x) + \cos x\frac{d}{dx}(\sin x) + \frac{d}{dx}(2)\right]$$

$$= \frac{1}{2}(\sin x \cos x + 2)^{-1/2}[\sin x(-\sin x) + \cos x(\cos x) + 0]$$

$$= \frac{1}{2}(\sin x \cos x + 2)^{-1/2}(\cos^2 x - \sin^2 x)$$

EXAMPLE 4-35: Differentiate $f(x) = \sqrt{x^2 + 3x + 1}\,\sqrt[3]{x^3 - 1}$.

Solution: Rewriting with fractional exponents,

$$f(x) = (x^2 + 3x + 1)^{1/2}(x^3 - 1)^{1/3}$$

you see that $f(x)$ is a product. Your first step then is an application of the product rule:

$$f'(x) = (x^2 + 3x + 1)^{1/2}\frac{d}{dx}(x^3 - 1)^{1/3} + (x^3 - 1)^{1/3}\frac{d}{dx}(x^2 + 3x + 1)^{1/2}$$

Each of the terms to be differentiated now requires the chian rule:

$$f'(x) = (x^2 + 3x + 1)^{1/2}\left[\frac{1}{3}(x^3 - 1)^{-2/3}\frac{d}{dx}(x^3 - 1)\right]$$

$$+ (x^3 - 1)^{1/3}\left[\frac{1}{2}(x^2 + 3x + 1)^{-1/2}\frac{d}{dx}(x^2 + 3x + 1)\right]$$

$$= \frac{1}{3}(x^2 + 3x + 1)^{1/2}(x^3 - 1)^{-2/3}(3x^2)$$

$$+ \frac{1}{2}(x^3 - 1)^{1/3}(x^2 + 3x + 1)^{-1/2}(2x + 3)$$

4-6. The Differentiation Formulas: Using the Chain Rule

A. Rename the inner function.

If $y = f(g(x))$, then the chain rule says

$$\frac{dy}{dx} = (\text{derivative of } f\text{, evaluated at } g)\,(\text{derivative of } g)$$

For notational purposes, rename the inner function $g(x) = u$. Then $y = f(u)$ and the chain rule becomes

$$\frac{dy}{dx} = f'(u)\frac{du}{dx}$$

EXAMPLE 4-36: Differentiate $f(x) = \sin\sqrt{x}$.

Solution: Let $u = \sqrt{x} = x^{1/2}$, the inner function. Then

$$\frac{du}{dx} = \frac{1}{2}x^{-1/2}$$

$$f(x) = \sin u$$

$$f'(x) = \cos u\frac{du}{dx} = \cos\sqrt{x}\left(\frac{1}{2}x^{-1/2}\right)$$

B. The chain rule applied to the basic functions

Reexamine the list of derivatives in Section 4-3. You will find these formulas following, written in a different notation. Each function is a composition of the given outer function with an inner function u.

1. $\frac{d}{dx}(u^n) = nu^{n-1}\frac{du}{dx}$ for n any real number
2. $\frac{d}{dx}(\sin u) = \cos u\frac{du}{dx}$
3. $\frac{d}{dx}(\cos u) = -\sin u\frac{du}{dx}$
4. $\frac{d}{dx}(\tan u) = \sec^2 u\frac{du}{dx}$
5. $\frac{d}{dx}(\sec u) = \sec u\tan u\frac{du}{dx}$
6. $\frac{d}{dx}(\csc u) = -\csc u\cot u\frac{du}{dx}$
7. $\frac{d}{dx}(\cot u) = -\csc^2 u\frac{du}{dx}$
8. $\frac{d}{dx}(a^u) = a^u\ln a\frac{du}{dx}$ for any number $a > 0$
9. $\frac{d}{dx}(e^u) = e^u\frac{du}{dx}$
10. $\frac{d}{dx}(\log_a u) = \left(\frac{1}{u}\right)\frac{1}{\ln a}\frac{du}{dx}$ for $a > 0, a \neq 1$
11. $\frac{d}{dx}(\ln u) = \frac{1}{u}\frac{du}{dx}$
12. $\frac{d}{dx}(\text{arc sin } u) = \frac{1}{\sqrt{1-u^2}}\frac{du}{dx}$
13. $\frac{d}{dx}(\text{arc cos } u) = \frac{-1}{\sqrt{1-u^2}}\frac{du}{dx}$
14. $\frac{d}{dx}(\text{arc tan } u) = \frac{1}{1+u^2}\frac{du}{dx}$
15. $\frac{d}{dx}(\text{arc sec } u) = \frac{1}{|u|\sqrt{u^2-1}}\frac{du}{dx}$
16. $\frac{d}{dx}(\text{arc csc } u) = \frac{-1}{|u|\sqrt{u^2-1}}\frac{du}{dx}$
17. $\frac{d}{dx}(\text{arc cot } u) = \frac{-1}{1+u^2}\frac{du}{dx}$
18. $\frac{d}{dx}(\sinh u) = \cosh u\frac{du}{dx}$
19. $\frac{d}{dx}(\cosh u) = \sinh u\frac{du}{dx}$
20. $\frac{d}{dx}(\tanh u) = \text{sech}^2 u\frac{du}{dx}$
21. $\frac{d}{dx}(\text{sech } u) = -\text{sech } u\tanh u\frac{du}{dx}$
22. $\frac{d}{dx}(\text{csch } u) = -\text{csch } u\coth u\frac{du}{dx}$
23. $\frac{d}{dx}(\coth u) = -\text{csch}^2 u\frac{du}{dx}$

These formulas are not an added complication. They are simply the same formulas you learned in Section 4-3, with the chain rule "built in." Remember that u represents any differentiable function of x. If, for example, $u = x$, then $du/dx = 1$ and these formulas become those you learned in that earlier section.

EXAMPLE 4-37: Differentiate $y = \sin(1/x)$.

Solution: Formula 2 tells us that $\frac{d}{dx}(\sin u) = \cos u \frac{du}{dx}$. So, with $u = 1/x$

$$\frac{d}{dx}\sin\frac{1}{x} = \cos\frac{1}{x}\frac{d}{dx}\frac{1}{x} = \cos\frac{1}{x}\left(\frac{-1}{x^2}\right)$$

EXAMPLE 4-38: Find $\frac{d}{dx}(\tanh xe^x)$.

Solution: Apply formula 20, where the inside function is $u = xe^x$:

$$\begin{aligned}\frac{d}{dx}(\tanh xe^x) &= \frac{d}{dx}(\tanh u) = \text{sech}^2 u\frac{du}{dx}\\ &= \text{sech}^2(xe^x)\frac{d}{dx}(xe^x)\\ &= \text{sech}^2(xe^x)(xe^x + e^x)\end{aligned}$$

Note that you needed the product rule to find $(d/dx)\, xe^x$.

4-7. Higher Order Derivatives

The **second derivative** of a function $y = f(x)$ is the derivative of the first derivative of f. The second derivative of $y = f(x)$ is denoted in a variety of ways:

$$f''(x) \qquad D^2 f \qquad D_x^2 f \qquad \frac{d^2y}{dx^2} \qquad \frac{d^2 f}{dx^2} \qquad \frac{d^2}{dx^2}(f(x)) \qquad y'' \qquad \text{or} \qquad \ddot{y}$$

EXAMPLE 4-39: Find the second derivative of $f(x) = 2x^5 - 4x^2 + 7x + 6$.

Solution:

$$\begin{aligned}f'(x) &= 10x^4 - 8x + 7\\ f''(x) &= \frac{d}{dx}(f'(x)) = \frac{d}{dx}(10x^4 - 8x + 7) = 40x^3 - 8\end{aligned}$$

The ***n*th derivative** of a function $y = f(x)$ (where n is a positive integer) is the derivative of the $(n - 1)$st derivative of f. The nth derivative of f is most often denoted $f^{(n)}(x)$ to avoid the use of many primes.

EXAMPLE 4-40: Find the sixth derivative of $f(x) = 2x^5 - 4x^2 + 7x + 6$.

Solution:

$$\begin{aligned}f'(x) &= 10x^4 - 8x + 7\\ f''(x) &= 40x^3 - 8\\ f^{(3)}(x) &= 120x^2\\ f^{(4)}(x) &= 240x\\ f^{(5)}(x) &= 240\\ f^{(6)}(x) &= 0\end{aligned}$$

EXAMPLE 4-41: Find the higher derivatives of $f(x) = \sin x$.

Solution:

$$\begin{aligned}f'(x) &= \cos x\\ f''(x) &= -\sin x\\ f^{(3)}(x) &= -\cos x\\ f^{(4)}(x) &= \sin x\end{aligned}$$

So the pattern continues.

SUMMARY

1. The derivative of a function is another function, which is defined as the limit of a difference quotient.
2. In practice, you differentiate the basic functions by using the formulas in Section 4-3. It is important that you know these formulas.
3. You may have to express the function in another form to use the formulas. If a function involves radicals, express them as fractional exponents.
4. Use the product, quotient, or chain rules to differentiate functions involving products, quotients or compositions.
5. You must correctly identify composite functions to use the chain rule.
6. You can differentiate a complicated function one step at a time. The notation will guide you.
7. To calculate the nth derivative of a function, differentiate n times.

SOLVED PROBLEMS

PROBLEM 4-1 Use the definition of the derivative to find $f'(x)$, where $f(x) = 3x - 4$.

Solution: First find the difference quotient and simplify:

$$\frac{f(x + \Delta x) - f(x)}{\Delta x} = \frac{[3(x + \Delta x) - 4] - [3x - 4]}{\Delta x}$$

$$= \frac{3x + 3\Delta x - 4 - 3x + 4}{\Delta x} = \frac{3\Delta x}{\Delta x}$$

To find the derivative, take the limit as Δx approaches zero:

$$f'(x) = \lim_{\Delta x \to 0} \left(\frac{3\Delta x}{\Delta x}\right) = 3$$

[See Section 4-2.]

PROBLEM 4-2 Use the definition of the derivative to find $f'(x)$, where $f(x) = 3 - 2x - 2x^2$.

Solution: Find the difference quotient and simplify:

$$\frac{f(x + \Delta x) - f(x)}{\Delta x} = \frac{[3 - 2(x + \Delta x) - 2(x + \Delta x)^2] - [3 - 2x - 2x^2]}{\Delta x}$$

$$= \frac{3 - 2x - 2\Delta x - 2x^2 - 4x\Delta x - 2(\Delta x)^2 - 3 + 2x + 2x^2}{\Delta x}$$

$$= \frac{-2\Delta x - 4x\Delta x - 2(\Delta x)^2}{\Delta x}$$

Apply the limit to find $f'(x)$:

$$f'(x) = \lim_{\Delta x \to 0} \left(\frac{-2\Delta x - 4x\Delta x - 2(\Delta x)^2}{\Delta x}\right)$$

$$= \lim_{\Delta x \to 0} \frac{\Delta x(-2 - 4x - 2\Delta x)}{\Delta x} = -2 - 4x$$

[See Section 4-2.]

PROBLEM 4-3 Use the definition of the derivative to find $f'(x)$, where $f(x) = 3/(2x - 1)$.

Solution: Find the difference quotient:

$$\frac{f(x + \Delta x) - f(x)}{\Delta x} = \frac{\dfrac{3}{2(x + \Delta x) - 1} - \dfrac{3}{2x - 1}}{\Delta x}$$

Find a common denominator and simplify:

$$\frac{f(x+\Delta x)-f(x)}{\Delta x}=\frac{3(2x-1)-3(2x+2\Delta x-1)}{\Delta x(2x-1)(2x+2\Delta x-1)}$$

$$=\frac{-6\Delta x}{\Delta x(2x-1)(2x+2\Delta x-1)}$$

Apply the limit:

$$f'(x)=\lim_{\Delta x\to 0}\frac{f(x+\Delta x)-f(x)}{\Delta x}=\lim_{\Delta x\to 0}\frac{-6\Delta x}{\Delta x(2x-1)(2x+2\Delta x-1)}$$

$$=\frac{-6}{(2x-1)^2}$$

[See Section 4-2.]

PROBLEM 4-4 Find the derivative of $f(x)=1/\sqrt{x}$, using the definition of the derivative.

Solution: Find the difference quotient and rationalize the numerator:

$$\frac{f(x+\Delta x)-f(x)}{\Delta x}=\frac{(1/\sqrt{x+\Delta x})-(1/\sqrt{x})}{\Delta x}$$

$$=\frac{\sqrt{x}-\sqrt{x+\Delta x}}{\Delta x\sqrt{x}\sqrt{x+\Delta x}}$$

$$=\frac{(\sqrt{x}-\sqrt{x+\Delta x})(\sqrt{x}+\sqrt{x+\Delta x})}{(\Delta x\sqrt{x}\sqrt{x+\Delta x})(\sqrt{x}+\sqrt{x+\Delta x})}$$

$$=\frac{x-(x+\Delta x)}{\Delta x\sqrt{x}\sqrt{x+\Delta x}(\sqrt{x}+\sqrt{x+\Delta x})}$$

$$=\frac{-\Delta x}{\Delta x\sqrt{x}\sqrt{x+\Delta x}(\sqrt{x}+\sqrt{x+\Delta x})}$$

Apply the limit:

$$f'(x)=\lim_{\Delta x\to 0}\frac{-\Delta x}{\Delta x\sqrt{x}\sqrt{x+\Delta x}(\sqrt{x}+\sqrt{x+\Delta x})}$$

$$=\lim_{\Delta x\to 0}\frac{-1}{\sqrt{x}\sqrt{x+\Delta x}(\sqrt{x}+\sqrt{x+\Delta x})}$$

$$=\frac{-1}{\sqrt{x}\sqrt{x}(\sqrt{x}+\sqrt{x})}=\frac{-1}{x(2\sqrt{x})}$$

$$=-\frac{1}{2}x^{-3/2}$$

[See Section 4-2.]

PROBLEM 4-5 Use the definition of the derivative to find $f'(x)$, where

$$f(x)=2x^3+x^2-5x+3.$$

Solution:

$$\frac{f(x+\Delta x)-f(x)}{\Delta x}=\frac{[2(x+\Delta x)^3+(x+\Delta x)^2-5(x+\Delta x)+3]-(2x^3+x^2-5x+3)}{\Delta x}$$

$$=\frac{2[(x+\Delta x)^3-x^3]+[(x+\Delta x)^2-x^2]-5[(x+\Delta x)-x]}{\Delta x}$$

$$=\frac{2[3x^2\Delta x+3x(\Delta x)^2+(\Delta x)^3]+[2x\Delta x+(\Delta x)^2]-5\Delta x}{\Delta x}$$

$$=\frac{\Delta x[6x^2+6x\Delta x+2(\Delta x)^2+2x+\Delta x-5]}{\Delta x}$$

$$\begin{aligned} f'(x) &= \lim_{\Delta x \to 0} \frac{f(x + \Delta x) - f(x)}{\Delta x} \\ &= \lim_{\Delta x \to 0} \frac{\Delta x[6x^2 + 6x\Delta x + 2(\Delta x)^2 + 2x + \Delta x - 5]}{\Delta x} \\ &= \lim_{\Delta x \to 0} [6x^2 + 6x\Delta x + 2(\Delta x)^2 + 2x + \Delta x - 5] \\ &= 6x^2 + 2x - 5 \end{aligned}$$

[See Section 4-2.]

PROBLEM 4-6 Find the derivative of $f(x) = 3x^4 + x^3 - 2x^2 - 4x + 5$.

Solution: The derivative of a sum is the sum of the derivatives:

$$f'(x) = \frac{d}{dx}(3x^4) + \frac{d}{dx}(x^3) - \frac{d}{dx}(2x^2) - \frac{d}{dx}(4x) + \frac{d}{dx}(5)$$

and the derivative of a constant times a function is the constant times the derivative of the function:

$$\begin{aligned} f'(x) &= 3\frac{d}{dx}(x^4) + \frac{d}{dx}(x^3) - 2\frac{d}{dx}(x^2) - 4\frac{d}{dx}(x) + \frac{d}{dx}(5) \\ &= 3(4x^3) + 3x^2 - 2(2x) - 4(1) + 0 \\ &= 12x^3 + 3x^2 - 4x - 4 \end{aligned}$$

[See Section 4-4.]

PROBLEM 4-7 Find dy/dx where $y = -2x^7 + 12x^4 - 11x^3 + 3x + 2$.

Solution: You proceed as in the last problem:

$$\begin{aligned} \frac{dy}{dx} &= -2\frac{d}{dx}(x^7) + 12\frac{d}{dx}(x^4) - 11\frac{d}{dx}(x^3) + 3\frac{d}{dx}(x) + \frac{d}{dx}(2) \\ &= -2(7x^6) + 12(4x^3) - 11(3x^2) + 3(1) + 0 \\ &= -14x^6 + 48x^3 - 33x^2 + 3 \end{aligned}$$

[See Section 4-4.]

PROBLEM 4-8 Differentiate $y = (2x^3/7) + (5x^2/9) + (2x/5) - 1$.

Solution: Because the denominators are constants, you don't need the quotient rule:

$$\begin{aligned} \frac{dy}{dx} &= \left(\frac{2}{7}\right)\frac{d}{dx}(x^3) + \left(\frac{5}{9}\right)\frac{d}{dx}(x^2) + \left(\frac{2}{5}\right)\frac{d}{dx}(x) - \frac{d}{dx}(1) \\ &= \left(\frac{2}{7}\right)3x^2 + \left(\frac{5}{9}\right)2x + \frac{2}{5} - 0 = \frac{6}{7}x^2 + \frac{10}{9}x + \frac{2}{5} \end{aligned}$$

[See Section 4-4.]

PROBLEM 4-9 Differentiate $f(x) = 8x^{5/3} + (2/x) - (3/x^2)$.

Solution: First write each of the terms as a power of x:

$$\begin{aligned} f(x) &= 8x^{5/3} + 2x^{-1} - 3x^{-2} \\ f'(x) &= 8\left(\frac{5}{3}x^{(5/3)-1}\right) + 2(-1)x^{-1-1} - 3(-2)x^{-2-1} \\ &= \frac{40}{3}x^{2/3} - 2x^{-2} + 6x^{-3} \end{aligned}$$

[See Section 4-4.]

PROBLEM 4-10 Differentiate $y = \sqrt[3]{x^5}/\sqrt[5]{x^3}$.

Solution: This function can be written as a power of x:

$$y = \frac{x^{5/3}}{x^{3/5}} = x^{5/3-3/5} = x^{16/15}$$

$$\frac{dy}{dx} = \frac{16}{15}x^{1/15}$$

[See Section 4-4.]

PROBLEM 4-11 Differentiate $f(x) = (x^3 + x^{-3})(x^2 - 3)$.

Solution: Because this is a product of two functions, you should use the product rule:

$$f'(x) = (x^3 + x^{-3})\frac{d}{dx}(x^2 - 3) + (x^2 - 3)\frac{d}{dx}(x^3 + x^{-3})$$

$$= (x^3 + x^{-3})(2x) + (x^2 - 3)(3x^2 - 3x^{-4})$$

[See Section 4-4.]

PROBLEM 4-12 Differentiate $f(x) = x^3e^x$.

Solution: Use the product rule:

$$f'(x) = x^3\frac{d}{dx}(e^x) + e^x\frac{d}{dx}(x^3) = x^3e^x + e^x(3x^2)$$

[See Section 4-4.]

PROBLEM 4-13 Differentiate $f(t) = t \arctan t$.

Solution: Use the product rule:

$$f'(x) = t\frac{d}{dt}(\arctan t) + \arctan t\frac{d}{dt}(t)$$

$$= \frac{t}{1 + t^2} + \arctan t$$

[See Section 4-4.]

PROBLEM 4-14 Differentiate $y = (2x + 5)/(x - 3)$.

Solution: Because this is a quotient of two functions, you use the quotient rule:

$$\frac{dy}{dx} = \frac{(x - 3)\frac{d}{dx}(2x + 5) - (2x + 5)\frac{d}{dx}(x - 3)}{(x - 3)^2}$$

$$= \frac{(x - 3)(2) - (2x + 5)(1)}{(x - 3)^2}$$

and finally simplify:

$$\frac{dy}{dx} = \frac{-11}{(x - 3)^2}$$

As an alternative, you can write y as a product and use the product rule:

$$\frac{dy}{dx} = \frac{d}{dx}[(2x + 5)(x - 3)^{-1}]$$

$$= (2x + 5)\frac{d}{dx}(x - 3)^{-1} + (x - 3)^{-1}\frac{d}{dx}(2x + 5)$$

To differentiate $(x - 3)^{-1}$, you use the chain rule:

$$\frac{dy}{dx} = (2x + 5)\left[-1(x - 3)^{-2}\frac{d}{dx}(x - 3)\right] + (x - 3)^{-1}(2)$$

$$= (2x + 5)[-(x - 3)^{-2}] + 2(x - 3)^{-1}$$

$$= \frac{-(2x + 5)}{(x - 3)^2} + \frac{2}{x - 3} = \frac{-2x - 5 + 2(x - 3)}{(x - 3)^2}$$

$$= \frac{-11}{(x - 3)^2}$$

You can use either of these methods to differentiate a quotient—it's up to you. [See Section 4-4.]

PROBLEM 4-15 Differentiate $y = (1 + \tan x)/\sin x$.

Solution: Use the quotient rule:

$$\frac{dy}{dx} = \frac{\sin x \dfrac{d}{dx}(1 + \tan x) - (1 + \tan x)\dfrac{d}{dx}(\sin x)}{(\sin x)^2}$$

$$= \frac{\sin x(0 + \sec^2 x) - (1 + \tan x)\cos x}{\sin^2 x}$$

$$= \frac{\sin x \sec^2 x - \cos x - \tan x \cos x}{\sin^2 x}$$

[See Section 4-4.]

PROBLEM 4-16 Differentiate $f(x) = (x^5 - 3x + 2)^{17}$.

Solution: Because this function is the composition of two functions, $x^5 - 3x + 2$, the inner function, with the seventeenth power function (x^{17}), the outer function, you can use the chain rule: $f'(x)$ is the derivative of x^{17}, evaluated at $x^5 - 3x + 2$, times the derivative of $x^5 - 3x + 2$.

$$f'(x) = 17(x^5 - 3x + 2)^{16}\frac{d}{dx}(x^5 - 3x + 2)$$

$$= 17(x^5 - 3x + 2)^{16}(5x^4 - 3)$$

[See Section 4-4.]

PROBLEM 4-17 Differentiate $y = \sin^3 x$.

Solution: You'll recognize y as a composite function if you rewrite it as $y = (\sin x)^3$. Now you see that y is a composition of $\sin x$, the inner function, with x^3, the outer function. Use the chain rule:

$$\frac{dy}{dx} = 3(\sin x)^2\frac{d}{dx}(\sin x) = 3\sin^2 x \cos x$$

[See Section 4-4.]

PROBLEM 4-18 Differentiate $f(x) = e^{x^2+3x+1}$.

Solution: This is a composition of the exponential function, the outer function, with $x^2 + 3x + 1$, the inner function, so

$$f'(x) = e^{x^2+3x+1}\frac{d}{dx}(x^2 + 3x + 1) = e^{x^2+3x+1}(2x + 3)$$

[See Section 4-4.]

PROBLEM 4-19 Differentiate $y = \log_3(x^2 + 2x + 3)$.

Solution: Use the chain rule:

$$\frac{dy}{dx} = \frac{1}{x^2 + 2x + 3}\left(\frac{1}{\ln 3}\right)\frac{d}{dx}(x^2 + 2x + 3)$$

$$= \frac{1}{x^2 + 2x + 3}\left(\frac{1}{\ln 3}\right)(2x + 2) = \frac{2x + 2}{(x^2 + 2x + 3)\ln 3}$$

[See Section 4-4.]

PROBLEM 4-20 Differentiate $f(x) = 3/(x^2 + 1)$.

Solution: The easiest way to differentiate this function is to rewrite it as $f(x) = 3(x^2 + 1)^{-1}$ and use the chain rule:

$$f'(x) = 3(-1)(x^2 + 1)^{-2}\frac{d}{dx}(x^2 + 1)$$

$$= -3(x^2 + 1)^{-2}(2x) = \frac{-6x}{(x^2 + 1)^2}$$

You could also differentiate f using the quotient rule:

$$f'(x) = \frac{(x^2 + 1)\dfrac{d}{dx}(3) - 3\dfrac{d}{dx}(x^2 + 1)}{(x^2 + 1)^2}$$

$$= \frac{0 - 3(2x)}{(x^2 + 1)^2} = \frac{-6x}{(x^2 + 1)^2}$$

[See Section 4-4.]

PROBLEM 4-21 Differentiate $f(x) = \ln(1 + \sin^2 x)$.

Solution: Use the chain rule:

$$f'(x) = \frac{1}{1 + \sin^2 x}\left[\frac{d}{dx}(1 + \sin^2 x)\right]$$

$$= \frac{1}{1 + \sin^2 x}(2 \sin x \cos x)$$

[See Section 4-4.]

PROBLEM 4-22 Differentiate $y = (x^2 + x - 3)^{12}(2x^3 - 5x^2 + 9x + 1)^4$.

Solution: Use the product rule:

$$\frac{dy}{dx} = (x^2 + x - 3)^{12}\frac{d}{dx}(2x^3 - 5x^2 + 9x + 1)^4 + (2x^3 - 5x^2 + 9x + 1)\frac{d}{dx}(x^2 + x - 3)^{12}$$

$$= (x^2 + x - 3)^{12}4(2x^3 - 5x^2 + 9x + 1)^3\frac{d}{dx}(2x^3 - 5x^2 + 9x + 1)$$

$$+ (2x^3 - 5x^2 + 9x + 1)^4 12(x^2 + x - 3)^{11}\frac{d}{dx}(x^2 + x - 3)$$

$$= (x^2 + x - 3)^{12}4(2x^3 - 5x^2 + 9x + 1)^3(6x^2 - 10x + 9)$$

$$+ (2x^3 - 5x^2 + 9x + 1)^4 12(x^2 + x - 3)^{11}(2x + 1)$$

You may want to simplify this expression:

$$\frac{dy}{dx} = 4(x^2 + x - 3)^{11}(2x^3 - 5x^2 + 9x + 1)^3$$

$$\times [(x^2 + x - 3)(6x^2 - 10x + 9) + 3(2x^3 - 5x^2 + 9x + 1)(2x + 1)]$$

$$= 4(x^2 + x - 3)^{11}(2x^3 - 5x^2 + 9x + 1)^3(18x^4 - 28x^3 + 20x^2 + 72x - 24)$$

$$= 8(x^2 + x - 3)^{11}(2x^3 - 5x^2 + 9x + 1)^3(9x^4 - 14x^3 + 10x^2 + 36x - 12)$$

[See Section 4-5.]

PROBLEM 4-23 Differentiate $f(x) = \ln(\log_{10} x)$.

Solution: Use the chain rule:

$$f'(x) = \frac{1}{\log_{10} x}\left[\frac{d}{dx}(\log_{10} x)\right]$$

$$= \frac{1}{\log_{10} x}\left(\frac{1}{x}\right)\frac{1}{\ln 10}$$

[See Section 4-4.]

PROBLEM 4-24 Differentiate $f(x) = x^2\sqrt{x^2 - 4}$.

Solution: Use the product rule:

$$f'(x) = x^2\frac{d}{dx}(x^2 - 4)^{1/2} + (x^2 - 4)^{1/2}\frac{d}{dx}(x^2)$$

$$= x^2\left[\frac{1}{2}(x^2 - 4)^{-1/2}\frac{d}{dx}(x^2 - 4)\right] + (x^2 - 4)^{1/2}(2x)$$

$$= \frac{1}{2}x^2(x^2 - 4)^{-1/2}(2x) + 2x(x^2 - 4)^{1/2}$$

Simplify:

$$f'(x) = \frac{x^3}{\sqrt{x^2 - 4}} + \frac{2x(x^2 - 4)}{\sqrt{x^2 - 4}} = \frac{3x^3 - 8x}{\sqrt{x^2 - 4}}$$

[See Section 4-5.]

PROBLEM 4-25 Differentiate $f(x) = e^{\sin(x^2 + 1)}$.

Solution: Use the chain rule:

$$f'(x) = e^{\sin(x^2+1)} \frac{d}{dx}(\sin(x^2 + 1))$$

$$= e^{\sin(x^2+1)}\cos(x^2 + 1)\frac{d}{dx}(x^2 + 1)$$

$$= e^{\sin(x^2+1)}[\cos(x^2 + 1)](2x)$$

[See Section 4-5.]

PROBLEM 4-26 Differentiate $f(x) = \sqrt{(x^2 + 1)/(x - 5)}$.

Solution: You can differentiate $f(x) = [(x^2 + 1)/(x - 5)]^{1/2}$ using the chain and quotient rules:

$$f'(x) = \frac{1}{2}\left(\frac{x^2 + 1}{x - 5}\right)^{-1/2} \frac{d}{dx}\left(\frac{x^2 + 1}{x - 5}\right)$$

$$= \frac{1}{2}\left(\frac{x^2 + 1}{x - 5}\right)^{-1/2}\left(\frac{(x - 5)\frac{d}{dx}(x^2 + 1) - (x^2 + 1)\frac{d}{dx}(x - 5)}{(x - 5)^2}\right)$$

$$= \frac{1}{2}\left(\frac{x^2 + 1}{x - 5}\right)^{-1/2}\left[\frac{(x - 5)(2x) - (x^2 + 1)}{(x - 5)^2}\right] = \frac{1}{2}\left(\frac{x^2 + 1}{x - 5}\right)^{-1/2}\frac{x^2 - 10x - 1}{(x - 5)^2}$$

You can also differentiate f by rewriting it in the form,

$$f(x) = (x^2 + 1)^{1/2}(x - 5)^{-1/2}$$

and using the product formula.

[See Section 4-5.]

PROBLEM 4-27 Differentiate $f(x) = \sec(\pi\sqrt{x^2 - 1})$.

Solution: Use the chain rule:

$$f'(x) = \sec(\pi\sqrt{x^2 - 1})\tan(\pi\sqrt{x^2 - 1})\frac{d}{dx}(\pi\sqrt{x^2 - 1})$$

$$= \sec(\pi\sqrt{x^2 - 1})\tan(\pi\sqrt{x^2 - 1})\pi\left[\frac{1}{2}(x^2 - 1)^{-1/2}\frac{d}{dx}(x^2 - 1)\right]$$

$$= \sec(\pi\sqrt{x^2 - 1})\tan(\pi\sqrt{x^2 - 1})\frac{\pi x}{\sqrt{x^2 - 1}}$$

[See Section 4-5.]

PROBLEM 4-28 Differentiate $f(x) = \cosh 2x \sinh x$.

Solution: Use the product rule:

$$f'(x) = \cosh 2x \frac{d}{dx}(\sinh x) + \sinh x \frac{d}{dx}(\cosh 2x)$$

$$= \cosh 2x \cosh x + 2 \sinh x \sinh 2x$$

[See Section 4-5.]

PROBLEM 4-29 Find the third derivative of $f(x) = 4x^{3/2} - 3x^4 + x^2 - 1$.

Solution: Differentiate three times:

$$f'(x) = 6x^{1/2} - 12x^3 + 2x$$

$$f''(x) = 3x^{-1/2} - 36x^2 + 2$$

$$f'''(x) = -\frac{3}{2}x^{-3/2} - 72x$$

[See Section 4-7.]

PROBLEM 4-30 Find the second derivative of $y = \sec^2 x$.

Solution: Differentiate twice:

$$\frac{dy}{dx} = 2 \sec x \frac{d}{dx}(\sec x) = 2 \sec x \sec x \tan x = 2 \sec^2 x \tan x$$

$$\frac{d^2y}{dx^2} = 2 \sec^2 x \frac{d}{dx}(\tan x) + 2 \tan x \frac{d}{dx}(\sec^2 x)$$

$$= 2 \sec^2 x \sec^2 x + (2 \tan x)(2 \sec x)(\sec x \tan x)$$

$$= 2 \sec^4 x + 4 \sec^2 x \tan^2 x$$

[See Section 4-7.]

PROBLEM 4-31 Find the sixth derivative of $f(x) = \ln(1 + x)$.

Solution: Differentiate six times:

$$f'(x) = \frac{1}{1+x} = (1 + x)^{-1}$$

$$f''(x) = -(1 + x)^{-2}$$

$$f'''(x) = 2(1 + x)^{-3}$$

$$f^{(4)}(x) = -6(1 + x)^{-4}$$

$$f^{(5)}(x) = 24(1 + x)^{-5}$$

$$f^{(6)}(x) = -120(1 + x)^{-6}$$

[See Section 4-7.]

Supplementary Exercises

In Problems 4-32 through 4-41 use the definition of the derivative to find the derivative of $f(x)$.

4-32 $f(x) = 5x + 4$

4-33 $f(x) = 2x^2 + x - 1$

4-34 $f(x) = 3x^2 - 12x + 1$

4-35 $f(x) = 1/(x + 1)$

4-36 $f(x) = \sqrt{2x - 1}$

4-37 $f(x) = x^3 + x^2 - 4$

4-38 $f(x) = x + (1/x)$

4-39 $f(x) = \sqrt{x} - 2x$

4-40 $f(x) = [3/(x - 1)] + \sqrt{x - 1}$

4-41 $f(x) = 2/(x^2 + 1)$

In Problems 4-42 through 4-127 find the derivative of the given function.

4-42 $f(x) = 3x^3 + 5x^2 - 7x + 4$

4-43 $f(x) = x - (1/x)$

4-44 $f(x) = 5x^2 - 13x + 6$

4-45 $y = x^4 + 2x^3 + 5x^2 + 6x + 5$

4-46 $y = 4/\sqrt[5]{x^2}$

4-47 $y = \sqrt{x^3}/\sqrt[3]{x^2}$

4-48 $f(x) = (x^3\sqrt{x^5})/\sqrt{x}$

4-49 $f(x) = x^2 + 2^x$

4-50 $f(x) = 3 \sin x + 4 \tan x$

4-51 $y = x \sin x$

4-52 $y = x^2e^x + 3xe^{-x} + 4$

4-53 $y = \arcsin x + \arccos x$

4-54 $f(x) = (x^2 + 1)/(x^2 - 1)$

4-55 $y = (x + \sqrt{x})/(x - \sqrt{x})$

4-56 $y = x/\ln x$

4-57 $f(x) = (\arcsin x)/x$

4-58 $f(x) = (\ln 2)/\ln x$

4-59 $f(x) = (x^3 - 3x + 7) \times (5 - 2\sqrt{x} + 3x)$

4-60 $y = (x^2 + x - 1)/(3 - x)$

4-61 $y = 1/(1 + \cos x)$

4-62 $y = \sin x/(\sin x + \cos x)$

4-63 $f(x) = (x^3 + 1)^{11}$

4-64 $f(x) = x^2 + \sqrt{x - 4}$

4-65 $f(x) = (1/\sqrt{x}) + \sqrt[3]{x + 2}$

4-66 $f(x) = \sqrt{\sin x}$

4-67 $y = \ln(x^2 + 2)$

4-68 $y = \tan^3 x$

4-69 $y = e^{\arctan x}$

4-70 $y = \sqrt{1 + e^x}$

4-71 $y = \sqrt[3]{x^3 + x - 5}$

4-72 $y = \cosh(\sinh x)$

4-73 $y = e^{\sqrt{2}}$

4-74 $f(x) = 2^{\sin x}$

4-75 $f(x) = \sin(2^x)$

4-76 $f(x) = \sin^x 2$

4-77 $f(x) = 5/(x^2 - 2)$

4-78 $y = \sqrt{\sin x + \cos x}$

4-79 $y = 2^{1/x}$

4-80 $y = \cot(1/x)$

4-81 $y = (x^2 - 4x^{7/3} + 3)^{-2}$

4-82 $y = \operatorname{arc\,sec} 4x$

4-83 $y = \sin xe^x$

4-84 $y = \sqrt{\sin \sqrt{x}}$

4-85 $f(x) = x^2 \ln \sin x$

4-86 $f(x) = e^{\sqrt{x}}/(1 + \sqrt{x})$

4-87 $f(x) = 2x^{3/2} - 5x^{1/2} + 7x^{-1/2}$

4-88 $y = \dfrac{1}{e^{\sin x}}$

4-89 $y = (2x^2 + 3x + 4)/[(x - 5)^2]$

4-90 $y = \sqrt{(x - 1)/(x + 1)}$

4-91 $y = 1/(2x - \sqrt{x^2 - 1})$

4-92 $f(x) = \cos \sqrt{1 - x^2}$

4-93 $f(x) = (2x^3 - 11)(x^2 - 3)$

4-94 $f(x) = \ln(x + \sqrt{1 + x^2})$

4-95 $f(x) = (e^{2x} - e^{-2x})/(e^{2x} + e^{-2x})$

4-96 $y = [\ln(2x - 1)]^2$

4-97 $y = \ln \ln(x + e^x)$

4-98 $y = \log_5 (\sqrt{x} + x^2)$

4-99 $y = 10^x \log_{10} x$

4-100 $y = \sqrt[3]{x^2 - 1}\sqrt{1 + \sqrt{x}}$

4-101 $y = [(\sin x)/x]^5$

4-102 $y = (\ln x)\ln(x + 1)$

4-103 $y = \ln[x \ln(x + 1)]$

4-104 $y = \tan^3(\ln x)$

4-105 $y = \csc e^x + e^{\csc x}$

4-106 $y = \ln^3(2 + \sin x)$

4-107 $y = \ln[(x^3 \sin x)/(x^2 + 1)]$

4-108 $y = e^{e^x}$

4-109 $y = e^{x^e}$

4-110 $f(x) = \sin(\sin x) + \sin^2 x$

4-111 $f(x) = \ln(2 + e^3)$

4-112 $y = \tanh(3 - 5x)$

4-113 $y = \sqrt{\log_3 x}$

4-114 $f(x) = \sqrt{x - 1}/\sqrt[3]{x + 1}$

4-115 $f(x) = (5/6)\sqrt{1 + \sin^2 x}$

4-116 $y = \sqrt{x}\sqrt{x+1}$

4-117 $y = x \sin x \sinh x$

4-118 $y = 5/[(1 + \tanh x)^2]$

4-119 $y = (e^{2x})/(1 + e^{3x})$

4-120 $y = \sqrt{x}e^{\sqrt{x}}$

4-121 $y = \sec^4 x \tan x$

4-122 $y = \ln(e^{3x})$

4-123 $y = 5^{\cosh x}$

4-124 $y = \sqrt{e^x - (3/x)}$

4-125 $y = \sin 5$

4-126 $y = \tan^2(3x)$

4-127 $y = \sqrt{\sin(3x - 4)}$

In Problems 4-128 through 4-139 find the second derivative:

4-128 $f(x) = x^5 - 3x^3 + 2x - 1$

4-129 $f(x) = x^3 - 2x^{-1}$

4-130 $y = \sqrt{x} + (1/\sqrt{x})$

4-131 $y = \sqrt[3]{x^7} - x$

4-132 $y = e^{4x-1}$

4-133 $y = (x + 1)/(x - 1)$

4-134 $f(x) = \tan x$

4-135 $f(x) = 2^x$

4-136 $f(x) = \arctan x$

4-137 $y = xe^x$

4-138 $y = \ln(1 + 2x)$

4-139 $y = 3^{x^2}$

In Problems 4-140 through 4-150 find the fourth derivative:

4-140 $f(x) = x^5 - x^4 + x^3 + 2x - 1$

4-141 $f(x) = 3e^{2x}$

4-142 $f(x) = \ln x$

4-143 $f(x) = \sin 3x + \cos 2x$

4-144 $f(x) = 1/(1 - x)$

4-145 $f(x) = \cosh(1 - 2x)$

4-146 $f(x) = \ln(\cos x)$

4-147 $f(x) = e^{x^2}$

4-148 $f(x) = 3^{2x+5}$

4-149 $f(x) = x \arctan x - \frac{1}{2}\ln(1 + x^2)$

4-150 $f(x) = 5x^{8/3}$

Solutions to Supplementary Exercises

(4-32) $$f'(x) = \lim_{\Delta x \to 0} \frac{[5(x + \Delta x) + 4] - (5x + 4)}{\Delta x} = 5$$

(4-33) $$f'(x) = \lim_{\Delta x \to 0} \frac{[2(x + \Delta x)^2 + (x + \Delta x) - 1] - (2x^2 + x - 1)}{\Delta x} = 4x + 1$$

(4-34) $$f'(x) = \lim_{\Delta x \to 0} \frac{[3(x + \Delta x)^2 - 12(x + \Delta x) + 1] - (3x^2 - 12x + 1)}{\Delta x} = 6x - 12$$

(4-35) $$f'(x) = \lim_{\Delta x \to 0} \frac{\dfrac{1}{x + \Delta x + 1} - \dfrac{1}{x + 1}}{\Delta x} = \lim_{\Delta x \to 0} \frac{(x + 1) - (x + \Delta x + 1)}{\Delta x(x + 1)(x + \Delta x + 1)}$$

$$= \frac{-1}{(x + 1)^2}$$

(4-36) $f'(x) = \lim_{\Delta x \to 0} \dfrac{\sqrt{2(x+\Delta x)-1} - \sqrt{2x-1}}{\Delta x}$

$= \lim_{\Delta x \to 0} \dfrac{2(x+\Delta x) - 1 - (2x-1)}{\Delta x[\sqrt{2(x+\Delta x)-1} + \sqrt{2x-1}]} = \dfrac{1}{\sqrt{2x-1}}$

(4-37) $f'(x) = \lim_{\Delta x \to 0} \dfrac{[(x+\Delta x)^3 + (x+\Delta x)^2 - 4] - (x^3 + x^2 - 4)}{\Delta x} = 3x^2 + 2x$

(4-38) $f'(x) = \lim_{\Delta x \to 0} \dfrac{[(x+\Delta x) + 1/(x+\Delta x)] - (x + 1/x)}{\Delta x}$

$= \lim_{\Delta x \to 0} \dfrac{\Delta x + \dfrac{x - (x+\Delta x)}{x(x+\Delta x)}}{\Delta x} = 1 - \dfrac{1}{x^2}$

(4-39) $f'(x) = \lim_{\Delta x \to 0} \dfrac{[\sqrt{x+\Delta x} - 2(x+\Delta x)] - (\sqrt{x} - 2x)}{\Delta x}$

$= \lim_{\Delta x \to 0} \left[\dfrac{x+\Delta x - x}{\Delta x(\sqrt{x+\Delta x} + \sqrt{x})} - \dfrac{2\Delta x}{\Delta x}\right] = \dfrac{1}{2\sqrt{x}} - 2$

(4-40) $f'(x) = \lim_{\Delta x \to 0} \dfrac{\left[\dfrac{3}{x+\Delta x-1} + \sqrt{x+\Delta x-1}\right] - \left[\dfrac{3}{x-1} + \sqrt{x-1}\right]}{\Delta x}$

$= \lim_{\Delta x \to 0} \left[\dfrac{3(x-1) - 3(x+\Delta x-1)}{\Delta x(x-1)(x+\Delta x-1)} + \dfrac{(x+\Delta x-1) - (x-1)}{\Delta x(\sqrt{x+\Delta x-1} + \sqrt{x-1})}\right]$

$= \dfrac{-3}{(x-1)^2} + \dfrac{1}{2\sqrt{x-1}}$

(4-41) $f'(x) = \lim_{\Delta x \to 0} \dfrac{\dfrac{2}{(x+\Delta x)^2+1} - \dfrac{2}{x^2+1}}{\Delta x}$

$= \lim_{\Delta x \to 0} \dfrac{2(x^2+1) - 2[(x+\Delta x)^2+1]}{\Delta x(x^2+1)[(x+\Delta x)^2+1]} = \dfrac{-4x}{(x^2+1)^2}$

(4-42) $f'(x) = 9x^2 + 10x - 7$

(4-43) $f'(x) = 1 + (1/x^2)$

(4-44) $f'(x) = 10x - 13$

(4-45) $y' = 4x^3 + 6x^2 + 10x + 6$

(4-46) $y' = (-8/5)x^{-7/5}$

(4-47) $y' = (5/6)x^{-1/6}$

(4-48) $y' = 5x^4$

(4-49) $f'(x) = 2x + (\ln 2)2^x$

(4-50) $f'(x) = 3 \cos x + 4 \sec^2 x$

(4-51) $y' = \sin x + x \cos x$

(4-52) $y' = 2xe^x + x^2e^x + 3e^{-x} - 3xe^{-x}$

(4-53) $y' = 0$

(4-54) $f'(x) = -4x/[(x^2 - 1)^2]$

(4-55) $y' = -\sqrt{x}/[(x - \sqrt{x})^2]$

(4-56) $y' = (\ln x - 1)/(\ln^2 x)$

(4-57) $f'(x) = (1/x^2)[(x/\sqrt{1 - x^2}) - \text{arc sin } x]$

(4-58) $f'(x) = (-\ln 2)/(x \ln^2 x)$

(4-59) $f'(x) = (x^3 - 3x + 7)(-x^{-1/2} + 3) + (5 - 2x^{1/2} + 3x)(3x^2 - 3)$

(4-60) $y' = \dfrac{(3 - x)(2x + 1) + (x^2 + x - 1)}{(3 - x)^2} = \dfrac{-x^2 + 6x + 2}{(3 - x)^2}$

(4-61) $y' = (\sin x)/[(1 + \cos x)^2]$

(4-62) $y' = \dfrac{(\sin x + \cos x)\cos x - \sin x(\cos x - \sin x)}{(\sin x + \cos x)^2} = \dfrac{1}{(\sin x + \cos x)^2}$

(4-63) $f'(x) = 33x^2(x^3 + 1)^{10}$

(4-64) $f'(x) = 2x + \frac{1}{2}(x - 4)^{-1/2}$

(4-65) $f'(x) = -\frac{1}{2}x^{-3/2} + \frac{1}{3}(x + 2)^{-2/3}$

(4-66) $f'(x) = \frac{1}{2}\sin^{-1/2} x \cos x$

(4-67) $y' = 2x/(x^2 + 2)$

(4-68) $y' = 3 \tan^2 x \sec^2 x$

(4-69) $y' = [1/(1 + x^2)]\, e^{\text{arc tan } x}$

(4-70) $y' = \frac{1}{2}(1 + e^x)^{-1/2} e^x$

(4-71) $y' = \frac{1}{3}(x^3 + x - 5)^{-2/3}(3x^2 + 1)$

(4-72) $y' = \sinh(\sinh x)\cosh x$

(4-73) $y' = 0$

(4-74) $f'(x) = \ln 2(\cos x)2^{\sin x}$

(4-75) $f'(x) = (\ln 2)2^x \cos 2^x$

(4-76) $f'(x) = (\ln \sin 2)\sin^x 2$

(4-77) $f'(x) = -10x/[(x^2 - 2)^2]$

(4-78) $y' = \frac{1}{2}(\sin x + \cos x)^{-1/2}(\cos x - \sin x)$

(4-79) $y' = [-\ln 2/(x^2)]2^{1/x}$

(4-80) $y' = [\csc^2(1/x)]/(x^2)$

(4-81) $y' = -2(x^2 - 4x^{7/3} + 3)^{-3}[2x - (28/3)x^{4/3}]$

(4-82) $y' = 1/(|x|\sqrt{16x^2 - 1})$

(4-83) $y' = \cos(xe^x)(e^x + xe^x)$

(4-84) $y' = \frac{1}{2}(\sin x^{1/2})^{-1/2}\cos(x^{1/2})\frac{1}{2}x^{-1/2} = (\cos \sqrt{x})/(4\sqrt{x \sin\sqrt{x}})$

(4-85) $f'(x) = 2x \ln \sin x + x^2[(\cos x)/(\sin x)]$

(4-86) $f'(x) = e^{\sqrt{x}}/[2(1 + \sqrt{x})^2]$

(4-87) $f'(x) = 3x^{1/2} - (5/2)x^{-1/2} - (7/2)x^{-3/2}$

(4-88) $y' = (-\cos x)/(e^{\sin x})$

(4-89) $y' = -23[(x + 1)/(x - 5)^3]$

(4-90) $y' = (x + 1)^{-2}[(x - 1)/(x + 1)]^{-1/2} = 1/[(x + 1)^2\sqrt{x^2 - 1}]$

(4-91) $f'(x) = [-1/(2x - \sqrt{x^2 - 1})^2][2 - (x/\sqrt{x^2 - 1})]$

(4-92) $f'(x) = (x \sin\sqrt{1 - x^2})/(\sqrt{1 - x^2})$

(4-93) $f'(x) = 2x(2x^3 - 11) + 6x^2(x^2 - 3) = 10x^4 - 18x^2 - 22x$

(4-94) $f'(x) = [1/(x + \sqrt{1 + x^2})][1 + (x/\sqrt{1 + x^2})] = 1/\sqrt{1 + x^2}$

(4-95) $f'(x) = 8/[(e^{2x} + e^{-2x})^2]$

(4-96) $y' = [4 \ln(2x - 1)]/(2x - 1)$

(4-97) $y' = (1 + e^x)/[(x + e^x)\ln(x + e^x)]$

(4-98) $y' = (\frac{1}{2}x^{-1/2} + 2x)/[\ln 5(x^{1/2} + x^2)]$

(4-99) $y' = 10^x \ln 10 \log_{10} x + [10^x/(x \ln 10)]$

(4-100) $y' = \frac{2}{3}x(x^2 - 1)^{-2/3}(1 + x^{1/2})^{1/2} + \frac{1}{4}(x^2 - 1)^{1/3}(1 + x^{1/2})^{-1/2}x^{-1/2}$

(4-101) $y' = 5[(\sin x)/x]^4[(x \cos x - \sin x)/x^2]$

(4-102) $y' = (\ln x)/(x + 1) + [\ln(x + 1)]/x$

(4-103) $y' = (1/x) + [1/(x + 1)\ln(x + 1)]$

(4-104) $y' = [3 \tan^2(\ln x) \sec^2 (\ln x)]/x$

(4-105) $y' = -e^x \csc e^x \cot e^x - \csc x \cot x e^{\csc x}$

(4-106) $y' = 3 \ln^2(2 + \sin x)[(\cos x)/(2 + \sin x)]$

(4-107) $y' = (3/x) + [(\cos x)/(\sin x)] - [2x/(x^2 + 1)]$

(4-108) $y' = e^x e^{e^x}$

(4-109) $y' = ex^{e-1}e^{x^e}$

(4-110) $f'(x) = \cos x \cos(\sin x) + 2 \sin x \cos x$

(4-111) $f'(x) = 0$

(4-112) $y' = -5 \operatorname{sech}^2(3 - 5x)$

(4-113) $y' = \frac{1}{2} \log_3^{-1/2} x(1/\ln 3)(1/x) = 1/(2x\sqrt{\log_3 x} \ln 3)$

(4-114) $f'(x) = [\frac{1}{2}(x + 1)^{1/3}(x - 1)^{-1/2} - (1/3)(x - 1)^{1/2}(x + 1)^{-2/3}]/(x + 1)^{2/3}$

(4-115) $f'(x) = (5/6)(1 + \sin^2 x)^{-1/2} \sin x \cos x$

(4-116) $y' = \frac{1}{2}\sqrt{(x + 1)/x} + \frac{1}{2}\sqrt{x/(x + 1)}$

(4-117) $y' = \sin x \sinh x + x \cos x \sinh x + x \sin x \cosh x$

(4-118) $y' = (-10 \operatorname{sech}^2 x)/[(1 + \tanh x)^3]$

(4-119) $y' = (2e^{2x} - e^{5x})/[(1 + e^{3x})^2]$

(4-120) $y' = \frac{1}{2}e^{\sqrt{x}}[1 + (1/\sqrt{x})]$

(4-121) $y' = 4 \sec^4 x \tan^2 x + \sec^6 x$

(4-122) $y' = 3$

(4-123) $y' = \ln 5 \sinh x\ 5^{\cosh x}$

(4-124) $y' = \frac{1}{2}(e^x - 3x^{-1})^{-1/2}(e^x + 3x^{-2})$

(4-125) $y' = 0$

(4-126) $y' = 6 \tan 3x \sec^2 3x$

(4-127) $y' = [3 \cos(3x - 4)]/[2\sqrt{\sin(3x - 4)}]$

(4-128) $f''(x) = 20x^3 - 18x$

(4-129) $f''(x) = 6x - 4x^{-3}$

(4-130) $y'' = -\frac{1}{4}x^{-3/2} + \frac{3}{4}x^{-5/2}$

(4-131) $y'' = (28/9)\sqrt[3]{x}$

(4-132) $y'' = 16e^{4x-1}$

(4-133) $y'' = 4/[(x - 1)^3]$

(4-134) $f''(x) = 2 \sec^2 x \tan x$

(4-135) $f''(x) = (\ln^2 2)2^x$

(4-136) $f''(x) = -2x/[(1 + x^2)^2]$

(4-137) $y'' = 2e^x + xe^x$

(4-138) $y'' = -4/[(1 + 2x)^2]$

(4-139) $y'' = 2(\ln 3)3^{x^2} + 4x^2(\ln^2 3)3^{x^2}$

(4-140) $f^{(4)}(x) = 120x - 24$

(4-141) $f^{(4)}(x) = 48e^{2x}$

(4-142) $f^{(4)}(x) = -6x^{-4}$

(4-143) $f^{(4)}(x) = 81 \sin 3x + 16 \cos 2x$

(4-144) $f^{(4)}(x) = 24(1 - x)^{-5}$

(4-145) $f^{(4)}(x) = 16 \cosh(1 - 2x)$

(4-146) $f^{(4)}(x) = -4 \sec^2 x \tan^2 x - 2 \sec^4 x$

(4-147) $f^{(4)}(x) = 4e^{x^2}(3 + 12x^2 + 4x^4)$

(4-148) $f^{(4)}(x) = 16 (\ln^4 3)3^{2x+5}$

(4-149) $f^{(4)}(x) = (6x^2 - 2)/[(1 + x^2)^3]$

(4-150) $f^{(4)}(x) = (-400/81)x^{-4/3}$

EXAM 1 (CHAPTERS 1–4)

1. Find the equation of the line through (1, 2) perpendicular to the line through (2, 5) and (3, −4) in slope-intercept form.
2. Find the partial fraction decomposition of

$$\frac{x^2 - 2}{x^2(x^2 + 1)}$$

3. Use the identity for $\cos 2\theta$ to find $\cos(2 \arctan (1/4))$.
4. Identify, put in standard form, and graph:

$$x^2 - 4y^2 - 2x - 8y - 7 = 0$$

5. Find the following limits:

$$\textbf{(a)} \lim_{x \to 1} \frac{x - (1/x)}{x - 1} \qquad \textbf{(b)} \lim_{x \to \pi^-} \frac{\sin x}{1 + \cos x} \qquad \textbf{(c)} \lim_{x \to \infty} \frac{3x^2 + 2x - 5}{1 - 3x - x^2}$$

6. Use the definition to find the derivative of $f(x) = 1/(x + 1)$.
7. Determine whether $f(x)$ has any discontinuities:

$$f(x) = \begin{cases} \frac{2}{3}x + 3 & \text{for } x < 1 \\ 2 & \text{for } 1 \leqslant x < 5 \\ 7 - x & \text{for } x \geqslant 5 \end{cases}$$

In Problems 8 through 15, find the derivatives:

8. $y = \dfrac{x + 3}{x^2 - 5}$
9. $u = 2t^3\sqrt{5 - t}$
10. $f(x) = (x^2 + x - 4)^5$
11. $G(x) = \sin^3 x$
12. $Q(t) = e^{\sqrt{1 - t^2}}$
13. $w = \ln(z + \sqrt{z})$
14. $h(x) = \sec \sqrt{x}$
15. $v = \arctan \left(\dfrac{1}{x}\right)$

SOLUTIONS TO EXAM 1

1. The line joining (2, 5) and (3, −4) has slope

$$m = \frac{5 - (-4)}{2 - 3} = -9$$

A line perpendicular to this line has slope 1/9. Thus you want the line through (1, 2) with slope 1/9:

$$y - 2 = \frac{1}{9}(x - 1)$$

$$y = \frac{1}{9}x - \frac{1}{9} + 2 = \frac{1}{9}x + \frac{17}{9}$$

2. The form for the partial fraction decomposition is

$$\frac{x^2 - 2}{x^2(x^2 + 1)} = \frac{A}{x} + \frac{B}{x^2} + \frac{Cx + D}{x^2 + 1}$$

Clear the denominators:

$$\begin{aligned} x^2 - 2 &= Ax(x^2 + 1) + B(x^2 + 1) + (Cx + D)x^2 \\ &= x^3(A + C) + x^2(B + D) + xA + B \end{aligned}$$

and equate the coefficients:

$$\begin{aligned} 0 &= A + C \\ 1 &= B + D \\ 0 &= A \\ -2 &= B \end{aligned}$$

Solve using $A = 0$, $B = -2$, $C = 0$, $D = 3$:

$$\frac{x^2 - 2}{x^2(x^2 + 1)} = \frac{-2}{x^2} + \frac{3}{x^2 + 1}$$

3. The identity for $\cos 2\theta$ is $\cos 2\theta = 2\cos^2\theta - 1$. Let $\theta = \arctan(1/4)$. Draw a triangle with an angle θ, as shown in the figure. Then $\tan\theta = (1/4)$ and $\cos\theta = 4/\sqrt{17}$. Finally,

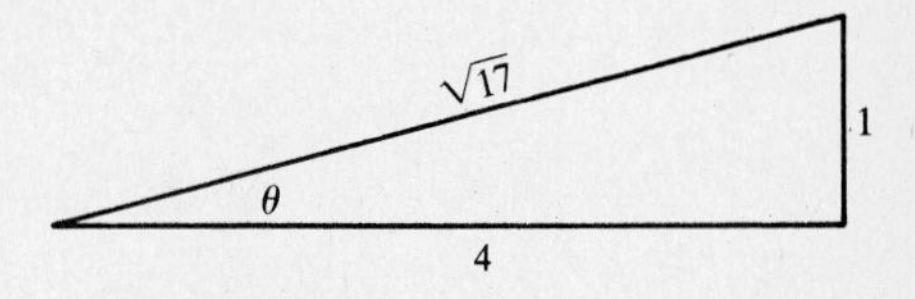

$$\cos\left(2 \arctan\frac{1}{4}\right) = \cos(2\theta) = 2\cos^2\theta - 1 = 2\left(\frac{4}{\sqrt{17}}\right)^2 - 1 = \frac{15}{17}$$

4. Because the coefficients of x^2 and y^2 are opposite in sign, the graph is a hyperbola. Put the equation in standard form by completing the square in x and y:

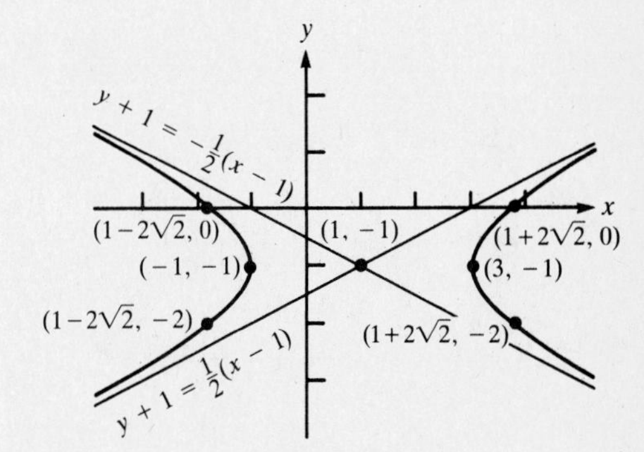

$$\begin{aligned} x^2 - 2x + 1 - 4(y^2 + 2y + 1) &= 7 + 1 - 4 \\ (x - 1)^2 - 4(y + 1)^2 &= 4 \\ \frac{(x - 1)^2}{4} - \frac{(y + 1)^2}{1} &= 1 \end{aligned}$$

The center is $(1, -1)$, with $a = 2$, $b = 1$. The asymptotes are $y + 1 = \pm(1/2)(x - 1)$, as in the figure.

5. **(a)**

$$\begin{aligned} \lim_{x\to 1} \frac{x - (1/x)}{x} &= \lim_{x\to 1} \frac{x^2 - 1}{x(x - 1)} = \lim_{x\to 1} \frac{(x - 1)(x + 1)}{x(x - 1)} \\ &= \lim_{x\to 1} \frac{x + 1}{x} = \frac{2}{1} = 2 \end{aligned}$$

(b)

$$\begin{aligned} \lim_{x\to\pi^-} \frac{\sin x}{1 + \cos x} &= \lim_{x\to\pi^-} \frac{\sin x}{1 + \cos x} \cdot \frac{1 - \cos x}{1 - \cos x} = \lim_{x\to\pi^-} \frac{\sin x(1 - \cos x)}{1 - \cos^2 x} \\ &= \lim_{x\to\pi^-} \frac{\sin x(1 - \cos x)}{\sin^2 x} = \lim_{x\to\pi^-} \frac{1 - \cos x}{\sin x} = \infty \end{aligned}$$

(c) $$\lim_{x\to\infty} \frac{3x^2 + 2x - 5}{1 - 3x - x^2} = \frac{3}{-1} = -3$$

6. $$\lim_{\Delta x \to 0} \frac{f(x + \Delta x) - f(x)}{\Delta x} = \lim_{\Delta x \to 0} \frac{\dfrac{1}{x + \Delta x + 1} - \dfrac{1}{x + 1}}{\Delta x}$$

$$= \lim_{\Delta x \to 0} \frac{(x + 1) - (x + \Delta x + 1)}{(x + 1)(x + \Delta x + 1)\Delta x}$$

$$= \lim_{\Delta x \to 0} \frac{-\Delta x}{(x + 1)(x + \Delta x + 1)\Delta x}$$

$$= \lim_{\Delta x \to 0} \frac{-1}{(x + 1)(x + \Delta x + 1)} = \frac{-1}{(x + 1)^2}$$

7. The only possible discontinuities are at $x = 1$ and $x = 5$. Look at the one-sided limits at $x = 1$ and $x = 5$:

$$\lim_{x \to 1^-} f(x) = \lim_{x \to 1^-} \left(\frac{2}{3}x + 3\right) = \frac{2}{3} + 3 = \frac{11}{3}$$

$$\lim_{x \to 1^+} f(x) = \lim_{x \to 1^+} 2 = 2$$

Because $\lim_{x \to 1^-} f(x) \neq \lim_{x \to 1^+} f(x)$, the limit $\lim_{x \to 1} f(x)$ does not exist and $f(x)$ is not continuous at $x = 1$.

$$\lim_{x \to 5^-} f(x) = \lim_{x \to 5^-} 2 = 2$$

$$\lim_{x \to 5^+} f(x) = \lim_{x \to 5^+} (7 - x) = 2$$

Thus, $\lim_{x \to 5^-} f(x) = \lim_{x \to 5^+} f(x) = f(5)$, and so $f(x)$ is continuous at $x = 5$.

8. Use the quotient rule:

$$\frac{dy}{dx} = \frac{(x^2 - 5)\dfrac{d}{dx}(x + 3) - (x + 3)\dfrac{d}{dx}(x^2 - 5)}{(x^2 - 5)^2}$$

$$= \frac{(x^2 - 5) - (x + 3)(2x)}{(x^2 - 5)^2}$$

$$= \frac{-x^2 - 6x - 5}{(x^2 - 5)^2}$$

9. Use the product rule:

$$\frac{du}{dt} = 2t^3 \frac{d}{dt}(\sqrt{5 - t}) + \sqrt{5 - t}\frac{d}{dt}(2t^3) = 2t^3\left(\frac{-1}{2\sqrt{5 - t}}\right) + \sqrt{5 - t}\,(6t^2)$$

$$= \frac{-t^3}{\sqrt{5 - t}} + 6t^2\sqrt{5 - t}$$

10. Use the chain rule:

$$f'(x) = 5(x^2 + x - 4)^4 \frac{d}{dx}(x^2 + x - 4) = 5(x^2 + x - 4)^4(2x + 1)$$

11. $$G'(x) = 3 \sin^2 x \cos x$$

12. $$Q'(t) = e^{\sqrt{1-t^2}} \frac{d}{dt}(\sqrt{1 - t^2}) = e^{\sqrt{1-t^2}} \frac{1}{2}(1 - t^2)^{-1/2} \frac{d}{dt}(1 - t^2)$$

$$= e^{\sqrt{1-t^2}} \frac{1}{2}(1 - t^2)^{-1/2}(-2t) = \frac{-te^{\sqrt{1-t^2}}}{\sqrt{1 - t^2}}$$

13. $$\frac{dw}{dz} = \frac{1}{z + \sqrt{z}} \cdot \frac{d}{dz}(z + \sqrt{z}) = \frac{1}{z + \sqrt{z}}\left(1 + \frac{1}{2}z^{-1/2}\right) = \frac{2\sqrt{z} + 1}{2\sqrt{z}(z + \sqrt{z})}$$

14. $$h'(x) = \sec\sqrt{x}\tan\sqrt{x}\,\frac{d}{dx}(\sqrt{x}) = \frac{\sec\sqrt{x}\tan\sqrt{x}}{2\sqrt{x}}$$

15. $$\frac{dv}{dx} = \frac{1}{1 + (1/x)^2} \cdot \frac{d}{dx}\left(\frac{1}{x}\right) = \frac{1}{1 + (1/x)^2}\left(\frac{-1}{x^2}\right) = \frac{-1}{x^2 + 1}$$

5 BASIC APPLICATIONS OF THE DERIVATIVE

THIS CHAPTER IS ABOUT

☑ **Tangent Lines**
☑ **Velocity**
☑ **General Rates of Change**
☑ **Tangent Line Approximations and the Differential**

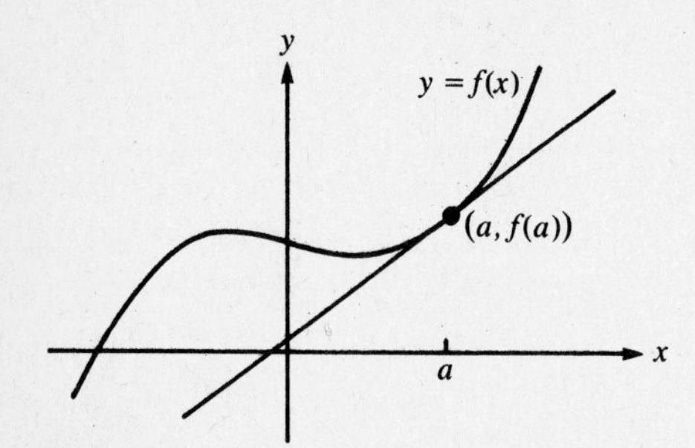

Figure 5-1
The tangent line to the graph of $f(x)$ at $(a, f(a))$.

5-1. Tangent Lines

The **tangent line** to the graph of the function $f(x)$ at the point $(a, f(a))$ is the line through $(a, f(a))$ that best approximates the graph of $f(x)$ in the vicinity of the point $(a, f(a))$ (see Figure 5-1).

A. The slope of the tangent line at $(a, f(a))$ is $f'(a)$.

You can obtain information about the tangent line at a point $(a, f(a))$ from $f'(x)$. Suppose that you want to estimate the slope of the tangent line. You could approximate that slope by finding the slope of a line through the point $(a, f(a))$ and a nearby point on the graph, say $(b, f(b))$, as in Figure 5-2. The slope of this approximating line is $(f(b) - f(a))/(b - a)$. Let Δx be the quantity $b - a$. Then $b = a + \Delta x$. Thus, the slope of the approximating line is

$$\frac{f(a + \Delta x) - f(a)}{\Delta x}$$

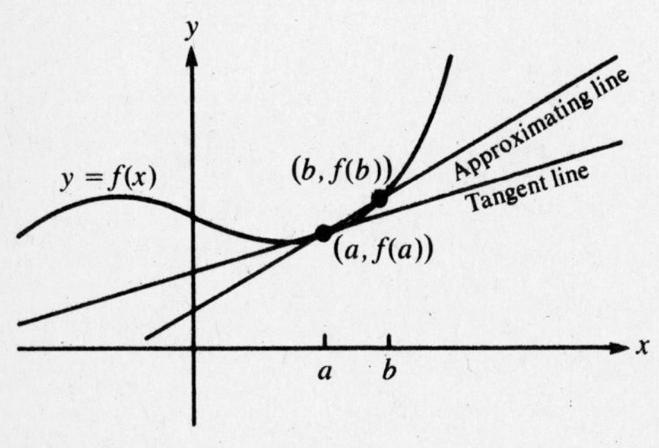

Figure 5-2

Examining Figure 5-2, you see that you obtain a better approximation by choosing b closer to a, that is, Δx closer to zero. As Δx approaches zero, you find that the slope of the approximating line approaches the slope of the tangent line:

SLOPE OF TANGENT LINE

$$\lim_{\Delta x \to 0} \frac{f(a + \Delta x) - f(a)}{\Delta x} \qquad \textbf{(5-1)}$$

is the slope of the tangent line. Recall from the definition of the derivative that this limit is just $f'(a)$. Thus, $f'(a)$ is the slope of the tangent line to the graph of $f(x)$ at the point $(a, f(a))$.

EXAMPLE 5-1: Find the slope of the tangent line to the graph of $f(x) = x^2$ at the point $(2, f(2)) = (2, 4)$.

Solution: In other words you must find $f'(x)$ at $x = 2$:

$$f'(x) = 2x$$

$$f'(2) = 4$$

Thus, the slope of the tangent line at $(2, 4)$ is 4.

EXAMPLE 5-2: Find the slope of the tangent line to the graph of $f(x) = x^3 + 2x$ at the point $(-1, -3)$.

Solution: Differentiating, you find

$$f'(x) = 3x^2 + 2$$

$$f'(-1) = 5$$

The slope of the tangent line at $(-1, -3)$ is 5.

B. You can find the equation of the tangent line.

You can now find the equation of the line tangent to the graph of $f(x)$ at $(a, f(a))$. You know the slope of the tangent line, $f'(a)$, and a point on the tangent line, $(a, f(a))$. In **point-slope** form the equation of the tangent line is

POINT-SLOPE $$y - f(a) = f'(a)(x - a) \qquad \textbf{(5-2)}$$

EXAMPLE 5-3: Find the equation of the line tangent to the graph of $f(x) = 2x^3 - 7x + 1$ at $(2, 3)$.

Solution: Differentiate to find the slope:

$$f'(x) = 6x^2 - 7$$

$$f'(2) = 17$$

Thus, the tangent line is

$$y - 3 = 17(x - 2)$$

or in slope-intercept form

$$y = 17x - 31$$

EXAMPLE 5-4: Find the equation of the tangent line to the graph of $f(x) = x - x^3$ at $x = -2$.

Solution: This is another way of asking for the tangent line at $(-2, f(-2)) = (-2, 6)$. Differentiate to find the slope:

$$f'(x) = 1 - 3x^2$$

$$f'(-2) = -11$$

So the tangent line has slope -11 and passes through the point $(-2, 6)$. The equation of the tangent line is

$$y - 6 = -11[x - (-2)]$$

$$y = -11x - 16$$

EXAMPLE 5-5: Find the equation of the line tangent to the graph of $f(x) = \sin x$ at $x = \pi/4$.

Solution: The tangent line passes through the point $(\pi/4, f(\pi/4)) = (\pi/4, \sqrt{2}/2)$ and has slope $f'(\pi/4)$:

$$f'(x) = \cos x$$

$$f'(\pi/4) = \sqrt{2}/2$$

The equation of the tangent line is

$$y - \frac{\sqrt{2}}{2} = \frac{\sqrt{2}}{2}\left(x - \frac{\pi}{4}\right)$$

$$y = \frac{\sqrt{2}}{2}x + \frac{\sqrt{2}}{2}\left(1 - \frac{\pi}{4}\right)$$

The graph of the function is shown in Figure 5-3.

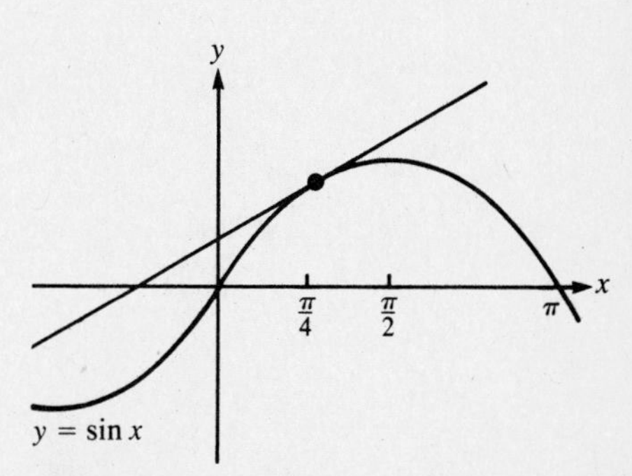

Figure 5-3
The tangent line to the graph of $f(x) = \sin x$ at $x = \pi/4$.

EXAMPLE 5-6: Find the point on the graph of $f(x) = x^2 + 2x + 5$ where the tangent line is parallel to the line $y = 3x - 1$.

Solution: Recall that two lines are parallel if they have the same slope. Thus, you are searching for the point on the graph of $f(x) = x^2 + 2x + 5$ where the slope of the tangent line is 3 (the slope of $y = 3x - 1$ is 3). You want the point $(a, f(a))$ where $f'(a) = 3$:

$$f'(x) = 2x + 2$$

So $f'(a) = 3$ when $2a + 2 = 3$, i.e., $a = \frac{1}{2}$. Thus the desired point is $(1/2, f(1/2)) = (1/2, 25/4)$.

EXAMPLE 5-7: Find another point on the graph of $f(x) = x^3 - 2$ where the tangent line is parallel to the tangent line at $(1, -1)$.

Solution: The slope of the tangent line at $(1, -1)$ is $f'(1)$:

$$f'(x) = 3x^2$$

$$f'(1) = 3$$

The problem asks you to find another point on the graph where the tangent line has slope 3, so you seek a value of a such that $f'(a) = 3$:

$$f'(a) = 3a^2 = 3$$

$$a = 1 \quad \text{or} \quad a = -1$$

The value $a = 1$ is the original point. You find the desired point by setting $a = -1$:

$$(-1, f(-1)) = (-1, -3)$$

5-2. Velocity

Suppose that a train leaves the station at time $t = 0$ and that after t hours the train has traveled $f(t)$ kilometers. How might you estimate the velocity (velocity = distance/time) of the train at time $t = 1$ hour? One method is to find the average velocity of the train between time $t = 1$ hour and $t = 2$ hours: $f(2) - f(1)$ is the distance traveled by the train between $t = 1$ hour and $t = 2$ hours. Thus,

$$\frac{f(2) - f(1)}{1 \text{ hour}}$$

is the average velocity of the train between $t = 1$ and $t = 2$ hours. A better approximation would be obtained from the average velocity of the train between $t = 1$ hour and $t = 1 + \Delta t$ hours, where Δt is smaller than 1: $f(1 + \Delta t) - f(1)$ is the distance traveled by the train during the Δt hours following $t = 1$. Thus, the average velocity of the train between $t = 1$ hour and $t = 1 + \Delta t$ hours is

$$\frac{\text{distance traveled}}{\text{time elapsed}} = \frac{f(1 + \Delta t) - f(1)}{\Delta t}$$

As Δt approaches zero, this quantity approaches the instantaneous velocity of the train at $t = 1$. So you see that

$$f'(1) = \lim_{\Delta t \to 0} \frac{f(1 + \Delta t) - f(1)}{\Delta t}$$

is the instantaneous velocity of the train at $t = 1$ hour.

In general if $f(t)$ describes the position of (or distance traveled by) an object at time t, then $f'(t)$ describes the *velocity* of that object at time t.

EXAMPLE 5-8: If the train has traveled $f(t) = 30t + t^2$ kilometers after t hours, how fast is the train traveling after 1 hour? After 2 hours?

Solution: You find that

$$f'(t) = 30 + 2t$$

$$f'(1) = 32 \qquad f'(2) = 34$$

Thus after 1 hour the train is traveling at 32 kilometers per hour and after 2 hours, 34 kilometers per hour.

Note: The units (e.g., kilometers per hour) of the derivative are determined by the units of the function (e.g., kilometers) and of the variable (e.g., hours).

EXAMPLE 5-9: A ball is thrown upward. After t seconds the ball is $p(t) = 64t - 16t^2$ feet above the thrower. What is the velocity of the ball after 1 second? After 2 seconds? After 3 seconds?

Solution: You find that $p'(t) = 64 - 32t$

$$p'(1) = 32 \qquad p'(2) = 0 \qquad p'(3) = -32$$

You interpret these values as follows: After 1 second, the ball is traveling upward at 32 ft/s. At 2 seconds the velocity of the ball is zero and the ball is at the highest point of its flight. At 3 seconds the ball is falling (notice the minus sign) at 32 ft/s.

5-3. General Rates of Change

EXAMPLE 5-10: Suppose that air is pumped into a spherical balloon so that after t seconds the diameter of the balloon is $f(t)$ inches. Find a function that describes the rate at which the diameter is growing at any time.

Solution: You can approximate the rate at which the diameter is growing at time $t = t_0$ by using the average rate of growth of the diameter between time $t = t_0$ and time $t = t_0 + \Delta t$. Of course a small value of Δt will give you a better approximation. The average rate of growth of the diameter during this time is

$$\frac{\text{change in diameter}}{\text{change in time}} = \frac{f(t_0 + \Delta t) - f(t_0)}{\Delta t}$$

As Δt approaches zero, this quantity approaches the instantaneous rate of growth of diameter at time t_0. Thus, $f'(t_0)$ is the instantaneous rate of growth of diameter at time t_0. If $f(t) = 4t^{1/3}$ inches, then $f'(t) = (4/3)t^{-2/3}$ in./s. So, for example, after 8 seconds the diameter is growing at a rate of $f'(8) = (4/3)8^{-2/3} = 1/3$ in./s.

In general, if f is a function of x, then the *rate of change* of f with respect to x is $f'(x)$. As you have seen in Section 5-2, the rate of change of position with respect to time is velocity. In Example 5-10 the rate of change of diameter with respect to time is the derivative of diameter as a function of time. If $P(t)$ describes the population of a bacterial culture at time t, then $P'(t)$ gives the rate of growth of the population at time t. If $f(x)$ describes the profits of a company as a function of x, the production level, then $f'(x)$ gives the rate of change of profits with respect to production level.

EXAMPLE 5-11: A ski rental facility calculates that if the depth of snow at their facility is x in., then they will make $f(x) = 20x - 50 - (1/20)x^2$ dollars/day. Find and interpret $f'(50)$.

Solution: Differentiate to find

$$f'(x) = 20 - \frac{x}{10}$$

You can interpret $f'(x)$ as the rate of change of income with respect to snow

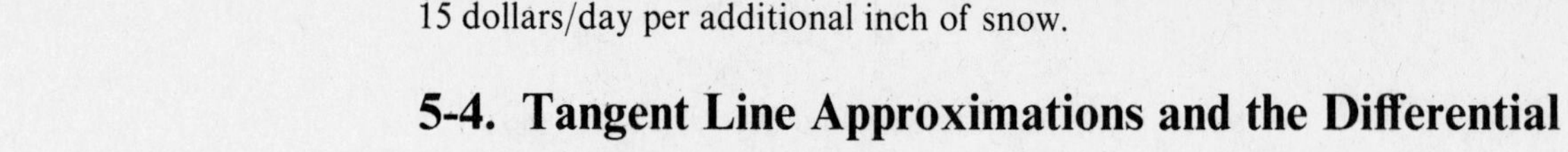

depth. For example, $f'(50) = 15$, which means that if there are 50 in. of snow on the ground, additional snow will increase the facility's receipts at a rate of 15 dollars/day per additional inch of snow.

5-4. Tangent Line Approximations and the Differential

A. Tangent line approximations

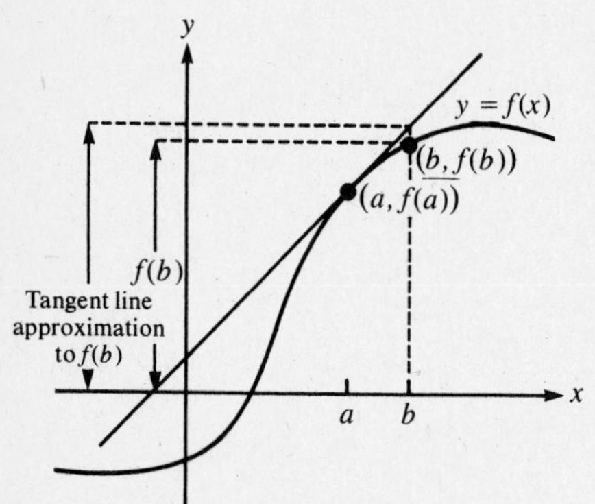

Figure 5-4
A tangent line approximation to $f(b)$.

You can use tangent lines to approximate the value of a function $f(x)$ at $x = b$ when you know the values of $f(a)$ and $f'(a)$ for b close to a. Recall that the point-slope equation of the line tangent to the graph of $f(x)$ at $(a, f(a))$ is

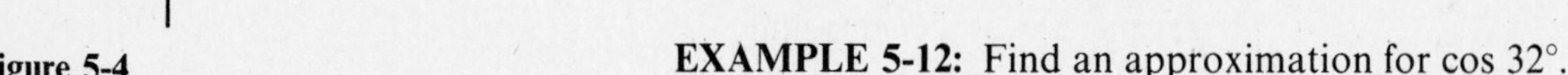

$$y - f(a) = f'(a)(x - a)$$

As you can see in Figure 5-4, the graph of the tangent line to $y = f(x)$ at $(a, f(a))$ is a reasonable approximation to the graph of $f(x)$ in the vicinity of a. Because b is close to a, the tangent line approximation to $f(b)$ is close to $f(b)$. You can use this observation to solve a variety of problems.

EXAMPLE 5-12: Find an approximation for cos 32°.

Solution: First notice that 32° is near 30° and that cos 30° is easily found. The tangent line to $f(x) = \cos x$ at $x = 30°$ should give a good approximation to cos 32°. Before you proceed, change to radians: $30° = \pi/6$ radians. Then

$$f(x) = \cos x$$
$$f'(x) = -\sin x$$
$$f'\left(\frac{\pi}{6}\right) = -\frac{1}{2}$$

The tangent line at $(\pi/6, \sqrt{3}/2)$ thus has equation

$$y - \frac{\sqrt{3}}{2} = -\frac{1}{2}\left(x - \frac{\pi}{6}\right)$$

At $x = 32° = 32\pi/180$ radians $= 8\pi/45$:

$$y = \frac{\sqrt{3}}{2} - \frac{1}{2}\left(\frac{8\pi}{45} - \frac{\pi}{6}\right) = \frac{\sqrt{3}}{2} - \frac{1}{2}\cdot\frac{\pi}{90} \approx 0.848\,57$$

EXAMPLE 5-13: Estimate $\sqrt{4.1}$ in decimal form.

Solution: View $\sqrt{4.1}$ as the value of $f(x) = \sqrt{x}$ at $x = 4.1$. You can use the tangent line to $f(x) = \sqrt{x}$ at $x = 4$ to estimate $\sqrt{4.1}$:

$$f(x) = \sqrt{x}$$
$$f'(x) = \tfrac{1}{2}x^{-1/2}$$
$$f'(4) = \tfrac{1}{4}$$

The tangent line to $f(x) = \sqrt{x}$ at (4, 2) is $y = \frac{1}{4}x + 1$. Thus, your estimate of $\sqrt{4.1}$ is

$$y = \tfrac{1}{4}(4.1) + 1 = 2.025$$

Use the following steps to estimate using tangent line approximations:

(1) Identify the function f (so that the value being estimated is $f(b)$).
(2) Find a point $x = a$ that is close to b so that $f(a)$ and $f'(a)$ are easily found. (Typically, if $f(a)$ is easily found, then $f'(a)$ will be easily found.)
(3) Find the equation of the tangent line and evaluate at $x = b$.

EXAMPLE 5-14: Estimate $(1.99)^{-5}$.

Solution: You recognize $(1.99)^{-5}$ as $f(b)$, where $f(x) = x^{-5}$ and $b = 1.99$. You can easily evaluate $f(x)$ at $a = 2$, which is close to 1.99. So you must find the tangent line to $f(x)$ at $x = 2$:

$$f'(x) = -5x^{-6}$$

$$f'(2) = \frac{-5}{64}$$

The tangent line is

$$y - \frac{1}{32} = \frac{-5}{64}(x - 2)$$

Evaluate at $x = 1.99$:

$$y = \frac{1}{32} - \frac{5}{64}(1.99 - 2) = 0.032\,031\,205$$

A decimal approximation to $(1.99)^{-5}$ is 0.032 031 205.

This method of approximation is given more as a matter of interest than of utility. It is obvious that if you have your calculator, and if your batteries are charged, you can find a much better approximation to cos 32° or $(1.99)^{-5}$ by pushing the right buttons!

B. Differentials: Another viewpoint

The **differential** dy of a differentiable function $y = f(x)$ is defined by

$$dy = f'(x)\,dx$$

where dx is defined as the function $dx = \Delta x$.

Recall that the derivative $f'(x)$ is defined by

$$f'(x) = \lim_{\Delta x \to 0} \frac{\Delta y}{\Delta x}$$

and so, for small Δx, $\dfrac{\Delta y}{\Delta x} \approx f'(x)$ and $\Delta y \approx f'(x)\,\Delta x$.

Compare that approximation for Δy with the definition of dy, and notice that $\Delta y \approx dy$; that is, you can use dy to approximate a change in y (Δy) produced by a change in x (Δx).

EXAMPLE 5-15: Use the differential to approximate $\sqrt{4.1}$.

Solution: If $y = f(x) = \sqrt{x}$, then $f(4) = \sqrt{4} = 2$. If we can find dy (thus approximating Δy) corresponding to $\Delta x = 0.1$ (x changes from $x = 4$ to $x = 4.1$), then

$$f(4.1) \approx 2 + dy$$

By definition:

$$dy = f'(x)\,dx$$
$$= \frac{1}{2\sqrt{x}}\Delta x$$

Setting $x = 4$, $\Delta x = 0.1$, you have

$$dy = \frac{1}{2\sqrt{4}}(0.1) = 0.025$$

and so

$$\sqrt{4.1} \approx 2.025$$

SUMMARY

1. The tangent line to the graph of $f(x)$ at $(a, f(a))$ has slope $f'(a)$.
2. The equation of the tangent line to the graph of $f(x)$ at $(a, f(a))$ can be computed using the slope, $f'(a)$, a point on the line $(a, f(a))$, and the point-slope formula for a line: $y - f(a) = f'(a)(x - a)$.
3. Velocity is the rate of change of position. The velocity function is the derivative of the position function with respect to time.
4. The derivative of the function $f(x)$ with respect to x yields the rate of change of f with respect to x.
5. You can find a tangent line approximation to $f(b)$ by using $f(b) \approx f'(a)(b - a) + f(a)$ for a close to b.
6. When you approximate a value of a function using differentials, you are really using a tangent line approximation.

SOLVED PROBLEMS

PROBLEM 5-1 Find the equation of the tangent line to the graph of $f(x) = x^4 - 3x + 1$ at $(1, -1)$.

Solution: The slope of this tangent line is $f'(1)$:

$$f'(x) = 4x^3 - 3$$

$$f'(1) = 1$$

So the tangent line has slope 1 and passes through $(1, -1)$. Using the point-slope formula for a line, you find that this tangent line has equation

$$y - (-1) = 1(x - 1) \quad \text{or} \quad y = x - 2$$

[See Section 5-1.]

PROBLEM 5-2 Find the equation of the tangent line to the graph of $f(x) = (x^2 + 1)/(x^2 - 1)$ at $x = 2$.

Solution: The tangent line has equation

$$y = f(2) + f'(2)(x - 2)$$

Since

$$f'(x) = \frac{(x^2 - 1)(2x) - (x^2 + 1)(2x)}{(x^2 - 1)^2} = \frac{-4x}{(x^2 - 1)^2}$$

$$f'(2) = -\frac{8}{9}$$

So the tangent line is

$$y = \frac{5}{3} + \left(\frac{-8}{9}\right)(x - 2) \quad \text{or} \quad y = \frac{-8}{9}x + \frac{31}{9}$$

[See Section 5-1.]

PROBLEM 5-3 Find the equation of the tangent line to the graph of $f(x) = e^{2x}$ at $x = 0$.

Solution: You find that

$$f'(x) = 2e^{2x}$$

The tangent line is

$$y = f(0) + f'(0)(x - 0)$$

$$y = 1 + 2x$$

[See Section 5-1.]

PROBLEM 5-4 Find the y intercept of the tangent line to the graph of $f(x) = (x - 3)(x^2 + 1)$ at its x intercept.

Solution: The x intercept of the graph of f is $x = 3$. Since

$$f'(x) = (x^2 + 1) + (x - 3)(2x)$$

$$f'(3) = 10$$

the tangent line is

$$y = f(3) + f'(3)(x - 3)$$

$$y = 0 + 10(x - 3) = 10x - 30$$

So the y intercept of this tangent line is $(0, -30)$. [See Section 5-1.]

PROBLEM 5-5 Find all points on the curve $y = x^5 - 20x^2$ where the tangent line is horizontal.

Solution: The slope of the tangent line is

$$\frac{dy}{dx} = 5x^4 - 40x = 5x(x^3 - 8)$$

The tangent line at x is horizontal whenever $dy/dx = 0$, i.e., at $x = 0$ and $x = 2$. The points of interest are $(0, f(0)) = (0, 0)$ and $(2, f(2)) = (2, -48)$. [See Section 5-1.]

PROBLEM 5-6 Find all points on the curve $y = (x + 1)/(x - 1)$ where the tangent line is parallel to the line $y = -\frac{1}{2}x + 5$.

Solution: The tangent line at x has slope

$$\frac{dy}{dx} = \frac{-2}{(x - 1)^2}$$

You are searching for values of x such that

$$\frac{-2}{(x - 1)^2} = -\frac{1}{2}$$

$$(x - 1)^2 = 4$$

$$x = -1, 3$$

So the points are $(-1, f(-1)) = (-1, 0)$ and $(3, f(3)) = (3, 2)$. [See Section 5-1.]

PROBLEM 5-7 Find all points on the graph of $f(x) = x^4 - 2x^2 + 1$ where the tangent line is parallel to the tangent line at $x = 1$.

Solution: Since

$$f'(x) = 4x^3 - 4x$$

$$f'(1) = 0$$

So the slope of the tangent line at $x = 1$ is 0. Two lines are parallel precisely when their slopes are equal. You search for all other values of x for which the tangent line at x has slope 0 (i.e., $f'(x) = 0$):

$$0 = 4x^3 - 4x = 4x(x^2 - 1)$$

So the points of interest are $x = 0, \pm 1$, i.e., $(0, 1), (1, 0), (-1, 0)$. [See Section 5-1.]

PROBLEM 5-8 Find all points on the graph of $f(x) = (x^2 + 1)/(x - 1)$ where the tangent line is parallel to the tangent line at $x = 0$.

Solution: Since

$$f'(x) = \frac{x^2 - 2x - 1}{(x - 1)^2}$$

$$f'(0) = -1$$

You want to find all values of x where $f'(x) = -1$:

$$-1 = \frac{x^2 - 2x - 1}{(x - 1)^2}$$

$$-(x - 1)^2 = x^2 - 2x - 1$$

$$2x^2 - 4x = 0$$

$$x = 0, 2$$

So the points of interest are $(0, f(0)) = (0, -1)$ and $(2, f(2)) = (2, 5)$. [See Section 5-1.]

PROBLEM 5-9 Find the point on the graph of the function $f(x) = x^4$ where the x intercept of the tangent line is (2, 0).

Solution: The tangent line at $x = a$ is

$$y = f(a) + f'(a)(x - a)$$

$$y = a^4 + 4a^3(x - a) = 4a^3x - 3a^4$$

The x intercept of this line is found by setting y equal to zero:

$$0 = 4a^3x - 3a^4$$

$$x = \frac{3a}{4}$$

So the x intercept is 2 when $2 = 3a/4$, that is, when $a = 8/3$. The point of interest is thus $(a, f(a)) = (8/3, f(8/3)) = (8/3, 4096/81)$. [See Section 5-1.]

PROBLEM 5-10 If an object is dropped from a tower 400 feet above the ground, its distance above the ground after t seconds is given by $400 - 16t^2$ feet. Find the velocity of the object at $t = 1$, $t = 3$, and on impact.

Solution: If $A(t)$ is the altitude at time t, then $A'(t)$ is the rate of change of altitude with respect to time, i.e., velocity:

$$A'(t) = -32t$$

So $A'(1) = -32$ ft/s and $A'(3) = -96$ ft/s. The minus sign indicates that the object is falling (its altitude is decreasing).

The object strikes the ground when $A(t) = 0$:

$$0 = 400 - 16t^2$$

$$t = \pm 5$$

But $t = -5$ is not a time under consideration, so $t = 5$ is the time when the object strikes the ground. The velocity when $t = 5$ is $A'(5) = -160$ ft/s. [See Section 5-2.]

PROBLEM 5-11 The population of a city at time t years is given by $P(t) = 50\,000(t^{1/2} + 3)$. At what rate is the population growing when $t = 4$?

Solution: The rate of change of population with respect to time is given by the derivative of the population function with respect to time:

$$P'(t) = 25\,000t^{-1/2}$$

$$P'(4) = 25\,000(4)^{-1/2} = 12\,500 \text{ people per year}$$ [See Section 5-3.]

PROBLEM 5-12 For a certain dosage of a drug, d (given in cubic centimeters), the resultant temperature change (in degrees Fahrenheit) of the person taking the drug is $T(d) = 0.05d - 0.3d^2$. Find the rate of change of temperature with respect to dosage.

Solution: $T'(d) = 0.05 - 0.6d$ degrees per cubic centimeter [See Section 5-3.]

PROBLEM 5-13 The position of a point moving on a horizontal line at time t is given by $s(t) = t^3 - 8t^2 + 5t + 1$. Determine the times at which the point is stationary.

Solution: The velocity function is $s'(t) = 3t^2 - 16t + 5$. You want to know when $s'(t) = 0$:

$$0 = 3t^2 - 16t + 5 = (3t - 1)(t - 5)$$

So the point is stationary when $t = \frac{1}{3}$ and $t = 5$. [See Section 5-2.]

PROBLEM 5-14 Sales of a product and the amount spent on advertising the product are related by $S = -a^2 + 100a + 6$, where a is advertising dollars (in thousands) and S is number of sales. At what rate are sales changing with respect to advertising dollars when $a = 40$?

Solution: Since

$$S'(a) = -2a + 100$$

$$S'(40) = 20$$

When \$40 000 are spent on advertising, additional expenditure on advertising will increase sales at a rate of 20 sales per thousand dollars. [See Section 5-3.]

PROBLEM 5-15 The total discharge of a water faucet t minutes after opening is $3t^2 - 5t + 4$ liters. At what rate is the water leaving the faucet when $t = 2$?

Solution: Let $f(t) = 3t^2 - 5t + 4$. Then

$$f'(t) = 6t - 5$$

$$f'(2) = 7 \text{ liters/min.}$$

[See Section 5-3.]

PROBLEM 5-16 Find an approximate value for $(3.97)^3$ using a tangent line approximation.

Solution: You can use a tangent line approximation to the graph of the function $f(x) = x^3$ at $x = 4$:

$$f(4) = 4^3 = 64$$

$$f'(x) = 3x^2$$

$$f'(4) = 48$$

The tangent line at $x = 4$ is

$$y = f(4) + f'(4)(x - 4) = 64 + 48(x - 4)$$

So an estimate for $(3.97)^3$ is

$$(3.97)^3 \approx 64 + 48(3.97 - 4) = 62.56$$

[See Section 5-4.]

PROBLEM 5-17 Find an approximation for $\sqrt[3]{8.03}$.

Solution: Let $f(x) = x^{1/3}$. Then

$$f'(x) = \tfrac{1}{3}x^{-2/3}$$

You can find the tangent line to f at $x = 8$:

$$f(8) = 2$$

$$f'(8) = \frac{1}{12}$$

The tangent line is

$$y = f(8) + f'(8)(x - 8) = 2 + \frac{1}{12}(x - 8)$$

$$\sqrt[3]{8.03} \approx 2 + \frac{1}{12}(8.03 - 8) = 2.0025$$

[See Section 5-4.]

PROBLEM 5-18 Find an approximate value for sin 42°.

Solution: First find the tangent line to the graph of $f(x) = \sin x$ at $x = 45° = \pi/4$ radians:

$$y = f\left(\frac{\pi}{4}\right) + f'\left(\frac{\pi}{4}\right)\left(x - \frac{\pi}{4}\right)$$

$$= \sin\frac{\pi}{4} + \cos\frac{\pi}{4}\left(x - \frac{\pi}{4}\right)$$

$$= \frac{\sqrt{2}}{2} + \frac{\sqrt{2}}{2}\left(x - \frac{\pi}{4}\right)$$

$$\sin 42° = \sin\frac{7\pi}{30} \approx \frac{\sqrt{2}}{2} + \frac{\sqrt{2}}{2}\left(\frac{7\pi}{30} - \frac{\pi}{4}\right)$$

Recall that $\sqrt{2} \approx 1.414$ and $\pi \approx 3.141\,59$, and obtain

$$\sin 42° \approx 0.670\,0827$$

[See Section 5-4.]

PROBLEM 5-19 Use the differential to approximate $8.99/(19 + (8.99)^2)$.

Solution: Let $y = f(x) = x/(19 + x^2)$, with $x = 9$ and $dx = \Delta x = -0.01$. Then

$$f(8.99) = f(9) + \Delta y$$

$$\approx f(9) + dy$$

Now

$$dy = f'(x)\,dx = \frac{19 - x^2}{(19 + x^2)^2}\,dx$$

$$= \frac{19 - 81}{(19 + 81)^2}(-0.01) = \frac{0.62}{100^2} = 0.000\,062$$

$$f(9) = \frac{9}{19 + 81} = 0.09$$

You have

$$f(8.99) \approx 0.09 + 0.000\,062 = 0.090\,062$$

[See Section 5-4.]

PROBLEM 5-20 Use the differential to approximate $(9.02)^{3/2} + [1/(9.02)^{1/2}] - 26$.

Solution: Choose $y = f(x) = x^{3/2} + \dfrac{1}{x^{1/2}} - 26$, with $x = 9.00$, $\Delta x = dx = 0.02$. Then

$$f(9) = 9^{3/2} + \frac{1}{9^{1/2}} - 26 = \frac{4}{3}$$

and

$$dy = \left(\frac{3}{2}x^{1/2} - \frac{1}{2x^{3/2}}\right)dx$$

$$= \left[\frac{3}{2}(9)^{1/2} - \frac{1}{2(9)^{3/2}}\right](0.02)$$

$$= \left(\frac{9}{2} - \frac{1}{54}\right)(0.02) = \frac{2.42}{27}$$

Thus

$$f(9.02) \approx \frac{4}{3} + \frac{2.42}{27} = \frac{38.42}{27} \approx 1.423$$

[See Section 5-4.]

Supplementary Exercises

In Problems 5-21 through 5-35 find the equation of the line that is tangent to the given curve at the given point:

5-21 $f(x) = x^5 - 3x^3 - 3$ at $(2, 5)$

5-22 $f(x) = \sqrt{x} - x$ at $(4, -2)$

5-23 $f(x) = (x^2 - 1)/(2x + 3)$ at $(-2, -3)$

5-24 $f(x) = e^x - e^{x^2}$ at $(0, 0)$

5-25 $f(x) = \sin[\frac{1}{2}(x^2 - 1)\pi]$ at $(2, -1)$

5-26 $f(x) = x^7 - 3x^4 + x^2 - 1$ at $x = 1$

5-27 $f(x) = x^2 - (9/\sqrt{x}) + 2$ at $x = 9$

5-28 $f(x) = (x - 3)/(x^2 - 2x)$ at $x = 1$

5-29 $f(x) = (\sqrt{x} + 1)/(\sqrt{x} - 1)$ at $x = 4$

5-30 $f(x) = \cos x - 2 \sin x$ at $x = \pi/2$

5-31 $f(x) = \ln(x^2 + e)$ at $x = \sqrt{e}$

5-32 $f(x) = (1 - \sin x)/(x + 1)$ at $x = 0$

5-33 $f(x) = (x^{1/3} - 1)/(x + 4)$ at $x = -8$

5-34 $f(x) = x(\sqrt{x^3} - 3x + 2)$ at $x = 2$

5-35 $f(x) = 2^x - 3^x$ at $x = -1$

In Problems 5-36 through 5-40 find all points on the graph of f where the tangent line is horizontal:

5-36 $f(x) = x^2 - 4\sqrt{x} + 1$

5-37 $f(x) = x^5 - 5x^3 - 20x + 7$

5-38 $f(x) = 3e^{-x^2}$

5-39 $f(x) = \sqrt{x^4 + x^2}$

5-40 $f(x) = e^{x^2 - \cos x}$

In Problems 5-41 through 5-45 find all points on the graph of f where the tangent line is parallel to the given line:

5-41 $f(x) = x^3 - 6x + 2$ $\quad y = 6x - 7$

5-42 $f(x) = \sqrt{x} + x$ $\quad y = 2x + 4$

5-43 $f(x) = (x + 1)/(x - 2)$ $\quad y = -3x + \pi$

5-44 $f(x) = e^x - e^{2x} + 1$ $\quad 2x + 2y = 5$

5-45 $f(x) = \tan x$ $\quad y = 2x + 3$

In Problems 5-46 through 5-50 find all points where the tangent line is parallel to the tangent line at the given point:

5-46 $f(x) = x^4 - 2x^2 + 7$ at $x = 1$

5-47 $f(x) = x^3 - x + 2$ at $x = -1$

5-48 $f(x) = (x - 2)/(x + 7)$ at $x = -6$

5-49 $f(x) = e^{x+1} - e^{-x+1}$ at $x = -1$

5-50 $f(x) = \ln(x - 1) + x^2$ at $x = 2$

5-51 A line is tangent to $y = x^3$ at (2, 8). At which point does it again meet the curve?

5-52 Find all points on the curve $y = x - (1/x)$ where the tangent line has y intercept $\frac{1}{2}$.

5-53 Find all points on the curve $y = x^2$ where the tangent line passes through (4, 12).

5-54 Find all points on the curve $y = x^3 - x$ where the tangent line has x intercept 1.

5-55 Find all points on the curve $y = x^4 + 1$ where the tangent line passes through (1, 1).

5-56 An automobile's odometer reads $60t - 30 \ln(t + \frac{1}{2})$ miles t hours after leaving on a trip. At what rate is the car traveling after 1 hour?

5-57 A brushfire spreads so that after t hours $80t - 20t^2$ acres are burning. What is the rate of growth of the acreage that is burning after 90 minutes?

5-58 If the distance l (in yards) that a ball rolls down a hill is related to the weight w (in pounds) of the ball by $l = 2w^2 + w + 10$, find the rate of change of distance with respect to weight for a ball that weighs 4 pounds.

5-59 The velocity of a ball is $80 - 32t$ ft/s t seconds after being thrown. If acceleration is defined as the rate of change of velocity with respect to time, find the acceleration of the ball.

5-60 A gardener discovers that if she uses x pounds of fertilizer on her garden, it will yield $(-12/5)x^2 + 24x + 60$ pints of fruit. What is the rate of change of yield with respect to amount of fertilizer for her garden when she uses 3 pounds of fertilizer?

In Problems 5-61 through 5-70 find an approximate value for the given quantity using a tangent line approximation:

5-61 $\sqrt{37}$

5-62 $(2.013)^4$

5-63 $(28)^{2/3}$

5-64 $\dfrac{(2.97)^2 - 1}{(2.97)^2 + 1}$

5-65 $\tan 47°$

5-66 $e^{0.01}$

5-67 $\dfrac{\sqrt{3.96} + 1}{\sqrt{3.96} - 1}$

5-68 $\ln 1.023$

5-69 $(9.95)^{-4}$

5-70 $(1.013)^3 + (1.013)^{-2}$

Solutions to Supplementary Exercises

(5-21) $y = 44x - 83$

(5-22) $y = -\frac{3}{4}x + 1$

(5-23) $y = -2x - 7$

(5-24) $y = x$

(5-25) $y = -1$

(5-26) $y = -3x + 1$

(5-27) $y = (109/6)x - 167/2$

(5-28) $y = -x + 3$

(5-29) $y = -\frac{1}{2}x + 5$

(5-30) $y = -x + (\pi/2) - 2$

(5-31) $y = x/\sqrt{e} + \ln 2$

(5-32) $y = -2x + 1$

(5-33) $y = (1/6)x + 25/12$

(5-34) $y = (13/2)x - 9$

(5-35) $y = \dfrac{1}{6} + \left(\dfrac{\ln 2}{2} - \dfrac{\ln 3}{3}\right)(x + 1)$

(5-36) $(1, -2)$

(5-37) $(2, -41)$ and $(-2, 55)$

(5-38) $(0, 3)$

(5-39) none

(5-40) $(0, 1/e)$

(5-41) $(2, -2)$ and $(-2, 6)$

(5-42) $(\frac{1}{4}, \frac{3}{4})$

(5-43) $(1, -2)$ and $(3, 4)$

(5-44) $(0, 1)$

(5-45) $\left(\frac{\pi}{4} + k\pi, 1\right)$ and $\left(-\frac{\pi}{4} + k\pi, -1\right)$ for any integer k

(5-46) $(1, 6), (-1, 6),$ and $(0, 7)$

(5-47) $(-1, 2)$ and $(1, 2)$

(5-48) $(-6, -8)$ and $(-8, 10)$

(5-49) $(-1, 1 - e^2)$ and $(1, e^2 - 1)$

(5-50) $(2, 4)$ and $(3/2, (9/4) - \ln 2)$

(5-51) $(-4, -64)$

(5-52) $(-4, -15/4)$

(5-53) $(2, 4)$ and $(6, 36)$

(5-54) $(1, 0)$ and $(-\frac{1}{2}, \frac{3}{8})$

(5-55) $(0, 1)$ and $(\frac{4}{3}, \frac{337}{81})$

(5-56) 40 m/h

(5-57) 20 acres/h

(5-58) 17 yards/pound

(5-59) -32 feet/sec^2

(5-60) 48/5 pints/pound

(5-61) 6.0833

(5-62) 16.416

(5-63) 9.222 22

(5-64) 0.7964

(5-65) 1.069 8132

(5-66) 1.01

(5-67) 3.02

(5-68) 0.023

(5-69) 0.000 102

(5-70) 2.013

6 IMPLICIT AND LOGARITHMIC DIFFERENTIATION

THIS CHAPTER IS ABOUT

☑ **Implicit Differentiation**
☑ **Logarithmic Differentiation**

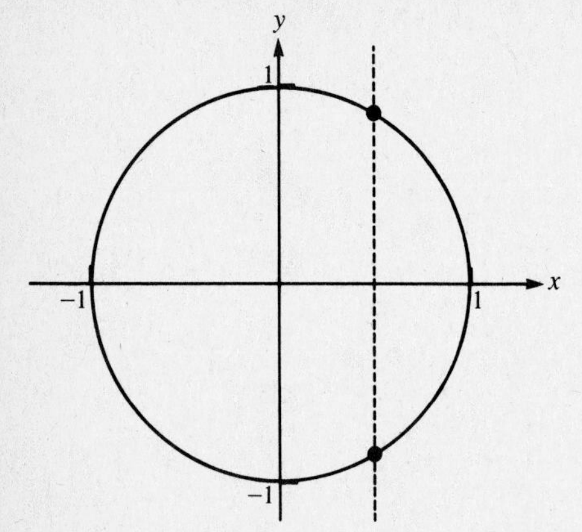

Figure 6-1
$x^2 + y^2 = 1$.

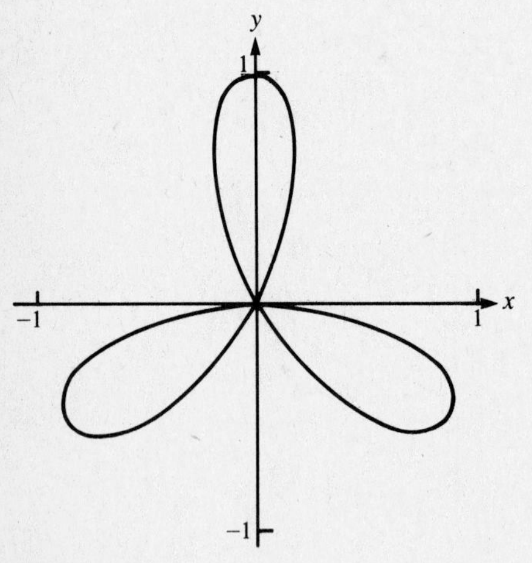

Figure 6-2
$(x^2 + y^2)^2 + 3x^2y - y^3 = 0$.

6-1. Implicit Differentiation

A. Deciding when to use implicit differentiation

You can use implicit differentiation to find dy/dx when y is not given as a function of x. An equation relating two variables, x and y, describes a relationship between the two variables that can be graphed. This does not necessarily define one variable as a function of the other.

EXAMPLE 6-1: Consider the equation $x^2 + y^2 = 1$. When you solve for y, you find two values of y for each value of x between -1 and 1:

$$y = \pm\sqrt{1 - x^2}$$

Figure 6-1 shows the graph of $x^2 + y^2 = 1$. Since you can draw a vertical line that touches the graph at more than one point, y is not a function of x.

Sometimes you will find an equation relating x to y where it is difficult or impossible to solve for y, regardless of whether y actually is a function of x.

EXAMPLE 6-2: Consider the equation $(x^2 + y^2)^2 + 3x^2y - y^3 = 0$. This is a fourth degree equation in y, and can't easily be solved for y. The graph of this equation is shown in Figure 6-2. Again you can see y is not a function of x.

EXAMPLE 6-3: Consider $x^2e^x + y^2 - e^y = 2$. This is a transcendental equation (it mixes algebraic functions with exponential functions) and can't be solved for y. Nevertheless, if you graphed all ordered pairs (x, y) that satisfy this equation, you would find that they form a curve in the xy plane.

Because the relationships between x and y discussed above define curves, you can reasonably speak of the slope of the tangent line to these curves at all but a few points on the curve (some points may have vertical tangents, for example). Although y may not be a function of x, you treat it as if it were. Technically, y is a function of x in some small region containing x if you ignore other branches of the graph.

B. How to differentiate implicitly

To do **implicit differentiation**, you treat y as a function of x, and differentiate both sides of the equation.

EXAMPLE 6-4: Find the slope of the curve described by $x^2 + y^2 = 1$ at $(1/2, \sqrt{3}/2)$.

Solution: Differentiating both sides with respect to x,

$$\frac{d}{dx}(x^2 + y^2) = \frac{d}{dx}(1)$$

$$\frac{d}{dx}(x^2) + \frac{d}{dx}(y^2) = 0$$

$$2x + 2y\frac{dy}{dx} = 0$$

From the chain rule, $\frac{d}{dx}[y(x)]^2 = 2y(x)\left(\frac{dy}{dx}\right)$. Here we use the notation $y = y(x)$ to emphasize that we are treating y as a function of x. Now, to find the slope, dy/dx, at the point $(\frac{1}{2}, \sqrt{3}/2)$, insert $\frac{1}{2}$ for x and $\sqrt{3}/2$ for y, obtaining,

$$2 \cdot \frac{1}{2} + 2\frac{\sqrt{3}}{2}\frac{dy}{dx} = 0$$

Solving for dy/dx, you obtain $dy/dx = -\sqrt{3}/3$.

We will adopt the notation $\left.\frac{dy}{dx}\right|_{(1/2, \sqrt{3}/2)}$ to denote the slope at the point $(1/2, \sqrt{3}/2)$. You couldn't merely say $\frac{dy}{dx}(\frac{1}{2})$, because there are two points on the curve at $x = \frac{1}{2}$, namely $(1/2, \sqrt{3}/2)$ and $(1/2, -\sqrt{3}/2)$.

C. Exercise caution in using the rules for differentiation!

You must apply the chain, product, and quotient rules carefully when you differentiate implicitly. You must practice.

EXAMPLE 6-5: Differentiate the expression xy with respect to x.

Solution: Treat y as a function of x and use the product rule:

$$\frac{d}{dx}(xy) = \left(\frac{d}{dx}x\right)y + x\left(\frac{d}{dx}y\right) = y + x\frac{dy}{dx}$$

Note: Differentiating with respect to x is entirely different from differentiating with respect to y.

EXAMPLE 6-6: Differentiate the expression $y^{1/2}$.

Solution: Using the chain rule,

$$\frac{d}{dx}(y^{1/2}) = \frac{1}{2}y^{-1/2}\left(\frac{dy}{dx}\right)$$

EXAMPLE 6-7: Differentiate the expression $\sin(x^2y^2)$.

Solution: First use the chain rule:

$$\frac{d}{dx}[\sin(x^2y^2)] = \cos(x^2y^2)\left[\frac{d}{dx}(x^2y^2)\right]$$

and then the product rule:

$$= \cos(x^2y^2)\left[\left(\frac{d}{dx}x^2\right)y^2 + x^2\left(\frac{d}{dx}y^2\right)\right]$$

$$= \cos(x^2y^2)\left[2xy^2 + x^2\left(2y\frac{dy}{dx}\right)\right]$$

EXAMPLE 6-8: On the curve $xy + y^{1/2} = x + y$, find an expression for dy/dx, and calculate $\left.\dfrac{dy}{dx}\right|_{(\frac{2}{3},4)}$

Solution: Differentiate both sides with respect to x:

$$\frac{d}{dx}(xy) + \frac{d}{dx}(y^{1/2}) = \frac{d}{dx}x + \frac{d}{dx}y$$

$$\left(\frac{d}{dx}x\right)y + x\left(\frac{dy}{dx}\right) + \frac{1}{2}y^{-1/2}\left(\frac{dy}{dx}\right) = 1 + \frac{dy}{dx}$$

$$y + x\left(\frac{dy}{dx}\right) + \frac{1}{2}y^{-1/2}\left(\frac{dy}{dx}\right) = 1 + \frac{dy}{dx}$$

Solving for dy/dx,

$$\left(x + \frac{1}{2}y^{-1/2} - 1\right)\frac{dy}{dx} = 1 - y$$

$$\frac{dy}{dx} = \frac{1 - y}{x + \frac{1}{2}y^{-1/2} - 1}$$

Before you find $\left.\dfrac{dy}{dx}\right|_{(\frac{2}{3},4)}$, you should verify that $(\frac{2}{3}, 4)$ actually lies on the curve. Substituting $(\frac{2}{3}, 4)$ into the original equation, you get

$$\tfrac{2}{3} \cdot 4 + 4^{1/2} = \tfrac{2}{3} + 4$$

so $(\frac{2}{3}, 4)$ does satisfy the equation defining the curve. Finally, calculate

$$\left.\frac{dy}{dx}\right|_{(\frac{2}{3},4)} = \frac{1 - 4}{[(\frac{2}{3}) + \frac{1}{2}4^{-1/2} - 1]} = 36$$

EXAMPLE 6-9: On the curve $y^2 + e^{xy} = \sin(x^2y^2) + 2$, find $\left.\dfrac{dy}{dx}\right|_{(0,1)}$.

Solution: First, verify that (0, 1) does lie on the curve:

$$1^2 + e^{1 \cdot 0} = \sin(0^2 1^2) + 2$$

$$2 = 2$$

Differentiate both sides with respect to x:

$$2y\frac{dy}{dx} + e^{xy}\left[\frac{d}{dx}(xy)\right] = \cos(x^2y^2)\left[\frac{d}{dx}(x^2y^2)\right] + 0$$

$$2y\frac{dy}{dx} + e^{xy}\left(y + x\frac{dy}{dx}\right) = \cos(x^2y^2)\left(2xy^2 + 2x^2y\frac{dy}{dx}\right)$$

Rather than solve for dy/dx and then evaluate at (0, 1), you can save some effort by first evaluating at (0, 1):

$$2\frac{dy}{dx} + 1\left(1 + 0\frac{dy}{dx}\right) = \cos(0)(0 + 0)$$

$$2\frac{dy}{dx} + 1 = 0$$

$$\frac{dy}{dx} = -\frac{1}{2}$$

Thus,

$$\left.\frac{dy}{dx}\right|_{(0,1)} = -\frac{1}{2}.$$

EXAMPLE 6-10: On the curve $1 + (x^2y)/(x + y) = x^2 + y^2$, find dy/dx.

Solution: Applying d/dx to both sides of the equation, you get

$$\frac{d}{dx}(1) + \frac{d}{dx}\left(\frac{x^2y}{x+y}\right) = \frac{d}{dx}(x^2) + \frac{d}{dx}(y^2)$$

$$0 + \frac{(x+y)\dfrac{d}{dx}(x^2y) - x^2y\dfrac{d}{dx}(x+y)}{(x+y)^2} = 2x + 2y\frac{dy}{dx}$$

$$\frac{(x+y)\left[\left(\dfrac{d}{dx}x^2\right)y + x^2\left(\dfrac{d}{dx}y\right)\right] - x^2y\left(1 + \dfrac{dy}{dx}\right)}{(x+y)^2} = 2x + 2y\frac{dy}{dx}$$

$$\frac{(x+y)\left(2xy + x^2\cdot\dfrac{dy}{dx}\right) - x^2y\left(1 + \dfrac{dy}{dx}\right)}{(x+y)^2} = 2x + 2y\frac{dy}{dx}$$

Because this is a linear equation in dy/dx, you can solve for dy/dx:

$$(x+y)2xy + (x+y)x^2\frac{dy}{dx} - x^2y - x^2y\frac{dy}{dx}$$

$$= (x+y)^2\cdot 2x + (x+y)^2\cdot 2y\frac{dy}{dx}$$

$$[(x+y)x^2 - x^2y - 2y(x+y)^2]\frac{dy}{dx} = 2x(x+y)^2 - (x+y)2xy + x^2y$$

$$[x^3 - 2y(x+y)^2]\frac{dy}{dx} = 3x^2y + 2x^3$$

$$\frac{dy}{dx} = \frac{3x^2y + 2x^3}{x^3 - 2y(x+y)^2}$$

EXAMPLE 6-11: On the curve $x^2 + 2xy - y^2 + 2y + 4 = 0$, find all points where the tangent line is parallel to the tangent line at $(-2, 2)$.

Solution: Applying d/dx, you get

$$2x + 2x\frac{dy}{dx} + 2y - 2y\frac{dy}{dx} + 2\frac{dy}{dx} = 0$$

$$\frac{dy}{dx} = \frac{-(x+y)}{x-y+1}$$

The slope of the tangent line at $(-2, 2)$ is

$$\left.\frac{dy}{dx}\right|_{(-2,2)} = \frac{-(-2+2)}{-2-2+1} = 0$$

The tangent line at (x, y) will be parallel to the tangent line at $(-2, 2)$ only when their slopes are equal, that is, when $\left.\dfrac{dy}{dx}\right|_{(x,y)} = 0$. So,

$$\frac{-(x+y)}{x-y+1} = 0$$

$$y = -x$$

You need to find all points on the curve $x^2 + 2xy - y^2 + 2y + 4 = 0$ where $y = -x$. Substituting $-x$ for y in the original expression, you get

$$x^2 + 2x(-x) - (-x)^2 + 2(-x) + 4 = 0$$

$$x^2 + x - 2 = 0$$

$$(x-1)(x+2) = 0$$

The tangent line has slope 0 when $x = 1$ and $x = -2$, that is, at $(1, -1)$ and $(-2, 2)$. So $(-1, 1)$ is the only other point on the curve where the tangent line is parallel to the tangent line at $(-2, 2)$.

D. Derivatives of inverse functions

When you know the derivative of a function, you can use implicit differentiation to find the derivative of its inverse function.

EXAMPLE 6-12: Find the derivative of $y = \ln x$.

Solution: You can use the fact that the exponential function is the inverse of the natural logarithm function. Solve for x in terms of y:

$$e^y = x$$

Using the chain rule, differentiate both sides with respect to x, obtaining

$$e^y \cdot \frac{dy}{dx} = 1$$

Solving for dy/dx, you find

$$\frac{dy}{dx} = \frac{1}{e^y} = \frac{1}{x}$$

Thus, $(d/dx)(\ln x) = 1/x$. This is the formula that you learned in Chapter 4.

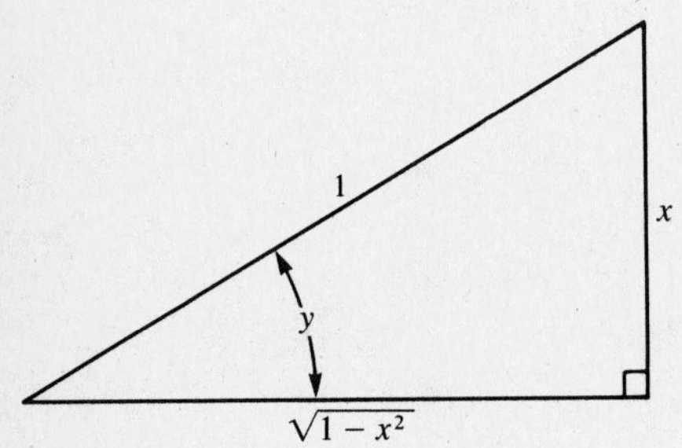

Figure 6-3
$y = \text{arc sin } x$.

EXAMPLE 6-13: For $y = \text{arc sin } x$, find dy/dx.

Solution: You can use the inverse function, $\sin y = x$. Differentiating both sides with respect to x, you get

$$\cos y \frac{dy}{dx} = 1$$

$$\frac{dy}{dx} = \frac{1}{\cos y} = \frac{1}{\cos(\text{arc sin } x)}$$

From the triangle in Figure 6-3, you see that $\cos(\text{arc sin } x) = \sqrt{1 - x^2}$. Thus, $(d/dx)(\text{arc sin } x) = 1/\sqrt{1 - x^2}$.

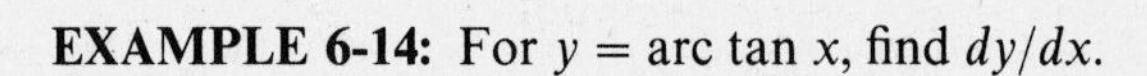

EXAMPLE 6-14: For $y = \text{arc tan } x$, find dy/dx.

Solution: Use the inverse function, $\tan y = x$, and differentiate:

$$\sec^2 y \frac{dy}{dx} = 1$$

$$\frac{dy}{dx} = \frac{1}{\sec^2 y} = \cos^2 y = \cos^2(\text{arc tan } x)$$

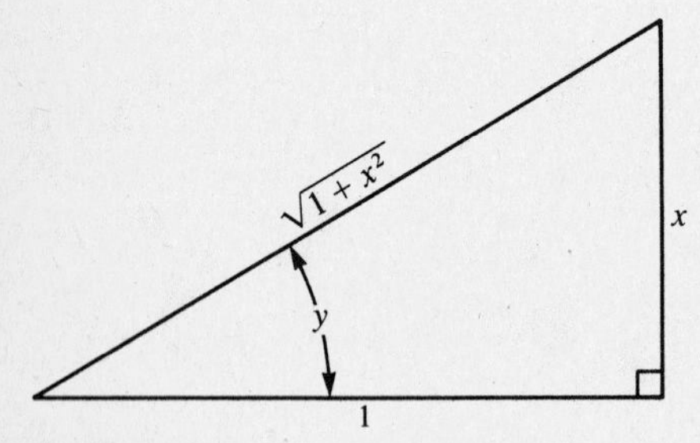

Figure 6-4
$y = \text{arc tan } x$.

Figure 6-4 shows that $\cos(\text{arc tan } x) = 1/\sqrt{1 + x^2}$, so

$$\frac{dy}{dx} = \frac{1}{1 + x^2}$$

6-2. Logarithmic Differentiation

Logarithmic differentiation is primarily a labor-saving technique for differentiating functions that might otherwise involve repeated use of the product or quotient rules.

A. The formula for logarithmic differentiation

Because $\dfrac{d}{dx}(\ln f(x)) = \dfrac{f'(x)}{f(x)}$, you can see that

LOGARITHMIC DIFFERENTIATION

$$f'(x) = f(x)\cdot\frac{d}{dx}(\ln f(x)). \tag{6-1}$$

Logarithmic differentiation is the process of applying this formula.

EXAMPLE 6-15: Calculate the derivative of $f(x) = x^3 (\sin x)\, e^x$.

Solution: You'll need to complete the following steps:
(1) Find $\ln f(x)$.
(2) Simplify $\ln f(x)$ using the following properties of the logarithm:

- $\log ab = \log a + \log b$
- $\log a^m = m \log a$
- $\log (a/b) = \log a - \log b$

(3) Differentiate.
(4) Multiply the result by $f(x)$.

In this case you have

$$\begin{aligned}
&\textbf{(1)}\ \ln f(x) = \ln\left(x^3(\sin x)e^x\right)\\
&\textbf{(2)}\qquad\quad\ = \ln x^3 + \ln(\sin x) + \ln e^x\\
&\qquad\qquad\quad\ = 3\ln x + \ln(\sin x) + x\ln e\\
&\qquad\qquad\quad\ = 3\ln x + \ln(\sin x) + x
\end{aligned}$$

$$\textbf{(3)}\ \frac{d}{dx}(\ln f(x)) = 3\cdot\frac{1}{x} + \frac{\cos x}{\sin x} + 1$$

$$\begin{aligned}
\textbf{(4)}\ \frac{d}{dx} f(x) &= f(x)\frac{d}{dx}\ln f(x)\\
&= x^3(\sin x)e^x\left(\frac{3}{x} + \cot x + 1\right)
\end{aligned}$$

You would usually use the product rule twice to calculate

$$\frac{d}{dx}[x^3(\sin x)e^x].$$

By using logarithmic differentiation, you didn't need the product rule at all.

There is a slightly different method of performing logarithmic differentiation that you can try: Let $y = f(x)$, take the natural logarithm of both sides, differentiate implicitly, and solve for dy/dx.

EXAMPLE 6-16: Calculate the derivative of $f(x) = \dfrac{\sqrt{x}(x^3 - 1)^2}{x - 5}$.

Solution: Let $y = f(x)$, so

$$\begin{aligned}
\ln y &= \ln\left[\frac{\sqrt{x}(x^3 - 1)^2}{x - 5}\right]\\
&= \ln\sqrt{x} + \ln(x^3 - 1)^2 - \ln(x - 5)\\
&= \tfrac{1}{2}\ln x + 2\ln(x^3 - 1) - \ln(x - 5)
\end{aligned}$$

Differentiating implicity,

$$\frac{1}{y}\cdot\frac{dy}{dx} = \frac{1}{2}\cdot\frac{1}{x} + 2\cdot\frac{3x^2}{x^3 - 1} - \frac{1}{x - 5}$$

Solving for dy/dx,

$$\frac{dy}{dx} = y\left(\frac{1}{2x} + \frac{6x^2}{x^3 - 1} - \frac{1}{x - 5}\right) = \frac{\sqrt{x}(x^3 - 1)^2}{x - 5}\left(\frac{1}{2x} + \frac{6x^2}{x^3 - 1} - \frac{1}{x - 5}\right)$$

B. Use logarithmic differentiation to avoid repeated use of the product and quotient rules.

Your greatest advantage in using logarithmic differentiation occurs when you are differentiating a function that consists of many products and quotients.

EXAMPLE 6-17: Differentiate $f(x) = (x^3\sqrt{1+x}(3-x^2)^4)/((x+1)^7\sqrt{x+2})$.

Solution: Following the procedure of Example 6-15,

$$\ln f(x) = 3\ln x + \tfrac{1}{2}\ln(1+x) + 4\ln(3-x^2) - 7\ln(x+1) - \tfrac{1}{2}\ln(x+2)$$

$$\frac{d}{dx}\ln f(x) = \frac{3}{x} + \frac{1}{2}\cdot\frac{1}{1+x} + 4\cdot\frac{-2x}{3-x^2} - 7\cdot\frac{1}{x+1} - \frac{1}{2}\cdot\frac{1}{x+2}$$

$$= \frac{3}{x} + \frac{1}{2(1+x)} - \frac{8x}{3-x^2} - \frac{7}{x+1} - \frac{1}{2(x+2)}$$

$$f'(x) = f(x)\cdot\frac{d}{dx}\ln f(x)$$

$$= \frac{x^3\sqrt{1+x}(3-x^2)^4}{(x+1)^7\sqrt{x+2}}$$

$$\times\left(\frac{3}{x} + \frac{1}{2(1+x)} - \frac{8x}{3-x^2} - \frac{7}{x+1} - \frac{1}{2(x+2)}\right)$$

EXAMPLE 6-18: For $f(x) = (\sin x\cos x\sin(x+1))/(\sin(x+3))$, find $f'(x)$.

Solution: Following our procedures,

$$\ln f(x) = \ln\sin x + \ln\cos x + \ln\sin(x+1) - \ln\sin(x+3)$$

$$\frac{d}{dx}\ln f(x) = \frac{\cos x}{\sin x} + \frac{-\sin x}{\cos x} + \frac{\cos(x+1)}{\sin(x+1)} - \frac{\cos(x+3)}{\sin(x+3)}$$

$$= \cot x - \tan x + \cot(x+1) - \cot(x+3)$$

$$f'(x) = f(x)\cdot\frac{d}{dx}\ln f(x)$$

$$= \frac{\sin x\cos x\sin(x+1)}{\sin(x+3)}[\cot x - \tan x + \cot(x+1) - \cot(x+3)]$$

You need only try to differentiate any of the functions in the preceding examples in the usual way to appreciate the power of logarithmic differentiation.

C. Use logarithmic differentiation for functions with the variable in an exponent.

Functions in which the variable appears in an exponent (for example, x^x or $(x^2+1)^{\sin x}$) may require logarithmic differentiation.

EXAMPLE 6-19: Differentiate $f(x) = x^x$.

Solution: Differentiating logarithmically,

$$\ln f(x) = \ln(x^x) = x\ln x$$

$$\frac{d}{dx}\ln f(x) = \ln x + 1$$

$$f'(x) = f(x)\frac{d}{dx}\ln f(x)$$

$$= x^x(\ln x + 1)$$

EXAMPLE 6-20: Differentiate $f(x) = (x^2 + 1)^{\sin x}$.

Solution: Following our procedures,

$$\ln f(x) = \ln[(x^2 + 1)^{\sin x}] = (\sin x)\ln(x^2 + 1)$$

$$\frac{d}{dx}\ln f(x) = (\cos x)\ln(x^2 + 1) + (\sin x)\cdot\frac{2x}{x^2 + 1}$$

$$f'(x) = f(x)\frac{d}{dx}\ln f(x)$$

$$= (x^2 + 1)^{\sin x}\left((\cos x)\ln(x^2 + 1) + (\sin x)\cdot\frac{2x}{x^2 + 1}\right)$$

SUMMARY

1. Use implicit differentiation to find dy/dx when the equation relating x and y is not easily solved for y.
2. To differentiate implicitly, differentiate both sides of the equation with respect to x, treating y as a function of x.
3. Implicit differentiation requires careful application of the chain, product, and quotient rules.
4. You can use implicit differentiation to find tangent lines to curves.
5. You can use implicit differentiation to find derivatives of inverse functions.
6. Use logarithmic differentiation to avoid repeated use of the product and quotient rules.
7. Use logarithmic differentiation to differentiate a function of x raised to an exponent that is a function of x.

SOLVED PROBLEMS

For all problems that ask you to find $\left.\frac{dy}{dx}\right|_{(a,b)}$ for a curve, (a, b) is on the curve.

PROBLEM 6-1 On the curve $x^2 + y^2 = 4$, find $\left.\frac{dy}{dx}\right|_{(1,\sqrt{3})}$ and $\left.\frac{dy}{dx}\right|_{(1,-\sqrt{3})}$.

Solution: Differentiating with respect to x,

$$\frac{d}{dx}(x^2 + y^2) = \frac{d}{dx}(4)$$

$$2x + 2y\frac{dy}{dx} = 0$$

$$\frac{dy}{dx} = \frac{-x}{y}$$

At $x = 1$ and $y = \sqrt{3}$, you have

$$\frac{dy}{dx} = \frac{-1}{\sqrt{3}}$$

At $x = 1$ and $y = -\sqrt{3}$, you have

$$\frac{dy}{dx} = \frac{-1}{-\sqrt{3}} = \frac{1}{\sqrt{3}}$$

[See Section 6-1.]

PROBLEM 6-2 If $\dfrac{x+y}{x-y} = \sqrt{x+y+6}$, find $\left.\dfrac{dy}{dx}\right|_{(2,1)}$.

Solution: Differentiating with respect to x,

$$\frac{d}{dx}\left(\frac{x+y}{x-y}\right) = \frac{d}{dx}(x+y+6)^{1/2}$$

$$\frac{(x-y)\left(1+\frac{dy}{dx}\right) - (x+y)\left(1-\frac{dy}{dx}\right)}{(x-y)^2} = \frac{1}{2}(x+y+6)^{-1/2}\left(1+\frac{dy}{dx}\right)$$

At $x = 2$ and $y = 1$, you have

$$\left(1+\frac{dy}{dx}\right) - 3\left(1-\frac{dy}{dx}\right) = \frac{1}{2}\cdot\frac{1}{3}\left(1+\frac{dy}{dx}\right)$$

$$\frac{dy}{dx}\left(1+3-\frac{1}{6}\right) = -1+3+\frac{1}{6}$$

$$\left.\frac{dy}{dx}\right|_{(2,1)} = \frac{13/6}{23/6} = \frac{13}{23}$$

[See Section 6-1.]

PROBLEM 6-3 Find dy/dx if y is defined implicitly by $e^{x+y} = (x^2 - y^2)^{1/3} + 1$.

Solution: Differentiate both sides of the equation:

$$e^{x+y}\cdot\frac{d}{dx}(x+y) = \frac{1}{3}(x^2-y^2)^{-2/3}\cdot\frac{d}{dx}(x^2-y^2)$$

$$e^{x+y}\left(1+\frac{dy}{dx}\right) = \frac{1}{3}(x^2-y^2)^{-2/3}\left(2x - 2y\frac{dy}{dx}\right)$$

Solve for dy/dx:

$$e^{x+y}\cdot\frac{dy}{dx} + \frac{2}{3}(x^2-y^2)^{-2/3}y\frac{dy}{dx} = \frac{2}{3}x(x^2-y^2)^{-2/3} - e^{x+y}$$

$$\frac{dy}{dx} = \frac{2x(x^2-y^2)^{-2/3} - 3e^{x+y}}{2y(x^2-y^2)^{-2/3} + 3e^{x+y}}$$

PROBLEM 6-4 If $(x^2+y^2)^2 + 3x^2y - y^3 = 0$, find $\left.\dfrac{dy}{dx}\right|_{(\frac{1}{2},-\frac{1}{2})}$.

Solution: Differentiate with respect to x:

$$2(x^2+y^2)\frac{d}{dx}(x^2+y^2) + 3\frac{d}{dx}(x^2y) - 3y^2\cdot\frac{dy}{dx} = 0$$

$$2(x^2+y^2)\left(2x+2y\frac{dy}{dx}\right) + 3\left(x^2\cdot\frac{dy}{dx} + 2xy\right) - 3y^2\cdot\frac{dy}{dx} = 0$$

Substitute $x = \frac{1}{2}$, $y = -\frac{1}{2}$:

$$2\left(\frac{1}{4}+\frac{1}{4}\right)\left(1-\frac{dy}{dx}\right) + 3\left(\frac{1}{4}\frac{dy}{dx} - 2\cdot\frac{1}{4}\right) - \frac{3}{4}\cdot\frac{dy}{dx} = 0$$

$$1 - \frac{dy}{dx} + \frac{3}{4}\cdot\frac{dy}{dx} - \frac{3}{2} - \frac{3}{4}\cdot\frac{dy}{dx} = 0$$

$$\left.\frac{dy}{dx}\right|_{(\frac{1}{2},-\frac{1}{2})} = -\frac{1}{2}$$

[See Section 6-1.]

PROBLEM 6-5 Find the points on the graph of $x^2 + xy + y^2 = 4$ where the tangent line is horizontal.

Solution: You want those points where $dy/dx = 0$. Differentiating both sides of the equation with respect to x, you get:

$$2x + x\frac{dy}{dx} + y + 2y\frac{dy}{dx} = 0$$

$$\frac{dy}{dx}(x + 2y) = -(2x + y)$$

$$\frac{dy}{dx} = -\frac{2x + y}{x + 2y}$$

So $dy/dx = 0$ when $2x + y = 0$. You now have two conditions on the desired point (or points):

(1) They are points on the graph of $x^2 + xy + y^2 = 4$.
(2) Their coordinates satisfy $2x + y = 0$.

From the second condition, you get $y = -2x$. Substitute for y in the original equation:

$$x^2 + x(-2x) + (-2x)^2 = 4$$

$$x^2 - 2x^2 + 4x^2 = 4$$

$$x^2 = \frac{4}{3}$$

$$x = \frac{2\sqrt{3}}{3} \quad \text{or} \quad x = \frac{-2\sqrt{3}}{3}$$

When $x = 2\sqrt{3}/3$, $y = -2(2\sqrt{3}/3) = -4\sqrt{3}/3$; when $x = -2\sqrt{3}/3$, $y = -2(-2\sqrt{3}/3) = 4\sqrt{3}/3$. You conclude that the graph of $x^2 + xy + y^2 = 4$ has horizontal tangent lines at two points: $(2\sqrt{3}/3, -4\sqrt{3}/3)$ and $(-2\sqrt{3}/3, 4\sqrt{3}/3)$. [See Section 6-1.]

PROBLEM 6-6 If $y^3 - \left(\frac{x}{y}\right) + x^2 = xy + y$, find $\left.\frac{dy}{dx}\right|_{(2,1)}$.

Solution: You could differentiate both sides with respect to x, but, because of the x/y term, you can simplify your work by first multiplying both sides of the equation by y. You can do this because $y \neq 0$ at (2, 1), the point of interest. So,

$$y^4 - x + x^2y = xy^2 + y^2$$

Now, differentiate:

$$4y^3 \cdot \frac{dy}{dx} - 1 + x^2 \cdot \frac{dy}{dx} + 2xy = 2xy\frac{dy}{dx} + y^2 + 2y\frac{dy}{dx}$$

and substitute $x = 2$, $y = 1$:

$$4\frac{dy}{dx} - 1 + 4\frac{dy}{dx} + 4 = 4\frac{dy}{dx} + 1 + 2\frac{dy}{dx}$$

$$2\frac{dy}{dx} = -2$$

$$\left.\frac{dy}{dx}\right|_{(2,1)} = -1$$

[See Section 6-1.]

PROBLEM 6-7 Let y be defined implicitly by $\sin(x + y) = x \cos y^2$. Find dy/dx at the point (0, 0).

Solution: Differentiating both sides:

$$\cos(x + y)\frac{d}{dx}(x + y) = -x\sin(y^2)\frac{d}{dx}(y^2) + \cos y^2$$

$$\cos(x + y)\left(1 + \frac{dy}{dx}\right) = -2xy\sin(y^2)\frac{dy}{dx} + \cos y^2$$

When $x = 0$ and $y = 0$, you obtain,

$$1 + \frac{dy}{dx} = 1$$

$$\left.\frac{dy}{dx}\right|_{(0,0)} = 0$$

[See Section 6-1.]

PROBLEM 6-8 Find the equation of the line that is tangent to the curve $(x^2 - y^2)^{1/2} = x + y - 6$ at (5, 4).

Solution: First, find the slope of the tangent line, $\left.\frac{dy}{dx}\right|_{(5,4)}$:

$$\frac{1}{2}(x^2 - y^2)^{-1/2}\left(2x - 2y\frac{dy}{dx}\right) = 1 + \frac{dy}{dx}$$

At (5, 4), you have

$$\frac{1}{2}(9)^{-1/2}\left(10 - 8\frac{dy}{dx}\right) = 1 + \frac{dy}{dx}$$

$$\frac{dy}{dx} = \frac{2}{7}$$

The tangent line has slope $\frac{2}{7}$ and passes through the point (5, 4). You can write the equation of the line in point-slope form:

$$y - 4 = \frac{2}{7}(x - 5)$$

[See Section 6-1.]

PROBLEM 6-9 Find the equation of the tangent line to the graph of $\cos xy = x \sin xy$ at $(1, \pi/4)$.

Solution: If you find $\left.\frac{dy}{dx}\right|_{(1,\pi/4)}$, you'll have the information necessary to write the equation of the tangent line in point-slope form. Differentiating, you find

$$(-\sin xy)\left(x\frac{dy}{dx} + y\right) = (x \cos xy)\left(x\frac{dy}{dx} + y\right) + \sin xy$$

When $x = 1$ and $y = \pi/4$, $xy = \pi/4$, so $\sin xy = \cos xy = \sqrt{2}/2$. The simplified equation is

$$-\frac{\sqrt{2}}{2}\left(\frac{dy}{dx} + \frac{\pi}{4}\right) = \frac{\sqrt{2}}{2}\left(\frac{dy}{dx} + \frac{\pi}{4}\right) + \frac{\sqrt{2}}{2}$$

$$-\left(\frac{dy}{dx} + \frac{\pi}{4}\right) = \frac{dy}{dx} + \frac{\pi}{4} + 1$$

$$2\frac{dy}{dx} = -\left(1 + \frac{\pi}{2}\right)$$

$$\left.\frac{dy}{dx}\right|_{(1,\pi/4)} = -\frac{2 + \pi}{4}$$

and the equation of the tangent line is

$$y - \frac{\pi}{4} = -\frac{2 + \pi}{4}(x - 1)$$

[See Section 6-1.]

PROBLEM 6-10 Show that the graphs of $y = x^{4/3} + x$ and $x = y + y^5$ have a common tangent line at the origin.

Solution: Find $\left.\dfrac{dy}{dx}\right|_{(0,0)}$ for both curves. For $y = x^{4/3} + x$,

$$\frac{dy}{dx} = \frac{4}{3}x^{1/3} + 1$$

$$\left.\frac{dy}{dx}\right|_{(0,0)} = 1$$

For $x = y + y^5$,

$$1 = \frac{dy}{dx} + 5y^4 \cdot \frac{dy}{dx}$$

$$\left.\frac{dy}{dx}\right|_{(0,0)} = 1$$

Both tangent lines have slope 1 and pass through (0, 0). Therefore, they are the same line.

[See Section 6-1.]

PROBLEM 6-11 Show that the graphs of $xy = \sqrt{2}$ and $x^2 - y^2 = 1$ have perpendicular tangent lines at their points of intersection.

Solution: Find the points of intersection by solving $xy = \sqrt{2}$ for y,

$$y = \frac{\sqrt{2}}{x}$$

and substitute the value of y into the second equation:

$$x^2 - \left(\frac{\sqrt{2}}{x}\right)^2 = 1$$

$$x^4 - 2 = x^2$$

$$(x^2 + 1)(x^2 - 2) = 0$$

$$x^2 - 2 = 0$$

$$x = \pm\sqrt{2}$$

The curves meet at $(\sqrt{2}, 1)$ and $(-\sqrt{2}, -1)$. For the curve $xy = \sqrt{2}$,

$$x\frac{dy}{dx} + y = 0$$

$$\frac{dy}{dx} = -\frac{y}{x}$$

$$\left.\frac{dy}{dx}\right|_{(\sqrt{2},1)} = \left.\frac{dy}{dx}\right|_{(-\sqrt{2},-1)} = \frac{-1}{\sqrt{2}}$$

For the curve $x^2 - y^2 = 1$,

$$2x - 2y\frac{dy}{dx} = 0$$

$$\frac{dy}{dx} = \frac{x}{y}$$

$$\left.\frac{dy}{dx}\right|_{(\sqrt{2},1)} = \left.\frac{dy}{dx}\right|_{(-\sqrt{2},-1)} = \sqrt{2}$$

Because the slopes are negative reciprocals, the tangent lines are perpendicular. [See Section 6-1.]

PROBLEM 6-12 For the curve $xy + x^2y^2 = 1$, find any points where the tangent line has slope -1.

Solution: Check to see if the equation $dy/dx = -1$ has any solution among the points of the graph. First, find dy/dx:

$$x\frac{dy}{dx} + y + x^2 2y\frac{dy}{dx} + 2xy^2 = 0$$

$$\frac{dy}{dx}(x + 2x^2y) = -(2xy^2 + y)$$

$$\frac{dy}{dx} = -\frac{y + 2xy^2}{x + 2x^2y}$$

Setting dy/dx equal to -1, you get

$$y + 2xy^2 = x + 2x^2y$$

$$2xy^2 - 2x^2y + y - x = 0$$

$$2xy(y - x) + (y - x) = 0$$

$$(2xy + 1)(y - x) = 0$$

There are two possible conditions. Either $2xy + 1 = 0$ (that is, $xy = -\frac{1}{2}$), or $y - x = 0$. The first condition is not satisfied by any points on the curve:

$$xy + (xy)^2 = -\tfrac{1}{2} + (-\tfrac{1}{2})^2 = -\tfrac{1}{4} \neq 1$$

But there are points on the curve that satisfy the second condition:

$$x^2 + x^2x^2 = 1$$

$$x^4 + x^2 - 1 = 0$$

From the quadratic formula, $x^2 = \dfrac{-1 \pm \sqrt{5}}{2}$. The negative value is discarded, as x^2 can't be negative. Thus you have two points on the curve where the tangent line has a slope of -1:

$$\left(\sqrt{\frac{\sqrt{5} - 1}{2}}, \sqrt{\frac{\sqrt{5} - 1}{2}}\right) \text{ and } \left(-\sqrt{\frac{\sqrt{5} - 1}{2}}, -\sqrt{\frac{\sqrt{5} - 1}{2}}\right).$$

[See Section 6-1.]

PROBLEM 6-13 For $x^{1/2} + y^{1/2} = a^{1/2}$, find d^2y/dx^2.

Solution: Differentiating implicitly,

$$\frac{1}{2}x^{-1/2} + \frac{1}{2}y^{-1/2} \cdot \frac{dy}{dx} = 0$$

$$\frac{dy}{dx} = -\frac{y^{1/2}}{x^{1/2}}$$

Now, differentiate a second time, using the quotient rule on the right:

$$\frac{d^2y}{dx^2} = -\frac{x^{1/2}\frac{1}{2}y^{-1/2} \cdot \frac{dy}{dx} - y^{1/2}\frac{1}{2}x^{-1/2}}{x}$$

$$= \frac{\left(\frac{y}{x}\right)^{1/2} - \left(\frac{x}{y}\right)^{1/2} \cdot \frac{dy}{dx}}{2x}$$

Since $dy/dx = -(y/x)^{1/2}$,

$$\frac{d^2y}{dx^2} = \frac{\left(\frac{y}{x}\right)^{1/2} + 1}{2x} = \frac{y^{1/2} + x^{1/2}}{2x^{3/2}}$$

Finally, note that $x^{1/2} + y^{1/2} = a^{1/2}$, so you can write the answer in its simplest form as

$$\frac{d^2y}{dx^2} = \frac{a^{1/2}}{2x^{3/2}}$$

Alternatively, you could perform a second implicit differentiation immediately following the first one. However, you must exercise caution in the handling of the dy/dx term. After division by $\frac{1}{2}$ to remove the coefficients,

$$x^{-1/2} + y^{-1/2} \cdot \frac{dy}{dx} = 0$$

Differentiate again, using the product rule on the second factor:

$$-\frac{1}{2}x^{-3/2} + y^{-1/2}\frac{d}{dx}\left(\frac{dy}{dx}\right) + \frac{d}{dx}(y^{-1/2})\frac{dy}{dx} = 0$$

$$-\frac{1}{2}x^{-3/2} + y^{-1/2} \cdot \frac{d^2y}{dx^2} - \frac{1}{2}y^{-3/2} \cdot \frac{dy}{dx} \cdot \frac{dy}{dx} = 0$$

Substitute $-(y/x)^{1/2}$ for dy/dx, and solve for d^2y/dx^2:

$$y^{-1/2} \cdot \frac{d^2y}{dx^2} = \frac{1}{2}\left[y^{-3/2}\left(-\left(\frac{y}{x}\right)^{1/2}\right)^2 + x^{-3/2}\right]$$

$$\frac{d^2y}{dx^2} = \frac{1}{2}y^{1/2}\left[\frac{y^{-1/2}}{x} + x^{-3/2}\right]$$

$$= \frac{1}{2}\left[\left(\frac{1}{x}\right) + \left(\frac{y^{1/2}}{x^{3/2}}\right)\right] = \frac{x^{1/2} + y^{1/2}}{2x^{3/2}} = \frac{a^{1/2}}{2x^{3/2}}$$

[See Section 6-1.]

PROBLEM 6-14 On the curve $x^3 + y^3 = 16$, find d^2y/dx^2.

Solution: Differentiating implicitly,

$$3x^2 + 3y^2 \cdot \frac{dy}{dx} = 0$$

$$\frac{dy}{dx} = \frac{-x^2}{y^2}$$

$$\frac{d^2y}{dx^2} = -\frac{2xy - 2x^2 \cdot dy/dx}{y^3} = -\frac{2xy + (2x^4/y^2)}{y^3}$$

$$= -\frac{2x(y^3 + x^3)}{y^5} = \frac{-32x}{y^5}$$

[See Section 6-1.]

PROBLEM 6-15 For $y = \text{arc cos } x$, use implicit differentiation to find dy/dx.

Solution: You use the inverse function, $\cos y = x$. Differentiate both sides with respect to x:

$$-(\sin y)\frac{dy}{dx} = 1$$

$$\frac{dy}{dx} = \frac{-1}{\sin y} = \frac{-1}{\sin(\text{arc cos } x)}$$

To simplify this expression, construct a triangle in which $\cos y = x$ (see Figure 6-5). You conclude that $\sin(\text{arc cos } x) = \sqrt{1 - x^2}$, so

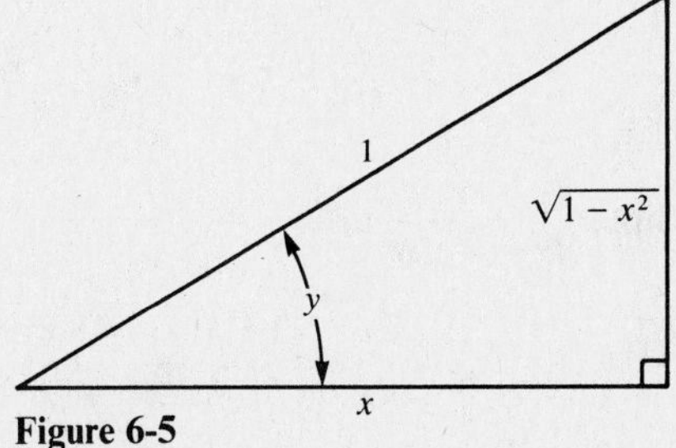

Figure 6-5
$y = \text{arc cos } x$.

$$\frac{dy}{dx} = \frac{-1}{\sqrt{1 - x^2}}$$

You already know this formula for the derivative of the arc cos function.

[See Section 6-1.]

PROBLEM 6-16 For $y = (x - 1)(x - 2)^2(x - 3)^3(x - 4)^4$, find dy/dx.

Solution: Use logarithmic differentiation to avoid the somewhat complicated process for finding the derivative of a product of four distinct factors. Using the following two properties of the logarithm,

$$\log ab = \log a + \log b \quad \text{and} \quad \log a^m = m \log a$$

you obtain

$$\ln y = \ln(x - 1) + 2\ln(x - 2) + 3\ln(x - 3) + 4\ln(x - 4)$$

Differentiating both sides,

$$\frac{1}{y}\cdot\frac{dy}{dx} = \frac{1}{x-1} + \frac{2}{x-2} + \frac{3}{x-3} + \frac{4}{x-4}$$

Now, solve for dy/dx, using y as originally given:

$$\frac{dy}{dx} = (x - 1)(x - 2)^2(x - 3)^3(x - 4)^4\left(\frac{1}{x-1} + \frac{2}{x-2} + \frac{3}{x-3} + \frac{4}{x-4}\right)$$

[See Section 6-2.]

PROBLEM 6-17 For $y = [(x + 9)/(x - 9)]^{1/6}$, find dy/dx.

Solution: Again, simplify the problem by taking advantage of the properties of logarithms:

$$\ln y = \frac{1}{6}\ln\left(\frac{x+9}{x-9}\right)$$

$$= \frac{1}{6}[\ln(x + 9) - \ln(x - 9)]$$

$$\frac{d}{dx}\ln y = \frac{1}{6}\left(\frac{1}{x+9} - \frac{1}{x-9}\right)$$

$$\frac{dy}{dx} = y\left(\frac{d}{dx}\ln y\right)$$

$$= \frac{1}{6}\left(\frac{x+9}{x-9}\right)^{1/6}\left(\frac{1}{x+9} - \frac{1}{x-9}\right)$$

[See Section 6-2.]

PROBLEM 6-18 For $y = \sqrt[5]{\dfrac{(x + 4)(x^2 - 1)^3}{x^2 + x + 1}}$, find $\dfrac{dy}{dx}$.

Solution: From the properties of the logarithm,

$$\ln y = \frac{1}{5}[\ln(x + 4) + 3\ln(x^2 - 1) - \ln(x^2 + x + 1)]$$

Of course, you resist the temptation to "simplify" the log of a sum!
Differentiating,

$$\frac{d}{dx}\ln y = \frac{1}{y}\cdot\frac{dy}{dx} = \frac{1}{5}\left[\frac{1}{x+4} + \frac{3}{x^2-1}\cdot 2x - \frac{2x+1}{x^2+x+1}\right]$$

Since $\dfrac{dy}{dx} = y\dfrac{d}{dx}\ln y$,

$$\frac{dy}{dx} = \frac{1}{5}\sqrt[5]{\frac{(x + 4)(x^2 - 1)^3}{x^2 + x + 1}}\left[\frac{1}{x+4} + \frac{6x}{x^2-1} - \frac{2x+1}{x^2+x+1}\right]$$

[See Section 6-2.]

PROBLEM 6-19 For $y = x^{\ln x}$, find dy/dx.

Solution: This function is neither a constant to a variable power (e.g., 3^x) nor a variable to a constant power (e.g., x^4). So, differentiate logarithmically:

$$\ln y = \ln(x^{\ln x}) = (\ln x)(\ln x) = (\ln x)^2$$

$$\frac{d}{dx}\ln y = \frac{1}{y}\cdot\frac{dy}{dx} = 2\ln x\left(\frac{1}{x}\right)$$

$$\frac{dy}{dx} = y\frac{d}{dx}\ln y = x^{\ln x}\cdot\frac{2\ln x}{x}$$

$$= \frac{2x^{\ln x}\cdot\ln x}{x}$$

[See Section 6-2.]

PROBLEM 6-20 For $y = x^{(x^x)}$, find $\left.\dfrac{dy}{dx}\right|_{(2,16)}$.

Solution: As you might suspect, this problem is similar to the preceding one, but it is a bit more complex. You will need to apply the logarithm twice:

$$\ln y = \ln[x^{(x^x)}] = x^x\cdot\ln x$$

$$\ln(\ln y) = \ln(x^x\cdot\ln x) = x\ln x + \ln(\ln x)$$

Now, differentiate carefully:

$$\frac{1}{\ln y}\cdot\frac{1}{y}\cdot\frac{dy}{dx} = x\cdot\frac{1}{x} + \ln x + \frac{1}{\ln x}\cdot\frac{1}{x}$$

Substituting $x = 2$ and $y = 16$, you obtain:

$$\frac{1}{\ln(16)}\cdot\frac{1}{16}\cdot\frac{dy}{dx} = 1 + \ln 2 + \frac{1}{2\ln 2}$$

$$\left.\frac{dy}{dx}\right|_{(2,16)} = 16\ln(16)\left(1 + \ln 2 + \frac{1}{2\ln 2}\right)$$

[See Section 6-2.]

PROBLEM 6-21 For $y = [(x-1)/(2x+1)]^3(4x+3)$, find d^2y/dx^2.

Solution: Take the natural logarithm of both sides:

$$\ln y = 3\ln(x-1) - 3\ln(2x+1) + \ln(4x+3)$$

Differentiate:

$$\frac{1}{y}\cdot\frac{dy}{dx} = \frac{3}{x-1} - \frac{6}{2x+1} + \frac{4}{4x+3}$$

$$\frac{dy}{dx} = y\left(\frac{3}{x-1} - \frac{6}{2x+1} + \frac{4}{4x+3}\right)$$

Differentiate again, using the product rule on the right:

$$\frac{d^2y}{dx^2} = y\left[\frac{-3}{(x-1)^2} + \frac{12}{(2x+1)^2} - \frac{16}{(4x+3)^2}\right] + \frac{dy}{dx}\left(\frac{3}{x-1} - \frac{6}{2x+1} + \frac{4}{4x+3}\right)$$

Finally, substitute for y and dy/dx and simplify:

$$\frac{d^2y}{dx^2} = \left(\frac{x-1}{2x+1}\right)^3(4x+3)$$

$$\times\left[\frac{-3}{(x-1)^2} + \frac{12}{(2x+1)^2} - \frac{16}{(4x+3)^2} + \left(\frac{3}{x-1} - \frac{6}{2x+1} + \frac{4}{4x+3}\right)^2\right]$$

[See Section 6-2.]

Supplementary Exercises

In Problems 6-22 through 6-35 use implicit differentiation to find dy/dx.

6-22 $4x^2 + 9y^2 = 36$

6-23 $xy + x^2y^2 = 1$

6-24 $y + \sqrt{y} = 2x^2$

6-25 $x^{2/3} + y^{2/3} = 1$

6-26 $xy^{3/2} = 2x + y$

6-27 $(x + y)^4 = x^4 + y^4$

6-28 $x^2 + 3y^2 - x = 6$

6-29 $ye^y = x$

6-30 $\sin(y - x) = x \cos(x + y)$

6-31 $xy + \ln(xy) = 0$

6-32 $y^5 = x^2y^3 + x^3y^2 + 1$

6-33 $2^y = y^2 - x^2$

6-34 $x \ln x = \ln y + 1$

6-35 $(x + y)/(x - y) = \sin y$

In Problems 6-36 through 6-45 find the slope of the tangent line to the given curve at the given point $\left(\left.\frac{dy}{dx}\right|_{(x,y)}\right)$.

6-36 $(1/x) + (1/y) = 1$ at $(2, 2)$

6-37 $y^3 - 3y^2 + 3y = 3(x + 1)$ at $(2, 3)$

6-38 $\sqrt{xy} + xy^4 = 12$ at $(9, 1)$

6-39 $x^2y^2 = x^2 + y^2$ at $(\sqrt{2}, -\sqrt{2})$

6-40 $x(1 + y)^{1/2} + y(1 + x)^{1/2} = 4$ at $(4, 0)$

6-41 $\sqrt{(y/x)} + \sqrt{(x/y)} = \frac{5}{2}$ at $(1, 4)$

6-42 $e^{x+y} + x + y = 1$ at $(\ln 2, \ln \frac{1}{2})$

6-43 $4 \sin^2 x \cos^2 y = 3$ at $(\pi/2, \pi/6)$

6-44 $xe^y = ye^x$ at $(0, 0)$

6-45 $(x - y)/(x + y) + xy = 4$ at $(2, 2)$

In Problems 6-46 through 6-50 find the equation of the tangent line to the given curve at the given point. [See Problems 6-8 and 6-9.]

6-46 $x^3y + x^2y^2 + xy^3 + 1 = 0$ at $(1, -1)$

6-47 $x^2 - xy + y^2 = 9$ at $(3, 0)$

6-48 $x^3 + y^3 = 2$ at $(1, 1)$

6-49 $y^3 + x + y + x^3 = 20$ at $(-2, 3)$

6-50 $5y^3 - 4y^5 = x$ at $(1, 1)$

6-51 For the curve $x = (2y - 1)/(3y + 1)$, find the equation of the tangent line at the x intercept.

6-52 On the curve $y^3 - xy + 2x^2 = 0$, find all points where the tangent line is horizontal. [See Problem 6-5.]

6-53 On the curve $x^2y^3 - x^3 - y^3 = 1$, find all points where the tangent line is horizontal.

6-54 On the curve $2xy = x + 3y^2$, find all points where the tangent line is parallel to the tangent line at (4, 2).

6-55 The slant height of a cone is given by $S = \pi r\sqrt{r^2 + h^2}$. Assume that r and h are changing in such a way that S remains constant. Use implicit differentiation to find dr/dh.

6-56 On the curve $e^x + e^y = x + y + 2$, find all points where the tangent line is parallel to the line $y = x + 3$.

6-57 On the curve $x^4 - x^2y^2 + y^4 = 1$, find all points where the tangent line at (0, 1) intersects the curve.

In Problems 6-58 through 6-62 find d^2y/dx^2. [See Problems 6-13 and 6-14.]

6-58 $\sqrt{x} + \sqrt{y} = 4$

6-59 $xy + y^2 = 1$

6-60 $x = y^3 - 2y^5$

6-61 $x^2 + y^2 = 10$

6-62 $x^2 = y^2 - y$ at $(\sqrt{2}, 2)$

In Problems 6-63 through 6-75, use logarithmic differentiation to find dy/dx.

6-63 $y = (2x + 3)(x - 5)^2$

6-64 $y = (x - 1)(x - 2)^2(x - 3)^3(x - 4)^4$

6-65 $y = \sqrt{(x + 1)/(x - 1)}$

6-66 $y = \sqrt[4]{\dfrac{x^2(x^2 + 3)^3}{x^2 - 2x + 5}}$

6-67 $y = \dfrac{(x^2 + 1)\sqrt[6]{2x^4 - 7x^3 + 9}}{x + 3}$

6-68 $y = (x^2 + 5)^{6x^2+7}$

6-69 $y = (\sin x)^{\cos x}$

6-70 $y = x^{(x^2)}$

6-71 $y = x^{(2^x)}$

6-72 $y = 2^{(2^x)}$

6-73 $y = x^{\sin x}$

6-74 $y = x^{\sqrt{x+1}}$

6-75 $y = \left(\dfrac{x^2 \sin x}{x + 1}\right)^x$

In Problems 6-76 through 6-80 use logarithmic differentiation and implicit differentiation to find d^2y/dx^2. [See Problem 6-21.]

6-76 $y = (\ln x)^x$

6-77 $y = x^x$

6-78 $y = \dfrac{(x + 3)^2}{\sqrt{x - 1}}$

6-79 $y = \sqrt{(2x - 1)(x + 2)}$

6-80 $y = (\sin x)^x$

Solutions to Supplementary Exercises

(6-22) $y' = -4x/9y$

(6-23) $y' = -y/x$

(6-24) $y' = \dfrac{8x\sqrt{y}}{2\sqrt{y} + 1}$

(6-25) $y' = -(y/x)^{1/3}$

(6-26) $y' = \dfrac{2(y^{3/2} - 2)}{2 - 3xy^{1/2}}$

(6-27) $y' = \dfrac{x^3 - (x + y)^3}{(x + y)^3 - y^3}$

(6-28) $y' = (1 - 2x)/(6y)$

(6-29) $y' = 1/[(y + 1)e^y]$

(6-30) $y' = \dfrac{\cos(y - x) + \cos(x + y) - x\sin(x + y)}{\cos(y - x) + x\sin(x + y)}$

(6-31) $y' = -y/x$

(6-32) $y' = \dfrac{2xy^2 + 3x^2y}{5y^3 - 3x^2y - 2x^3}$

(6-33) $y' = (2x)/(2y - 2^y \ln 2)$

(6-34) $y' = y(1 + \ln x)$

(6-35) $y' = (2y)/(2x - (x - y)^2 \cos y)$

(6-36) -1

(6-37) $\frac{1}{4}$

(6-38) $-7/225$

(6-39) 1

(6-40) $2 - \sqrt{5}$

(6-41) 4

(6-42) -1

(6-43) 0

(6-44) 1

(6-45) $-9/7$

(6-46) $y + 1 = x - 1$

(6-47) $y = 2(x - 3)$

(6-48) $y - 1 = -(x - 1)$

(6-49) $y - 3 = (-13/28)(x + 2)$

(6-50) $y - 1 = (-1/5)(x - 1)$

(6-51) $y = (1/5)(x + 1)$

(6-52) $(0, 0)$ and $(1/32, 1/8)$

(6-53) $(0, -1)$ and $(2, \sqrt[3]{3})$

(6-54) $(-1, -1)$

(6-55) $dr/dh = (-rh)/(2r^2 + h^2)$

(6-56) $(0, 0)$

(6-57) $(1, 1), (-1, 1),$ and $(0, 1)$

(6-58) $d^2y/dx^2 = 2/(x^{3/2})$

(6-59) $\dfrac{d^2y}{dx^2} = \dfrac{2y(x + y)}{(x + 2y)^3}$

(6-60) $\dfrac{d^2y}{dx^2} = \dfrac{2(20y^2 - 3)}{y^5(3 - 10y^2)^3}$

(6-61) $d^2y/dx^2 = -10/(y^3)$

(6-62) $\left.\dfrac{d^2y}{dx^2}\right|_{(\sqrt{2}, 2)} = 2/27$

(6-63) $y' = (2x + 3)(x - 5)^2\left(\dfrac{2}{2x + 3} + \dfrac{2}{x - 5}\right)$

(6-64) $y' = (x-1)(x-2)^2(x-3)^3(x-4)^4\left[\frac{1}{x-1} + \frac{2}{x-2} + \frac{3}{x-3} + \frac{4}{x-4}\right]$

(6-65) $y' = \frac{1}{2}\sqrt{\frac{x+1}{x-1}}\left(\frac{1}{x+1} - \frac{1}{x-1}\right)$

(6-66) $y' = \frac{1}{4}\sqrt[4]{\frac{x^2(x^2+3)^3}{x^2-2x+5}}\left(\frac{2}{x} + \frac{6x}{x^2+3} - \frac{2x-2}{x^2-2x+5}\right)$

(6-67) $y' = \frac{(x^2+1)\sqrt[6]{2x^4-7x^3+9}}{x+3}\left[\frac{2x}{x^2+1} + \frac{8x^3-21x^2}{6(2x^4-7x^3+9)} - \frac{1}{x+3}\right]$

(6-68) $y' = (x^2+5)^{6x^2+7}\left[(6x^2+7)\cdot\frac{2x}{x^2+5} + 12x\ln(x^2+5)\right]$

(6-69) $y' = (\sin x)^{\cos x}[(\cos x)(\cos x/\sin x) - (\sin x)\ln \sin x]$

(6-70) $y' = x^{(x^2)}(x + 2x\ln x)$

(6-71) $y' = x^{(2^x)}[(\ln 2)2^x(\ln x) + (2^x)/x]$

(6-72) $y' = 2^{(2^x)}2^x(\ln 2)^2$

(6-73) $y' = x^{\sin x}[(\sin x)/x + (\cos x)\ln x]$

(6-74) $y' = x^{\sqrt{x+1}}\left(\frac{\sqrt{x+1}}{x} + \frac{\ln x}{2\sqrt{x+1}}\right)$

(6-75) $y' = \left(\frac{x^2\sin x}{x+1}\right)^x\left[2\ln x + \ln\sin x - \ln(x+1) + x\left(\frac{2}{x} + \frac{\cos x}{\sin x} - \frac{1}{x+1}\right)\right]$

(6-76) $y'' = (\ln x)^x\left[\frac{-1}{x(\ln x)^2} + \frac{1}{x\ln x} + \left(\frac{1}{\ln x} + \ln(\ln x)\right)^2\right]$

(6-77) $y'' = x^x[(1/x) + (1 + \ln x)^2]$

(6-78) $y'' = \frac{(x+3)^2}{\sqrt{x-1}}\left[\frac{-2}{(x+3)^2} + \frac{1}{2(x-1)^2} + \left(\frac{2}{x+3} - \frac{1}{2(x-1)}\right)^2\right]$

(6-79) $y'' = \frac{1}{2}\sqrt{(2x-1)(x+2)}\left[\frac{-4}{(2x-1)^2} - \frac{1}{(x+2)^2} + \left(\frac{2}{2x-1} + \frac{1}{x+2}\right)^2\right]$

(6-80) $y'' = (\sin x)^x[2\cot x - x\csc^2 x + (x\cot x + \ln\sin x)^2]$

EXAM 2 (CHAPTERS 5 AND 6)

1. A particle moves along a straight line so that its position at any time $t > 0$ seconds is $s = 5t + 20/(t + 1)$ units. Find the position and the acceleration of the particle when it comes to rest.

2. The tangent line and the normal line to the curve $y = 6x - x^2$ at (5, 5) form a triangle with the x axis. Find the area of that triangle.

In Problems 3 through 6, find the derivative of y with respect to x:

3. $x^2y + y^2x = 6$

4. $e^x \sin y + e^y \sin x = 1$

5. $y = \left(\dfrac{1}{x}\right)^{1/x}$

6. $y = \dfrac{\sqrt[3]{(x^2 + 1)(x - 1)}}{(2x + 1)^2(3x - 2)^4}$

7. Use a tangent line approximation to estimate $1/\sqrt{10}$.

SOLUTIONS TO EXAM 2

1. The particle comes to rest when its velocity $s'(t)$ is zero:

$$v = s'(t) = 5 - \frac{20}{(t + 1)^2}$$

$$0 = 5 - \frac{20}{(t + 1)^2}$$

$$(t + 1)^2 = 4$$

Because $t > 0, t + 1 = 2$ or $t = 1$. The particle comes to rest at $t = 1$. The acceleration at time t is

$$a = s''(t) = \frac{40}{(t + 1)^3}$$

The position at $t = 1$ is $s(1) = 15$ units and the acceleration is $s''(1) = 5$ units/s^2.

2. Find the slope of the tangent line:

$$y' = 6 - 2x$$

$$y'(5) = 6 - 2(5) = -4$$

The tangent line at (5, 5) has slope -4, and so the normal line at (5, 5) has slope 1/4. The tangent line is

$$y - 5 = -4(x - 5)$$

$$y = -4x + 25$$

which intersects the x axis at (25/4, 0). The normal line is

$$y - 5 = \frac{1}{4}(x - 5)$$

$$y = \frac{1}{4}x + \frac{15}{4}$$

which intersects the x axis at $(-15, 0)$. Thus, triangle ABC has vertices $(5, 5)$, $(25/4, 0)$ and $(-15, 0)$ as shown in the figure. This triangle has base $25/4 + 15 = 85/4$ and height 5:

$$\text{area} = \frac{1}{2}\left(\frac{85}{4}\right)(5) = \frac{425}{8}$$

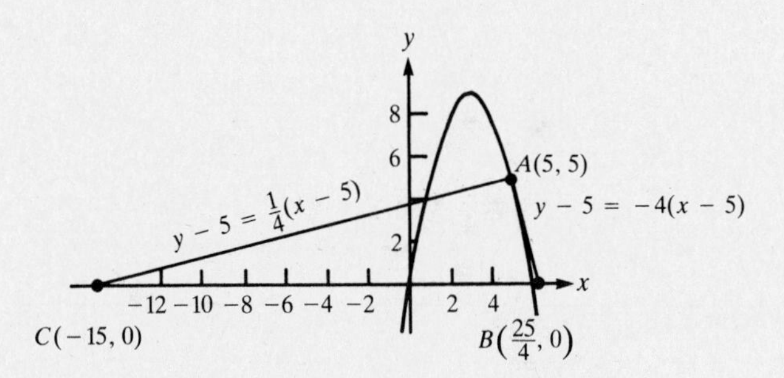

3. Differentiate implicitly:

$$x^2\frac{dy}{dx} + 2xy + y^2 + 2y\frac{dy}{dx}x = 0$$

$$\frac{dy}{dx}(x^2 + 2xy) = -(2xy + y^2)$$

$$\frac{dy}{dx} = -\frac{2xy + y^2}{x^2 + 2xy}$$

4. $e^x\sin y + e^x\cos y\dfrac{dy}{dx} + e^y\dfrac{dy}{dx}\sin x + e^y\cos x = 0$

$$\frac{dy}{dx}(e^x\cos y + e^y\sin x) = -(e^x\sin y + e^y\cos x)$$

$$\frac{dy}{dx} = -\frac{e^x\sin y + e^y\cos x}{e^x\cos y + e^y\sin x}$$

5. Use logarithmic differentiation by taking the natural log of both sides and differentiating implicitly:

$$\ln y = \ln\left[\left(\frac{1}{x}\right)^{1/x}\right] = \left(\frac{1}{x}\right)\ln\left(\frac{1}{x}\right) = \frac{-1}{x}\ln x$$

$$\frac{1}{y}\cdot\frac{dy}{dx} = \left(\frac{-1}{x}\right)\left(\frac{1}{x}\right) + \left(\frac{1}{x^2}\right)\ln x = \frac{\ln x - 1}{x^2}$$

$$\frac{dy}{dx} = (y)\frac{\ln x - 1}{x^2} = \left(\frac{1}{x}\right)^{1/x}\left(\frac{\ln x - 1}{x^2}\right)$$

6. Use logarithmic differentiation:

$$\ln y = \frac{1}{3}\ln(x^2 + 1) + \frac{1}{3}\ln(x - 1) - 2\ln(2x + 1) - 4\ln(3x - 2)$$

$$\frac{1}{y}\cdot\frac{dy}{dx} = \frac{2x}{3(x^2 + 1)} + \frac{1}{3(x - 1)} - \frac{4}{2x + 1} - \frac{12}{3x - 2}$$

$$\frac{dy}{dx} = y\left[\frac{2x}{3(x^2 + 1)} + \frac{1}{3(x - 1)} - \frac{4}{2x + 1} - \frac{12}{3x - 2}\right]$$

$$= \sqrt[3]{\frac{(x^2 + 1)(x - 1)}{(2x + 1)^2(3x - 2)^4}}\left[\frac{2x}{3(x^2 + 1)} + \frac{1}{3(x - 1)} - \frac{4}{2x + 1} - \frac{12}{3x - 2}\right]$$

7. Find the tangent line to $f(x) = 1/\sqrt{x}$ at $x = 9$.

$$f'(x) = \frac{-1}{2x^{3/2}}$$

$$f'(9) = \frac{-1}{2(9)^{3/2}} = \frac{-1}{54}$$

The tangent line at $x = 9$ has slope $-1/54$ and passes through $(9, f(9)) = (9, 1/3)$.

$$y - \frac{1}{3} = \frac{-1}{54}(x - 9)$$

$$y = \frac{-1}{54}x + \frac{9}{54} + \frac{1}{3} = \frac{-1}{54}x + \frac{1}{2}$$

At $x = 10$, you have

$$y = \frac{-1}{54}(10) + \frac{1}{2} = \frac{17}{54}$$

So the tangent line approximation is $1/\sqrt{10} \approx 17/54$.

7 BASIC THEOREMS OF CALCULUS

THIS CHAPTER IS ABOUT

☑ **The Intermediate Value Theorem**
☑ **The Extreme Value Theorem**
☑ **Rolle's Theorem**
☑ **The Mean Value Theorem**

7-1. The Intermediate Value Theorem

Intermediate Value Theorem: If f is a continuous function at all points a through b (i.e., f is a continuous function on the closed interval $[a, b]$), then f attains all values between $f(a)$ and $f(b)$.

EXAMPLE 7-1: Consider $f(x) = 3\sin\left(\frac{\pi}{2}x\right) + 1$, which is a continuous function everywhere. Because $f(1) = 4$ and $f(3) = -2$, we can assert that f takes on all values between -2 and 4. In particular, there is at least one number x between 1 and 3 such that $f(x) = \sqrt{5}$ (because $-2 \leqslant \sqrt{5} \leqslant 4$).

A. View the Intermediate Value Theorem graphically.

Recall that a function is continuous if it can be graphed without lifting the pencil from the paper. This theorem says that if the pencil goes from here to there, then it goes everywhere in between.

In Figure 7-1, c is an arbitrary number between $f(a)$ and $f(b)$. You see that it is possible to find a number x between a and b such that $f(x) = c$. This is the Intermediate Value Theorem.

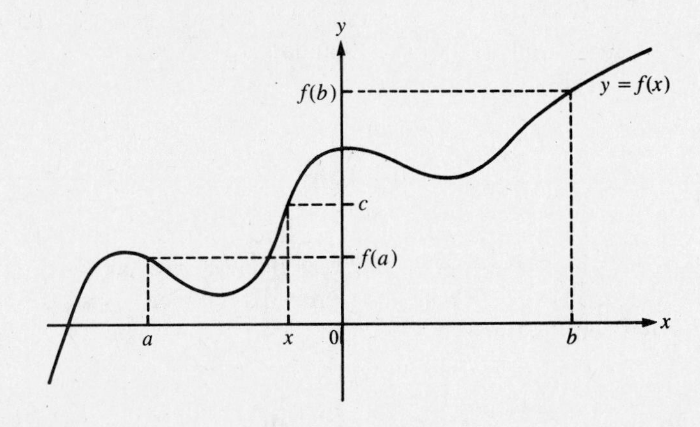

Figure 7-1
An arbitrary number between $f(a)$ and $f(b)$ and the corresponding value x between a and b for which $f(x) = c$.

B. Verify the continuity hypothesis.

You must be sure that the function is continuous on the interval in order to apply the theorem.

EXAMPLE 7-2: Consider the function $f(x) = 2/x$ for $-1 \leqslant x \leqslant 1$. Because $f(-1) = -2$ and $f(1) = 2$, you might expect that there would be a value x

between -1 and 1 for which $f(x) = 0$ (because 0 is between -2 and 2). This is not true ($2/x \neq 0$), of course, because f is not continuous between -1 and 1. There is a discontinuity at $x = 0$.

C. Applications of the Intermediate Value Theorem.

1. You can use the theorem to *check for roots of polynomials.*

EXAMPLE 7-3: Consider the polynomial $f(x) = 3x^7 + x^2 + 2x - 1$. Although the polynomial is difficult to factor, you can use the theorem to assert that it has a **zero** (a number x such that $f(x) = 0$). A polynomial function is a continuous function.

$$f(0) = -1 < 0 \qquad f(1) = 5 > 0$$

Let $c = 0$. Because $-1 \leqslant c \leqslant 5$, you know that there is a number x ($0 \leqslant x \leqslant 1$) such that $f(x) = c$; i.e., $3x^7 + x^2 + 2x - 1 = 0$.

2. You can use the theorem to *establish the existence of roots for functions* other than polynomial functions.

EXAMPLE 7-4: It is difficult to solve the equation

$$2^x = x^2$$

But if you consider the continuous function $f(x) = 2^x - x^2$ on the interval $[1, 3]$, then you can assert that there is a solution to the equation:

$$f(1) = 1 \qquad f(3) = -1$$

So there is a number x such that $f(x) = 0$ (because $-1 \leqslant 0 \leqslant 1$). Although you can't easily find x, you do know that it is between 1 and 3.

3. You can use the theorem in diverse *word problems.*

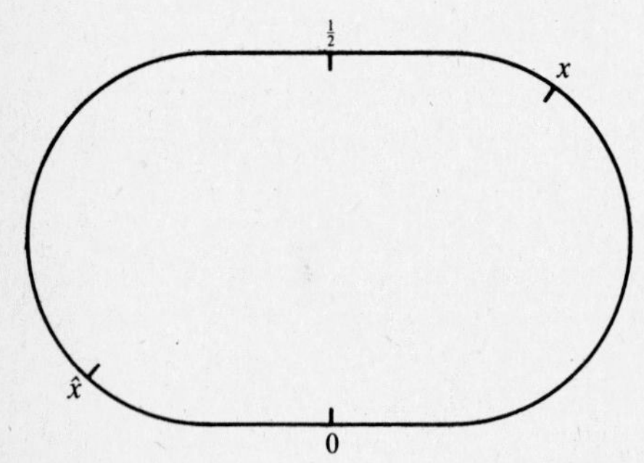

Figure 7-2
A running track.

EXAMPLE 7-5: Consider the following situation. A runner competes in a race involving one lap of a track. The race begins and ends at the same point. The runner begins and ends the race with the same velocity. Define a continuous function as follows: The variable will be x, the position on the track (x is between 0 and $\frac{1}{2}$ lap). Use $\hat{x}$ to denote the point opposite x on the track (see Figure 7-2) and $v(x)$ to denote the velocity of the runner when she is at position x. Now let $f(x) = v(x) - v(\hat{x})$. That is, $f(x)$ gives the difference between the runner's velocity at position x and her velocity $\frac{1}{2}$ lap later. Then you have

$$f(0) = v(0) - v(\hat{0}) = v(0) - v(\tfrac{1}{2})$$

and

$$f(\tfrac{1}{2}) = v(\tfrac{1}{2}) - v(\hat{\tfrac{1}{2}}) = v(\tfrac{1}{2}) - v(0) = -f(0)$$

because $\hat{\frac{1}{2}} = 0$. The fact that $f(\frac{1}{2}) = -f(0)$ ensures that either $f(0) \leqslant 0 \leqslant f(\frac{1}{2})$ or $f(\frac{1}{2}) \leqslant 0 \leqslant f(0)$. Thus, there is an x between 0 and $\frac{1}{2}$ for which $f(x) = 0$. But if $f(x) = 0$, then $v(x) = v(\hat{x})$. You have proved that there must be a pair of opposite points on the track at which the runner had exactly the same speed. (Hard to believe, isn't it!)

7-2. The Extreme Value Theorem

Extreme Value Theorem: If f is a continuous function on the closed interval $[a, b]$, then f attains a maximum and a minimum value on $[a, b]$.

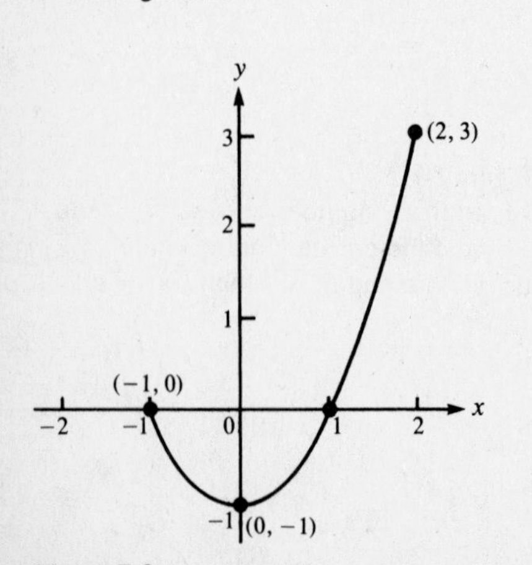

Figure 7-3
$f(x) = x^2 - 1$.

EXAMPLE 7-6: Let $f(x) = x^2 - 1$. The function is continuous on the interval $[-1, 2]$. On this interval, f attains a maximum value of 3 at $x = 2$ and a minimum value of -1 at $x = 0$. These points are seen as the high point and low point on the graph (see Figure 7-3).

A. The function must be continuous on a closed interval.

1. A function that fails to be continuous at every point of the interval may have no maximum or minimum value.

EXAMPLE 7-7: Consider the function $f(x) = 1/x^2$ on $[-3, 3]$. On this interval, f does not attain a maximum value because f is not continuous on $[-3, 3]$ (see Figure 7-4). Indeed, f is not even defined at $x = 0$.

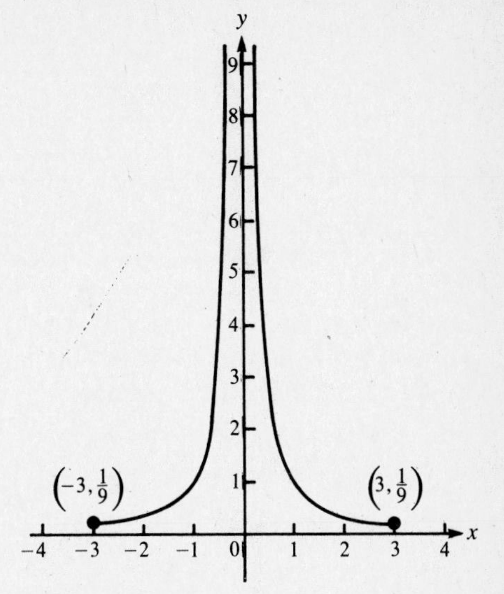

Figure 7-4
$f(x) = 1/x^2$ on $[-3, 3]$.

2. The interval must be closed. Otherwise, the maximum or minimum value might not be attained.

EXAMPLE 7-8: Consider the function $f(x) = 1/x$ on $(0, 3]$. On this interval, $f(x)$ does not attain a maximum value (see Figure 7-5). The difficulty is that f has been defined on an interval that is not closed.

EXAMPLE 7-9: Consider the function $f(x) = x + 1$ on $(2, 5]$. The function does not attain a minimum on this interval. Although $f(2) = 3$, the function never quite reaches a value of 3 on the interval $(2, 5]$ because 2 is not included in the interval.

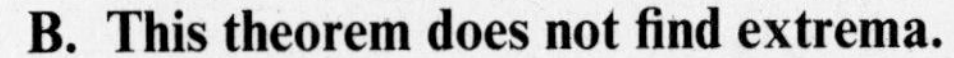

B. This theorem does not find extrema.

Although this theorem ensures the *existence* of **extrema** (minima and maxima), it does *not locate* the extrema for us. Rather, we must rely upon the methods described in Chapter 9.

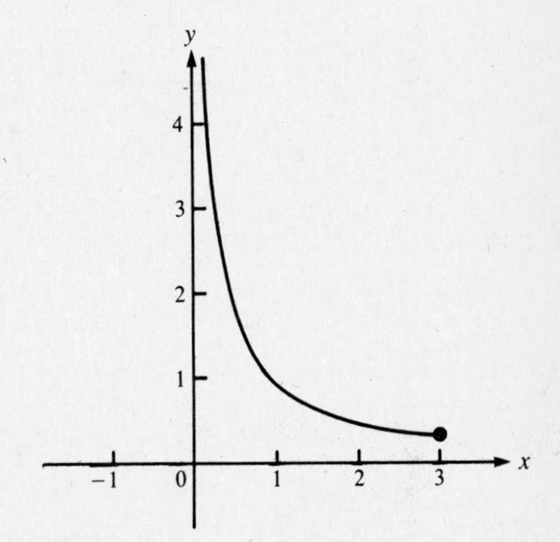

Figure 7-5
$f(x) = 1/x$ on $(0, 3]$.

EXAMPLE 7-10: An alchemist induces a certain chemical reaction in order to produce gold. He finds that the temperature of the reaction vessel affects the amount of gold produced in a continuous fashion. He can keep the vessel at any temperature between 0 and 100°C. The Extreme Value Theorem then tells him that there is a temperature x $(0 \leqslant x \leqslant 100)$ for the vessel that will produce a maximum amount of gold.

If the alchemist can determine the formula for the amount of gold produced as a function of the vessel temperature, then the methods of Chapter 9 will enable him to determine the temperature at which the maximum occurs.

7-3. Rolle's Theorem

Rolle's Theorem: If f is a continuous function on the closed interval $[a, b]$, and if

(1) $f(a) = f(b) = 0$, and
(2) f is differentiable on the interval (a, b)

then there is some point c $(a < c < b)$ such that $f'(c) = 0$.

There may be many points where $f'(c) = 0$. Figure 7-6 shows several values of c for which $f'(c) = 0$.

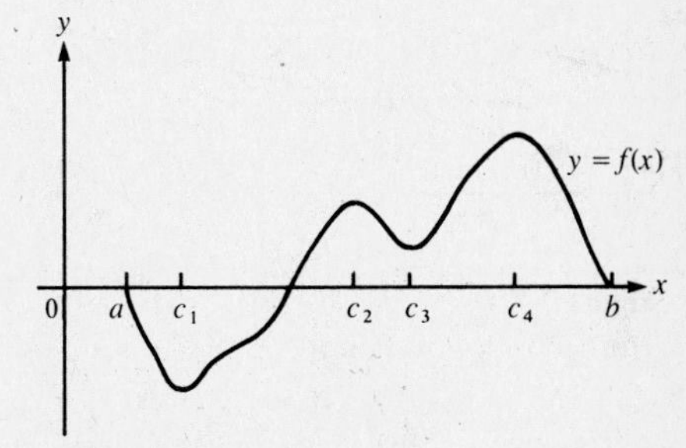

Figure 7-6
Several values of c (c_1, c_2, c_3, c_4) where $f'(c) = 0$.

EXAMPLE 7-11: Consider the function $f(x) = x^2 - 4x + 3$ on $[1, 3]$. You first verify that $f(1) = f(3) = 0$. Then f is indeed continuous on $[1, 3]$ and differentiable on $(1, 3)$:

$$f'(x) = 2x - 4$$

So $f'(2) = 0$. Thus, $c = 2$ is the point whose existence is ensured by Rolle's Theorem.

EXAMPLE 7-12: Consider the function $f(x) = x^3 + x^2 - 2x$ for $-2 \leqslant x \leqslant 0$. The function is continuous on $[-2, 0]$ and differentiable on

$(-2, 0)$. Since

$$f(-2) = f(0) = 0 \qquad \text{and}$$
$$f'(x) = 3x^2 + 2x - 2$$

you know that a number c exists between -2 and 0 where $f'(c) = 0$. To find c, set $f'(x)$ equal to zero:

$$3x^2 + 2x - 2 = 0$$

This expression is not easily factored, so use the quadratic formula:

$$c = \frac{-2 \pm \sqrt{2^2 - 4 \cdot 3 \cdot (-2)}}{2 \cdot 3} = \frac{-2 \pm \sqrt{28}}{2 \cdot 3} = \frac{-1 \pm \sqrt{7}}{3}$$

So the value of c between -2 and 0 is $(-1 - \sqrt{7})/3$.

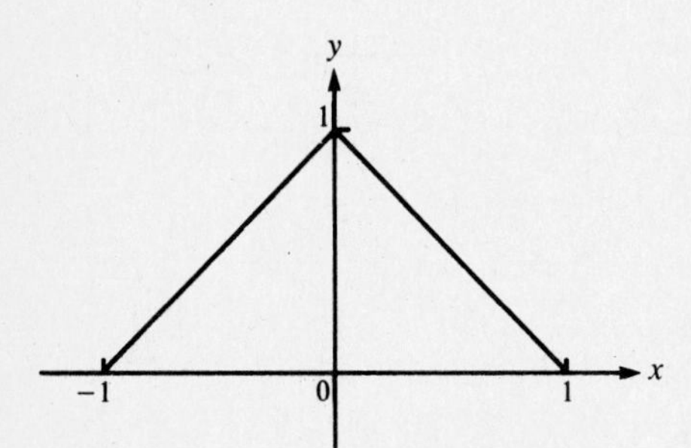

Figure 7-7
$$f(x) = \begin{cases} 1 + x & \text{for } -1 \le x \le 0 \\ 1 - x & \text{for } 0 \le x \le 1 \end{cases}$$

A. The hypotheses must be verified.

The function must be differentiable on (a, b).

EXAMPLE 7-13: Consider the function

$$f(x) = \begin{cases} 1 + x & \text{for} \quad -1 \leqslant x \leqslant 0 \\ 1 - x & \text{for} \quad 0 \leqslant x \leqslant 1 \end{cases}$$

(see Figure 7-7). Here you have $f(-1) = f(1) = 0$, and f is continuous on $[-1, 1]$. But, no point c exists between -1 and 1 for which $f'(c) = 0$. The failure to obtain such a point is a result of the failure of f to be differentiable on $(-1, 1)$. Indeed, $f'(x)$ is not defined at $x = 0$. Note that the Extreme Value Theorem still holds: f attains a maximum at zero.

7-4. The Mean Value Theorem

Mean Value Theorem: If f is a continuous function on the closed interval $[a, b]$ and differentiable on the open interval (a, b), then there is at least one point c between a and b such that

$$f'(c) = \frac{f(b) - f(a)}{b - a}$$

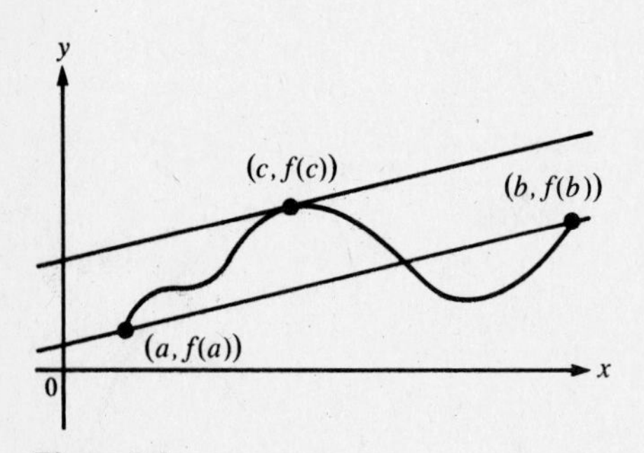

Figure 7-8
The Mean Value Theorem.

A. The Mean Value Theorem is a "tilted" Rolle's Theorem.

The quantity $\dfrac{f(b) - f(a)}{b - a}$ is the slope of the line through the points $(a, f(a))$ and $(b, f(b))$. The slope of the tangent line to the graph of $f(x)$ at $x = c$ is $f'(c)$. Thus, the Mean Value Theorem says that the tangent line at $x = c$ is parallel to the line through the points at $x = a$ and $x = b$ (see Figure 7-8).

You can therefore view the Mean Value Theorem as a "tilted" Rolle's Theorem. In Rolle's Theorem the slope of the line through $(a, f(a))$ and $(b, f(b))$ is zero because $f(a) = f(b) = 0$.

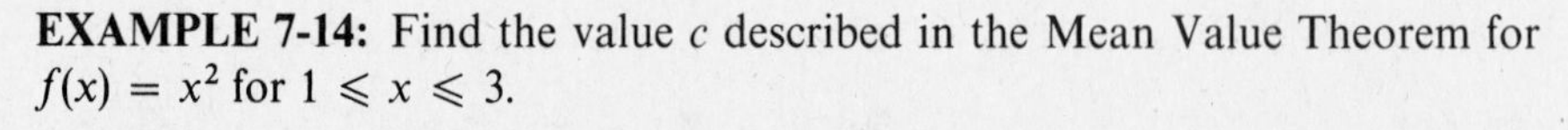

EXAMPLE 7-14: Find the value c described in the Mean Value Theorem for $f(x) = x^2$ for $1 \leqslant x \leqslant 3$.

Solution: You must find c such that

$$f'(c) = \frac{f(3) - f(1)}{3 - 1} = \frac{9 - 1}{3 - 1} = 4$$

However, $f'(x) = 2x$. Because $f'(2) = 4$, you see that $c = 2$ is the desired value.

EXAMPLE 7-15: Consider $f(x) = x^3$ on $[1, 3]$. You want

$$f'(c) = \frac{f(3) - f(1)}{3 - 1} = 13$$

But $f'(x) = 3x^2$. So you must have $3c^2 = 13$. The desired value of c between 1 and 3 is $\sqrt{13/3}$. That is, the line tangent to $f(x)$ at $x = \sqrt{13/3}$ is parallel to the line through $(1, f(1))$ and $(3, f(3))$.

EXAMPLE 7-16: Consider $f(x) = 3/(x + 2)$ on $[-1, 1]$.

$$f'(c) = \frac{f(1) - f(-1)}{1 - (-1)} = \frac{1 - 3}{2} = -1$$

But $f'(x) = -3/(x + 2)^2$, so you must solve

$$\frac{-3}{(x + 2)^2} = -1$$

$$(x + 2)^2 = 3$$

$$x + 2 = \pm\sqrt{3}$$

Thus, $x = -2 \pm \sqrt{3}$. The value that lies between -1 and 1 is $c = -2 + \sqrt{3}$.

B. The function must be differentiable on (a, b).

EXAMPLE 7-17: Consider $f(x) = |x|$ on $[-1, 2]$ (see Figure 7-9). If the Mean Value Theorem were applicable in this instance, it would imply the existence of a number c between -1 and 2 such that

$$f'(c) = \frac{f(2) - f(-1)}{2 - (-1)} = \frac{1}{3}$$

However,

$$f'(x) = \begin{cases} -1 & \text{for } x < 0 \\ 1 & \text{for } x > 0 \end{cases}$$

So no such number c exists. This is because $f'(0)$ does not exist.

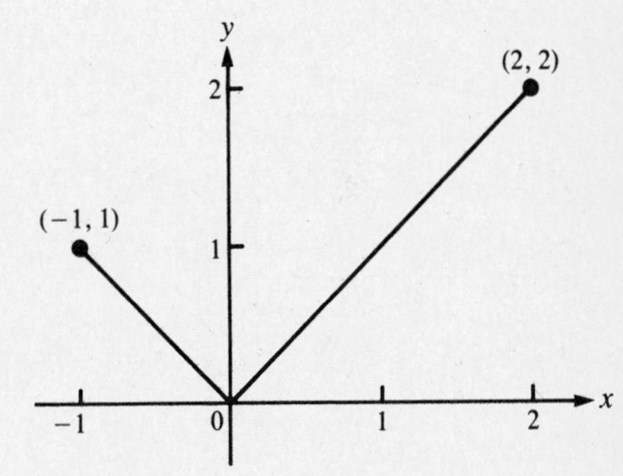

Figure 7-9 $f(x) = |x|$ on $[-1, 2]$.

SUMMARY

1. The Intermediate Value Theorem is valid for continuous functions.
2. You can sometimes use the Intermediate Value Theorem to show that a polynomial equation has real roots.
3. You can sometimes use the Intermediate Value Theorem to show that complicated algebraic equations have solutions.
4. Use the Extreme Value Theorem to establish that continuous functions attain maximum and minimum values on closed intervals.
5. You must check all hypotheses of Rolle's Theorem before you apply it.
6. Rolle's Theorem implies the existence of points where the first derivative is zero.
7. The Mean Value Theorem is valid for continuous, differentiable functions.
8. The Mean Value Theorem is analogous to Rolle's Theorem, viewed at a slant.

SOLVED PROBLEMS

PROBLEM 7-1 Show that the polynomial $x^5 - 2x^2 + 5x - 1$ has a real zero.

Solution: You search for both positive and negative values of the function $f(x) = x^5 - 2x^2 + 5x - 1$. Because $f(0) = -1 < 0$ and $f(1) = 3 > 0$, you know that $f(x) = 0$ for some x between 0 and 1.

[See Section 7-1.]

PROBLEM 7-2 Show that the equation $4^x = x + 5$ has a solution.

Solution: You consider the function $f(x) = 4^x - x - 5$ because, as you know, zeros of f (numbers x such that $f(x) = 0$) are solutions to the equation in question. Observe that $f(1) = -2 < 0$ and $f(2) = 9 > 0$. Thus, the function f has a zero (and in turn, the equation has a solution) x where $1 < x < 2$. [See Section 7-1.]

PROBLEM 7-3 A 22-in. child is born, weighing 8 lb. When fully grown, he is 5 ft 6 in. tall and weighs 145 lb. Show that there is some point in his life when his weight (in pounds) and his height (in inches) are equal.

Solution: Construct a function $f(x)$ that gives the child's height minus his weight at age x. Because his height and weight are continuous functions of his age, f must be a continuous function. If $x = b$ represents an age at which he is fully grown, then

$$f(b) = 66 - 145 = -79$$

But at age 0, you obtain

$$f(0) = 22 - 8 = 14$$

Now, apply the Intermediate Value Theorem. Because $-79 < 0 < 14$, there must be an age x between 0 and fully grown for which $f(x) = 0$; i.e., his height and weight at age x are equal. [See Section 7-1.]

PROBLEM 7-4 Consider the function $f(x) = (x^3 - 1)/x$. Although $f(1) = 0$ and $f(-1) = 2$, there is no number x between -1 and 1 for which $f(x) = 1$. Explain why this is not a contradiction of the Intermediate Value Theorem.

Solution: The Intermediate Value Theorem can't be applied because f is discontinuous at $x = 0$. [See Section 7-1.]

PROBLEM 7-5 Show that the polynomial $f(x) = x^6 + 3x^3 + 2x + 1$ has a real zero.

Solution: You search for both positive and negative values of the function $f(x)$. Because $f(0) = 1$ and $f(-1) = -3$, you know that $f(x) = 0$ for some x between -1 and 0. Thus, the polynomial has a zero somewhere in the interval $(-1, 0)$. Further observe that because $f(-2) = 37 > 0$, there must be an x between -2 and -1 for which $f(x) = 0$. Thus, you are certain that this polynomial has at least two real zeros. [See Section 7-1.]

PROBLEM 7-6 Show that the equation $\sin x = x + 1$ has a solution.

Solution: Considering the function $f(x) = \sin x - x - 1$,

$$f(0) = \sin 0 - 0 - 1 = -1 < 0$$

$$f(-\pi) = \sin(-\pi) - (-\pi) - 1 = 0 + \pi - 1 > 0$$

Thus, $f(x)$ has a zero (and the equation has a solution) for some x between $-\pi$ and 0. [See Section 7-1.]

PROBLEM 7-7 Suppose that $f(x)$ is a continuous function on $[a, b]$ and that $a \leqslant f(x) \leqslant b$ for all x in $[a, b]$. Show that $f(x) = x$ for some x where $a \leqslant x \leqslant b$.

Solution: Let $g(x) = f(x) - x$, which is another continuous function on $[a, b]$. Then

$$g(a) = f(a) - a \geqslant 0$$

$$g(b) = f(b) - b \leqslant 0$$

Thus, g goes from positive to negative values as x goes from a to b. So there is an x between a and b for which $g(x) = 0$. This implies $f(x) = x$. [See Section 7-1.]

PROBLEM 7-8 Show that at any time there is a pair of points that are directly opposite each other on the earth's equator and that have exactly the same temperature.

Solution: Draw a diagram of the earth, showing the equator (Figure 7-10). Let p denote some point on the earth's equator. Then let x denote the position x miles east of p on the equator. Further, let $\hat{x}$ denote the position directly opposite x on the equator; i.e., $\hat{x} = x + \frac{1}{2}$ circumference. Let $f(x)$ = temperature at x − temperature at $\hat{x}$. Assume that temperature varies continuously as position varies, and thus conclude that f is a continuous function. You easily verify that $f(p) = -f(\hat{p})$:

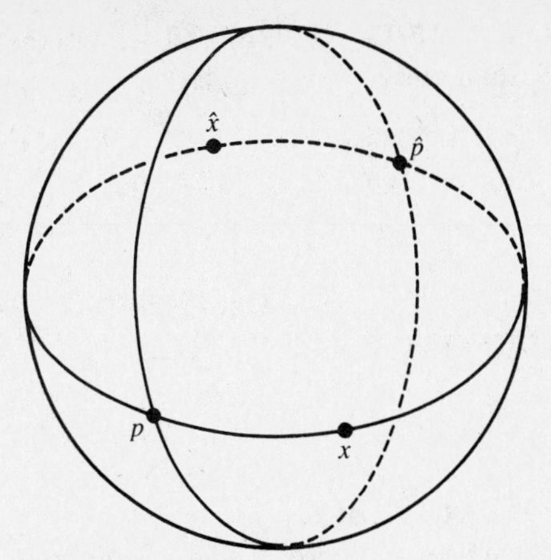

Figure 7-10 Problem 7-8.

$$\begin{aligned} f(p) &= \text{temp at } p - \text{temp at } \hat{p} \\ &= -(\text{temp at } \hat{p} - \text{temp at } p) \\ &= -(\text{temp at } \hat{p} - \text{temp at } \hat{\hat{p}}) \\ &= -f(\hat{p}) \end{aligned}$$

Now you have two possibilities: Either $f(p) = f(\hat{p}) = 0$ or 0 lies between $f(p)$ and $f(\hat{p})$. In the first case, p and $\hat{p}$ are the points you want. In the second case, you apply the Intermediate Value Theorem. You can assert that there is a point on the equator (between p and $\hat{p}$) for which $f(x) = 0$. But then x and $\hat{x}$ are the points you want.

[See Section 7-1.]

PROBLEM 7-9 Show that the equation $x = \ln x + 2$ has a solution.

Solution: Let $f(x) = x - \ln x - 2$. Then

$$f(1) = 1 - \ln 1 - 2 = 1 - 0 - 2 = -1 < 0$$

$$f(4) = 2 - \ln 4 > 0$$

Thus, $f(x) = 0$ for some x between 1 and 4. So x provides a solution to equation.

[See Section 7-1.]

PROBLEM 7-10 A bicyclist leaves her house at 8:00 AM and rides to her friend's house. The next morning she leaves her friend's house at 8:00 and rides home along the same route. Show that there is a time of day when she was in the same place on both days.

Solution: Let $f(x)$ = (position of the cyclist at time x on the first day) minus (position of cyclist at time x on the second day). Then f is a continuous function. If b represents the time of day when she reaches her friend's house, then f(8:00 AM) and $f(b)$ have opposite signs. Thus, there is an x between 8:00 AM and b for which $f(x) = 0$. That is, there is a time x at which the bicyclist was in the same position on both days.

[See Section 7-1.]

PROBLEM 7-11 Let $f(x) = (x^2 - 1)^2$ on the interval $[-1, 1]$. Find the value of c whose existence is guaranteed by Rolle's Theorem.

Solution: Since f is a polynomial, it is continuous and differentiable everywhere. Because

$$f(-1) = f(1) = 0$$

there is a point c $(-1 < c < 1)$ such that $f'(c) = 0$.

$$f'(x) = 2(x^2 - 1)(2x) = 4x(x^2 - 1)$$

So $f'(x) = 0$ when $x = \pm 1, 0$. The value whose existence is guaranteed by Rolle's Theorem is $c = 0$.

[See Section 7-3.]

PROBLEM 7-12 Let $f(x) = x^{2/3} - 1$ on $[-1, 1]$. Determine whether Rolle's Theorem applies, and if so, find the value c for which $f'(c) = 0$.

Solution: The function is continuous on $[-1, 1]$. However, $f'(x) = \frac{2}{3}x^{-1/3}$, which is undefined at $x = 0$. Thus, f is not differentiable at $x = 0$, and Rolle's Theorem does not apply.

[See Section 7-3.]

PROBLEM 7-13 Use Rolle's Theorem to show that the polynomial $x^3 - 12x + D$ cannot have more than one zero in the interval $[-2, 2]$ (regardless of what number D is).

Solution: Let $f(x) = x^3 - 12x + D$. The function is continuous and differentiable everywhere. If there are two zeros, say a and b, in $[-2, 2]$, then

$$f(a) = f(b) = 0$$

But then Rolle's Theorem says that there must be a number c between a and b (in particular, c must satisfy $-2 < c < 2$) such that $f'(c) = 0$. However,

$$\begin{aligned} f'(x) &= 3x^2 - 12x \\ &= 3(x - 2)(x + 2) \end{aligned}$$

so $f'(x) = 0$ only at $x = -2$ and $x = 2$. Thus, no such c $(-2 < c < 2)$ can exist, and there must not have been two zeros a and b as described above. [See Section 7-3.]

PROBLEM 7-14 A ball has position (height) $f(t) = -16t^2 + 32t + 40$, t seconds after being thrown upward. Note that $f(2) = f(0)$, and so after 2 seconds the ball returns to its initial height. What does Rolle's Theorem imply about the velocity of the ball? Find the value of t guaranteed by the theorem.

Solution: Let $g(t) = f(t) - 40 = -16t^2 + 32t$. Then $g(0) = g(2) = 0$. Because it is a polynomial, g is continuous and differentiable. Rolle's Theorem implies that $g'(c) = 0$ for some c between 0 and 2. But $g'(c) = f'(c)$. If $f(t)$ is the position at time t, then $f'(t)$ is the velocity at time t. So Rolle's Theorem says that the ball has zero velocity at some time between 0 and 2 seconds. Because $f'(t) = -32t + 32$, you see that $f'(t) = 0$ when $t = 1$. [See Section 7-3.]

PROBLEM 7-15 Show that the polynomial $7x^6 - 5x^4 - 2x + 1$ has at least one zero between 0 and 1.

Solution: Let $f(x) = 7x^6 - 5x^4 - 2x + 1$. Because $f(0) = f(1) = 1$, you can't use the Intermediate Value Theorem. But another trick will work. Let $g(x) = x^7 - x^5 - x^2 + x$. Because it is a polynomial, g is continuous and differentiable. Note that $g(0) = 0$ and $g(1) = 0$, so Rolle's Theorem applies. There is a number c between 0 and 1 for which $g'(c) = 0$. But g was chosen so that

$$g'(x) = 7x^6 - 5x^4 - 2x + 1 = f(x)$$

So $0 = g'(c) = f(c)$, and c is a zero of the original polynomial. [See Section 7-3.]

PROBLEM 7-16 Consider the function $f(x) = (2x^2 - 1)/(1 - x^2)$ on $[-\sqrt{2}/2, \sqrt{2}/2]$. Determine whether or not Rolle's Theorem applies. If it does apply, find the value c $(-\sqrt{2}/2 < c < \sqrt{2}/2)$ for which $f'(c) = 0$.

Solution: The function is continuous and differentiable on $[-\sqrt{2}/2, \sqrt{2}/2]$. The only discontinuities of f are at $x = -1$ and $x = 1$, which lie outside the interval of interest. So Rolle's Theorem does apply. So,

$$f'(x) = 2x(1 - x^2)^{-2}$$

and $f'(x) = 0$ when $x = 0$. Thus, $c = 0$ is the desired quantity. [See Section 7-3.]

PROBLEM 7-17 Let $f(x) = (x^2 - 1)/x^2$ on $[-1, 1]$. Show that there is no value x in $[-1, 1]$ for which $f'(x) = 0$. Explain why this does not contradict Rolle's Theorem.

Solution: Since

$$f'(x) = 2x^{-3}$$

$f'(x)$ is never zero. However, $f(x)$ is not continuous on $[-1, 1]$ (discontinuous at $x = 0$), so Rolle's Theorem does not apply. [See Section 7-3.]

PROBLEM 7-18 Let $f(x) = x^3 - x + 1$ on $[-1, 1]$. Observe that $f(-1) = f(1)$. Show that an application of Rolle's Theorem to $g(x) = f(x) - 1$ will establish that $f'(x) = 0$ for some x between -1 and 1. Find all values c whose existence is guaranteed by the theorem.

Solution: Notice that

$$f(-1) = f(1) = 1$$

Choose $g(x) = f(x) - 1$, so that $g(1) = g(-1) = 0$. Moreover, $g(x)$ is continuous and differentiable everywhere; hence, Rolle's Theorem applies: $g'(c) = 0$ for at least one c between -1 and 1. Solving for x,

$$g'(x) = f'(x) = 3x^2 - 1 = 0$$

$$x^2 = \frac{1}{3}$$

$$x = \pm\frac{\sqrt{3}}{3}$$

Both values are contained in the interval $[-1, 1]$. [See Section 7-3.]

PROBLEM 7-19 Show that the polynomial $3x^3 + 5x + 2$ has exactly one real zero.

Solution: Let $f(x) = 3x^3 + 5x + 2$. Then f is continuous with $f(-2) = -32 < 0$ and $f(2) = 36 > 0$.

The Intermediate Value Theorem says that $f(r) = 0$ for some r between -2 and 2; i.e., the polynomial has a zero between -2 and 2. To show that there are no other zeros, assume that s is another zero. Then $f(r) = f(s) = 0$. Because f is a polynomial function, it is continuous and differentiable everywhere. Rolle's Theorem then asserts that there is a real number c between r and s for which $f'(c) = 0$. But $f'(x) = 9x^2 + 5$, which is always positive. Thus, $f'(x)$ is never equal to zero, which contradicts the assumption that the polynomial $3x^3 + 5x + 2$ has two distinct real zeros. So $3x^3 + 5x + 2$ has exactly one real zero. [See Sections 7-1 and 7-3.]

PROBLEM 7-20 Let $f(x) = x^3 - 3x^2 + 1$ on the interval $[1, 4]$. Find the value c whose existence is guaranteed by the Mean Value Theorem.

Solution: Because f is a polynomial, it is continuous and differentiable everywhere. You find

$$\frac{f(b) - f(a)}{b - a} = \frac{f(4) - f(1)}{4 - 1} = \frac{17 - (-1)}{3} = 6$$

$$f'(x) = 3x^2 - 6x$$

You search for a number c that satisfies $f'(c) = 6$:

$$3c^2 - 6c = 6$$

$$c^2 - 2c - 2 = 0$$

$$c = \frac{2 \pm \sqrt{(-2)^2 - 4(1)(-2)}}{2(1)} = \frac{2 \pm \sqrt{12}}{2} = 1 \pm \sqrt{3}$$

So $c = 1 + \sqrt{3}$ is the number you want. [See Section 7-4.]

PROBLEM 7-21 Suppose that you average 25 miles per hour on a trip in your car. Show that at some instant in the trip you must travel exactly 25 miles per hour.

Solution: Let $s(t)$ = distance traveled at time t. The speed, or velocity, at any time t is given by $v(t) = s'(t)$. Physical considerations lead you to believe that $s(t)$ is differentiable. If a is the starting time and b is the ending time of the trip, then the Mean Value Theorem implies that

$$\frac{s(b) - s(a)}{b - a} = s'(c)$$

for some c between a and b. However,

$$\frac{s(b) - s(a)}{b - a} = 25$$

because the average speed for the entire trip is 25 mi/h. So $v(c) = s'(c) = 25$ for some time c; i.e., there is a time c at which the velocity is 25 mi/h. [See Section 7-4.]

PROBLEM 7-22 Let $f(x) = x - x^{2/3}$ on $[-1, 1]$. Determine whether or not the Mean Value Theorem applies. If it does, find the value(s) c where the mean value is attained.

Solution: The function f is continuous on $[-1, 1]$, with

$$f'(x) = 1 - \tfrac{2}{3}x^{-1/3}$$

So $f'(x)$ is discontinuous at $x = 0$. Thus, the Mean Value Theorem does not apply.

[See Section 7-4.]

PROBLEM 7-23 Use the Mean Value Theorem to show that $\sqrt{26} - \sqrt{25} < 0.1$.

Solution: Let $f(x) = \sqrt{x}$ on the interval $[25, 26]$. The function is continuous and differentiable on this interval, with

$$\frac{f(b) - f(a)}{b - a} = \frac{f(26) - f(25)}{26 - 25} = \sqrt{26} - \sqrt{25}$$

$$f'(x) = \tfrac{1}{2}x^{-1/2}$$

So, by the Mean Value Theorem, a number c exists between 25 and 26 for which $f'(c) = \sqrt{26} - \sqrt{25}$:

$$\sqrt{26} - \sqrt{25} = \frac{1}{2}c^{-1/2}$$

But because $c > 25$,

$$\frac{1}{2}c^{-1/2} < \frac{1}{2}\cdot\frac{1}{5} = \frac{1}{10}$$

So

$$\sqrt{26} - \sqrt{25} = \frac{1}{2}c^{-1/2} < \frac{1}{10}$$

[See Section 7-4.]

PROBLEM 7-24 Let $f(x) = x + \sin x$ on the interval $[\pi/2, \pi]$. Find the value of c whose existence is guaranteed by the Mean Value Theorem.

Solution: Since $f(x)$ is continuous and differentiable everywhere, the theorem applies. With $a = \pi/2$, $b = \pi$, you have

$$\frac{f(\pi) - f(\pi/2)}{\pi - (\pi/2)} = \frac{\pi - (\pi/2 + 1)}{\pi/2} = 1 - \frac{2}{\pi}$$

Moreover, $f'(x) = 1 + \cos x$. Hence,

$$1 + \cos x = 1 - \frac{2}{\pi}$$

$$\cos x = \frac{-2}{\pi}$$

$$x = \arccos\frac{-2}{\pi}$$

Notice that $\arccos(-2/\pi) \cong 2.26$ radians, so $\pi/2 < \arccos(-2/\pi) < \pi$, as assured by the theorem. [See Section 7-4.]

PROBLEM 7-25 Use the Mean Value Theorem to establish the following: If $f(x)$ is continuous on $[a, b]$ and differentiable on (a, b), and further, $f'(x) = 0$ for every x between a and b, then $f(x) = k$ (a constant).

Solution: The function $f(x)$ satisfies the hypotheses of the Mean Value Theorem. Hence, for any x_1 and x_2 such that $a \leqslant x_1 < x_2 \leqslant b$, there exists a number c, with $x_1 < c < x_2$ and

$$\frac{f(x_2) - f(x_1)}{x_2 - x_1} = f'(c)$$

But $f'(c) = 0$. Hence $f(x_2) - f(x_1) = 0$ and so $f(x_2) = f(x_1) = k$ for some constant k.

[See Section 7-4.]

PROBLEM 7-26 Use the result of Problem 7-25 to show that if two differentiable functions have the property that $f'(x) = g'(x)$ for every x in some open interval (a, b), then $f(x) - g(x) = k$ (a constant).

Solution: Let $h(x) = f(x) - g(x)$. Then h is differentiable and $h'(x) = f'(x) - g'(x) = 0$ for every x between a and b. By the results of Problem 7-25, $h(x) = k$. Hence, $f(x) - g(x) = k$.

[See Section 7-4.]

PROBLEM 7-27 Let $f(x) = 1/x$ on $[-2, -\frac{1}{2}]$. Determine whether or not the Mean Value Theorem applies. If it does apply, find the value(s) c where the mean value is attained.

Solution: Since f is continuous and differentiable everywhere except at $x = 0$, the Mean Value Theorem applies on the interval $[-2, -\frac{1}{2}]$. So,

$$\frac{f(b) - f(a)}{b - a} = \frac{f(-2) - f(-\frac{1}{2})}{-2 - (-\frac{1}{2})} = -1$$

$$f'(x) = -x^{-2}$$

You are searching for solutions to

$$-x^{-2} = -1$$

$$x = \pm 1$$

So $c = -1$ is the desired value.

[See Section 7-4.]

PROBLEM 7-28 Show that $1/6 < \ln(6/5) < 1/5$.

Solution: Consider the function $f(x) = \ln x$ on $[1, 6/5]$. Applying the Mean Value Theorem,

$$\frac{f(6/5) - f(1)}{(6/5) - 1} = \frac{\ln(6/5) - \ln(1)}{(1/5)} = 5\ln(6/5)$$

$$f'(x) = \frac{1}{x}$$

You are searching for solutions to

$$\frac{1}{x} = 5\ln(6/5)$$

$$x = \frac{1}{5\ln(6/5)}$$

So $c = 1/(5\ln(6/5))$ is the solution. The theorem also tells you that $1 < c < 6/5$. So,

$$1 < \frac{1}{5\ln(6/5)} < \frac{6}{5}$$

$$5 < \frac{1}{\ln(6/5)} < 6$$

Inverting, you obtain

$$\frac{1}{5} > \ln(6/5) > \frac{1}{6}$$

[See Section 7-4.]

Supplementary Exercises

For each of the following functions on the indicated intervals (Problems 7-29 through 7-48), determine whether or not Rolle's Theorem applies. If it does apply, find the value(s) in the interval for which $f'(c) = 0$.

7-29 $f(x) = x^4 - 2x^2$ on $[-2, 2]$

7-30 $f(x) = x - \sqrt{x}$ on $[0, 1]$

7-31 $f(x) = |x| - 2$ on $[-2, 2]$

7-32 $f(x) = x^3 - 9x$ on $[-3, 3]$

7-33 $f(x) = \ln(\sin x - \frac{1}{2})$ on $[\pi/6, 5\pi/6]$

7-34 $f(x) = (x + 1)/(x + 2)$ on $[-3, 0]$

7-35 $f(x) = \sqrt{1 - x^2}$ on $[-1, 1]$

7-36 $f(x) = x\sqrt{2x + 1}$ on $[-\frac{1}{2}, 0]$

7-37 $f(x) = (x + 1)(x - 2)^2$ on $[-1, 2]$

7-38 $f(x) = x^2 - x^{3/2}$ on $[0, 1]$

7-39 $f(x) = (x^2 - 1)(x - 2)^{-1}$ on $[-1, 1]$

7-40 $f(x) = -x + \sin\left(\frac{\pi}{2} \cdot x\right)$ on $[0, 1]$

7-41 $f(x) = 2x^6 - 9x^4 + 12x^2 - 32$ on $[-2, 2]$

7-42 $f(x) = \sqrt{x^2 - x^4} - \frac{1}{2}$ on $\left[\frac{-\sqrt{2}}{2}, \frac{\sqrt{2}}{2}\right]$

7-43 $f(x) = (x^2 - 1)e^x$ on $[-2, 2]$

7-44 $f(x) = x^2 - x^3$ on $[0, 1]$

7-45 $f(x) = (4 - x^2)^{2/3}$ on $[-1, 1]$

7-46 $f(x) = \sqrt{1 - x^{2/3}}$ on $[-1, 1]$

7-47 $f(x) = (x - 1)^2(x + 2)^2$ on $[-2, 1]$

7-48 $f(x) = \dfrac{\sin x}{1 + \cos x}$ on $[0, \pi]$

For each of the following functions on the indicated intervals (Problems 7-49 through 7-68), determine whether or not the Mean Value Theorem applies. If it does apply, find the value(s) c where the mean value is attained.

7-49 $f(x) = x^4$ on $[0, 2]$

7-50 $f(x) = e^x$ on $[0, 1]$

7-51 $f(x) = \sqrt{1 - x^2}$ on $[0, 1]$

7-52 $f(x) = \sin x$ on $[0, \pi/2]$

7-53 $f(x) = x\sqrt{2x + 1}$ on $[0, 4]$

7-54 $f(x) = x^2 - x^{2/5}$ on $[-1, 1]$

7-55 $f(x) = x \ln x - x$ on $[1, e]$

7-56 $f(x) = x^{2/3}$ on $[0, 1]$

7-57 $f(x) = \ln x$ on $[1, e]$

7-58 $f(x) = \dfrac{x - 1}{x - 2}$ on $[3, 4]$

7-59 $f(x) = x^3 - 3x^2$ on $[-1, 3]$

7-60 $f(x) = x^{4/3}$ on $[8, 27]$

7-61 $f(x) = \sqrt{x + 3}$ on $[1, 6]$

7-62 $f(x) = x - (1/x)$ on $[1, 2]$

7-63 $f(x) = x/(x + 1)$ on $[-2, 0]$

7-64 $f(x) = x - \sqrt{x}$ on $[1, 4]$

7-65 $f(x) = x^3$ on $[-2, 2]$

7-66 $f(x) = 1/(\sqrt{x + 1})$ on $[0, 3]$

7-67 $f(x) = x/(x - 1)$ on $[2, 4]$

7-68 $f(x) = \dfrac{\sin x}{1 + \cos x}$ on $[0, \pi/2]$

Solutions to Supplementary Exercises

(7-29) $c = 0, \pm 1$

(7-30) $c = \frac{1}{4}$

(7-31) The theorem does not apply (not differentiable at $x = 0$).

(7-32) $c = \pm\sqrt{3}$

(7-33) The theorem does not apply (not continuous at endpoints).

(7-34) The theorem does not apply (not continuous at $x = -2$).

(7-35) $c = 0$

(7-36) $c = -\frac{1}{3}$

(7-37) $c = 0$

(7-38) $c = 9/16$

(7-39) $c = 2 - \sqrt{3}$

(7-40) $c = \dfrac{2}{\pi} \arccos \dfrac{2}{\pi}$

(7-41) $c = 0, \pm 1, \pm\sqrt{2}$

(7-42) The theorem does not apply (not differentiable at $x = 0$).

(7-43) The theorem does not apply $(f(-2) \neq f(2))$.

(7-44) $c = \frac{2}{3}$

(7-45) $c = 0$

(7-46) The theorem does not apply (not differentiable at $x = 0$).

(7-47) $c = -\frac{1}{2}$

(7-48) The theorem does not apply (not continuous at $x = \pi$).

(7-49) $c = \sqrt[3]{2}$

(7-50) $c = \ln(e - 1)$

(7-51) $c = \frac{1}{2}\sqrt{2}$

(7-52) $c = \arccos(2/\pi)$

(7-53) $c = (2 + 2\sqrt{3})/3$

(7-54) The theorem does not apply (not differentiable at $x = 0$).

(7-55) $c = e^{1/(e-1)}$

(7-56) $c = 8/27$

(7-57) $c = e - 1$

(7-58) $c = 2 + \sqrt{2}$

(7-59) $c = (3 \pm 2\sqrt{3})/3$

(7-60) $c = (195/76)^3$

(7-61) $c = 13/4$

(7-62) $c = \sqrt{2}$

(7-63) The theorem does not apply (not continuous at $x = -1$).

(7-64) $c = 9/4$

(7-65) $c = \pm(2\sqrt{3})/3$

(7-66) $c = 3^{2/3} - 1$

(7-67) $c = 1 + \sqrt{3}$

(7-68) $c = \arccos\left(\dfrac{\pi}{2} - 1\right)$

8 GRAPHING

THIS CHAPTER IS ABOUT

☑ **Intercepts and Symmetry**
☑ **Asymptotes**
☑ **The Sign of $f(x)$**
☑ **The Sign of $f'(x)$ and $f''(x)$**
☑ **The First and Second Derivative Tests**
☑ **Graphing a Function**

8-1. Intercepts and Symmetry

There are several basic tools available to you for graphing functions. Among them are the concepts of intercepts and symmetry.

A. Intercepts

The **intercepts** of the graph of a function are the points at which the graph crosses the x or y axis, that is, where y or x equals zero.

EXAMPLE 8-1: Find the intercepts of $f(x) = x^2 - 1$.

Solution: For $x = 0, y = -1$; for $y = 0, x = \pm 1$. The graph of $f(x)$ crosses the y axis at $(0, -1)$ and crosses the x axis at $(+1, 0)$ and $(-1, 0)$.

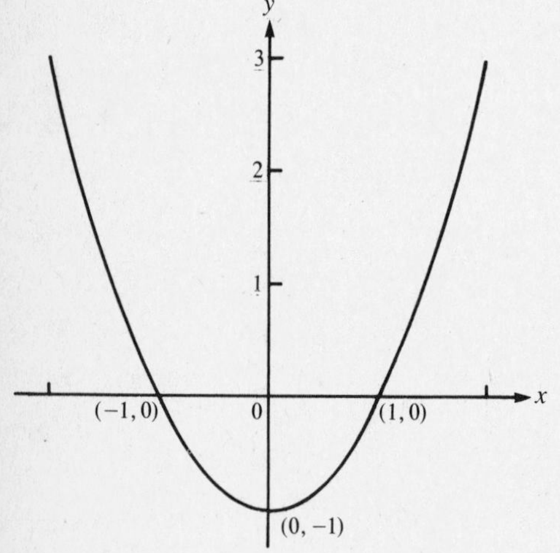

Figure 8-1
$f(x) = x^2 - 1$: A function whose graph is symmetric with respect to the y axis.

B. Symmetry

You can determine the **symmetry** of a graph with respect to the y axis or with respect to the origin as follows: The graph of an *even function* (i.e., $f(-x) = f(x)$) is symmetric with respect to the y axis; the graph of an *odd function* (i.e., $f(-x) = -f(x)$) is symmetric with respect to the origin.

EXAMPLE 8-2: Does the graph of $f(x) = x^2 - 1$ display either type of symmetry?

Solution: Because $f(-x) = f(x)$, the function is even: If you folded its graph along the y axis, the portion of the curve for $x < 0$ would coincide with the portion for which $x > 0$. (See Figure 8-1.)

EXAMPLE 8-3: Does the graph of $f(x) = x^3$ display either type of symmetry?

Solution: Because $f(-x) = -f(x)$, the function is odd: The graph would remain unchanged if you were to rotate it 180° about the origin. (See Figure 8-2.)

EXAMPLE 8-4: Does the graph of $f(x) = x^2 - x$ display either type of symmetry?

Solution: The function is neither odd nor even: Its graph displays neither of the symmetries of the previous examples. (See Figure 8-3.)

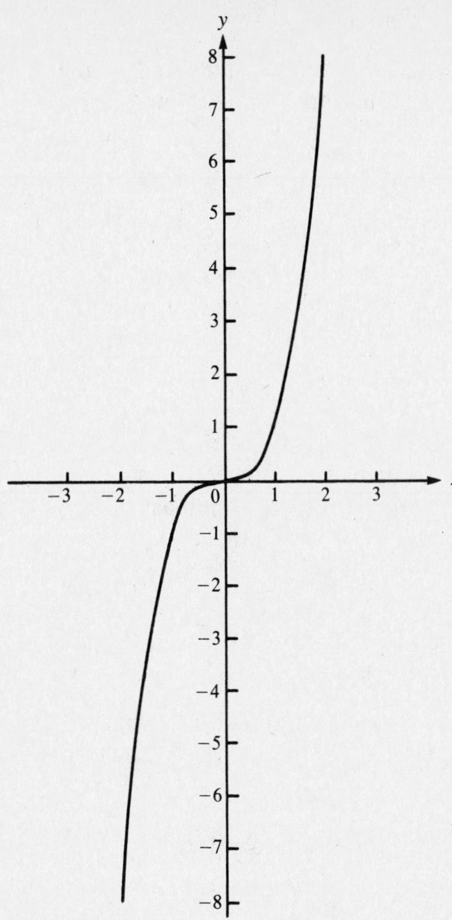

Figure 8-2
$f(x) = x^3$: A function whose graph is symmetric with respect to the origin.

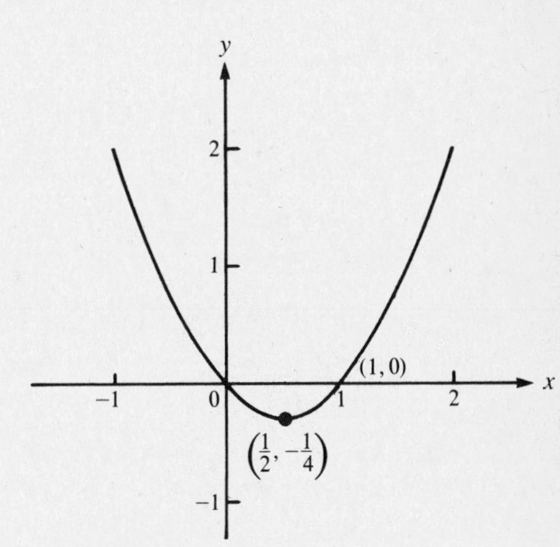

Figure 8-3
$f(x) = x^2 - x$.

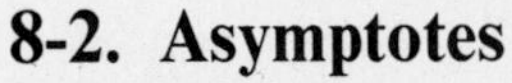

8-2. Asymptotes

The graphs of some functions seem to "approach" some vertical or horizontal line. Such lines are called **asymptotes**.

A. Horizontal asymptotes

To find a *horizontal asymptote*, consider the limit of $f(x)$ as x becomes numerically large. In Figure 8-4 the graph of $y = f(x)$ has a horizontal asymptote at $y = c$:

$$\lim_{x \to \infty} f(x) = c$$

The graph approaches the line $y = c$ as x gets larger.

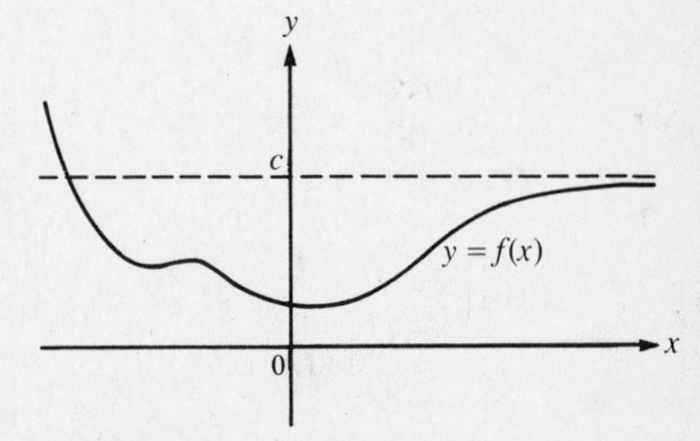

Figure 8-4
The graph of a function f such that $\lim_{x\to\infty} f(x) = c$.

EXAMPLE 8-5: Does the graph of $f(x) = 3 - 1/(x^2 + 1)$ have a horizontal asymptote?

Solution: Evaluating the limit of $f(x)$,

$$\lim_{x \to \infty} 3 - \frac{1}{x^2 + 1} = 3$$

Thus, $f(x)$ has a horizontal asymptote at $y = 3$, as shown in Figure 8-5.

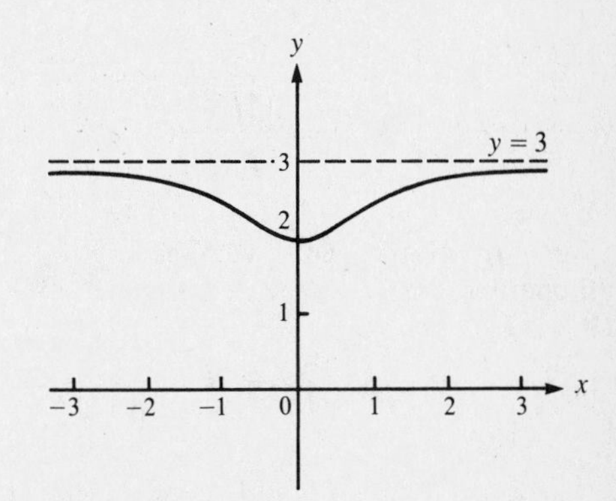

Figure 8-5
$f(x) = 3 - 1/(x^2 + 1)$; the graph has horizontal asymptote $y = 3$.

For rational functions, use the techniques of Section 3-7 on infinite limits to find the horizontal asymptotes. For

$$f(x) = \frac{a_n x^n + a_{n-1} x^{n-1} + \cdots + a_1 x + a_0}{b_m x^m + b_{m-1} x^{m-1} + \cdots + b_1 x + b_0}$$

there are three possibilities:

(1) If $n > m$, then $f(x)$ has no horizontal asymptote.
(2) If $n < m$, then $f(x)$ has horizontal asymptote $y = 0$.
(3) If $n = m$, then $f(x)$ has horizontal asymptote $y = a_n/b_m$.

EXAMPLE 8-6: Does the graph of $f(x) = (4x^3 - 5x + 1)/(7x^3 + x^2 + 2)$ have any horizontal asymptotes?

Solution: The numerator and denominator have the same degree ($n = m$) and the leading coefficients are 4 and 7. This function has horizontal asymptote $y = 4/7$.

EXAMPLE 8-7: Does the graph of $f(x) = (4x^5 - 5x + 1)/(7x^3 + x^2 + 2)$ have a horizontal asymptote?

Solution: Because the degree of the numerator is greater than the degree of the denominator, this function has no horizontal asymptote.

EXAMPLE 8-8: Does the graph of $f(x) = (7x^3 + x^2 + 2)/(4x^5 - 5x + 1)$ have a horizontal asymptote?

Solution: Because the degree of the numerator is less than the degree of the denominator, this function has horizontal asymptote $y = 0$ (the x axis).

For functions that are not rational, use whatever techniques are appropriate for finding the limit as x becomes infinite.

EXAMPLE 8-9: Let

$$f(x) = \frac{1}{3\cos(1/x)} + \frac{1}{x + e^x}$$

then

$$\lim_{x \to \infty} f(x) = \frac{1}{3} + 0 = \frac{1}{3}$$

The function has horizontal asymptote $y = 1/3$.

B. Vertical asymptotes

To find a *vertical asymptote*, you first find values of x for which y is not defined. If $f(x)$ is not defined for $x = c$, then the graph of f will not cross the vertical line $x = c$. Since x may assume values "near" c (values less than c and values greater than c), the graph will appear to approach the vertical line, as in Figure 8-6. To look at the behavior of y for x near c, use the one-sided limit techniques introduced in Section 3-5. If

$$\lim_{x \to a^-} f(x) = \pm\infty \qquad \text{or} \qquad \lim_{x \to a^+} f(x) = \pm\infty$$

then $f(x)$ will have a vertical asymptote at $x = a$.

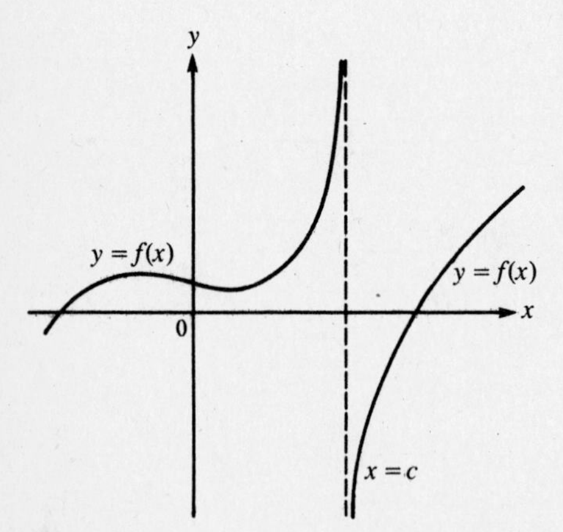

Figure 8-6
The graph of a function f with the properties $\lim_{x \to c^-} f(x) = \infty$ and $\lim_{x \to c^+} f(x) = -\infty$.

EXAMPLE 8-10: Consider the function $f(x) = 1/x$. The function has a vertical asymptote at $x = 0$: $f(0)$ is not defined. Since $f(x) > 0$ when $x > 0$ and $f(x) < 0$ when $x < 0$, it follows that

$$\lim_{x \to 0^+} f(x) = +\infty \qquad \text{and} \qquad \lim_{x \to 0^-} f(x) = -\infty$$

EXAMPLE 8-11: Consider the function $f(x) = 8/(x^2 - 16)$. The function is undefined at $x = 4$ and at $x = -4$. If $x^2 > 16$ (i.e., when $x > 4$ or $x < -4$), then $f(x) > 0$. If $x^2 < 16$ (when $-4 < x < 4$), then $f(x) < 0$. Thus, the one-sided limits are

$$\lim_{x \to -4^-} f(x) = \infty \qquad \lim_{x \to -4^+} f(x) = -\infty$$

$$\lim_{x \to 4^-} f(x) = -\infty \qquad \lim_{x \to 4^+} f(x) = \infty$$

Hence, $f(x)$ has vertical asymptotes at $x = -4$ and $x = 4$ (Figure 8-7). The sign of the one-sided limits (i.e., $+\infty$ or $-\infty$) will help you when graphing the function.

Caution: A function $f(x)$ may be undefined at $x = c$ and yet have no vertical asymptote at $x = c$.

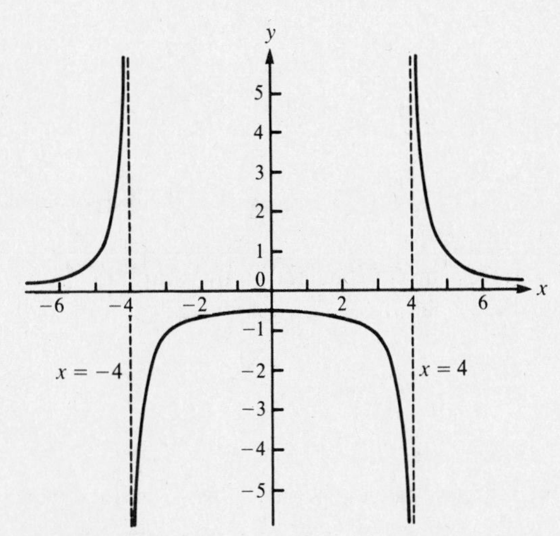

Figure 8-7
$f(x) = 8/(x^2 - 16)$.

EXAMPLE 8-12: Consider the function $f(x) = x^3/x$. The function is not defined at $x = 0$, but the graph has no vertical asymptote. The graph is identical to that of $y = x^2$, with the origin "punched out" (see Figure 8-8):

$$\lim_{x \to 0^-} f(x) = \lim_{x \to 0^+} f(x) = 0$$

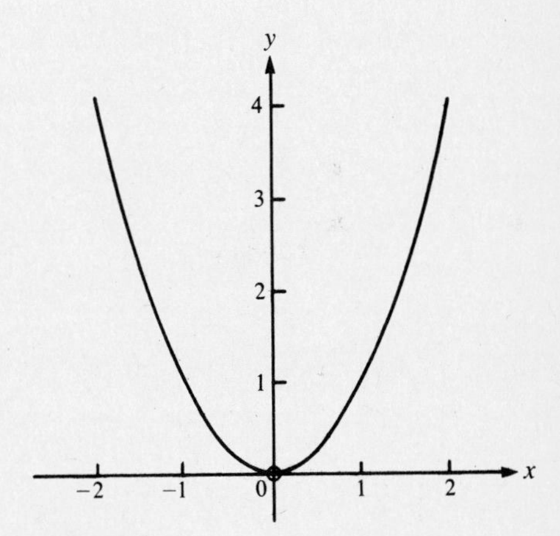

Figure 8-8
$f(x) = x^3/x$.

8-3. The Sign of $f(x)$

From the Intermediate Value Theorem you know that if a continuous function goes from negative to positive values (or from positive to negative), then it must go through zero. There is an immediate (and graphically useful!) consequence of this theorem: If $f(x) \neq 0$ at every point of an interval on which f is continuous, then $f(x)$ is either always positive or always negative on that interval. This means that $f(x)$ can change sign only at those values of x for which $f(x) = 0$ or for which f is discontinuous.

EXAMPLE 8-13: Consider the function $f(x) = (x - 1)(2x + 3)(x - 4)$. You can see that $f(x) = 0$ when $x = 1$, $x = -3/2$, or $x = 4$. Only for these three values of x can $f(x)$ change sign. If you exclude those numbers from the x axis, you are left with four open intervals:

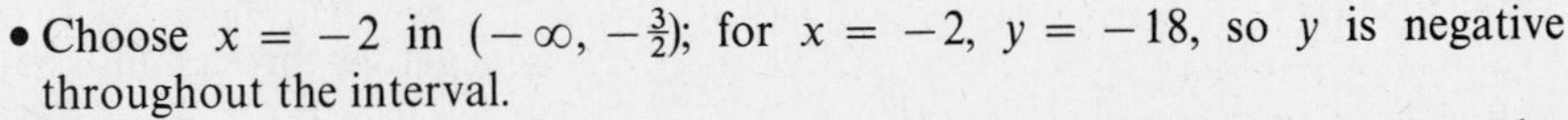

$$(-\infty, -\tfrac{3}{2}) \qquad (-\tfrac{3}{2}, 1) \qquad (1, 4) \qquad (4, \infty)$$

To determine the sign of $f(x)$ in each interval, calculate $y = f(x)$ for one arbitrarily chosen value of x in each interval.

- Choose $x = -2$ in $(-\infty, -\frac{3}{2})$; for $x = -2$, $y = -18$, so y is negative throughout the interval.
- Choose $x = 0$ in $(-\frac{3}{2}, 1)$; for $x = 0$, $y = 12$, so y is positive throughout the interval.
- Choose $x = 2$ in $(1, 4)$; for $x = 2$, $y = -14$, so y is negative throughout the interval.
- Choose $x = 5$ in $(4, \infty)$; for $x = 5$, $y = 52$, so y is positive throughout the interval.

We can now say with certainty that the graph of y lies below the x axis for $x < -\frac{3}{2}$ and for $1 < x < 4$, and above the x axis for $-\frac{3}{2} < x < 1$ and for $x > 4$.

The above observations apply only to continuous functions.

Note: Functions may also change signs at points of discontinuity. For example, see the graph of $f(x) = 8/(x^2 - 16)$ in Figure 8-7. At $x = -4$ and again at $x = 4$ (where $f(x)$ is undefined), y changes sign.

8-4. The Sign of $f'(x)$ and $f''(x)$

The method outlined in Section 8-3 is also useful when applied to the first and second derivatives when they are continuous.

A. Use the sign of f' to determine where the graph rises.

In Section 4-1 you saw that you could interpret the derivative of $y = f(x)$ at $x = a$ as the slope of the tangent line to the graph of $f(x)$ at $x = a$. Hence, you can relate the behavior of the graph to the sign of the derivative as follows (see Figure 8-9):

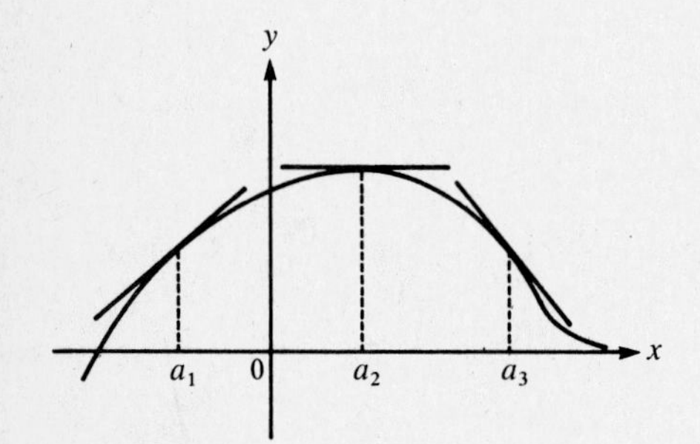

Figure 8-9
Lines tangent to graph at $x = a_1, a_2$, and a_3.

$f'(a_1) > 0$	Tangent line rises (from left to right) at $x = a_1$.	f is increasing at $x = a_1$.
$f'(a_2) = 0$	Tangent line is horizontal at $x = a_2$.	f is neither increasing nor decreasing at $x = a_2$.
$f'(a_3) < 0$	Tangent line falls (from left to right) at $x = a_3$.	f is decreasing at $x = a_3$.

A point of the domain of f at which $f'(x)$ changes sign is called a **critical point**. Notice that if $x = c$ is a critical point for $f(x)$, then either $f'(c) = 0$, or $f'(c)$ is not defined (see Figure 8-10). At a critical point the graph has a horizontal tangent line, as at $x = a_1$ and $x = a_2$, or no (uniquely defined) tangent line, as at $x = a_3$.

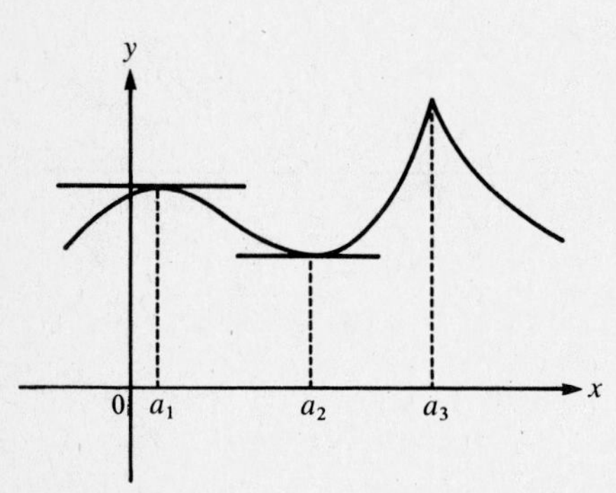

Figure 8-10
Horizontal tangent lines at $x = a_1, a_2$.
No tangent line at $x = a_3$.

Finding where a graph is rising and falling is similar to determining where a graph lies above (or below) the x axis. First, find where $f'(x) = 0$ and where $f'(x)$ is discontinuous. Use these values to define intervals on the x axis. Since $f'(x)$ can change sign only at those points, $f'(x)$ must be of constant sign over each interval. Hence, the graph is always rising, or always falling, over the interval.

EXAMPLE 8-14: Find where the graph of $f(x) = (x^3/3) - x - 1$ rises, where it falls, and where the tangent line is horizontal.

Solution: Solving $f'(x) = 0$ for x, you get

$$f'(x) = x^2 - 1 = 0 \quad \text{for} \quad x = \pm 1$$

The defined intervals are then $(-\infty, -1)$, $(-1, 1)$, and $(1, \infty)$. Following the procedure in Example 8-13, you calculate the values of $f'(x)$ (not $f(x)$, because you want to know where f' is positive and negative) at one arbitrarily chosen x in each interval.

- Choose $x = -2$ in $(-\infty, -1)$; $f'(-2) = 3$, so $f'(x)$ is positive throughout the interval.
- Choose $x = 0$ in $(-1, 1)$; $f'(0) = -1$, so $f'(x)$ is negative throughout the interval.
- Choose $x = 2$ in $(1, \infty)$; $f'(2) = 3$, so $f'(x)$ is positive throughout the interval.

You conclude that the graph of $y = (x^3/3) - x - 1$ has horizontal tangent lines at $x = -1$ and at $x = 1$. The graph rises for $x < -1$ and for $x > 1$, and falls for $-1 < x < 1$.

B. The sign of f'': Concavity

The sign of the second derivative tells whether a graph is concave upward or concave downward.

If $f''(x) > 0$, then $f'(x)$ is increasing. This follows from the discussion in Section 8-4A because $f''(x)$ is the derivative of $f'(x)$. When $f''(x) > 0$, the slope of the tangent line increases from left to right. This means that the graph is "curving upward."

Increasing slope does not rule out the possibility that the slope is negative (see Figure 8-11). For example, the slope might increase from -1 to $-\frac{1}{2}$ to 0 as x increases from $x = a_1$ to $x = a_2$ to $x = a_3$. Such a graph would belong to a function $f(x)$ with $f''(x) > 0$ for $a_1 < x < a_3$.

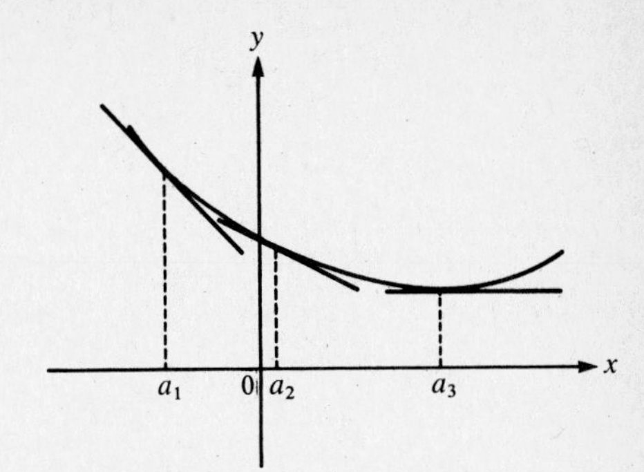

Figure 8-11
Graph of f such that $f''(x) > 0$ and $f'(x) < 0$ for $a_1 \leqslant x \leqslant a_3$.

If $f''(x) > 0$ at $x = a$, we say that the graph of f is *concave upward* at $x = a$. If $f''(x) < 0$ at $x = b$, then we conclude (by a similar line of reasoning) that the graph is curving downward, and we say that the graph is *concave downward* at $x = b$. In Figure 8-12 the graph is concave upward for $x < x_1$ and for $x_2 < x < x_3$, and concave downward for $x_1 < x < x_2$ and for $x_3 < x$.

A point in the domain of $f(x)$ where the graph changes from concave upward to concave downward, or vice versa, is called an **inflection point**. In Figure 8-12, x_1, x_2, and x_3 are inflection points.

You determine the concavity of the graph by considering f'' and its sign changes. As in the case of f and f', f'' can change sign only where $f''(x) = 0$ and where $f''(x)$ is discontinuous. Hence, $f''(x)$ will be of constant sign on each interval determined by the zeros and discontinuities, and the graph will always be concave upward, or always concave downward, on each interval.

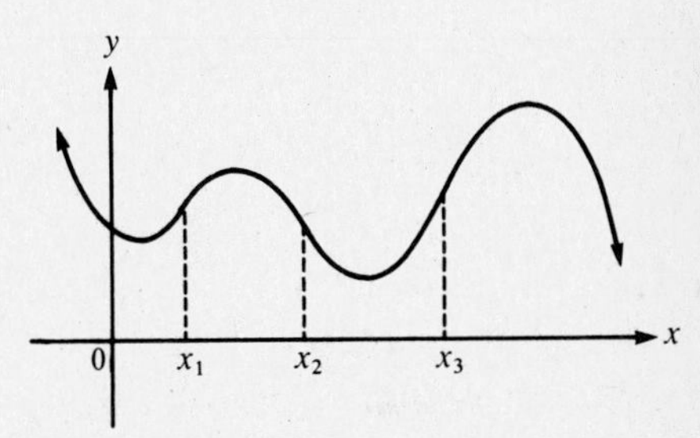

Figure 8-12
Inflection points at $x = x_1$, x_2, and x_3.

EXAMPLE 8-15: Determine where the graph of $f(x) = x^4 - 6x^2 - 3x - 5$ is concave upward and where it is concave downward.

Solution: Solve $f''(x) = 0$ for x:

$$f'(x) = 4x^3 - 12x - 3$$

$$f''(x) = 12x^2 - 12$$

When $x = \pm 1$, $f''(x) = 0$. Using the procedures established in Examples 8-13 and 8-14, you find values of $f''(x)$ for one x in each of the intervals $(-\infty, -1)$, $(-1, 1)$, and $(1, \infty)$. You use the values of $f''(x)$, not $f(x)$ or $f'(x)$, because you want to determine the concavity of the graph.

- Choose $x = -2$ in $(-\infty, 1)$; $f''(-2) = 36$, so $f''(x)$ is positive throughout the interval.
- Choose $x = 0$ in $(-1, 1)$; $f''(0) = -12$, so $f''(x)$ is negative throughout the interval.
- Choose $x = 2$ in $(1, \infty)$; $f''(2) = 36$, and $f''(x)$ is positive throughout the interval.

You conclude that the graph is concave upward for $x < -1$ and for $x > 1$, and concave downward for $-1 < x < 1$. Thus, $x = -1$ and $x = 1$ are inflection points.

8-5. The First and Second Derivative Tests

You may need to determine the location of "peaks" and "valleys" on a graph. You can conclude from the theorems of Chapter 7 that the graph of a continuous function over a closed interval will have a highest point and a lowest point. Discontinuous functions, or continuous functions graphed over an open interval, do not necessarily have this property.

A brief inspection of Figure 8-12 suggests that high points and low points on a graph occur at critical points and at endpoints of intervals. These special points and methods for finding them follow. In Chapter 9 we will return to the subject in greater detail.

A. Maxima and minima

If, for a given f, there exists a number a in the domain of f such that $f(x) \leqslant f(a)$ for all x in the domain of f, then f has a **global** (or absolute) **maximum** at $x = a$. There are no restrictions on the domain; it might be any interval, open or closed, including $(-\infty, \infty)$. The idea of **local** (or relative) **maximum** is similar. The function f is said to have a local maximum at $x = c$ if, for a given f, there exists an open interval (a, b) about c (i.e., $a < c < b$) such that $f(x) \leqslant f(c)$ for all x in the interval. *Global* (absolute) and *local minima* may be similarly defined.

Look at Figure 8-13. This function (defined on $a \leqslant x \leqslant b$) has a global maximum at $x = x_3$ and local maxima at $x = x_1$ and $x = x_5$. There is a global minimum at $x = a$ and local minima at $x = x_2$, $x = x_4$, and $x = b$.

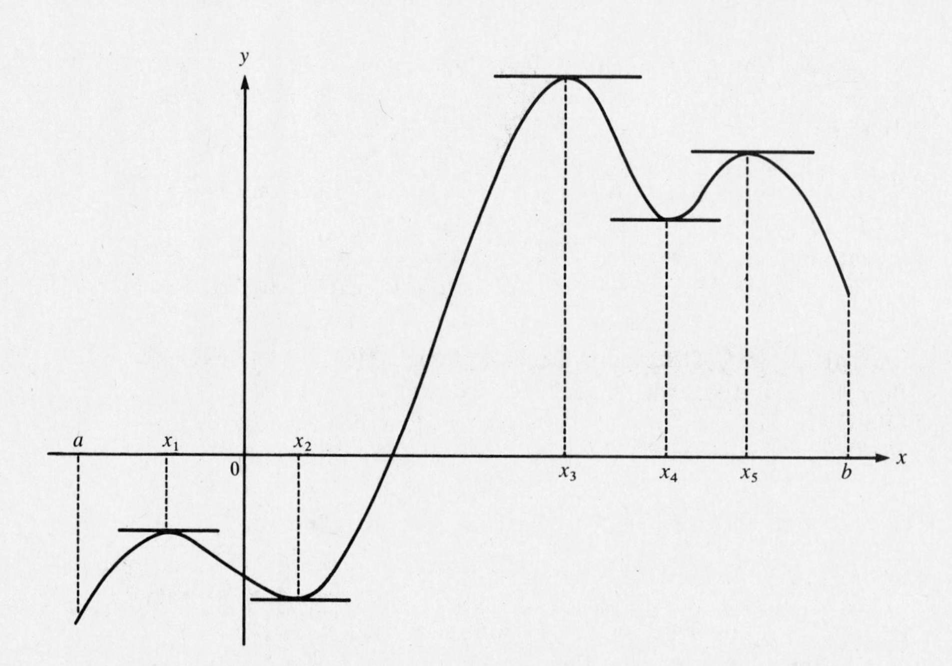

Figure 8-13
Local extrema at $x = a, x_1, x_2, x_3, x_4, x_5$, and b.

B. The first derivative test

Each local maximum occurs at a point at which the first derivative changes from positive (f increasing) to negative (f decreasing). Each local minimum occurs at a point at which f' changes from negative (f decreasing) to positive (f increasing). These observations constitute the **first derivative test**.

EXAMPLE 8-16: Find all local maxima and minima of

$$f(x) = 4x^3 - 9x^2 - 30x + 1.$$

Solution: You must find all critical points (since f' is continuous, find where $f'(x) = 0$):

$$f'(x) = 12x^2 - 18x - 30$$

Set $f'(x)$ equal to zero:

$$12x^2 - 18x - 30 = 0$$

$$6(2x - 5)(x + 1) = 0$$

Thus, the critical points are at $x = \frac{5}{2}$ and $x = -1$. Following our earlier procedure, you find that $f'(x)$ changes from positive to negative at $x = -1$, so f

has a local maximum at $(-1, f(-1)) = (-1, 18)$. Moreover, $f'(x)$ changes from negative to positive at $x = \frac{5}{2}$, so f has a local minimum at $(5/2, f(5/2)) = (5/2, -271/4)$.

EXAMPLE 8-17: Consider the function $f(x) = x^3$ and its first derivative $f'(x) = 3x^2$. Notice that $f'(x) \geq 0$ for all x: Even though $f'(x) = 0$ at $x = 0$, $f'(x)$ does not change sign. Hence, you can conclude that $f(x)$ has no relative maxima or minima (see Figure 8-2).

C. The second derivative test

You can often use a different test to determine whether or not a given critical point is a maximum or minimum. Referring again to Figure 8-13, notice that at a local maximum at which $f'(x) = 0$, the graph is concave downward, and at a local minimum at which $f'(x) = 0$, the graph is concave upward. This leads to the **second derivative test**: If $f''(c) < 0$, then f has a local maximum at a critical point $x = c$. If $f''(c) > 0$, then f has a local minimum at $x = c$.

EXAMPLE 8-18: Find the local maxima and minima of the function $f(x) = 4x^3 - 9x^2 - 30x + 1$ from Example 8-16, this time using the second derivative test.

Solution:

$$f'(x) = 12x^2 - 18x - 30$$

$$f''(x) = 24x - 18 = 6(4x - 3)$$

Substituting the critical values $x = -1$ and $x = \frac{5}{2}$ into $f''(x)$ you find,

$$f''(-1) = 6(-4 - 3) < 0$$

so there is a local maximum at $x = -1$. At $\frac{5}{2}$ you find

$$f''(5/2) = 6(10 - 3) > 0$$

so there is a local minimum at $x = \frac{5}{2}$.

Caution: (1) The second derivative test is used only at points at which $f'(x) = 0$.
(2) If there is a critical point at $x = a$ and $f''(a) = 0$, then the second derivative test is of no help in determining whether $x = a$ is a local maximum or minimum or neither.

8-6. Graphing a Function

When graphing a function, use the procedures described in Sections 8-1 through 8-5 to obtain the relevant information on

(1) intercepts
(2) symmetry
(3) asymptotes
(4) critical points
(5) slope of the tangent line
(6) concavity

The procedures will often provide a duplication of information. This serves as a double check.

Keep in mind that

(1) Values of $f(x)$ give you information about the location of points on the graph.
(2) Values of $f'(x)$ give you information about the tangent line to the graph.
(3) Values of $f''(x)$ give you information about the concavity of the graph.

EXAMPLE 8-19: Graph $f(x) = (x + 2)(x - 1)^2$.

Solution: You need to determine the following:

(1) Intercepts:

$$y \text{ intercept} \quad (0, f(0)) = (0, 2)$$
$$x \text{ intercepts} \quad 0 = (x + 2)(x - 1)^2$$
$$x = -2, 1$$

(2) Symmetry: $f(-x) = (-x + 2)(-x - 1)^2$ which is neither $f(x)$ nor $-f(x)$. Hence, the graph is not symmetric about either the y axis or the origin.

(3) Asymptotes: Because f is defined for all x, there are no vertical asymptotes. Because $\lim_{x\to\infty} f(x) = \infty$, there are no horizontal asymptotes. In general, polynomial functions have no asymptotes.

(4) Critical points:

$$\begin{aligned} f'(x) &= (x - 1)^2 + 2(x + 2)(x - 1) \\ &= (x - 1)[(x - 1) + 2(x + 2)] \\ &= (x - 1)(3x + 3) = 3(x - 1)(x + 1) \end{aligned}$$

So, $f'(x) = 0$ when $x = 1$ or $x = -1$. The critical points of the graph are $(1, f(1)) = (1, 0)$ and $(-1, f(-1)) = (-1, 4)$.

(5) Slope of tangent line: Choosing an x in each of the intervals determined by the critical points, you discover that

- $f'(x) > 0$ (f increasing) to the left of $x = -1$, and $f'(x) < 0$ (f decreasing) to the right of $x = -1$, so f has a local maximum at $x = -1$.
- $f'(x) < 0$ to the left of $x = 1$, and $f'(x) > 0$ to the right of $x = 1$, so f has a local minimum at $x = 1$.

(6) Concavity and inflection points: $f''(x) = (d/dx)[3(x^2 - 1)] = 6x$. Thus, $f''(x) = 0$ at $x = 0$, with $f''(x) > 0$ for $x > 0$, and $f''(x) < 0$ for $x < 0$. The graph is concave downward for $x < 0$ and concave upward for $x > 0$. Consequently, there is an inflection point at $x = 0$ (i.e., at $(0, f(0)) = (0, 2)$).

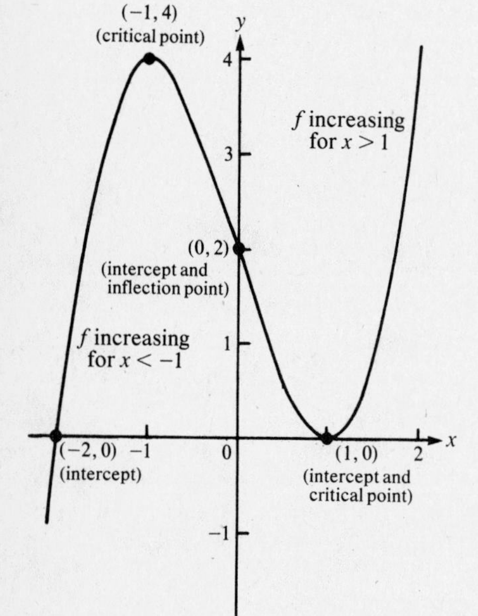

Figure 8-14
$f(x) = (x + 2)(x - 1)^2$.

To sketch the graph, you first locate the specific points found: the intercepts, critical points, and inflection points. If you had found an asymptote, you would sketch it. At each critical point where $f'(x) = 0$, you draw a short horizontal segment to indicate the tangent line at the point. Then enter the information relevant to each point: $(-1, 4)$ is a local maximum and $(1, 0)$ is a local minimum. Carefully draw a smooth curve through the points, making sure your graph is consistent with the information you have entered on the graph. Note that the curve has concavity as predicted by $f''(x)$: concave downward for negative values of x and concave upward for positives values of x.

If you have any doubt about the accuracy of the graph, a small amount of arithmetic will reassure you. If $x = 2$, then $y = 4$, and $(2, 4)$ fits on the graph. If $x = -3$, then $y = -16$, and $(-3, -16)$ fits on the graph as well. See Figure 8-14.

EXAMPLE 8-20: Graph $f(x) = (x^2 + 1)/x$.

Solution: Again you find the following information:

(1) Intercepts: y is undefined for $x = 0$, so there is no y intercept; $f(x)$ is never zero, so there is no x intercept. Thus, the graph never crosses either coordinate axis.

(2) Symmetry:

$$f(-x) = \frac{(-x)^2 + 1}{-x} = -\frac{x^2 + 1}{x} = -f(x)$$

Hence, the graph is symmetric with respect to the origin.

(3) Asymptotes: f is not defined at $x = 0$;

$$\lim_{x\to 0^-} f(x) = -\infty \qquad \lim_{x\to 0^+} f(x) = +\infty$$

There is a vertical asymptote at $x = 0$ (the y axis). There is no horizontal asymptote because f is a rational function with the degree of the numerator greater than the degree of the denominator.

(4) Critical points:

$$f'(x) = \frac{x(2x) - (x^2 + 1)}{x^2} = \frac{x^2 - 1}{x^2}$$

So $f'(x) = 0$ for $x = \pm 1$; $f'(x)$ is defined everywhere except at $x = 0$, so the only critical points are $(1, f(1)) = (1, 2)$ and $(-1, f(-1)) = (-1, -2)$.

(5) Slope of the tangent line: Choose an x in each interval determined by the critical points and the discontinuity of f' (i.e., at $x = -1, 0$, and 1). For $x < -1$, $f'(x) > 0$; for $-1 < x < 0$, $f'(x) < 0$; for $0 < x < 1$, $f'(x) < 0$; and for $x > 1$, $f'(x) > 0$. Thus, by the first derivative test you see that $f(x)$ has a local maximum at $x = -1$ (i.e., at $(-1, -2)$), and a local minimum at $x = 1$ (i.e., $(1, 2)$).

(6) Concavity and inflection points:

$$f''(x) = \frac{x^2(2x) - (x^2 - 1)2x}{x^4} = \frac{2}{x^3}$$

For $x > 0$, $f''(x) > 0$ (graph is concave upward) and for $x < 0$, $f''(x) < 0$ (graph is concave downward). Notice that there are no inflection points because f'' changes sign only at $x = 0$, where f is not defined.

To sketch the graph, you first label the y axis as a vertical asymptote; this should remind you that the graph will not cross the y axis. Then locate the critical points $(-1, -2)$ (local maximum) and $(1, 2)$ (local minimum) and draw short horizontal segments through each point. You can see from the information on asymptotes that as x approaches zero from the left, y becomes very large negatively. Moreover, for $-1 < x < 0$, f is decreasing. You have little choice as to the behavior of the graph for $x < 0$, so draw in a smooth curve. If you want, you can calculate two or three points. Notice that the concavity is correct: For $x < 0$ the graph is concave downward. Now use the symmetry (with respect to the origin) to draw in the rest of the graph.

Note: We have not mentioned asymptotes that are neither horizontal nor vertical, but this function furnishes an example of such behavior. Notice that $(x^2 + 1)/x = x + (1/x)$. So

$$\lim_{x \to \infty} [x + (1/x)] = \lim_{x \to \infty} (x) + \lim_{x \to \infty} (1/x)$$

Because $\lim_{x \to \infty} (1/x) = 0$, the points of the graph of f are getting closer to the line $y = x$; that is, $y = x$ is an asymptote for $f(x)$. See Figure 8-15.

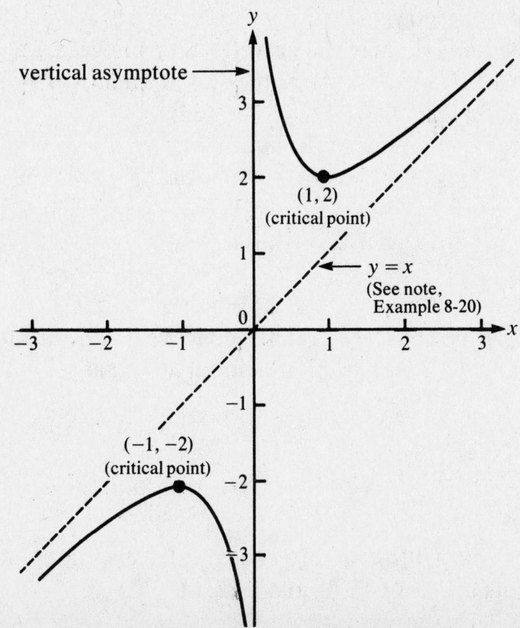

Figure 8-15
$f(x) = (x^2 + 1)/x$.

SUMMARY

1. The intercepts of a graph are the points where the graph crosses the coordinate axes.
2. The graph of an even (or odd) function will be symmetric with respect to the y axis (or the origin).
3. Calculate $\lim_{x\to\infty} f(x)$ or $\lim_{x\to-\infty} f(x)$ to find any horizontal asymptotes.
4. Vertical asymptotes occur at points at which the function is not defined. One-sided limits verify that the line is a vertical asymptote and help you to graph the function near the asymptote.
5. You need only evaluate $f(x)$ at a few values of x to determine exactly where f is positive and negative.
6. The first derivative of $f(x)$ gives the slope of the graph of $f(x)$. In particular, the sign of $f'(x)$ tells you whether the graph of f is rising or falling.
7. The second derivative of $f(x)$ gives the concavity of $f(x)$. In particular, the sign of $f''(x)$ tells you whether the graph of f is concave upward or downward.
8. The first and second derivatives are used to determine whether critical points are local maxima or local minima.
9. You graph a function by first finding the intercepts, symmetry, asymptotes, critical points, slope of a tangent line, and concavity. All of this information must then be faithfully represented in the graph.

SOLVED PROBLEMS

PROBLEM 8-1 Graph $f(x) = x^3 - 6x^2$.

Solution:

(1) Intercepts: The y intercept is $(0, f(0)) = (0, 0)$. To find the x intercept, you solve

$$0 = x^3 - 6x^2$$

$$0 = x^2(x - 6)$$

The x intercepts are (0, 0) and (6, 0).

(2) Symmetry: $f(-x) = (-x)^3 - 6(-x)^2 = -x^3 - 6x^2$, which is neither $f(x)$ nor $-f(x)$. So f does not display either kind of symmetry.

(3) Asymptotes: Because f is a polynomial, there are no vertical or horizontal asymptotes.

(4) Critical points: $f'(x) = 3x^2 - 12x = 3x(x - 4)$. So $f'(x) = 0$ when $x = 0$ and when $x = 4$. Thus, you find critical points at (0, 0) and $(4, f(4)) = (4, -32)$.

(5) Slope of the tangent line: Choose an x in each of the intervals determined by the critical points: $(-\infty, 0)$, $(0, 4)$, and $(4, \infty)$, for example, $x = -1$, 1, and 5. For these values of x, $f'(-1) = 15 > 0$, $f'(1) = -9 < 0$, $f'(5) = 15 > 0$. For all $x < 0$ and all $x > 4$, f is increasing. For $0 < x < 4$, f is decreasing. Thus, f has a local maximum at (0, 0) and a local minimum at $(4, -32)$.

(6) Concavity and inflection points: $f''(x) = 6x - 12$. When $x = 2$, $f''(x) = 0$. For $x > 2$, $f''(x) > 0$, and for $x < 2$, $f''(x) < 0$. Thus, the graph is concave downward for all $x < 2$ and concave upward for all $x > 2$. There is an inflection point at $x = 2$: at $(2, f(2)) = (2, -16)$.

[See Figure 8-16.]

PROBLEM 8-2 Graph $f(x) = x(x^2 - 9)$.

Solution:

(1) Intercepts: The y intercept is $(0, f(0)) = (0, 0)$. The x intercepts are found by solving $0 = x(x^2 - 9)$. The x intercepts are (0, 0), (3, 0), and $(-3, 0)$.

Figure 8-16
$f(x) = x^3 - 6x^2$.

(2) Symmetry: $f(-x) = (-x)((-x)^2 - 9) = -x(x^2 - 9) = -f(x)$. So f is symmetric with respect to the origin.
(3) Asymptotes: Because f is a polynomial, there are no horizontal or vertical asymptotes.
(4) Critical points: $f'(x) = 3x^2 - 9 = 3(x^2 - 3)$. So $f'(x) = 0$ when $x = \pm\sqrt{3}$. The critical points are $(\sqrt{3}, f(\sqrt{3})) = (\sqrt{3}, -6\sqrt{3})$, and $(-\sqrt{3}, f(-\sqrt{3})) = (-\sqrt{3}, 6\sqrt{3})$.
(5) Slope of tangent line: The critical points determine three intervals: $(-\infty, -\sqrt{3})$, $(-\sqrt{3}, \sqrt{3})$, and $(\sqrt{3}, \infty)$. Choose test points in each interval, for example, -2, 0, and 2. For these values of x, $f'(-2) = 3 > 0$, $f'(0) = -9 < 0$, and $f'(2) = 3 > 0$. For all $x < -\sqrt{3}$ and all $x > \sqrt{3}$, f is increasing. For $-\sqrt{3} < x < \sqrt{3}$, f is decreasing. So, $(-\sqrt{3}, 6\sqrt{3})$ is a local maximum and $(\sqrt{3}, -6\sqrt{3})$ is a local minimum.
(6) Concavity and inflection points: $f''(x) = 6x$. When $x > 0$, $f''(x) > 0$; when $x < 0$, $f''(x) < 0$. Thus, the graph is concave upward for $x > 0$ and concave downward for $x < 0$. There is an inflection point at $x = 0$: i.e., at (0, 0). [See Figure 8-17.]

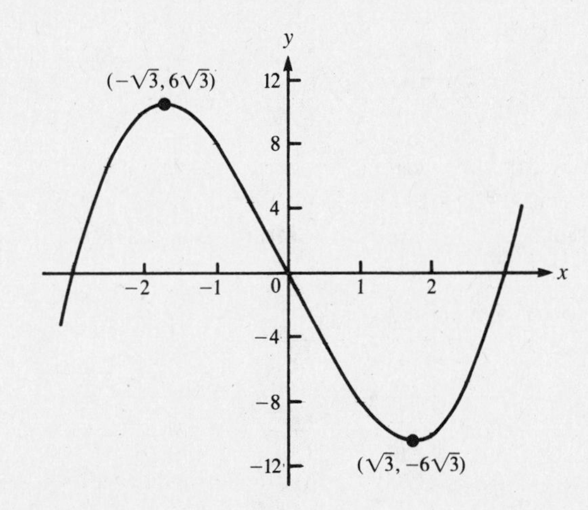

Figure 8-17
$f(x) = x(x^2 - 9)$.

PROBLEM 8-3 Graph $f(x) = x^4 - 2x^3$.

Solution:

(1) Intercepts: The y intercept is $(0, f(0)) = (0, 0)$. Set $f(x)$ equal to 0: $x^4 - 2x^3 = 0$, $x^3(x - 2) = 0$. When $x = 0$ and when $x = 2$, $f(x) = 0$, so the x intercepts are (0, 0) and (2, 0).
(2) Symmetry: $f(-x) = x^4 + 2x^3$, which is neither $f(x)$ nor $-f(x)$. So f exhibits neither form of symmetry.
(3) Asymptotes: Because f is a polynomial, there are no asymptotes.
(4) Critical points: $f'(x) = 4x^3 - 6x^2 = 2x^2(2x - 3)$. It follows that $f'(x) = 0$ when $x = 0$ and when $x = \frac{3}{2}$. The critical points are (0, 0) and $(\frac{3}{2}, f(\frac{3}{2})) = (3/2, -27/16)$.
(5) Slope of tangent line: Choosing $x = -1$, 1, and 2 as test points, $f'(-1) = -10 < 0$, $f'(1) = -2 < 0$, and $f'(2) = 8 > 0$. For $x > \frac{3}{2}$, f is increasing; for $x < 0$ and for $0 < x < \frac{3}{2}$, f is decreasing. Thus, there is a local minimum at $(3/2, -27/16)$; (0, 0) is neither a local minimum nor a local maximum.
(6) Concavity and inflection points: $f''(x) = 12x^2 - 12x$. When $x = 0$ and $x = 1$, $f''(x) = 0$. Selecting -1, $\frac{1}{2}$, and 2 as test points, $f''(-1) = 24 > 0$, $f''(\frac{1}{2}) = -3 < 0$, and $f''(2) = 24 > 0$. When $x < 0$ and when $x > 1$, the graph is concave upward; when $0 < x < 1$, the graph is concave downward. There are inflection points at (0, 0) and $(1, f(1)) = (1, -1)$.
[See Figure 8-18.]

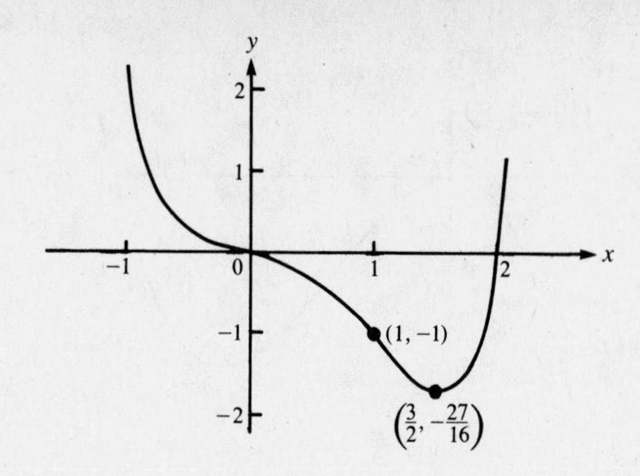

Figure 8-18
$f(x) = x^4 - 2x^3$.

PROBLEM 8-4 Graph $f(x) = (x^2 - 1)/(x^2 + 1)$.

Solution:

(1) Intercepts: The y intercept is $(0, f(0)) = (0, -1)$. When $x = \pm 1, f(x) = 0$, so $(1, 0)$ and $(-1, 0)$ are the x intercepts.

(2) Symmetry: $f(-x) = f(x)$, so f is symmetric about the y axis.

(3) Asymptotes:

$$\lim_{x\to\infty} \frac{x^2 - 1}{x^2 + 1} = 1 = \lim_{x\to-\infty} \frac{x^2 - 1}{x^2 + 1}$$

so the horizontal asymptote is $y = 1$. The denominator, $x^2 + 1$, is never zero; thus, there are no vertical asymptotes.

(4) Critical points:

$$f'(x) = \frac{(x^2 + 1)\cdot 2x - (x^2 - 1)\cdot 2x}{(x^2 + 1)^2} = \frac{4x}{(x^2 + 1)^2}$$

So, $f'(x) = 0$ when $x = 0$. Thus, the only critical point is $(0, -1)$.

(5) Slope of tangent line: Observe that when $x > 0, f'(x) > 0$, and when $x < 0, f'(x) < 0$. Thus, f is increasing for $x > 0$ and decreasing for $x < 0$. The point $(0, -1)$ is a local minimum.

(6) Concavity and inflection points:

$$f''(x) = \frac{(x^2 + 1)^2 \cdot 4 - 4x \cdot 2(x^2 + 1)\cdot 2x}{(x^2 + 1)^4}$$

$$= \frac{4(x^2 + 1)[(x^2 + 1) - 2x\cdot 2x]}{(x^2 + 1)^4}$$

$$= \frac{4(x^2 + 1)(1 - 3x^2)}{(x^2 + 1)^4} = \frac{4 - 12x^2}{(x^2 + 1)^3}$$

Set $f''(x) = 0$ to obtain $4 - 12x^2 = 0$ or $x = \pm\sqrt{\frac{1}{3}}$. Choosing -1, 0, and 1 as test points, $f''(-1) = -1 < 0$, $f''(0) = 4 > 0$, and $f''(1) = -1 < 0$. For $x < -\sqrt{\frac{1}{3}}$ and for $x > \sqrt{\frac{1}{3}}$, the graph is concave downward; for $-\sqrt{\frac{1}{3}} < x < \sqrt{\frac{1}{3}}$, the graph is concave upward. You find inflection points at $(\sqrt{\frac{1}{3}}, f(\sqrt{\frac{1}{3}})) = (\sqrt{\frac{1}{3}}, -\frac{1}{2})$ and $(-\sqrt{\frac{1}{3}}, f(-\sqrt{\frac{1}{3}})) = (-\sqrt{\frac{1}{3}}, -\frac{1}{2})$.

[See Figure 8-19.]

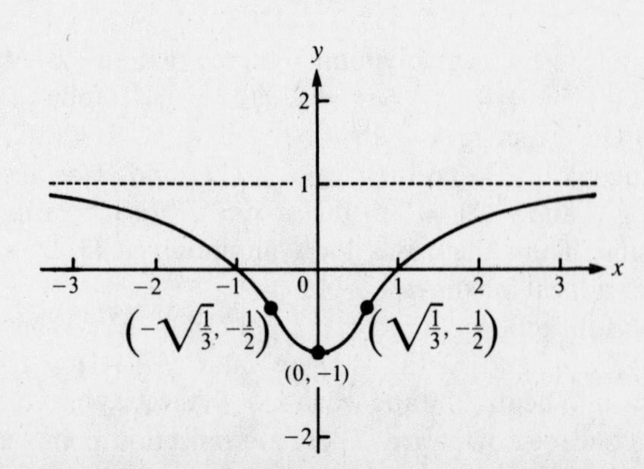

Figure 8-19
$f(x) = (x^2 - 1)/(x^2 + 1)$.

PROBLEM 8-5 Graph $f(x) = (x^3 - 16)/x$.

Solution:

(1) Intercepts: Because $f(x)$ is not defined at $x = 0$, there is no y intercept. When $x = \sqrt[3]{16} = 2\sqrt[3]{2}$, $f(x) = 0$. So the x intercept is $(2\sqrt[3]{2}, 0)$.

(2) Symmetry:

$$f(-x) = \frac{-x^3 - 16}{-x} = \frac{x^3 + 16}{x}$$

so f is not symmetric with respect to the origin or the y axis.

(3) Asymptotes:

$$\lim_{x \to \infty} \frac{x^3 - 16}{x} = \infty \qquad \lim_{x \to -\infty} \frac{x^3 - 16}{x} = \infty$$

So, there is no horizontal asymptote.

$$\lim_{x \to 0^-} \frac{x^3 - 16}{x} = \infty \qquad \lim_{x \to 0^+} \frac{x^3 - 16}{x} = -\infty$$

The line $x = 0$ is a vertical asymptote.

(4) Critical points:

$$f'(x) = \frac{x(3x^2) - (x^3 - 16)}{x^2} = \frac{2x^3 + 16}{x^2}$$

So $f'(x) = 0$ when $2x^3 + 16 = 0$, that is, when $x = -2$. There is a critical point at $(-2, f(-2)) = (-2, 12)$.

(5) Slope of tangent line: Choosing -3, -1, and 1 as test points, $f'(-3) = \frac{-38}{9} < 0$, $f'(-1) = 14 > 0$, and $f'(1) = 18 > 0$. For $-2 < x < 0$ and for $x > 0$, f is increasing; for $x < -2$, f is decreasing. There is a local minimum at $(-2, 12)$.

(6) Concavity and inflection points:

$$f''(x) = \frac{d}{dx}\left(\frac{2x^3 + 16}{x^2}\right) = \frac{d}{dx}(2x + 16x^{-2}) = 2 - 32x^{-3} = \frac{2x^3 - 32}{x^3}$$

When $2x^3 - 32 = 0$, at $x = \sqrt[3]{16} = 2\sqrt[3]{2}$. At $x = 0$, $f''(x)$ is undefined. To find where f'' is positive and negative, you examine the regions determined by the zeros of f'' and the discontinuities of $f''(2\sqrt[3]{2}$ and $0)$: $(-\infty, 0)$, $(0, 2\sqrt[3]{2})$, and $(2\sqrt[3]{2}, \infty)$. Choosing -1, 1, and 4 as test points, $f''(-1) = 34 > 0$, $f''(1) = -30 < 0$, and $f''(4) = \frac{3}{2} > 0$. For $x < 0$ and for $x > 2\sqrt[3]{2}$, the graph is concave upward; for $0 < x < 2\sqrt[3]{2}$, the graph is concave downward. There is an inflection point at $(2\sqrt[3]{2}, 0)$. [See Figure 8-20.]

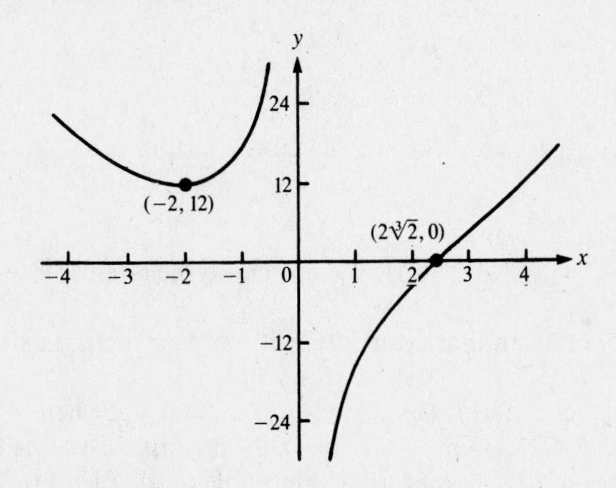

Figure 8-20
$f(x) = (x^3 - 16)/x$.

PROBLEM 8-6 Graph $f(x) = 4x/(x - 1)^2$.

Solution:

(1) Intercepts: (0, 0) is the x intercept and the y intercept.

(2) Symmetry: none

(3) Asymptotes:

$$\lim_{x\to\infty}\frac{4x}{(x-1)^2}=\lim_{x\to\infty}\frac{4x}{x^2-2x+1}=0 \qquad \lim_{x\to-\infty}\frac{4x}{(x-1)^2}=0$$

The degree of the denominator is greater than the degree of the numerator, so the horizontal asymptote is the line $y = 0$ (the x axis). The vertical asymptote is $x = 1$:

$$\lim_{x\to1^+}\frac{4x}{(x-1)^2}=\infty \qquad \lim_{x\to1^-}\frac{4x}{(x-1)^2}=\infty$$

(4) Critical points:

$$f'(x)=\frac{(x-1)^2\cdot4-4x\cdot2(x-1)}{(x-1)^4}=\frac{-4(x+1)}{(x-1)^3}$$

So $f'(x) = 0$ when $x = -1$. You find a critical point at $(-1, f(-1)) = (-1, -1)$.

(5) Slope of tangent line: Choosing -2, 0, and 2 as test points, $f'(-2) = -4/27 < 0$, $f'(0) = 4 > 0$, and $f'(2) = -12 < 0$. For $-1 < x < 1$, f is increasing; for $x < -1$ and $x > 1$, f is decreasing. The point $(-1, -1)$ is a local minimum.

(6) Concavity and inflection points:

$$f''(x)=\frac{(x-1)^3(-4)-(-4(x+1))\cdot3\cdot(x-1)^2}{(x-1)^6}=\frac{8x+16}{(x-1)^4}$$

When $x = -2$, $f''(x) = 0$; when $x < -2$, $f''(x) < 0$; and when $x > -2$, $f''(x) > 0$. The graph is concave downward for $x < -2$, and concave upward for all $x > -2$ (except $x = 1$, of course). There is an inflection point at $(-2, f(-2)) = (-2, -8/9)$. [See Figure 8-21.]

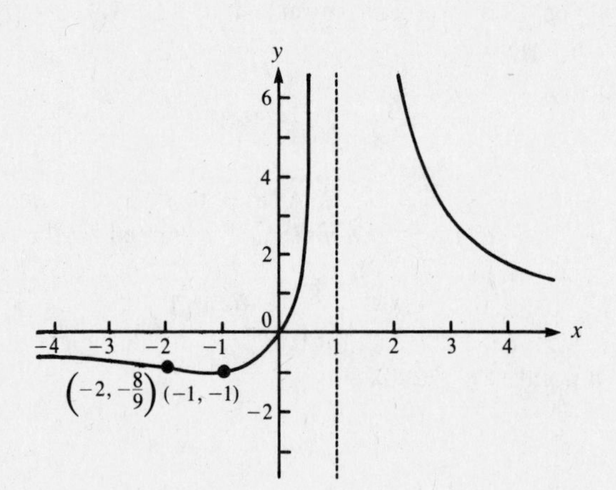

Figure 8-21
$f(x) = 4x/(x-1)^2$.

PROBLEM 8-7 Graph $f(x) = \sin^2 x$.

Solution:

(1) Intercepts: $\sin^2 x = 0$ when $x = k\pi$, where k is any integer. So the intercepts are $(k\pi, 0)$, k any integer.

(2) Symmetry: $f(-x) = f(x)$, so the graph is symmetric with respect to the y axis.

(3) Asymptotes: none.

(4) Critical points: $f'(x) = 2\sin x\cos x$; so $f'(x) = 0$ when $\sin x = 0$ and when $\cos x = 0$. When $x = k\pi$, $\sin x = 0$. When $x = (2k-1)\pi/2$, $\cos x = 0$. So the critical points occur at every integer multiple of π (where $y = 0$) and at every odd multiple of $\pi/2$ (where $y = 1$).

(5) Slope of tangent line: When $\sin x$ and $\cos x$ agree in sign (i.e., when x is in the first or third quadrant), $f'(x) > 0$. When x is in the second or fourth quadrant, $f'(x) < 0$.

(6) Concavity and inflection points: $f''(x) = -2\sin^2 x + 2\cos^2 x$. When $\sin x = \pm\cos x$ (i.e., when $x = (2k-1)\pi/4$, k any integer), $f''(x) = 0$. Thus you have inflection points at $((2k-1)\pi/4, 1/2)$, for every integer k. [See Figure 8-22.]

Figure 8-22
$f(x) = \sin^2 x$.

PROBLEM 8-8 Graph $f(x) = x^3(x - 7)^4$.

Solution:

(1) Intercepts: The y intercept is $(0, 0)$. The x intercepts are $(0, 0)$ and $(7, 0)$.
(2) Symmetry: none
(3) Asymptotes: none
(4) Critical points:

$$\begin{aligned} f'(x) &= 3x^2(x - 7)^4 + x^3 4(x - 7)^3 \\ &= x^2(x - 7)^3[3(x - 7) + 4x] \\ &= 7x^2(x - 7)^3(x - 3) \end{aligned}$$

So $f'(x) = 0$ when $x = 0, 3$, or 7. The critical points are $(0, 0)$, $(3, 6912)$, and $(7, 0)$.

(5) Slope of tangent line: Selecting -1, 1, 4, and 8 as test points, $f'(-1) = 14\,336 > 0$, $f'(1) = 3024 > 0$, $f'(4) = -3024 < 0$, and $f'(8) = 2240 > 0$. For $0 < x < 3$, $x < 0$, and $x > 7$, f is increasing; for $3 < x < 7$, f is decreasing. Thus, $(3, 6912)$ is a local maximum, $(7, 0)$ is a local minimum, and $(0, 0)$ is neither.

(6) Concavity and inflection points: $f''(x) = 42x(x - 7)^2(x^2 - 6x + 7)$. When $x = 0, 7, 3 - \sqrt{2}$, or $3 + \sqrt{2}$, $f''(x) = 0$. You find that the graph is concave upward for $0 < x < 3 - \sqrt{2}$ and for $x > 3 + \sqrt{2}$; the graph is concave downward for $x < 0$ and for $3 - \sqrt{2} < x < 3 + \sqrt{2}$. There are inflection points at $(0, 0)$, $(3 - \sqrt{2}, f(3 - \sqrt{2}))$, and $(3 + \sqrt{2}, f(3 + \sqrt{2}))$.

[See Figure 8-23.]

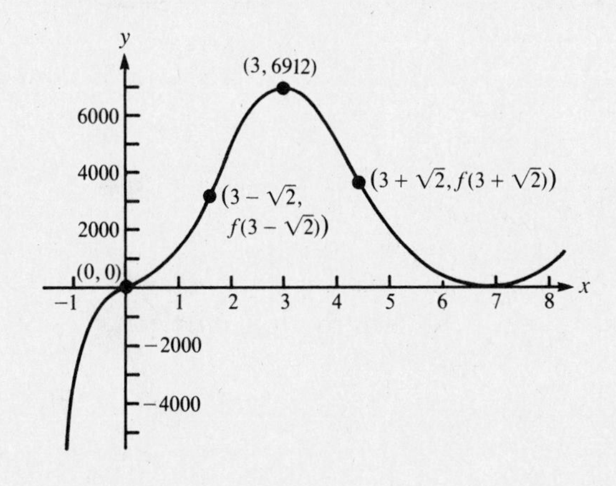

Figure 8-23
$f(x) = x^3(x - 7)^4$.

PROBLEM 8-9 Graph $f(x) = x^2/(1 - x^2)$.

Solution:

(1) Intercepts: The only intercept (x or y) is at $(0, 0)$.
(2) Symmetry:

$$f(-x) = \frac{(-x)^2}{1 - (-x)^2} = \frac{x^2}{1 - x^2} = f(x)$$

so f is symmetric about the y axis.

(3) Asymptotes:

$$\lim_{x \to \infty} \frac{x^2}{1 - x^2} = -1 = \lim_{x \to -\infty} \frac{x^2}{1 - x^2}$$

so f has horizontal asymptote $y = -1$. The vertical asymptotes are $x = 1$ and $x = -1$:

$$\lim_{x \to -1^-} \frac{x^2}{1 - x^2} = -\infty, \qquad \lim_{x \to -1^+} \frac{x^2}{1 - x^2} = \infty,$$

$$\lim_{x \to 1^-} \frac{x^2}{1 - x^2} = \infty, \qquad \lim_{x \to 1^+} \frac{x^2}{1 - x^2} = -\infty$$

(4) Critical points:

$$f'(x) = \frac{(1 - x^2) \cdot 2x - x^2(-2x)}{(1 - x^2)^2} = \frac{2x}{(1 - x^2)^2}$$

so there is a critical point at (0, 0).

(5) Slope of tangent line: Selecting -2, $-\frac{1}{2}$, $\frac{1}{2}$, and 2 as test points, $f'(-2) = -4/9 < 0$, $f'(-1/2) = -16/9 < 0$, $f'(1/2) = 16/9 > 0$, and $f'(2) = 4/9 > 0$. For $0 < x < 1$ and $x > 1$, f is increasing; for $-1 < x < 0$ and $x < -1$, f is decreasing. The point (0, 0) is a local minimum.

(6) Concavity and inflection points:

$$f''(x) = \frac{(1 - x^2)^2 \cdot 2 - 2x \cdot 2(1 - x^2)(-2x)}{(1 - x^2)^4} = \frac{2 + 6x^2}{(1 - x^2)^3}$$

so $f''(x)$ is never zero. You find that the graph is concave downward for $x < -1$ and for $x > 1$, and concave upward for $-1 < x < 1$. There are no inflection points. [See Figure 8-24.]

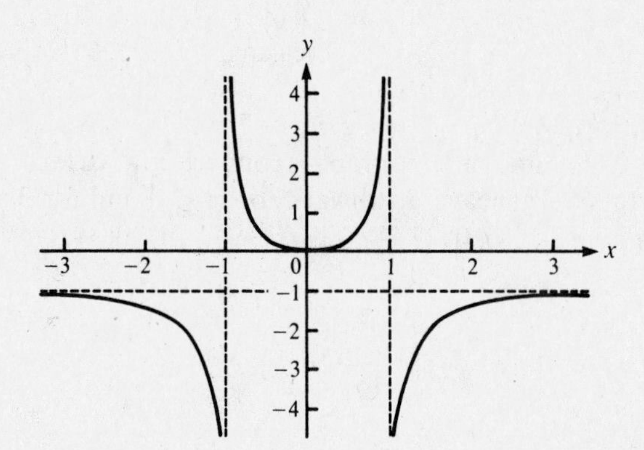

Figure 8-24
$f(x) = x^2/(1 - x^2)$.

PROBLEM 8-10 Graph $f(x) = 3x^5 - 10x^3$.

Solution:

(1) Intercepts: The y intercept is (0, 0). Solving for the x intercepts,

$$3x^5 - 10x^3 = 0$$

$$x^3(3x^2 - 10) = 0$$

$$x = 0, -\sqrt{\frac{10}{3}}, \sqrt{\frac{10}{3}}$$

So the x intercepts are $(0, 0)$, $(-\sqrt{10/3}, 0)$, and $(\sqrt{10/3}, 0)$.

(2) Symmetry: $f(-x) = -f(x)$ so f is symmetric about the origin.

(3) Asymptotes: none

(4) Critical points: $f'(x) = 15x^4 - 30x^2 = 15x^2(x^2 - 2)$. So $f'(x) = 0$ when $x = 0, -\sqrt{2}$, or $\sqrt{2}$. The critical points are $(0, 0)$, $(-\sqrt{2}, 8\sqrt{2})$, and $(\sqrt{2}, -8\sqrt{2})$.

(5) Slope of tangent line: Choosing -2, -1, 1, and 2 as test points, $f'(-2) = 120 > 0$, $f'(-1) = -15 < 0$, $f'(1) = -15 < 0$, and $f'(2) = 120 > 0$. For $x < -\sqrt{2}$ and $x > \sqrt{2}$, f is increasing; for $-\sqrt{2} < x < 0$ and $0 < x < \sqrt{2}$, f is decreasing. Thus, $(-\sqrt{2}, 8\sqrt{2})$ is a local maximum, and $(\sqrt{2}, -8\sqrt{2})$ is a local minimum.

(6) Concavity and inflection points: $f''(x) = 60x^3 - 60x = 60x(x^2 - 1)$. When $x = -1, 0$, or 1, $f''(x) = 0$. Choosing -2, $-\frac{1}{2}$, $\frac{1}{2}$, and 2 as test points, $f''(-2) = -360 < 0$, $f''(-1/2) = 45/2 > 0$, $f''(1/2) = -45/2 < 0$, and $f''(2) = 360 > 0$. Thus, the graph is concave downward for $x < -1$ and $0 < x < 1$, and concave upward for $-1 < x < 0$ and $x > 1$. There are inflection points at $(-1, f(-1)) = (-1, 7)$, $(0, 0)$, and $(1, -7)$. [See Figure 8-25.]

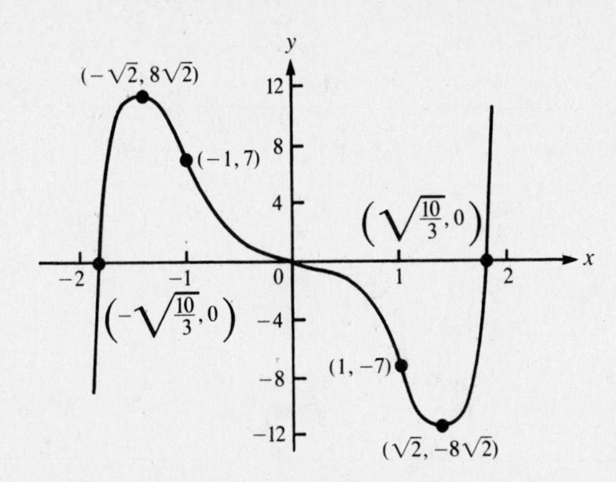

Figure 8-25
$f(x) = 3x^5 - 10x^3$.

PROBLEM 8-11 Graph $f(x) = 1/(1 + \sin x)$.

Solution:

(1) Intercepts: The y intercept is $(0, 1)$. There is no x intercept.
(2) Symmetry: none
(3) Asymptotes: There is a vertical asymptote when $\sin x = -1$ (i.e., at $x = 3\pi/2 + 2k\pi = (4k + 3)\pi/2$, k any integer). Because $f(x)$ is always positive, $\lim_{x\to 3\pi/2^+} f(x) = \lim_{x\to 3\pi/2^-} f(x) = \infty$.
(4) Critical points:

$$f'(x) = \frac{-\cos x}{(1 + \sin x)^2}$$

So $f'(x) = 0$ when $\cos x = 0$ and $\sin x \neq -1$ (i.e., $x = (4k + 1)\pi/2$, k any integer).
(5) Slope of tangent line: When $\cos x < 0$, that is, when x is in the second or third quadrant, $f'(x) > 0$. When $\cos x > 0$, that is, when x is in the first or fourth quadrant, $f'(x) < 0$.
(6) Concavity and inflection points:

$$\begin{aligned} f''(x) &= \frac{\sin x}{(1 + \sin x)^2} + \frac{2\cos^2 x}{(1 + \sin x)^3} \\ &= \frac{\sin x}{(1 + \sin x)^2} + \frac{2(1 + \sin x)(1 - \sin x)}{(1 + \sin x)^3} \\ &= \frac{2 - \sin x}{(1 + \sin x)^2} > 0 \text{ for all } x \end{aligned}$$

So the graph is always concave upward. [See Figure 8-26.]

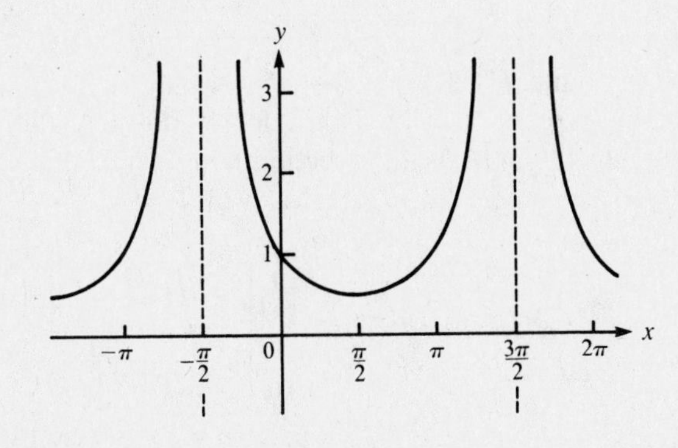

Figure 8-26
$f(x) = 1/(1 + \sin x)$.

PROBLEM 8-12 Graph $f(x) = (4x - 4)/x^2$.

Solution:

(1) Intercepts: There is no y intercept. The x intercept is $(1, 0)$.
(2) Symmetry: none

(3) Asymptotes: The horizontal asymptote is $y = 0$; the vertical asymptote is $x = 0$:

$$\lim_{x\to 0^-} \frac{4x-4}{x^2} = -\infty \qquad \lim_{x\to 0^+} \frac{4x-4}{x^2} = -\infty$$

(4) Critical points: $f'(x) = (-4x + 8)/x^3$, so there is a critical point at (2, 1).

(5) Slope of tangent line: Choosing $-1, 1$, and 3 as test points, $f'(-1) = -12 < 0$, $f'(1) = 4 > 0$, and $f'(3) = -4/27 < 0$. For $0 < x < 2$, f is increasing; for $x < 0$ and $x > 2$, f is decreasing. A local maximum occurs at (2, 1).

(6) Concavity and inflection points:

$$f''(x) = \frac{8x-24}{x^4}$$

When $x = 3$, $f''(x) = 0$, and when $x = 0$, f'' is undefined. Choosing $-1, 1$, and 4 as test points, $f''(-1) = -32 < 0$, $f''(1) = -16 < 0$, and $f''(4) = 1/32 > 0$. So the graph is concave downward for $x < 0$ and for $0 < x < 3$, and concave upward for $x > 3$. There is an inflection point $(3, \frac{8}{9})$.
[See Figure 8-27.]

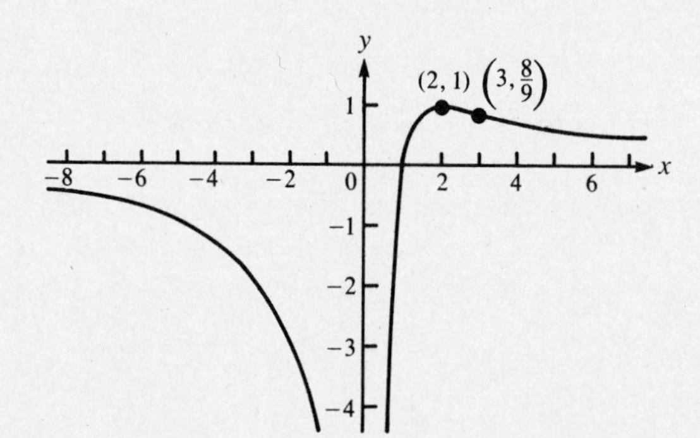

Figure 8-27
$f(x) = (4x - 4)/x^2$.

PROBLEM 8-13 Graph $f(x) = e^{2x} - e^x$.

Solution:

(1) Intercepts: The only intercept is (0, 0).

(2) Symmetry: none

(3) Asymptotes:

$$\lim_{x\to\infty} (e^{2x} - e^x) = \lim_{x\to\infty} (e^x(e^x - 1)) = \infty \qquad \lim_{x\to-\infty} (e^{2x} - e^x) = 0$$

so $y = 0$ is a horizontal asymptote (to the left). There is no vertical asymptote.

(4) Critical points: $f'(x) = 2e^{2x} - e^x = e^x(2e^x - 1)$. So $f'(x) = 0$ when $2e^x - 1 = 0$, that is, when $x = \ln\frac{1}{2} = -\ln 2$. There is one critical point: $(-\ln 2, -\frac{1}{4})$.

(5) Slope of tangent line: You find that for $x < -\ln 2$, $f'(x) < 0$, and for $x > -\ln 2$, $f'(x) > 0$. So f is decreasing for $x < -\ln 2$ and increasing for $x > -\ln 2$. The point $(-\ln 2, -\frac{1}{4})$ is a local minimum.

(6) Concavity and inflection points: $f''(x) = 4e^{2x} - e^x = e^x(4e^x - 1)$. When $x = \ln\frac{1}{4} = -\ln 4$, $f''(x) = 0$. So the graph is concave downward for $x < -\ln 4$ and concave upward for $x > -\ln 4$. The point $(-\ln 4, -3/16)$ is an inflection point.
[See Figure 8-28.]

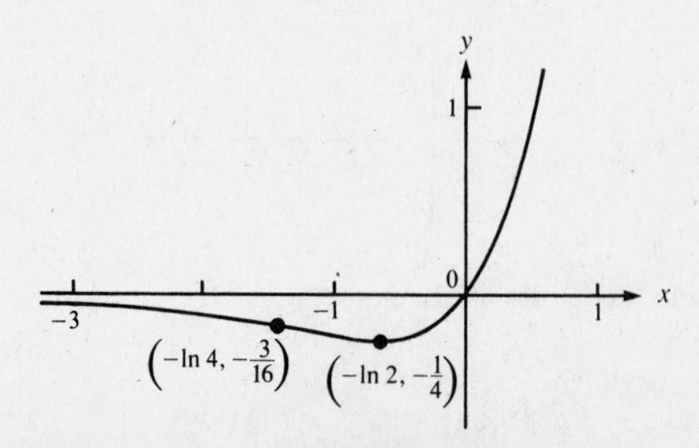

Figure 8-28
$f(x) = e^{2x} - e^x$.

PROBLEM 8-14 Graph $f(x) = e^{-x^2}$.

Solution:

(1) Intercepts: The y intercept is $(0, 1)$. There is no x intercept.
(2) Symmetry: $f(-x) = f(x)$, so f is symmetric about the y axis.
(3) Asymptotes:

$$\lim_{x\to\infty} e^{-x^2} = 0 = \lim_{x\to-\infty} e^{-x^2}$$

So $y = 0$ is the horizontal asymptote. There is no vertical asymptote.
(4) Critical points: $f'(x) = -2xe^{-x^2}$, so the only critical point is $(0, 1)$.
(5) Slope of tangent line: You find that f is increasing for $x < 0$ and decreasing for $x > 0$, and so $(0, 1)$ is a local maximum.
(6) Concavity and inflection points: $f''(x) = (4x^2 - 2)e^{-x^2}$. When $x = \pm\sqrt{1/2} = \pm\sqrt{2}/2$, $f''(x) = 0$. Choosing $-1, 0$, and 1 as test points, $f''(-1) = 2e^{-1} > 0, f''(0) = -2 < 0$, and $f''(1) = 2e^{-1} > 0$. So the graph is concave upward for $x < -\sqrt{2}/2$ and $x > \sqrt{2}/2$; the graph is concave downward for $-\sqrt{2}/2 < x < \sqrt{2}/2$. The inflection points are $(-\sqrt{2}/2, e^{-1/2})$ and $(\sqrt{2}/2, e^{-1/2})$. [See Figure 8-29.]

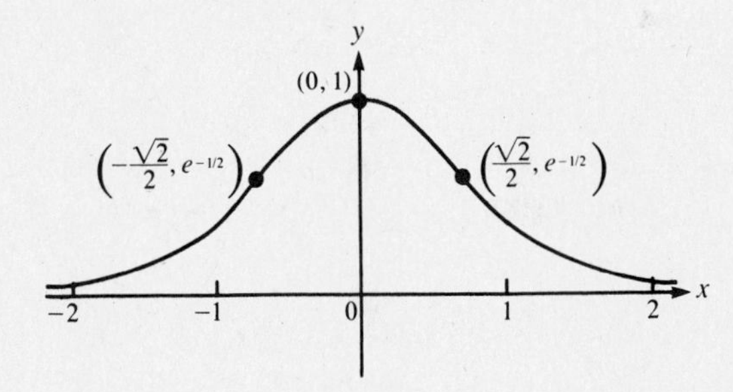

Figure 8-29
$f(x) = e^{-x^2}$.

PROBLEM 8-15 Graph $f(x) = (x^2 - 1)/x^3$.

Solution:

(1) Intercepts: There is no y intercept. The x intercepts are $(1, 0)$ and $(-1, 0)$.
(2) Symmetry: $f(-x) = -f(x)$, so f is symmetric with respect to the origin.
(3) Asymptotes:

$$\lim_{x\to\infty} \frac{x^2 - 1}{x^3} = 0 = \lim_{x\to-\infty} \frac{x^2 - 1}{x^3}$$

so $y = 0$ is the horizontal asymptote. The vertical asymptote is $x = 0$:

$$\lim_{x\to0^-} \frac{x^2 - 1}{x^3} = \infty \qquad \lim_{x\to0^+} \frac{x^2 - 1}{x^3} = -\infty$$

(4) Critical points: $f'(x) = (3 - x^2)/x^4$, so $f'(x) = 0$ when $x = \pm\sqrt{3}$. The critical points are $(-\sqrt{3}, -2\sqrt{3}/9)$ and $(\sqrt{3}, 2\sqrt{3}/9)$.
(5) Slope of tangent line: Choosing $-2, -1, 1$, and 2 as test points, $f'(-2) = f'(2) = -1/16 < 0$, and $f'(-1) = f'(1) = 2 > 0$. For $-\sqrt{3} < x < 0$ and $0 < x < \sqrt{3}$, f is increasing; for $x < -\sqrt{3}$ and for $x > \sqrt{3}$, f is decreasing. The point $(-\sqrt{3}, -2\sqrt{3}/9)$ is a local minimum and the point $(\sqrt{3}, 2\sqrt{3}/9)$ is a local maximum.
(6) Concavity and inflection points:

$$f''(x) = \frac{2(x^2 - 6)}{x^5}$$

so when $x = \pm\sqrt{6}, f''(x) = 0$; $f''(x)$ is undefined at $x = 0$. Choosing $-3, -1, 1$, and 3 as test points, $f''(-3) = -2/81 < 0, f''(-1) = 10 > 0, f''(1) = -10 < 0$, and $f''(3) = 2/81 > 0$. For $-\sqrt{6} < x < 0$ and for $x > \sqrt{6}$, the graph is concave upward; for $x < -\sqrt{6}$ and for $0 < x < \sqrt{6}$, the graph is concave downward. There are inflection points at $(-\sqrt{6}, -5\sqrt{6}/36)$ and $(\sqrt{6}, 5\sqrt{6}/36)$. [See Figure 8-30.]

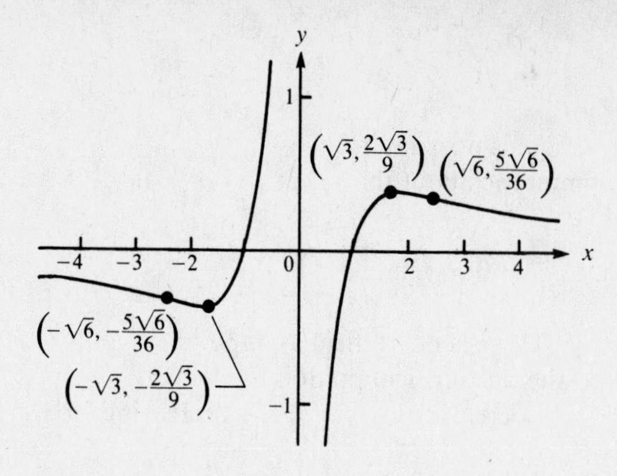

Figure 8-30
$f(x) = (x^2 - 1)/x^3$.

PROBLEM 8-16 Graph $f(x) = x - 3x^{1/3}$.

Solution:

(1) Intercepts: The y intercept is (0, 0). Solving for the x intercept,

$$0 = x - 3x^{1/3}$$
$$0 = x(1 - 3x^{-2/3})$$

When $x = 0$ and when $1 = 3x^{-2/3}$, that is, when $x = \pm 3^{3/2} = \pm 3\sqrt{3}$, $f(x) = 0$. Thus, the x intercepts are (0, 0), $(3\sqrt{3}, 0)$ and $(-3\sqrt{3}, 0)$.

(2) Symmetry: $f(-x) = (-x) - 3(-x)^{1/3} = -x + 3x^{1/3} = -f(x)$, so f is symmetric with respect to the origin.

(3) Asymptotes: none

(4) Critical points: $f'(x) = 1 - x^{-2/3}$, so $f'(x) = 0$ when $x = \pm 1$. The critical points are $(-1, 2)$ and $(1, -2)$.

(5) Slope of tangent line: Choosing $-2, -\frac{1}{2}, \frac{1}{2}$, and 2 as test points,

$$f'(-2) = f'(2) = \frac{2 - \sqrt[3]{2}}{2} > 0 \qquad f'\left(-\frac{1}{2}\right) = f'\left(\frac{1}{2}\right) = 1 - \sqrt[3]{4} < 0$$

For $x < -1$ and $x > 1$, f is increasing; for $-1 < x < 0$ and $0 < x < 1$, f is decreasing. The point $(-1, 2)$ is a local maximum; $(1, -2)$ is a local minimum.

(6) Concavity and inflection points: $f''(x) = (2/3)x^{-5/3}$, so $f''(x)$ is never zero, although it is undefined at $x = 0$. For $x > 0$, the graph is concave upward; for $x < 0$, the graph is concave downward.

[See Figure 8-31.]

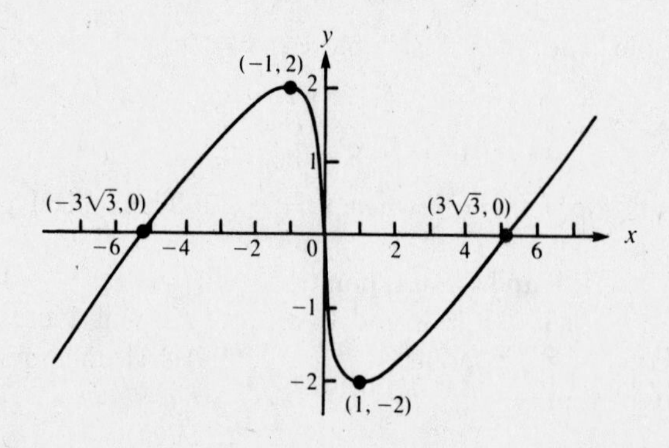

Figure 8-31
$f(x) = x - 3x^{1/3}$.

PROBLEM 8-17 Graph $f(x) = x\sqrt{1 - x^2}$.

Solution:

(1) Intercepts: The y intercept is (0, 0); the x intercepts are $(-1, 0)$, (0, 0), and (1, 0).

(2) Symmetry: $f(-x) = -f(x)$, so f is symmetric with respect to the origin. Note that f is defined only for $-1 \leqslant x \leqslant 1$.

(3) Asymptotes: none

(4) Critical points:

$$f'(x) = (1 - x^2)^{1/2} + \frac{1}{2}x(1 - x^2)^{-1/2}(-2x)$$
$$= (1 - 2x^2)(1 - x^2)^{-1/2}$$

so $f'(x) = 0$ when $x = \pm\sqrt{2}/2$. The critical points are $(-\sqrt{2}/2, -1/2)$ and $(\sqrt{2}/2, 1/2)$. At $x = -1$ and $x = 1$, f' is discontinuous.

(5) Slope of tangent line: Choosing $-\frac{3}{4}$, 0, and $\frac{3}{4}$ as test points, $f'(3/4) = f'(-3/4) = -32/49 < 0$, and $f'(0) = 1 > 0$. Notice that you can't select points $x < -1$ or $x > 1$ because they are outside the domain of f. For $-\sqrt{2}/2 < x < \sqrt{2}/2$, f is increasing; for $-1 < x < -\sqrt{2}/2$ and for $\sqrt{2}/2 < x < 1$, f is decreasing. There is a local minimum at $(-\sqrt{2}/2, -1/2)$ and a local maximum at $(\sqrt{2}/2, 1/2)$.

(6) Concavity and inflection points:

$$f''(x) = \frac{x(2x^2 - 3)}{(1 - x^2)^{3/2}}$$

When $x = 0$, or $\pm\sqrt{3/2}$, $f''(x) = 0$. But $x = \pm\sqrt{3/2}$ is not in the domain of f. For $-1 < x < 0$, f is concave upward; for $0 < x < 1$, f is concave downward. The origin is an inflection point.

[See Figure 8-32.]

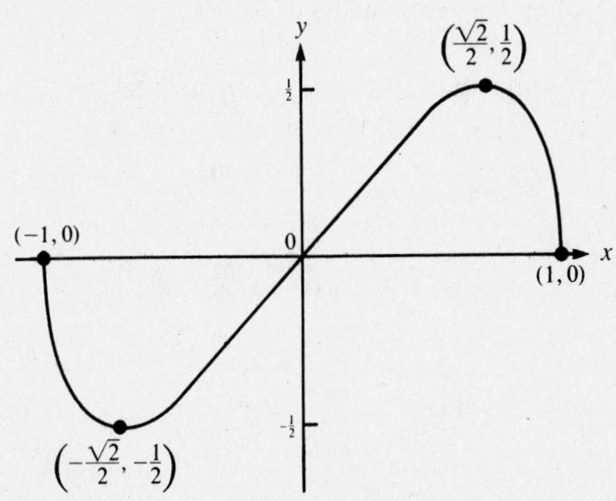

Figure 8-32
$f(x) = x\sqrt{1 - x^2}$.

PROBLEM 8-18 Graph $f(x) = 2x/(x^2 + 4)$.

Solution:

(1) Intercepts: The only intercept is (0, 0).

(2) Symmetry: $f(-x) = -2x/(x^2 + 4) = -f(x)$, so f is symmetric with respect to the origin.

(3) Asymptotes:

$$\lim_{x\to\infty} \frac{2x}{x^2 + 4} = 0 = \lim_{x\to-\infty} \frac{2x}{x^2 + 4}$$

so $y = 0$ is the horizontal asymptote. Since $x^2 + 4$ is never zero, there is no vertical asymptote.

(4) Critical points:

$$f'(x) = \frac{(x^2 + 4)\cdot 2 - (2x)(2x)}{(x^2 + 4)^2} = \frac{-2x^2 + 8}{(x^2 + 4)^2}$$

so $f'(x) = 0$ when $x = \pm 2$. You find critical points at $(-2, f(-2)) = (-2, -\frac{1}{2})$ and $(2, f(2)) = (2, \frac{1}{2})$.

(5) Slope of tangent line: Selecting -3, 0, and 3 as test points, $f'(-3) = -10/169 < 0$, $f'(0) = 1/2 > 0$, and $f'(3) = -10/169 < 0$. Thus, f is increasing for $-2 < x < 2$ and decreasing for $x < -2$ and $x > 2$. There is a local minimum at $(-2, -\frac{1}{2})$ and a local maximum at $(2, \frac{1}{2})$.

(6) Concavity and inflection points:

$$f''(x) = \frac{4x(x^2 - 12)}{(x^2 + 4)^3}$$

When $x = 0$, $-2\sqrt{3}$, or $+2\sqrt{3}$, $f''(x) = 0$. Selecting $-4, -1, 1$, and 4 as test points, $f''(-4) = -30/19^3 < 0$, $f''(-1) = 5/8 > 0$, $f''(1) = -5/8 < 0$, and $f''(4) = 30/19^3 > 0$. So f is concave upward for $-2\sqrt{3} < x < 0$ and for $x > 2\sqrt{3}$; f is concave downward for $x < -2\sqrt{3}$ and for $0 < x < 2\sqrt{3}$. There are inflection points at $(2\sqrt{3}, f(2\sqrt{3})) = (2\sqrt{3}, \sqrt{3}/4)$ and $(-2\sqrt{3}, f(-2\sqrt{3})) = (-2\sqrt{3}, -\sqrt{3}/4)$. [See Figure 8-33.]

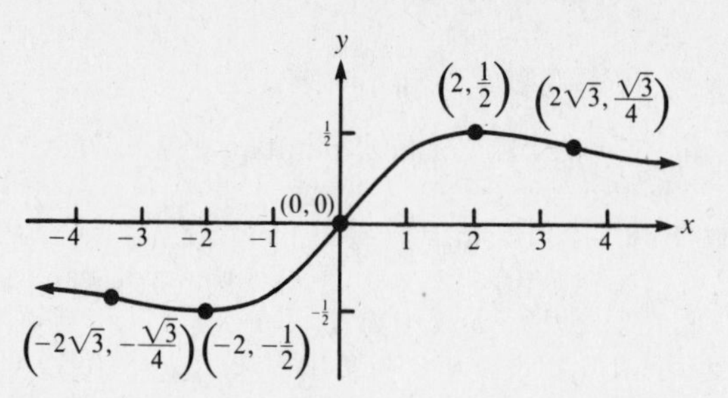

Figure 8-33
$f(x) = 2x/(x^2 + 4)$.

PROBLEM 8-19 Graph $f(x) = x^3/(x^2 - 3)$.

Solution:

(1) Intercepts: The only intercept is (0, 0).
(2) Symmetry: $f(-x) = -f(x)$ so f is symmetric about the origin.
(3) Asymptotes:

$$\lim_{x\to\infty} \frac{x^3}{x^2 - 3} = \infty \qquad \lim_{x\to-\infty} \frac{x^3}{x^2 - 3} = -\infty$$

so there is no horizontal asymptote. The vertical asymptotes are $x = -\sqrt{3}$ and $x = \sqrt{3}$:

$$\lim_{x\to-\sqrt{3}^-} \frac{x^3}{x^2 - 3} = -\infty \qquad \lim_{x\to-\sqrt{3}^+} \frac{x^3}{x^2 - 3} = \infty$$

$$\lim_{x\to\sqrt{3}^-} \frac{x^3}{x^2 - 3} = -\infty \qquad \lim_{x\to\sqrt{3}^+} \frac{x^3}{x^2 - 3} = \infty$$

(4) Critical points:

$$f'(x) = \frac{(x^2 - 3)3x^2 - x^3(2x)}{(x^2 - 3)^2} = \frac{x^4 - 9x^2}{(x^2 - 3)^2} = \frac{x^2(x^2 - 9)}{(x^2 - 3)^2}$$

so $f'(x) = 0$ when $x = 0, -3$, or 3. There are critical points at (0, 0), $(-3, -9/2)$, and $(3, 9/2)$ and discontinuities at $x = -\sqrt{3}$ and $x = \sqrt{3}$.
(5) Slope of tangent line: Choosing $-4, -2, -1, 1, 2$, and 4 as test points, $f'(-4) = f'(4) = 112/169 > 0$, $f'(-2) = f'(2) = -20 < 0$, and $f'(-1) = f'(1) = -2 < 0$. So, f is increasing, for $x < -3$ and $x > 3$, and decreasing for x between $x = -3$ and $x = 3$ (except, of course, at $x = \pm\sqrt{3}$). There is a local maximum at $(-3, -9/2)$ and a local minimum at $(3, 9/2)$.
(6) Concavity and inflection points:

$$f''(x) = \frac{6x(x^2 + 9)}{(x^2 - 3)^3}$$

so $f''(x) = 0$ when $x = 0$. You find that the graph is concave downward for $x < -\sqrt{3}$ and $0 < x < \sqrt{3}$; the graph is concave upward for $-\sqrt{3} < x < 0$ and $x > \sqrt{3}$. There is an inflection point at (0, 0).
[See Figure 8-34.]

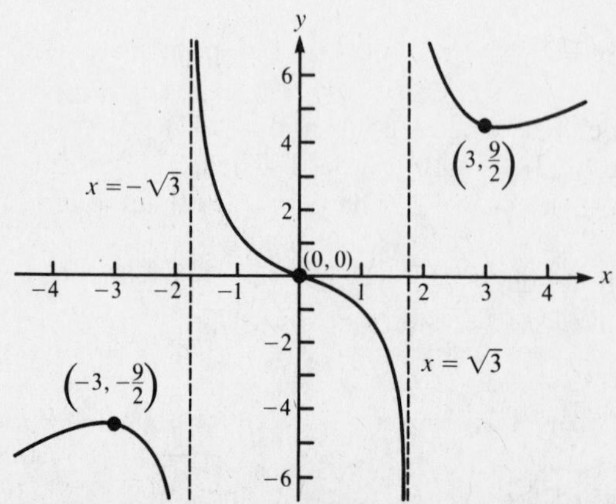

Figure 8-34
$f(x) = x^3/(x^2 - 3)$.

PROBLEM 8-20 Graph $f(x) = \sin x/(1 + \cos x)$.

Solution:

(1) Intercepts: $(2k\pi, 0)$, k any integer.
(2) Symmetry: $f(-x) = -f(x)$, so f is symmetric with respect to the origin.
(3) Asymptotes: There is a vertical asymptote when $\cos x = -1$, that is, when $x = (2k + 1)\pi$, k any integer. As x approaches $(2k + 1)\pi$ from the left, $f(x)$ approaches ∞. As x approaches $(2k + 1)\pi$ from the right, $f(x)$ approaches $-\infty$.
(4) Critical points:

$$f'(x) = \frac{(1 + \cos x)\cos x + \sin^2 x}{(1 + \cos x)^2} = \frac{1}{1 + \cos x}$$

so $f'(x)$ is never zero.
(5) Slope of tangent line: $f'(x) > 0$ for all x, except $x = (2k + 1)\pi$.
(6) Concavity and inflection points:

$$f''(x) = \frac{\sin x}{(1 + \cos x)^2}$$

so $f''(x) = 0$ when $x = 2k\pi$, k any integer. For x in the first or second quadrant, the graph is concave upward; for x in the third or fourth quadrant, the graph is concave downward. There are inflection points at $(2k\pi, 0)$, k any integer. [See Figure 8-35.]

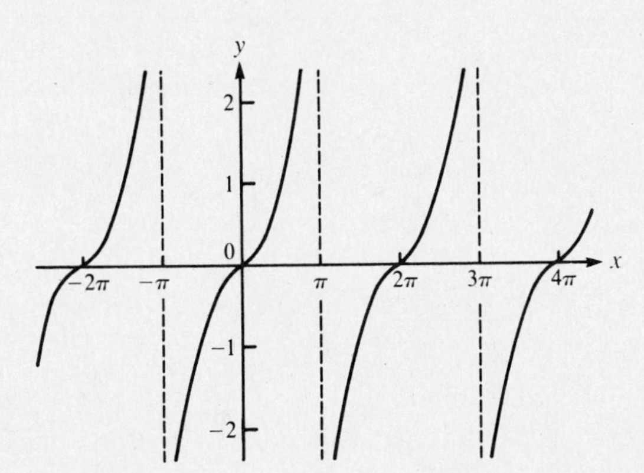

Figure 8-35
$f(x) = \sin x/(1 + \cos x)$.

Supplementary Exercises

Graph the following functions:

8-21 $f(x) = 2x^3 - 9x^2 + 12x - 2$

8-22 $f(x) = x^2/(x^2 + 1)$

8-23 $f(x) = 3x^4 + 4x^3 + 1$

8-24 $f(x) = (x - 1)/(x + 1)$

8-25 $f(x) = x^{2/3}$

8-26 $f(x) = x - \sqrt{1 - x^2}$

8-27 $f(x) = \sin x/(1 + \sin x)$

8-28 $f(x) = (x^2 - 9)/(1 - x^2)$

8-29 $f(x) = x^2 + (1/x^2)$

8-30 $f(x) = \sqrt{x^2 - x^4}$

8-31 $f(x) = x^4 - 2x^2 + 2$

8-32 $f(x) = 2x/(x^2 + 1)$

8-33 $f(x) = \cos^2 x$

8-34 $f(x) = (x^2 + 2x - 3)/x^2$

8-35 $f(x) = xe^{-x}$

8-36 $f(x) = (x^2 - 1)^2$

8-37 $f(x) = (x + 1)/\sqrt{x}$

8-38 $f(x) = x - 2\sqrt{x}$

8-39 $f(x) = x^3/(x^3 + 1)$

8-40 $f(x) = 1/(1 - \cos x)$

8-41 $f(x) = \ln(1 + x^2)$

Solutions to Supplementary Exercises

(**8-21**) Fig. 8-36

(**8-22**) Fig. 8-37

(**8-23**) Fig. 8-38

(**8-24**) Fig. 8-39

(**8-25**) Fig. 8-40

(**8-26**) Fig. 8-41

(**8-27**) Fig. 8-42

(**8-28**) Fig. 8-43

(**8-29**) Fig. 8-44

(**8-30**) Fig. 8-45

(**8-31**) Fig. 8-46

(**8-32**) Fig. 8-47

(**8-33**) Fig. 8-48

(**8-34**) Fig. 8-49

(**8-35**) Fig. 8-50

(**8-36**) Fig. 8-51

(**8-37**) Fig. 8-52

(**8-38**) Fig. 8-53

(**8-39**) Fig. 8-54

(**8-40**) Fig. 8-55

(**8-41**) Fig. 8-56

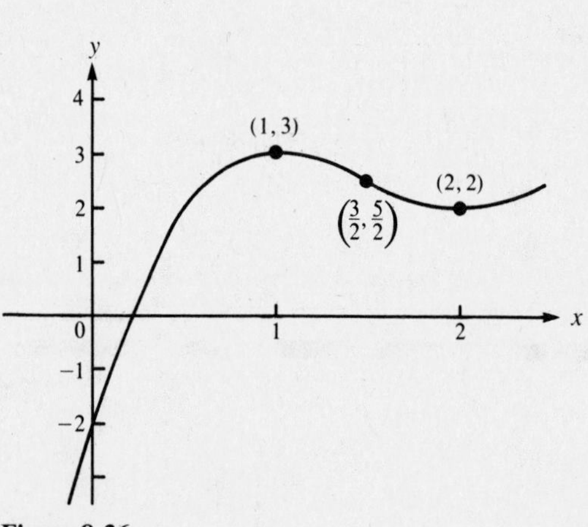

Figure 8-36
$f(x) = 2x^3 - 9x^2 + 12x - 2.$

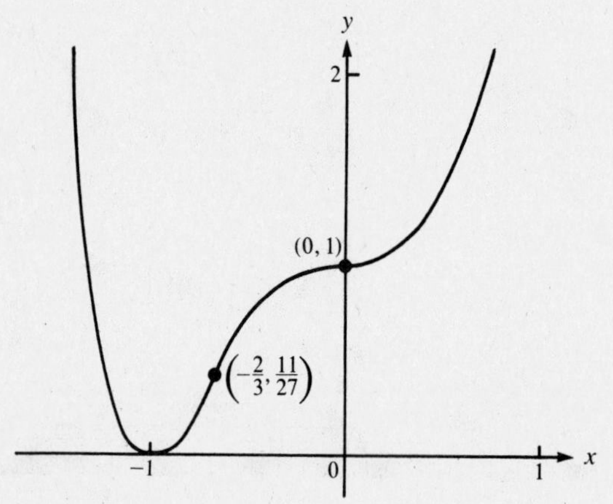

Figure 8-38
$f(x) = 3x^4 + 4x^3 + 1.$

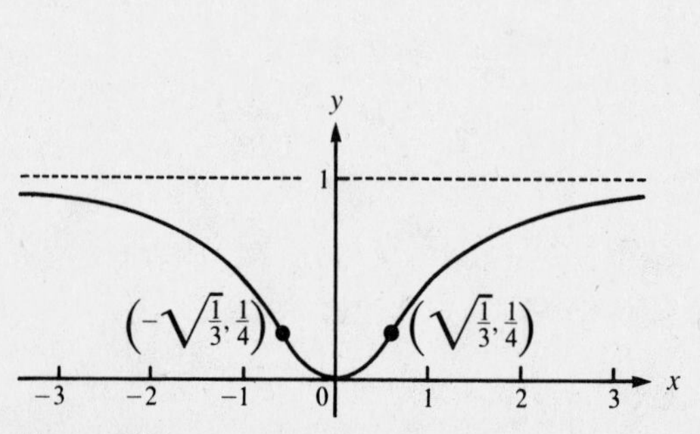

Figure 8-37
$f(x) = x^2/(x^2 + 1).$

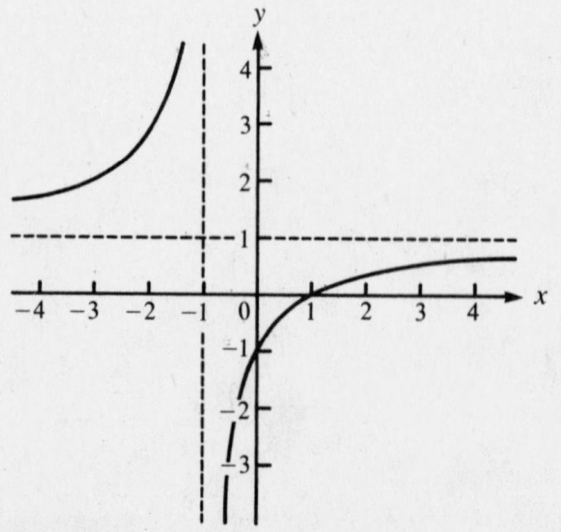

Figure 8-39
$f(x) = (x - 1)/(x + 1).$

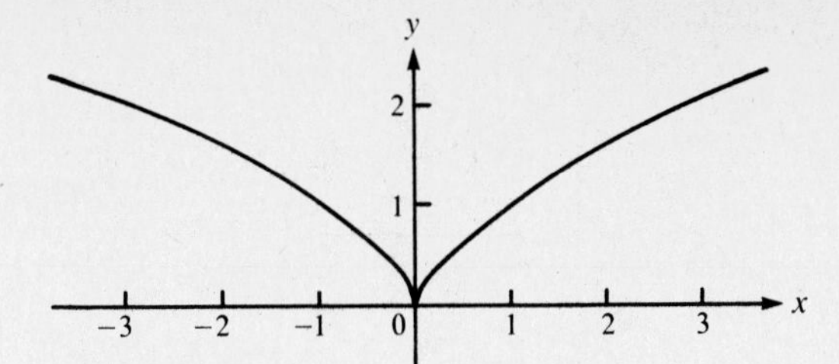

Figure 8-40
$f(x) = x^{2/3}$.

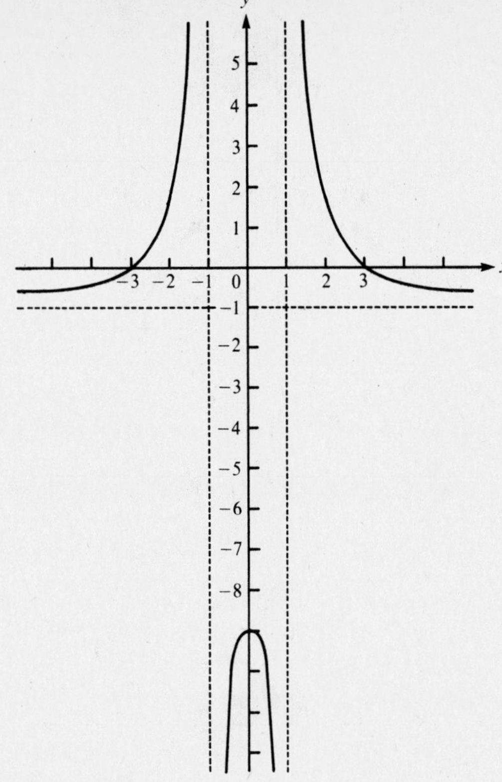

Figure 8-43
$f(x) = (x^2 - 9)/(1 - x^2)$.

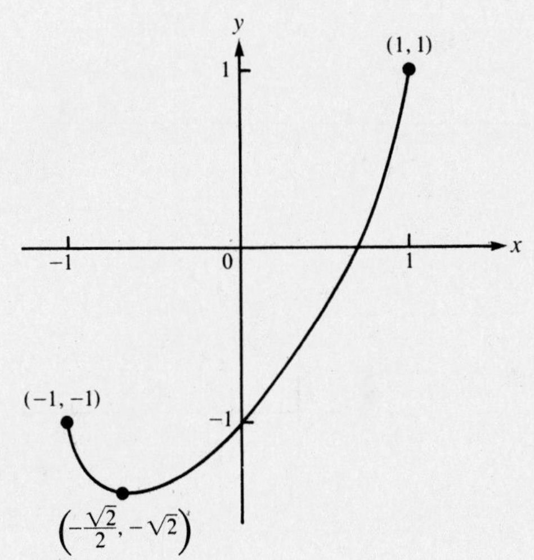

Figure 8-41
$f(x) = x - \sqrt{1 - x^2}$.

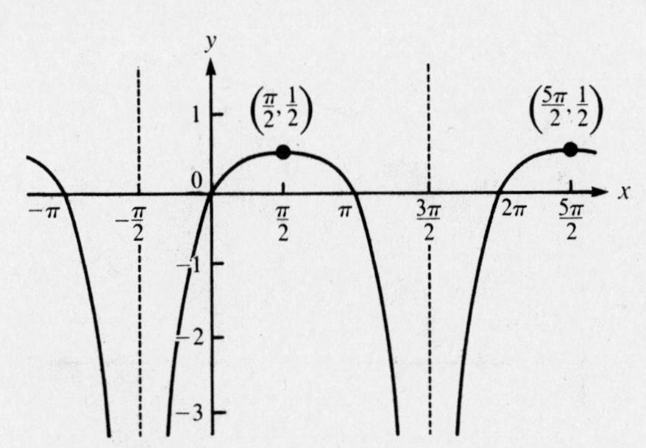

Figure 8-42
$f(x) = \sin x/(1 + \sin x)$.

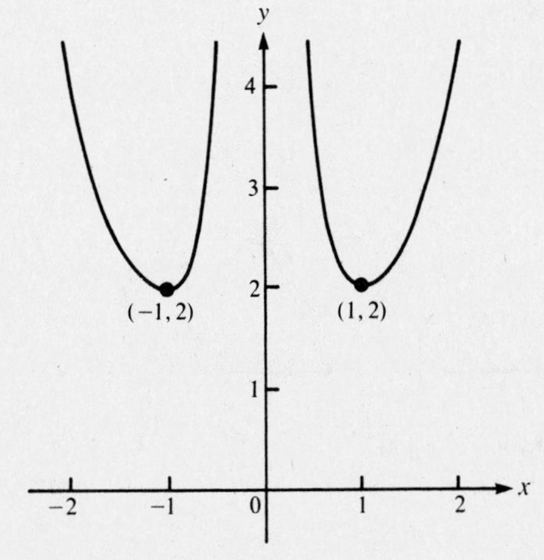

Figure 8-44
$f(x) = x^2 + (1/x^2)$.

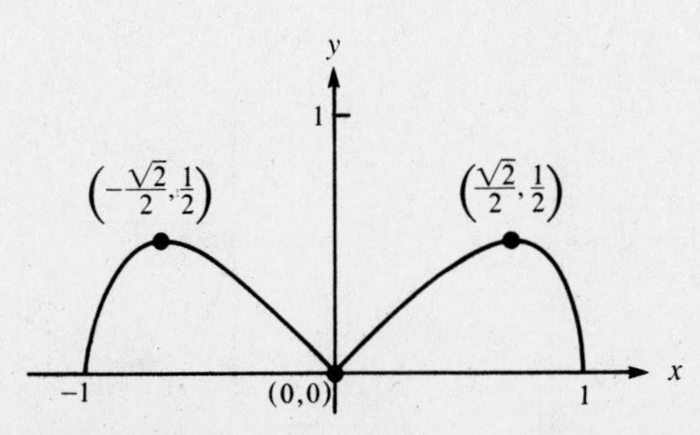

Figure 8-45
$f(x) = \sqrt{x^2 - x^4}$.

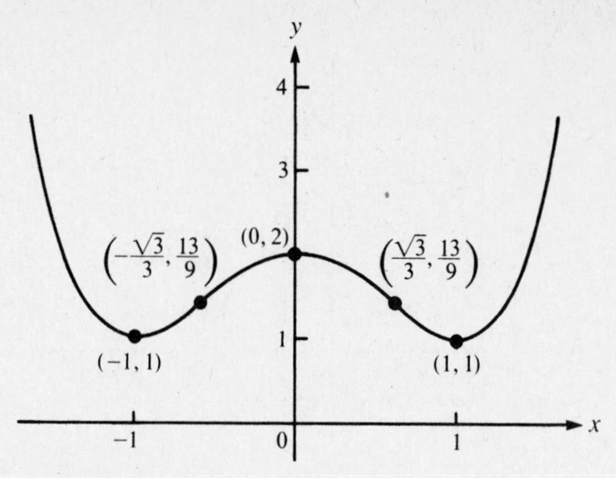

Figure 8-46
$f(x) = x^4 - 2x^2 + 2.$

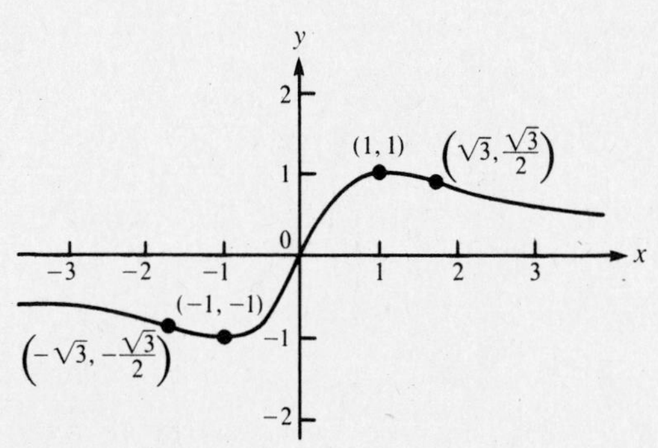

Figure 8-47
$f(x) = 2x/(x^2 + 1).$

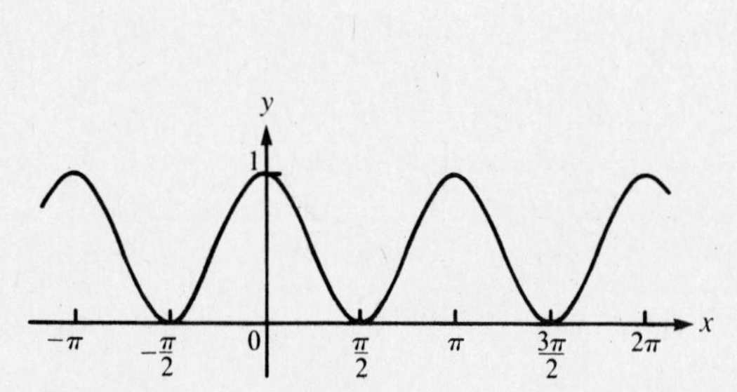

Figure 8-48
$f(x) = \cos^2 x.$

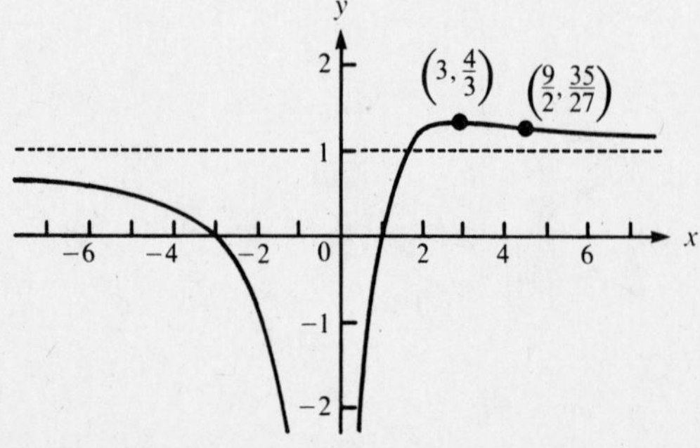

Figure 8-49
$f(x) = (x^2 + 2x - 3)/x^2.$

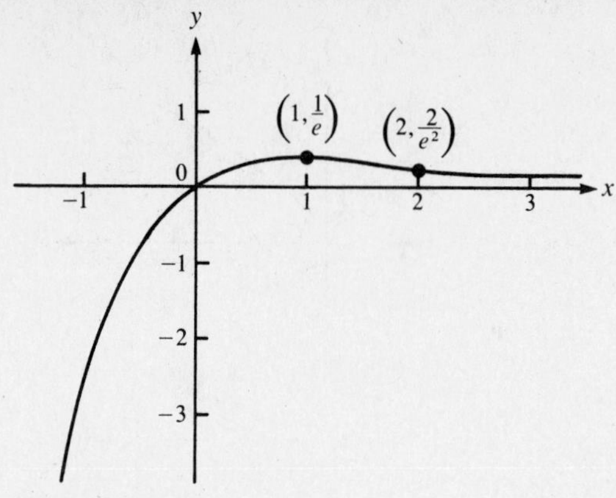

Figure 8-50
$f(x) = xe^{-x}.$

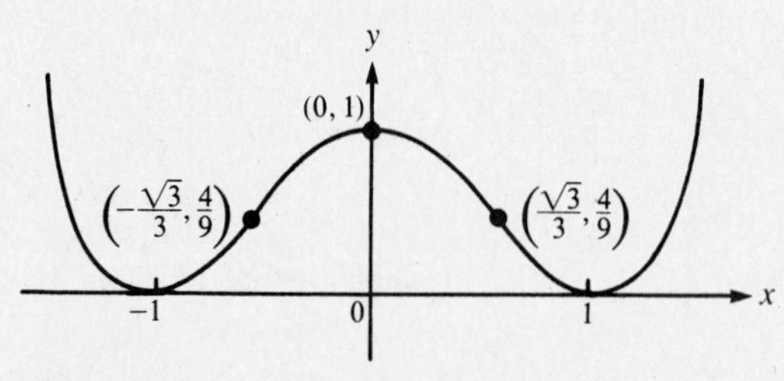

Figure 8-51
$f(x) = (x^2 - 1)^2.$

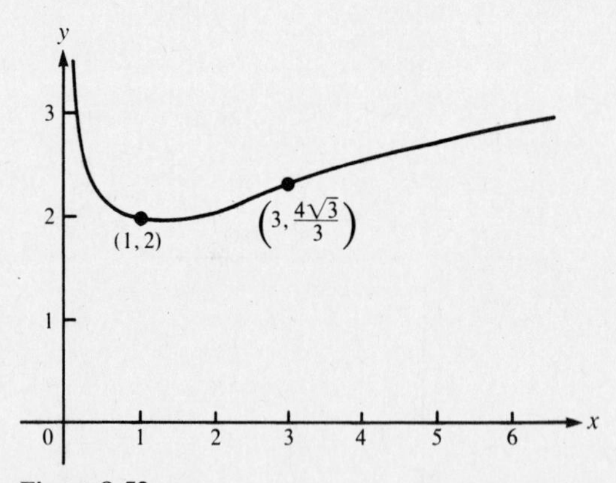

Figure 8-52
$f(x) = (x + 1)/\sqrt{x}.$

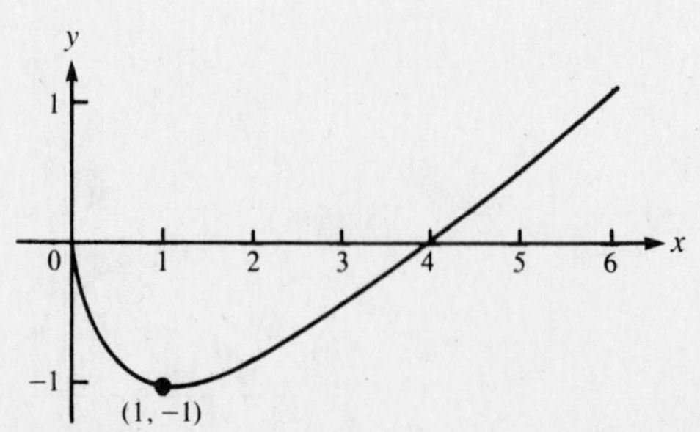

Figure 8-53
$f(x) = x - 2\sqrt{x}.$

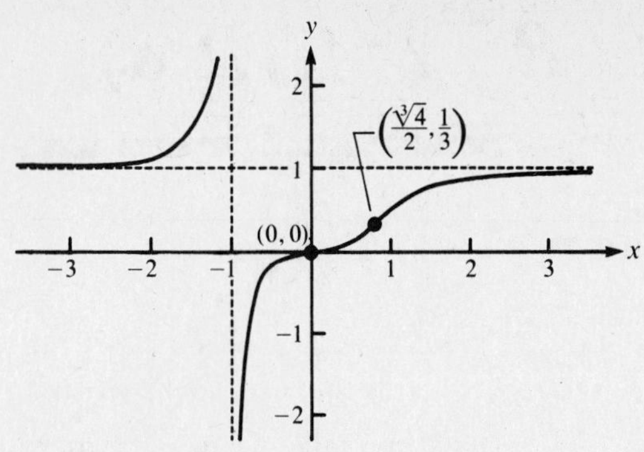

Figure 8-54
$f(x) = x^3/(x^3 + 1)$.

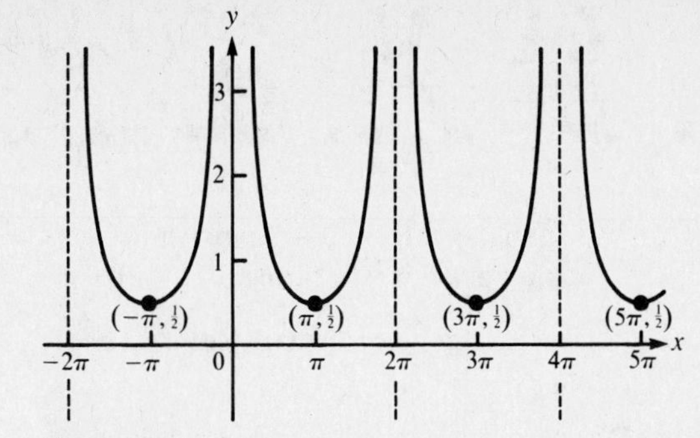

Figure 8-55
$f(x) = 1/(1 - \cos x)$.

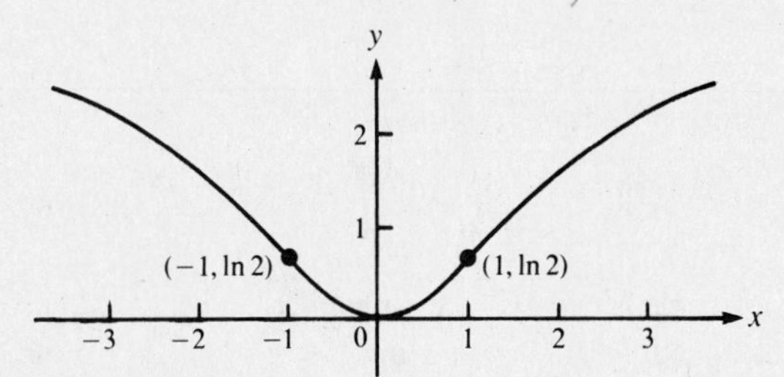

Figure 8-56
$f(x) = \ln(1 + x^2)$.

EXAM 3 (CHAPTERS 7 AND 8)

1. The Intermediate Value Theorem allows you to determine whether $2x^3 + x - 1 = 0$ has a solution between $x = 0$ and $x = 1$. Explain.

2. Let $f(x) = x^3 - 3x$. Verify Rolle's Theorem by finding all x such that $f(x) = 0$, and all x for which $f'(x) = 0$.

3. Does $g(x) = x + (1/x)$ have a minimum value on $(0, 1)$? Is this a contradiction to the Extreme Value Theorem?

In Problems 4 through 7, determine which conclusions follow from the given statement.

4. Given $f'(a) > 0$, then: **(a)** The graph of f is concave upward at $x = a$; **(b)** f is increasing at $x = a$; **(c)** f has a local maximum at $x = a$.

5. Given $g'(a)$ is defined, then: **(a)** $g(x)$ is continuous at $x = a$; **(b)** the graph of $g(x)$ has a tangent line at $x = a$; **(c)** no conclusion can be drawn about $g(x)$.

6. Given $h'(a) = 0$, then: **(a)** $h(x)$ has a maximum at $x = a$; **(b)** $h(x)$ may have a minimum at $x = a$; **(c)** $h(x)$ may have an inflection point at $x = a$.

7. Given $F'(a) = 0$ and $F''(a) > 0$, then: **(a)** F has a maximum at $x = a$; **(b)** F has a minimum at $x = a$; **(c)** the graph of F has a horizontal tangent line at $x = a$.

In Problems 8 through 10, graph the curve, giving the coordinates of the intercepts and of all local maxima, local minima, and inflection points.

8. $y = xe^{-x}$
9. $y = \dfrac{1}{x} + \dfrac{4}{1 - x}$
10. $y = \dfrac{1}{36}x^4 - x^2$

SOLUTIONS TO EXAM 3

1. The Intermediate Value Theorem allows you to conclude that if $f(x)$ is continuous on $[0, 1]$, with $f(0)$ and $f(1)$ of opposite signs, then $f(x) = 0$ for some x between $x = 0$ and $x = 1$. Because $f(x) = 2x^3 + x - 1$ is continuous for all x, with $f(0) = -1$ and $f(1) = 2$, you can conclude that $2x^3 + x - 1 = 0$ has at least one solution between 0 and 1.

2. Rolle's Theorem says that if $f(x)$ is continuous on $[a, b]$ and differentiable on (a, b), with $f(a) = f(b) = 0$, then $f'(x) = 0$ for at least one value of x between a and b.

$$f(x) = x^3 - 3x = x(x^2 - 3)$$
$$0 = x(x^2 - 3)$$
$$x = 0, -\sqrt{3}, \sqrt{3}$$

Thus you would expect to find a zero of $f'(x)$ in $(-\sqrt{3}, 0)$ and one in $(0, \sqrt{3})$. (Because $f(x)$ is a polynomial, it is continuous and differentiable for all x.)

$$f'(x) = 3x^2 - 3$$
$$0 = 3x^2 - 3$$
$$x = -1, 1$$
$$-\sqrt{3} < -1 < 0 \qquad 0 < 1 < \sqrt{3}$$

This confirms Rolle's Theorem in this case.

3. Differentiate $g(x)$:

$$g'(x) = 1 - \frac{1}{x^2}$$

So $g'(x) = 0$ for $x = -1$ and $x = 1$. Neither -1 nor 1 is in the interval (0, 1), yet $g(x)$ is differentiable for all x in (0, 1). You see that $g(x)$ has neither maximum nor minimum in the given interval. The Extreme Value Theorem assures the existence of a maximum and a minimum for every function that is continuous on a closed interval. But (0, 1) is not a closed interval, so this example is not a contradiction to the Extreme Value Theorem.

4. If $f'(a) > 0$, then: **(b)** f is increasing at $x = a$.

5. If $g'(a)$ is defined then: **(a)** $g(x)$ is continuous at $x = a$, and **(b)** the graph of $g(x)$ has a tangent line at $x = a$.

6. If $h'(a) = 0$, then: **(b)** $h(x)$ may have a minimum at $x = a$, and **(c)** $h(x)$ may have an inflection point at $x = a$.

7. If $F'(a) = 0$ and $F''(a) > 0$, then: **(b)** $F(x)$ has a minimum at $x = a$, and **(c)** the graph of F has a horizontal tangent line at $x = a$.

8. Intercepts: (0, 0)

$$y' = -xe^{-x} + e^{-x} = e^{-x}(1 - x)$$

Thus $y' = 0$ for $x = 1$, $y' > 0$ for $x < 1$, and $y' < 0$ for $x > 1$. You conclude that there is a maximum at $(1, 1/e)$, and that y is increasing for $x < 1$ and decreasing for $x > 1$.

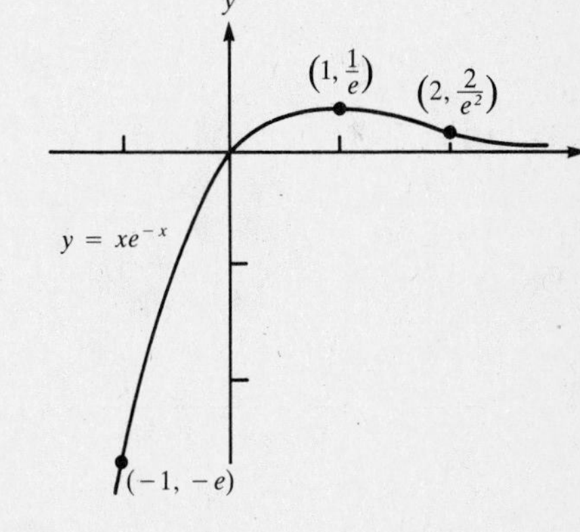

$$y'' = xe^{-x} - e^{-x} - e^{-x} = e^{-x}(x - 2)$$

The second derivative is negative for $x < 2$, so the graph is concave downward for $x < 2$. The graph is concave upward for $x > 2$, and $(2, 2/e^2)$ is an inflection point, as shown in the figure.

9. Find the intercepts:

$$y = \frac{1}{x} + \frac{4}{1 - x} = \frac{1 + 3x}{x(1 - x)}$$

Thus, $y = 0$ when $x = -1/3$ and there is no y intercept. There are vertical asymptotes at $x = 0$ and $x = 1$. The horizontal asymptote

$$\lim_{x \to \infty} \frac{1 + 3x}{x(1 - x)} = 0$$

is $y = 0$. Differentiate:

$$y' = \frac{-1}{x^2} + \frac{4}{(1 - x)^2} = \frac{-(1 - 2x + x^2) + 4x^2}{x^2(1 - x)^2} = \frac{3x^2 + 2x - 1}{x^2(1 - x)^2}$$

$$= \frac{(3x - 1)(x + 1)}{x^2(1 - x)^2}$$

$$y' = 0 \quad \text{at} \quad x = \frac{1}{3} \qquad \text{and} \qquad x = -1$$

$$y' > 0 \quad \text{for} \quad x < -1 \qquad \text{and for} \quad x > \frac{1}{3}, x \neq 1$$

$$y' < 0 \quad \text{for} \quad -1 < x < \frac{1}{3}, \qquad x \neq 0$$

$$y'' = \frac{2}{x^3} + \frac{8}{(1 - x)^3}$$

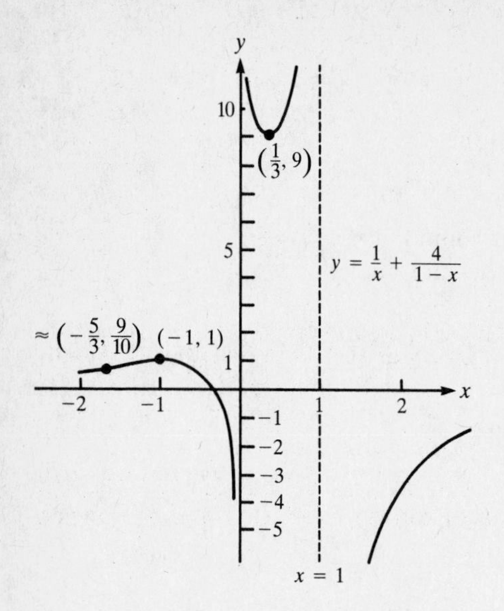

At $x = 1/3$, $y'' > 0$, so there is a local minimum at $(1/3, 9)$. At $x = -1$, $y'' < 0$, so there is a local maximum at $(-1, 1)$. Find where $y'' = 0$:

$$0 = \frac{2}{x^3} + \frac{8}{(1-x)^3}$$

$$\left(\frac{1-x}{x}\right)^3 = -4$$

$$\frac{1-x}{x} = -\sqrt[3]{4}$$

$$1 - x = -x\sqrt[3]{4}$$

$$x = \frac{1}{1 - \sqrt[3]{4}}$$

There is an inflection point at $x = 1/(1 - \sqrt[3]{4}) \approx -5/3$, as shown in the figure.

10. Find the intercepts: When $x = 0$, $y = 0$; when $y = 0$, $x = 0, \pm 6$. The graph is symmetric with respect to the y axis. Differentiate:

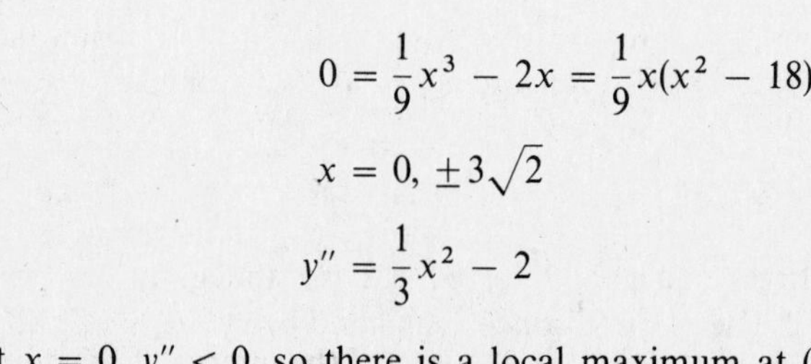

$$y' = \frac{1}{9}x^3 - 2x$$

$$0 = \frac{1}{9}x^3 - 2x = \frac{1}{9}x(x^2 - 18)$$

$$x = 0, \pm 3\sqrt{2}$$

$$y'' = \frac{1}{3}x^2 - 2$$

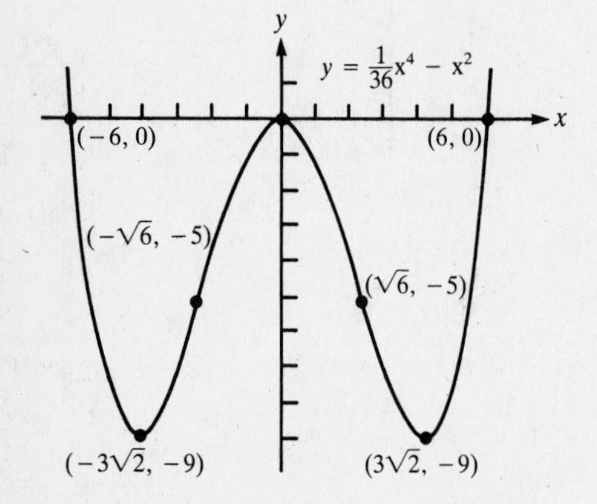

At $x = 0$, $y'' < 0$, so there is a local maximum at $(0, 0)$. At $x = \pm 3\sqrt{2}$, $y'' > 0$, so there are local minima at $x = \pm 3\sqrt{2}$. Find inflection points:

$$\frac{1}{3}x^2 - 2 = 0$$

$$x = \pm\sqrt{6}$$

The graph is concave downward for $-\sqrt{6} < x < \sqrt{6}$ and concave upward for $x < -\sqrt{6}$ and $x > \sqrt{6}$, as shown in the figure.

9 EXTREMA

THIS CHAPTER IS ABOUT

☑ How to Find Extrema: The Method
☑ How to Solve Word Problems Involving Maxima and Minima

9-1. How to Find Extrema: The Method

You will often be given a function and asked to find the maximum or minimum value of the function. Sometimes the extremum is found among a restricted set of values of the variable. Other times the mathematical problem is hidden in a word problem.

A. Find the critical points.

To find the maximum or minimum value of a function $f(x)$, you can simply graph the function. However, this is more effort than is necessary merely to find an extremum. Instead, you can *find the critical points first*, ignoring those that involve values of the variable outside the interval of interest.

B. Determine the global extrema.

If f has no vertical asymptotes in the interval of interest, you simply evaluate f at each critical point (and at the endpoint(s) of the interval if the domain is restricted) and select the largest value and the smallest value. If the values of the variable do not include a right endpoint b, you consider $\lim_{x\to b^-} f(x)$ (or $\lim_{x\to\infty} f(x)$ if x is unbounded on the right) along with the values of f at the critical points. If $\lim_{x\to b^-} f(x)$ or $\lim_{x\to\infty} f(x)$ is the largest (or smallest) of these values, then f does not attain a maximum (or minimum). In particular, if $\lim_{x\to b^-} f(x) = \infty$ or $-\infty$ (or if $\lim_{x\to\infty} f(x) = \infty$ or $-\infty$), then f does not attain a maximum (or minimum). Similarly, if the values of x under consideration do not include a left endpoint a, then $\lim_{x\to a^+} f(x)$ (or $\lim_{x\to -\infty} f(x)$) must be considered.

If f does have a vertical asymptote at $x = c$ in the interval of interest, you have to determine the behavior of f in the vicinity of $x = c$. If $\lim_{x\to c^+} f(x) = \infty$ or $\lim_{x\to c^-} f(x) = \infty$, then f does not have a maximum value. If $\lim_{x\to c^+} f(x) = -\infty$ or $\lim_{x\to c^-} f(x) = -\infty$, then f has no minimum value.

EXAMPLE 9-1: Find the maximum and minimum values of $f(x) = 3x^4 - 4x^3 + 1$ for $-1 \leqslant x \leqslant 2$.

Solution: First, find the critical points:

$$f'(x) = 12x^3 - 12x^2$$

Set $f'(x)$ equal to zero:

$$12x^3 - 12x^2 = 0$$

$$12x^2(x - 1) = 0$$

Thus, there are critical points at $x = 0$ and $x = 1$.

Next, find the *global* (*absolute*) *extrema*: Now you evaluate f at the critical points and at the endpoints (-1 and 2). Evaluating f at 0, 1, -1, and 2,

$$f(0) = 1 \qquad f(1) = 0 \qquad f(-1) = 8 \qquad f(2) = 17$$

So the minimum value of f on $[-1, 2]$ is 0, which is attained at $x = 1$. The maximum value is 17, which is attained at $x = 2$.

EXAMPLE 9-2: Find the maximum and minimum values of $f(x) = (x^2 - x + 1)/(x^2 + 1)$ on the interval $0 \leqslant x \leqslant 3$.

Solution:

$$f'(x) = \frac{(x^2 + 1)(2x - 1) - (x^2 - x + 1)(2x)}{(x^2 + 1)^2} = \frac{x^2 - 1}{(x^2 + 1)^2}$$

So $f'(x) = 0$ when $x^2 - 1 = 0$, i.e., $x = \pm 1$. Thus, the critical points are at -1 and 1. Because -1 is not in the interval of interest, remove it from consideration. Evaluating f at 1 and at 0 and 3 (endpoints), you get

$$f(1) = \frac{1}{2} \qquad f(0) = 1 \qquad f(3) = \frac{7}{10}$$

Thus, the minimum value is $\frac{1}{2}$ and the maximum value is 1.

EXAMPLE 9-3: Find the extreme values of $f(x) = (1 - x)^{-2}$ on $[0, 3]$.

Solution: Since $f(1)$ is not defined and

$$\lim_{x \to 1^+} (1 - x)^{-2} = \infty \qquad \lim_{x \to 1^-} (1 - x)^{-2} = \infty$$

it follows that f does not attain a maximum value on $[0, 3]$ (i.e., $f(x)$ gets arbitrarily large). Differentiating,

$$f'(x) = 2(1 - x)^{-3}$$

so $f'(x)$ is never zero. The only candidates for points at which the minimum is attained are 0 and 3, the endpoints:

$$f(0) = 1 \qquad f(3) = \tfrac{1}{4}$$

Thus, the minimum value is $\frac{1}{4}$, which is attained at $x = 3$.

EXAMPLE 9-4: Find the extreme values of $f(x) = (x^2 - 3)/(x^2 + 1)$ for $x \geqslant -1$.

Solution: You find that

$$f'(x) = 8x(x^2 + 1)^{-2}$$

and $f'(x) = 0$ when $x = 0$; f has no vertical asymptotes. Evaluate f at 0 and at the endpoint -1:

$$f(0) = -3 \qquad f(-1) = -1$$

Because the interval is unbounded on the right, you must evaluate $\lim_{x \to \infty} f(x)$:

$$\lim_{x \to \infty} f(x) = \lim_{x \to \infty} \frac{x^2 - 3}{x^2 + 1} = \lim_{x \to \infty} \frac{(x^2 - 3)(1/x^2)}{(x^2 + 1)(1/x^2)}$$

$$= \lim_{x \to \infty} \frac{1 - (3/x^2)}{1 + (1/x^2)} = \frac{1 - 0}{1 + 0} = 1$$

The minimum value of f is -3, which is attained at $x = 0$. Among the three values under consideration for the maximum (-3, -1, and 1), $y = 1$ is the largest. However, the maximum is never attained as $y = 1$ is only approached as a horizontal asymptote (see Figure 9-1).

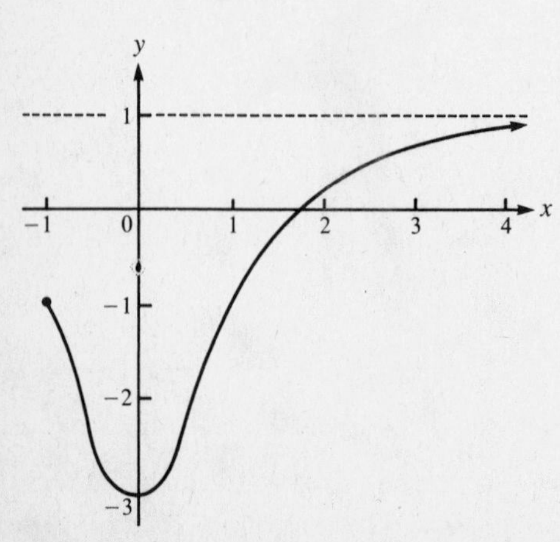

Figure 9-1
$f(x) = (x^2 - 3)/(x^2 + 1)$ for $x \geqslant -1$.

EXAMPLE 9-5: Find the extreme points of $f(x) = 4/(x - 2)$ for $2 < x < 4$.

Solution: You find that

$$f'(x) = -4(x - 2)^{-2}$$

so $f'(x)$ is never zero. Note that 2 and 4 are endpoints of the interval under consideration but are not included in the interval. Thus, $\lim_{x\to 2^+} f(x)$ and $\lim_{x\to 4^-} f(x)$ must be considered:

$$\lim_{x\to 2^+} \frac{4}{x - 2} = \infty \qquad \lim_{x\to 4^-} \frac{4}{x - 2} = 2$$

Thus, $f(x)$ does not attain a maximum and $f(x)$ does not attain a minimum; even though $f(4) = 2$, f never quite reaches 2 on this interval.

9-2. How to Solve Word Problems Involving Maxima and Minima

A. Translate to a mathematical problem.

When you have a word problem that asks for the maximum or minimum value of a certain quantity, you have to translate the problem to a purely mathematical question of the kind discussed in Section 9-1.

Typically, you can follow a three-step procedure:

(1) *Draw a picture.* Label the picture with the quantities given in the problem and with as many unknowns as you need.

(2) *Find an expression for the quantity to be maximized (or minimized).* This expression will usually involve two or more variables. Using the picture, find equations relating these variables to each other. Use these equations to eliminate all but one variable in the expression to be maximized.

(3) *Note any restrictions on this variable that are imposed by the problem.*

Now the problem is entirely translated to a mathematical problem, and you can proceed as outlined in Section 9-1.

- Usually the translation process is the most difficult task.

B. Use this shortcut.

Suppose you wish to find the maximum (or minimum) value of a differentiable function $f(x)$ on a certain region. Find where $f'(x) = 0$ on the interval of interest. If there is only *one* such point $x = a$ and if f has no vertical asymptotes, then

- If it is a local maximum (or minimum), it is a global maximum (or minimum).

You determine whether it is a local maximum by examining $f'(x)$ on both sides of $x = a$. Or, you might use the second derivative test: If $f''(a) < 0$ (or > 0), then a is a local maximum (or minimum).

C. Answer the question asked in the problem.

For example, one problem might ask for the maximum area of a given figure, while another problem might ask for a dimension of the figure which produces the maximum area.

EXAMPLE 9-6: A cone with a slant height of 6 in. is to be constructed. What is the largest possible volume of such a cone?

Solution:

(1) *Draw the picture*: Draw a cone, labeling the height h, the radius r, and the slant height 6 in. (see Figure 9-2).

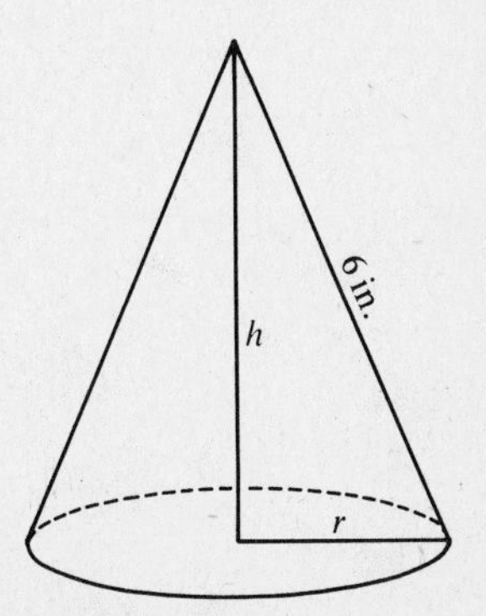

Figure 9-2

(2) *Find the quantity to be maximized* and express it as a function of one variable. The volume of a cone is given by

$$V = \tfrac{1}{3}\pi r^2 h$$

an expression that involves two variables. From the picture, you see that h and r are related by the Pythagorean theorem:

$$h^2 + r^2 = 36$$

and hence $r^2 = 36 - h^2$. You can now substitute for r^2, and write V as a function of h only:

$$V = \frac{\pi}{3}(36 - h^2)h$$

where $0 < h < 6$.

(3) Now you can *find the critical points*:

$$V(h) = \frac{\pi}{3}(36h - h^3)$$

$$V'(h) = \frac{\pi}{3}(36 - 3h^2)$$

So $V'(h) = 0$ when

$$36 - 3h^2 = 0$$

$$h^2 = 12$$

$$h = 2\sqrt{3}$$

(Notice that $h \neq -2\sqrt{3}$, since $h > 0$.)

The second derivative test tells you that $V(h)$ is a maximum when $h = 2\sqrt{3}$:

$$V''(h) = \frac{\pi}{3}(-6h) = -2\pi h$$

and $V''(2\sqrt{3}) < 0$. Thus, the maximum volume is

$$V(2\sqrt{3}) = \frac{\pi}{3}(36(2\sqrt{3}) - (2\sqrt{3})^3)$$

$$= \frac{\pi}{3}(72\sqrt{3} - 24\sqrt{3}) = 16\pi\sqrt{3}\text{ in}^3.$$

EXAMPLE 9-7: Find the dimensions of the rectangle of largest area that can be inscribed in a circle of radius 4.

Solution:

(1) *Draw the picture*: Draw the circle with a rectangle inscribed, as shown in Figure 9-3. The only quantity given in the problem is the radius, which is 4. Label this on the picture. Also label the dimensions of the rectangle, x and y—they will be necessary to express the area of the rectangle.

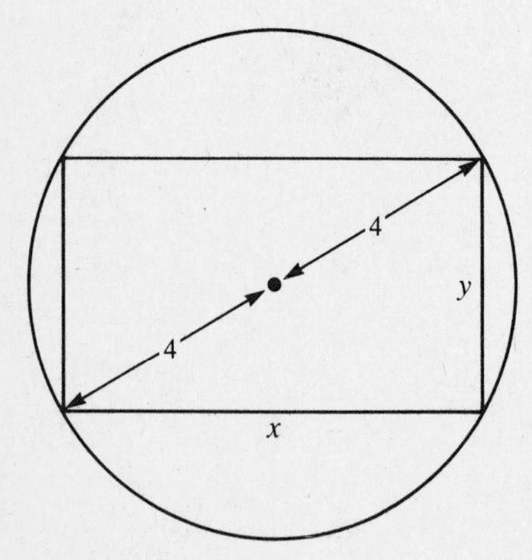

Figure 9-3

(2) *Find the expression to be maximized*, as a function of one variable. The expression

$$A = xy$$

involves two variables, x and y. Search for an equation relating x and y. There is a right triangle pictured, with hypotenuse 8 ($2r$). Thus, by the Pythagorean Theorem,

$$x^2 + y^2 = 64$$

You can use this, solving for one variable in terms of the other, to eliminate a variable from the expression for A:

$$y = (64 - x^2)^{1/2}$$

$$A = xy = x(64 - x^2)^{1/2}$$

(3) *Note restrictions*: The remaining variable is x. The restrictions imposed by the picture are $0 \leqslant x \leqslant 8$.

Now you have reduced the problem to one of finding the maximum value of $A = x(64 - x^2)^{1/2}$ on the interval [0, 8]. Proceed by finding the critical points:

$$A'(x) = x(\tfrac{1}{2})(64 - x^2)^{-1/2}(-2x) + (64 - x^2)^{1/2} = \frac{64 - 2x^2}{\sqrt{64 - x^2}}$$

You find that $A'(x) = 0$ when $64 - 2x^2 = 0$ (i.e., $x = \pm 4\sqrt{2}$). The critical point in the interval of interest is at $x = 4\sqrt{2}$. Because $A'(x) > 0$ when $0 < x < 4\sqrt{2}$ and $A'(x) < 0$ when $4\sqrt{2} < x < 8$, you conclude that $4\sqrt{2}$ is a local maximum. As it is the only critical point of a continuous function, you conclude that it is the global maximum of A. The problem requires that you find the values of x and y that yield this maximum value of A. From $x = 4\sqrt{2}$ and $x^2 + y^2 = 64$, you get

$$(4\sqrt{2})^2 + y^2 = 64$$
$$32 + y^2 = 64$$

So $y = 4\sqrt{2}$. Thus, the rectangle is a square whose sides have length $4\sqrt{2}$.

EXAMPLE 9-8: A cylindrical can is constructed using 100 in.^{2} of tin. This amount includes a top and bottom for the can. What is the largest volume such a can might contain?

Solution: First, draw a picture and label it, as in Figure 9-4. The quantity to be maximized is the volume V:

$$V = \pi r^2 h$$

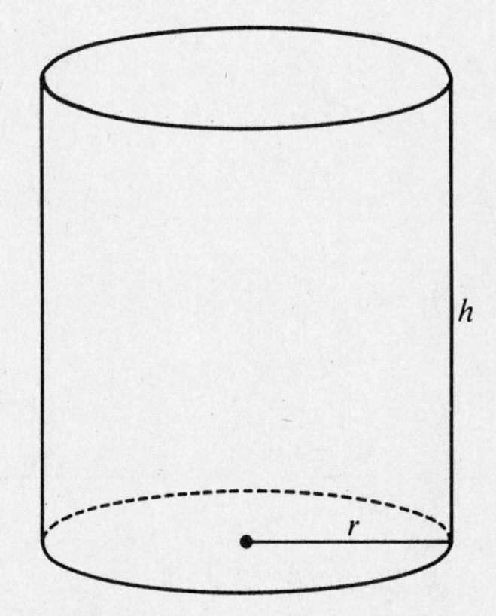

Figure 9-4

You need an equation relating radius r and height h. The fact that the area of tin in the can is 100 means that

$$100 = (\text{area of sides of can}) + (\text{area of top and bottom}) = 2\pi rh + 2(\pi r^2)$$

Use this equation to eliminate a variable from the expression for V.

$$h = \frac{100 - 2\pi r^2}{2\pi r} = \frac{50}{\pi r} - r$$

Thus, V becomes

$$V = \pi r^2\left(\frac{50}{\pi r} - r\right) = 50r - \pi r^3$$

The restrictions on r are $0 \leqslant r \leqslant \sqrt{50/\pi}$. (The exact value of the upper bound on r will not be important.) Proceed to maximize $V(r) = 50r - \pi r^3$ on the interval $[0, \sqrt{50/\pi}]$:

$$V'(r) = 50 - 3\pi r^2$$

So $V'(r) = 0$ when $r = \pm 5\sqrt{2/(3\pi)}$. Thus, $r = 5\sqrt{2/(3\pi)}$ is your critical point. Because this is the only critical point, you need to verify that it yields a local maximum in order to be certain that it yields the global maximum:

$$V''(r) = -6\pi r$$

$$V''\left(5\sqrt{\frac{2}{3\pi}}\right) = -30\pi\sqrt{\frac{2}{3\pi}} < 0$$

So there is a local maximum at $r = 5\sqrt{2/(3\pi)}$, and hence a global maximum. The largest volume the can might contain is then

$$V(r) = r(50 - \pi r^2)$$

$$V\left(5\sqrt{\frac{2}{3\pi}}\right) = 5\sqrt{\frac{2}{3\pi}}\left(50 - \frac{\pi \cdot 50}{3\pi}\right)$$

$$= 250\sqrt{\frac{2}{3\pi}}\left(1 - \frac{1}{3}\right) = \frac{500}{3}\sqrt{\frac{2}{3\pi}}$$

EXAMPLE 9-9: Find the distance x that an observer should stand away from a 12-ft-tall sign, mounted 4 ft above eye-level, so that the angle subtended by the sign at his eye is a maximum.

Solution: The quantity to be maximized is the angle θ, as shown by Figure 9-5. Find an expression for θ:

$$\tan(\theta + \alpha) = 16/x$$

$$\theta + \alpha = \arctan(16/x)$$

$$\theta = \arctan(16/x) - \alpha$$

To express θ in terms of one variable, you need an equation relating x and α:

$$\tan \alpha = 4/x$$

$$\alpha = \arctan(4/x)$$

Thus, θ becomes simply

$$\theta = \arctan(16/x) - \arctan(4/x)$$

The restriction on x is $x \geqslant 0$. You search for critical points:

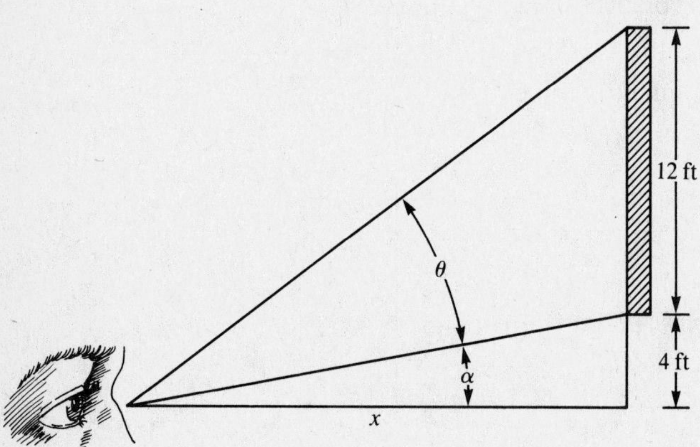

Figure 9-5

$$\theta'(x) = \frac{1}{1 + (16/x)^2}\left(\frac{-16}{x^2}\right) - \frac{1}{1 + 16/x^2}\left(\frac{-4}{x^2}\right)$$

$$= \frac{-16}{x^2 + 16^2} - \frac{-4}{x^2 + 16}$$

$$= \frac{-16(x^2 + 16) + 4(x^2 + 16^2)}{(x^2 + 16^2)(x^2 + 16)}$$

$$= \frac{4(16^2 - 64 - 3x^2)}{(x^2 + 16^2)(x^2 + 16)}$$

$$= \frac{12(64 - x^2)}{(x^2 + 16^2)(x^2 + 16)}$$

So $\theta'(x) = 0$ when $64 - x^2 = 0$, i.e., when $x = \pm 8$. Of course, $x > 0$, so $x = 8$ is the only critical point. If $0 < x < 8$, then $\theta'(x) > 0$. If $x > 8$, then $\theta'(x) < 0$. Thus, $x = 8$ is the location of a local maximum. Because it is the only critical point, it is the global maximum. The observer should stand 8 ft away from the sign.

SUMMARY

1. Continuous functions on closed intervals achieve extreme values at endpoints and critical points.
2. When the region is unbounded on the left or right, you must consider $\lim_{x \to -\infty} f(x)$ or $\lim_{x \to \infty} f(x)$.
3. When the function is defined on an open interval (a, b), you must consider $\lim_{x \to a^+} f(x)$ and $\lim_{x \to b^-} f(x)$.
4. When the function has a vertical asymptote at $x = a$, you must consider $\lim_{x \to a^-} f(x)$ and $\lim_{x \to a^+} f(x)$.
5. Translate word problems to purely mathematical problems.
6. Careful drawings are essential.
7. Express the quantity to be maximized or minimized in terms of one variable. (Sometimes this requires that you find more than one relationship in order to eliminate the extra variables by substitution.)
8. Once you find the critical point, there are a variety of methods available to determine whether the point is a minimum or a maximum. For example, you might use the second derivative test or you might determine where f is increasing and decreasing.

SOLVED PROBLEMS

PROBLEM 9-1 Find the maximum and minimum values of $f(x) = x - x^3$ on $[-1, 5]$.

Solution: You find that

$$f'(x) = 1 - 3x^2$$

so $f'(x) = 0$ when $x = \pm\sqrt{3}/3$. Since f has no vertical asymptotes, you evaluate f at $\pm\sqrt{3}/3$ and at -1 and 5, the endpoints:

$$f(-1) = 0 \qquad f\left(-\frac{\sqrt{3}}{3}\right) = \frac{-2}{9}\sqrt{3} \qquad f\left(\frac{\sqrt{3}}{3}\right) = \frac{2}{9}\sqrt{3} \qquad f(5) = -120$$

The maximum value is $2\sqrt{3}/9$, which is attained at $x = \sqrt{3}/3$. The minimum value is -120, attained at $x = 5$. [See Section 9-1.]

PROBLEM 9-2 Find the extrema of the function $f(x) = x^5 - 5x^3 + 10x - 7$ on $[0, 2]$.

Solution: You find that

$$f'(x) = 5x^4 - 15x^2 + 10 = 5(x^2 - 2)(x^2 - 1)$$

so $f'(x) = 0$ when $x = \pm\sqrt{2}, \pm 1$. The points in the interval of interest are $x = 1$ and $x = \sqrt{2}$. Evaluating f at 0, 1, $\sqrt{2}$, and 2,

$$f(0) = -7 \qquad f(1) = -1 \qquad f(\sqrt{2}) = 4\sqrt{2} - 7 \qquad f(2) = 5$$

You find that the minimum occurs at $(0, -7)$ and the maximum at $(2, 5)$. [See Section 9-1.]

PROBLEM 9-3 Find the extrema of $f(x) = \dfrac{\sin x}{\cos x + \sqrt{2}}$ on $[0, 2\pi]$.

Solution: You find that

$$f'(x) = \frac{1 + \sqrt{2}\cos x}{(\cos x + \sqrt{2})^2}$$

so $f'(x) = 0$ when $\cos x = -\sqrt{2}/2$ (i.e., $x = 3\pi/4$ and $x = 5\pi/4$). Since f has no vertical asymptotes, you evaluate f at $3\pi/4$, $5\pi/4$, 0, and 2π:

$$f\left(\frac{3\pi}{4}\right) = 1 \qquad f\left(\frac{5\pi}{4}\right) = -1 \qquad f(0) = 0 \qquad f(2\pi) = 0$$

So the maximum occurs at $(3\pi/4, 1)$ and the minimum occurs at $(5\pi/4, -1)$. [See Section 9-1.]

PROBLEM 9-4 Find the extrema of $f(x) = 1 + x^{-2}$ on $[-1, 2]$.

Solution: You find that

$$f'(x) = -2x^{-3}$$

so $f'(x)$ is never zero. However, there is a vertical asymptote at $x = 0$ (which lies within $[-1, 2]$), so you must examine the behavior of $f(x)$ in the vicinity of $x = 0$. Since

$$\lim_{x \to 0^-} (1 + x^{-2}) = \infty \qquad \lim_{x \to 0^+} (1 + x^{-2}) = \infty$$

there is no maximum. To find the minimum, evaluate f at -1 and 2:

$$f(-1) = 2 \qquad f(2) = 5/4$$

So the minimum occurs at $(2, 5/4)$. [See Section 9-1.]

PROBLEM 9-5 Find the extreme values of $f(x) = x/(x^2 + 1)$ for $x \geqslant \frac{1}{2}$.

Solution: You find that

$$f'(x) = (1 - x^2)(x^2 + 1)^{-2}$$

so $f'(x) = 0$ when $x = \pm 1$. Consider the values of f at 1 and $\frac{1}{2}$:

$$f(1) = \frac{1}{2} \qquad f\left(\frac{1}{2}\right) = \frac{2}{5}$$

Because x is unbounded on the right, you must also consider $\lim_{x \to \infty} f(x)$:

$$\lim_{x \to \infty} x(x^2 + 1)^{-1} = 0$$

The maximum value is $\frac{1}{2}$, attained at $x = 1$. Because zero, the smallest of the three numbers under consideration, occurs in the limit only, there is no minimum value. [See Section 9-1.]

PROBLEM 9-6 Find the maximum and minimum values of $f(x) = x^{1/3} + 1$ on $[-3, 3]$.

Solution: You find that

$$f'(x) = \tfrac{1}{3}x^{-2/3}$$

so $f'(x)$ is never zero. Although $f'(x)$ is undefined at $x = 0$, this is not a critical point because f' does not change sign at $x = 0$. Evaluate f at -3 and 3:

$$f(-3) = 1 - \sqrt[3]{3} \qquad f(3) = 1 + \sqrt[3]{3}$$

So the minimum value is $1 - \sqrt[3]{3}$ and the maximum value is $1 + \sqrt[3]{3}$. [See Section 9-1.]

PROBLEM 9-7 Find the extrema of $f(x) = x^4 - 6x^2 + 3$.

Solution: You find that

$$f'(x) = 4x^3 - 12x = 4x(x^2 - 3)$$

so $f'(x) = 0$ when $x = 0, \pm\sqrt{3}$. Because the variable is unrestricted, you must consider

$$\lim_{x \to \infty} (x^4 - 6x^2 + 3) = \infty \qquad \lim_{x \to -\infty} (x^4 - 6x^2 + 3) = \infty$$

$$f(0) = 3 \qquad f(\sqrt{3}) = f(-\sqrt{3}) = -6$$

Thus, there is no maximum, and there are two points at which the minimum is attained: $(-\sqrt{3}, -6)$ and $(\sqrt{3}, -6)$. [See Section 9-1.]

PROBLEM 9-8 Find the extrema of $f(x) = x(x^2 - 1)^{-1}$ on $[0, 3]$.

Solution: You find that

$$f'(x) = -(x^2 + 1)(x^2 - 1)^{-2}$$

so $f'(x)$ is never zero. You must consider the vertical asymptote at $x = 1$:

$$\lim_{x \to 1^-} x(x^2 - 1)^{-1} = -\infty \qquad \lim_{x \to 1^+} x(x^2 - 1)^{-1} = \infty$$

Thus, there are no extrema. [See Section 9-1.]

PROBLEM 9-9 Find the extrema of $f(x) = \sin^2 x + 2 \cos x$ on $[0, 3\pi/2]$.

Solution: You find that

$$\begin{aligned} f'(x) &= 2 \sin x \cos x - 2 \sin x \\ &= 2 \sin x(\cos x - 1) \end{aligned}$$

so $f'(x) = 0$ when $\sin x = 0$ and when $\cos x = 1$. In the interval $[0, 3\pi/2]$ the critical points are at $x = 0$ and $x = \pi$. You must evaluate f at 0, π, and $3\pi/2$:

$$f(0) = 2 \qquad f(\pi) = -2 \qquad f\left(\frac{3\pi}{2}\right) = 1$$

Thus, the maximum occurs at $(0, 2)$ and the minimum occurs at $(\pi, -2)$. [See Section 9-1.]

PROBLEM 9-10 Find the extrema of $f(x) = x^2/(x^2 + 1)$ for $x \geqslant 1$.

Solution: Differentiating, you find

$$f'(x) = \frac{2x}{(x^2 + 1)^2}$$

For $x \geqslant 1$, $f'(x) \geqslant 0$. Thus, on $[1, \infty) f(x)$ is increasing: The minimum occurs at $(1, \frac{1}{2})$ and there is no maximum.
[See Section 9-1.]

PROBLEM 9-11 Find the extreme points of $f(x) = 4x/(x - 1)^2$ for $-2 < x < 1$.

Solution: If you rewrite $f(x) = 4x(x - 1)^{-2}$, then

$$\begin{aligned} f'(x) &= 4x(-2)(x - 1)^{-3} + 4(x - 1)^{-2} \\ &= 4(x - 1)^{-3}[-2x + x - 1] \\ &= -4(x + 1)(x - 1)^{-3} \end{aligned}$$

so $f'(x) = 0$ when $x = -1$. In addition to $f(-1) = -1$, you must consider $\lim_{x \to -2^+} f(x)$ and $\lim_{x \to 1^-} f(x)$:

$$\lim_{x \to -2^+} \frac{4x}{(x - 1)^2} = -\frac{8}{9} \qquad \lim_{x \to 1^-} \frac{4x}{(x - 1)^2} = \infty$$

So f does not attain a maximum; f attains a minimum at $(-1, -1)$.
[See Section 9-1.]

PROBLEM 9-12 Find the extreme points of the graph of $x^2 + xy + y^2 = 3$.

Solution: Differentiating implicitly, you have

$$2x + (xy' + y) + 2yy' = 0$$

Set $y' = 0$:

$$2x + y = 0$$

You should interpret this as an additional condition on the extreme points you are seeking; that is, you solve $x^2 + xy + y^2 = 3$ and $2x + y = 0$ simultaneously.
Substituting $y = -2x$ into the first equation, you get

$$x^2 - 2x^2 + 4x^2 = 3$$

or

$$3x^2 = 3$$

$$x = \pm 1$$

So the tangent lines to the curve at $(1, -2)$ and $(-1, 2)$ are horizontal.

You may now use the second derivative test to determine whether or not either point is a maximum or a minimum and to properly label each one. Returning to your first differentiation and differentiating a second time, you have

$$2x + xy' + y + 2yy' = 0$$

$$2 + (xy'' + y') + y' + 2(yy'' + (y')^2) = 0$$

At $(1, -2)$, with $y' = 0$:

$$2 + y'' - 4y'' = 0$$

$$y'' = \frac{2}{3} > 0$$

You have a minimum at $(1, -2)$. At $(-1, 2)$, with $y' = 0$:

$$2 - y'' + 4y'' = 0$$

$$y'' = -\frac{2}{3} < 0$$

You have a maximum at $(-1, 2)$.

PROBLEM 9-13 What are the dimensions of the cone with surface area 10π that encloses the largest volume? [*Hint*: Surface area $= \pi r(h^2 + r^2)^{1/2}$; volume $= \frac{1}{3}\pi r^2 h$.]

Solution:

(1) Draw the picture and label the unknown quantities h and r, as in Figure 9-6:
(2) The quantity to be maximized is the volume V:

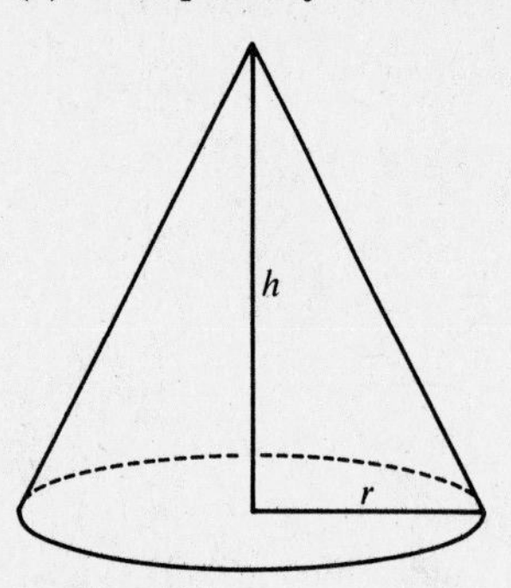

Figure 9-6
Problem 9-13

$$V = \tfrac{1}{3}\pi r^2 h$$

The equation that relates r to h arises from the condition that the surface area is 10π:

$$10\pi = \pi r(h^2 + r^2)^{1/2}$$

Solve for h in terms of r:

$$h = \left(\frac{100}{r^2} - r^2\right)^{1/2}$$

Use this expression to eliminate h from the expression for V:

$$V = \frac{1}{3}\pi r^2\left(\frac{100}{r^2} - r^2\right)^{1/2} = \frac{1}{3}\pi r(100 - r^4)^{1/2}$$

(3) The restriction on the variable is $r \geqslant 0$. Also, the largest possible value of r is $\sqrt{10}$ (otherwise the surface area would exceed 10π). (This observation is not necessary because such an endpoint will lead to a volume of zero, which certainly is not the maximum.) Next, compute $V'(r)$:

$$V'(r) = \tfrac{1}{3}\pi(100 - 3r^4)(100 - r^4)^{-1/2}$$

So $V'(r) = 0$ when $r = (100/3)^{1/4}$ in the interval of interest. You deduce that this value for r yields a maximum because it is the only point where $V'(r) = 0$ and V is zero at both endpoints while positive between.
(4) Finally, you can answer the question:

$$r = \left(\frac{100}{3}\right)^{1/4} \quad \text{and} \quad r^2 = \left(\frac{100}{3}\right)^{1/2} = \frac{10\sqrt{3}}{3}$$

$$h = \left(\frac{100}{r^2} - r^2\right)^{1/2} = \left(\frac{100 \cdot 3}{10\sqrt{3}} - \frac{10\sqrt{3}}{3}\right)^{1/2} = \left[10\sqrt{3}\left(1 - \frac{1}{3}\right)\right]^{1/2}$$

$$= \left(\frac{20\sqrt{3}}{3}\right)^{1/2} = \frac{2}{3}\sqrt{5}(27)^{1/4}$$

[See Section 9-2.]

PROBLEM 9-14 A silo consists of a cylinder with a hemispherical top, as shown in Figure 9-7. Find the dimensions of the silo with fixed volume $V = 40\pi/3$ that has the least surface area. Include the floor.

Solution: You'll need the equations for the volumes of a hemisphere and of a cylinder, and for the surface areas of each. A sphere of radius r has volume $V_s = \frac{4}{3}\pi r^3$ and surface area $A_s = 4\pi r^2$. A cylinder of radius r and height h has volume $V_c = \pi r^2 h$ and surface area $A_c = 2\pi rh + \pi r^2$ (including one base). Hence, the volume of the silo is given by

$$V = \tfrac{2}{3}\pi r^3 + \pi r^2 h$$

and the surface area by

$$A = 2\pi r^2 + 2\pi rh + \pi r^2 = \pi(3r^2 + 2rh)$$

You want to minimize A. The expression for A involves two variables, r and h; but because $V = 40\pi/3$, you have an equation relating r and h:

$$\frac{40\pi}{3} = \frac{2}{3}\pi r^3 + \pi r^2 h$$

Solving this equation for h (because it is easier than solving for r):

$$h = \frac{40}{3}r^{-2} - \frac{2}{3}r$$

Now, substitute into the expression for A:

$$A = \pi\left(3r^2 + 2r\left(\frac{40}{3}r^{-2} - \frac{2}{3}r\right)\right)$$

$$= \pi\left(3r^2 + \frac{80}{3}r^{-1} - \frac{4}{3}r^2\right)$$

$$= \pi\left(\frac{5}{3}r^2 + \frac{80}{3}r^{-1}\right)$$

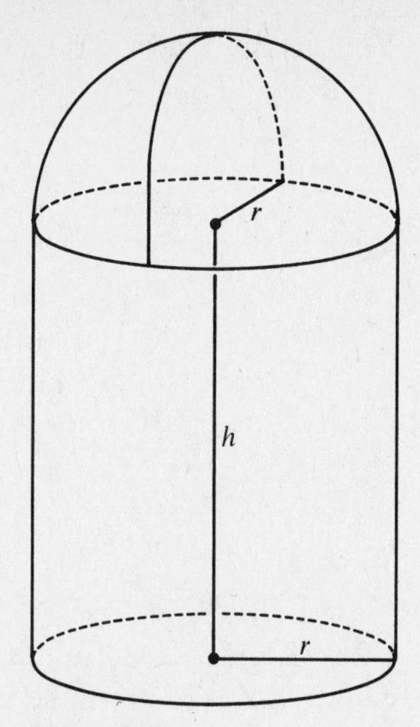

Figure 9-7
Problem 9-14

Now that you have expressed A as a function of one variable, $r (r > 0)$, search for critical points.

$$A' = \pi\left(\frac{10}{3}r - \frac{80}{3}r^{-2}\right)$$

So $A' = 0$ when

$$\frac{10}{3}r - \frac{80}{3}r^{-2} = 0$$

$$\frac{10}{3}r^3 - \frac{80}{3} = 0$$

$$r^3 = 8$$

$$r = 2$$

Note that

$$A'' = \pi\left(\frac{10}{3} + \frac{160}{3}r^{-3}\right)$$

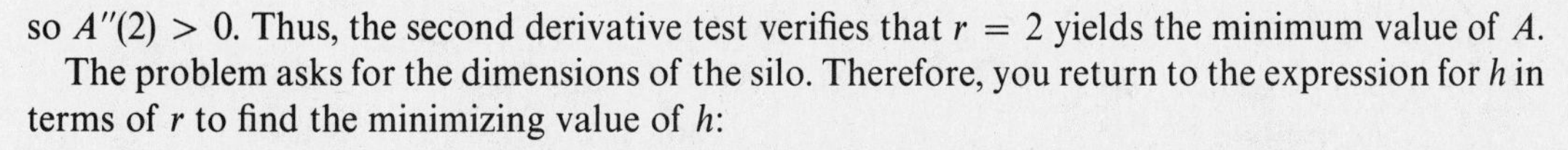

so $A''(2) > 0$. Thus, the second derivative test verifies that $r = 2$ yields the minimum value of A.

The problem asks for the dimensions of the silo. Therefore, you return to the expression for h in terms of r to find the minimizing value of h:

$$h = \frac{40}{3}r^{-2} - \frac{2}{3}r \quad = \frac{40}{3}(2)^{-2} - \frac{2}{3}(2) = 2$$

Therefore, the silo (with floor) with volume $40\pi/3$ that has the least surface area has radius $r = 2$ and height (of cylinder) $h = 2$. [See Section 9-2.]

PROBLEM 9-15 The open cylindrical bucket shown in Figure 9-8 is to be built with a volume of 1 ft^3. Find the dimensions that minimize the area of material used in its construction.

Solution: The area of material to be used is

$$M = \text{(area of sides)} + \text{(area of bottom)}$$
$$= 2\pi rh + \pi r^2$$

However, r and h are related by $\pi r^2 h = 1$, so

$$h = \frac{1}{\pi r^2}$$

Thus, M can be expressed as

$$M = 2\pi r\left(\frac{1}{\pi r^2}\right) + \pi r^2 = \frac{2}{r} + \pi r^2$$

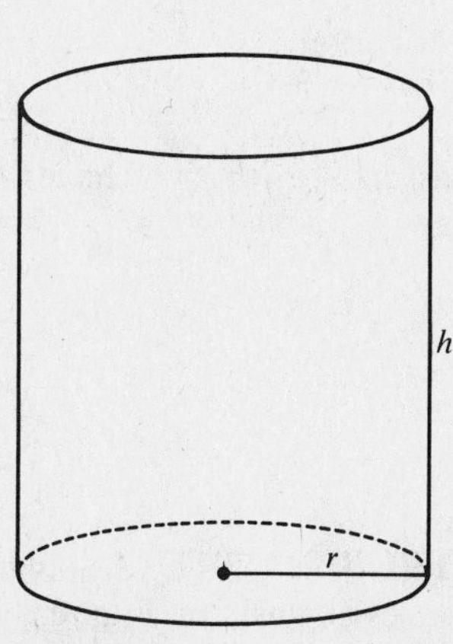

Figure 9-8
Problem 9-15

Your restriction on r is $r \geqslant 0$. Now differentiate:

$$M'(r) = \frac{-2}{r^2} + 2\pi r$$
$$= 2\left(\frac{\pi r^3 - 1}{r^2}\right)$$

So $M'(r) = 0$ when $r = (\pi)^{-1/3}$. You find that $M'(r) < 0$ for $0 < r < (\pi)^{-1/3}$ and $M'(r) > 0$ for $r > (\pi)^{-1/3}$. Thus, the minimal amount of material is used when

$$r = (\pi)^{-1/3}\text{ ft}$$
$$h = \frac{1}{\pi(\pi^{-1/3})^2} = (\pi)^{-1/3}\text{ ft}$$

[See Section 9-2.]

PROBLEM 9-16 Find the dimensions of the right circular cone of maximum surface area that can be inscribed in a sphere of radius $r = 1$.

Solution: The surface area of a cone of base radius r, height h, slant height s (Figure 9-9) is given by

$$A = \pi rs = \pi r(h^2 + r^2)^{1/2}$$

Use the fact that the cone is inscribed in a sphere of radius 1 to relate the variables:

$$h = 1 + x = 1 + (1 - r^2)^{1/2}$$

This allows you to express A as a function of r only. But notice the "messiness" of the result:

$$A = \pi r((1 + (1 - r^2)^{1/2})^2 + r^2)^{1/2}$$

You might do better by choosing as the single variable the quantity labeled "x." Then $h = 1 + x$ and $r = (1 - x^2)^{1/2}$. If, furthermore, you note that maximizing A^2 is equivalent to maximizing A, you can reduce the problem to simple terms:

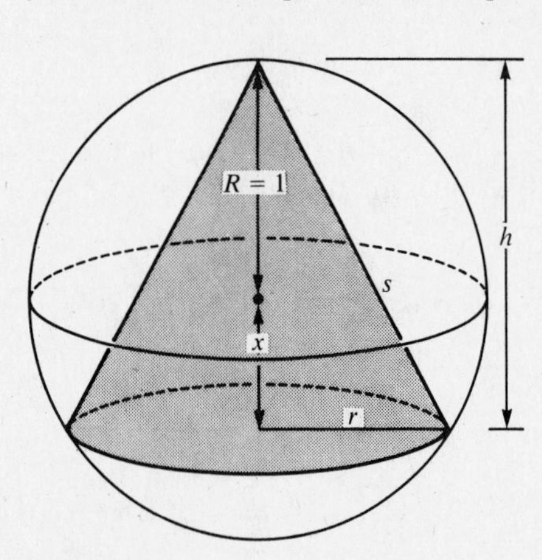

Figure 9-9
Problem 9-16

$$A^2 = \pi^2 r^2(h^2 + r^2)$$
$$= \pi^2(1 - x^2)((1 + x)^2 + (1 - x^2))$$
$$= 2\pi^2(1 + x - x^2 - x^3)$$

You now find $\frac{d}{dx}(A^2)$ and set it equal to zero:

$$\frac{d}{dx}(A^2) = 2\pi^2(1 - 2x - 3x^2) = 0$$
$$(1 - 3x)(1 + x) = 0$$

The only possible solution is $x = \frac{1}{3}$. A quick look at the second derivative

$$\left(\frac{d^2}{dx^2}(A^2) = -2x - 6x < 0 \quad \text{at} \quad x = \frac{1}{3}\right)$$

assures you that you have the desired solution. The cone of maximum surface area has dimensions

$$h = 1 + \frac{1}{3} = \frac{4}{3}$$
$$r = \left(1 - \frac{1}{9}\right)^{1/2} = \frac{2\sqrt{2}}{3}$$

[See Section 9-2.]

PROBLEM 9-17 A man in a boat is 24 miles from a straight shore and wishes to reach a point 20 miles downshore (Figure 9-10). He can travel 5 miles per hour in the boat and 13 mi/h on land. At what point should he land the boat in order to minimize the time required to get to his desired destination?

Solution: Let x be the number of miles from point P (the point on shore nearest to the man's present position) where the boat should land. Then the distance to be traveled at 5 mi/h is

$$d_1 = (24^2 + x^2)^{1/2}$$

so that the time required (using the basic relationship rate × time = distance) is

$$t_1 = \frac{(24^2 + x^2)^{1/2}}{5}$$

The distance to be traveled along the shore is

$$d_2 = 20 - x$$

so that

$$t_2 = \frac{20 - x}{13}$$

This allows you to write the expression for total time:

$$T = t_1 + t_2 = \frac{(24^2 + x^2)^{1/2}}{5} + \frac{20 - x}{13}$$

Differentiating and setting equal to zero yields

$$T' = \frac{1}{5}x(24^2 + x^2)^{-1/2} - \frac{1}{13} = 0$$

$$13x = 5(24^2 + x^2)^{1/2}$$

$$169x^2 = 25(24^2 + x^2)$$

$$x^2 = \frac{(25)24^2}{144}$$

$$x = 10 \text{ miles}$$

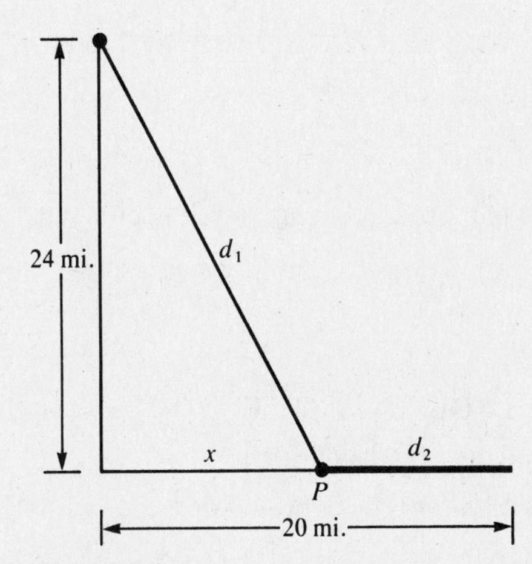

Figure 9-10
Problem 9-17

Note that $T' < 0$ for $x < 10$ and $T' > 0$ for $x > 10$, so this is the minimum. [See Section 9-2.]

PROBLEM 9-18 A poster is to contain a printed area of 150 in.^{2}, with clear margins of 3 in., top and bottom, and 2 in. on each side (Figure 9-11). Find minimum total area.

Solution: Let the dimensions of the printed area of the poster be x and y, so that $xy = 150$ or $y = 150/x$. Then the dimensions of the poster are $x + 4$ and $y + 6$, so that the total area is given by

$$A = (x + 4)(y + 6) = (x + 4)\left(\frac{150}{x} + 6\right)$$

$$= 174 + 6x + 600x^{-1}$$

Differentiating and setting equal to zero:

$$A' = 6 - 600x^{-2} = 0$$

$$x^2 = 100$$

$$x = 10$$

Notice that $A'' = 1200x^{-3} > 0$ for $x = 10$. Hence, the poster is of minimum area when $x = 10$ and $y = 150/10 = 15$, i.e., when the poster measures 14 by 21 in. [See Section 9-2.]

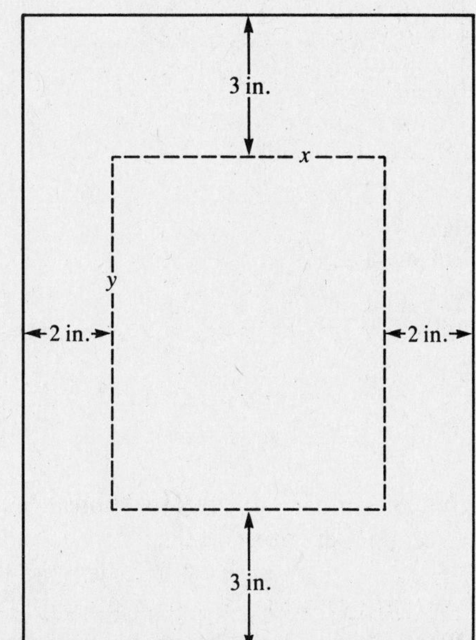

Figure 9-11
Problem 9-18

PROBLEM 9-19 A square piece of cardboard 1 foot on each side (Figure 9-12) is to be cut and folded into a box without a top (small squares will be cut from each corner). Find the dimensions of the box that will enclose the greatest volume.

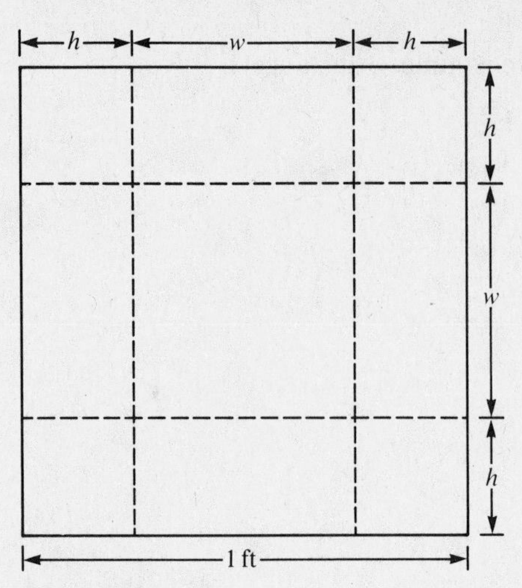

Figure 9-12
Problem 9-19

Solution: The quantity to be maximized is the volume V:

$$V = hw^2$$

The variables, h and w, are related by the equation:

$$2h + w = 1 \quad \text{or} \quad h = \frac{1 - w}{2}$$

Thus, you can express the volume as

$$V = \frac{1 - w}{2} w^2 = \frac{1}{2} w^2 - \frac{1}{2} w^3$$

The restriction on w is $0 \leqslant w \leqslant 1$. You find that

$$V'(w) = w - \frac{3}{2} \cdot w^2$$

so $V'(w) = 0$ when $w = 0$ and $w = 2/3$; $V'(w) > 0$ for $0 < w < 2/3$; and $V'(w) < 0$ for $w > 2/3$.

Thus, the maximum for V occurs when

$$w = 2/3 \text{ ft}$$

$$h = 1/6 \text{ ft}$$

[See Section 9-2.]

PROBLEM 9-20 Find the point on the graph of $y = x^2 + 1$ (Figure 9-13) that is closest to the point $(8, \frac{3}{2})$.

Solution: Let $P(x, y) = (x, x^2 + 1)$ be a point of the parabola. Then the distance formula

$$s = \sqrt{(x_2 - x_1)^2 + (y_2 - y_1)^2}$$

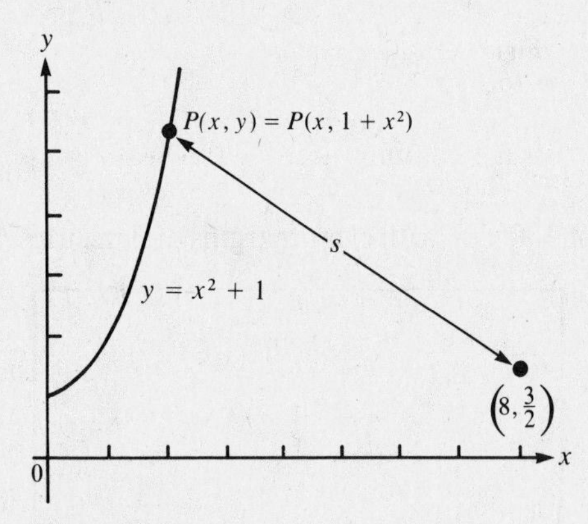

Figure 9-13
Problem 9-20

gives an expression for the distance from P to $(8, \frac{3}{2})$:

$$s = [(x - 8)^2 + (x^2 + 1 - \tfrac{3}{2})^2]^{1/2}$$
$$= [(x - 8)^2 + (x^2 - \tfrac{1}{2})^2]^{1/2}$$

and s is the quantity to be minimized. Note that you can save yourself some algebraic labor by choosing to minimize s^2 rather than s:

$$s^2 = (x - 8)^2 + (x^2 - \tfrac{1}{2})^2$$

Differentiate both sides of this equation and find the x for which the derivative is zero:

$$\frac{d}{dx}(s^2) = 2(x - 8) + 2\left(x^2 - \frac{1}{2}\right)(2x) = 0$$

$$2x - 16 + 4x^3 - 2x = 0$$

$$x^3 = 4$$

This equation has one real solution $x = \sqrt[3]{4}$. A quick application of the second derivative test gives a second derivative of $8x^2 > 0$ for all x—in particular, for $x = \sqrt[3]{4}$. Hence, the point of $y = x^2 + 1$ closest to $(8, \frac{3}{2})$ is the point $(\sqrt[3]{4}, 1 + \sqrt[3]{16}) = (\sqrt[3]{4}, 1 + 2\sqrt[3]{2})$. [See Section 9-2.]

PROBLEM 9-21 The girth of a shipping carton is the perimeter of an end (Figure 9-14). Shipping restrictions require that the sum of the girth and length not exceed 100 in. Find the dimensions of the package with a square end that has the largest volume.

Solution: Let the end be a square x in. on a side. Let the length be l in. Then the shipping restrictions require that $l + 4x \leqslant 100$. Assuming the equality (in order to obtain maximum volume), you have

$$l + 4x = 100 \qquad \text{with} \quad 0 < x < 25$$

so that $l = 100 - 4x$. The volume (to be maximized) is

$$V = x^2 l = x^2(100 - 4x) = 100x^2 - 4x^3$$

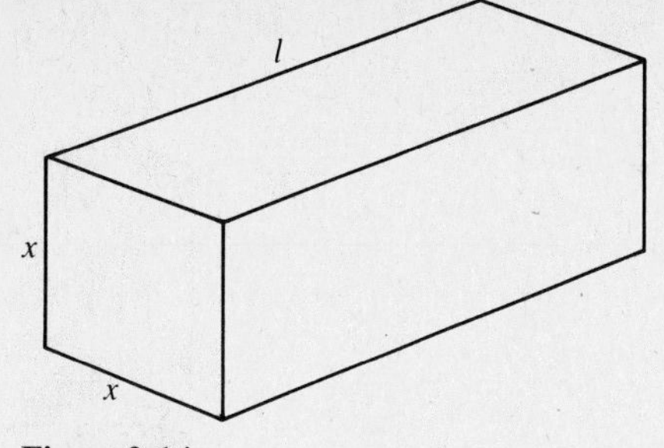

Figure 9-14
Problem 9-21

Differentiating, you obtain

$$V' = 200x - 12x^2 = 0$$

$$4x(50 - 3x) = 0$$

But $x \neq 0$, so $50 - 3x = 0$ and $x = 50/3$. The second derivative test gives $V'' = 200 - 24x$, which is less than zero for $x = 50/3$. Hence, V is a maximum when $x = 50/3$ in., $l = 100 - 200/3 = 100/3$ in. [See Section 9-2.]

PROBLEM 9-22 For the shipping carton of Problem 9-21, assume that the package is cylindrical (i.e., the end is a circle, as in Figure 9-15) and find the dimensions that yield the largest surface area.

Solution: Assume radius r and length l; then the perimeter of the end is the circumference $2\pi r$. The shipping restriction on size then becomes (again, choosing the equality)

$$2\pi r + l = 100 \qquad 0 < r < \frac{100}{2\pi}$$

$$l = 100 - 2\pi r$$

The surface area (to be maximized) of the cylinder is

$$S = 2\pi r l + 2\pi r^2$$

and, in terms of r only

$$S = 2\pi r(100 - 2\pi r) + 2\pi r^2$$

$$= 200\pi r - 4\pi^2 r^2 + 2\pi r^2$$

Differentiating and setting equal to zero gives

$$S' = 200\pi - 8\pi^2 r + 4\pi r = 0$$

$$4\pi r(1 - 2\pi) + 200\pi = 0$$

$$r = \frac{50}{2\pi - 1}$$

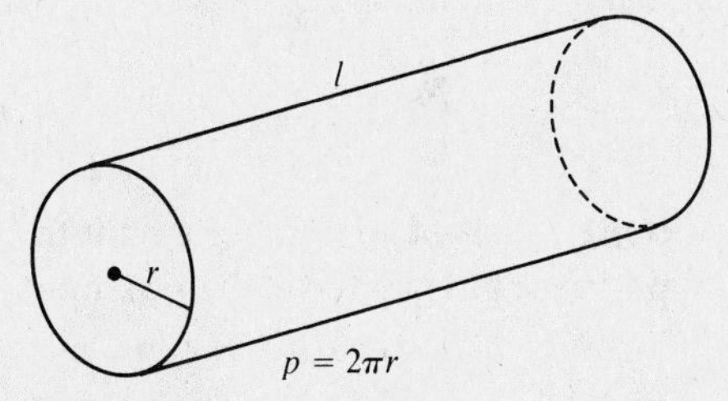

Figure 9-15
Problem 9-22

The second derivative test ($S'' = 4\pi - 8\pi^2 < 0$) assures us that this value of r does in fact maximize S. Hence, the dimensions for maximum surface area are

$$r = \frac{50}{2\pi - 1} \text{in.}$$

$$l = 100 - \frac{2\pi(50)}{2\pi - 1} = \frac{100(\pi - 1)}{2\pi - 1} \text{in.}$$

[See Section 9-2.]

PROBLEM 9-23 Find the dimensions of the cylinder of largest volume that will fit inside a cone of radius 3 and height 5. Assume that the axis of the cylinder lies on the axis of the cone. (See Figure 9-16.)

Figure 9-16
Problem 9-23

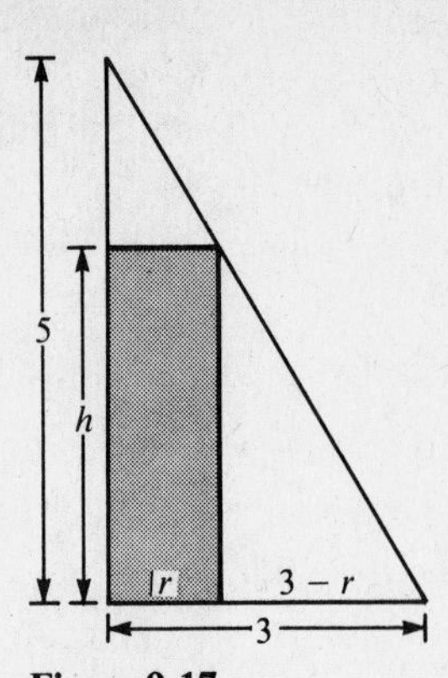

Figure 9-17
Problem 9-23

Solution: The quantity to be maximized is V, the volume of the cylinder:

$$V = \pi r^2 h$$

The condition that relates h and r is shown in Figure 9-17, which depicts the silhouette of half of Figure 9-16. From similar triangles you know that

$$\frac{h}{5} = \frac{3 - r}{3}$$

so

$$h = 5 - \frac{5r}{3}$$

Now you can express volume solely in terms of r:

$$V = \pi r^2 \left(5 - \frac{5r}{3}\right) = 5\pi r^2 - \frac{5}{3}\pi r^3$$

where $0 \leqslant r \leqslant 3$. Differentiating,

$$V'(r) = 10\pi r - 5\pi r^2 = 5\pi r(2 - r)$$

so $V'(r) = 0$ when $r = 0$ and when $r = 2$. When $r = 0$ and $r = 3$, $V = 0$. Thus, $r = 2$ must be the global maximum. So the largest cylinder will have dimensions

$$r = 2 \qquad h = \frac{5}{3}$$

[See Section 9-2.]

PROBLEM 9-24 Consider a right triangle with two sides lying on the coordinate axes, with the hypotenuse passing through (4, 3). Find the minimum area such a triangle could enclose (see Figure 9-18).

Solution: To find the area, you must first find the length of the base. The line through $(0, b)$ and $(4, 3)$ is

$$y = \frac{3 - b}{4}x + b$$

This line has x intercept $(4b/(b - 3), 0)$. Thus, the area of the triangle is

$$A = \tfrac{1}{2}(\text{base})(\text{height})$$

$$= \frac{1}{2}\left(\frac{4b}{b - 3}\right) b$$

$$= \frac{2b^2}{b - 3}$$

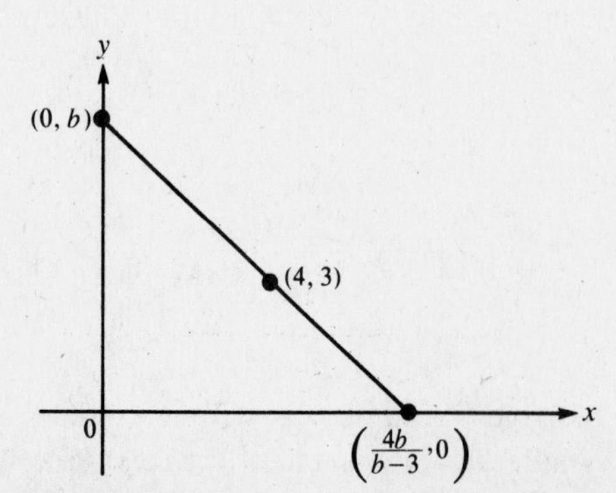

Figure 9-18
Problem 9-24

The variable b is restricted to $b \geqslant 3$. You find that

$$A'(b) = \frac{2b(b-6)}{(b-3)^2}$$

so $A'(b) = 0$ when $b = 0$ or $b = 6$. Thus, $b = 6$ is the point of interest. Evidently, A increases for $b > 6$, and A decreases for $3 < b < 6$. Thus, $b = 6$ must yield a minimum for A. The smallest area for such a triangle is then $A(6) = 24$. [See Section 9-2.]

PROBLEM 9-25 Consider circles that have centers on the positive x axis and pass through the point $(0, a)$ where $a > 0$. Among such circles, what is the center $(x, 0)$ of the one that maximizes the ratio of x to the area of the circle (Figure 9-19)?

Solution: The quantity to be maximized is the ratio R:

$$R = \frac{x}{\pi r^2}$$

But x and r are related by the equation

$$a^2 + x^2 = r^2$$

Thus,

$$R = \frac{x}{\pi(a^2 + x^2)}$$

The only restriction on x is $x > 0$. Searching for critical points,

$$R'(x) = \frac{a^2 - x^2}{\pi(a^2 + x^2)^2}$$

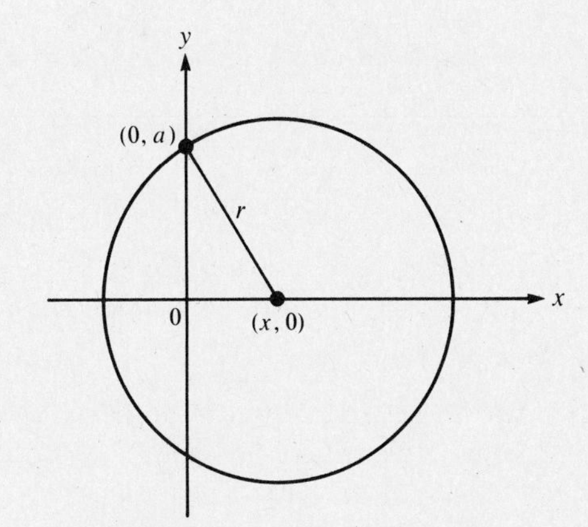

Figure 9-19
Problem 9-25

so $R'(x) = 0$ when $x = a$. Because $R'(x) > 0$ for $0 < x < a$ and $R'(x) < 0$ for $x > a$, you verify that $x = a$ yields the maximum value for R. [See Section 9-2.]

PROBLEM 9-26 What positive number gives the least sum when added to its reciprocal?

Solution: You want to minimize the sum

$$S = x + (1/x)$$

where $x \geqslant 0$. Differentiating,

$$S'(x) = 1 - (1/x^2)$$

so $S'(x) = 0$ when $x = \pm 1$. The value of interest in the interval is $x = 1$. Since S is decreasing for $0 < x < 1$ and increasing for $x > 1$; $x = 1$ is the minimizing number. [See Section 9-2.]

PROBLEM 9-27 Find the dimensions of the isosceles triangle of largest area that can be inscribed in a circle of radius a (Figure 9-20).

Solution: The area of the triangle is given by

$$A = \tfrac{1}{2}(\text{height})(\text{base})$$

with $h = a + y = a + \sqrt{a^2 - x^2}$, $0 < x < a$, and $b = 2x$. You want to maximize

$$A = x(a + \sqrt{a^2 - x^2}) = ax + x\sqrt{a^2 - x^2}$$

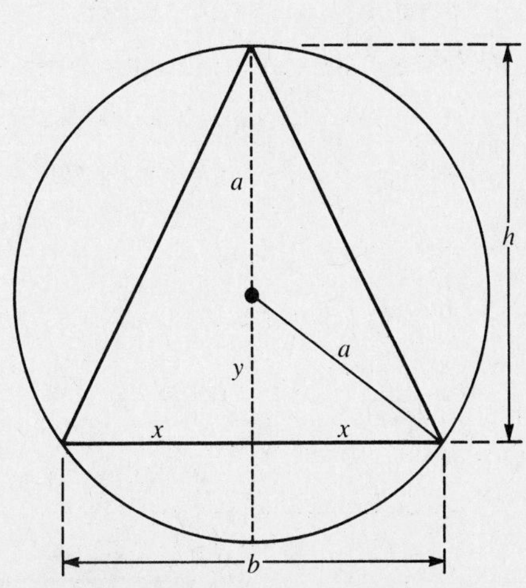

Figure 9-20
Problem 9-27

Differentiating, setting equal to zero, and performing the necessary algebra, you obtain

$$A' = a + (a^2 - x^2)^{1/2} - x^2(a^2 - x^2)^{-1/2} = 0$$

$$a(a^2 - x^2)^{1/2} + a^2 - x^2 - x^2 = 0$$

$$a(a^2 - x^2)^{1/2} = 2x^2 - a^2$$

$$a^2(a^2 - x^2) = 4x^4 - 4a^2x^2 + a^4$$

$$4x^4 - 3a^2x^2 = 0$$

$$x^2(4x^2 - 3a^2) = 0$$

But $x \neq 0$, so $4x^2 = 3a^2$ and $x = a\sqrt{3}/2$. A check of the expression for A' will show that $A' > 0$ for $x < a\sqrt{3}/2$, and $A' < 0$ for $x > a\sqrt{3}/2$, so you do indeed have maximum area when $x = a\sqrt{3}/2$; i.e., when

$$h = a + \left(a^2 - \frac{3a^2}{4}\right)^{1/2} = \frac{3a}{2} \qquad b = 2\left(\frac{\sqrt{3}}{2}a\right) = a\sqrt{3}$$

[See Section 9-2.]

PROBLEM 9-28 A wire of length 100 in. is to be cut into two pieces. One piece is to be bent into a square, the other into a circle. (See Figure 9-21.) Where should the cut be made if the sum of the two areas is to be a minimum?

Solution: Let x be the length of the piece of wire to be bent into a circle of radius r. Then $100 - x$ is the length to be bent into a square of side s. Note that this means that the perimeter of each figure, and hence the area, can be expressed in terms of x:

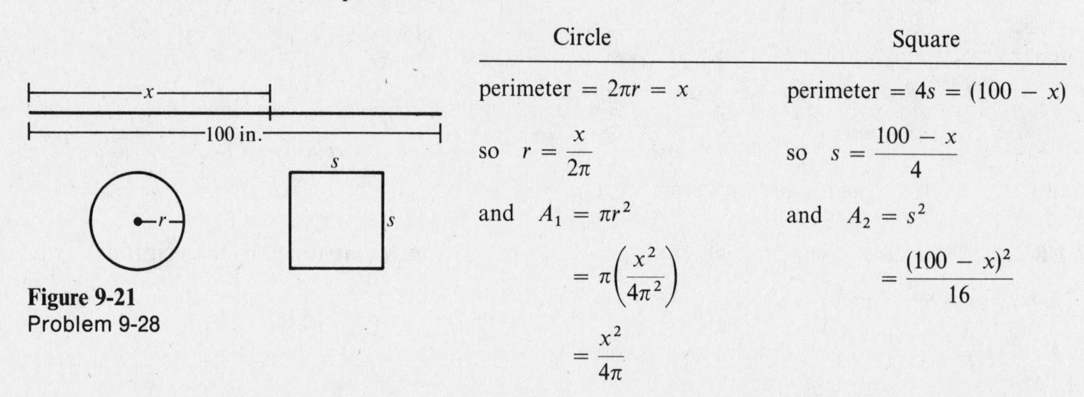

Circle	Square
perimeter $= 2\pi r = x$	perimeter $= 4s = (100 - x)$
so $r = \dfrac{x}{2\pi}$	so $s = \dfrac{100 - x}{4}$
and $A_1 = \pi r^2$	and $A_2 = s^2$
$= \pi\left(\dfrac{x^2}{4\pi^2}\right)$	$= \dfrac{(100 - x)^2}{16}$
$= \dfrac{x^2}{4\pi}$	

Figure 9-21
Problem 9-28

Total area, then, is given by $A = A_1 + A_2$:

$$A = \frac{x^2}{4\pi} + \frac{(100 - x)^2}{16}$$

You differentiate, set equal to zero, and solve for x:

$$A' = \frac{2x}{4\pi} - \frac{2(100 - x)}{16} = 0$$

$$4x - \pi(100 - x) = 0$$

$$x(4 + \pi) = 100\pi$$

$$x = \frac{100\pi}{4 + \pi}$$

If A' is simplified and rewritten, you get

$$A' = \frac{x}{2\pi} - \frac{100 - x}{8} = \frac{4x - 100\pi + \pi x}{8\pi}$$

$$= \frac{x(4 + \pi) - 100\pi}{8\pi} = \left(\frac{4 + \pi}{8\pi}\right)\left(x - \frac{100\pi}{4 + \pi}\right)$$

It is easy to see that A in fact has a minimum at $x = 100\pi/(4 + \pi)$: $A' < 0$ for $x < 100\pi/(4 + \pi)$, $A' > 0$ for $x > 100\pi/(4 + \pi)$.

[See Section 9-2.]

PROBLEM 9-29 A rectangular warehouse that is to have two rectangular rooms separated by one interior wall must have 5000 square feet of floor space (Figure 9-22). The cost of exterior walls is \$150 per linear foot, and the cost of interior walls is \$90 per linear foot. Find the dimensions of the least expensive warehouse.

Solution: Let the length be x ft and width be y ft, with the interior wall x ft long. Then $xy = 5000$ (total area), and the quantity to be minimized is

$$C = 150(2x + 2y) + 90x$$

or in terms of x only

$$C = 390x + 300\left(\frac{5000}{x}\right)$$

From this you have

$$C' = 390 - 1\,500\,000x^{-2} = 0$$

$$x^2 = (1\,500\,000/390)$$

so $x = 100\sqrt{5/13}$ and hence $y = 50\sqrt{13/5}$.

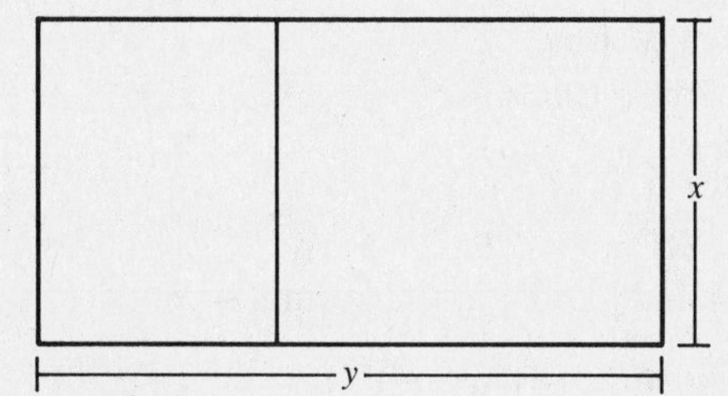

Figure 9-22
Problem 9-29

[See Section 9-2.]

PROBLEM 9-30 A rectangular building is located on an intersection of two streets, as shown in Figure 9-23. The city wants to build a straight-line sidewalk from one street to the other, so that the sidewalk just touches the corner of the building. Show that the angle θ, which minimizes the distance d along the walk, is given by $\tan\theta = 4/3$.

Solution: From Figure 9-23 you have

$$d = d_1 + d_2 = 64 \csc\theta + 27 \sec\theta$$

$$\frac{d}{d\theta}(d) = -64 \csc\theta \cot\theta + 27 \sec\theta \tan\theta = 0$$

Using the identities $\csc\theta = \dfrac{1}{\sin\theta}$; $\cot\theta = \dfrac{\cos\theta}{\sin\theta}$; $\sec\theta = \dfrac{1}{\cos\theta}$; $\tan\theta = \dfrac{\sin\theta}{\cos\theta}$; you have

$$\frac{64\cos\theta}{\sin^2\theta} = \frac{27\sin\theta}{\cos^2\theta}$$

$$64\cos^3\theta = 27\sin^3\theta$$

$$\tan^3\theta = 64/27$$

so that $\tan\theta = 4/3$.

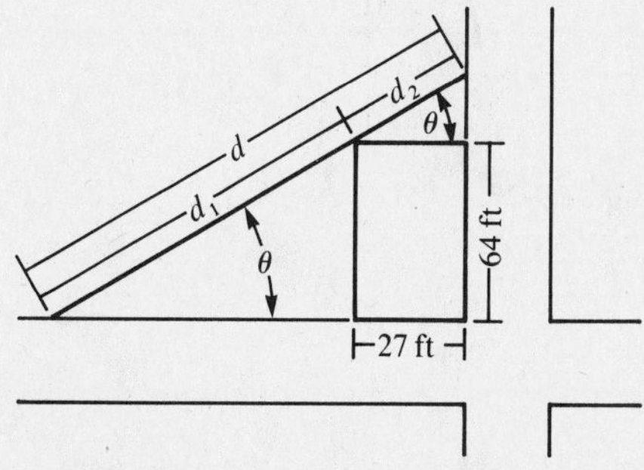

Figure 9-23
Problem 9-30

[See Section 9-2.]

PROBLEM 9-31 For what angle α does the isosceles trapezoid in Figure 9-24 possess a maximum area?

Solution: The area of a trapezoid is given by

$$A = \tfrac{1}{2}h(b_1 + b_2)$$

where b_1 and b_2 are the lengths of the parallel sides, and h is the distance between them. Then in the trapezoid of Figure 9-24 you have $h = a \sin\alpha$, $b_1 = a$, and $b_2 = a + 2a\cos\alpha$, so that

$$A = \tfrac{1}{2}a \sin\alpha(a + a + 2a\cos\alpha)$$

$$= a^2(\sin\alpha + \sin\alpha\cos\alpha)$$

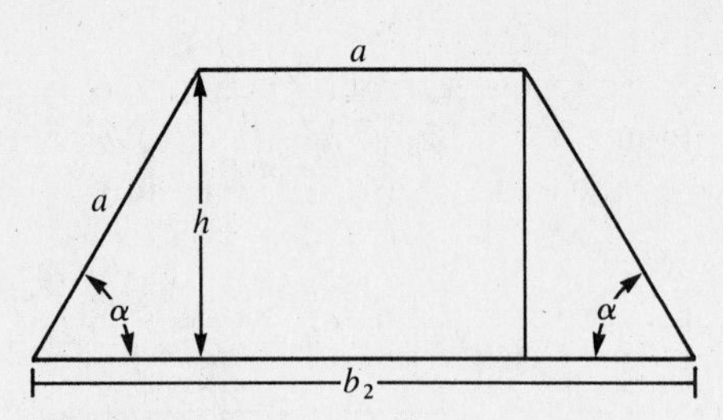

Figure 9-24
Problem 9-31

To maximize A, set $A' = 0$:

$$A' = a^2(\cos\alpha - \sin^2\alpha + \cos^2\alpha) = 0$$

Setting $\sin^2\alpha = 1 - \cos^2\alpha$:

$$\cos\alpha - 1 + \cos^2\alpha + \cos^2\alpha = 0$$

$$(2\cos\alpha - 1)(\cos\alpha + 1) = 0$$

$$\cos\alpha = \tfrac{1}{2} \qquad \cos\alpha = -1$$

Clearly, $\cos\alpha = -1$ yields $\alpha = \pi$ radians, or 180°, which would give $h = 0$ and $A = 0$. Maximum area then comes from $\cos\alpha = \frac{1}{2}$, or $\alpha = 60°$.

[See Section 9-2.]

PROBLEM 9-32 On $y = x^3$, find two points whose abscissas differ by two, such that the line joining them has minimum slope.

Solution: Let $P_1 = (x_1, y_1)$ and $P_2 = (x_2, y_2)$ be two points on the graph of $y = x^3$ (Figure 9-25). Then

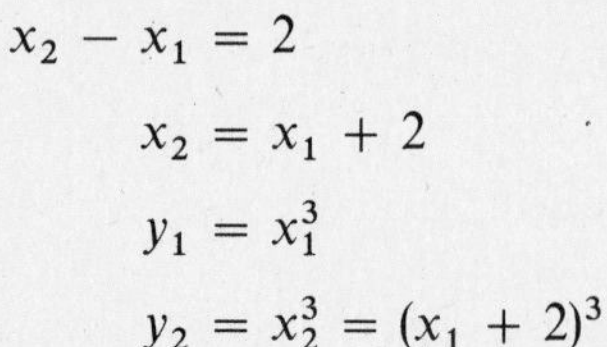

$$x_2 - x_1 = 2$$

$$x_2 = x_1 + 2$$

$$y_1 = x_1^3$$

$$y_2 = x_2^3 = (x_1 + 2)^3$$

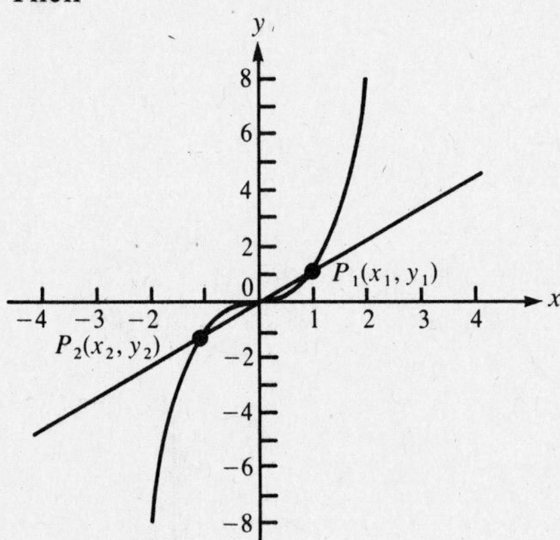

Figure 9-25
Problem 9-32

and the slope m of the line joining the two points is given by

$$m = \frac{y_2 - y_1}{x_2 - x_1} = \frac{(x_1 + 2)^3 - x_1^3}{2}$$

$$= \frac{1}{2}(6x_1^2 + 12x_1 + 8)$$

To find the smallest m, we set dm/dx_1 equal to zero and solve:

$$\frac{dm}{dx_1} = \frac{1}{2}(12x_1 + 12) = 0$$

so $x_1 = -1$. This gives the minimum because the second derivative is positive. The two points are then $(-1, -1)$ and $(1, 1)$.

[See Section 9-2.]

PROBLEM 9-33 What is the maximum possible area of a rectangle whose base lies on the x axis, with its two upper vertices on the graph of $y = 4 - x^2$ (Figure 9-26)?

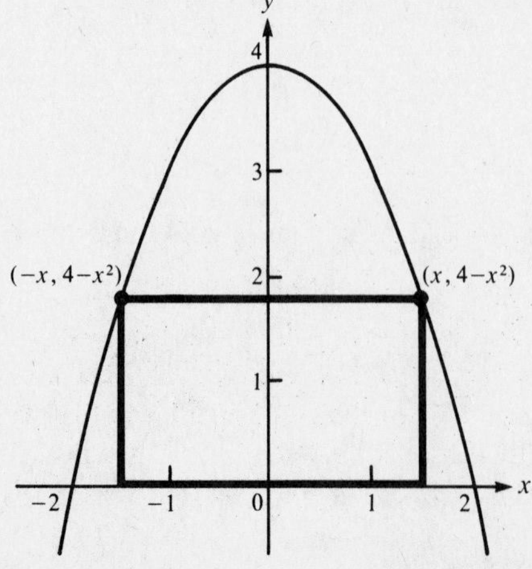

Figure 9-26
Problem 9-33

Solution: Let the two upper vertices be given by $(x, 4 - x^2)$ and $(-x, 4 - x^2)$. Then the area A is

$$A = 2x(4 - x^2) = 8x - 2x^3$$

where $0 < x < 2$. Maximum area will occur when $A' = 0$:

$$A' = 8 - 6x^2$$

so $A' = 0$ when $x = 2\sqrt{3}/3$. Thus, $y = 4 - x^2 = 4 - 4/3 = 8/3$. Since $A'' < 0$, you have maximum area

$$A = 2 \cdot \frac{2\sqrt{3}}{3} \cdot \frac{8}{3} = \frac{32\sqrt{3}}{9}$$

[See Section 9-2.]

PROBLEM 9-34 Find the dimensions of the rectangle of greatest area that can be inscribed in a semicircle of radius 2 (Figure 9-27).

Solution: The quantity to be maximized is the area A:

$$A = hl$$

The variables, h and l, are related by the Pythagorean Theorem:

$$(l/2)^2 + h^2 = 4$$

which becomes

$$l = 2(4 - h^2)^{1/2}$$

Figure 9-27
Problem 9-34

So A can be expressed as

$$A = 2h(4 - h^2)^{1/2}$$

where $0 \leqslant h \leqslant 2$. Now differentiate:

$$A'(h) = (8 - 4h^2)(4 - h^2)^{-1/2}$$

so $A'(h) = 0$ when $h = \sqrt{2}$. Evidently, this yields a maximum value for A, so the dimensions are

$$h = \sqrt{2} \qquad l = 2\sqrt{2}$$

[See Section 9-2.]

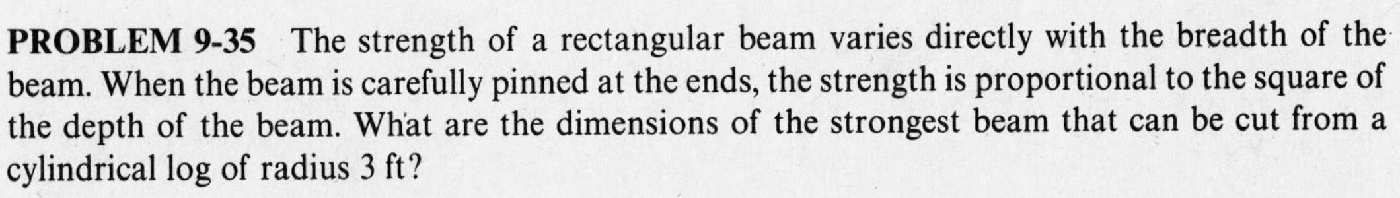

PROBLEM 9-35 The strength of a rectangular beam varies directly with the breadth of the beam. When the beam is carefully pinned at the ends, the strength is proportional to the square of the depth of the beam. What are the dimensions of the strongest beam that can be cut from a cylindrical log of radius 3 ft?

Solution: Look at the log in cross-section (as in Figure 9-28); the quantity to be maximized is the strength S:

$$S = kxy^2$$

where k is the constant of proportionality. Since x and y are related by

$$(x/2)^2 + (y/2)^2 = 9$$

you have

$$y^2 = 36 - x^2$$

and

$$S = kx(36 - x^2) = 36kx - kx^3$$

for $0 \leqslant x \leqslant 6$. Differentiate to find

$$S'(x) = 36k - 3kx^2$$

Figure 9-28
Problem 9-35

so $S'(x) = 0$ when $x = 2\sqrt{3}$. You can easily verify that this is indeed the maximum:

$$x = 2\sqrt{3}\text{ ft} \qquad y = 2\sqrt{6}\text{ ft}$$

[See Section 9-2.]

PROBLEM 9-36 A ray of light travels from point A to point B, where A and B are in different media. Suppose that the common boundary of the two media is a plane. Fermat's principle in optics states that the light will travel along the path for which the time of travel is a minimum. Show that if v_1 and v_2 are the velocities of light in media 1 and 2, respectively, then the light will travel a path that crosses the boundary in accordance with Snell's law:

$$\frac{\sin \theta_1}{\sin \theta_2} = \frac{v_1}{v_2}$$

where θ_1 and θ_2 are the angles noted in Figure 9-29.

Solution: Assume the positions of A and B as defined by the constants a, b, and c, as in Figure 9-29. Then the quantity to be minimized is T, the time of travel of the light ray:

$$\begin{aligned} T &= (1/v_1) \times (\text{distance traveled in medium 1}) \\ &\quad + (1/v_2) \times (\text{distance traveled in medium 2}) \\ &= (1/v_1)(a^2 + x^2)^{1/2} + (1/v_2)(b^2 + (c - x)^2)^{1/2} \end{aligned}$$

where x is in $[0, c]$. Then

$$T'(x) = \frac{x}{v_1}(a^2 + x^2)^{-1/2} - \frac{(c - x)}{v_2}(b^2 + (c - x)^2)^{-1/2}$$

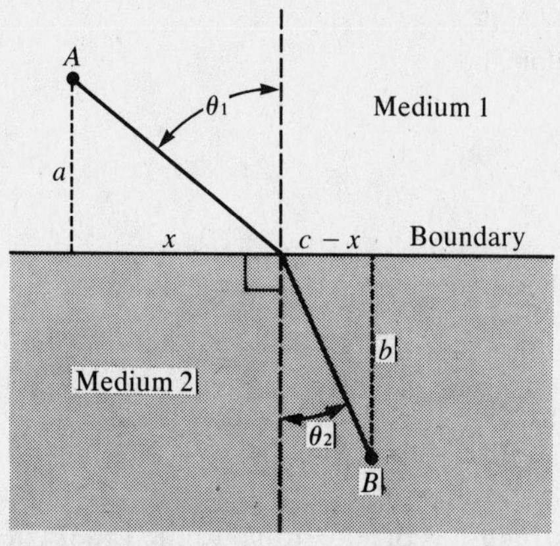

Figure 9-29
Problem 9-36

so $T'(x) = 0$ when

$$\frac{x}{v_1}(a^2 + x^2)^{-1/2} = \frac{(c - x)}{v_2}(b^2 + (c - x)^2)^{-1/2}$$

But an examination of the diagram reveals that

$$x(a^2 + x^2)^{-1/2} = \sin\theta_1$$

$$(c - x)(b^2 + (c - x)^2)^{-1/2} = \sin\theta_2$$

Thus, you see that $T'(x) = 0$ when

$$\frac{\sin\theta_1}{v_1} = \frac{\sin\theta_2}{v_2}$$

or

$$\frac{v_1}{v_2} = \frac{\sin\theta_1}{\sin\theta_2}$$

Apparently, this gives the minimum value for T.

[See Section 9-2.]

Supplementary Exercises

For each of the following functions (Problems 9-37 through 9-87), find the points (if they exist) at which the maximum and minimum values are attained on the given interval:

9-37 $f(x) = 2x^3 - 9x^2 + 12x + 12$ on $[0, 2]$

9-38 $f(x) = x^2 + 54x^{-1}$ for $x \geqslant 1$

9-39 $f(x) = \sqrt{x + 1} - \sqrt{x}$ on $[0, 1]$

9-40 $f(x) = x^2(x^2 + 1)^{-1}$ for $x \leqslant 2$

9-41 $f(x) = x^2(x^2 - 1)^{-1}$ for $-1 < x < 1$

9-42 $f(x) = xe^x - e^x$ for $x \geqslant -1$

9-43 $f(x) = 2x^6 - 9x^4 + 12x^2 - 1$ on $[-2, 2]$

9-44 $f(x) = \sin x - \cos x$ on $[0, 2\pi]$

9-45 $f(x) = \sqrt{4 - x^2}$ on $[-2, 2]$

9-46 $f(x) = x^{2/3}$ on $[-1, 1]$

9-47 $f(x) = (x - 1)/(x - 2)$ on $[0, 3]$

9-48 $f(x) = (x^3 - 1)(x^3 + 1)^{-1}$ for $x \geqslant 0$

9-49 $f(x) = (x + 1)(x - 2)^2$ on $[-1, 3]$

9-50 $f(x) = x - \sqrt{1 - x^2}$ on $[-1, 1]$

9-51 $f(x) = 3x^4 + 4x^3 + 1$ on $[-2, 1]$

9-52 $f(x) = (x^2 - 1)(x^2 + 1)^{-1}$ for all x

9-53 $f(x) = e^{2x} - e^x$ for all x

9-54 $f(x) = (x^2 - 1)x^{-3}$ on $[-1, 1]$

9-55 $f(x) = x(1 + 3x)^{-1/3}$ on $[-1, 0]$

9-56 $f(x) = x^{-2} \ln x$ for $x \geqslant 1$

9-57 $f(x) = (x - 1)/(x + 1)$ for $x \geqslant 0$

9-58 $f(x) = x^2\sqrt{4x + 3}$ on $[-\frac{3}{4}, 0]$

9-59 $f(x) = (x^2 - 9)(1 - x^2)^{-1}$ for $-1 < x < 1$

9-60 $f(x) = 2x - \tan x$ for all x

9-61 $f(x) = x^2 + 16x^{-2}$ on $[-3, 3]$

9-62 $f(x) = x^4 - 2x^2 + 2$ on $[-2, 2]$

9-63 $f(x) = \sqrt{x} - \sqrt[3]{x}$ on $[0, 1]$

9-64 $f(x) = (3x + 1)(x^2 + x + 3)^{-1}$ on $[-3, 0]$

9-65 $f(x) = xe^{-x}$ for all x

9-66 $f(x) = x^{1/2} + x^{-1/2}$ for $x \geqslant \frac{1}{2}$

9-67 $f(x) = 4x^5 - 5x^4 + 1$ on $[-1, 1]$

9-68 $f(x) = x^2 - x^{3/2}$ on $[0, 4]$

9-69 $f(x) = 2 \sin x - x$ on $[0, \pi/2]$

9-70 $f(x) = x(x^2 + 1)^{-1}$ for all x

9-71 $f(x) = \sqrt{16 - (x - 2)^2}$ on $[-1, 6]$

9-72 $f(x) = x - x^{-2}$ on $[-2, \frac{1}{2}]$

9-73 $f(x) = x \ln x - x$ for $x \geqslant \frac{1}{2}$

9-74 $f(x) = (x - 1)(x + 1)^2$ for all x

9-75 $f(x) = 3 - (x + 1)^4$ for all x

9-76 $f(x) = (x^2 + 2x - 3)x^{-2}$ for $x \geqslant 1$

9-77 $f(x) = e^x + e^{-x}$ on $[-1, 1]$

9-78 $f(x) = x^4 - 2x^3$ on $[-1, 2]$

9-79 $f(x) = \sec x + 2 \cos x$ for $0 < x < \pi/2$

9-80 $f(x) = x + (1/x)$ for $x > 0$

9-81 $f(x) = (3x + 1)\sqrt{4x + 3}$ on $[-\frac{3}{4}, 1]$

9-82 $f(x) = \sqrt{x^2 - x^4}$ on $[-1, 1]$

9-83 $f(x) = x(1 + 3x)^{-2/3}$ on $[-2, -\frac{1}{4}]$

9-84 $f(x) = e^x \sin^2 x$ on $[0, 2\pi]$

9-85 $f(x) = (x - 1)x^{-2}$ for $x \geqslant \frac{1}{2}$

9-86 $f(x) = x + 2 \cos x - 2 \sin x - \cos 2x - 1$ for $-\pi/2 < x < \pi/2$

9-87 $f(x) = \sqrt{4x + 3} - x$ for $x \geqslant 0$

Word Problems

9-88 Find the minimum distance between the point (5, 1) and the parabola $y = -x^2$.

[See Problem 9-20.]

9-89 The strength of a rectangular beam varies directly with the breadth of the beam. When the beam is carelessly pinned at the ends, the strength varies directly with the 3/2 power of the depth of the beam. What are the dimensions of the strongest rectangular beam that can be hewn from a cylindrical log of radius 20 in.?

[See Problem 9-35.]

9-90 A rectangular piece of cardboard 2 ft by 1 ft is to be cut and folded into a box without a top. What should the dimensions be in order that the box enclose the greatest volume?

[See Problem 9-19.]

9-91 The sum of two numbers is 16. Find the numbers if the sum of their cubes is a minimum.

[See Problem 9-26.]

9-92 What are the dimensions of the cone with the largest volume that can be circumscribed about a cylinder of radius 4 and height 5? (The cone with radius r and height h has volume $\frac{1}{3}\pi r^2 h$.)

[See Problem 9-23.]

9-93 A rancher wishes to construct a rectangular corral to enclose 900 ft^2. The corral is to be divided into two parts by a fence parallel to two of the sides. What should the outer dimensions of the corral be if he is trying to conserve fence material? [See Problem 9-29.]

9-94 In a certain town lots are taxed at a rate of \$1 per front foot (i.e., \$1 for every foot of width of the lot) and 50¢ per foot of depth. What are the dimensions of the lot with the greatest area that yields the town \$200 in taxes?

9-95 A certain athlete can swim at a speed 5/13 times her running speed. She must travel from a point on one shore of a canal (no current) to a point 1000 yards upstream on the opposite shore. The canal is straight and 24 yards wide at all points. If she begins her trip by swimming and wants to minimize her total time, where should she emerge from the water? [See Problem 9-17.]

9-96 Find the straight line through the point (8, 18) with positive intercepts such that the sum of the intercepts is a minimum. [See Problem 9-24.]

9-97 What is the largest perimeter of a rectangle that can be inscribed in a semicircle of radius 5? [See Problem 9-34.]

9-98 What number most exceeds its square?

9-99 A rectangle is inscribed in an isosceles triangle whose sides have lengths 5, 5, and 6. One side of the rectangle lies along the base (the unequal side) of the triangle. What is the greatest area that such a rectangle can enclose?

9-100 A track with perimeter 400 meters encircles a rectangular region with semicircular regions attached on two opposite sides. What is the largest area the rectangular region might contain?

9-101 For what x is the slope of the tangent line at x to the curve $y = x^{1/2} + x^{-1/2}$ a maximum?

9-102 You want to inscribe a cone inside another cone. The outer cone has height 6 and radius (of the base) 4. The inner cone is inscribed so that its apex lies on the base of the outer cone. The base of the inner cone is parallel to the base of the outer cone. The axes of the two cones are collinear. What should the height of the inner cone be in order that it contain the largest possible volume? (The volume of a cone with height h and radius r is $\frac{1}{3}\pi r^2 h$.)

9-103 A silo consists of a cylinder with a hemispherical top. Find the dimensions of the silo with fixed volume $V = 18\pi$ that has minimal surface area. Exclude the floor. [See Problem 9-14.]

9-104 A Norman window consists of a rectangle surmounted by a semicircle. Find the dimensions of the window of perimeter 10 that has the largest area.

9-105 Find the dimensions of the right circular cone of maximal volume that can be inscribed in a sphere of radius 1. [See Problem 9-16.]

9-106 Find the dimensions of the isosceles triangle with fixed perimeter $p = 6$ that has maximal area.

9-107 A poster is to contain a printed area of 150 in.^{2} with a clear 3-in. border on the top and bottom and a clear 2-in. border on the sides. Find the dimensions of the poster with the smallest perimeter. [See Problem 9-18.]

9-108 The girth of a shipping carton is the perimeter of an end. Shipping restrictions require that the sum of the girth and length be no more than 100 in. Find the dimensions of the package with square ends that has the greatest surface area. [See Problem 9-21.]

9-109 In the box of Problem 9-108 assume a cylindrical package (end is a circle). Find the dimensions that will give the largest volume. [See Problem 9-22.]

9-110 Find two positive real numbers whose sum is 40 and whose product is a maximum.

9-111 What is the shape of the rectangle with fixed perimeter p that encloses the greatest area?

Solutions to Supplementary Exercises

(9-37) minimum at $(0, 12)$; maximum at $(1, 17)$

(9-38) min. at $(3, 27)$; no max.

(9-39) min. $(1, \sqrt{2} - 1)$; max. $(0, 1)$

(9-40) min. $(0, 0)$; no max.

(9-41) no min.; max. $(0, 0)$

(9-42) min. $(0, -1)$; no max.

(9-43) min. $(0, -1)$; max. $(-2, 31)$ and $(2, 31)$

(9-44) min. $(7\pi/4, -\sqrt{2})$; max. $(3\pi/4, \sqrt{2})$

(9-45) min. $(-2, 0)$ and $(2, 0)$; max. $(0, 2)$

(9-46) min. $(0, 0)$; max. $(-1, 1)$ and $(1, 1)$

(9-47) no min.; no max.

(9-48) min. $(0, -\frac{1}{3})$; no max.

(9-49) min. $(-1, 0)$ and $(2, 0)$; max. $(0, 4)$ and $(3, 4)$

(9-50) min. $(-\frac{1}{2}\sqrt{2}, -\sqrt{2})$; max. $(1, 1)$

(9-51) min. $(-1, 0)$; max. $(-2, 17)$

(9-52) min. $(0, -1)$; no max.

(9-53) min. $(-\ln 2, -\frac{1}{4})$; no max.

(9-54) no min.; no max.

(9-55) no min.; no max.

(9-56) min. $(1, 0)$; max. $(\sqrt{e}, 1/(2e))$

(9-57) min. $(0, -1)$; no max.

(9-58) min. $\left(\frac{-3}{4}, 0\right)$ and $(0, 0)$; max. $\left(\frac{-3}{5}, \frac{9\sqrt{3}}{125}\right)$

(9-59) no min.; max. $(0, -9)$

(9-60) no min.; no max.

(9-61) min. $(2, 8)$ and $(-2, 8)$; no max.

(9-62) min. $(1, 1)$ and $(-1, 1)$; max. $(2, 10)$ and $(-2, 10)$

(9-63) min. $((2/3)^6, -4/27)$; max. $(0, 0)$ and $(1, 0)$

(9-64) min. $(-2, -1)$; max. $(0, \frac{1}{3})$

(9-65) no min.; max. $(1, 1/e)$

(9-66) min. $(1, 2)$; no max.

(9-67) min. $(-1, -8)$; max. $(0, 1)$

(9-68) min. $(9/16, -27/256)$; max. $(4, 8)$

(9-69) min. $(0, 0)$; max. $(\pi/3, \sqrt{3} - (\pi/3))$

(9-70) min. $(-1, -\frac{1}{2})$; max. $(1, \frac{1}{2})$

(9-71) min. $(6, 0)$; max. $(2, 4)$

(9-72) no min.; max. $(-\sqrt[3]{2}, -\frac{3}{2}\sqrt[3]{2})$

(9-73) min. $(1, -1)$; no max.

(9-74) no min.; no max.

(9-75) no min.; max. $(-1, 3)$

(9-76) min. $(1, 0)$; max. $(3, \frac{4}{3})$

(9-77) min. $(0, 2)$; max. $(-1, e + (1/e))$ and $(1, e + (1/e))$

(9-78) min. $(3/2, -27/16)$; max. $(-1, 3)$

(9-79) min. $(\pi/4, 2\sqrt{2})$; no max.

(9-80) min. $(1, 2)$; no max.

(9-81) min. $\left(\frac{-11}{18}, \frac{-5\sqrt{5}}{18}\right)$; max. $(1, 4\sqrt{7})$

(9-82) min. $(-1, 0)$ and $(0, 0)$ and $(1, 0)$; max. $(\frac{1}{2}\sqrt{2}, \frac{1}{2})$ and $(-\frac{1}{2}\sqrt{2}, \frac{1}{2})$

(9-83) no min.; max. $(-1, -\frac{1}{2}\sqrt[3]{2})$ and $(-\frac{1}{4}, -\frac{1}{2}\sqrt[3]{2})$

(9-84) min. $(0, 0)$ and $(\pi, 0)$ and $(2\pi, 0)$; max. $(2\pi + \arctan(-2), \frac{4}{5}e^{2\pi + \arctan(-2)})$

(9-85) min. $(\frac{1}{2}, -2)$; max. $(2, \frac{1}{4})$

(9-86) no min.; max. $\left(\frac{-\pi}{3}, \frac{1}{2} + \sqrt{3} - \frac{\pi}{3}\right)$

(9-87) no min.; max. $\left(\frac{1}{4}, \frac{7}{4}\right)$

(9-88) $2\sqrt{5}$ (at $(1, -1)$)

(9-89) breadth $= 8\sqrt{10}$ in.; depth $= 8\sqrt{15}$ in.

(9-90) depth $= \frac{1}{2} - (\sqrt{3}/6)$ ft; base $= \sqrt{3}/3$ ft by $1 + (\sqrt{3}/3)$ ft

(9-91) 8 and 8

(9-92) radius $= 6$; height $= 15$

(9-93) $10\sqrt{6}$ ft by $15\sqrt{6}$ ft

(9-94) 100 ft wide by 200 ft deep

(9-95) 10 yards upstream

(9-96) $y = 30 - (3/2)x$

(9-97) $10\sqrt{5}$

(9-98) $\frac{1}{2}$

(9-99) 6

(9-100) $20\,000/\pi$ m^2

(9-101) $x = 3$

(9-102) $h = 2$

(9-103) $r = 3$; $h = 0$ (i.e., a hemisphere of radius 3)

(9-104) $r = 10/(4 + \pi)$; $h = 10/(4 + \pi)$

(9-105) $r = (2\sqrt{2})/3$; $h = 4/3$

(9-106) equilateral triangle with each side of length 2

(9-107) width $= 5\sqrt{6} + 4$ in.; length $= 5\sqrt{6} + 6$ in.

(9-108) 100/7 in. by 100/7 in. by 300/7 in.

(9-109) radius $r = 100/3\pi$ in.; length $l = 100/3$ in.

(9-110) 20 and 20

(9-111) square (each side has length $p/4$)

10 RELATED RATES

THIS CHAPTER IS ABOUT

☑ **Solving Related Rates Problems**
☑ **Solving Variants of Related Rates Problems**
☑ **Three Important Steps to Remember**

In this chapter, you'll learn to solve **related rates** problems. These problems examine the relationships among the rates of change of various quantities.

10-1. Solving Related Rates Problems

A. Search for an equation that relates the quantities!

In a basic related rates problem, you'll be asked to find the rate of change of one quantity that is linked to the rate of change of some other quantity. You can express the relationship between these quantities in an equation. You will often find it helpful to draw a picture of the situation. You can then develop an equation that relates the dimensions in the picture.

B. Differentiate your equation!

To obtain an equation relating the rates of change (derivatives) of the quantities, differentiate your equation with respect to time. You will use implicit differentiation, so apply the chain rule carefully. Your new equation may also involve the original variables. You can now substitute the values given in the problem into your equation and solve for the rate of change that answers the original question.

EXAMPLE 10-1: A snowball is melting at a rate of 2 ft^3 per hour. If it remains spherical, at what rate is the radius changing when the radius of the snowball is 20 in?

Solution: First, identify the two quantities whose rates of change are related. In this case you are asked for the rate of change of the radius r and you are given the rate of change of the volume V (note that the units, ft^3 per hour, tell you that this is the rate of change of volume). You want to find an equation that relates these quantities. Because these are dimensions of a sphere, you know that

$$V = \frac{4}{3}\pi r^3$$

(See Figure 10-1.) When you differentiate this equation relating V and r, you'll produce an equation that relates their rates of change. Differentiating with respect to time t, you get

$$\frac{dV}{dt} = 4\pi r^2 \cdot \frac{dr}{dt}$$

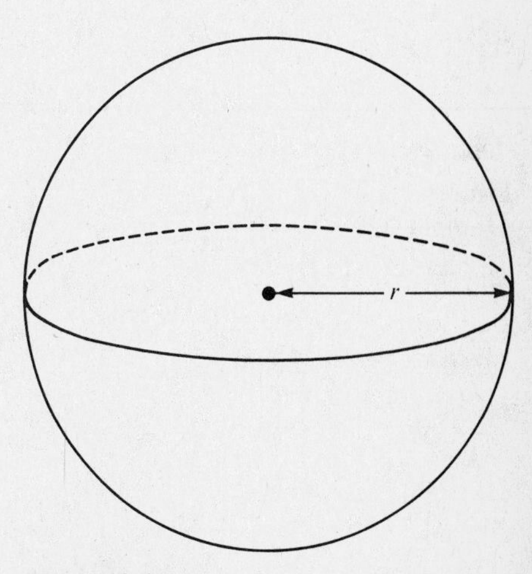

Figure 10-1
Example 10-1

You want to determine the value of dr/dt when $r = 20$ in. (5/3 ft). To find this value, solve the equation for dr/dt when $r = 5/3$ and $dV/dt = -2$ (the minus sign is required because the volume is decreasing):

$$-2 = 4\pi\left(\frac{5}{3}\right)^2 \cdot \frac{dr}{dt}$$

$$\frac{dr}{dt} = \frac{-9}{50\pi}$$

The radius is changing at the rate of $-9/(50\pi)$ feet per hour.

Before you leave this example, look at it one more time with your main focus on units. In the initial equation

$$V = (\text{constant}) \times r^3$$

and the units should be clear:

$$\text{ft}^3 = \text{ft}^3$$

Remember that when you differentiate, you are dividing by Δt (a measure of time) and taking a limit as Δt approaches zero. That means that the units have become ft^3/h on the left; what happened on the right? You have

$$\frac{dV}{dt}\frac{\text{ft}^3}{\text{h}} = (\text{constant}) \times r^2 \cdot \frac{dr}{dt}$$

The units of r^2 are ft^2, and the units of dr/dt are ft/h. Thus, everything checks:

$$\frac{\text{ft}^3}{\text{h}} = \text{ft}^2 \cdot \frac{\text{ft}}{\text{h}} = \frac{\text{ft}^3}{\text{h}}$$

This amount of detail is not necessary in every problem, but your understanding of the examples will increase if you stay aware of the units.

EXAMPLE 10-2: A child is standing still and flying a kite. The kite remains at an altitude of 30 ft above the child's hands while traveling parallel to the ground at a rate of 10 ft/s. When the kite is 50 ft away from the child, how fast is the kite string leaving the child's hand?

Solution: You are asked to find the rate at which the kite string leaves the child's hand and you are given the lateral velocity of the kite. First, draw a picture (see Figure 10-2). The side labeled y must not be labeled 50, otherwise the variables will not vary. Indeed, you are asked to find dy/dt when $y = 50$. Note that the rate of change of x is the lateral velocity of the kite (10 ft/s), and the rate of change of y is the rate at which the kite string leaves the child's hand.

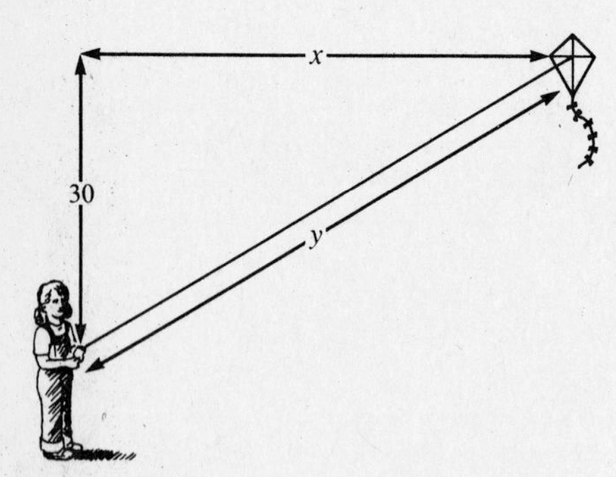

Figure 10-2
Example 10-2

From the Pythagorean theorem, you can write an equation relating x and y:

$$x^2 + 900 = y^2$$

Differentiating with respect to time t, you get

$$2x\frac{dx}{dt} = 2y\frac{dy}{dt}$$

$$x\frac{dx}{dt} = y\frac{dy}{dt}$$

This equation will allow you to solve for dy/dt: y is given (50 ft), dx/dt is given (10 ft/s), and you can find x from the original equation:

$$x^2 + 900 = 50^2$$

$$x^2 = 1600$$

$$x = 40$$

Now substitute:

$$(40)(10) = 50\frac{dy}{dt}$$

$$\frac{dy}{dt} = 8\frac{\text{ft}}{\text{s}}$$

The kite string leaves the child's hand at 8 ft/s. You should verify the units.

10-2. Solving Variants of Related Rates Problems

Most related rates problems are more complicated than those covered in Section 10-1. You will discover that they all have the following in common:

(1) You can write one or more equations relating the quantities.
(2) You can differentiate to find the relationships among the rates of change of the quantities.

A. Quantities related by more than one equation

You will encounter related rates problems where the quantities are related by several equations. You may be able to manipulate these equations to obtain a single equation, but this is often unnecessary. You should differentiate all of the equations, and then solve for the quantity of interest.

EXAMPLE 10-3: An ice cube is melting. When the volume of the cube is 8 cm^3, it is melting at a rate of 4 cm^3/s. Find the rate of change of the surface area of the cube at that moment.

Solution: You are given the rate of change of the volume of the cube and are asked to find the rate of change of the surface area. Accordingly, you look for equations relating the surface area to the volume. The volume V of a cube with sides of length x is x^3. Because there are six sides, each with area x^2, the surface area S of a cube is $S = 6x^2$. Rather than solve to find a single equation relating S and V by eliminating the variable x, differentiate both equations with respect to time:

$$\frac{dV}{dt} = 3x^2 \cdot \frac{dx}{dt}$$

$$\frac{dS}{dt} = 12x\frac{dx}{dt}$$

When the volume of the cube is 8 cm^3, x must be 2 cm. You want to find dS/dt when $x = 2$, given that $dV/dt = -4$ at that time. So,

$$-4 = 3(2)^2 \cdot \frac{dx}{dt}$$

When $x = 2$, $dx/dt = -1/3$. Thus,

$$\frac{dS}{dt} = 12(2)\left(-\frac{1}{3}\right) = -8\frac{\text{cm}^2}{\text{s}}$$

B. Problems that involve more than two related quantities

The rate of change you seek may depend on the rates of change of several quantities. As before, you find the equation(s) relating these quantities and differentiate.

EXAMPLE 10-4: A car is 30 miles north of town, heading north at 25 miles per hour. At the same time, a truck is 40 miles east of town, traveling east at 50 mi/h. At what rate is the distance between the two vehicles changing?

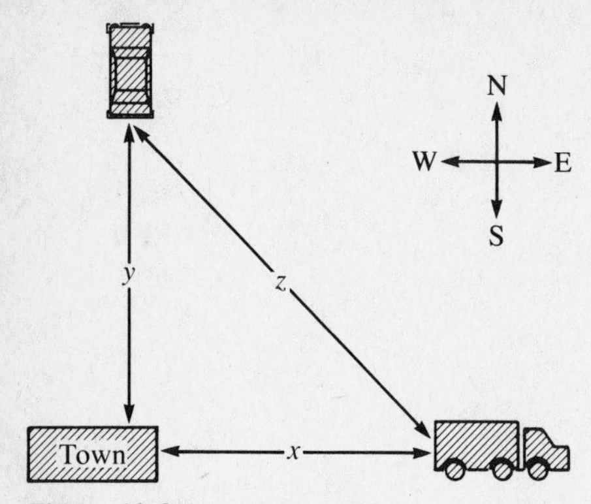

Figure 10-3
Example 10-4

Solution: Begin by drawing a diagram (see Figure 10-3). From the Pythagorean Theorem, you know that

$$x^2 + y^2 = z^2$$

Differentiating with respect to time,

$$2x\frac{dx}{dt} + 2y\frac{dy}{dt} = 2z\frac{dz}{dt}$$

or simply

$$x\frac{dx}{dt} + y\frac{dy}{dt} = z\frac{dz}{dt}$$

You want to find dz/dt when $x = 40$, $y = 30$, $dx/dt = 50$, and $dy/dt = 25$. At that time,

$$z = \sqrt{x^2 + y^2} = \sqrt{40^2 + 30^2} = 50$$

The equation relating the rates of change then becomes

$$(40)(50) + (30)(25) = (50)\frac{dz}{dt}$$

so $dz/dt = 55$. The vehicles are separating at a rate of 55 mi/h.

10-3. Three Important Steps to Remember

A. Draw the picture carefully!

Draw pictures that accurately include all of the variables in the problem.

B. Don't label variables as constants!

Some dimensions given in the problem remain fixed as time passes. Label these as constants in the diagram. Other information defines the point in time at which you are to calculate the rate of change. Don't label these dimensions as constants, as they vary with time.

C. Translate to a mathematical problem!

Once you've drawn the diagram, translate the information given in the problem to statements about the variables in the diagram. Carefully pose the question asked by the problem as a question about the variables or their rates of change.

D. Determine where to evaluate the equation relating the rates of change!

You must give careful consideration to your equation before you can decide where to evaluate.

EXAMPLE 10-5: A worker holds one end of a rope that is 36 feet long with a weight attached to the other end. The rope runs through a pulley 20 feet directly above the worker's hand. If the worker walks away from the pulley at 5 ft/s, how fast is the weight rising when it is 10 feet above its original position?

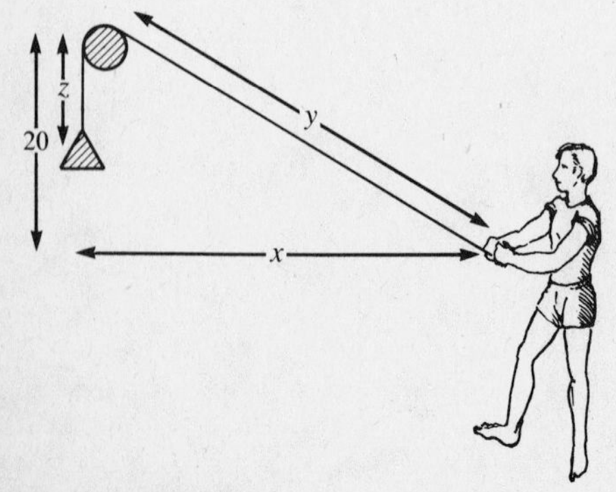

Figure 10-4
Example 10-5

Solution: First, draw a picture (see Figure 10-4). You want to find dz/dt, given that $dx/dt = 5$. From the diagram, you see that

$$x^2 + 400 = y^2$$

You want an equation relating x to z so you can find an equation relating dx/dt to dz/dt. Since the rope is 36 feet long, $y + z = 36$. So,

$$x^2 + 400 = (36 - z)^2$$

Differentiating,

$$2x\frac{dx}{dt} = -2(36 - z)\frac{dz}{dt}$$

$$x\frac{dx}{dt} = -(36 - z)\frac{dz}{dt}$$

Now, you want to find dz/dt when z is 10 feet shorter than it was initially, that is, when $z = 6$. At that time,

$$x^2 + 400 = (36 - 6)^2$$

so, $x = \sqrt{500}$. Finally, find dz/dt when $z = 6$ and $x = \sqrt{500}$:

$$\sqrt{500}(5) = -(36 - 6)\frac{dz}{dt}$$

$$\frac{dz}{dt} = \frac{-5\sqrt{500}}{30} = \frac{-5\sqrt{5}}{3}$$

The weight is rising at a rate of $(5\sqrt{5})/3$ ft/s.

SUMMARY

1. In a basic related rates problem, you are asked to find the rate of change of one quantity, given the rate of change of a related quantity.
2. You can express the relationship among the quantities by equations.
3. Differentiate these equations with respect to time to find the relationship among the rates of change of the quantities.
4. Solve the equation(s) to answer the question. You may need to refer to the original equation to correctly substitute into the final equation.
5. To obtain the correct equations, an accurate diagram is essential.

SOLVED PROBLEMS

PROBLEM 10-1 When the depth of liquid in a specific bowl is h in., the volume of liquid in the bowl is h^3 in.^{3}. When $h = 3$ in., the depth of liquid in the bowl is increasing at a rate of 2 in./min. Find the rate at which liquid is entering the bowl at that time.

Solution: You are given that $dh/dt = 2$ and are asked to find dV/dt, when $h = 3$. Find the equation relating h to V, $V = h^3$, and differentiate with respect to time t:

$$\frac{dV}{dt} = 3h^2 \cdot \frac{dh}{dt}$$

When $h = 3$,

$$\frac{dV}{dt} = 3(3)^2 \cdot 2 = 54$$

When $h = 3$, the liquid is entering the bowl at 54 in.^{3}/min. [See Section 10-1.]

PROBLEM 10-2 A car is traveling west on the interstate. A highway patrolman is parked 90 feet north of the interstate. The patrolman takes a radar reading and finds that the car is 150 feet from his position, and that the distance separating them is increasing at the rate of 72 ft/s. Find the speed of the car at that moment.

Solution: First, draw a diagram (see Figure 10-5). You are told that, when $x = 150$, $dx/dt = 72$; you want to find dy/dt. From the Pythagorean Theorem,

$$y^2 + 90^2 = x^2$$

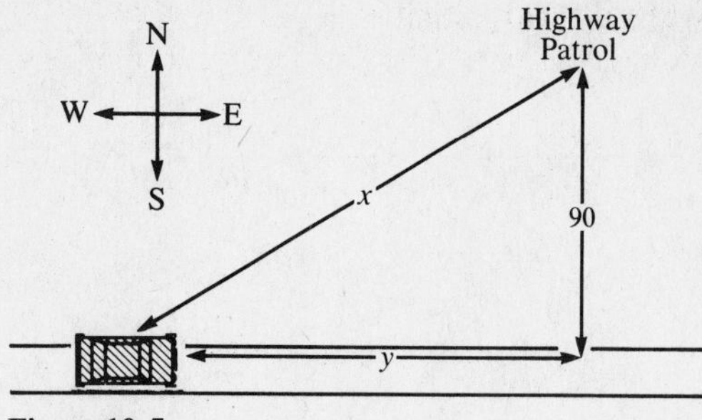

Figure 10-5
Problem 10-2

Differentiating with respect to time t:

$$2y\frac{dy}{dt} = 2x\frac{dx}{dt}$$

$$y\frac{dy}{dt} = \frac{dx}{dt}$$

When $x = 150$,

$$y^2 + 90^2 = 150^2$$

so $y = 120$. You must find dy/dt when $x = 150$, $y = 120$, and $dx/dt = 72$. Substituting into your second equation, you get

$$120\frac{dy}{dt} = 150(72)$$

When $x = 150$, $dy/dt = 90$ ft/s.

[See Section 10-1.]

PROBLEM 10-3 If $y = -x^2$ and $dx/dt = 4$ for all time t, find dy/dt and d^2y/dt^2 when $x = 2$.

Solution: Differentiate the equation relating y and x with respect to time:

$$\frac{dy}{dt} = -2x\frac{dx}{dt}$$

Because $dx/dt = 4$ for all time t,

$$\frac{dy}{dt} = -2x \cdot 4 = -8x$$

When $x = 2$, $dy/dt = -16$. Again, differentiating with respect to time,

$$\frac{d^2y}{dt^2} = -8\frac{dx}{dt} = -8 \cdot 4 = -32$$

[See Section 10-2.]

PROBLEM 10-4 A man who is 5 ft tall walks away from a lamppost at 7 ft/s. The lamppost is 20 ft tall. When he is 8 ft from the lamppost, find the rate at which the tip of his shadow is moving.

Solution: Draw a diagram (see Figure 10-6). In terms of x, y, and z, you are told that $dx/dt = 7$, and you are asked to find dz/dt. The equation relating x to y comes from the geometry of similar triangles. The triangle with vertices at the person's head, feet, and tip of shadow is similar to the triangle with vertices at the top of lamppost, bottom of lamppost, and tip of shadow. Thus,

$$z/20 = (z - x)/5$$

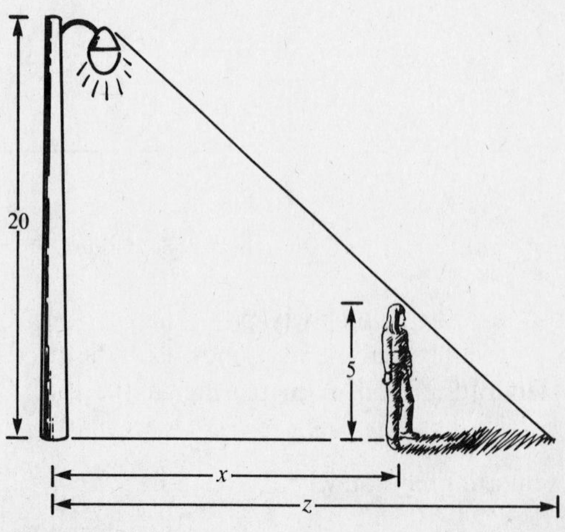

Figure 10-6
Problem 10-4

So, $3z = 4x$ and $3(dz/dt) = 4(dx/dt)$. You know that $dx/dt = 7$, so

$$3\frac{dz}{dt} = 4 \cdot 7$$

and $dz/dt = 28/3$ ft/s.

[See Section 10-1.]

PROBLEM 10-5 Each side of a square baseball diamond is 90 ft long. If a ball is hit down the third base line at a speed of 100 ft/s, how fast is the distance from the ball to first base changing when the ball is halfway to third base?

Solution: The diagram in Figure 10-7 illustrates the situation. You are told that $dx/dt = 100$, and are asked to find dy/dt when $x = 45$. From the Pythagorean Theorem, x and y are related by

$$x^2 + 90^2 = y^2$$

Differentiate with respect to time t:

$$2x\frac{dx}{dt} = 2y\frac{dy}{dt}$$

When $x = 45$, $45^2 + 90^2 = y^2$, so $y = 45\sqrt{5}$. At that point:

$$2(45)(100) = 2(45\sqrt{5})\frac{dy}{dt}$$

When $x = 45$, $dy/dt = 20\sqrt{5}$ ft/s.

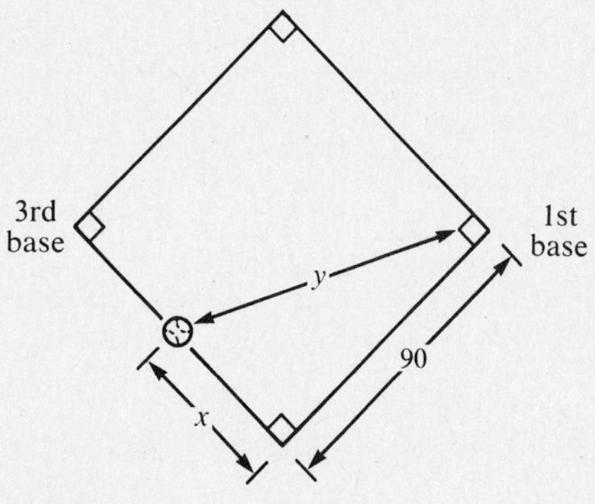

Figure 10-7
Problem 10-5

[See Section 10-1.]

PROBLEM 10-6 The velocity of a particle moving along the x axis is given by the equation $V = 6x^{2/3}$. Find the acceleration of the particle when it is at $x = 27$.

Solution: You want to find the acceleration of the particle, that is, the derivative with respect to time of its velocity: $a = dV/dt$. Differentiate the velocity equation with respect to time t:

$$\frac{dV}{dt} = 6\left(\frac{2}{3}\right)x^{-1/3}\cdot\frac{dx}{dt} = 4x^{-1/3}\cdot\frac{dx}{dt}$$

Since dx/dt is the rate of change of position, or velocity, you can substitute the original expression for V into the second equation:

$$\frac{dV}{dt} = 4x^{-1/3}V = 4x^{-1/3}(6x^{2/3}) = 24x^{1/3}$$

Thus, when $x = 27$, $dV/dt = 24(27)^{1/3} = 72$.

[See Section 10-2.]

PROBLEM 10-7 A rectangular sign, 24 feet wide and of negligible thickness, turns at a rate of 5 revolutions per minute about a vertical axis through its center. A distant observer sees the sign as a rectangle of variable width. How fast is the apparent width changing when the sign is 12 ft wide, as viewed by the observer, and is increasing in width?

Solution: Draw a diagram of the sign as viewed from above (see Figure 10-8). In the diagram w is the apparent width of the sign. You are told that the sign makes 5 revolutions per minute. Hence, $d\theta/dt = 10\pi$ radians/min. You are looking for the relationship between w and θ. From trigonometry,

$$w = 24\sin\theta$$

Differentiating with respect to time t,

$$\frac{dw}{dt} = 24\cos\theta\cdot\frac{d\theta}{dt}$$

When $w = 12$, $\sin\theta = \frac{1}{2}$. Since the width of the sign is increasing, θ must lie between $\pi/2$ and 0, so $\theta = \pi/6$. Thus,

$$\frac{dw}{dt} = 24\left(\cos\frac{\pi}{6}\right)(10\pi) = 120\pi\sqrt{3}\text{ ft/min.}$$

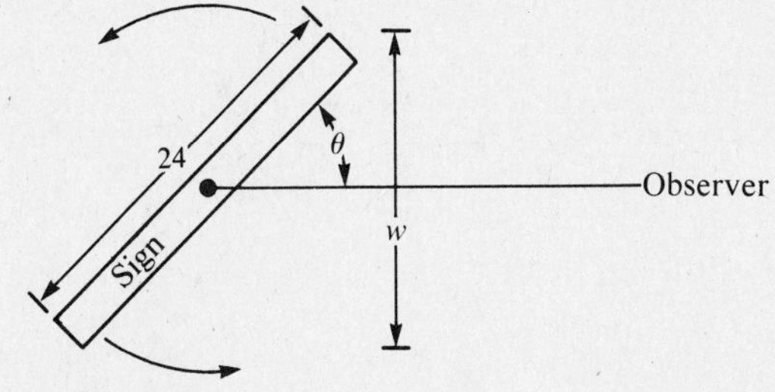

Figure 10-8
Problem 10-7

[See Section 10-2.]

PROBLEM 10-8 Sand is poured on a conical pile at a rate of 20 m^3/min. The height of the pile is always equal to the radius of the base of the pile. When the pile is 3 meters high, how fast is the height of the pile increasing?

Solution: Draw a picture (see Figure 10-9). The volume of the cone is $\frac{1}{3}\pi r^2 h$. Since $r = h$, the volume is

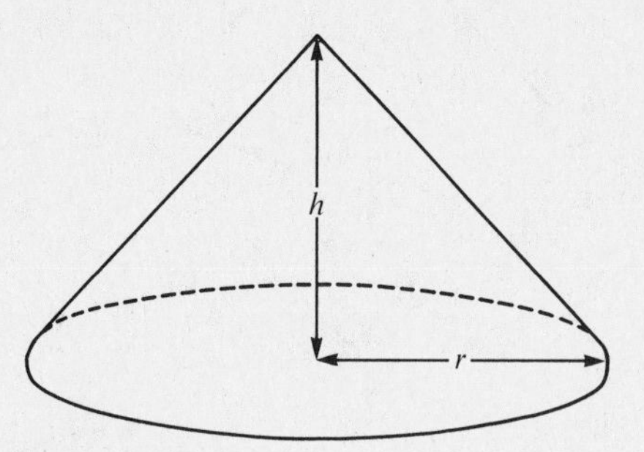

Figure 10-9
Problem 10-8

$$V = \tfrac{1}{3}\pi h^2 h = \tfrac{1}{3}\pi h^3$$

You differentiate to find

$$\frac{dV}{dt} = \pi h^2 \cdot \frac{dh}{dt}$$

You are told that $dV/dt = 20$. When $h = 3$,

$$20 = \pi 3^2 \cdot \frac{dh}{dt}$$

$$\frac{dh}{dt} = \frac{20}{9\pi} \text{ m/min.}$$

[See Section 10-1.]

PROBLEM 10-9 Consider a variable right triangle ABC in a rectangular coordinate system. Vertex A is the origin, the right angle is at vertex B on the y axis, and vertex C is on the parabola $y = (7/4)x^2 + 1$. If point B starts at (0, 1) and moves upward at a constant rate of 2 units/s, how fast is the area of the triangle increasing when $t = 7/2$ seconds?

Solution: First, draw a diagram (see Figure 10-10). When B is at $(0, y)$, C must be at the point (x, y), where $(7/4)x^2 + 1 = y$. That is,

$$x = \left[\frac{4}{7}(y-1)\right]^{1/2}$$

The area of the triangle is

$$T = \frac{1}{2}y\left[\frac{4}{7}(y-1)\right]^{1/2}$$

Differentiating with respect to time t,

$$\frac{dT}{dt} = \frac{1}{2}y \cdot \frac{d}{dt}\left[\frac{4}{7}(y-1)\right]^{1/2} + \left[\frac{4}{7}(y-1)\right]^{1/2} \cdot \frac{d}{dt}\left(\frac{1}{2}y\right)$$

$$= \frac{1}{2}y\frac{1}{2}\left[\frac{4}{7}(y-1)\right]^{-1/2}\left(\frac{4}{7}\right)\left(\frac{dy}{dt}\right) + \left[\frac{4}{7}(y-1)\right]^{1/2}\left(\frac{1}{2}\right)\left(\frac{dy}{dt}\right)$$

$$= \frac{y}{7}\left[\frac{4}{7}(y-1)\right]^{-1/2}\left(\frac{dy}{dt}\right) + \frac{1}{2}\left[\frac{4}{7}(y-1)\right]^{1/2}\frac{dy}{dt}$$

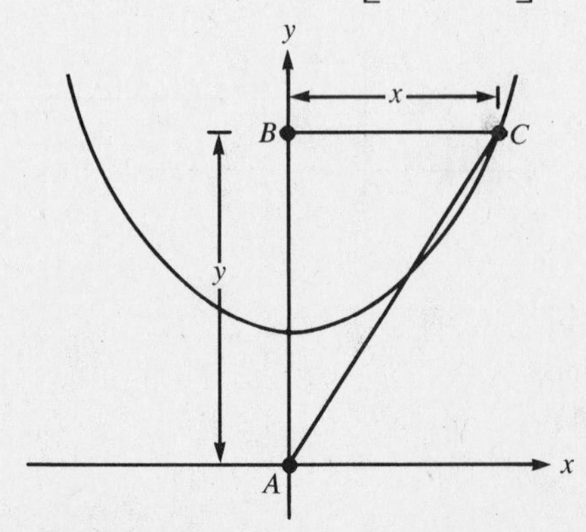

Figure 10-10
Problem 10-9

Because B starts at (0, 1) and moves upward at a constant rate of 2 units/s ($dy/dt = 2$), when $t = 7/2$, $y = 1 + (7/2)2 = 8$. At that time,

$$\frac{dT}{dt} = \frac{8}{7}\left[\frac{4}{7}(8-1)\right]^{-1/2}(2) + \frac{1}{2}\left[\frac{4}{7}(8-1)\right]^{1/2}(2)$$

$$= 22/7 \text{ units/s}$$

[See Section 10-2.]

PROBLEM 10-10 A particle is moving along the parabola $y = x^2$. At what point of its path are the abcissa and the ordinate of the particle changing at the same rate?

Solution: You want to know when $dx/dt = dy/dt$. Differentiating the equation relating y to x:

$$\frac{dy}{dt} = 2x\frac{dx}{dt}$$

So, $dy/dt = dx/dt$ when $2x = 1$. This occurs when $x = \frac{1}{2}$; that is, at $(\frac{1}{2}, \frac{1}{4})$. [See Section 10-2.]

PROBLEM 10-11 Consider a rubber washer that is being compressed. At a certain time, the following measurements are obtained: The outer diameter of the washer is found to be 3 cm; the inner diameter of the washer is 1 cm; the thickness of the washer is decreasing at a rate of $\frac{1}{4}$ cm/min; and the outer diameter is increasing at a rate of $\frac{1}{2}$ cm/min. If the volume of the washer remains $\pi\ \text{cm}^3$ at all times, at what rate is the inner diameter changing at the time the measurements are made?

Solution: Figure 10-11 shows the washer. The volume V, thickness T, inner diameter H, and outer diameter D of the washer are related by

$$V = T\left[\pi\left(\frac{D}{2}\right)^2 - \pi\left(\frac{H}{2}\right)^2\right] = \frac{\pi T}{4}(D^2 - H^2)$$

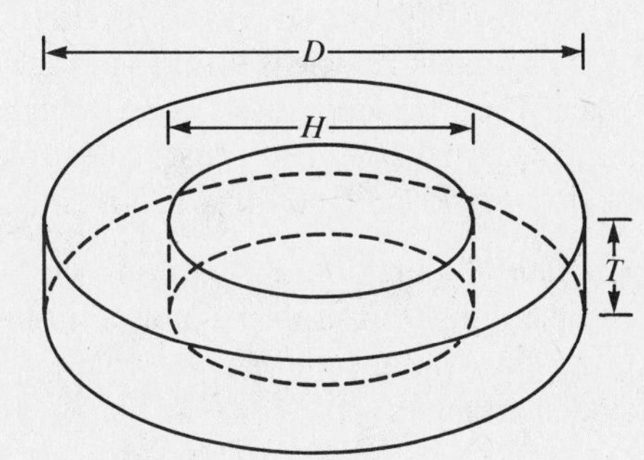

Figure 10-11
Problem 10-11

Differentiating with respect to time t,

$$\frac{dV}{dt} = \frac{\pi}{4}T\cdot\frac{d}{dt}(D^2 - H^2) + \frac{\pi}{4}(D^2 - H^2)\cdot\frac{dT}{dt}$$

$$= \frac{\pi}{4}T\left(2D\frac{dD}{dt} - 2H\frac{dH}{dt}\right) + \frac{\pi}{4}(D^2 - H^2)\frac{dT}{dt}$$

At the time of interest, $dT/dt = -\frac{1}{4}$, $dD/dt = \frac{1}{2}$, $D = 3$, $H = 1$. You need to find T at that time, so,

$$\pi = V = \frac{\pi}{4}T(D^2 - H^2) = \frac{\pi}{4}T(3^2 - 1^2)$$

and so $T = \frac{1}{2}$.
Because the volume is always π, $dV/dt = 0$. At the time of interest,

$$0 = \frac{\pi}{4}\left(\frac{1}{2}\right)\left[2(3)\left(\frac{1}{2}\right) - 2(1)\frac{dH}{dt}\right] + \frac{\pi}{4}(3^2 - 1^2)\left(-\frac{1}{4}\right)$$

$$= \frac{\pi}{8}\left(3 - 2\frac{dH}{dt}\right) - \frac{\pi}{2}$$

and $dH/dt = -\frac{1}{2}$ cm/min. [See Section 10-2.]

PROBLEM 10-12 Let A, D, C, and r be the area, diameter, circumference, and radius of a circle, respectively. At a certain instant, $r = 6$ and $dr/dt = 3$ in./s. Find the rate of change of A with respect to **(a)** r, **(b)** D, **(c)** C, and **(d)** t.

Solution: A and r are related by $A = \pi r^2$.

(a) Differentiating with respect to r,

$$\frac{dA}{dr} = 2\pi r$$

When $r = 6$, $dA/dr = 12\pi\ \text{in.}^2/\text{in.}$

(b) You know that $2r = D$, so

$$A = \pi\left(\frac{D}{2}\right)^2 = \frac{1}{4}\pi D^2$$

$$\frac{dA}{dD} = \frac{1}{2}\pi D = \frac{1}{2}\pi(12) = 6\pi\ \text{in.}^2/\text{in.}$$

(c) You know that $2\pi r = C$, so differentiate both this equation and $A = \pi r^2$ with respect to C:

$$\frac{dA}{dC} = 2\pi r \frac{dr}{dC}$$

$$2\pi \frac{dr}{dC} = 1$$

So

$$\frac{dA}{dC} = 2\pi r\left(\frac{1}{2\pi}\right) = r$$

When $r = 6$, $dA/dC = 6$ in.^{2}/in.

(d) Differentiate $A = \pi r^2$ with respect to time t,

$$\frac{dA}{dt} = 2\pi r \frac{dr}{dt}$$

When $r = 6$ and $dr/dt = 3$, $dA/dt = 36\pi$ in.^{2}/s. [See Section 10-2.]

PROBLEM 10-13 Two motorcycles are approaching each other at night on a straight, two-lane highway. Each vehicle is traveling in the center of its lane, and the centers of the lanes are 10 yds apart. The westbound cycle is traveling at 25 yds/s. The eastbound cycle is traveling at a rate of 30 yds/s, and its headlight casts a shadow of the second cycle onto a fence, 20 yds from the center of the westbound lane. How fast is the shadow of the westbound cycle moving on the fence?

Solution: Figure 10-12 illustrates the situation. A good sketch is your key to the successful completion of this problem. You want to find dx/dt. Without investigation, it isn't clear that dx/dt is independent of time. You are given that $dz/dt = -30$ and $dy/dt = 25$. From the geometry of similar triangles,

$$\frac{z - x}{30} = \frac{y - x}{20}$$

$$2z = 3y - x$$

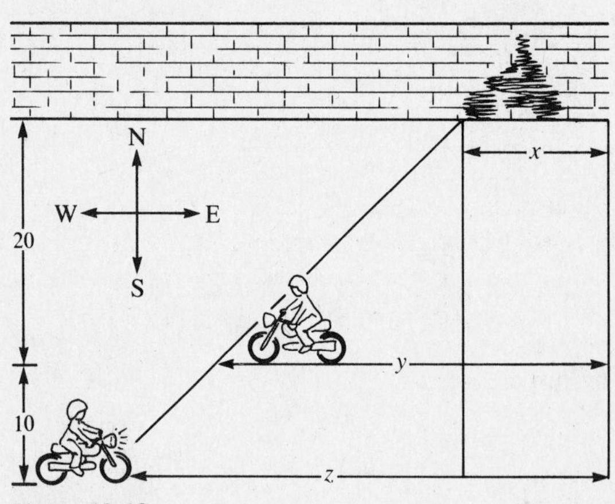

Figure 10-12
Problem 10-13

Differentiating and substituting for the known quantities, you get

$$2\frac{dz}{dt} = 3\frac{dy}{dt} - \frac{dx}{dt}$$

$$2(-30) = 3(25) - \frac{dx}{dt}$$

$$\frac{dx}{dt} = 135$$

So, the shadow moves at a rate of 135 yds/sec. [See Section 10-2]

PROBLEM 10-14 A streetlight, 20 ft in height, stands 5 ft from a sidewalk. If a policeman, 6 ft in height, walks along the sidewalk at 4 ft/s, at what rate is the length of his shadow changing when he is 13 ft from the base of the streetlight?

Solution: Carefully draw a diagram (see Figure 10-13). You want to find dz/dt, when $x = 13$. You are given that $dy/dt = 4$. You find that y and z are related by two equations, the first a result of the Pythagorean theorem:

$$y^2 + 25 = x^2$$

and the second a result of similar triangles:

$$\frac{z}{6} = \frac{x + z}{20}$$

$$7z = 3x$$

Combine these equations to obtain:

$$y^2 + 25 = \frac{49}{9} z^2$$

Differentiate:

$$2y \frac{dy}{dt} = \frac{98}{9} z \frac{dz}{dt}$$

You need to find y and z, when $x = 13$:

$$y^2 + 25 = 13^2$$

$$y = 12$$

$$7z = 3(13)$$

$$z = \frac{39}{7}$$

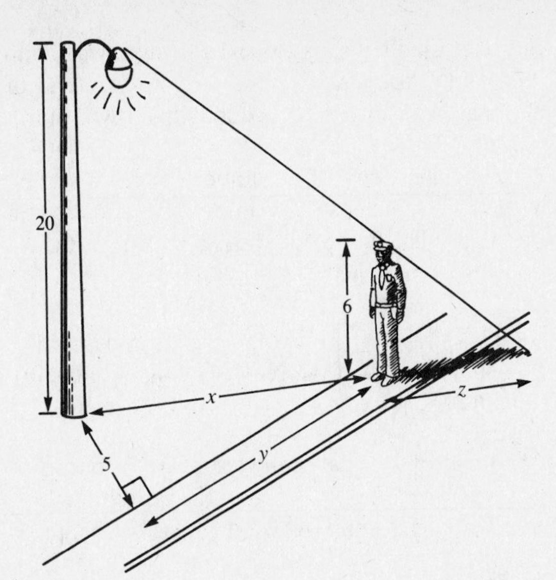

Figure 10-13
Problem 10-14

Finally, then, you use $y = 12$, $z = 39/7$, and $dy/dt = 4$:

$$2(12)(4) = \frac{98}{9} \cdot \frac{39}{7} \cdot \frac{dz}{dt}$$

$$\frac{dz}{dt} = \frac{144}{91} \text{ ft/s}$$

[See Section 10-2.]

Supplementary Exercises

10-15 Suppose that the height of a certain tree is $40D^{3/2}$ in., where D is the diameter of the trunk of the tree. If the diameter of the trunk grows at a constant rate of $\frac{1}{4}$ in. per year, at what rate is the height of the tree changing when the diameter is 4 in.?

10-16 An angler has hooked a fish. The fish is traveling in an east-west direction along a line 30 ft north of the angler. If the fishing line is leaving the reel at a rate of 6 ft/s when the fish is 50 ft from the angler, how fast is the fish traveling?

10-17 A boat is pulled toward a pier by means of a taut cable. If the boat is 20 ft below the level of the pier and the cable is pulled in at a rate of 36 ft/min, how fast is the boat moving when it is 48 ft from the base of the pier?

10-18 A car, traveling at 40 ft/s, crosses a bridge over a canal 10 s before a boat traveling at 20 ft/s passes under the bridge. The canal and the road are straight and at right angles to each other. At what rate are the car and boat separating 10 s after the boat passes under the bridge?

10-19 A certain pine tree maintains the shape of a cone. When the base of the tree is 28 ft in diameter, the diameter of the base is growing at the rate of 2 ft/year. At the same time, the tree is 60 ft tall and its height is growing at the rate of 4 ft/year. At what rate is the volume of the tree changing at that time? (The volume V of a cone with radius r and height h is $V = \frac{1}{3}\pi r^2 h$.)

10-20 A spherical balloon is inflated at the rate of 4 ft^3/min. What is the volume of the balloon when the radius is increasing at the rate of 6 in./min?

10-21 A stone dropped in a still pond creates a circular ripple whose radius increases at a constant rate of 3 ft/s. At what rate is the area enclosed by the ripple increasing 8 s after the stone strikes the pond?

10-22 A ladder 15 ft in length leans against a vertical wall, with the bottom of the ladder 5 ft from the wall on a horizontal floor. If at that time the bottom end of the ladder is being pulled away at the rate of 2 ft/s, at what rate does the top of the ladder slip down the wall?

10-23 A ladder leans against a vertical wall, with the bottom of the ladder 8 ft from the wall on a horizontal floor. At that time the bottom end of the ladder is being pulled away at the rate of 3 ft/s and the top of the ladder slips down the wall at the rate of 4 ft/s. How long is the ladder?

10-24 The radius of a cylinder increases at a constant rate. Its height is a linear function of its radius and increases three times as fast as the radius. When the radius is 1 ft, the height is 6 ft. When the radius is 6 ft, the volume is increasing at the rate of 1 ft^3/s. Find the rate at which the volume increases when the radius is 36 ft.

10-25 A satellite is moving in an elliptical orbit about a planet. The equation of its planar orbit is $3x^2 + 4y^2 = 20$. If the velocity of the satellite in the y direction is 10 when the y coordinate of the satellite is 2, what is the velocity in the x direction at that time?

10-26 A cylindrical tank with a radius of 5 ft and a height of 20 ft is filled with a certain liquid chemical. A hole is punched in the bottom. At that moment the chemical drains out of the tank at the rate of 2 ft^3/min. At what rate is the height of liquid in the tank changing?

10-27 A point moves on the graph of $y = x^3 - x$ so that, when the point is at $(x, x^3 - x)$, the rate of change of x with respect to time is $1/x$. Find the rate of change of y with respect to time when $y = 6$.

10-28 Ships A and B leave the same port. Ship A sails west at 20 knots (nautical miles per hour) and ship B sails south at 15 knots. At what rate is the distance between them changing at 2 PM if:

(a) A and B both leave at noon?
(b) A leaves at noon and B leaves at 1 PM?

10-29 A balloon rises vertically at the rate of 10 ft/s. A person on the ground 100 ft away from the spot below the rising balloon watches the balloon ascend; at what rate is the distance between balloon and observer changing when the balloon is 100 ft above ground?

10-30 A balloon rises vertically at the rate of 10 ft/s. A person watches the balloon ascend from a point on the ground 100 ft away from the spot below the rising balloon. At what rate (radians/s) is the observer's eye rotating upward to follow the balloon when the balloon is 50 ft above the level of the observer's eye?

10-31 A balloon rises vertically at the rate of 10 ft/s. A lamppost stands 20 ft from the spot below the balloon. The lamppost is 25 ft high. At what rate is the shadow of the balloon moving when the balloon is 15 ft above the ground?

10-32 A balloon rises vertically at the rate of 10 ft/s. A lamppost stands 20 ft from the spot below the rising balloon. The lamppost is 25 ft high. On the other side of the balloon stands a vertical wall, 10 ft from the balloon (30 ft from the lamppost). At what rate is the shadow moving on the wall when the balloon is 15 ft above the ground?

Solutions to Supplementary Exercises

(10-15) 30 in./year

(10-16) 15/2 ft/s

(10-17) 39 ft/min.

(10-18) $\dfrac{180\sqrt{17}}{17}$ ft/s

(10-19) $2464\pi/3$ ft^3/year

(10-20) $\dfrac{8}{3}\sqrt{\dfrac{2}{\pi}}$ ft^3

(10-21) 144π ft^2/s

(10-22) $-\sqrt{2}/2$ ft/s

(10-23) 10 ft

(10-24) 33 ft^3/s

(10-25) $\pm 40\sqrt{3}/3$

(10-26) $2/25\pi$ ft/min

(10-27) 11/2

(10-28) (a) 25 knots
(b) $\dfrac{205\sqrt{73}}{73}$ knots

(10-29) $5\sqrt{2}$ ft/s

(10-30) 2/25 radians/s

(10-31) 50 ft/s

(10-32) 15 ft/s

EXAM 4 (CHAPTERS 9 AND 10)

1. Find the maximum and minimum values of each of the following functions over the given interval:

   (a) $f(x) = \dfrac{x^2 + x + 4}{x + 1}$ on $[0, 2]$ **(b)** $g(x) = \dfrac{2x}{\sqrt{3x - 5}}$ on $[2, 5]$.

2. Find the area of the largest rectangle that can be drawn with its base on the x axis and with two vertices on the graph of $y = 8/(x^2 + 4)$.

3. Let V_1 be the volume of a right circular cylinder with fixed lateral surface area 4π ft^2. Let V_2 be the volume of a hemisphere whose radius is equal to that of the cylinder. Find the dimensions of the cylinder that will maximize $V_1 - V_2$.

4. At a certain instant the length of a rectangle is 3 in. and is increasing at the rate of 1 in./min, and the width is 2 in. and decreasing at the rate of 1/2 in./min. Is the area increasing or decreasing at that instant? At what rate?

5. A point $P(x, y)$ moves along the upper half of the right-hand branch of the hyperbola $x^2 - y^2 = 1$. Let $A(0, y)$ and $B(x, 0)$ be the projection of P on the y and x axes, respectively. If x is changing at the rate of 2 cm/s when $x = \sqrt{13}$ cm, find the rate at which the length of $\overline{AB}$ is changing.

SOLUTIONS TO EXAM 4

1. Find the critical points and examine the values of the function at these points and at the endpoints of the given interval.

(a)

$$f'(x) = \frac{(x + 1)(2x + 1) - (x^2 + x + 4)}{(x + 1)^2} = \frac{x^2 + 2x - 3}{(x + 1)^2} = \frac{(x + 3)(x - 1)}{(x + 1)^2}$$

$$f'(x) = 0 \quad \text{for } x = -3 \quad \text{and for } x = 1$$

Because -3 is not in $[0, 2]$, examine:

$$f(0) = 4 \qquad f(1) = 3 \qquad f(2) = \frac{10}{3}$$

On the interval $[0, 2]$, f has a maximum value of 4 at $x = 0$ and a minimum value of 3 at $x = 1$.

(b)

$$g'(x) = \frac{\sqrt{3x - 5}(2) - (2x)3/(2\sqrt{3x - 5})}{3x - 5} = \frac{2(3x - 5) - 3x}{(3x - 5)^{3/2}} = \frac{3x - 10}{(3x - 5)^{3/2}}$$

$$g'(x) = 0 \quad \text{at } x = \frac{10}{3}$$

$$g(2) = 4 \qquad g\left(\frac{10}{3}\right) = \frac{4\sqrt{5}}{3} \qquad g(5) = \sqrt{10}$$

On the interval $[2, 5]$, g has a maximum value of 4 at $x = 2$ and a minimum value of $4\sqrt{5}/3$ at $x = 10/3$ $(4\sqrt{5}/3 \approx 2.98 < 3.16 \approx \sqrt{10})$.

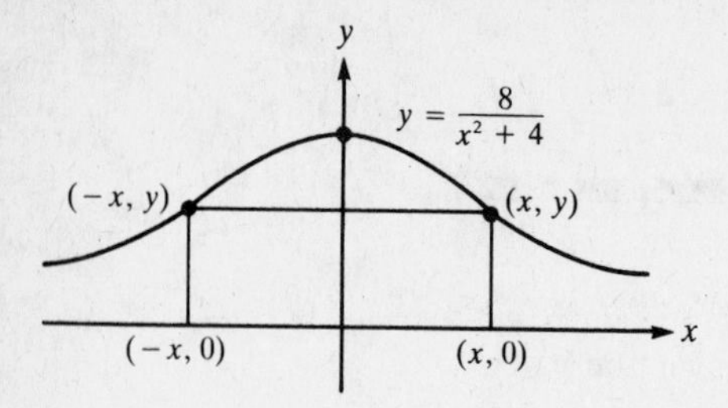

2. Let (x, y) be a point on $y = 8/(x^2 + 4)$ with $x > 0$ and let (x, y), $(x, 0)$, $(-x, 0)$, and $(-x, y)$ be the vertices of the rectangle as shown in the figure. The area of the rectangle is

$$A = 2xy = \frac{16x}{x^2 + 4}$$

Differentiate:

$$A'(x) = \frac{(x^2 + 4)16 - 16x(2x)}{(x^2 + 4)^2} = \frac{64 - 16x^2}{(x^2 + 4)^2} = \frac{16(4 - x^2)}{(x^2 + 4)^2}$$

So $A'(x) = 0$ when $x = 2$. If $0 < x < 2$, then $A'(x) > 0$. If $x > 2$, then $A'(x) < 0$. Thus by the first derivative test, $A(x)$ is a maximum when $x = 2$ and

$$A(2) = \frac{32}{4 + 4} = 4 \text{ units}^2$$

3. The lateral surface area of a cylinder is $A = 2\pi rh$; the volume of a cylinder is $V_1 = \pi r^2 h$, and the volume of a hemisphere is $V_2 = (2/3)\pi r^3$. It is given that $A = 2\pi rh = 4\pi$. Thus $rh = 2$, or $h = 2/r$.

$$V_1 - V_2 = \pi r^2 h - \frac{2}{3}\pi r^3 = \pi\left(2r - \frac{2}{3}r^3\right)$$

$$\frac{d}{dr}(V_1 - V_2) = \pi(2 - 2r^2)$$

$$0 = \pi(2 - 2r^2)$$

$$r = \pm 1$$

But $r < 0$ makes no sense, so examine $r = 1$: $\frac{d}{dr}(V_1 - V_2) > 0$ for $0 < r < 1$ and $\frac{d}{dh}(V_1 - V_2) < 0$ for $r > 1$. Thus $V_1 - V_2$ is a maximum when $r = 1$ (and when $r = 1, h = 2$).

4. Let x be the length and y the width of the rectangle. Then $A = xy$, and

$$\frac{dA}{dt} = x\frac{dy}{dt} + y\frac{dx}{dt}$$

Substitute $x = 3$, $dx/dt = 1$, $y = 2$, $dy/dt = -1/2$:

$$\frac{dA}{dt} = 3\left(-\frac{1}{2}\right) + 2(1) = \frac{1}{2}$$

The area is increasing at the rate of $1/2$ in.^{2}/min.

5. The distance $\overline{AB} = s$ is given by

$$s = \sqrt{x^2 + y^2} = \sqrt{x^2 + x^2 - 1}$$

because $y^2 = x^2 - 1$. Differentiate:

$$s = \sqrt{2x^2 - 1}$$

$$\frac{ds}{dt} = \frac{4x}{2\sqrt{2x^2 - 1}} \cdot \frac{dx}{dt} = \frac{4\sqrt{13}}{2\sqrt{25}} \cdot 2 = \frac{4\sqrt{13}}{5} \text{ cm/s}$$

FINAL EXAM (Chapters 1–10)

In Problems 1 through 10, find the derivative:

1. $f(t) = (t + \sqrt{t+1})^5$
2. $g(x) = (2x + 3)/(x^2 + x + 1)$
3. $y = e^x(\sin x + \cos x)$
4. $F(x) = \ln(x + \sqrt{x^2 - 1})$
5. $w = 1/(\sin \sqrt{x+1})$
6. $G(x) = x^{\ln x}$
7. $v = e^{\sin x}$
8. $H(y) = \tan \sqrt{y}$
9. $y = (x \ln x)^{1/4}$
10. $u = (x^2 + 4)^3 e^x$
11. Use implicit differentiation to find dy/dx and d^2y/dx^2:

$$x^2 + 2xy - 3y = 5$$

12. Find $\lim_{h\to 0} (\sqrt{x+h} - \sqrt{x})/h$.
13. Find the points on the graph of $y = x^3/3 - 2x^2 + 1$ where the tangent line is perpendicular to the line $x + 5y = 7$.
14. Show that the graphs of $y = 1/x$ and $y = \sqrt{2 - x^2}$ have a common tangent line at their point of intersection, and find the equation of that line.
15. Find the point on the graph of $2x - 3y = 6$ that is nearest to the origin.
16. Let ABC be a right triangle with fixed hypotenuse $c = 10$ in. If α (the angle at vertex A) is increasing at the rate of 4π radians/min, find the rate at which a (the side opposite A) is increasing at the instant that $\alpha = \pi/3$ radians.
17. Let f be defined as follows:

$$f(x) = \begin{cases} 2x + 1 & \text{for } x < 1 \\ 3 & \text{for } x = 1 \\ 4 - x & \text{for } x > 1 \end{cases}$$

Use the definition of continuity of $f(x)$ at $x = a$ to determine whether f is continuous at $x = 1$.

18. Let $g(x)$ be a continuous function such that:

$$g(3) = 0$$

$$\lim_{x\to 3^-} g'(x) = -1$$

$$\lim_{x\to 3^+} g'(x) = 1$$

$$g''(x) < 0 \quad \text{for all } x \neq 3$$

Draw a graph that might represent $g(x)$ for x near 3.

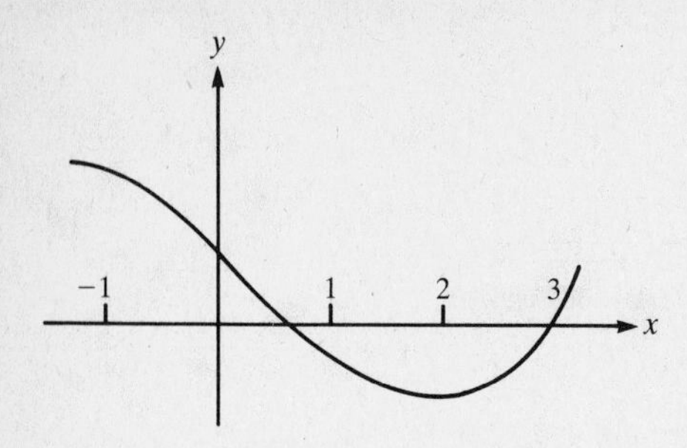

19. Examine the given graph (shown in the figure) of a function $F(x)$, and determine whether each of the following is positive, negative, or zero:

(a) $F'(-1)$ **(c)** $F(1)$ **(e)** $F(3)$
(b) $F''(0)$ **(d)** $F'(2)$ **(f)** $F''(3)$

20. Graph $y = x/(x^2 - 9)$, showing coordinates of all intercepts, relative maxima and minima, and points of inflection, and equations of asymptotes.

SOLUTIONS TO FINAL EXAM

1. Use the chain rule:

$$f'(t) = 5(t + \sqrt{t+1})^4 \frac{d}{dt}(t + \sqrt{t+1}) = 5(t + \sqrt{t+1})^4 \left(1 + \frac{1}{2\sqrt{t+1}}\right)$$

2. Use the quotient rule:

$$g'(x) = \frac{\left[\frac{d}{dx}(2x+3)\right](x^2+x+1) - (2x+3)\frac{d}{dx}(x^2+x+1)}{(x^2+x+1)^2}$$

$$= \frac{2(x^2+x+1) - (2x+3)(2x+1)}{(x^2+x+1)^2} = \frac{-2x^2 - 6x - 1}{(x^2+x+1)^2}$$

3. Use the product rule:

$$\frac{dy}{dx} = e^x \frac{d}{dx}(\sin x + \cos x) + \left[\frac{d}{dx}(e^x)\right](\sin x + \cos x)$$

$$= e^x(\cos x - \sin x) + e^x(\sin x + \cos x)$$

$$= 2e^x \cos x$$

4. $$F'(x) = \frac{1}{x + \sqrt{x^2-1}} \cdot \frac{d}{dx}(x + \sqrt{x^2-1}) = \frac{1}{x+\sqrt{x^2-1}} \cdot \left(1 + \frac{2x}{2\sqrt{x^2-1}}\right)$$

$$= \frac{1}{x+\sqrt{x^2-1}} \cdot \frac{\sqrt{x^2-1} + x}{\sqrt{x^2-1}} = \frac{1}{\sqrt{x^2-1}}$$

5. Use the chain rule:

$$\frac{dw}{dx} = \frac{d}{dx}(\sin\sqrt{x+1})^{-1} = -(\sin\sqrt{x+1})^{-2}\frac{d}{dx}(\sin\sqrt{x+1})$$

$$= -(\sin\sqrt{x+1})^{-2}\left[\cos\sqrt{x+1}\,\frac{d}{dx}(\sqrt{x+1})\right]$$

$$= -(\sin\sqrt{x+1})^{-2}\cos\sqrt{x+1}\,\frac{1}{2\sqrt{x+1}} = \frac{-\cos\sqrt{x+1}}{2\sqrt{x+1}\,\sin^2\sqrt{x+1}}$$

6. Use logarithmic differentiation. Take the natural log of both sides and differentiate implicitly:

$$\ln G(x) = \ln(x^{\ln x}) = (\ln x)(\ln x) = (\ln x)^2$$

$$\frac{G'(x)}{G(x)} = 2 \ln x \left(\frac{1}{x}\right)$$

$$G'(x) = G(x)\frac{2 \ln x}{x} = \frac{x^{\ln x}\, 2 \ln x}{x}$$

7.
$$\frac{dv}{dx} = e^{\sin x} \frac{d}{dx}(\sin x) = e^{\sin x}\cos x$$

8.
$$H'(y) = \sec^2\sqrt{y}\,\frac{d}{dy}\sqrt{y} = \frac{\sec^2\sqrt{y}}{2\sqrt{y}}$$

9.
$$\frac{dy}{dx} = \frac{1}{4}(x \ln x)^{-3/4}\frac{d}{dx}(x \ln x) = \frac{1}{4}(x \ln x)^{-3/4}\left(x \cdot \frac{1}{x} + \ln x\right)$$

$$= \frac{1 + \ln x}{4(x \ln x)^{3/4}}$$

10.
$$\frac{du}{dx} = (x^2 + 4)^3 e^x + e^x 3(x^2 + 4)^2(2x)$$

$$= e^x(x^2 + 4)^3 + 6xe^x(x^2 + 4)^2 = e^x(x^2 + 4)^2(x^2 + 6x + 4)$$

11. Differentiate with respect to x, treating y as a function of x:

$$2x + 2\left(x\frac{dy}{dx} + y\right) - 3\frac{dy}{dx} = 0$$

Solve for $\dfrac{dy}{dx}$:

$$\frac{dy}{dx}(2x - 3) = -(2x + 2y) \qquad (1)$$

$$\frac{dy}{dx} = \frac{-(2x + 2y)}{2x - 3}$$

To find $\dfrac{d^2y}{dx^2}$, differentiate (1) again:

$$\frac{d^2y}{dx^2}(2x - 3) + \frac{dy}{dx}(2) = -2 - 2\frac{dy}{dx}$$

$$\frac{d^2y}{dx^2}(2x - 3) = -2 - 4\frac{dy}{dx}$$

Substitute for $\dfrac{dy}{dx}$:

$$\frac{d^2y}{dx^2}(2x - 3) = -2 - 4\left[\frac{-(2x + 2y)}{2x - 3}\right] = -2 + \frac{8x + 8y}{2x - 3} = \frac{4x + 8y + 6}{2x - 3}$$

and solve for d^2y/dx^2:

$$\frac{d^2y}{dx^2} = \frac{4x + 8y + 6}{(2x - 3)^2}$$

12.
$$\frac{\sqrt{x + h} - \sqrt{x}}{h} \cdot \frac{\sqrt{x + h} + \sqrt{x}}{\sqrt{x + h} + \sqrt{x}} = \frac{x + h - x}{h(\sqrt{x + h} + \sqrt{x})} = \frac{h}{h(\sqrt{x + h} + \sqrt{x})}$$

$$\lim_{h \to 0} \frac{\sqrt{x + h} - \sqrt{x}}{h} = \lim_{h \to 0} \frac{h}{h(\sqrt{x + h} + \sqrt{x})} = \lim_{h \to 0} \frac{1}{\sqrt{x + h} + \sqrt{x}}$$

$$= \frac{1}{\sqrt{x} + \sqrt{x}} = \frac{1}{2\sqrt{x}}$$

13. The line $x + 5y = 7$ has slope $m = -1/5$. Any line perpendicular to this line has slope $m = 5$. The slope of the tangent line to the graph of $f(x)$ at x is $m = f'(x)$. So, you want to find a so that $f'(a) = 5$.

$$f(x) = \frac{1}{3}x^3 - 2x^2 + 1$$

$$f'(x) = x^2 - 4x$$

$$f'(a) = a^2 - 4a = 5$$

$$a^2 - 4a - 5 = 0$$

$$(a - 5)(a + 1) = 0$$

$$a = 5 \qquad a = -1$$

$$f(5) = \frac{1}{3}125 - 2(25) + 1 = \frac{-22}{3}$$

$$f(-1) = \frac{-1}{3} - 2 + 1 = \frac{-4}{3}$$

At $(5, -22/3)$ and $(-1, -4/3)$ the tangent lines have slope 5 and hence are perpendicular to $x + 5y = 7$.

14. The graphs of $y = 1/x$ and $y = \sqrt{2 - x^2}$ intersect when

$$\frac{1}{x} = \sqrt{2 - x^2}$$

$$\frac{1}{x^2} = 2 - x^2$$

$$x^4 - 2x^2 + 1 = 0$$

$$(x^2 - 1)^2 = 0$$

$$x = 1 \quad \text{or} \quad x = -1$$

However, $x = -1$ does not give a point of intersection because $\sqrt{2 - 1} \neq -1$. Thus the point of intersection is $(1, 1)$. The slope of the tangent line to $y = 1/x$ at $x = 1$ is:

$$y' = \left.\frac{-1}{x^2}\right|_{x=1} = -1$$

The slope of the tangent line to $y = \sqrt{2 - x^2}$ at $x = 1$ is

$$y' = \left.\frac{-x}{\sqrt{2 - x^2}}\right|_{x=1} = -1$$

Thus, the tangent lines are the same. The line has slope -1 and passes through $(1, 1)$:

$$y - 1 = -1(x - 1)$$

15. If (x, y) is a point on $2x - 3y = 6$, then $y = (2/3)(x - 3)$.
The distance s from $(0, 0)$ to (x, y) is

$$s = \sqrt{(x - 0)^2 + (y - 0)^2} = \sqrt{x^2 + \frac{4}{9}(x - 3)^2}$$

You want to find the minimum value of s, or more simply, s^2. Find the derivative of s^2, set it equal to zero, and solve for x:

$$s^2 = x^2 + \frac{4}{9}(x - 3)^2$$

$$\frac{d}{dx}(s^2) = 2x + \frac{8}{9}(x - 3) = 0$$

$$26x - 24 = 0$$

$$x = \frac{12}{13}$$

$$y = \frac{2}{3}\left(\frac{12}{13} - 3\right) = \frac{-18}{13}$$

For $x > 12/13, \frac{d}{dx}s^2 > 0$; for $x < 12/13, \frac{d}{dx}s^2 < 0$. Thus, by the first derivative test, s^2 has a minimum at $(12/3, -18/13)$. That is, $(12/13, -18/13)$ is the point on $2x - 3y = 6$ that is closest to the origin.

16. Draw the triangle, as shown in the figure. You are given that $c = 10$ in. and $d\alpha/dt = 4\pi$ rad/min. You want to find da/dt when $\alpha = \pi/3$. Find the equation that relates a to α.

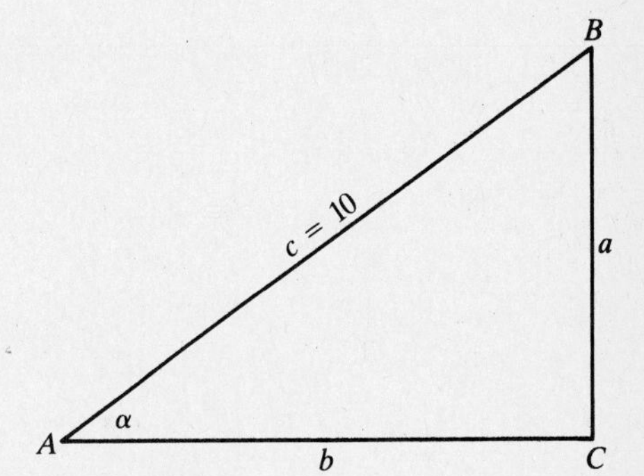

$$\sin \alpha = \frac{a}{c}$$

$$a = c \sin \alpha = 10 \sin \alpha$$

Differentiate with respect to t

$$\frac{da}{dt} = 10 \cos \alpha \frac{d\alpha}{dt}$$

When $\alpha = \pi/3$ and $\frac{dx}{dt} = 4\pi$, you have

$$\frac{da}{dt} = 10 \cos (\pi/3)(4\pi) = 10\left(\frac{1}{2}\right)(4\pi) = 20\pi \text{ in./min}$$

17. The function $f(x)$ is continuous at $x = a$ if $f(a)$ is defined and $\lim_{x\to a} f(x) = f(a)$. The given function is defined at $x = 1$ (you are given that $f(1) = 3$). To determine $\lim_{x\to 1} f(x)$, you must use one sided limits, because $f(x)$ is defined differently for $x < 1$ and for $x > 1$.

$$\lim_{x\to 1^-} f(x) = \lim_{x\to 1^-} (2x + 1) = 2(1) + 1 = 3$$

$$\lim_{x\to 1^+} f(x) = \lim_{x\to 1^+} (4 - x) = 4 - 1 = 3$$

Therefore: $\lim_{x\to 1} f(x) = 3 = f(1)$, and $f(x)$ is continuous at $x = 1$.

18. The point $(3, 0)$ lies on the graph because $g(3) = 0$. Because $g(x)$ is continuous and $\lim_{x\to 3^-} g'(x) = -1$, the graph must approach $(3, 0)$ from the left, tangent to the $-45°$ line. Because $\lim_{x\to 3^+} g'(x) = 1$, you also know that the graph must approach $(3, 0)$ from the right, tangent to the $45°$ line. Because $g''(x) < 0$ for all $x \neq 3$, the graph is concave downward, as shown in the figure.

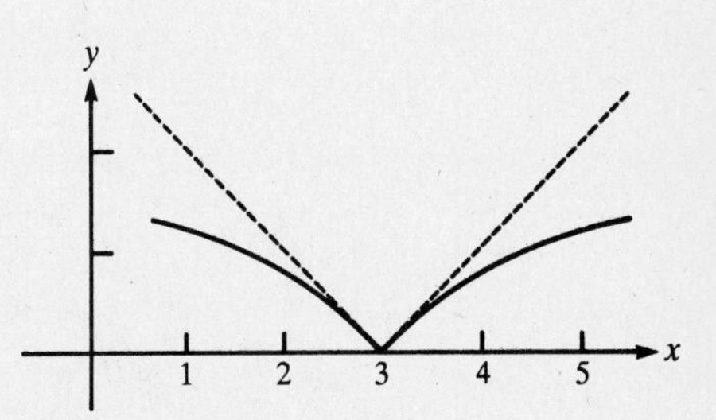

19. **(a)** The graph is falling from left to right at $x = -1$, so $F'(1) < 0$. **(b)** The curve changes from concave upward to concave downward at $x = 0$, an inflection point, so $F''(0) = 0$. **(c)** The curve is below the x axis at $x = 1$, so $F(1) < 0$. **(d)** The tangent line at $x = 2$ is horizontal, so $F'(2) = 0$. **(e)** The graph crosses the x axis at $x = 3$, so $F(3) = 0$. **(f)** The curve is concave upward at $x = 3$, so $F''(3) > 0$.

20. Intercepts: $(0, 0)$

Symmetry: $f(-x) = \frac{-x}{(-x)^2 - 9} = -f(x)$, so it is symmetric with respect to the origin.

Asymptotes: The degree of the numerator is less than the degree of the denominator, so $y = 0$ is the horizontal asymptote. There are vertical asymptotes at $x = \pm 3$.

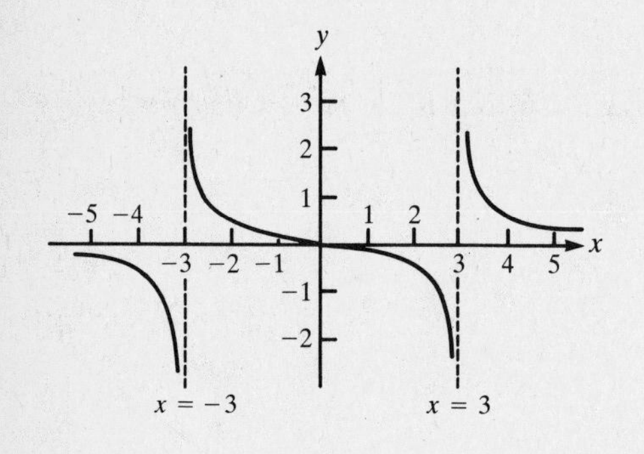

Critical points:

$$y' = \frac{(x^2 - 9) - x(2x)}{(x^2 - 9)^2} = -\frac{x^2 + 9}{(x^2 - 9)^2}$$

Notice that $y' < 0$ for all $x \neq \pm 3$. Thus there are no local maxima or minima. The curve is decreasing for all $x \neq \pm 3$.

Concavity:

$$y'' = -\frac{(x^2 - 9)^2 2x - (x^2 + 9)2(x^2 - 9)(2x)}{(x^2 - 9)^4}$$

$$= -\frac{2x(x^2 - 9) - 4x(x^2 + 9)}{(x^2 - 9)^3} = \frac{2x^3 + 54x}{(x^2 - 9)^3} = \frac{2x(x^2 + 27)}{(x^2 - 9)^3}$$

At $x = 0$, $y'' = 0$. For $x < -3$ and $0 < x < 3$, $y'' < 0$. For $x > 3$ and $-3 < x < 0$, $y'' > 0$.

The graph is shown in the figure.

11 THE ANTIDERIVATIVE

THIS CHAPTER IS ABOUT

☑ **Definition of the Antiderivative**
☑ **Antidifferentiation Formulas**

11-1. Definition of the Antiderivative

An **antiderivative** of a function $f(x)$ is another function $g(x)$ whose derivative is $f(x)$. In some texts the antiderivative of f is called the **indefinite integral** of f. **Antidifferentiation** is the reverse process to differentiation.

EXAMPLE 11-1: Find an antiderivative of $f(x) = 4x^3$.

Solution: Because $\frac{d}{dx}(x^4) = 4x^3$, $g(x) = x^4$ is an antiderivative of $f(x) = 4x^3$. Notice that $h(x) = x^4 + 8$ and $k(x) = x^4 - 3$ are also antiderivatives of $f(x)$ because $h'(x) = f(x)$ and $k'(x) = f(x)$. In general, $x^4 + C$ is an antiderivative of $4x^3$ for any constant C.

The antiderivative of a function is a family of functions, any two of which differ by a constant. The most general antiderivative of the function $f(x)$ is denoted

ANTIDERIVATIVE OF f(x) $$\int f(x)\,dx$$

The function you are antidifferentiating, in this case $f(x)$, is called the **integrand**.

EXAMPLE 11-2: Calculate $\int (1 + x)\,dx$.

Solution: Because $\frac{d}{dx}(x + \frac{1}{2}x^2 + C) = 1 + x$, you see that $\int (1 + x)\,dx = x + \frac{1}{2}x^2 + C$. Don't forget to include the $+C$, the so-called **constant of antidifferentiation** or **constant of integration**. The significance of the constant of antidifferentiation is that no matter what value C might take, $x + \frac{1}{2}x^2 + C$ is still an antiderivative of $1 + x$.

EXAMPLE 11-3: Calculate $\int t^2\,dt$.

Solution: Because $\frac{d}{dt}(t^3) = 3t^2$, you should notice that $\frac{d}{dt}(\frac{1}{3}t^3) = t^2$, and so $\int t^2\,dt = \frac{1}{3}t^3 + C$.

EXAMPLE 11-4: Calculate $\int 2x^{10}\,dx$.

Solution: First notice that $\frac{d}{dx}(x^{11}) = 11x^{10}$.

Thus you recognize that $\frac{d}{dx}\left(\frac{2}{11}x^{11}\right) = 2x^{10}$, and so $\int 2x^{10}\,dx = \frac{2}{11}x^{11} + C$.

EXAMPLE 11-5: Calculate $\int x^n\,dx$ where n is a real number not equal to -1.

Solution: From the previous examples you can now guess the formula:

$$\int x^n\,dx = \frac{1}{n+1}x^{n+1} + C$$

because

$$\frac{d}{dx}\left(\frac{1}{n+1}x^{n+1}\right) = x^n$$

EXAMPLE 11-6: Calculate $\int e^{2x}\,dx$.

Solution: The derivative of e^{2x} is $2e^{2x}$, so $\frac{d}{dx}(\frac{1}{2}e^{2x}) = e^{2x}$. Thus

$$\int e^{2x}\,dx = \frac{1}{2}e^{2x} + C$$

At this stage you are using clever guesswork to find antiderivatives. In Chapters 13 and 14 you will learn the techniques of antidifferentiation. Even when you master these techniques, there will be functions that you will be unable to antidifferentiate. Antidifferentiation is inherently more difficult than differentiation. There is no set of rules (as there is for differentiation) that enables you to antidifferentiate any complicated function one step at a time. For example, there is no "quotient rule" for antidifferentiating the quotient of two functions in terms of the antiderivatives of the two functions.

Note:

$$\int f(x)g(x)\,dx \neq \int f(x)\,dx \cdot \int g(x)\,dx$$

$$\int f(x)g(x)\,dx \neq f(x)\int g(x)\,dx$$

and

$$\int \frac{f(x)}{g(x)}\,dx \neq \frac{\int f(x)\,dx}{\int g(x)\,dx}$$

11-2. Antidifferentiation Formulas

The formulas for the derivatives of the basic functions (see Chapter 4) can be reversed to form the antidifferentiation formulas. To be successful in antidifferentiation, you must first be familiar with these formulas. If you already know the differentiation formulas, then these will be easy to learn.

1. $\int Cf(x)\,dx = C\int f(x)\,dx$

for any constant C

2. $\int [f(x) + g(x)]\,dx$

$= \int f(x)\,dx + \int g(x)\,dx$

3. $\int x^n\,dx = \frac{1}{n+1}x^{n+1} + C$

for $n \neq -1$

3a. $\int a\,dx = ax$

4. $\int \cos x\,dx = \sin x + C$

5. $\int \sin x\,dx = -\cos x + C$

6. $\int \sec^2 x\,dx = \tan x + C$

7. $\int \sec x \tan x\,dx = \sec x + C$

8. $\int \csc x \cot x\,dx = -\csc x + C$

9. $\int \csc^2 x\,dx = -\cot x + C$

10. $\int a^x\,dx = \frac{1}{\ln a}a^x + C$

for $a > 0$, $a \neq 1$

11. $\int e^x\,dx = e^x + C$

12. $\int \frac{1}{x}\,dx = \ln|x| + C$

13. $\int \frac{1}{\sqrt{1-x^2}}\,dx = \text{arc sin } x + C$

14. $\int \frac{1}{1+x^2}\,dx = \text{arc tan } x + C$

15. $\int \frac{1}{x\sqrt{x^2-1}}\,dx =$ arc sec $|x| + C$

16. $\int \cosh x\,dx = \sinh x + C$

17. $\int \sinh x\,dx = \cosh x + C$

18. $\int \text{sech}^2 x\,dx = \tanh x + C$

19. $\int \text{sech } x \tanh x\,dx = -\text{sech } x + C$

20. $\int \text{csch } x \coth x\,dx = -\text{csch } x + C$

21. $\int \text{csch}^2 x\,dx = -\coth x + C$

EXAMPLE 11-7: Calculate $\int (5x^2 - 2x + 3)\,dx$.

Solution: Use the rules for the antiderivative to reduce the problem to one of antidifferentiating basic functions:

$$\begin{aligned}\int (5x^2 - 2x + 3)\,dx &= \int 5x^2\,dx + \int (-2x)\,dx + \int 3\,dx\\ &= 5\int x^2\,dx - 2\int x\,dx + 3\int 1\,dx\\ &= 5\int x^2\,dx - 2\int x^1\,dx + 3\int x^0\,dx\\ &= 5\cdot\frac{1}{2+1}x^{2+1} - 2\cdot\frac{1}{1+1}x^{1+1}\\ &\quad + 3\cdot\frac{1}{0+1}x^{0+1} + C\\ &= \frac{5}{3}x^3 - x^2 + 3x + C\end{aligned}$$

If you want to check your answer, you need only differentiate:

$$\frac{d}{dx}\left(\frac{5}{3}x^3 - x^2 + 3x + C\right) = 5x^2 - 2x + 3$$

EXAMPLE 11-8: Calculate $\int [(5/x) - 2\sqrt[3]{x^2}]\,dx$.

Solution: When you see a radical symbol in an integrand, rewrite it using an exponent (e.g., $\sqrt[3]{x^2} = x^{2/3}$):

$$\begin{aligned}\int\left(\frac{5}{x} - 2\sqrt[3]{x^2}\right)dx &= 5\int\frac{1}{x}\,dx - 2\int x^{2/3}\,dx\\ &= 5\ln|x| - 2\left[\frac{1}{(\frac{2}{3})+1}x^{(2/3)+1}\right] + C\\ &= 5\ln|x| - \frac{6}{5}x^{5/3} + C\end{aligned}$$

EXAMPLE 11-9: Calculate $\int \frac{x^5 + 3x - 2}{x^3}\,dx$.

Solution: Simplify the integrand.

$$\int \frac{x^5 + 3x - 2}{x^3}\,dx = \int \left(\frac{x^5}{x^3} + \frac{3x}{x^3} - \frac{2}{x^3}\right) dx$$
$$= \int \left(x^2 + \frac{3}{x^2} - \frac{2}{x^3}\right) dx$$
$$= \int (x^2 + 3x^{-2} - 2x^{-3})\,dx$$
$$= \frac{1}{2+1}x^{2+1} + 3\frac{1}{-2+1}x^{-2+1} - 2\frac{1}{-3+1}x^{-3+1} + C$$
$$= \frac{1}{3}x^3 - 3x^{-1} + x^{-2} + C$$

Caution: $\int \frac{3}{x^2}\,dx \neq \frac{3}{x^3/3} + C.$

EXAMPLE 11-10: Find the function $f(x)$ such that $f'(x) = x^4$ and $f(0) = 2$.

Solution: If $f'(x) = x^4$, then f is an antiderivative of x^4.

$$\int x^4\,dx = \frac{1}{5}x^5 + C$$

Thus $f(x) = \frac{1}{5}x^5 + C$ for some constant C. You can find C because you know $f(0) = 2$:

$$2 = f(0) = \tfrac{1}{5}0^5 + C = C$$

So $C = 2$ and $f(x) = \frac{1}{5}x^5 + 2$.

EXAMPLE 11-11: Find the function $f(x)$ such that $f'(x) = 1/\sqrt{x}$ and $f(4) = 1$.

Solution: Because $f'(x) = x^{-1/2}$, $f(x)$ must be an antiderivative of $x^{-1/2}$:

$$\int x^{-1/2}\,dx = \frac{1}{-\frac{1}{2}+1}x^{(-1/2)+1} + C = 2x^{1/2} + C$$

So $f(x) = 2x^{1/2} + C$ for some constant C. However:

$$1 = f(4) = 2(4)^{1/2} + C = 4 + C$$
$$C = -3$$

Thus $f(x) = 2x^{1/2} - 3$.

SUMMARY

1. To antidifferentiate a function f, search for a function whose derivative is f.
2. Antidifferentiation can't be broken down into a step-by-step procedure. However, the antiderivative of the sum of two functions is the sum of the antiderivatives of the two functions. Also, the antiderivative of a constant times a function is the constant times the antiderivative of the function.
3. The antiderivative of a function is a family of functions, any two of which differ by a constant.
4. You should express radicals in the integrand as exponents in order to use the antidifferentiation formulas.

SOLVED PROBLEMS

PROBLEM 11-1 Calculate $\int (2x^7 - x^3 + 5)\,dx$.

Solution: First split the integrand into its components:

$$\int (2x^7 - x^3 + 5)\,dx = \int 2x^7\,dx + \int -x^3\,dx + \int 5\,dx$$

and then recall that the antiderivative of a constant times a function is the constant times the antiderivative of the function.

$$\int 2x^7\,dx + \int -x^3\,dx + \int 5\,dx = 2\int x^7\,dx - \int x^3\,dx + 5\int 1\,dx$$

$$= 2\frac{x^8}{8} - \frac{x^4}{4} + 5x + C$$

$$= \frac{x^8}{4} - \frac{x^4}{4} + 5x + C$$

[See Section 11-2.]

PROBLEM 11-2 Calculate $\displaystyle\int (2x^{5/2} + x^3)\,dx$.

Solution: You find that

$$\int (2x^{5/2} + x^3)\,dx = 2\int x^{5/2}\,dx + \int x^3\,dx$$

$$= 2\frac{x^{7/2}}{7/2} + \frac{x^4}{4} + C$$

$$= \frac{4}{7}x^{7/2} + \frac{1}{4}x^4 + C$$

[See Section 11-2.]

PROBLEM 11-3 Calculate $\displaystyle\int [e^x - (5/\sqrt{1 - x^2}) + 2\sin x]\,dx$.

Solution: You find that

$$\int \left(e^x - \frac{5}{\sqrt{1 - x^2}} + 2\sin x\right) dx = \int e^x\,dx - 5\int \frac{dx}{\sqrt{1 - x^2}} + 2\int \sin x\,dx$$

$$= e^x - 5\arcsin x - 2\cos x + C$$

[See Section 11-2.]

PROBLEM 11-4 Calculate $\displaystyle\int \left(\frac{1}{x^3} + \frac{1}{x^2} + \frac{1}{x}\right) dx$.

Solution: You find that

$$\int \left(\frac{1}{x^3} + \frac{1}{x^2} + \frac{1}{x}\right) dx = \int (x^{-3} + x^{-2} + x^{-1})\,dx$$

$$= \frac{1}{-3 + 1}x^{-3+1} + \frac{1}{-2 + 1}x^{-2+1} + \ln|x| + C$$

$$= -\frac{1}{2}x^{-2} - x^{-1} + \ln|x| + C$$

[See Section 11-2.]

PROBLEM 11-5 Calculate $\int [\sqrt{x^3} + (8/x^4) - 1]\, dx$.

Solution: First make all radicals and powers in denominators into powers of x, and then split the integrand into manageable pieces and integrate:

$$\begin{aligned}
\int\left(\sqrt{x^3} + \frac{8}{x^4} - 1\right) dx &= \int (x^{3/2} + 8x^{-4} - 1)\, dx \\
&= \int x^{3/2}\, dx + 8\int x^{-4}\, dx - \int 1\, dx \\
&= \frac{x^{(3/2)+1}}{(3/2)+1} + 8\frac{x^{-4+1}}{-4+1} - x + C \\
&= \frac{2}{5}x^{5/2} - \frac{8}{3}x^{-3} - x + C
\end{aligned}$$

[See Section 11-2.]

PROBLEM 11-6 Calculate $\int (\sin x + 3\sec^2 x)\, dx$.

Solution: You find that

$$\int (\sin x + 3\sec^2 x)\, dx = \int \sin x\, dx + 3\int \sec^2 x\, dx = -\cos x + 3\tan x + C$$

[See Section 11-2.]

PROBLEM 11-7 Calculate $\int (x^e + e^x)\, dx$.

Solution: The integrand consists of the sum of two functions, x^e and e^x, which you antidifferentiate by different formulas.

$$\int (x^e + e^x)\, dx = \int x^e\, dx + \int e^x\, dx = \frac{1}{e+1}x^{e+1} + e^x + C$$

[See Section 11-2.]

PROBLEM 11-8 Calculate $\int \frac{x^3 + x - 2}{x^2}\, dx$.

Solution: First express the integrand in a form that allows you to use the antidifferentiation formulas:

$$\begin{aligned}
\int \frac{x^3 + x - 2}{x^2}\, dx &= \int \frac{x^3}{x^2}\, dx + \int \frac{x}{x^2}\, dx - 2\int \frac{1}{x^2}\, dx \\
&= \int x\, dx + \int \frac{1}{x}\, dx - 2\int x^{-2}\, dx \\
&= \frac{1}{2}x^2 + \ln|x| + 2x^{-1} + C
\end{aligned}$$

[See Section 11-2.]

PROBLEM 11-9 Calculate $\int \frac{\sqrt{t} - t + t^3}{\sqrt[3]{t}}\, dt$.

Solution: You find that

$$\begin{aligned}
\int \frac{\sqrt{t} - t + t^3}{\sqrt[3]{t}}\, dt &= \int (t^{1/6} - t^{2/3} + t^{8/3})\, dt \\
&= \frac{t^{7/6}}{7/6} - \frac{t^{5/3}}{5/3} + \frac{t^{11/3}}{11/3} + C \\
&= \frac{6}{7}t^{7/6} - \frac{3}{5}t^{5/3} + \frac{3}{11}t^{11/3} + C
\end{aligned}$$

[See Section 11-2.]

PROBLEM 11-10 Calculate $\int \frac{\cos^3 t + 1}{\cos^2 t}\, dt$.

Solution: You find that

$$\int \frac{\cos^3 t + 1}{\cos^2 t}\, dt = \int \left(\cos t + \frac{1}{\cos^2 t}\right) dt$$

$$= \int (\cos t + \sec^2 t)\, dt$$

$$= \sin t + \tan t + C$$

[See Section 11-2.]

PROBLEM 11-11 Calculate $\int \frac{u^2 - 1}{u - 1}\, du$.

Solution: You find that

$$\int \frac{u^2 - 1}{u - 1}\, du = \int \frac{(u - 1)(u + 1)}{u - 1}\, du = \int (u + 1)\, du = \frac{1}{2}u^2 + u + C$$

[See Section 11-2.]

PROBLEM 11-12 Given that $\frac{d}{dx}\left(\frac{x^2 - 1}{x + 2}\right) = \frac{x^2 + 4x + 1}{(x + 2)^2}$, calculate $\int \frac{3x^2 + 12x + 3}{(x + 2)^2}\, dx$.

Solution: You find that

$$\int \frac{3x^2 + 12x + 3}{(x + 2)^2}\, dx = 3\int \frac{x^2 + 4x + 1}{(x + 2)^2}\, dx$$

But you know that $(x^2 - 1)/(x + 2)$ is an antiderivative of $(x^2 + 4x + 1)/[(x + 2)^2]$, so:

$$3\int \frac{x^2 + 4x + 1}{(x + 2)^2}\, dx = 3\left(\frac{x^2 - 1}{x + 2}\right) + C$$

[See Section 11-2.]

PROBLEM 11-13 Calculate $\frac{d}{dx}\int \frac{x^2 + 7}{x^2 + 10}\, dx$.

Solution: There is no need to antidifferentiate $(x^2 + 7)/(x^2 + 10)$. The definition of the antiderivative assures you that the derivative of the antiderivative of a function f is the function f, so:

$$\frac{d}{dx}\int \frac{x^2 + 7}{x^2 + 10}\, dx = \frac{x^2 + 7}{x^2 + 10}$$

[See Section 11-2.]

PROBLEM 11-14 If $f'(x) = 3x^2 + 2$ and $f(0) = 5$, find $f(x)$.

Solution: Because $f'(x) = 3x^2 + 2$, you see that $f(x)$ is an antiderivative of $3x^2 + 2$:

$$\int (3x^2 + 2)\, dx = x^3 + 2x + C$$

So $f(x) = x^3 + 2x + C$ for some constant C. You can use the initial condition $f(0) = 5$ to find C:

$$5 = f(0) = 0^3 + 2(0) + C$$

Thus $C = 5$ and $f(x) = x^3 + 2x + 5$.

[See Section 11-2.]

PROBLEM 11-15 If $f'(x) = 3\sqrt{x}$ and $f(1) = -1$, find $f(x)$.

Solution: The function $f(x)$ is an antiderivative of $3x^{1/2}$;

$$\int 3x^{1/2}\, dx = 3\left(\frac{x^{3/2}}{3/2}\right) + C = 2x^{3/2} + C$$

So $f(x) = 2x^{3/2} + C$ for some constant C:

$$-1 = f(1) = 2(1)^{3/2} + C = 2 + C$$

Thus $C = -3$ and $f(x) = 2x^{3/2} - 3$. [See Section 11-2.]

PROBLEM 11-16 If $f'(x) = e^x + 1 + 2\sin x$ and $f(0) = 4$, find $f(x)$.

Solution: Because f is an antiderivative of f',

$$f(x) = e^x + x - 2\cos x + C$$

$$4 = f(0) = e^0 + 0 - 2\cos 0 + C = -1 + C$$

Thus $C = 5$ and $f(x) = e^x + x - 2\cos x + 5$. [See Section 11-2.]

PROBLEM 11-17 If $f'(x) = 7\sqrt[3]{x^4} + x$ and $f(1) = \frac{1}{2}$, find $f(8)$.

Solution: You find that

$$\int (7\sqrt[3]{x^4} + x)\,dx = 7\int x^{4/3}\,dx + \int x\,dx = 3x^{7/3} + \frac{1}{2}x^2 + C$$

$$f(x) = 3x^{7/3} + \frac{1}{2}x^2 + C$$

$$\frac{1}{2} = f(1) = 3(1)^{7/3} + \frac{1}{2}(1)^2 + C = \frac{7}{2} + C$$

So $C = -3$ and

$$f(x) = 3x^{7/3} + \frac{1}{2}x^2 - 3$$

$$f(8) = 3(8)^{7/3} + \frac{1}{2}(8)^2 - 3 = 413$$

[See Section 11-2.]

PROBLEM 11-18 If $f'(x) = 1/(1 + x^2) + (1/x)$ and $f(1) = \pi/2$, find $f(-\sqrt{3})$.

Solution: You find that

$$f(x) = \arctan x + \ln|x| + C$$

$$\frac{\pi}{2} = f(1) = \arctan 1 + \ln 1 + C = \frac{\pi}{4} + C$$

So $C = \pi/4$ and

$$f(x) = \arctan x + \ln|x| + \frac{\pi}{4}$$

$$f(-\sqrt{3}) = \arctan(-\sqrt{3}) + \ln|-\sqrt{3}| + \frac{\pi}{4} = \frac{-\pi}{3} + \ln\sqrt{3} + \frac{\pi}{4}$$

$$= \frac{-\pi}{12} + \frac{1}{2}\ln 3$$

[See Section 11-2.]

PROBLEM 11-19 If $f'(x) = (x^2 + 1)/x$, find $f(3) - f(1)$.

Solution: You find that

$$\int \frac{x^2 + 1}{x}\,dx = \int \left(x + \frac{1}{x}\right)dx = \frac{1}{2}x^2 + \ln|x| + C$$

$$f(x) = \frac{1}{2}x^2 + \ln|x| + C$$

$$f(3) - f(1) = \left(\frac{3^2}{2} + \ln 3 + C\right) - \left(\frac{1}{2} + \ln 1 + C\right) = 4 + \ln 3$$

[See Section 11-2.]

PROBLEM 11-20 If $f'(x) = e^x + \cos x$, find $f(1) - f(-1)$.

Solution: You find that

$$\int (e^x + \cos x)\,dx = e^x + \sin x + C$$

$$f(x) = e^x + \sin x + C$$

$$f(1) - f(-1) = (e^1 + \sin 1 + C) - (e^{-1} + \sin(-1) + C)$$

$$= e + \sin 1 - e^{-1} + \sin 1 = e - e^{-1} + 2 \sin 1$$

[See Section 11-2.]

Supplementary Exercises

In Problems 11-21 through 11-50 calculate the antiderivative using the formulas in Section 11-2:

11-21 $\int (4x^2 - 7x + 1)\,dx$

11-22 $\int (2 - x^{15})\,dx$

11-23 $\int (x + x^{-1})\,dx$

11-24 $\int (x^{1/2} + x^{-1/2})\,dx$

11-25 $\int (6x^{3/2} + 5x^{-2})\,dx$

11-26 $\int (-x^{\pi} + 3x^{\sqrt{2}} + 3)\,dx$

11-27 $\int \left(\frac{4}{1 + x^2} - 6 \cosh x\right) dx$

11-28 $\int (\sec x \tan x + \sec^2 x)\,dx$

11-29 $\int (e/x)\,dx$

11-30 $\int e^{\sqrt{2}}\,dx$

11-31 $\int (\sqrt{x^3} + \sqrt[3]{x^2})\,dx$

11-32 $\int (4/\sqrt[5]{x^3})\,dx$

11-33 $\int (\sqrt{x^3}/\sqrt[3]{x})\,dx$

11-34 $\int ((x^2 + 3x - 1)/\sqrt{x})\,dx$

11-35 $\int (x + 2)(x - 2)\,dx$

11-36 $\int (2x + 1)^2\,dx$

11-37 $\int ((x^3 - 1)/(x - 1))\,dx$

11-38 $\int \left\{\frac{d}{dx}\left[\frac{x^2 + 3x - 9}{e^x + \sin x}\right]\right\} dx$

11-39 $\int \sqrt[3]{x}(\sqrt{x} + 1)\,dx$

11-40 $\int ((x^2 - 5x + 6)/(x - 2))\,dx$

11-41 $\int (d/dx)[(x^4 - 1)\sqrt{\sqrt{x} + \sqrt[3]{x}}]\,dx$

11-42 $\int (x^{0.99} + x^{1.01})\,dx$

11-43 $\int e\,dx$

11-44 $\int \left[\frac{17}{\sqrt{1 - x^2}} + \sqrt{(x^2 + 1)^2}\right] dx$

11-45 $\int (2 \operatorname{sech} x \tanh x - x)\,dx$

11-46 $\int ((x \sin x + 1)/x)\,dx$

11-47 $\int (5^x - 4^x)\,dx$

11-48 $\int \sqrt{\pi}\,dx$

11-49 $\int (x^{-0.99} + x^{-1})\,dx$

11-50 $\int (\tan x/(\sin^2 x \sec x + \cos x))\,dx$

In Problems 11-51 through 11-70 find the function $f(x)$ that satisfies the given condition:

11-51 $f'(x) = x^2 \quad f(0) = 2$

11-52 $f'(x) = 4x^5 \quad f(1) = 1$

11-53 $f'(x) = 1/x \quad f(e) = 2$

11-54 $f'(x) = 9\sqrt{x^7} \quad f(4) = 0$

11-55 $f'(x) = \sin x - \cos x \quad f(0) = 3$

11-56 $f'(x) = 3x^2 - 4x + 3x^{-2} \quad f(2) = \frac{1}{2}$

11-57 $f'(x) = (7x^3 + \sqrt{x})/\sqrt[3]{x^5} \quad f(1) = 1$

11-58 $f'(x) = (x - 1)(x^2 + 3) \quad f(0) = 3$

11-59 $f'(x) = e^x + x \quad f(0) = 2$

11-60 $f'(x) = (1/\sqrt{1 - x^2}) + 1 - x^2 \quad f(\frac{1}{2}) = 11/24$

11-61 $f'(x) = 3 \cosh x - \sinh x \quad f(0) = -7$

11-62 $f'(x) = (7/x) + (7/x^2) \quad f(-1) = 6$

11-63 $f'(x) = (x^2 - 3x + 2)/(x - 2) \quad f(-2) = 0$

11-64 $f'(x) = 2^x + x^2 \quad f(3) = -1$

11-65 $f'(x) = 5^2 \quad f(2) = 20$

11-66 $f'(x) = \pi \sec^2 x \quad f(\pi/3) = 0$

11-67 $f'(x) = \sqrt{x}[x - (1/\sqrt[3]{x})] \quad f(0) = -3$

11-68 $f'(x) = (3 - 2x)^2 \quad f(3) = 2$

11-69 $f'(x) = (x^2 - 3x + 1)/e \quad f(1) = 0$

11-70 $f'(x) = (x^2 - 3x + 1)/x \quad f(e) = 1$

11-71 If $f'(x) = 4x^3 - 1$ and $f(0) = 3$, find $f(1)$.

11-72 If $f'(x) = x^2 - 2x + 1$ and $f(0) = 2$, find $f(3)$.

11-73 If $f'(x) = x^3 + 3x - 1$ and $f(2) = 6$, find $f(0)$.

11-74 If $f'(x) = 3 - (2/x)$ and $f(1) = 2$, find $f(e)$.

11-75 If $f'(x) = 3x^2 + 3x^{-2}$ and $f(3) = 11$, find $f(1)$.

11-76 If $f'(x) = \cos x + \sin x$ and $f(0) = 3$, find $f(\pi/2)$.

11-77 If $f'(x) = \sqrt{x} + e^x$ and $f(0) = -4$, find $f(9)$.

11-78 If $f'(x) = 2/(1 + x^2)$ and $f(1) = \pi/4$, find $f(0)$.

11-79 If $f'(x) = (\sqrt{x} - 1)/\sqrt{x}$ and $f(9) = 1$, find $f(4)$.

11-80 If $f'(x) = 2\,\text{sech}^2\,x$ and $f(0) = 3$, find $f(\ln 2)$.

11-81 If $f'(x) = x^2 + 1$, find $f(3) - f(0)$.

11-82 If $f'(x) = \sqrt[3]{x}$, find $f(27) - f(8)$.

11-83 If $f'(x) = (x + 1)/x$, find $f(2e) - f(e)$.

11-84 If $f'(x) = 2^x$, find $f(1) - f(0)$.

11-85 If $f'(x) = \sec x \tan x$, find $f(\pi/4) - f(0)$.

Solutions to Supplementary Exercises

(11-21) $(4/3)x^3 - (7/2)x^2 + x + C$

(11-22) $2x - (x^{16}/16) + C$

(11-23) $\frac{1}{2}x^2 + \ln|x| + C$

(11-24) $\frac{2}{3}x^{3/2} + 2x^{1/2} + C$

(11-25) $(12/5)x^{5/2} - 5x^{-1} + C$

(11-26) $\dfrac{-1}{\pi + 1}x^{\pi+1} + \dfrac{3}{\sqrt{2} + 1}x^{\sqrt{2}+1} + 3x + C$

(11-27) $4 \arcsin$ — $4 \arctan x - 6 \sinh x + C$

(11-28) $\sec x + \tan x + C$

(11-29) $e \ln|x| + C$

(11-30) $e^{\sqrt{2}}x + C$

(11-31) $(2/5)x^{5/2} + (3/5)x^{5/3} + C$

(11-32) $10x^{2/5} + C$

(11-33) $(6/13)x^{13/6} + C$

(11-34) $(2/5)x^{5/2} + 2x^{3/2} - 2x^{1/2} + C$

(11-35) $(x^3/3) - 4x + C$

(11-36) $(4/3)x^3 + 2x^2 + x + C$

(11-37) $(x^3/3) + (x^2/2) + x + C$

(11-38) $(x^2 + 3x - 9)/(e^x + \sin x) + C$

(11-39) $(6/11)x^{11/6} + (3/4)x^{4/3} + C$

(11-40) $\frac{1}{2}x^2 - 3x + C$

(11-41) $(x^4 - 1)\sqrt{\sqrt{x} + \sqrt[3]{x}} + C$

(11-42) $(x^{1.99}/1.99) + (x^{2.01}/2.01) + C$

(11-43) $ex + C$

(11-44) $17 \arcsin x + (x^3/3) + x + C$

(11-45) $-2\,\text{sech}\,x - \frac{1}{2}x^2 + C$

(11-46) $-\cos x + \ln|x| + C$

(11-47) $[(1/\ln 5)5^x] - [(1/\ln 4)4^x] + C$

(11-48) $\sqrt{\pi}x + C$

(11-49) $100x^{0.01} + \ln|x| + C$

(11-50) $-\cos x + C$

(11-51) $f(x) = (x^3/3) + 2$

(11-52) $f(x) = (2/3)x^6 + (1/3)$

(11-53) $f(x) = \ln|x| + 1$

(11-54) $f(x) = 2x^{9/2} - 1024$

(11-55) $f(x) = -\cos x - \sin x + 4$

(11-56) $f(x) = x^3 - 2x^2 - 3x^{-1} + 2$

(11-57) $f(x) = 3x^{7/3} - 6x^{-1/6} + 4$

(11-58) $f(x) = (1/4)x^4 - (1/3)x^3 + (3/2)x^2 - 3x + 3$

(11-59) $f(x) = e^x + \frac{1}{2}x^2 + 1$

(11-60) $f(x) = \arcsin x + x - (x^3/3) - (\pi/6)$

(11-61) $f(x) = 3 \sinh x - \cosh x - 6$

(11-62) $f(x) = 7 \ln|x| - 7x^{-1} - 1$

(11-63) $f(x) = \frac{1}{2}x^2 - x - 4$

(11-64) $f(x) = (2^x/\ln 2) + (x^3/3) - 10 - (8/\ln 2)$

(11-65) $f(x) = 25x - 30$

(11-66) $f(x) = \pi \tan x - \pi\sqrt{3}$

(11-67) $f(x) = (2/5)x^{5/2} - (6/7)x^{7/6} - 3$

(11-68) $f(x) = (4/3)x^3 - 6x^2 + 9x - 7$

(11-69) $f(x) = (x^3/3e) - (3x^2/2e) + (x/e) + (1/6e)$

(11-70) $f(x) = \frac{1}{2}x^2 - 3x + \ln|x| + 3e - \frac{1}{2}e^2$

(11-71) 3

(11-72) 5

(11-73) -2

(11-74) $3e - 3$

(11-75) -17

(11-76) 5

(11-77) $13 + e^9$

(11-78) $-\pi/4$

(11-79) -2

(11-80) 21/5

(11-81) 12

(11-82) 195/4

(11-83) $e + \ln 2$

(11-84) $1/\ln 2$

(11-85) $\sqrt{2} - 1$

12 THE INTEGRAL

THIS CHAPTER IS ABOUT

☑ **The Integral**
☑ **The Fundamental Theorem of Calculus**
☑ **Riemann Sums**

12-1. The Integral

The **integral** (sometimes called the **definite integral**) of a function $f(x)$ between $x = a$ and $x = b$ is denoted

$$\int_a^b f(x)\,dx$$

and may be interpreted as the area of the region bounded by the graph of $y = f(x)$, the x axis, and the vertical lines $x = a$ and $x = b (a < b)$ (Figure 12-1). Any area above the x axis is given a positive value; any area below the x axis is given a negative value. The function $f(x)$ is called the **integrand**. The values $x = a$ and $x = b$ are called the **limits of integration**.

The notation for an integral resembles the notation for an antiderivative. This is because you can calculate integrals using antiderivatives (see Section 12-2). Let's examine three examples that illustrate the concept of the integral as an area.

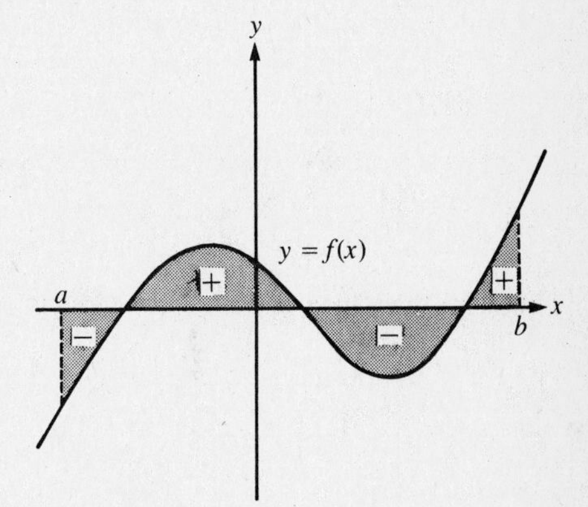

Figure 12-1

EXAMPLE 12-1: Evaluate $\int_1^3 (x - 1)\,dx$ by finding the area of the appropriate region.

Solution: You must examine the region bordered by the graphs of

$y = f(x) = x - 1$ (the integrand)

the x axis

$x = 1$ (the lower limit of integration)

$x = 3$ (the upper limit of integration)

This triangular region is shown in Figure 12-2. The area of this triangle is

$$\tfrac{1}{2}(\text{base}) \times (\text{height}) = \tfrac{1}{2}(2)(2) = 2$$

Because all of this region lies above the x axis, you'll give this area a positive value, $+2$. So you have

$$\int_1^3 (x - 1)\,dx = 2$$

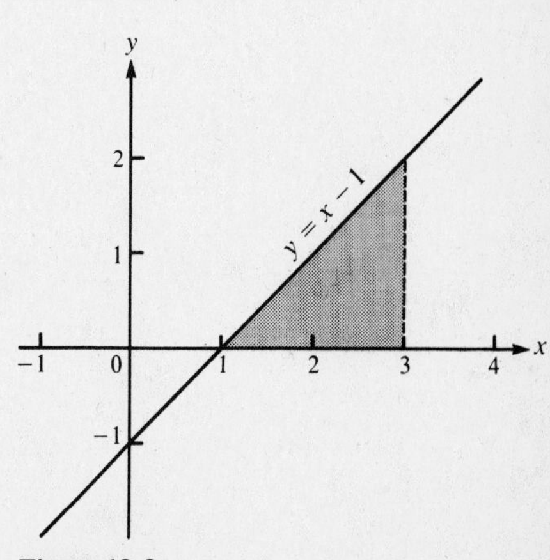

Figure 12-2

EXAMPLE 12-2: Calculate $\int_2^3 (1 - x)\,dx$ by finding the area of the appropriate region.

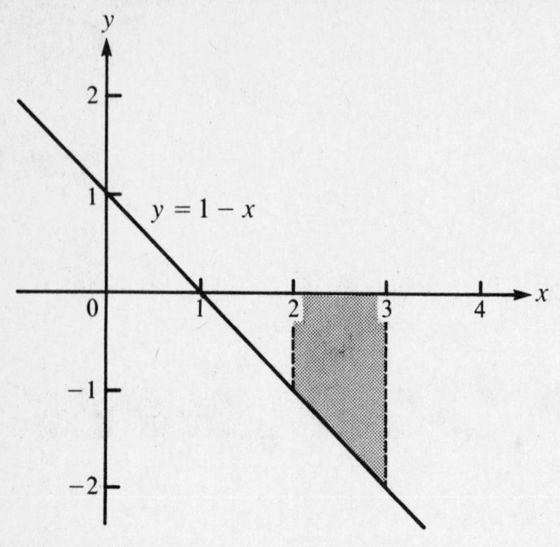

Figure 12-3

Solution: First examine the region bounded by the graphs of $y = 1 - x$, the x axis, $x = 2$, and $x = 3$, as shown in Figure 12-3. You can view the region as a rectangle and a triangle, so you have

$$\text{area of triangle} + \text{area of rectangle} = \tfrac{1}{2}(1)(1) + (1)(1) = \tfrac{3}{2}$$

Because this area lies below the x axis, you'll give this area a negative value:

$$\int_2^3 (1 - x)\,dx = \frac{-3}{2}$$

EXAMPLE 12-3: Calculate $\int_{-1}^{2} (1 - x)\,dx$ by finding the area of the appropriate region.

Solution: Examine the region bordered by the graphs of $y = 1 - x$, the x axis, $x = -1$, and $x = 2$, as shown in Figure 12-4. The triangle above the x axis has area 2; the triangle below the x axis has area $\frac{1}{2}$. The integral is the area above the x axis minus the area below the x axis:

$$\int_{-1}^{2} (1 - x)\,dx = 2 - \frac{1}{2} = \frac{3}{2}$$

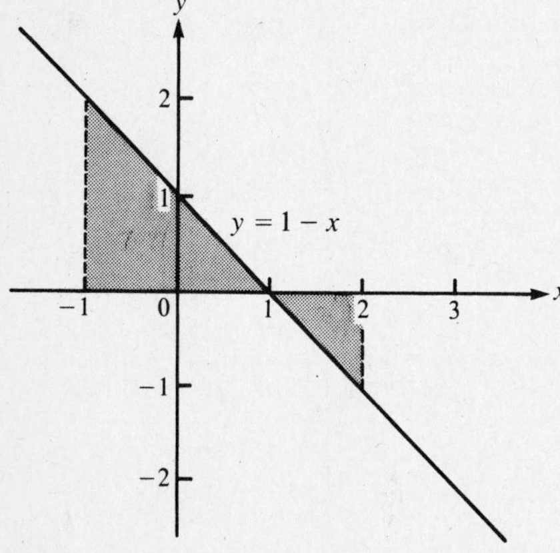

Figure 12-4

You'll undoubtedly appreciate that using such techniques for complicated integrals would be tedious at best and sometimes impossible. Fortunately, you can evaluate integrals using the Fundamental Theorem of Calculus, not by finding areas of regions.

12-2. The Fundamental Theorem of Calculus

The **Fundamental Theorem of Calculus** enables you to evaluate integrals using antiderivatives. Let $f(x)$ be a continuous function on $[a, b]$ and let $F(x)$ be any antiderivative of $f(x)$. Then:

FUNDAMENTAL THEOREM OF CALCULUS

$$\int_a^b f(x)\,dx = F(b) - F(a) \qquad \textbf{(12-1)}$$

To evaluate $\int_a^b f(x)\,dx$ using the Fundamental Theorem, you must first find $F(x)$, that is, any antiderivative of $f(x)$. You may recognize the antiderivative immediately, or you may need one or more of the techniques of integration from Chapters 13 and 14. Then you evaluate $F(x)$ at $x = a$ and $x = b$ and find the difference $F(b) - F(a)$.

EXAMPLE 12-4: Evaluate $\displaystyle\int_1^3 x\,dx$.

Solution: The function $F(x) = \frac{1}{2}x^2$ is an antiderivative of $f(x) = x$. From the Fundamental Theorem of Calculus,

$$\int_1^3 x\,dx = F(3) - F(1) = \frac{1}{2}(3)^2 - \frac{1}{2}(1)^2 = 4$$

However, you could have used any antiderivative of $f(x) = x$. For example, $F(x) = \frac{1}{2}x^2 + 17$ is an antiderivative of $f(x) = x$:

$$\int_1^3 x\,dx = F(3) - F(1) = \left[\frac{1}{2}(3)^2 + 17\right] - \left[\frac{1}{2}(1)^2 + 17\right] = 4$$

Because any antiderivative of $f(x)$ is of the form $\frac{1}{2}x^2 + C$ for some constant C, you might as well use the simplest antiderivative, $F(x) = \frac{1}{2}x^2$.

EXAMPLE 12-5: Evaluate $\displaystyle\int_{-1}^{0} x^3\,dx$.

Solution: The simplest antiderivative of $f(x) = x^3$ is $F(x) = \frac{1}{4}x^4$. Thus you have

$$\int_{-1}^{0} x^3\,dx = F(0) - F(-1) = \frac{1}{4}(0)^4 - \frac{1}{4}(-1)^4 = -\frac{1}{4}$$

Because you'll calculate $F(b) - F(a)$ so often, use the notation

$$F(x)\Big|_a^b = F(b) - F(a)$$

EXAMPLE 12-6: Evaluate $\int_2^4 (3x^2 - 4)\,dx$.

Solution: An antiderivative of $f(x) = 3x^2 - 4$ is $F(x) = x^3 - 4x$. You have

$$\int_2^4 (3x^2 - 4)\,dx = (x^3 - 4x)\Big|_2^4 = [(4)^3 - 4(4)] - [(2)^3 - 4(2)] = 48$$

EXAMPLE 12-7: Evaluate $\int_0^1 \left(3e^x + \frac{1}{1+x^2}\right)dx$.

Solution: An antiderivative of $f(x) = 3e^x + [1/(1 + x^2)]$ is $F(x) = 3e^x +$ arc tan x:

$$\int_0^1 \left(3e^x + \frac{1}{1+x^2}\right)dx = (3e^x + \text{arc tan } x)\Big|_0^1$$

$$= (3e^1 + \text{arc tan } 1) - (3e^0 + \text{arc tan } 0)$$

$$= \left(3e + \frac{\pi}{4}\right) - (3 + 0) = 3e + \frac{\pi}{4} - 3$$

12-3. Riemann Sums

Since you can think of the integral as an area, you can approximate the value of an integral by approximating the area. A **Riemann sum** is a sum of areas of rectangles and is one method of approximating the value of an integral.

To estimate $\int_a^b f(x)\,dx$, where $a < b$, first divide the interval $[a, b]$ into n subintervals, using $x_0, x_1, \ldots, x_n$ as endpoints for the subintervals:

$$a = x_0 < x_1 < \cdots < x_{n-1} < x_n = b$$

The subintervals are then

$$[x_0, x_1], [x_1, x_2], \ldots, [x_{n-1}, x_n]$$

For simplicity, let the subintervals have the same length, $\Delta x = (b - a)/n$. You then have $x_i = a + i\Delta x$.

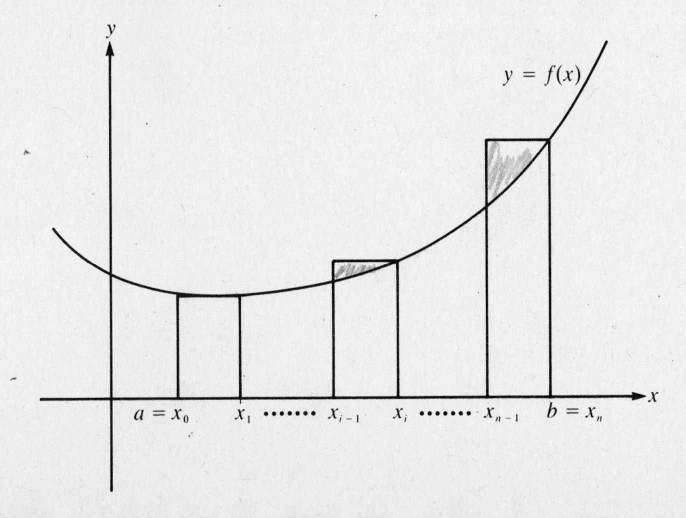

Figure 12-5

Choose a point c in each subinterval and construct a rectangle above the subinterval with height $f(c)$. For simplicity, above the ith subinterval $[x_{i-1}, x_i]$ make the height of the rectangle $f(x_i)$. This is shown in Figure 12-5. Then the sum of the areas of the rectangles is an approximation to $\int_a^b f(x)\,dx$. The ith rectangle has width Δx and height $f(x_i)$, and hence area $A_i = f(x_i)\,\Delta x$. Thus:

$$\int_a^b f(x)\,dx \approx A_1 + A_2 + \cdots + A_n = f(x_1)\,\Delta x + f(x_2)\,\Delta x + \cdots + f(x_n)\,\Delta x$$

$$= \sum_{i=1}^{n} f(x_i)\,\Delta x$$

This is a Riemann sum for the integral $\int_a^b f(x)\,dx$. (Refer to Chapter 1 for a review of summation notation if necessary.) In general, if you increase n, you'll improve your approximation. For all continuous functions f on a closed boundary interval $[a, b]$,

$$\lim_{n \to \infty} \sum_{i=1}^{n} f(x_i)\Delta x = \int_a^b f(x)\,dx$$

Note: Definite integrals may exist for discontinuous functions, depending on the type of discontinuity.

A. Calculating a limit of Riemann sums

For very simple functions f (e.g., polynomials of low degree) you can explicitly calculate the limit of the Riemann sums without antidifferentiating. You may need the following formulas:

$$\sum_{i=1}^{n} i = \frac{n(n+1)}{2}$$

$$\sum_{i=1}^{n} i^2 = \frac{n(n+1)(2n+1)}{6} \qquad \textbf{(12-3)}$$

$$\sum_{i=1}^{n} i^3 = \frac{n^2(n+1)^2}{4}$$

EXAMPLE 12-8: Find the area of the region bounded by the graphs of $f(x) = 2x$, $x = 0$, $x = 1$, and the x axis by calculating the limit of the Riemann sums.

Solution: First divide the interval $[0, 1]$ into n subintervals of equal length:

$$\Delta x = \frac{b-a}{n} = \frac{1-0}{n} = \frac{1}{n}$$

$$x_i = a + i\,\Delta x = 0 + i\frac{1}{n} = \frac{i}{n}$$

The nth Riemann sum is

$$\sum_{i=1}^{n} f(x_i)\,\Delta x = \sum_{i=1}^{n} f\left(\frac{i}{n}\right)\Delta x = \sum_{i=1}^{n} 2\left(\frac{i}{n}\right)\Delta x = \sum_{i=1}^{n} 2\left(\frac{i}{n}\right)\left(\frac{1}{n}\right)$$

$$= \sum_{i=1}^{n} \frac{2}{n^2} i$$

Inside of this summation, $2/n^2$ doesn't involve i, and so you can pull it outside the summation:

$$\sum_{i=1}^{n} \frac{2}{n^2} i = \frac{2}{n^2} \sum_{i=1}^{n} i = \frac{2}{n^2}\left[\frac{n(n+1)}{2}\right] = \frac{2n^2 + 2n}{2n^2} = 1 + \frac{1}{n}$$

Finally, you can find the limit of the Riemann sums:

$$\lim_{n\to\infty} \sum_{i=1}^{n} f(x_i)\,\Delta x = \lim_{n\to\infty}\left(1 + \frac{1}{n}\right) = 1$$

You can verify this result by evaluating the integral

$$\int_0^1 2x\,dx = 1$$

or by recognizing that the area represented by this integral is the area of a triangle with base 1 and height 2.

EXAMPLE 12-9: Find the area of the region bounded by the graphs of $f(x) = x^3 - 2x^2 + 1, x = 1, x = 3$, and the x axis by calculating the limit of the Riemann sums.

Solution: First divide [1, 3] into n subintervals of equal length:

$$\Delta x = \frac{3-1}{n} = \frac{2}{n}$$

$$x_i = a + i\,\Delta x = 1 + \frac{2i}{n}$$

The nth Riemann sum is

$$\begin{aligned}
\sum_{i=1}^{n} f(x_i)\Delta x &= \sum_{i=1}^{n}\left[\left(1+\frac{2i}{n}\right)^3 - 2\left(1+\frac{2i}{n}\right)^2 + 1\right]\frac{2}{n}\\
&= \sum_{i=1}^{n}\left\{1 + 3\frac{2i}{n} + 3\left(\frac{2i}{n}\right)^2 + \left(\frac{2i}{n}\right)^3 - 2\left[1 + \frac{4i}{n} + \frac{4i^2}{n^2}\right] + 1\right\}\frac{2}{n}\\
&= \sum_{i=1}^{n}\left(\frac{8}{n^3}i^3 + \frac{4}{n^2}i^2 - \frac{2}{n}i\right)\frac{2}{n}\\
&= \frac{16}{n^4}\sum_{i=1}^{n} i^3 + \frac{8}{n^3}\sum_{i=1}^{n} i^2 - \frac{4}{n^2}\sum_{i=1}^{n} i\\
&= \frac{16}{n^4}\left[\frac{n^2(n+1)^2}{4}\right] + \frac{8}{n^3}\left[\frac{n(n+1)(2n+1)}{6}\right] - \frac{4}{n^2}\left[\frac{n(n+1)}{2}\right]\\
&= \frac{4(n+1)^2}{n^2} + \frac{4(n+1)(2n+1)}{3n^2} - \frac{2(n+1)}{n}
\end{aligned}$$

and so

$$\begin{aligned}
\lim_{n\to\infty}\sum_{i=1}^{n} f(x_i)\Delta x &= \lim_{n\to\infty}\left[\frac{4(n+1)^2}{n^2} + \frac{4(n+1)(2n+1)}{3n^2} - \frac{2(n+1)}{n}\right]\\
&= 4 + \frac{8}{3} - 2 = \frac{14}{3}
\end{aligned}$$

You can verify this by evaluating the integral

$$\int_1^3 (x^3 - 2x^2 + 1)\,dx = \frac{14}{3}$$

B. Recognizing a limit of Riemann sums as an integral

If $f(x)$ is at all complicated, you'll find Riemann sums of little help in evaluating $\int_a^b f(x)\,dx$. However, you can use them in applications by "going the other way"; many physical problems have certain sums as approximations to their solutions. If you can recognize these sums as Riemann sums for a particular function $f(x)$ over an interval, then the solution to the problem is an integral. You'll find numerous examples in Chapters 18 and 19.

To recognize which integral a Riemann sum represents, you translate as follows: Δx becomes dx, x_i becomes x, and the limits of integration are $a = x_0$, $b = x_n$ from the original problem.

EXAMPLE 12-10: Evaluate $\lim_{n\to\infty} \sum_{i=1}^{n} x_i^3 \Delta x$, where $x_0 = 0, \ldots, x_n = 2$.

Solution: Use the rules for translation: Δx becomes dx, x_i^3 becomes x^3, and the limits of integration are from zero to two. You find

$$\lim_{n\to\infty} \sum_{i=1}^{n} x_i^3 \Delta x = \int_0^2 x^3\,dx = \frac{1}{4}x^4\bigg|_0^2 = \frac{1}{4}(2)^4 - \frac{1}{4}(0)^4 = 4$$

EXAMPLE 12-11: Evaluate $\lim_{n\to\infty} \sum_{i=1}^{n} (1/x_i)\Delta x$, where $x_0 = 1, \ldots, x_n = 5$.

Solution: You have

$$\lim_{n\to\infty} \sum_{i=1}^{n} \frac{1}{x_i}\Delta x = \int_1^5 \frac{1}{x}\,dx = \ln|x|\bigg|_1^5 = \ln|5| - \ln|1| = \ln 5$$

EXAMPLE 12-12: Evaluate $\lim_{n\to\infty} \sum_{i=1}^{n} (x_i - x_{i-1})/(x_i + x_{i-1})$, where $x_0 = 1, \ldots, x_n = 2$.

Solution: Recall that Δx is the width of a subinterval, so $x_i - x_{i-1} = \Delta x$. Translate both x_i and x_{i-1} as x in the integral:

$$\lim_{n\to\infty} \sum_{i=1}^{n} \frac{x_i - x_{i-1}}{x_i + x_{i-1}} = \lim_{n\to\infty} \sum_{i=1}^{n} \frac{\Delta x}{x_i + x_{i-1}} = \int_1^2 \frac{dx}{x + x}$$

$$= \int_1^2 \frac{1}{2x}\,dx = \frac{1}{2}\int_1^2 \frac{1}{x}\,dx$$

$$= \frac{1}{2}\ln|x|\bigg|_1^2 = \frac{1}{2}\ln|2| - \frac{1}{2}\ln|1| = \frac{1}{2}\ln 2$$

SUMMARY

1. You can interpret an integral as an area.
2. Use the Fundamental Theorem of Calculus to evaluate integrals.
3. To integrate, first antidifferentiate.
4. A Riemann sum approximates the value of an integral.
5. For simple functions you can evaluate the limit of the Riemann sums directly.
6. You can evaluate the limit of a Riemann sum by evaluating the appropriate integral.

SOLVED PROBLEMS

PROBLEM 12-1 Evaluate $\int_0^2 x^3\,dx$.

Solution: First find an antiderivative of $f(x) = x^3$:

$$\int x^3\,dx = \frac{1}{4}x^4 + C$$

So $F(x) = \frac{1}{4}x^4$ is an antiderivative of $f(x)$. Finally, you have

$$\int_0^2 x^3\,dx = F(2) - F(0) = \frac{1}{4}(2)^4 - \frac{1}{4}(0)^4 = 4$$

[See Section 12-2.]

PROBLEM 12-2 Evaluate $\int_0^2 (2x^7 - x^3 + 5)\,dx$.

Solution: First find an antiderivative of $f(x) = 2x^7 - x^3 + 5$:

$$\int (2x^7 - x^3 + 5)\,dx = 2\int x^7\,dx - \int x^3\,dx + 5\int dx$$

$$= 2\left(\frac{x^8}{8}\right) - \frac{x^4}{4} + 5x + C$$

So $F(x) = \frac{1}{4}x^8 - \frac{1}{4}x^4 + 5x$ is an antiderivative of $f(x)$. Thus

$$\int_0^2 (2x^7 - x^3 + 5)\,dx = F(2) - F(0)$$

$$= \left[\frac{1}{4}(2)^8 - \frac{1}{4}(2)^4 + 5(2)\right] - \left[\frac{1}{4}(0)^8 - \frac{1}{4}(0)^4 + 5(0)\right]$$

$$= (64 - 4 + 10) - (0) = 70$$

[See Section 12-2.]

PROBLEM 12-3 Evaluate $\int_1^4 (2x^{5/2} + x^3)\,dx$.

Solution: Following the previous examples,

$$\int_1^4 (2x^{5/2} + x^3)\,dx = \left[2\left(\frac{2}{7}x^{7/2}\right) + \frac{1}{4}x^4\right]\Bigg|_1^4$$

$$= \left[\frac{4}{7}(4)^{7/2} + \frac{1}{4}(4)^4\right] - \left[\frac{4}{7}(1)^{7/2} + \frac{1}{4}(1)^4\right]$$

$$= \left[\frac{4}{7}(128) + 64\right] - \left[\frac{4}{7} + \frac{1}{4}\right] = \frac{3817}{28}$$

[See Section 12-2.]

PROBLEM 12-4 Evaluate $\int_0^{1/2} (5/\sqrt{1 - x^2})\,dx$.

Solution: You find that

$$\int_0^{1/2} \frac{5}{\sqrt{1 - x^2}}\,dx = 5\int_0^{1/2} \frac{dx}{\sqrt{1 - x^2}} = 5 \arcsin x\,\Bigg|_0^{1/2}$$

$$= 5 \arcsin \frac{1}{2} - 5 \arcsin 0$$

$$= 5\left(\frac{\pi}{6}\right) - 0 = \frac{5\pi}{6}$$

[See Section 12-2.]

PROBLEM 12-5 Evaluate $\int_0^{\pi} (e^x + 2 \sin x)\,dx$.

Solution: You get

$$\int_0^{\pi} (e^x + 2 \sin x)\,dx = (e^x - 2 \cos x)\Bigg|_0^{\pi}$$

$$= (e^{\pi} - 2 \cos \pi) - (e^0 - 2 \cos 0)$$

$$= e^{\pi} - 2(-1) - (1 - 2) = 3 + e^{\pi}$$

[See Section 12-2.]

PROBLEM 12-6 Evaluate $\int_{-8}^{-1} 1/\sqrt[3]{x^2}\,dx$.

Solution: You find that

$$\int_{-8}^{-1} \frac{1}{\sqrt[3]{x^2}}\,dx = \int_{-8}^{-1} x^{-2/3}\,dx = 3x^{1/3}\Big|_{-8}^{-1} = 3(-1)^{1/3} - 3(-8)^{1/3}$$

$$= 3(-1) - 3(-2) = 3$$

[See Section 12-2.]

PROBLEM 12-7 Evaluate $\int_{\pi/6}^{\pi/4} \sec x(2\sec x + 3\tan x)\,dx$.

Solution: You find that

$$\int_{\pi/6}^{\pi/4} \sec x(2\sec x + 3\tan x)\,dx = \int_{\pi/6}^{\pi/4} (2\sec^2 x + 3\sec x\tan x)\,dx$$

$$= (2\tan x + 3\sec x)\Big|_{\pi/6}^{\pi/4}$$

$$= \left(2\tan\frac{\pi}{4} + 3\sec\frac{\pi}{4}\right) - \left(2\tan\frac{\pi}{6} + 3\sec\frac{\pi}{6}\right)$$

$$= [2(1) + 3\sqrt{2}] - \left[2\frac{\sqrt{3}}{3} + 3\left(\frac{2}{3}\sqrt{3}\right)\right]$$

$$= 2 + 3\sqrt{2} - \frac{8}{3}\sqrt{3}$$

[See Section 12-2.]

PROBLEM 12-8 Evaluate $\int_{1}^{5} 1/x\,dx$.

Solution: You find that

$$\int_{1}^{5} \frac{1}{x}\,dx = (\ln|x|)\Big|_{1}^{5} = \ln|5| - \ln|1| = \ln 5$$

[See Section 12-2.]

PROBLEM 12-9 Evaluate $\int_{-1}^{2} 3^x\,dx$.

Solution: You find that

$$\int_{-1}^{2} 3^x\,dx = \frac{1}{\ln 3}3^x\Big|_{-1}^{2} = \frac{1}{\ln 3}(3^2 - 3^{-1}) = \frac{26}{3\ln 3}$$

[See Section 12-2.]

PROBLEM 12-10 Evaluate $\int_{-\sqrt{3}}^{0} 2/(x^2 + 1)\,dx$.

Solution: You get

$$\int_{-\sqrt{3}}^{0} \frac{2}{x^2 + 1}\,dx = 2\arctan x\Big|_{-\sqrt{3}}^{0} = 2\arctan 0 - 2\arctan(-\sqrt{3})$$

$$= 2(0) - 2\left(\frac{-\pi}{3}\right) = \frac{2\pi}{3}$$

[See Section 12-2.]

PROBLEM 12-11 Evaluate $\int_{17}^{128} \ln 5\,dx$.

Solution: Because ln 5 is a constant, you have

$$\int_{17}^{128} \ln 5\,dx = \ln 5\int_{17}^{128} dx = (\ln 5)x\Big|_{17}^{128}$$

$$= (\ln 5)128 - (\ln 5)17 = 111\ln 5$$

[See Section 12-2.]

PROBLEM 12-12 Evaluate $\int_1^4 (x^2 + 1)/\sqrt{x}\, dx$.

Solution: You get

$$\int_1^4 \frac{x^2+1}{\sqrt{x}}\, dx = \int_1^4 \left(\frac{x^2}{\sqrt{x}} + \frac{1}{\sqrt{x}}\right) dx$$

$$= \int_1^4 (x^{3/2} + x^{-1/2})\, dx = \left(\frac{2}{5}x^{5/2} + 2x^{1/2}\right)\Bigg|_1^4$$

$$= \left[\frac{2}{5}(4)^{5/2} + 2(4)^{1/2}\right] - \left(\frac{2}{5} + 2\right) = \frac{72}{5}$$

[See Section 12-2.]

PROBLEM 12-13 Evaluate $\int_0^1 (2 \sinh x - 3 \cosh x)\, dx$.

Solution: You find that

$$\int_0^1 (2 \sinh x - 3 \cosh x)\, dx = (2 \cosh x - 3 \sinh x)\Big|_0^1$$

$$= (2 \cosh 1 - 3 \sinh 1) - (2 \cosh 0 - 3 \sinh 0)$$

$$= 2 \cosh 1 - 3 \sinh 1 - 2 = \frac{5}{2}e^{-1} - \frac{1}{2}e - 2$$

[See Section 12-2.]

PROBLEM 12-14 Given that $\frac{d}{dx}(\sqrt{x^2 - 1}) = x/\sqrt{x^2 - 1}$, evaluate $\int_{-3}^{-2} (x/\sqrt{x^2 - 1})\, dx$.

Solution: Because $F(x) = \sqrt{x^2 - 1}$ is an antiderivative of $f(x) = x/\sqrt{x^2 - 1}$, you find

$$\int_{-3}^{-2} \frac{x\, dx}{\sqrt{x^2 - 1}} = \sqrt{x^2 - 1}\,\Big|_{-3}^{-2} = \sqrt{4 - 1} - \sqrt{9 - 1} = \sqrt{3} - \sqrt{8}$$

[See Section 12-2.]

PROBLEM 12-15 Evaluate $\int_1^{64} (\sqrt[3]{x^2}/\sqrt{5x^3})\, dx$.

Solution: You find

$$\int_1^{64} \frac{\sqrt[3]{x^2}}{\sqrt{5x^3}}\, dx = \int_1^{64} \frac{1}{\sqrt{5}} x^{(2/3)-(3/2)}\, dx = \frac{\sqrt{5}}{5}\int_1^{64} x^{-5/6}\, dx$$

$$= \frac{6\sqrt{5}}{5} x^{1/6}\Big|_1^{64} = \frac{6\sqrt{5}}{5}[(64)^{1/6} - (1)^{1/6}]$$

$$= \frac{6\sqrt{5}}{5}$$

[See Section 12-2.]

PROBLEM 12-16 Find the area of the region bounded by the graphs of $f(x) = x^2$, $x = 0$, $x = 2$, and the x axis by calculating the limit of the Riemann sums.

Solution: First divide $[0, 2]$ into n subintervals of equal length:

$$\Delta x = \frac{2 - 0}{n} = \frac{2}{n}$$

$$x_i = a + i\,\Delta x = 0 + i\frac{2}{n} = \frac{2i}{n}$$

The nth Riemann sum is

$$\sum_{i=1}^{n} f(x_i)\,\Delta x = \sum_{i=1}^{n} f\left(\frac{2i}{n}\right)\left(\frac{2}{n}\right) = \sum_{i=1}^{n}\left(\frac{2i}{n}\right)^2\left(\frac{2}{n}\right)$$

$$= \sum_{i=1}^{n}\frac{8}{n^3}i^2 = \frac{8}{n^3}\sum_{i=1}^{n} i^2$$

$$= \frac{8}{n^3}\left[\frac{n(n+1)(2n+1)}{6}\right] = \frac{4(n+1)(2n+1)}{3n^2}$$

The area of the region is the limit of the Riemann sums:

$$\lim_{n\to\infty}\sum_{i=1}^{n} f(x_i)\,\Delta x = \lim_{n\to\infty}\frac{4(n+1)(2n+1)}{3n^2} = \frac{8}{3}$$

[See Section 12-3.]

PROBLEM 12-17 Find the area of the region bounded by the graphs of $f(x) = (x-1)^2 + 2$, $x = -1$, $x = 2$, and the x axis by finding the limit of the Riemann sums.

Solution: Divide $[-1, 2]$:

$$\Delta x = \frac{2-(-1)}{n} = \frac{3}{n}$$

$$x_i = a + i\Delta x = -1 + \frac{3i}{n}$$

The nth Riemann sum is

$$\sum_{i=1}^{n} f(x_i)\,\Delta x = \sum_{i=1}^{n} f\left(-1+\frac{3i}{n}\right)\frac{3}{n} = \sum_{i=1}^{n}\left[\left(-1+\frac{3i}{n}-1\right)^2+2\right]\frac{3}{n}$$

$$= \sum_{i=1}^{n}\left[\left(\frac{3i}{n}-2\right)^2+2\right]\frac{3}{n} = \sum_{i=1}^{n}\left(\frac{9i^2}{n^2}-\frac{12i}{n}+4+2\right)\frac{3}{n}$$

$$= \sum_{i=1}^{n}\left(\frac{27}{n^3}i^2-\frac{36}{n^2}i+\frac{18}{n}\right) = \frac{27}{n^3}\sum_{i=1}^{n} i^2 - \frac{36}{n^2}\sum_{i=1}^{n} i + \frac{18}{n}\sum_{i=1}^{n} 1$$

$$= \frac{27}{n^3}\left[\frac{n(n+1)(2n+1)}{6}\right] - \frac{36}{n^2}\left[\frac{n(n+1)}{2}\right] + \frac{18}{n}(n)$$

$$= \frac{9(n+1)(2n+1)}{2n^2} - \frac{18(n+1)}{n} + 18$$

The area is the limit of the Riemann sum:

$$\lim_{n\to\infty}\sum_{i=1}^{n} f(x_i)\,\Delta x = \lim_{n\to\infty}\left[\frac{9(n+1)(2n+1)}{2n^2} - \frac{18(n+1)}{n} + 18\right]$$

$$= 9 - 18 + 18 = 9$$

[See Section 12-3.]

PROBLEM 12-18 Find the area of the region bounded by the graphs of $f(x) = 2(x+2)^3$, $x = -2$, $x = 0$, and the x axis by calculating the limit of the Riemann sums.

Solution: Divide $[-2, 0]$:

$$\Delta x = \frac{2}{n} \qquad x_i = -2 + \frac{2i}{n}$$

The nth Riemann sum is

$$\sum_{i=1}^{n} f(x_i)\,\Delta x = \sum_{i=1}^{n} 2\left(-2+\frac{2i}{n}+2\right)^3\left(\frac{2}{n}\right) = \sum_{i=1}^{n}\frac{32}{n^4}i^3 = \frac{32}{n^4}\sum_{i=1}^{n} i^3$$

$$= \frac{32}{n^4}\left[\frac{n^2(n+1)^2}{4}\right] = \frac{8(n+1)^2}{n^2}$$

Find the limit:

$$\lim_{n\to\infty}\sum_{i=1}^{n} f(x_i)\,\Delta x = \lim_{n\to\infty}\frac{8(n+1)^2}{n^2} = 8$$

[See Section 12-3.]

PROBLEM 12-19 Evaluate $\lim_{n\to\infty}\sum_{i=1}^{n}(x_i^2 - 2x_i)\,\Delta x$, where

$$x_0 = 1, \qquad x_1 = 1 + \Delta x, \qquad \ldots, \qquad x_n = 3$$

by evaluating the appropriate integral.

Solution: You must translate this Riemann sum into an integral: Δx becomes dx, x_i becomes x, and the interval of integration is [1, 3].

$$\lim_{n\to\infty}\sum_{i=1}^{n}(x_i^2 - 2x_i)\,\Delta x = \int_1^3 (x^2 - 2x)\,dx = \left(\frac{x^3}{3} - x^2\right)\Bigg|_1^3$$

$$= \left(\frac{3^3}{3} - 3^2\right) - \left(\frac{1^3}{3} - 1^2\right) = \frac{2}{3}$$

[See Section 12-3.]

PROBLEM 12-20 Evaluate $\lim_{n\to\infty}\sum_{i=1}^{n}\sqrt{x_i}\,\Delta x$, where $x_0 = 0, \ldots, x_n = 4$.

Solution: Recognize the integral that this Riemann sum represents:

$$\lim_{n\to\infty}\sum_{i=1}^{n}\sqrt{x_i}\,\Delta x = \int_0^4 \sqrt{x}\,dx = \int_0^4 x^{1/2}\,dx = \frac{x^{1/2+1}}{\frac{1}{2}+1}\Bigg|_0^4$$

$$= \frac{2}{3}x^{3/2}\Bigg|_0^4 = \frac{2}{3}(4^{3/2} - 0^{3/2}) = \frac{16}{3}$$

[See Section 12-3.]

PROBLEM 12-21 Evaluate $\lim_{n\to\infty}\sum_{i=1}^{n}[e^{x_i} - (1/x_i)]\,\Delta x$, where $x_0 = 1, \ldots, x_n = 3$.

Solution: You find

$$\lim_{n\to\infty}\sum_{i=1}^{n}\left(e^{x_i} - \frac{1}{x_i}\right)\Delta x = \int_1^3\left(e^x - \frac{1}{x}\right)dx = (e^x - \ln|x|)\Bigg|_1^3$$

$$= (e^3 - \ln 3) - (e^1 - \ln 1) = e^3 - e - \ln 3$$

[See Section 12-3.]

PROBLEM 12-22 Evaluate $\lim_{n\to\infty}\sum_{i=1}^{n}(x_{i+1} - x_i)\cos x_i$, where $x_0 = 0, \ldots, x_n = \pi/6$.

Solution: Recognize that $x_{i+1} - x_i = \Delta x$ and obtain

$$\lim_{n\to\infty}\sum_{i=1}^{n}(x_{i+1} - x_i)\cos x = \lim_{n\to\infty}\sum_{i=1}^{n}\Delta x\cos x$$

$$= \int_0^{\pi/6}\cos x\,dx = \sin x\Bigg|_0^{\pi/6} = \frac{\sin\pi}{6} - \sin 0 = \frac{1}{2} - 0 = \frac{1}{2}$$

[See Section 12-3.]

PROBLEM 12-23 Evaluate $\lim_{n\to\infty}\sum_{i=1}^{n}(x_{i+1}^2 - x_i^2)$, where $x_0 = 3, \ldots, x_n = 5$.

Solution: You get

$$\lim_{n\to\infty}\sum_{i=1}^{n}(x_{i+1}^2 - x_i^2) = \lim_{n\to\infty}\sum_{i=1}^{n}(x_{i+1} + x_i)(x_{i+1} - x_i) = \lim_{n\to\infty}\sum_{i=1}^{n}(x_{i+1} + x_i)\,\Delta x$$

$$= \int_3^5 (x + x)\,dx = \int_3^5 2x\,dx = x^2\Bigg|_3^5 = 16$$

[See Section 12-3.]

Supplementary Exercises

In Problems 12-24 through 12-50 evaluate the integral:

12-24 $\int_{-1}^{2} x^3\,dx$

12-25 $\int_{2}^{3} (2 - x^2)\,dx$

12-26 $\int_{-2}^{2} (4x^2 - 7x + 1)\,dx$

12-27 $\int_{1}^{8} (2\sqrt[3]{x} + 5x^{-2})\,dx$

12-28 $\int_{1}^{e} (7/x)\,dx$

12-29 $\int_{-\pi/6}^{\pi/4} (\sec x \tan x + \sec^2 x)\,dx$

12-30 $\int_{2}^{4} (\sqrt{2} + \frac{1}{2}x^{-1/2})\,dx$

12-31 $\int_{1}^{32} (4/\sqrt[5]{x^3})\,dx$

12-32 $\int_{0}^{\pi} (\sin x + 2e^x)\,dx$

12-33 $\int_{0}^{1/2} (4/\sqrt{1 - x^2})\,dx$

12-34 $\int_{1}^{4} [(x^2 + 3x - 1)/\sqrt{x}]\,dx$

12-35 $\int_{-3}^{3} (x + 2)(x - 2)\,dx$

12-36 $\int_{3}^{5} e\,dx$

12-37 $\int_{0}^{2} (2^x + x^2)\,dx$

12-38 $\int_{1}^{\sqrt{3}} \left(\frac{5}{x^2 + 1} + \frac{5}{x}\right) dx$

12-39 $\int_{1}^{9} [(\sqrt{x} + 1)/\sqrt{x}]\,dx$

12-40 $\int_{0}^{1} [e^x/(\pi + 3)]\,dx$

12-41 $\int_{0}^{\pi} (\cos x + \cosh x)\,dx$

12-42 $\int_{0}^{\pi} (\sin x + \sinh x)\,dx$

12-43 $\int_{0}^{1} \operatorname{sech}^2 x\,dx$

12-44 $\int_{0}^{1} (x^{\sqrt{3}} + (\sqrt{3})^x)\,dx$

12-45 $\int_{1}^{2} [(x/2) + (2/x)]\,dx$

12-46 $\int_{1}^{16} [(\sqrt{x} + 2)/\sqrt[4]{x^3}]\,dx$

12-47 $\int_{-12}^{12} (x^7 - 3x^5 + x^3 - 8x)\,dx$

12-48 $\int_{1}^{4} [\sqrt{x} + (1/\sqrt{x})]\,dx$

12-49 $\int_{0}^{1} \left(\frac{2}{1 + x^2} - \cosh x\right) dx$

12-50 $\int_{1}^{2} (\sqrt{x^3}/\sqrt[3]{x^2})\,dx$

12-51 Given that $\frac{d}{dx}\left(\frac{x - 7}{x - 3}\right) = \frac{4}{(x - 3)^2}$, evaluate $\int_{1}^{2} \frac{4}{(x - 3)^2}\,dx.$

12-52 Given that $\frac{d}{dx}(\sin e^x) = e^x \cos e^x$, evaluate $\int_{\ln(\pi/4)}^{\ln(\pi/2)} e^x \cos e^x\,dx.$

12-53 Given that $\frac{d}{dx}(\sqrt{\sqrt{x} + 1}) = \frac{1}{4\sqrt{x}\sqrt{\sqrt{x} + 1}}$, evaluate $\int_{9}^{64} \frac{dx}{\sqrt{x}\sqrt{\sqrt{x} + 1}}.$

In Problems 12-54 through 12-60 find the area of the region bounded by the x axis and the given curves by calculating the limit of the Riemann sums:

12-54 $f(x) = 3x \qquad x = 0 \qquad x = 1$

12-55 $f(x) = x + 7 \qquad x = -2 \qquad x = 3$

12-56 $f(x) = 3x^2 \qquad x = -1 \qquad x = 1$

12-57 $f(x) = x^2 - 2 \qquad x = 2 \qquad x = 3$

12-58 $f(x) = 2x^3 + x \qquad x = 0 \qquad x = 1$

12-59 $f(x) = (x + 1)^2 \qquad x = -1 \qquad x = 3$

12-60 $f(x) = x^3 + x^2 + x + 1 \qquad x = 0 \qquad x = 1$

In Problems 12-61 through 12-70 recognize the limit of Riemann sums as an integral and evaluate:

12-61 $\lim_{n\to\infty} \sum_{i=1}^{n} x_i^4 \,\Delta x$, where $x_0 = 0, \ldots, x_n = 5$

12-62 $\lim_{n\to\infty} \sum_{i=1}^{n} (x_i^2 - 3x_i + 1)\,\Delta x$, where $x_0 = 0, \ldots, x_n = 3$

12-63 $\lim_{n\to\infty} \sum_{i=1}^{n} (x_{i+1} - x_i)/\sqrt{x_i}$, where $x_0 = 1, \ldots, x_n = 4$

12-64 $\lim_{n\to\infty} \sum_{i=1}^{n} \sec^2 x_i(x_{i+1} - x_i)$, where $x_0 = 0, \ldots, x_n = \pi/4$

12-65 $\lim_{n\to\infty} \sum_{i=1}^{n} (x_{i+1} - x_i)/(x_i^2 + 1)$, where $x_0 = 0, \ldots, x_n = 1$

12-66 $\lim_{n\to\infty} \sum_{i=1}^{n} (2x_{i+1} - 2x_i)/(x_{i+1} + 2x_i)$, where $x_0 = -3, \ldots, x_n = -2$

12-67 $\lim_{n\to\infty} \sum_{i=1}^{n} (x_i^2 - x_{i+1}^2)/(x_{i+1} + x_i)^2$, where $x_0 = 1, \ldots, x_n = 3$

12-68 $\lim_{n\to\infty} \sum_{i=1}^{n} 2^{x_i}(x_{i+1} - x_i)$, where $x_0 = 0, \ldots, x_n = 2$

12-69 $\lim_{n\to\infty} \sum_{i=1}^{n} \sec x_i \tan x_{i+1}(x_{i+1} - x_i)$, where $x_0 = 0, \ldots, x_n = \pi/4$

12-70 $\lim_{n\to\infty} \sum_{i=1}^{n} (x_i^2 - x_{i+1}x_i)/\sqrt{x}$, where $x_0 = 1, \ldots, x_n = 4$

Solutions to Supplementary Exercises

(12-24) $15/4$

(12-25) $-13/3$

(12-26) $76/3$

(12-27) $215/8$

(12-28) 7

(12-29) $1 + \sqrt{2} - (\sqrt{3}/3)$

(12-30) $2 + \sqrt{2}$

(12-31) 30

(**12-32**) $2e^{\pi}$

(**12-33**) $2\pi/3$

(**12-34**) $122/5$

(**12-35**) -6

(**12-36**) $2e$

(**12-37**) $(3/\ln 2) + (8/3)$

(**12-38**) $(5\pi/12) + (5/2)\ln 3$

(**12-39**) 12

(**12-40**) $(e - 1)/(\pi + 3)$

(**12-41**) $\sinh \pi$

(**12-42**) $1 + \cosh \pi$

(**12-43**) $\tanh 1$

(**12-44**) $[1/(1 + \sqrt{3})] + [(2\sqrt{3} - 2)/\ln 3]$

(**12-45**) $\frac{3}{4} + 2\ln 2$

(**12-46**) $52/3$

(**12-47**) 0

(**12-48**) $20/3$

(**12-49**) $\frac{1}{2}(\pi - e + e^{-1})$

(**12-50**) $(6/11)(2^{11/6} - 1)$

(**12-51**) 2

(**12-52**) $(2 - \sqrt{2})/2$

(**12-53**) 4

(**12-54**) $\lim_{n\to\infty} \sum_{i=1}^{n} 3\left(\frac{i}{n}\right)\frac{1}{n} = \lim_{n\to\infty} \frac{3}{n^2}\left[\frac{n(n+1)}{2}\right] = \frac{3}{2}$

(**12-55**) $\lim_{n\to\infty} \sum_{i=1}^{n} \left[\left(-2 + \frac{5i}{n}\right) + 7\right]\frac{5}{n} = \lim_{n\to\infty} \left[\frac{25}{n^2}\left(\frac{n(n+1)}{2}\right) + 25\right] = \frac{75}{2}$

(**12-56**) $\lim_{n\to\infty} \sum_{i=1}^{n} 3\left(-1 + \frac{2i}{n}\right)^2\left(\frac{2}{n}\right) = \lim_{n\to\infty} \left[\frac{24}{n^3}\left(\frac{n(n+1)(2n+1)}{6}\right) - \frac{24}{n^2}\left(\frac{n(n+1)}{2}\right) + 6\right] = 2$

(**12-57**) $\lim_{n\to\infty} \sum_{i=1}^{n} \left[\left(2 + \frac{i}{n}\right)^2 - 2\right]\frac{1}{n} = \lim_{n\to\infty} \left[\frac{1}{n^3}\left(\frac{n(n+1)(2n+1)}{6}\right) + \frac{4}{n^2}\left(\frac{n(n+1)}{2}\right) + 2\right] = \frac{13}{3}$

(**12-58**) $\lim_{n\to\infty} \sum_{i=1}^{n} \left[2\left(\frac{i}{n}\right)^3 + \frac{i}{n}\right]\frac{1}{n} = \lim_{n\to\infty} \left[\frac{2}{n^4}\left(\frac{n^2(n+1)^2}{4}\right) + \frac{1}{n^2}\left(\frac{n(n+1)}{2}\right)\right] = 1$

(**12-59**) $\lim_{n\to\infty} \sum_{i=1}^{n} \left(-1 + \frac{4i}{n} + 1\right)^2\left(\frac{4}{n}\right) = \lim_{n\to\infty} \frac{64}{n^3}\left[\frac{n(n+1)(2n+1)}{6}\right] = \frac{64}{3}$

(**12-60**) $\lim_{n\to\infty} \sum_{i=1}^{n} \left[\left(\frac{i}{n}\right)^3 + \left(\frac{i}{n}\right)^2 + \left(\frac{i}{n}\right) + 1\right]\frac{1}{n}$

$$= \lim_{n\to\infty} \left[\frac{1}{n^4}\left(\frac{n^2(n+1)^2}{4}\right) + \frac{1}{n^3}\left(\frac{n(n+1)(2n+1)}{6}\right) + \frac{1}{n^2}\left(\frac{n(n+1)}{2}\right) + 1\right] = \frac{25}{12}$$

(**12-61**) $\int_0^5 x^4\,dx = 625$

(**12-62**) $\int_0^3 (x^2 - 3x + 1)\,dx = -3/2$

(**12-63**) $\int_1^4 x^{-1/2}\,dx = 2$

(**12-64**) $\int_0^{\pi/4} \sec^2 x\,dx = 1$

(**12-65**) $\int_0^1 1/(x^2 + 1)\,dx = \pi/4$

(**12-66**) $\int_{-3}^{-2} 2/(3x)\,dx = \frac{2}{3}\ln\frac{2}{3}$

(**12-67**) $\int_1^3 -2x/(2x)^2\,dx = -\frac{1}{2}\ln 3$

(**12-68**) $\int_0^2 2^x\,dx = 3/(\ln 2)$

(**12-69**) $\int_0^{\pi/4} \sec x \tan x\,dx = \sqrt{2} - 1$

(**12-70**) $\int_1^4 (-x/\sqrt{x})\,dx = -14/3$

13 INTEGRATION BY SUBSTITUTION

THIS CHAPTER IS ABOUT

☑ The Chain Rule Revisited
☑ Simple Substitutions in Antidifferentiation
☑ More Complicated Substitutions
☑ Substitutions in Integrals
☑ Overview

13-1. The Chain Rule Revisited

Recall that if you can write $y = f(x)$ as a composite of functions $g(x)$ and $h(x)$, then you find the derivative of $f(x)$ by taking the product of the derivatives of g and h. That is, if

$$y = g(u) \quad \text{with} \quad u = h(x)$$

then

$$\frac{dy}{dx} = g'(u)\frac{du}{dx}$$

If you restate this equation in terms of differentials, you get

$$dy = g'(u)\,du$$

This brief look at the chain rule furnishes you with one of the most useful techniques for finding an antiderivative: substitution.

13-2. Simple Substitutions in Antidifferentiation

You can think of each differentiation formula as an antidifferentiation formula. Consider the following example:

EXAMPLE 13-1: Let $y = \frac{1}{2}\sin^2 x$. Then

$$\frac{dy}{dx} = \frac{1}{2}(2\sin x)\frac{d}{dx}(\sin x) = \sin x \cos x$$

So

$$dy = \sin x \cos x\,dx$$

The resulting antiderivative statement is

$$\int \sin x \cos x\,dx = \frac{1}{2}\sin^2 x + C$$

Look at that equation again, in a slightly different form:

$$\int \sin x\,d(\sin x) = \frac{\sin^2 x}{2} + C$$

or one more time, with $\sin x$ replaced by u and $d(\sin x)$ by du:

$$\int u\,du = \frac{u^2}{2} + C$$

This should look familiar to you. It's the antiderivative formula:

$$\int x^n \, dx = \frac{x^{n+1}}{n+1} + C$$

with x replaced by u, and $n = 1$.

EXAMPLE 13-2: Find $\int \sin x \cos x \, dx$.

Solution: Without Example 13-1 as a guide you would probably be at a loss as to how to proceed on this problem; it certainly bears no resemblance to any of the basic antidifferentiation formulas. But with Example 13-1 you should recognize that if you make the simple substitution $u = \sin x$, so that

$$\frac{du}{dx} = \cos x$$

and

$$du = \cos x \, dx$$

the given problem immediately transforms into one of the recognizable formulas:

$$\int \sin x \cos x \, dx = \int u \, du = \frac{u^2}{2} + C$$

The problem is finished when you retrace your substitution (replace u by $\sin x$) to obtain

$$\int \sin x \cos x \, dx = \frac{\sin^2 x}{2} + C$$

EXAMPLE 13-3: Find $\int \frac{x}{x^2 + 1} \, dx$.

Solution: This doesn't, as it stands, fit any of the known patterns. But the fact that

$$\frac{d}{dx}(x^2 + 1) = 2x$$

and so

$$d(x^2 + 1) = 2x \, dx$$

suggests the substitution (actual or mental):

$$x^2 + 1 = u$$

$$2x = \frac{du}{dx}$$

and

$$x \, dx = \frac{1}{2} \, du$$

The given problem changes form and becomes recognizable:

$$\begin{aligned} \int \frac{x \, dx}{x^2 + 1} &= \int \frac{\frac{1}{2} \, du}{u} \\ &= \frac{1}{2} \int \frac{du}{u} \\ &= \frac{1}{2} \ln|u| + C \end{aligned}$$

Returning to x's:

$$\int \frac{x\,dx}{x^2+1} = \frac{1}{2}\ln|x^2+1| + C$$

EXAMPLE 13-4: Find $\displaystyle\int \frac{\sec^2\sqrt{x}}{\sqrt{x}}\,dx$.

Solution: Once again, this doesn't immediately fit into one of the known formulas; it does, however, make you think of

$$\int \sec^2 u\,du = \tan u + C$$

With this as a hint, you try the substitution $u = \sqrt{x}$ with $du = 1/(2\sqrt{x})\,dx$, so that $2\,du = 1/\sqrt{x}\,dx$. Make this substitution in the given problem, and proceed to the answer:

$$\int (\sec^2\sqrt{x})\frac{1}{\sqrt{x}}\,dx = \int (\sec^2 u)\,2\,du = 2\int \sec^2 u\,du$$

$$= 2\tan u + C = 2\tan\sqrt{x} + C$$

Caution: When you make a substitution, substitute for every part of the integrand. In Example 13-4 you substitute not only for x but also for dx. (Otherwise your integral would look like $\sec^2 u(1/u)\,dx$, a mixture of symbols that would be difficult to interpret.)

The ability to spot an essentially obvious substitution comes with a sure knowledge of the basic antidifferentiation formulas and with practice. You must train your eye to separate the integrand into factors: functions (u) (or composite functions $f(u)$) and their differentials (du).

EXAMPLE 13-5: $\int 1/(x\ln x)\,dx = \int (1/\ln x)(1/x)\,dx$. Do you see that this has the form $\int (1/u)\,du$, with $u = \ln x$? You can write the answer immediately:

$$\int \frac{dx}{x\ln x} = \ln|u| + C = \ln|\ln x| + C$$

13-3. More Complicated Substitutions

Many integrals require more sophisticated techniques of substitution than the preceding examples.

EXAMPLE 13-6: Find $\displaystyle\int \sqrt{1+\sqrt{x}}\,dx$.

Solution: This one clearly fits no recognizable formula. The motivation in introducing a substitution is to change its form with the hope that something recognizable will emerge.

Try $u = \sqrt{1+\sqrt{x}}$ as a first guess. Notice, however, that in this form the correct substitution for dx isn't apparent. You can overcome this difficulty by solving for x:

$$u^2 = 1 + \sqrt{x}$$

$$\sqrt{x} = u^2 - 1$$

$$x = (u^2-1)^2$$

Now:

$$dx = 2(u^2-1)\,2u\,du$$

$$= 4u(u^2-1)\,du$$

and you are ready to explore what this substitution does to the given problem:

$$\int \sqrt{1+\sqrt{x}}\,dx = \int (u)4u(u^2-1)\,du = 4\int u^2(u^2-1)\,du$$
$$= 4\int (u^4-u^2)\,du = 4\left[\frac{u^5}{5}-\frac{u^3}{3}\right]+C$$

Aha! An apparently difficult problem in antidifferentiation has been changed, by a wisely chosen substitution, into an immediate combination of recognizable formulas. All that remains is to simplify and return to the x's:

$$\int \sqrt{1+\sqrt{x}}\,dx = 4\left(\frac{3u^5-5u^3}{15}\right)+C = \frac{4u^3}{15}(3u^2-5)+C$$

From the original substitution:

$$u^2 = 1+\sqrt{x}$$
$$u^3 = (1+\sqrt{x})^{3/2}$$

and so

$$\int \sqrt{1+\sqrt{x}}\,dx = \frac{4}{15}(1+\sqrt{x})^{3/2}[3(1+\sqrt{x})-5]+C$$
$$= \frac{4}{15}(1+\sqrt{x})^{3/2}[3\sqrt{x}-2]+C$$

Perhaps the use of expressions like "first guess" and "wisely chosen substitution" has you wondering if any other substitution would have worked. Suppose, for example, you try $u=\sqrt{x}$, with $x=u^2$ and $dx=2u\,du$. The problem becomes

$$\int \sqrt{1+\sqrt{x}}\,dx = \int (\sqrt{1+u})\,2u\,du$$

which is, in fact, somewhat simpler than the given integral. The critical consideration is that you still can't perform the antidifferentiation by using a basic formula. Further simplification is required.

Let $\sqrt{1+u}=v$, so $1+u=v^2$ and $du=2v\,dv$. Then the integral becomes

$$2\int \sqrt{1+u}\,(u)\,du = 2\int v(v^2-1)\,2v\,dv = 4\int (v^4-v^2)\,dv$$
$$= 4\left(\frac{v^5}{5}-\frac{v^3}{3}\right)+C = \frac{4}{15}v^3(3v^2-5)+C$$

Return first to u's ($v^2=1+u$):

$$2\int \sqrt{1+u}\,(u)\,du = \frac{4}{15}(1+u)^{3/2}(3u-2)+C$$

and then to x's ($u=\sqrt{x}$):

$$\int \sqrt{1+\sqrt{x}}\,dx = \frac{4}{15}(1+\sqrt{x})^{3/2}(3\sqrt{x}-2)+C$$

In general, then, it is not usually true that there is only one correct substitution; it is true that, most frequently, there is one best substitution.

EXAMPLE 13-7: Find $\int (2+\sqrt{x})/(1-\sqrt{x})\,dx$.

Solution: You could try $u=\sqrt{x}$ (with $x=u^2$ and $dx=2u\,du$) or $u=2+\sqrt{x}$ (with $x=(u-2)^2$ and $dx=2(u-2)\,du$ and $1-\sqrt{x}=3-u$). Each of these will lead to antiderivatives that must be further simplified by a long division. If, however, you let $u = 1 - \sqrt{x}$ (the denominator), the algebra will be less tedious. You have

$$\sqrt{x} = 1 - u$$
$$x = (1 - u)^2$$
$$dx = 2(1 - u)(-1)\,du = 2(u - 1)\,du$$

with

$$2 + \sqrt{x} = 2 + 1 - u = 3 - u$$

The problem becomes

$$\int \frac{2 + \sqrt{x}}{1 - \sqrt{x}}\,dx = \int \frac{3 - u}{u}\,2(u - 1)\,du$$

Do the algebra: Multiply $(3 - u)$ by $(u - 1)$ and divide every term by u. You have

$$\int \frac{2 + \sqrt{x}}{1 - \sqrt{x}}\,dx = 2\int \left(4 - u - \frac{3}{u}\right) du$$
$$= 2\left[4u - \frac{u^2}{2} - 3\ln|u|\right] + C$$

Returning to x's ($u = 1 - \sqrt{x}$), you have the answer:

$$\int \frac{2 + \sqrt{x}}{1 - \sqrt{x}}\,dx = 2\left[4(1 - \sqrt{x}) - \frac{(1 - \sqrt{x})^2}{2} - 3\ln|1 - \sqrt{x}|\right] + C$$

13-4. Substitutions in Integrals

The techniques of substitution that you use in the evaluation of an integral are no different from those discussed in the preceding examples. There is, however, one further aspect to the problem: where, how, and when are the limits of integration introduced? There are two basic approaches.

A. Use the Fundamental Theorem of Calculus.

The Fundamental Theorem of Calculus says

$$\int_a^b f(x)\,dx = F(b) - F(a)$$

where $F(x)$ is any antiderivative of $f(x)$. That is, you consider each integration problem as an antidifferentiation problem until you have an answer; then evaluate between the limits.

EXAMPLE 13-8: Evaluate $\int_0^{\pi/2} \sin^2 x \cos x\,dx$.

Solution: Is it clear to you that your immediate choice is $u = \sin x$, with $du = \cos x\,dx$? Then

$$\int \sin^2 x \cos x\,dx = \int u^2\,du = \frac{u^3}{3} + C$$
$$= \frac{\sin^3 x}{3} + C$$

Therefore:

$$\int_0^{\pi/2} \sin^2 x \cos x\,dx = \frac{\sin^3 x}{3}\bigg|_0^{\pi/2} = \frac{\sin^3(\pi/2)}{3} - \frac{\sin^3(0)}{3} = \frac{1}{3}$$

Note: When you rewrite the problem as an antidifferentiation problem, you omit the limits of integration. They reappear only after you have an antiderivative to evaluate.

EXAMPLE 13-9: Evaluate $\int_1^4 1/(1 + \sqrt{x})\, dx$.

Solution: To find an antiderivative, substitute $u = 1 + \sqrt{x}$, with $x = (u - 1)^2$ and $dx = 2(u - 1)\, du$.
Substituting:

$$\begin{aligned}\int \frac{dx}{1 + \sqrt{x}} &= 2\int \frac{u - 1}{u}\, du \\ &= 2\int \left(1 - \frac{1}{u}\right) du \\ &= 2[u - \ln|u|] + C \\ &= 2[(1 + \sqrt{x}) - \ln|1 + \sqrt{x}|] + C\end{aligned}$$

Therefore:

$$\begin{aligned}\int_1^4 \frac{dx}{1 + \sqrt{x}} &= 2[(1 + \sqrt{x}) - \ln|1 + \sqrt{x}|]\Big|_1^4 \\ &= 2[[(1 + \sqrt{4}) - \ln|1 + \sqrt{4}|] \\ &\quad - [(1 + \sqrt{1}) - \ln|1 + \sqrt{1}|]] \\ &= 2[3 - \ln 3 - 2 + \ln 2] \\ &= 2[1 + \ln 2 - \ln 3]\end{aligned}$$

B. Change limits when you substitute.

A second approach to this problem is for you to change the limits of integration in a manner that is determined by your substitution. Look again at the problems of the last two examples using a different approach.

EXAMPLE 13-10: Evaluate $\int_0^{\pi/2} \sin^2 x \cos x\, dx$.

Solution: Again you observe that the proper substitution is $u = \sin x$, with $du = \cos x\, dx$. This time, however, you note that the interval of integration is given by $0 \leqslant x \leqslant \pi/2$, and you ask what this implied for u. Since $u = \sin x$, it follows that when $x = 0$, $u = \sin 0 = 0$; when $x = \pi/2$, $u = \sin(\pi/2) = 1$. That is, if x lies between 0 and $\pi/2$, $u = \sin x$ lies between 0 and 1. Thus you change the complete integral:

$$\int_0^{\pi/2} \sin^2 x \cos x\, dx = \int_0^1 u^2\, du = \frac{u^3}{3}\Big|_0^1 = \frac{1}{3}$$

EXAMPLE 13-11: Evaluate $\int_1^4 1/(1 + \sqrt{x})\, dx$.

Solution: The substitution: $u = 1 + \sqrt{x}$, with $dx = 2(u - 1)\, du$. The interval of integration: $1 \leqslant x \leqslant 4$. Substitute these values of x to get the corresponding values of u:

$$\begin{aligned} x &= 1 \qquad u = 1 + \sqrt{1} = 2 \\ x &= 4 \qquad u = 1 + \sqrt{4} = 3\end{aligned}$$

Therefore:

$$\int_1^4 \frac{dx}{1+\sqrt{x}} = 2\int_2^3 \frac{u-1}{u}\,du = 2\int_2^3 \left(1 - \frac{1}{u}\right) du$$

$$= 2[u - \ln|u|]\Big|_2^3 = 2[(3 - \ln 3) - (2 - \ln 2)]$$

$$= 2(1 - \ln 3 + \ln 2)$$

You can see the advantages and the disadvantages of the two methods: The first is a straightforward application of the Fundamental Theorem, while the second cuts out the sometimes awkward and involved algebra required to "back up" through your substitution.

13-5. Overview

Substitution is a technique for changing the form of a given antiderivative (or integral) in such a way that it becomes recognizable as one (or a combination of two or more) of the basic formulas. It is a device for "undoing" the chain rule.

Before you can become skillful with this technique, you must know the derivative and the antiderivative formulas. For example, to recognize $\int (\sin x/\cos x)\,dx$ as being in the form $-\int 1/u\,du$, you must know that $d(\cos x) = -\sin x\,dx$; and having noted that, you must know that

$$-\int \frac{du}{u} = -\ln|u| + C$$

before you can get to the final answer.

Secondly, you must practice. As you work more and more problems, your eye will become adept at picking out combinations of *u*'s and *du*'s, and hence at making at least a good guess at a helpful substitution.

SUMMARY

1. Substitution is one of the techniques that will help you in the problem of finding the antiderivative (and hence the integral) of a function. You use it, in general, in two situations:
   - To help you recognize a "disguised" use of the chain rule.
   - To change the form of an integrand that doesn't appear to be any one of the recognizable antiderivative formulas.
2. It is important to remember that, in changing $f(x)\,dx$, you must substitute for dx as well as for $f(x)$; the new integrand must contain only u's and du (just as the original contained only x's and dx).
3. To evaluate an integral, you have a choice of two methods:
   - First find an antiderivative (*in* x's), and then use the Fundamental Theorem:

     $$\int_a^b f(x)\,dx = F(b) - F(a)$$

     where $F(x)$ is any antiderivative of $f(x)$. You must remember that the given limits of integration are limits on x, and not on u.
   - Use your substitution to determine new limits of integration that are limits on u. If your substitution is $u = g(x)$, with $[a, b]$ the interval of integration for x, then the interval for u would be $[g(a), g(b)]$.
4. You may need several substitutions to successfully antidifferentiate some functions.

SOLVED PROBLEMS

PROBLEM 13-1 Find $\int \sqrt{3x + 5}\, dx$.

Solution: Choose $u = 3x + 5$, so $du/dx = 3$ and $du = 3\, dx$ or $dx = \frac{1}{3}\, du$. Substituting,

$$\int \sqrt{3x+5}\, dx = \int \sqrt{u}\, \frac{1}{3}\, du = \frac{1}{3} \int u^{1/2}\, du$$

$$= \frac{1}{3} \frac{u^{3/2}}{3/2} + C = \frac{2}{9} u^{3/2} + C$$

Return to x's ($u = 3x + 5$):

$$\int \sqrt{3x+5}\, dx = \frac{2}{9}(3x+5)^{3/2} + C$$

[See Section 13-2.]

PROBLEM 13-2 Find $\int xe^{x^2-1}\, dx$.

Solution: Rewrite the problem: $\int e^{x^2-1}\, x\, dx$. (This is suggested by the fact that $(d/dx)(x^2 - 1) = 2x$, with $d(x^2 - 1) = 2x\, dx$ or $x\, dx = \frac{1}{2}d(x^2 - 1)$.) Then let $u = x^2 - 1$, with $\frac{1}{2}\, du = x\, dx$. Substituting,

$$\int xe^{x^2-1}\, dx = \int e^u \frac{1}{2}\, du = \frac{1}{2} \int e^u\, du = \frac{1}{2} e^u + C$$

Return to x's:

$$\int xe^{x^2-1}\, dx = \frac{1}{2} e^{x^2-1} + C$$

[See Section 13-2.]

PROBLEM 13-3 Find $\int e^x/(1 + e^x)\, dx$.

Solution: Let $u = 1 + e^x$, with $du/dx = e^x$ or $du = e^x\, dx$. Substituting,

$$\int \frac{du}{u} = \ln|u| + C$$

or in terms of x:

$$\int \frac{e^x\, dx}{1 + e^x} = \ln|1 + e^x| + C$$

[See Section 13-2.]

PROBLEM 13-4 Find $\int \tan x \sec^3 x\, dx$.

Solution: Recalling that $(d/dx)(\sec x) = \sec x \tan x$, you rewrite the problem:

$$\int \tan x \sec^3 x\, dx = \int \sec^2 x \sec x \tan x\, dx$$

Then the substitution is easy: $u = \sec x$, $du = \sec x \tan x\, dx$, and

$$\int \sec^2 x \sec x \tan x\, dx = \int u^2\, du = \frac{u^3}{3} + C$$

Finally:

$$\int \tan x \sec^3 x\, dx = \frac{\sec^3 x}{3} + C$$

[See Section 13-2.]

Note: As you work more problems, you will begin to make substitutions like those in Problems 1 through 4 in your head, instead of writing it all down. Initially, though, there's no such thing as writing "too much."

PROBLEM 13-5 Find $\int x/(x^2 + 4)^2\, dx$.

Solution: Let $u = x^2 + 4$, with $du = 2x\, dx$ or $x\, dx = \frac{1}{2}\, du$. Then,

$$\int \frac{x\, dx}{(x^2 + 4)^2} = \int \frac{\frac{1}{2}\, du}{u^2} = \frac{1}{2}\int u^{-2}\, du = -\frac{1}{2}u^{-1} + C$$

and returning to x's:

$$\int \frac{x\, dx}{(x^2 + 4)^2} = -\frac{1}{2}(x^2 + 4)^{-1} + C$$

[See Section 13-2.]

PROBLEM 13-6 Find $\int (\sin x/\sqrt{1 + 2\cos x})\, dx$.

Solution: Let $u = 1 + 2\cos x$, so that $du = -2\sin x\, dx$ or $\sin x\, dx = -\frac{1}{2}\, du$. Substituting,

$$\int \frac{\sin x\, dx}{\sqrt{1 + 2\cos x}} = \int \frac{-\frac{1}{2}\, du}{\sqrt{u}} = -\frac{1}{2}\int u^{-1/2}\, du$$

$$= -\frac{1}{2}\left(\frac{u^{1/2}}{\frac{1}{2}}\right) + C = -\sqrt{u} + C$$

and

$$\int \frac{\sin x\, dx}{\sqrt{1 + 2\cos x}} = -\sqrt{1 + 2\cos x} + C$$

[See Section 13-2.]

PROBLEM 13-7 Find $\int (\ln x)^4/x\, dx$.

Solution: Let $u = \ln x$, with $du = dx/x$. The problem becomes

$$\int \frac{(\ln x)^4}{x}\, dx = \int u^4\, du = \frac{u^5}{5} + C$$

In x's:

$$\int \frac{(\ln x)^4}{x}\, dx = \frac{1}{5}(\ln x)^5 + C$$

[See Section 13-2.]

PROBLEM 13-8 Find $\int x/\sqrt{x + 1}\, dx$.

Solution: Let $\sqrt{x + 1} = u$. Then $x + 1 = u^2$, with $x = u^2 - 1$ and $dx = 2u\, du$. Substituting,

$$\int \frac{x\, dx}{\sqrt{x + 1}} = \int \frac{(u^2 - 1)\, 2u\, du}{u} = 2\int (u^2 - 1)\, du$$

$$= 2\left(\frac{u^3}{3} - u\right) + C = \frac{2u}{3}(u^2 - 3) + C$$

Now return to x's ($u = \sqrt{x + 1}$, $u^2 = x + 1$):

$$\int \frac{x\, dx}{\sqrt{x + 1}} = \frac{2}{3}\sqrt{x + 1}\,(x + 1 - 3) + C$$

$$= \frac{2}{3}\sqrt{x + 1}\,(x - 2) + C$$

[See Section 13-3.]

PROBLEM 13-9 Find $\int x^2\sqrt{2-x}\,dx$.

Solution: Choose $u = \sqrt{2-x}$. Then $u^2 = 2 - x$, $x = 2 - u^2$, and $dx = -2u\,du$. Notice also that you need $x^2 = (2-u^2)^2$. Now substitute:

$$\int x^2\sqrt{2-x}\,dx = \int (2-u^2)^2(u)(-2u)\,du$$

$$= -2\int (4 - 4u^2 + u^4)u^2\,du$$

$$= -2\int (4u^2 - 4u^4 + u^6)\,du$$

$$= -2\left[\frac{4u^3}{3} - \frac{4u^5}{5} + \frac{u^7}{7}\right] + C$$

$$= -\frac{2u^3}{105}[140 - 84u^2 + 15u^4] + C$$

From the original substitution ($u = (2-x)^{1/2}$):

$$u^3 = (2-x)^{3/2}$$

$$u^2 = 2 - x$$

$$u^4 = (2-x)^2$$

The answer, then, in x's:

$$\int x^2\sqrt{2-x}\,dx = -\frac{2}{105}(2-x)^{3/2}[140 - 84(2-x) + 15(2-x)^2] + C$$

(Clearly, further algebraic simplification is possible if the need arises.) [See Section 13-3.]

PROBLEM 13-10 Find $\int (x^3/\sqrt{1-x^2})\,dx$.

Solution: This is another example of a problem in which you can save yourself some algebraic complications by noting that $d(1-x^2)$ involves the term $x\,dx$ and taking that as a hint to first rewrite the problem:

$$\int \frac{x^3\,dx}{\sqrt{1-x^2}} = \int \frac{(x^2)\,x\,dx}{\sqrt{1-x^2}}$$

Now choose $u = 1 - x^2$, with $x^2 = 1 - u$ and $2x\,dx = -du$ or $x\,dx = -\frac{1}{2}\,du$. Substituting,

$$\int \frac{(x^2)\,x\,dx}{\sqrt{1-x^2}} = \int \frac{(1-u)(-\frac{1}{2})\,du}{\sqrt{u}} = -\frac{1}{2}\int \frac{1-u}{\sqrt{u}}\,du$$

$$= -\frac{1}{2}\int (u^{-1/2} - u^{1/2})\,du = -\frac{1}{2}\left[\frac{u^{1/2}}{1/2} - \frac{u^{3/2}}{3/2}\right] + C$$

$$= -\frac{u^{1/2}}{2}\left(2 - \frac{2u}{3}\right) + C = \frac{u^{1/2}}{3}(u-3) + C$$

Returning to x's:

$$\int \frac{x^3\,dx}{\sqrt{1-x^2}} = \frac{1}{3}(1-x^2)^{1/2}(1 - x^2 - 3) + C$$

$$= -\frac{1}{3}(1-x^2)^{1/2}(x^2+2) + C$$

[See Section 13-3.]

PROBLEM 13-11 Find $\int (e^{2x}/(1+e^x)^{1/3})\,dx$.

Solution: Rewrite (see the introductory comment of Problem 13-10):

$$\int \frac{e^{2x}\,dx}{(1+e^x)^{1/3}} = \int \frac{(e^x)e^x\,dx}{(1+e^x)^{1/3}}$$

Let $u = (1+e^x)^{1/3}$, so $1 + e^x = u^3$ and $e^x\,dx = 3u^2\,du$. Substituting,

$$\int \frac{(e^x)e^x\,dx}{(1+e^x)^{1/3}} = \int \frac{(u^3-1)\,3u^2\,du}{u} = 3\int (u^4 - u)\,du$$

$$= 3\left(\frac{u^5}{5} - \frac{u^2}{2}\right) + C = \frac{3}{10}u^2(2u^3 - 5) + C$$

Returning to x's $\left(u^2 = (1+e^x)^{2/3}\right)$:

$$\int \frac{e^{2x}\,dx}{(1+e^x)^{1/3}} = \frac{3}{10}(1+e^x)^{2/3}[2(1+e^x) - 5] + C$$

$$= \frac{3}{10}(1+e^x)^{2/3}(2e^x - 3) + C$$

[See Section 13-3.]

PROBLEM 13-12 Evaluate $\int_{\pi/6}^{\pi} \sin x \cos x\,dx$.

Solution: To evaluate using the Fundamental Theorem, you first seek an antiderivative. Substitute $u = \sin x$ and $du = \cos x\,dx$. You have

$$\int \sin x \cos x\,dx = \int u\,du = \frac{u^2}{2} + C$$

$$= \frac{\sin^2 x}{2} + C$$

Note: The Fundamental Theorem calls for any antiderivative of the integrand. This allows you to choose $C = 0$. Therefore:

$$\int_{\pi/6}^{\pi} \sin x \cos x\,dx = \left.\frac{\sin^2 x}{2}\right|_{\pi/6}^{\pi} = \frac{1}{2}\left(\sin^2\pi - \sin^2\frac{\pi}{6}\right) = -\frac{1}{8}$$

[See Section 13-4.]

PROBLEM 13-13 Evaluate $\int_0^4 \frac{x\,dx}{\sqrt{x^2+9}}$.

Solution: To find an antiderivative, you substitute $u = x^2 + 9$ and $du = 2x\,dx$ or $x\,dx = \frac{1}{2}\,du$. You get

$$\int \frac{x\,dx}{\sqrt{x^2+9}} = \int \frac{\frac{1}{2}\,du}{\sqrt{u}} = \frac{1}{2}\int u^{-1/2}\,du$$

$$= \frac{1}{2}\frac{u^{1/2}}{\frac{1}{2}} + C = \sqrt{x^2+9} + C$$

Thus:

$$\int_0^4 \frac{x\,dx}{\sqrt{x^2+9}} = \left.\sqrt{x^2+9}\right|_0^4$$

$$= \sqrt{16+9} - \sqrt{0+9}$$

$$= 5 - 3 = 2$$

[See Section 13-4.]

PROBLEM 13-14 Evaluate $\int_0^1 \left(\sqrt{x}/(\sqrt{x}+1)\right)dx$ after finding the new limits of integration determined by your substitution.

Solution: Let $u = \sqrt{x} + 1$. Then $\sqrt{x} = u - 1$, $x = (u - 1)^2$, and $dx = 2(u - 1)\,du$.
The interval of integration, $0 \leqslant x \leqslant 1$, is transformed by the substitution:

$$x = 0 \qquad u = \sqrt{0} + 1 = 1$$
$$x = 1 \qquad u = \sqrt{1} + 1 = 2$$

Therefore,

$$\int_0^1 \frac{\sqrt{x}}{\sqrt{x}+1}\,dx = \int_1^2 \frac{u-1}{u}\,2(u-1)\,du = \int_1^2 \frac{2u^2 - 4u + 2}{u}\,du$$
$$= 2\int_1^2 \left(u - 2 + \frac{1}{u}\right) du = 2\left[\frac{u^2}{2} - 2u + \ln|u|\right]\Bigg|_1^2$$
$$= 2\left[\frac{u^2}{2}\Bigg|_1^2 - 2u\Bigg|_1^2 + \ln|u|\Bigg|_1^2\right]$$
$$= 2\left[\frac{1}{2}(4-1) - 2(2-1) + (\ln 2 - \ln 1)\right] = -1 + 2\ln 2$$

[See Section 13-4.]

PROBLEM 13-15 Evaluate $\int_{-1}^{0} (1/(4 - 5x))\,dx$ after finding the new limits of integration determined by your substitution.

Solution: Let $u = 4 - 5x$, so $du = -5\,dx$ and $dx = -(1/5)\,du$.
To change the limits:

$$x = -1 \qquad u = 4 - 5(-1) = 9$$
$$x = 0 \qquad u = 4 - 0 = 4$$

Substituting,

$$\int_{-1}^{0} \frac{dx}{4 - 5x} = \int_9^4 -\frac{(1/5)\,du}{u} = -\frac{1}{5}\int_9^4 \frac{du}{u}$$

Note: There are two things to notice here:

1. Normally, in $\int_a^b$, you expect $a < b$, so $\int_9^4$ looks strange. However, this choice of $a = 9$, $b = 4$ is dictated by the substitution: The lower limit on the new integral must correspond to the lower limit on the original integral, and similarly for the upper limit.
2. The minus sign in front of the integral, coupled with the "backward" order of a and b, suggests the use of a property of integrals:

$$\int_a^b = -\int_b^a$$

$$-\frac{1}{5}\int_9^4 \frac{du}{u} = \frac{1}{5}\int_4^9 \frac{du}{u}$$
$$= \frac{1}{5}\ln|u|\Bigg|_4^9 = \frac{1}{5}(\ln 9 - \ln 4)$$

[See Section 13-4.]

PROBLEM 13-16 Evaluate $\int_0^{1/2} (1/(4x^2 + 1))\,dx$ after changing the limits to correspond with your substitution.

Solution: A good substitution on this one becomes more apparent if you rewrite: $4x^2 + 1 = (2x)^2 + 1$. Then $u = 2x$ and $du = 2\,dx$ or $dx = \frac{1}{2}\,du$.
The limits are

$$x = 0 \qquad u = 2(0) = 0$$
$$x = \frac{1}{2} \qquad u = 2\left(\frac{1}{2}\right) = 1$$

Substituting:

$$\int_0^{1/2} \frac{dx}{(2x)^2 + 1} = \int_0^1 \frac{\frac{1}{2}\,du}{u^2 + 1} = \frac{1}{2}\int_0^1 \frac{du}{u^2 + 1} = \frac{1}{2}\tan^{-1} u \bigg|_0^1$$

$$= \frac{1}{2}(\tan^{-1} 1 - \tan^{-1} 0) = \frac{1}{2}\left(\frac{\pi}{4}\right) = \frac{\pi}{8}$$

[See Section 13-4.]

PROBLEM 13-17 Evaluate $\int_0^{\ln 3} \sqrt{1 + e^x}\, e^x\, dx$ after finding new limits of integration to correspond to your substitution.

Solution: Let $u = 1 + e^x$, so $du = e^x\, dx$.
The limits are

$$x = 0 \qquad u = 1 + 1 = 2$$

$$x = \ln 3 \qquad u = 1 + e^{\ln 3} = 1 + 3 = 4$$

Thus the integral becomes

$$\int_0^{\ln 3} \sqrt{1 + e^x}\, e^x\, dx = \int_2^4 \sqrt{u}\, du = \frac{u^{3/2}}{3/2}\bigg|_2^4$$

$$= \frac{2}{3}(4^{3/2} - 2^{3/2}) = \frac{2}{3}(8 - 2\sqrt{2})$$

[See Section 13-4.]

PROBLEM 13-18 Evaluate $\displaystyle\int_0^2 [x^3/(x + 1)^4]\, dx$.

Solution: Let $u = x + 1$, so $du = dx$.
Changing the limits, you get

$$x = 0 \qquad u = 0 + 1 = 1$$

$$x = 2 \qquad u = 2 + 1 = 3$$

Substituting:

$$\int_0^2 \frac{x^3}{(x + 1)^4}\, dx = \int_1^3 \frac{(u - 1)^3}{u^4}\, du = \int_1^3 \frac{u^3 - 3u^2 + 3u - 1}{u^4}\, du$$

$$= \int_1^3 \left(\frac{1}{u} - \frac{3}{u^2} + \frac{3}{u^3} - \frac{1}{u^4}\right) du = \left(\ln|u| + \frac{3}{u} - \frac{3}{2u^2} + \frac{1}{3u^3}\right)\bigg|_1^3$$

$$= (\ln 3 - \ln 1) + 3\left(\frac{1}{3} - \frac{1}{1}\right) - \frac{3}{2}\left(\frac{1}{9} - \frac{1}{1}\right) + \frac{1}{3}\left(\frac{1}{27} - \frac{1}{1}\right)$$

$$= \ln 3 - 2 + \frac{4}{3} - \frac{26}{81} = \ln 3 - \frac{80}{81}$$

Had you elected not to change limits, you would have had, using the same substitution ($u = x + 1$),

$$\int \frac{x^3}{(x + 1)^4}\, dx = \int \frac{(u - 1)^3}{u^4}\, du = \ln|u| + \frac{3}{u} - \frac{3}{2u^2} + \frac{1}{3u^3} + C$$

Returning to x's and the integral:

$$\int_0^2 \frac{x^3}{(x + 1)^4}\, dx = \left(\ln|x + 1| + \frac{3}{x + 1} - \frac{3}{2(x + 1)^2} + \frac{1}{3(x + 1)^3}\right)\bigg|_0^2$$

$$= (\ln 3 - \ln 1) + 3\left(\frac{1}{2 + 1} - \frac{1}{1}\right) - \frac{3}{2}\left(\frac{1}{(2 + 1)^2} - \frac{1}{1}\right) + \frac{1}{3}\left(\frac{1}{(2 + 1)^3} - 1\right)$$

$$= \ln 3 - 2 + \frac{4}{3} - \frac{26}{81} = \ln 3 - \frac{80}{81}$$

[See Section 13-4.]

PROBLEM 13-19 Evaluate $\int_0^3 x\sqrt{1+x}\,dx$.

Solution: Choose $u = \sqrt{1+x}$. This gives you $1 + x = u^2$ and $x = u^2 - 1$, with $dx = 2u\,du$. Changing limits gives

$$x = 0 \qquad u = \sqrt{1} = 1$$

$$x = 3 \qquad u = \sqrt{4} = 2$$

Substituting,

$$\int_0^3 x\sqrt{1+x}\,dx = \int_1^2 (u^2 - 1)u\,2u\,du = 2\int_1^2 (u^4 - u^2)\,du$$

$$= 2\left(\frac{u^5}{5} - \frac{u^3}{3}\right)\Bigg|_1^2 = 2\left[\frac{1}{5}(32 - 1) - \frac{1}{3}(8 - 1)\right]$$

$$= 2\left(\frac{31}{5} - \frac{7}{3}\right) = \frac{116}{15}$$

[See Section 13-4.]

PROBLEM 13-20 Evaluate $\int_{\pi/6}^{\pi/3} \tan^3 x \sec^2 x\,dx$.

Solution: If $u = \tan x$, then $du = \sec^2 x\,dx$.
Limits:

$$x = \frac{\pi}{6} \qquad u = \tan\frac{\pi}{6} = \frac{1}{\sqrt{3}}$$

$$x = \frac{\pi}{3} \qquad u = \tan\frac{\pi}{3} = \sqrt{3}$$

Therefore,

$$\int_{\pi/6}^{\pi/3} \tan^3 x \sec^2 x\,dx = \int_{1/\sqrt{3}}^{\sqrt{3}} u^3\,du$$

$$= \frac{u^4}{4}\Bigg|_{1/\sqrt{3}}^{\sqrt{3}} = \frac{1}{4}\left[9 - \frac{1}{9}\right] = \frac{20}{9}$$

[See Section 13-4.]

Supplementary Exercises

Find the following antiderivatives:

13-21 $\int e^{2x+5}\,dx$

13-22 $\int x/(x^2 + 4)\,dx$

13-23 $\int x^2/\sqrt{2x^3 + 5}\,dx$

13-24 $\int 2\sin x\cos^2 x\,dx$

13-25 $\int e^x\cos(e^x)\,dx$

13-26 $\int e^{\sin x}\cos x\,dx$

13-27 $\int [x/(4 - x^2)^{1/3}]\,dx$

13-28 $\int (\sec^2 x)/\tan x\,dx$

13-29 $\int x^2\sqrt{1+x}\,dx$

13-30 $\int (1 - \sin x)/(x + \cos x)\,dx$

13-31 $\int x(x^2 - 1)^3\,dx$

13-32 $\int 1/(4 - 3x)\,dx$

13-33 $\int e^x\sqrt{1 + 4e^x}\,dx$

13-34 $\int e^{2x}\sqrt{1 + 4e^x}\,dx$

(*Hint*: Rewrite the integrand using $e^{2x} = e^x e^x$ and $d(e^x) = e^x\,dx$.)

13-35 $\int x^3(x^2 + 1)^{1/3}\,dx$

13-36 $\int 1/(x + x^{5/6})\,dx$ (*Hint*: Let $u = x^{1/6}$.)

13-37 $\int [x/(1 + 2x)^2]\,dx$

13-38 $\int [2x/(2 - x)^{2/3}]\,dx$

13-39 $\int (x + 2)^2\sqrt{1 + x}\,dx$

13-40 $\int \sqrt{1 + \sqrt{x}}/\sqrt{x}\,dx$

Evaluate each of the following integrals by applying the Fundamental Theorem directly. (Find an antiderivative, and evaluate it between the given limits.)

13-41 $\int_0^1 \sqrt{7x + 2}\,dx$

13-42 $\int_0^{\sqrt{2}} x\sqrt{4 - x^2}\,dx$

13-43 $\int_0^2 x/\sqrt{x^2 + 1}\,dx$

13-44 $\int_0^1 x/(x^2 + 1)\,dx$

13-45 $\int_0^{\pi/2} (\sin x)^{1/2}\cos x\,dx$

13-46 $\int_0^{\sqrt{2}} 3xe^{x^2+1}\,dx$

Evaluate each of the following integrals after changing limits of integration to correspond to your substitution.

13-47 $\int_1^2 x\sqrt{1 + x}\,dx$

13-48 $\int_0^{\pi/4} (1 + \tan x)\sec^2 x\,dx$

13-49 $\int_0^1 e^{3x+2}\,dx$

13-50 $\int_0^{\pi/4} \sec x\tan x/(4 + \sec x)\,dx$

13-51 $\int_{\pi/4}^{\pi/2} (\cos x)/\sin x\,dx$

13-52 $\int_0^{13} 1/\sqrt[3]{1 + 2x}\,dx$

13-53 $\int_0^{1/2 \ln 3} e^x/(1 + e^{2x})\,dx$

13-54 $\int_1^3 [x/(2x - 1)^2]\,dx$

13-55 $\int_0^1 x^2/(x^3 + 4)\,dx$

13-56 $\int_{\pi^2/9}^{\pi^2/4} (\sin\sqrt{x})/\sqrt{x}\,dx$

13-57 $\int_0^4 x/(1 + \sqrt{x})\,dx$

13-58 $\int_0^{\pi/2} \cos x\sqrt{1 + \sin x}\,dx$

13-59 $\int_0^9 \sqrt{1 + \sqrt{x}}\,dx$

13-60 $\int_1^2 e^{-1/x}/x^2\,dx$

Solutions to Supplementary Exercises

(13-21) $u = 2x + 5, du = 2\,dx; \quad \int e^u \tfrac{1}{2}\,du = \tfrac{1}{2}e^{2x+5} + C$

(13-22) $u = x^2 + 4, du = 2x\,dx; \ \tfrac{1}{2}\int 1/u\,du = \tfrac{1}{2}\ln|x^2 + 4| + C$

(13-23) $u = 2x^3 + 5, du = 6x^2\,dx; \quad (1/6)\int u^{-1/2}\,du = (1/3)(2x^3 + 5)^{1/2} + C$

(13-24) $u = \cos x, du = -\sin x\,dx; \quad -2\int u^2\,du = -\tfrac{2}{3}\cos^3 x + C$

(13-25) $u = e^x, du = e^x\,dx; \quad \int \cos u\,du = \sin e^x + C$

(13-26) $u = \sin x, du = \cos x\,dx; \quad \int e^u\,du = e^{\sin x} + C$

(13-27) $u = 4 - x^2, du = -2x\,dx; \quad -\tfrac{1}{2}\int u^{-1/3}\,du = -\tfrac{3}{4}(4 - x^2)^{2/3} + C$

(13-28) $u = \tan x, du = \sec^2 x\,dx; \quad \int 1/u\,du = \ln|\tan x| + C$

(13-29) $u = \sqrt{1 + x}, dx = 2u\,du;$

$$\int (u^2 - 1)^2(u)\,2u\,du = (2/105)(1 + x)^{3/2}[15(1 + x)^2 - 42(1 + x) + 35] + C$$

(13-30) $u = x + \cos x, du = (1 - \sin x)\,dx; \quad \int 1/u\,du = \ln|x + \cos x| + C$

(13-31) $u = x^2 - 1, du = 2x\,dx; \quad (1/2)\int u^3\,du = (1/8)(x^2 - 1)^4 + C$

(13-32) $u = 4 - 3x, du = -3\,dx; \quad -(1/3)\int 1/u\,du = -(1/3)\ln|4 - 3x| + C$

(13-33) $u = 1 + 4e^x, du = 4e^x\,dx; \quad (1/4)\int u^{1/2}\,du = (1/6)(1 + 4e^x)^{3/2} + C$

(13-34) $u = 1 + 4e^x, du = 4e^x\,dx; \quad \int \frac{u - 1}{4}(u^{1/2})\frac{1}{4}\,du = \frac{(1 + 4e^x)^{3/2}}{120}(12e^x - 2) + C$

(13-35) $u = (x^2 + 1)^{1/3}, u^3 = x^2 + 1, 3u^2\,du = 2x\,dx;$

$$\int (u^3 - 1)(3/2)u^2\,du = (3/56)(x^2 + 1)^{4/3}(4x^2 - 3) + C$$

(13-36) $u = x^{1/6}, u^6 = x, dx = 6u^5\,du; \quad \int 6u^5/(u^6 + u^5)\,du = 6\int 1/(u + 1)\,du = 6\ln|x^{1/6} + 1| + C$

(13-37) $u = 1 + 2x, du = 2\,dx; \int \frac{[(u - 1)/2](1/2)\,du}{u^2} = \frac{1}{4}\left(\ln|1 + 2x| + \frac{1}{1 + 2x}\right) + C$

(13-38) $u = (2 - x)^{1/3}, u^3 = 2 - x, 3u^2\,du = -dx;$

$$\int \frac{2(2 - u^3)(-3u^2)\,du}{u^2} = -(3/2)(2 - x)^{1/3}(x + 6) + C$$

(13-39) $u = \sqrt{1 + x}, u^2 = 1 + x, 2u\,du = dx;$

$$\int (u^2 + 1)^2(u)\,2u\,du = (2/105)(1 + x)^{3/2}\,[15(1 + x)^2 + 42(1 + x) + 35] + C$$

(13-40) $u = \sqrt{x}, du = 1/(2\sqrt{x})\,dx; \quad \int (1 + u)^{1/2}\,2\,du = (4/3)(1 + \sqrt{x})^{3/2} + C$

(13-41) $u = 7x + 2, du = 7\,dx; \quad (1/7)\int u^{1/2}\,du = (2/21)\,u^{3/2} + C;$

$$\int_0^1 \sqrt{7x + 2}\,dx = (2/21)(7x + 2)^{3/2}\Big|_0^1 = (2/21)(27 - 2\sqrt{2})$$

(13-42) $u = 4 - x^2, du = -2x\,dx; \quad -\tfrac{1}{2}\int u^{1/2}\,du = -\tfrac{1}{3}u^{3/2} + C;$

$$\int_0^{\sqrt{2}} x\sqrt{4 - x^2}\,dx = -\tfrac{1}{3}(4 - x^2)^{3/2}\Big|_0^{\sqrt{2}} = -\tfrac{1}{3}(2\sqrt{2} - 8)$$

(13-43) $u = x^2 + 1, du = 2x\,dx; \quad \tfrac{1}{2}\int u^{-1/2}\,du = u^{1/2} + C;$

$$\int_0^2 x/(x^2 + 1)\,dx = (x^2 + 1)^{1/2}\Big|_0^2 = \sqrt{5} - 1$$

(13-44) $u = x^2 + 1, du = 2x\,dx; \quad \tfrac{1}{2}\int 1/u\,du = \tfrac{1}{2}\ln|u| + C;$

$$\int_0^1 x/(x^2 + 1)\,dx = \tfrac{1}{2}\ln|x^2 + 1|\Big|_0^1 = \tfrac{1}{2}\ln 2$$

(13-45) $u = \sin x, du = \cos x\,dx; \quad \int u^{1/2}\,du = \tfrac{2}{3}u^{3/2} + C;$

$$\int_0^{\pi/2} (\sin x)^{1/2}\cos x\,dx = \tfrac{2}{3}(\sin x)^{3/2}\Big|_0^{\pi/2} = \tfrac{2}{3}$$

(13-46) $u = x^2 + 1, du = 2x\,dx; \quad (3/2)\int e^u\,du = (3/2)e^u + C;$

$$\int_0^{\sqrt{2}} 3x\,e^{x^2+1}\,dx = (3/2)e^{x^2+1}\Big|_0^{\sqrt{2}} = (3/2)(e^3 - e)$$

(13-47) $u = \sqrt{1 + x}, u^2 = 1 + x, 2u\,du = dx;$ if $x = 1, u = \sqrt{2};$ if $x = 2, u = \sqrt{3};$

$$\int_{\sqrt{2}}^{\sqrt{3}} (u^2 - 1)(u)2u\,du = 2[(u^5/5) - (u^3/3)]\Big|_{\sqrt{2}}^{\sqrt{3}} = (4/15)(6\sqrt{3} - \sqrt{2})$$

(13-48) $u = 1 + \tan x, du = \sec^2 x\,dx;$ if $x = 0, u = 1;$ if $x = \pi/4, u = 2;$

$$\int_1^2 u\,du = u^2/2\Big|_1^2 = 3/2$$

(13-49) $u = 3x + 2, du = 3\,dx;$ if $x = 0, u = 2;$ if $x = 1, u = 5;$

$$\int_2^5 (e^u)\,du/3 = e^u/3\Big|_2^5 = (e^5 - e^2)/3$$

(13-50) $u = 4 + \sec x, du = \sec x\tan x\,dx;$ if $x = 0, u = 5;$ if $x = \pi/4, u = 4 + \sqrt{2};$

$$\int_5^{4+\sqrt{2}} 1/u\,du = \ln|u|\Big|_5^{4+\sqrt{2}} = \ln(4 + \sqrt{2}) - \ln 5$$

(13-51) $u = \sin x, du = \cos x\, dx$; if $x = \pi/4, u = \sqrt{2}/2$; if $x = \pi/2, u = 1$;

$$\int_{\sqrt{2}/2}^{1} 1/u\, du = \ln|u| \Big|_{\sqrt{2}/2}^{1} = \ln\sqrt{2}$$

(13-52) $u = 1 + 2x, du = 2\, dx$; if $x = 0, u = 1$; if $x = 13, u = 27$;

$$\int_{1}^{27} \frac{\frac{1}{2}\, du}{u^{1/3}} = \tfrac{3}{4}u^{2/3} \Big|_{1}^{27} = 6$$

(13-53) $u = e^x, du = e^x\, dx, e^{2x} = u^2$; if $x = 0, u = 1$; if $x = \frac{1}{2}\ln 3, u = \sqrt{3}$;

$$\int_{1}^{\sqrt{3}} 1/(1 + u^2)\, du = \tan^{-1} u \Big|_{1}^{3} = \pi/12$$

(13-54) $u = 2x - 1, du = 2\, dx$; if $x = 1, u = 1$; if $x = 3, u = 5$;

$$\int_{1}^{5} \frac{[(u + 1)/2]\frac{1}{2}\, du}{u^2} = \frac{1}{4}\left(\ln|u| - \frac{1}{u}\right)\Big|_{1}^{5} = \frac{1}{4}\left(\ln 5 + \frac{4}{5}\right)$$

(13-55) $u = x^3 + 4, du = 3x^2\, dx$; if $x = 0, u = 4$; if $x = 1, u = 5$;

$$\int_{4}^{5} \frac{\frac{1}{3}\, du}{u} = \tfrac{1}{3}\ln|u| \Big|_{4}^{5} = \tfrac{1}{3}(\ln 5 - \ln 4)$$

(13-56) $u = \sqrt{x}, du = 1/(2\sqrt{x})\, dx$; if $x = \pi^2/9, u = \pi/3$; if $x = \pi^2/4, u = \pi/2$;

$$\int_{\pi/3}^{\pi/2} (\sin u)\, 2\, du = -2\cos u \Big|_{\pi/3}^{\pi/2} = 1$$

(13-57) $u = 1 + \sqrt{x}, x = (u - 1)^2, dx = 2(u - 1)\, du$; if $x = 0, u = 1$; if $x = 4, u = 3$;

$$\int_{1}^{3} \frac{(u - 1)^2\, 2(u - 1)\, du}{u} = 2\left(\frac{u^3}{3} - \frac{3u^2}{2} + 3u - \ln|u|\right)\Big|_{1}^{3} = 2\left(\frac{8}{3} - \ln 3\right)$$

(13-58) $u = 1 + \sin x, du = \cos x\, dx$; if $x = 0, u = 1$; if $x = \pi/2, u = 2$;

$$\int_{1}^{2} u^{1/2}\, du = \tfrac{2}{3}u^{3/2} \Big|_{1}^{2} = \tfrac{2}{3}(2\sqrt{2} - 1)$$

(13-59) $u = \sqrt{1 + \sqrt{x}}, x = (u^2 - 1)^2, dx = 2(u^2 - 1)\, 2u\, du$; if $x = 0, u = 1$; if $x = 9, u = 2$;

$$\int_{1}^{2} u2(u^2 - 1)2u\, du = 4[(u^5/5) - (u^3/3)] \Big|_{1}^{2} = 232/15$$

(13-60) $u = 1/x, du = -1/x^2\, dx$; if $x = 1, u = 1$; if $x = 2, u = \frac{1}{2}$;

$$\int_{1}^{1/2} e^{-u}(-du) = e^{-u} \Big|_{1}^{1/2} = e^{-1/2} - e^{-1}$$

EXAM 5 (CHAPTERS 11–13)

1. Antidifferentiate

(a) $\int (3x^4 - 5\sqrt{x} + 1)\, dx$

(b) $\int \left(\cos x + 2e^x + \frac{3}{\sqrt[3]{x^2}}\right) dx$

2. If $f(4) = 3$ and $f'(x) = \sqrt{x} - (1/\sqrt{x})$, find $f(1)$.

3. Evaluate

(a) $\int_{-2}^{-1} \left(x^2 + \frac{1}{x^2}\right) dx$

(b) $\int_{\pi/6}^{\pi/3} \left(\frac{1}{\cos^2 x} - \frac{1}{\sin^2 x}\right) dx$

4. Find the area of the region bounded by the graphs of $f(x) = 3(x + 1)^2 + x$, $x = -1$, and $x = 3$ by finding the limit of the Riemann sums.

5. Use substitution to find

(a) $\int (5x - 1)^7 dx$

(b) $\int_0^{\pi/3} \frac{\sin x}{\sqrt{\cos x}}\, dx$

(c) $\int \frac{e^x}{1 + 4e^{2x}}\, dx$

(d) $\int \frac{e^{\tan x}}{\cos^2 x}\, dx$

(e) $\int_0^1 \sqrt{1 - \sqrt{x}}\, dx$

SOLUTIONS TO EXAM 5

1. **(a)** $\int (3x^4 - 5\sqrt{x} + 1)\, dx = 3\int x^4\, dx - 5\int x^{1/2}\, dx + \int dx$

$$= 3\frac{x^5}{5} - \frac{5x^{3/2}}{3/2} + x + C = \frac{3}{5}x^5 - \frac{10}{3}x^{3/2} + x + C$$

(b) $\int \left(\cos x + 2e^x + \frac{3}{\sqrt[3]{x^2}}\right) dx = \int (\cos x + 2e^x + 3x^{-2/3})\, dx$

$$= \sin x + 2e^x + \frac{3x^{1/3}}{1/3} + C = \sin x + 2e^x + 9x^{1/3} + C$$

2. If $f'(x) = \sqrt{x} - (1/\sqrt{x})$, then $f(x)$ is an antiderivative of $f'(x)$,

$$\int f'(x)\, dx = \int \left(\sqrt{x} - \frac{1}{\sqrt{x}}\right) dx = \int (x^{1/2} - x^{-1/2})\, dx = \frac{2}{3}x^{3/2} - 2x^{1/2} + C$$

so $f(x) = \frac{2}{3}x^{3/2} - 2x^{1/2} + C$. You can find C because you know $f(4) = 3$:

$$3 = f(4) = \frac{2}{3}(4)^{3/2} - 2(4)^{1/2} + C = \frac{2}{3}(8) - 2(2) + C = \frac{4}{3} + C$$

$$C = 3 - \frac{4}{3} = \frac{5}{3}$$

$$f(x) = \frac{2}{3}x^{3/2} - 2x^{1/2} + \frac{5}{3}$$

$$f(1) = \frac{2}{3}(1)^{3/2} - 2(1)^{1/2} + \frac{5}{3} = \frac{2}{3} - 2 + \frac{5}{3} = \frac{1}{3}$$

3. **(a)**

$$\int_{-2}^{-1}\left(x^2 + \frac{1}{x^2}\right)dx = \int_{-2}^{-1}(x^2 + x^{-2})\,dx = \left(\frac{x^3}{3} - x^{-1}\right)\Bigg|_{-2}^{-1}$$

$$= \left(\frac{-1}{3} + 1\right) - \left(\frac{-8}{3} + \frac{1}{2}\right) = \frac{17}{6}$$

(b)

$$\int_{\pi/6}^{\pi/3}\left(\frac{1}{\cos^2 x} - \frac{1}{\sin^2 x}\right)dx = \int_{\pi/6}^{\pi/3}(\sec^2 x - \csc^2 x)\,dx = (\tan x + \cot x)\Bigg|_{\pi/6}^{\pi/3}$$

$$= \left(\tan\frac{\pi}{3} + \cot\frac{\pi}{3}\right) - \left(\tan\frac{\pi}{6} + \cot\frac{\pi}{6}\right)$$

$$= \left(\sqrt{3} + \frac{\sqrt{3}}{3}\right) - \left(\frac{\sqrt{3}}{3} + \sqrt{3}\right) = 0$$

4. Divide $[-1, 3]$ into n subintervals:

$$\Delta x = \frac{3 - (-1)}{n} = \frac{4}{n}$$

$$x_i = a + i\Delta x = -1 + \frac{4i}{n}$$

The nth Riemann sum is

$$\sum_{i=1}^{n} f(x_i)\,\Delta x = \sum_{i=1}^{n} f\left(-1 + \frac{4i}{n}\right)\frac{4}{n} = \sum_{i=1}^{n}\left\{3\left[\left(-1 + \frac{4i}{n}\right) + 1\right]^2 + \left(-1 + \frac{4i}{n}\right)\right\}\frac{4}{n}$$

$$= \sum_{i=1}^{n}\left(3\cdot\frac{16i^2}{n^2} + \frac{4i}{n} - 1\right)\frac{4}{n}$$

$$= \sum_{i=1}^{n}\left(\frac{192i^2}{n^3} + \frac{16i}{n^2} - \frac{4}{n}\right) = \frac{192}{n^3}\sum_{i=1}^{n} i^2 + \frac{16}{n^2}\sum_{i=1}^{n} i - \frac{4}{n}\sum_{i=1}^{n} 1$$

$$= \frac{192}{n^3}\left[\frac{n(n+1)(2n+1)}{6}\right] + \frac{16}{n^2}\left[\frac{n(n+1)}{2}\right] - \frac{4}{n}(n)$$

$$= \frac{32n(n+1)(2n+1)}{n^3} + \frac{8n(n+1)}{n^2} - 4$$

The area is the limit of the Riemann sums:

$$\lim_{n\to\infty}\sum_{i=1}^{n} f(x_i)\,\Delta x = \lim_{n\to\infty}\left[\frac{32n(n+1)(2n+1)}{n^3} + \frac{8n(n+1)}{n^2} - 4\right]$$

$$= 64 + 8 - 4 = 68$$

5. **(a)** Let $u = 5x - 1$ with $du = 5dx$:

$$\int(5x-1)^7\,dx = \int u^7\left(\frac{1}{5}du\right) = \frac{1}{5}\left(\frac{u^8}{8}\right) + C = \frac{1}{40}(5x-1)^8 + C$$

(b) Let $u = \cos x$ with $du = -\sin x\, dx$. When $x = 0$, $u = \cos 0 = 1$; when $x = \pi/3$, $u = \cos \pi/3 = 1/2$.

$$\int_0^{\pi/3} \frac{\sin x}{\sqrt{\cos x}}\, dx = \int_1^{1/2} \frac{-du}{\sqrt{u}} = -2u^{1/2}\Big|_1^{1/2} = -2\sqrt{\frac{1}{2}} + 2 = 2 - \sqrt{2}$$

(c) Let $u = 2e^x$ with $du = 2e^x\, dx$:

$$\int \frac{e^x}{1 + 4e^{2x}}\, dx = \int \frac{\frac{1}{2}\, du}{1 + u^2} = \frac{1}{2} \operatorname{arc\,tan} u + C = \frac{1}{2} \operatorname{arc\,tan}(2e^x) + C$$

(d) Let $u = \tan x$ with $du = \sec^2 x\, dx$:

$$\int \frac{e^{\tan x}}{\cos^2 x}\, dx = \int e^{\tan x} \sec^2 x\, dx = \int e^u\, du = e^u + C = e^{\tan x} + C$$

(e) Let $u = 1 - \sqrt{x}$; you find $x = (1 - u)^2$ and so $dx = -2(1 - u)\, du$. When $x = 0$, $u = 1$; when $x = 1$, $u = 0$.

$$\int_0^1 \sqrt{1 - \sqrt{x}}\, dx = \int_1^0 \sqrt{u}[-2(1 - u)\, du] = 2\int_1^0 (u^{3/2} - u^{1/2})\, du$$

$$= 2\left(\frac{2}{5}u^{5/2} - \frac{2}{3}u^{3/2}\right)\Big|_1^0 = 2\left[(0) - \left(\frac{2}{5} - \frac{2}{3}\right)\right] = \frac{8}{15}$$

14 OTHER METHODS OF INTEGRATION

THIS CHAPTER IS ABOUT

- ☑ Integration by Parts
- ☑ Integration of Certain Trigonometric Functions
- ☑ Inverse Trigonometric Substitutions
- ☑ Rational Functions
- ☑ Rational Functions of sin x and cos x
- ☑ Integral Tables
- ☑ An Overview

In Chapters 16, 18, and 19 you'll see that the solutions of many physical problems require that you evaluate an integral. In Chapter 12 you learned that the principal method used to evaluate an integral is the Fundamental Theorem of Calculus. In other words, you can always evaluate $\int_a^b f(x)\,dx$ if you can find an antiderivative of $f(x)$.

This problem isn't always solvable: Not every function $f(x)$—not even every continuous function—has an antiderivative that can be easily described. But many of them do, and this chapter will show you a number of techniques that will help you solve the basic antidifferentiation problem.

14-1. Integration by Parts

This technique is frequently useful when the integrand is a product. It is derived from the formula for the differential of a product. Recall that, if u and v are functions of x, then

$$\frac{d}{dx}(uv) = u\frac{dv}{dx} + v\frac{du}{dx}$$

or, in terms of differentials:

$$d(uv) = u\,dv + v\,du$$

The resulting antiderivative statement is

$$uv = \int u\,dv + \int v\,du$$

or, (in a more useful form):

$$\int u\,dv = uv - \int v\,du \tag{14-1}$$

To use this technique of integration, you must be able to recognize a given integrand as a product of two factors: u (a differentiable function of x) and dv (an integrable function of x).

EXAMPLE 14-1: Find $\int x\,e^x\,dx$.

Solution: First, examine possible choices for u and for dv: $u = x$, with $dv = e^x\,dx$; or $u = e^x$ and $dv = x\,dx$. Neither of these is an obvious "best choice"; examine each.

(a) Let $u = x$, with $du = dx$; then $dv = e^x\,dx$ and $v = e^x$ (obtain v from dv by antidifferentiating).

Substitution into the equation gives

$$\int \underbrace{x}_{u}\,\underbrace{e^x\,dx}_{dv} = \underbrace{x}_{u}\,\underbrace{e^x}_{v} - \int \underbrace{e^x}_{v}\,\underbrace{dx}_{du}$$

so that

$$\int xe^x\,dx = xe^x - e^x + C$$

(b) Let $u = e^x$, with $du = e^x\,dx$; then $dv = x\,dx$ and $v = x^2/2$. Substitute:

$$\int xe^x\,dx = \int e^x x\,dx = \frac{e^x(x^2)}{2} - \int \frac{x^2}{2}\,e^x\,dx$$

Notice here that the new integration problem (i.e., $\int (x^2/2)e^x\,dx$) is more complicated than the one you started with. This is a clue that your choice of u and dv hasn't been wise, and you need to consider possible alternatives. In general, the "best choice" for u is the one that becomes "simpler" when differentiated.

EXAMPLE 14-2: Find $\int \ln x\,dx$.

Solution: Choose u and dv in the only possible way:

$$u = \ln x,\ du = \frac{dx}{x} \qquad dv = dx,\ v = x$$

Then

$$\begin{aligned}\int \ln x\,dx &= (\ln x)(x) - \int x\,\frac{dx}{x}\\ &= x\ln x - \int dx\\ &= x\ln x - x + C\end{aligned}$$

EXAMPLE 14-3: Find $\int e^x \sin x\,dx$.

Solution: Choose:

$$u = e^x,\ du = e^x\,dx \qquad dv = \sin x\,dx,\ v = -\cos x$$

Remember that you get v from dv by antidifferentiating. You have

$$\begin{aligned}\int e^x \sin x\,dx &= -e^x\cos x - \int (-\cos x)e^x\,dx\\ &= -e^x\cos x + \int e^x\cos x\,dx\end{aligned}$$

The new integral is still not recognizable as a basic formula, but it's no more complicated than the one you started with. Do a second integration by parts. Choose:

$$u = e^x,\ du = e^x\,dx \qquad dv = \cos x\,dx,\ v = \sin x$$

Substitute:

$$\int e^x \sin x\, dx = -e^x \cos x + \left[e^x \sin x - \int e^x \sin x\, dx\right]$$

The new integral is the same as the one in the original problem; add it to both sides of the equation:

$$2\int e^x \sin x\, dx = -e^x \cos x + e^x \sin x$$

Finally, divide through by two, add an arbitrary constant, and you have found your answer:

$$\int e^x \sin x\, dx = \frac{1}{2}(e^x \sin x - e^x \cos x) + C$$

Warning: When you integrate by parts twice, you must be careful not to undo what you have just done. In your second integration, choose u to be the du from the first integration. Look back at Example 14-3; you have (after one integration)

$$\int e^x \sin x\, dx = -e^x \cos x + \int e^x \cos x\, dx$$

The correct choice of u in the second integral is $u = e^x$. Suppose you choose $u = \cos x$ and $dv = e^x\, dx$. Then

$$\int e^x \cos x\, dx = e^x \cos x + \int e^x \sin x\, dx$$

Substituting back:

$$\int e^x \sin x\, dx = -e^x \cos x + e^x \cos x + \int e^x \sin x\, dx$$

You have gone in a circle; the second integration by parts "undid" the first.

EXAMPLE 14-4: Find $\int x\sqrt{x+1}\, dx$.

Solution: Choose:

$$u = x,\ du = dx \qquad dv = (1+x)^{1/2}\, dx,\ v = \frac{2}{3}(1+x)^{3/2}$$

Then

$$\begin{aligned}\int x\sqrt{1+x}\, dx &= \frac{2x}{3}(1+x)^{3/2} - \frac{2}{3}\int (1+x)^{3/2}\, dx \\ &= \frac{2x}{3}(1+x)^{3/2} - \frac{2}{3}\left[\frac{2}{5}(1+x)^{5/2}\right] + C \\ &= \frac{2x}{3}(1+x)^{3/2} - \frac{4}{15}(1+x)^{5/2} + C\end{aligned}$$

Notice that this problem could also have been done by a substitution. It's frequently true that you'll have a choice of techniques to do a particular integration. When this happens, you should choose the method that *you* find the easiest to use.

14-2. Integration of Certain Trigonometric Functions

A. Integration of $\sin^m x \cos^n x\, dx$

The method of finding $\sin^m x \cos^n x\, dx$ depends on whether m (or n) is even or odd.

1. If m or n is odd and positive, then you can put the integrand in the form of $u^n\, du$.

EXAMPLE 14-5: Find $\int \sin^4 x \cos^3 x\, dx$.

Solution: The power of $\cos x$ is odd; if you "break off" one factor of $\cos x$, the other factor is an even power of $\cos x$ and can therefore (using the identity $\cos^2 x = 1 - \sin^2 x$) be written as a power of $1 - \sin^2 x$:

$$\begin{aligned}\int \sin^4 x \cos^3 x\, dx &= \int \sin^4 x \cos^2 x \cos x\, dx \\ &= \int \sin^4 x(1 - \sin^2 x)\cos x\, dx \\ &= \int \sin^4 x \cos x\, dx - \int \sin^6 x \cos x\, dx \\ &= \frac{\sin^5 x}{5} - \frac{\sin^7 x}{7} + C\end{aligned}$$

EXAMPLE 14-6: Find $\int \sin^5 x \sqrt{\cos x}\, dx$.

Solution: The power of $\sin x$ is odd; you can use the same technique as in Example 14-5, with the roles of $\sin x$ and $\cos x$ reversed:

$$\begin{aligned}\int \sin^5 x \sqrt{\cos x}\, dx &= \int \sin^4 x \sqrt{\cos x} \sin x\, dx \\ &= \int (1 - \cos^2 x)^2 (\cos x)^{1/2} \sin x\, dx \\ &= \int (1 - 2\cos^2 x + \cos^4 x)(\cos x)^{1/2} \sin x\, dx \\ &= \int (\cos x)^{1/2} \sin x\, dx - \int 2\cos^{5/2} x \sin x\, dx \\ &\quad + \int \cos^{9/2} x \sin x\, dx \\ &= -\frac{2}{3}(\cos x)^{3/2} + \frac{4}{7}(\cos x)^{7/2} - \frac{2}{11}(\cos x)^{11/2} + C\end{aligned}$$

2. If both m and n are even (including m (or n) $= 0$), use the double-angle identities:

$$\sin^2 x = \frac{1}{2}(1 - \cos 2x) \qquad \textbf{(14-2)}$$

$$\cos^2 x = \frac{1}{2}(1 + \cos 2x) \qquad \textbf{(14-3)}$$

EXAMPLE 14-7: Find $\int \cos^4 x\, dx$.

Solution: The power of cos x is even, so

$$\int \cos^4 x\, dx = \int \left[\frac{1}{2}(1 + \cos 2x)\right]^2 dx$$
$$= \frac{1}{4}\int (1 + 2\cos 2x + \cos^2 2x)\, dx$$
$$= \frac{1}{4}\left[\int dx + 2\int \cos 2x\, dx + \int \cos^2 2x\, dx\right]$$

The first two integrations are easy. The third requires a second use of Equation 14-3, but in a different form:

$$\cos^2 2x = \frac{1}{2}(1 + \cos 4x)$$

You have

$$\int \cos^4 x\, dx = \frac{1}{4}\left[x + \sin 2x + \frac{1}{2}\int (1 + \cos 4x)\, dx\right]$$
$$= \frac{1}{4}x + \frac{1}{4}\sin 2x + \frac{1}{8}\left(x + \frac{\sin 4x}{4}\right) + C$$
$$= \frac{3}{8}x + \frac{1}{4}\sin 2x + \frac{1}{32}\sin 4x + C$$

EXAMPLE 14-8: Find $\int \sin^2 x \cos^4 x\, dx$.

Solution: Both m and n are even, so

$$\int \sin^2 x \cos^4 x\, dx = \int \frac{1}{2}(1 - \cos 2x)\left[\frac{1}{2}(1 + \cos 2x)\right]^2 dx$$
$$= \frac{1}{8}\int (1 - \cos 2x)(1 + 2\cos 2x + \cos^2 2x)\, dx$$
$$= \frac{1}{8}\int (1 + \cos 2x - \cos^2 2x - \cos^3 2x)\, dx$$
$$= \frac{1}{8}\left[\int dx + \int \cos 2x\, dx - \frac{1}{2}\int (1 + \cos 4x)\, dx\right.$$
$$\left. - \int (1 - \sin^2 2x)\cos 2x\, dx\right]$$

In the third integral you have the double-angle identity, and in the fourth (where the power of cos $2x$ was odd), you have the method of Part 1. Performing each integration, you have

$$\int \sin^2 x \cos^4 x = \frac{1}{8}\left[x + \frac{\sin 2x}{2} - \frac{x}{2} - \frac{\sin 4x}{8} - \frac{\sin 2x}{2} + \frac{\sin^3 2x}{6}\right] + C$$
$$= \frac{x}{16} - \frac{\sin 4x}{64} + \frac{\sin^3 2x}{48} + C$$

B. Integration of $\sec^m x \tan^n x\, dx$

The technique for finding $\int \sec^m x \tan^n x\, dx$ depends on whether m (or n) is even or odd.

1. If m is even (for any n), or if n is odd (for any m), you can put the integrand in the form $u^n\, du$. The technique is similar to that used in

Examples 14-5 and 14-6 and uses the fact that $\frac{d}{dx}(\tan x) = \sec^2 x$ and $\frac{d}{dx}(\sec x) = \sec x \tan x$.

EXAMPLE 14-9: Find $\int \sec^4 x \tan^2 x \, dx$.

Solution: Break off a factor of $\sec^2 x$, leaving $\sec^2 x$, which you can express as $1 + \tan^2 x$.

$$\begin{aligned}\int \sec^4 x \tan^2 x \, dx &= \int \sec^2 x \tan^2 x \sec^2 x \, dx \\ &= \int (1 + \tan^2 x) \tan^2 x \sec^2 x \, dx \\ &= \int \tan^2 x \sec^2 x \, dx + \int \tan^4 x \sec^2 x \, dx \\ &= \frac{\tan^3 x}{3} + \frac{\tan^5 x}{5} + C\end{aligned}$$

EXAMPLE 14-10: Find $\int \sec^3 x \tan^3 x \, dx$.

Solution: The power of $\tan x$ is odd; break off (from the whole integrand) a factor of $\sec x \tan x$, leaving an even power of $\tan x$ that you change to a power of $\sec x$:

$$\begin{aligned}\int \sec^3 x \tan^3 x \, dx &= \int \sec^2 x \tan^2 x \sec x \tan x \, dx \\ &= \int \sec^2 x(\sec^2 x - 1) \sec x \tan x \, dx \\ &= \int \sec^4 x \sec x \tan x \, dx - \int \sec^2 x \sec x \tan x \, dx \\ &= \frac{\sec^5 x}{5} - \frac{\sec^3 x}{3} + C\end{aligned}$$

2. If m (the power of $\sec x$) is odd and n (the power of $\tan x$) is even, you'll need a two-step approach: First integrate by parts, and then use the identity

$$1 + \tan^2 x = \sec^2 x \qquad \textbf{(14-4)}$$

EXAMPLE 14-11: Find $\int \sec^3 x \, dx$.

Solution: The power of $\sec x$ is odd; the power of $\tan x$ is even, so

$$\int \sec^3 x \, dx = \int \sec x \sec^2 x \, dx$$

Integrate by parts:

$$u = \sec x, \; du = \sec x \tan x \, dx \qquad dv = \sec^2 x \, dx, \; v = \tan x$$

$$\begin{aligned}\int \sec^3 x \, dx &= \sec x \tan x - \int \sec x \tan^2 x \, dx \\ &= \sec x \tan x - \int \sec x (\sec^2 x - 1) \, dx \\ &= \sec x \tan x - \int \sec^3 x \, dx + \int \sec x \, dx\end{aligned}$$

The first integral is the one you started with; add it to both sides. You should recognize the second integral:

$$2\int \sec^3 x\, dx = \sec x \tan x + \ln|\sec x + \tan x| + C$$

$$\int \sec^3 x\, dx = \frac{1}{2}[\sec x \tan x + \ln|\sec x + \tan x|] + C$$

14-3. Inverse Trigonometric Substitutions

You can use trigonometry in order to simplify some integrals by means of a substitution as follows:

If, for some constant a, the integrand contains	let $x =$	with $dx =$
$a^2 - x^2$	$a \sin\theta$	$a\cos\theta\, d\theta$
$a^2 + x^2$	$a \tan\theta$	$a\sec^2\theta\, d\theta$
$x^2 - a^2$	$a \sec\theta$	$a\sec\theta\tan\theta\, d\theta$

Notice that these are substitutions (in the spirit of Chapter 13) of the form $\theta = \text{arc sin}(x/a)$, $\theta = \text{arc tan}(x/a)$, or $\theta = \text{arc sec}(x/a)$.

EXAMPLE 14-12: Find $\int \frac{dx}{4 - x^2}$.

Solution: Let $x = 2\sin\theta$, with $dx = 2\cos\theta\, d\theta$. Then

$$\int \frac{dx}{4 - x^2} = \int \frac{2\cos\theta\, d\theta}{4 - 4\sin^2\theta}$$

$$= 2\int \frac{\cos\theta\, d\theta}{4(1 - \sin^2\theta)}$$

$$= \frac{1}{2}\int \frac{\cos\theta\, d\theta}{\cos^2\theta}$$

$$= \frac{1}{2}\int \sec\theta\, d\theta$$

$$= \frac{1}{2}\ln|\sec\theta + \tan\theta| + C$$

As in Chapter 13, the problem isn't completed until you retrace your substitution to return to the variable (x) in the original problem: You used $x = 2\sin\theta$ or $\theta = \text{arc sin}(x/2)$; you need $\sec\theta$ and $\tan\theta$. The easiest way to find expressions for these functions is to use the right triangle definitions of the trigonometic functions. If $\theta = \text{arc sin}(x/2)$, then θ may be described by the right triangle of Figure 14-1. By definition,

$$\tan\theta = \frac{x}{\sqrt{4 - x^2}} \quad \text{and} \quad \sec\theta = \frac{2}{\sqrt{4 - x^2}}$$

You have as a final answer for the given problem:

$$\int \frac{dx}{4 - x^2} = \frac{1}{2}\ln\left|\frac{2}{\sqrt{4 - x^2}} + \frac{x}{\sqrt{4 - x^2}}\right| + C$$

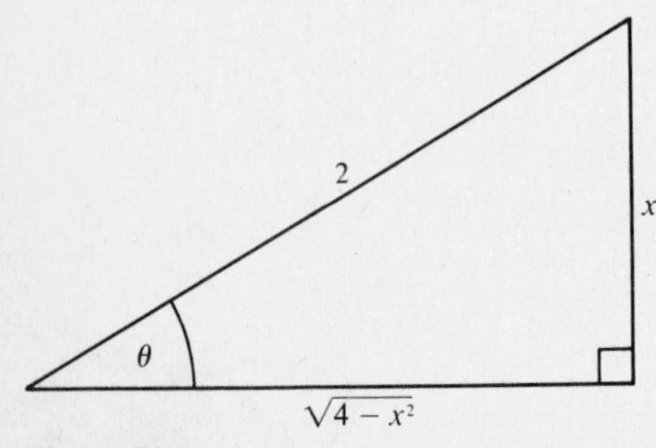

Figure 14-1
$\theta = \text{arc sin}\,\frac{x}{2}$.

However, you should notice that an algebraic manipulation, using properties of the logarithm function, allows you to write the answer in a different form:

$$\frac{1}{2}\ln\left|\frac{2 + x}{\sqrt{4 - x^2}}\right| = \frac{1}{2}\ln\left|\frac{2 + x}{\sqrt{2 + x}\sqrt{2 - x}}\right|$$

$$= \frac{1}{2}\ln\left|\frac{\sqrt{2 + x}}{\sqrt{2 - x}}\right| = \frac{1}{4}\ln\left|\frac{2 + x}{2 - x}\right|$$

EXAMPLE 14-13: Find $\int \sqrt{9 + x^2}\, dx$.

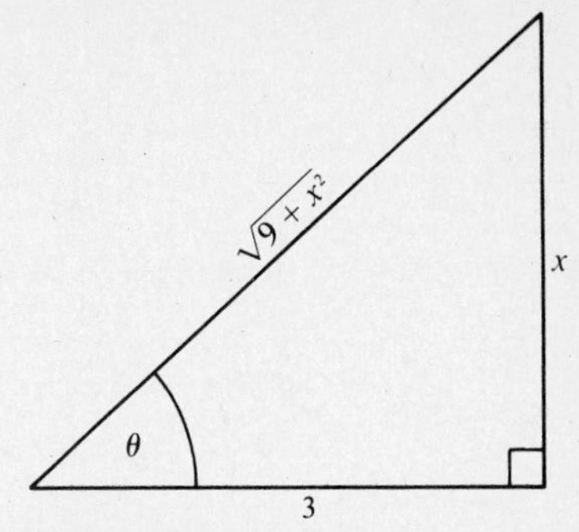

Figure 14-2
A right triangle with hypotenuse $\sqrt{9 + x^2}$.

Solution: Instead of memorizing a formula to help you find a correct substitution, you can use a right triangle and the Pythagorean theorem: The expression $\sqrt{9 + x^2}$ (the square root of the sum of two squares) suggests the length of the hypotenuse of a right triangle with sides of length x and 3. Draw such a triangle and label one of the angles θ (see Figure 14-2). Then the simplest relationship between θ and x gives you a substitution for x:

$$\frac{x}{3} = \tan\theta \quad \text{or} \quad x = 3\tan\theta$$

Use this and the triangle to find substitutions for dx and other quantities involving x:

$$dx = 3\sec^2\theta\, d\theta$$

$$\frac{\sqrt{x^2 + 9}}{3} = \sec\theta$$

$$\sqrt{x^2 + 9} = 3\sec\theta$$

Then,

$$\int \sqrt{9 + x^2}\, dx = \int 3\sec\theta\, 3\sec^2\theta\, d\theta = 9\int \sec^3\theta\, d\theta$$

Using the results of Example 14-11,

$$\int \sqrt{9 + x^2}\, dx = \frac{9}{2}\sec\theta\tan\theta + \frac{9}{2}\ln|\sec\theta + \tan\theta| + C$$

Use the triangle in Figure 14-2 to express this function of θ in terms of x:

$$\int \sqrt{9 + x^2}\, dx = \frac{9}{2}\frac{\sqrt{x^2 + 9}}{3}\left(\frac{x}{3}\right) + \frac{9}{2}\ln\left|\frac{\sqrt{x^2 + 9}}{3} + \frac{x}{3}\right| + C$$

$$= \frac{1}{2}x\sqrt{x^2 + 9} + \frac{9}{2}\ln|\sqrt{x^2 + 9} + x| + C'$$

(You've used the properties of logarithms to simplify:

$$\ln\left|\frac{(\sqrt{x^2 + 9} + x)}{3}\right| = \ln|\sqrt{x^2 + 9} + x| - \ln 3;$$

$C' = C - \ln 3$.)

EXAMPLE 14-14: Find $\int 1/\sqrt{x^2 - 1}\, dx$.

Solution: Let $x = \sec\theta$, with $dx = \sec\theta\tan\theta\, d\theta$ (see Figure 14-3 for an appropriately labeled triangle). Then,

$$\int \frac{dx}{\sqrt{x^2 - 1}} = \int \frac{\sec\theta\tan\theta\, d\theta}{\sqrt{\sec^2\theta - 1}}$$

$$= \int \frac{\sec\theta\tan\theta\, d\theta}{\tan\theta}$$

$$= \ln|\sec\theta + \tan\theta| + C$$

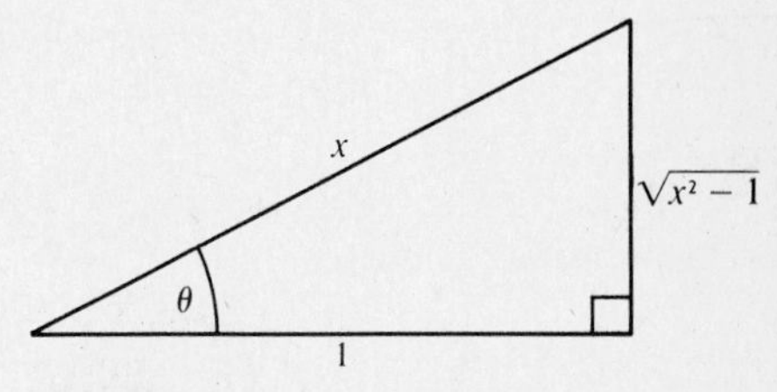

Figure 14-3
A right triangle in which $\sec\theta = x$.

Return to x's:

$$\int \frac{dx}{\sqrt{x^2 - 1}} = \ln|x + \sqrt{x^2 - 1}| + C$$

Note: Learning and using these substitutions will be easier if you remember the relationship between the substitutions and the Pythagorean identities.

When

$$x = \sin\theta,\quad 1 - x^2 = (1 - \sin^2\theta) = \cos^2\theta \tag{14-5}$$

$$x = \tan\theta,\quad 1 + x^2 = (1 + \tan^2\theta) = \sec^2\theta \tag{14-6}$$

$$x = \sec\theta,\quad x^2 - 1 = (\sec^2\theta - 1) = \tan^2\theta \tag{14-7}$$

14-4. Rational Functions

A rational function is the quotient of two polynomials. To integrate such a function, you should first be sure that the function is proper, i.e., the degree of the numerator is less than the degree of the denominator. If the function isn't proper, do the division to obtain a quotient, which is a polynomial, and a remainder, which is a proper rational function (see Chapter 1).

EXAMPLE 14-15: Write $(2x^3 + x^2 - 4x + 2)/(x^2 + 1)$ as the sum of a polynomial and a proper rational function.

Solution:

$$\frac{2x^3 + x^2 - 4x + 2}{x^2 + 1} = 2x + 1 + \frac{-6x + 1}{x^2 + 1}$$

A. In the form $f(x) = a/(bx + c)$

If $f(x) = a/(bx + c)$, then

$$\int f(x)\,dx = \int \frac{a}{bx + c}\,dx = \frac{a}{b}\int \frac{b}{bx + c}\,dx$$

$$= \frac{a}{b}\ln|bx + c| + C \tag{14-8}$$

EXAMPLE 14-16: Find $\int 3/(5x + 7)\,dx$.

Solution:

$$\int \frac{3}{5x + 7}\,dx = \frac{3}{5}\int \frac{5}{5x + 7}\,dx$$

$$= \frac{3}{5}\ln|5x + 7| + C$$

B. In the form $f(x) = 1/(a^2 \pm u^2)$

If you can write $f(x)$ in the form of $1/(a^2 \pm u^2)$, you use the methods of section 14-3 and get

$$\int \frac{du}{a^2 + u^2} = \frac{1}{a}\tan^{-1}\left(\frac{u}{a}\right) + C \tag{14-9}$$

$$\int \frac{du}{a^2 - u^2} = \frac{1}{2a}\ln\left|\frac{a + u}{a - u}\right| + C \tag{14-10}$$

(***Note:*** See Section 14-4D for an easier approach to $\int 1/(a^2 - u^2)\,du$.)

EXAMPLE 14-17: Find $\int 1/(x^2 + 4x + 13)\,dx$.

Solution: You can write $1/(x^2 + 4x + 13)$ in the form $1/(a^2 + u^2)$ by completing the square in the denominator:

$$x^2 + 4x + 13 = (x^2 + 4x + 4) + (13 - 4)$$

$$= (x + 2)^2 + 9$$

Thus

$$\int \frac{dx}{x^2 + 4x + 13} = \int \frac{dx}{3^2 + (x + 2)^2}$$

$$= \frac{1}{3} \tan^{-1}\left(\frac{x + 2}{3}\right) + C$$

using $u = x + 2$, with $du = dx$ and $a = 3$.

EXAMPLE 14-18: Find $\int 1/(3x^2 + 6x - 1)\,dx$.

Solution: Complete the square in the denominator (after factoring out the three):

$$3x^2 + 6x - 1 = 3\left(x^2 + 2x - \frac{1}{3}\right)$$

$$= 3\left(x^2 + 2x + 1 - \frac{1}{3} - 1\right)$$

$$= 3\left[(x + 1)^2 - \frac{4}{3}\right]$$

You have

$$\int \frac{dx}{3x^2 + 6x - 1} = \frac{1}{3}\int \frac{dx}{(x + 1)^2 - \frac{4}{3}}$$

$$= \frac{1}{3}\int \frac{dx}{(x + 1)^2 - \left(\frac{2}{\sqrt{3}}\right)^2}$$

If $u = x + 1$, with $du = dx$ and $a = 2/\sqrt{3}$, you get (from Equation 14-10)

$$\int \frac{dx}{3x^2 + 6x - 1} = \left(\frac{1}{3}\right)\frac{(-1)}{2\left(\frac{2}{\sqrt{3}}\right)} \ln \left|\frac{\frac{2}{\sqrt{3}} + x + 1}{\frac{2}{\sqrt{3}} - (x + 1)}\right| + C$$

$$= -\frac{\sqrt{3}}{12} \ln \left|\frac{2 + \sqrt{3}(x + 1)}{2 - \sqrt{3}(x + 1)}\right| + C$$

C. In the form $f(x) = (ax + b)/(px^2 + qx + r)$

If $f(x) = (ax + b)/(px^2 + qx + r)$, use an algebraic manipulation to write $f(x)$ as the sum of the two parts; one will be of the form du/u, and the other will be similar to the integrands of Examples 14-17 and 14-18. The details of the manipulation are best described with an example.

EXAMPLE 14-19: Find $\int (2x + 1)/(3x^2 + 4x + 2)\,dx$.

Solution: Notice that the derivative of the denominator is $6x + 4$. Thus, what you want to do is to find constants, C_1 and C_2, such that

$$\frac{2x + 1}{3x^2 + 4x + 2} = \frac{C_1(6x + 4)}{3x^2 + 4x + 2} + \frac{C_2}{3x^2 + 4x + 2}$$

Usually, C_1 is found by observation and C_2 is then easily computed. In this example $C_1 = 1/3$, and you have

$$2x + 1 = \frac{1}{3}(6x + 4) + C_2$$

$$= 2x + \frac{4}{3} + C_2$$

$$C_2 = -\frac{1}{3}$$

Now you can antidifferentiate:

$$\int \frac{2x + 1}{3x^2 + 4x + 2}\,dx = \frac{1}{3}\int \frac{6x + 4}{3x^2 + 4x + 2}\,dx - \frac{1}{3}\int \frac{1}{3x^2 + 4x + 2}\,dx$$

The first integrand is in the form du/u and yields a natural logarithm; the second is like that of Examples 14-17 and 14-18:

$$-\frac{1}{3}\int \frac{dx}{3x^2 + 4x + 2} = -\frac{1}{9}\int \frac{dx}{\left(x^2 + \frac{4}{3}x + \frac{4}{9}\right) + \left(\frac{2}{3} - \frac{4}{9}\right)}$$

$$= -\frac{1}{9}\int \frac{dx}{\left(x + \frac{2}{3}\right)^2 + \frac{2}{9}}$$

$$= -\frac{1}{9}\left(\frac{1}{\sqrt{2/3}}\right)\tan^{-1}\left(\frac{x + \frac{2}{3}}{\sqrt{2/3}}\right) + C$$

$$= -\frac{\sqrt{2}}{6}\tan^{-1}\left(\frac{3x + 2}{\sqrt{2}}\right) + C$$

Finally,

$$\int \frac{2x + 1}{3x^2 + 4x + 2}\,dx = \frac{1}{3}\ln|3x^2 + 4x + 2| - \frac{\sqrt{2}}{6}\tan^{-1}\left(\frac{3x + 2}{\sqrt{2}}\right) + C$$

Note: In each of the last examples the quadratic in the denominator was irreducible, i.e., nonfactorable. The rational function in which the denominator is a factorable quadratic is best treated with the method of partial fractions.

D. Method of partial fractions

If $f(x)$ is a proper rational function in which the denominator can be written as a product of linear and/or irreducible quadratic factors, use the method of partial fractions (see Section 1-3) to write $f(x)$ as a sum of terms that you antidifferentiate by the methods introduced above.

EXAMPLE 14-20: Find $\int 2x/(x^2 - 3x - 10)\,dx$.

Solution:

$$\frac{2x}{x^2 - 3x - 10} = \frac{2x}{(x - 5)(x + 2)} = \frac{C_1}{x - 5} + \frac{C_2}{x + 2}$$

$$2x = C_1(x + 2) + C_2(x - 5)$$

Setting $x = -2$,

$$-4 = -7C_2$$

$$C_2 = \frac{4}{7}$$

Setting $x = 5$,

$$10 = 7C_1$$

$$C_1 = \frac{10}{7}$$

You have, then:

$$\int \frac{2x}{x^2 - 3x - 10}\,dx = \frac{10}{7}\int \frac{dx}{x - 5} + \frac{4}{7}\int \frac{dx}{x + 2}$$

$$= \frac{10}{7}\ln|x - 5| + \frac{4}{7}\ln|x + 2| + C$$

EXAMPLE 14-21: Find $\displaystyle\int \frac{x^2 + 1}{(2x - 5)^2(x + 1)}\,dx.$

Solution:

$$\frac{x^2 + 1}{(2x - 5)^2(x + 1)} = \frac{C_1}{(2x - 5)^2} + \frac{C_2}{2x - 5} + \frac{C_3}{x + 1}$$

$$x^2 + 1 = C_1(x + 1) + C_2(x + 1)(2x - 5) + C_3(2x - 5)^2$$

As pointed out in Section 1-3, you have a choice of methods for solving for C_1, C_2, and C_3.

1. You may (as in Example 14-20) assign appropriate values of x—values that make the factors $x + 1$ and $2x - 5$ equal to zero—and any third value, chosen arbitrarily.

If $x = -1$,

$$(-1)^2 + 1 = C_1(0) + C_2(0) + C_3(-7)^2$$

$$C_3 = \frac{2}{49}$$

If $x = \frac{5}{2}$,

$$\left(\frac{5}{2}\right)^2 + 1 = C_1\left(\frac{5}{2} + 1\right) + C_2(0) + C_3(0)$$

$$C_1 = \frac{29/4}{7/2} = \frac{29}{14}$$

Choose $x = 0$ and substitute $C_1 = 29/14$ and $C_3 = 2/49$:

$$1 = \frac{29}{14}(0 + 1) + C_2(0 + 1)(0 - 5) + \frac{2}{49}(-5)^2$$

$$= \frac{29}{14} - 5C_2 + \frac{50}{49}$$

$$5C_2 = \frac{29}{14} + \frac{50}{49} - 1 = \frac{205}{98}$$

$$C_2 = \frac{41}{98}$$

2. You may perform the indicated multiplications on the right and regroup terms:

$$x^2 + 1 = C_1(x + 1) + C_2(2x^2 - 3x - 5) + C_3(4x^2 - 20x + 25)$$

$$= x^2(2C_2 + 4C_3) + x(C_1 - 3C_2 - 20C_3) + (C_1 - 5C_2 + 25C_3)$$

The coefficients of each power of x must be the same on both sides of the equation, so you equate the coefficients; that is, form equations by setting the coefficient of each power of x on the right side of the equation equal to the coefficient of the term of like degree on the left side:

$$2C_2 + 4C_3 = 1 \quad \text{(coefficient of } x^2\text{)}$$

$$C_1 - 3C_2 - 20C_3 = 0 \quad \text{(coefficient of } x\text{)}$$

$$C_1 - 5C_2 + 25C_3 = 1 \quad \text{(constant term)}$$

Solving, you get $C_1 = 29/14$, $C_2 = 41/98$, and $C_3 = 2/49$.

Now you can proceed to the problem of antidifferentiation:

$$\int \frac{x^2+1}{(2x-5)^2(x+1)}\,dx = \frac{29}{14}\int \frac{dx}{(2x-5)^2} + \frac{41}{98}\int \frac{dx}{2x-5} + \frac{2}{49}\int \frac{dx}{x+1}$$

The first integral (with $u = 2x - 5$, $du = dx$) may be put in the form $\int u^{-2}\,du$; the second and third easily become $\int 1/u\,du$. Thus:

$$\int \frac{x^2+1}{(2x-5)^2(x+1)}\,dx = -\frac{29}{28}\left(\frac{1}{2x-5}\right) + \frac{41}{196}\ln|2x-5| + \frac{2}{49}\ln|x+1| + C$$

EXAMPLE 14-22: Find $\displaystyle\int \frac{x\,dx}{(x-1)(x^2+1)^2}$.

Solution:

$$\frac{x}{(x-1)(x^2+1)^2} = \frac{C_1}{(x-1)} + \frac{C_2x + C_3}{(x^2+1)^2} + \frac{C_4x + C_5}{x^2+1}$$

$$x = C_1(x^2+1)^2 + (C_2x + C_3)(x-1) + (C_4x + C_5)(x-1)(x^2+1)$$

You can use a combination of techniques: Set $x = 1$ to solve for C_1, then perform indicated multiplications to find a system of equations whose solutions will give you values for the other constants. If $x = 1$,

$$1 = 4C_1$$

$$C_1 = \frac{1}{4}$$

Substituting for C_1, and performing all multiplications,

$$x = \frac{1}{4}(x^4 + 2x^2 + 1) + C_2x^2 + (C_3 - C_2)x - C_3 + C_4x^4 + (C_5 - C_4)x^3 + (C_4 - C_5)x^2 + (C_5 - C_4)x - C_5$$

$$= x^4\left(\frac{1}{4} + C_4\right) + x^3(C_5 - C_4) + x^2\left(\frac{1}{2} + C_2 + C_4 - C_5\right) + x(C_3 - C_2 + C_5 - C_4) + \left(\frac{1}{4} - C_3 - C_5\right)$$

Equating coefficients of terms of like degree, you have

$C_4 + \frac{1}{4} = 0, \quad C_4 = -\frac{1}{4}$

$C_5 - C_4 = 0, \quad C_5 = C_4 = -\frac{1}{4}$

$C_2 + C_4 - C_5 + \frac{1}{2} = 0, \quad C_2 = -\frac{1}{2}$

$C_3 - C_2 + C_5 - C_4 = 1, \quad C_3 = 1 - \frac{1}{2} = \frac{1}{2}$

Hence,

$$\int \frac{x\,dx}{(x-1)(x^2+1)^2} = \int \frac{\frac{1}{4}}{x-1}\,dx + \int \frac{-\frac{1}{2}x + \frac{1}{2}}{(x^2+1)^2}\,dx + \int \frac{-\frac{1}{4}x - \frac{1}{4}}{x^2+1}\,dx$$

The first of these integrals gives a natural logarithm:

$$\int \frac{\frac{1}{4}}{x-1}\,dx = \frac{1}{4}\int \frac{dx}{x-1} = \frac{1}{4}\ln|x-1| + C$$

The third integral is similar to Example 14-19:

$$\int \frac{-\frac{1}{4}x - \frac{1}{4}}{x^2+1}\,dx = -\frac{1}{4}\int \frac{x\,dx}{x^2+1} - \frac{1}{4}\int \frac{dx}{x^2+1}$$

$$= -\frac{1}{8}\ln|x^2+1| - \frac{1}{4}\tan^{-1}x + C$$

The second integral requires that you write it as a sum of two integrals:

$$\int \frac{-\frac{1}{2}x + \frac{1}{2}}{(x^2+1)^2}\,dx = -\frac{1}{2}\int \frac{x}{(x^2+1)^2}\,dx + \frac{1}{2}\int \frac{dx}{(x^2+1)^2}$$

Then

$$-\frac{1}{2}\int \frac{x}{(x^2+1)^2}\,dx = -\frac{1}{2}\left(\frac{1}{2}\right)\int (x^2+1)^{-2}(2x)\,dx$$

$$= \frac{1}{4}(x^2+1)^{-1} + C$$

You can best do the second part with an inverse trigonometric substitution. Let $x = \tan\theta$ (or $\theta = \tan^{-1}x$), with $dx = \sec^2\theta\,d\theta$:

$$\frac{1}{2}\int \frac{dx}{(x^2+1)^2} = \frac{1}{2}\int \frac{\sec^2\theta\,d\theta}{(1+\tan^2\theta)^2} = \frac{1}{2}\int \frac{\sec^2\theta\,d\theta}{\sec^4\theta}$$

$$= \frac{1}{2}\int \frac{d\theta}{\sec^2\theta} = \frac{1}{2}\int \cos^2\theta\,d\theta$$

$$= \frac{1}{4}\int (1+\cos 2\theta)\,d\theta = \frac{1}{4}\left(\theta + \frac{\sin 2\theta}{2}\right) + C$$

$$= \frac{1}{4}(\theta + \sin\theta\cos\theta) + C$$

(For the last equality, the double-angle identity $\sin 2\theta = 2\sin\theta\cos\theta$ was used.) Using a right triangle for $\theta = \tan^{-1}x$, you find $\sin\theta = x/\sqrt{1+x^2}$ and $\cos\theta = 1/\sqrt{1+x^2}$, so

$$\frac{1}{2}\int \frac{dx}{(x^2+1)^2} = \frac{1}{4}\left(\tan^{-1}x + \frac{x}{1+x^2}\right) + C$$

Finally, then, putting together the results of the three antidifferentiations:

$$\int \frac{x\,dx}{(x-1)(x^2+1)^2} = \frac{1}{4}\ln|x-1| + \frac{1}{4}(x^2+1)^{-1} + \frac{1}{4}\left(\tan^{-1}x + \frac{x}{1+x^2}\right)$$

$$- \frac{1}{8}\ln|x^2+1| - \frac{1}{4}\tan^{-1}x + C$$

$$= \frac{1}{4}\ln|x-1| + \frac{1}{4(x^2+1)} + \frac{x}{4(x^2+1)}$$

$$- \frac{1}{8}\ln|x^2+1| + C$$

As you can see, you may need many different techniques to integrate a rational function. In summary:

1. If the rational function is not proper, divide the denominator into the numerator to get the sum of a polynomial and a proper rational function.
2. Factor the denominator into irreducible polynomials (degree $\geqslant 1$).
3. Use partial fractions to obtain a sum of terms of the form

$$\frac{\text{constant}}{(\text{linear polynomial})^{\text{power}}} \quad \text{or} \quad \frac{\text{linear polynomial}}{(\text{irreducible quadratic})^{\text{power}}}$$

These can then be integrated by using the methods of Sections *A*, *B*, or *C*, and by inverse trigonometric substitutions.

14-5. Rational Functions of sin *x* and cos *x*

Integrals of rational functions of sin x and cos x may often be simplified by the use of a half-angle substitution:

$$u = \tan\frac{x}{2} \quad \text{or} \quad x = 2\tan^{-1}u \quad \text{and} \quad dx = \frac{2\,du}{1+u^2} \qquad \textbf{(14-11)}$$

Notice that this gives substitutions for sin x and cos x that result from the use of identities for sin 2α and cos 2α, respectively:

$$\begin{aligned}\sin x &= \sin 2(\tan^{-1}u)\\ &= 2\sin(\tan^{-1}u)\cos(\tan^{-1}u)\\ &= 2\left(\frac{u}{\sqrt{1+u^2}}\right)\frac{1}{\sqrt{1+u^2}}\end{aligned}$$

$$\sin x = \frac{2u}{1+u^2} \qquad \textbf{(14-12)}$$

$$\begin{aligned}\cos x &= \cos 2(\tan^{-1}u)\\ &= \cos^2(\tan^{-1}u) - \sin^2(\tan^{-1}u)\\ &= \frac{1}{1+u^2} - \frac{u^2}{1+u^2}\end{aligned}$$

$$\cos x = \frac{1-u^2}{1+u^2} \qquad \textbf{(14-13)}$$

EXAMPLE 14-23: Find $\int 1/(1 - \sin x)\,dx$.

Solution: Use Equations 14-11 and 14-12:

$$\begin{aligned}\int\frac{dx}{1-\sin x} &= \int\frac{2\,du}{(1+u^2)\left(1-\dfrac{2u}{1+u^2}\right)}\\ &= 2\int\frac{du}{1+u^2-2u}\\ &= 2\int\frac{du}{(1-u)^2}\\ &= \frac{2}{1-u} + C\\ &= \frac{2}{1-\tan\dfrac{x}{2}} + C\end{aligned}$$

14-6. Integral Tables

Integral tables are collections of formulas for the antiderivatives of functions commonly encountered in applications of calculus. The formulas were derived by the use of one or more of the techniques of Chapter 13 and the earlier sections of this chapter.

For the situations in which the integral (or the antiderivative) is more important than the method by which you find it, or when you don't have the time to work through the sometimes lengthy steps described in the preceding sections, you should know how to use the tables.

Typically, the formulas in the tables are grouped according to the integrand. In each formula u represents a function of x; you should be sure that (as in any substitution) you have the correct du.

Some typical formulas from a table of integrals follow, with examples.

A. Integrals involving $au + b$

$$\int u(au+b)^n\,du = \frac{(au+b)^{n+2}}{(n+2)a^2} - \frac{b(au+b)^{n+1}}{(n+1)a^2} + C, \qquad n \neq -1, -2 \tag{14-14}$$

EXAMPLE 14-24: Use Equation 14-14 to find $\int x(3x-4)^5\,dx$.

Solution: You can see that you have $u = x$, $a = 3$, $b = -4$ (it is important that you keep track of algebraic signs), and $n = 5$. Therefore:

$$\int x(3x-4)^5\,dx = \frac{(3x-4)^7}{7(9)} - \frac{(-4)(3x-4)^6}{6(9)} + C$$

$$= \frac{(3x-4)^7}{63} + \frac{4(3x-4)^6}{54} + C$$

EXAMPLE 14-25: Can you evaluate $\int x^2(2x^2-5)^4\,dx$ with Equation 14-14? To fit the formula, you must choose $u = x^2$, so that $x^2(2x^2-5)^4 = u(2u-5)^4$. However, $du = 2x\,dx$; the given integrand doesn't have the x factor that is needed, so this integration can't be done with Equation 14-14.

B. Integrals involving $\sqrt{au+b}$

1. $$\int \frac{du}{u\sqrt{au+b}} = \begin{cases} \dfrac{1}{\sqrt{b}} \ln\left|\dfrac{\sqrt{au+b}-\sqrt{b}}{\sqrt{au+b}+\sqrt{b}}\right| + C & \text{if } b > 0 \\[2ex] \dfrac{2}{\sqrt{-b}} \tan^{-1}\sqrt{\dfrac{au+b}{-b}} + C & \text{if } b < 0 \end{cases} \tag{14-15}$$

EXAMPLE 14-26: Use Equation 14-15 to find $\int 1/(t\sqrt{4t-1})\,dt$.

Solution: You have $u = t$, $a = 4$, and $b = -1$. Because $b < 0$, use the second of the two formulas:

$$\int \frac{dt}{t\sqrt{4t-1}} = \frac{2}{\sqrt{1}} \tan^{-1}\sqrt{\frac{4t-1}{1}} + C$$

2. $$\int \frac{\sqrt{au+b}}{u}\,du = 2\sqrt{au+b} + b\int \frac{du}{u\sqrt{au+b}} \tag{14-16}$$

Notice that you must also use Equation 14-15 when you use Equation 14-16.

EXAMPLE 14-27: Use Equation 14-16 to find $\int \dfrac{\sqrt{3x+2}}{x}\,dx$.

Solution: With $a = 3$, $b = 2$, you get

$$\int \frac{\sqrt{3x+2}}{x}\,dx = 2\sqrt{3x+2} + 2\int \frac{dx}{x\sqrt{3x+2}}$$

$$= 2\sqrt{3x+2} + \frac{2}{\sqrt{2}}\ln\left|\frac{\sqrt{3x+2}-\sqrt{2}}{\sqrt{3x+2}+\sqrt{2}}\right| + C$$

C. Integrals involving trigonometric functions

1. $$\int \sin^n au\,du = -\frac{\sin^{n-1}au\cos au}{an} + \frac{n-1}{n}\int \sin^{n-2}au\,du \qquad \textbf{(14-17)}$$

2. $$\int \cos^n au\,du = \frac{\sin au\cos^{n-1}au}{an} + \frac{n-1}{n}\int \cos^{n-2}au\,du \qquad \textbf{(14-18)}$$

3. $$\int \sin^m au\cos^n au\,du = \begin{cases} \dfrac{-\sin^{m-1}au\cos^{n+1}au}{a(m+n)} + \dfrac{m-1}{m+n}\displaystyle\int \sin^{m-2}au\cos^n au\,du & \textbf{(14-19a)} \\ \dfrac{\sin^{m+1}au\cos^{n-1}au}{a(m+n)} + \dfrac{n-1}{m+n}\displaystyle\int \sin^m au\cos^{n-2}au\,du & \textbf{(14-19b)} \end{cases}$$

You should notice three things about these formulas:

- They are designed for repeated use: each expresses the given integral in terms of an integral in which the power of sin *au* (or cos *au*) decreases. Repeated applications result eventually in an integrand in which the power of sin *au* (or cos *au*) is either zero or one.
- Equation 14-19 is best used when both m and n are even: Choose 14-19a for $m < n$ and 14-19b for $m > n$.
- The use of Equation 14-19 will sometimes require a subsequent use of either Equation 14-17 or 14-18.

EXAMPLE 14-28: Find $\int \sin^4 x\cos^6 x\,dx$.

Solution: You notice that $m = 4$, $n = 6$, and both m and n are even, with $n > m$. You choose Equation 14-19a, with $a = 1$ and $u = x$:

$$\int \sin^4 x\cos^6 x\,dx = \frac{-\sin^3 x\cos^7 x}{4+6} + \frac{3}{10}\int \sin^2 x\cos^6 x\,dx$$

$$= \frac{-\sin^3 x\cos^7 x}{10} + \frac{3}{10}\left[\frac{-\sin x\cos^7 x}{2+6} + \frac{1}{8}\int \cos^6 x\,dx\right]$$

(You've used the formula twice, reducing the power of sin x to zero.) Now use Equation 14-18 three times to find the last integral:

$$\int \cos^6 x\,dx = \frac{\sin x\cos^5 x}{6} + \frac{5}{6}\int \cos^4 x\,dx$$

$$= \frac{\sin x\cos^5 x}{6} + \frac{5}{6}\left(\frac{\sin x\cos^3 x}{4} + \frac{3}{4}\int \cos^2 x\,dx\right)$$

$$= \frac{\sin x\cos^5 x}{6} + \frac{5\sin x\cos^3 x}{24} + \frac{15}{24}\left(\frac{\sin x\cos x}{2} + \frac{1}{2}\int dx\right)$$

$$= \frac{\sin x\cos^5 x}{6} + \frac{5\sin x\cos^3 x}{24} + \frac{15\sin x\cos x}{48} + \frac{15}{48}x + C$$

Finally, then, you have

$$\int \sin^4 x \cos^6 x\,dx = \frac{-\sin^3 x \cos^7 x}{10} - \frac{3 \sin x \cos^7 x}{80} + \frac{3}{80}\left[\frac{\sin x \cos^5 x}{6} + \frac{5 \sin x \cos^3 x}{24} + \frac{15 \sin x \cos x}{48} + \frac{15}{48}x\right] + C$$

D. Integrals involving inverse trigonometric functions

$$\int u \sin^{-1}\frac{u}{a}\,du = \left(\frac{u^2}{2} - \frac{a^2}{4}\right)\sin^{-1}\frac{u}{a} + \frac{u\sqrt{a^2 - u^2}}{4} + C \qquad \textbf{(14-20)}$$

EXAMPLE 14-29: Use Equation 14-20 in the evaluation of

$$\int_0^{2/3} x \sin^{-1}(3x/2)\,dx.$$

Solution: With $u = 3x$ (and $du = 3dx$), $a = 2$, you have

$$\frac{1}{3}\int_0^{2/3} 3x \sin^{-1}\frac{3x}{2}\,dx = \frac{1}{3}\left[\left(\frac{9x^2}{2} - \frac{4}{4}\right)\sin^{-1}\frac{3x}{2} + \frac{3x\sqrt{4 - 9x^2}}{4}\right]\Bigg|_0^{2/3}$$

$$= \frac{1}{3}\left[\left(\frac{9(4/9)}{2} - 1\right)\sin^{-1}1 + \frac{3(2/3)\sqrt{4 - 9(4/9)}}{4}\right]$$

$$-\frac{1}{3}\left[(0 - 1)(0) + \frac{0\sqrt{4}}{4}\right]$$

$$= \frac{1}{3}(2 - 1)\frac{\pi}{2} + 0 - 0 = \frac{\pi}{6}$$

E. Integrals involving e^{au}

$$\int \frac{du}{p + qe^{au}} = \frac{u}{p} - \frac{1}{ap}\ln|p + qe^{au}| + C \qquad \textbf{(14-21)}$$

EXAMPLE 14-30: Use Equation 14-21 to find $\int_0^1 x/(4e^{x^2} - 2)\,dx$.

Solution: Choose $u = x^2$ (and $du = 2x\,dx$), $p = -2$, $q = 4$, $a = 1$:

$$\frac{1}{2}\int_0^1 \frac{2x\,dx}{-2 + 4e^{x^2}} = \frac{1}{2}\left[\frac{x^2}{-2} - \frac{1}{-2}\ln|-2 + 4e^{x^2}|\right]\Bigg|_0^1$$

$$= \frac{1}{2}\left[\left(-\frac{1}{2} + \frac{1}{2}\ln|-2 + 4e|\right) - \left(0 + \frac{1}{2}\ln|-2 + 4|\right)\right]$$

$$= -\frac{1}{4} + \frac{1}{4}\ln(4e - 2) - \frac{1}{4}\ln 2$$

14-7. An Overview

Perhaps the hardest part of an integration problem is the original decision: What technique of integration should you try first? Unfortunately, there is no formula or set of rules that answers that question for you.

There are a few suggestions that might be helpful; you have seen them before, scattered through Chapters 13 and 14, but they are brought together for emphasis:

1. Can you make the integrand, by a suitable substitution, take the form of a basic formula: $u^n\,du$, $(1/u)\,du$, $\sin u\,du$, $\sec^2 u\,du$, etc.?
2. Is there an algebraic manipulation that might simplify the integrand into two (or more) recognizable forms, for example,

$$\frac{\sin\theta + \cos\theta}{\cos\theta}\,d\theta = \frac{\sin\theta}{\cos\theta}\,d\theta + d\theta;$$

$$\frac{x^2+1}{x+1}\,dx = \left(x - 1 + \frac{2}{x+1}\right)dx$$

3. Is it possible to simplify the integrand by means of one or more trigonometric identities:

$$\tan^2\theta\,d\theta = (\sec^2\theta - 1)\,d\theta = \sec^2\theta\,d\theta - d\theta$$

$$\cos^2\theta\,d\theta = \frac{1}{2}(1 + \cos 2\theta)\,d\theta = \frac{1}{2}\,d\theta + \frac{1}{2}\cos 2\theta\,d\theta$$

4. Is the integrand a product of two distinct factors? Try an integration by parts.
5. Is the degree of the numerator less than the degree of the denominator? If not, divide it out, so you have the sum of a polynomial and a new rational function where the numerator does have lower degree than the denominator.
6. In a rational function is the denominator factorable? If so, you should probably try the method of partial fractions. If not, you should complete the square (in case of quadratics) and use an inverse trigonometric substitution.
7. When there is no obvious first step, try any reasonable substitution that will change the form of the integrand. Sometimes a new form will help you see what might have worked in the first place.

The most important factor in learning how to integrate is the element of experience. Work lots of problems. You will discover that as you do more and more integrations, they become easier and easier: The application of different techniques becomes less difficult, and, more importantly, that first decision becomes easier to make.

The supplementary exercises at the end of this chapter have been deliberately "mixed up." Before you start to work, go through the problems and make a tentative decision as to how you will begin each. You should begin to develop an eye for patterns and to make an association between pattern and "best guess" at an integration technique that not only will work, but will do the problem in the most efficient way.

SUMMARY

1. Integration by parts may be useful when the integrand is a product of functions.
2. Sometimes an integration that you can do by parts is simpler if you do it by substitution—and vice versa!
3. The technique that you use for integration $\sin^m x \cos^n x$, or $\sec^m x \tan^n x$, will depend on whether m and n are even or odd.
4. If your integral contains the expression $a^2 - x^2$, $a^2 + x^2$, or $x^2 - a^2$, you should consider using a properly chosen inverse trigonometric substitution.
5. When your integral contains a rational function, be sure that it is proper.
6. A knowledge of the organization of an integral table, and how to use the formulas, can be very helpful to you.

SOLVED PROBLEMS

Note: The examples in this chapter have illustrated the techniques of antidifferentiation. The following problems and supplementary exercises often call for the evaluation of integrals: you find an antiderivative, using one or more of the techniques of this chapter (and Chapter 13), and then use the Fundamental Theorem of Calculus to evaluate the integral.

PROBLEM 14-1 Find $\int x^2 \ln x\,dx$.

Solution: Integrate by parts:

$$u = \ln x,\ du = \frac{dx}{x} \qquad dv = x^2\,dx,\ v = \frac{x^3}{3}$$

$$\begin{aligned}\int x^2 \ln x\,dx &= (\ln x)\frac{x^3}{3} - \int \frac{x^3}{3}\left(\frac{dx}{x}\right)\\ &= \frac{1}{3}x^3 \ln x - \frac{1}{3}\int x^2\,dx\\ &= \frac{1}{3}x^3 \ln x - \frac{x^3}{9} + C\end{aligned}$$

[See Section 14-1.]

PROBLEM 14-2 Find $\int \arcsin x\,dx$.

Solution: Integrate by parts:

$$u = \arcsin x,\ du = \frac{dx}{\sqrt{1-x^2}} \qquad dv = dx,\ v = x$$

$$\begin{aligned}\int \arcsin x\,dx &= (\arcsin x)(x) - \int x \frac{dx}{\sqrt{1-x^2}}\\ &= x \arcsin x - \int (1-x^2)^{-1/2} x\,dx\\ &= x \arcsin x - \left(-\frac{1}{2}\right)\int (1-x^2)^{-1/2}(-2x)\,dx\\ &= x \arcsin x + \frac{1}{2}\frac{(1-x^2)^{1/2}}{1/2} + C\\ &= x \arcsin x + (1-x^2)^{1/2} + C\end{aligned}$$

[See Section 14-1.]

PROBLEM 14-3 Find $\int x^3 e^{2x}\,dx$.

Solution: Integrate by parts:

$$u = x^3,\ du = 3x^2\,dx \qquad dv = e^{2x}\,dx,\ v = \frac{1}{2}e^{2x}$$

$$\begin{aligned}\int x^3 e^{2x}\,dx &= (x^3)\left(\frac{1}{2}e^{2x}\right) - \frac{1}{2}\int e^{2x} 3x^2\,dx\\ &= \frac{1}{2}x^3 e^{2x} - \frac{3}{2}\int x^2 e^{2x}\,dx\end{aligned}$$

Again:

$$u = x^2,\ du = 2x\,dx \qquad dv = e^{2x}\,dx,\ v = \frac{e^{2x}}{2}$$

$$\int x^2 e^{2x}\,dx = \frac{1}{2}x^2 e^{2x} - \int x e^{2x}\,dx$$

And again:

$$u = x,\ du = dx \qquad dv = e^{2x}\,dx,\ v = \frac{e^{2x}}{2}$$

$$\int x e^{2x}\,dx = \frac{xe^{2x}}{2} - \frac{1}{2}\int e^{2x}\,dx$$

$$= \frac{xe^{2x}}{2} - \frac{1}{4}e^{2x} + C$$

Finally:

$$\int x^3 e^{2x}\,dx = \frac{1}{2}x^3 e^{2x} - \frac{3}{2}\left[\frac{1}{2}x^2 e^{2x} - \left(\frac{xe^{2x}}{2} - \frac{1}{4}e^{2x}\right)\right] + C$$

$$= \frac{1}{2}x^3 e^{2x} - \frac{3}{4}x^2 e^{2x} + \frac{3}{4}xe^{2x} - \frac{3}{8}e^{2x} + C$$ [See Section 14-1.]

PROBLEM 14-4 Find $\int e^{ax}\cos bx\,dx$.

Solution: Integrate by parts:

$$u = e^{ax},\ du = ae^{ax}\,dx \qquad dv = \cos bx\,dx,\ v = \frac{\sin bx}{b}$$

$$\int e^{ax}\cos bx\,dx = \frac{e^{ax}\sin bx}{b} - \frac{a}{b}\int e^{ax}\sin bx\,dx$$

Again:

$$u = e^{ax},\ du = ae^{ax}\,dx \qquad dv = \sin bx\,dx,\ v = \frac{-\cos bx}{b}$$

$$\int e^{ax}\sin bx\,dx = -\frac{e^{ax}\cos bx}{b} + \frac{a}{b}\int e^{ax}\cos bx\,dx$$

Thus:

$$\int e^{ax}\cos bx\,dx = \frac{e^{ax}\sin bx}{b} - \frac{a}{b}\left[-\frac{e^{ax}\cos bx}{b} + \frac{a}{b}\int e^{ax}\cos bx\,dx\right]$$

$$= \frac{e^{ax}\sin bx}{b} + \frac{a}{b^2}e^{ax}\cos bx - \frac{a^2}{b^2}\int e^{ax}\cos bx\,dx$$

Adding the last term to both sides:

$$\left(1 + \frac{a^2}{b^2}\right)\int e^{ax}\cos bx\,dx = \frac{e^{ax}}{b}\left(\sin bx + \frac{a}{b}\cos bx\right)$$

and

$$\int e^{ax}\cos bx\,dx = \frac{b^2}{a^2 + b^2}\left(\frac{e^{ax}}{b}\right)\left(\sin bx + \frac{a}{b}\cos bx\right)$$

$$= \frac{e^{ax}}{a^2 + b^2}(b\sin bx + a\cos bx) + C$$ [See Section 14-1.]

PROBLEM 14-5 Find $\int_0^{1/2} x^5/\sqrt{1-2x^3}\,dx$.

Solution: Integrate by parts:

$$u = x^3,\ du = 3x^2\,dx \qquad dv = \frac{x^2\,dx}{\sqrt{1-2x^3}},\ v = -\frac{1}{3}\sqrt{1-2x^3}$$

$$\int_0^{1/2} \frac{x^5\,dx}{1-2x^3} = -\frac{x^3}{3}\sqrt{1-2x^3}\Big|_0^{1/2} + \int_0^{1/2} x^2\sqrt{1-2x^3}\,dx$$

$$= \left[\frac{-x^3}{3}\sqrt{1-2x^3} - \frac{1}{6}\left(\frac{2}{3}\right)(1-2x^3)^{3/2}\right]\Bigg|_0^{1/2}$$

$$= \frac{-\sqrt{1-2x^3}}{3}\left[x^3 + \frac{1-2x^3}{3}\right]\Bigg|_0^{1/2}$$

$$= \frac{-\sqrt{1-2x^3}}{3}\left[\frac{x^3+1}{3}\right]\Bigg|_0^{1/2}$$

$$= -\frac{1}{9}\left[\sqrt{\frac{3}{4}}\left(\frac{9}{8}\right) - 1(1)\right]$$

$$= -\frac{1}{9}\left(\frac{9\sqrt{3}}{16} - 1\right) = \frac{1}{9} - \frac{\sqrt{3}}{16}$$

[See Section 14-1.]

PROBLEM 14-6 Find $\int_1^e \sin(\ln x)\,dx$.

Solution: Integrate by parts:

$$u = \sin(\ln x),\ du = \frac{\cos(\ln x)}{x} \qquad dv = dx,\ v = x$$

$$\int_1^e \sin(\ln x)\,dx = x\sin(\ln x)\Big|_1^e - \int_1^e \cos(\ln x)\,dx$$

Again:

$$u = \cos(\ln x),\ du = \frac{-\sin(\ln x)}{x} \qquad dv = dx,\ v = x$$

$$\int_1^e \cos(\ln x)\,dx = x\cos(\ln x)\Big|_1^e + \int_1^e \sin(\ln x)\,dx$$

Thus:

$$\int_1^e \sin(\ln x)\,dx = x\sin(\ln x)\Big|_1^e - x\cos(\ln x)\Big|_1^e - \int_1^e \sin(\ln x)\,dx$$

Adding the last term to both sides:

$$2\int_1^e \sin(\ln x)\,dx = x[\sin(\ln x) - \cos(\ln x)]\Big|_1^e$$

$$\int_1^e \sin(\ln x)\,dx = \frac{x}{2}[\sin(\ln x) - \cos(\ln x)]\Big|_1^e$$

$$= \frac{e}{2}[\sin 1 - \cos 1] - \frac{1}{2}[\sin 0 - \cos 0]$$

$$= \frac{e}{2}(\sin 1 - \cos 1) + \frac{1}{2}$$

[See Section 14-1.]

PROBLEM 14-7 Find $\int_0^{\pi/4} \sin^3 2x\cos^4 2x\,dx$.

Solution: Odd power of sin u ($u = 2x$, $du = 2\,dx$); if the x interval is $[0, \frac{1}{4}\pi]$, then the u interval is $[0, \frac{1}{2}\pi]$. So,

$$\int_0^{\pi/4} \sin^3 2x \cos^4 2x\,dx = \frac{1}{2}\int_0^{\pi/2} \sin^3 u \cos^4 u\,du = \frac{1}{2}\int_0^{\pi/2} \sin^2 u \cos^4 u \sin u\,du$$
$$= \frac{1}{2}\int_0^{\pi/2} (1 - \cos^2 u)\cos^4 u \sin u\,du$$
$$= \frac{1}{2}\int_0^{\pi/2} \cos^4 u \sin u\,du - \frac{1}{2}\int_0^{\pi/2} \cos^6 u \sin u\,du$$
$$= -\frac{1}{2}\frac{\cos^5 u}{5}\Bigg|_0^{\pi/2} + \frac{1}{2}\frac{\cos^7 u}{7}\Bigg|_0^{\pi/2}$$
$$= -\frac{1}{2}\left(0 - \frac{1}{5}\right) + \frac{1}{2}\left(0 - \frac{1}{7}\right) = \frac{1}{10} - \frac{1}{14} = \frac{1}{35}$$

[See Section 14-2.]

PROBLEM 14-8 Find $\int \cos^3 x \sin x/(2\sin^2 x + 5)\,dx$.

Solution: Substitute $u = \sin x$, $du = \cos x\,dx$, to get a rational function:

$$\int \frac{\cos^2 x \sin x}{2\sin^2 x + 5}\cos x\,dx = \int \frac{(1 - \sin^2 x)\sin x}{2\sin^2 x + 5}\cos x\,dx = \int \frac{(1-u^2)u}{2u^2+5}\,du$$
$$= \int\left(-\frac{1}{2}u + \frac{\frac{7}{2}u}{2u^2+5}\right)du = -\frac{1}{4}u^2 + \frac{7}{8}\ln|2u^2+5| + C$$
$$= -\frac{1}{4}\sin^2 x + \frac{7}{8}\ln|2\sin^2 x + 5| + C$$

[See Chapter 13 and Section 14-4.]

PROBLEM 14-9 Find $\int_0^2 \sqrt{4x - x^2}\,dx$.

Solution: Complete the square, then use an inverse trigonometric substitution:

$$\int_0^2 \sqrt{4x - x^2}\,dx = \int_0^2 \sqrt{4 - (4 - 4x + x^2)}\,dx$$
$$= \int_0^2 \sqrt{4 - (2 - x)^2}\,dx$$

The radicand is in the form $a^2 - u^2$; you substitute $u = a\sin\theta$:

$$2 - x = 2\sin\theta$$
$$dx = -2\cos\theta\,d\theta$$

To change limits: If $x = 0$, $\sin\theta = 1$ and $\theta = \pi/2$. If $x = 2$, $\sin\theta = 0$ and $\theta = 0$. You get

$$\int_0^2 \sqrt{4 - (2 - x)^2}\,dx = -\int_{\pi/2}^0 \sqrt{4 - 4\sin^2\theta}\,2\cos\theta\,d\theta = 4\int_0^{\pi/2} \cos^2\theta\,d\theta$$
$$= \frac{4}{2}\int_0^{\pi/2} (1 + \cos 2\theta)\,d\theta = 2\left[\theta + \frac{\sin 2\theta}{2}\right]\Bigg|_0^{\pi/2}$$
$$= 2\left[\frac{\pi}{2} + \frac{\sin\pi}{2}\right] = \pi$$

[See Section 14-3.]

PROBLEM 14-10 Find $\int 1/\sqrt{e^{2x} + e^x}\,dx$.

Solution: Substitute $e^x = u^2$, with $x = \ln u^2$ and $dx = 2(1/u)\,du$:

$$\int \frac{dx}{\sqrt{e^{2x} + e^x}} = 2\int \frac{du}{u\sqrt{u^4 + u^2}} = 2\int \frac{du}{u^2\sqrt{u^2 + 1}}$$

Now an inverse trigonometric substitution $u = \tan\theta$, $du = \sec^2\theta\,d\theta$:

$$2\int \frac{du}{u^2\sqrt{u^2+1}} = 2\int \frac{\sec^2\theta\,d\theta}{\tan^2\theta\sec\theta} = 2\int \frac{\cos\theta\,d\theta}{\sin^2\theta}$$

$$= 2\int \csc\theta\cot\theta\,d\theta = -2\csc\theta + C$$

$$= \frac{-2\sqrt{1+u^2}}{u} + C = \frac{-2\sqrt{1+e^x}}{\sqrt{e^x}} + C$$

[See Chapter 13 and Section 14-3.]

PROBLEM 14-11 . Find $\int_1^{e^\pi} \cos(\ln x)\,dx$.

Solution: Integrate by parts:

$$u = \cos(\ln x),\ du = -\sin(\ln x)\frac{1}{x}\,dx \qquad dv = dx,\ v = x$$

$$\int_1^{e^\pi} \cos(\ln x)\,dx = x\cos(\ln x)\Big|_1^{e^\pi} + \int_1^{e^\pi} x\sin(\ln x)\frac{dx}{x}$$

$$= e^\pi\cos(\ln e^\pi) - 1\cos(\ln 1) + \int_1^{e^\pi}\sin(\ln x)\,dx$$

Integrate by parts, again, with

$$u = \sin(\ln x),\ du = \frac{\cos(\ln x)}{x} \qquad dv = dx,\ v = x$$

$$\int_1^{e^\pi}\cos(\ln x)\,dx = e^\pi\cos\pi - 1\cos 0 + \left[x\sin(\ln x)\Big|_1^{e^\pi} - \int_1^{e^\pi}\cos(\ln x)\,dx\right]$$

Add $\int_1^{e^\pi}\cos(\ln x)\,dx$ to both sides and divide by two:

$$\int_1^{e^\pi}\cos(\ln x)\,dx = \frac{1}{2}[-e^\pi - 1 + (e^\pi\sin\pi - 1\sin 0)] = -\frac{1}{2}(e^\pi + 1)$$

[See Section 14-1.]

PROBLEM 14-12 Find $\int_0^{\pi/2} 1/(1 + \sin x + \cos x)\,dx$.

Solution: This is a rational function of $\sin x$ and $\cos x$. Use the substitution $u = \tan(x/2)$, or $x = 2\tan^{-1}u$, $dx = 2/(1+u^2)\,du$. You get $\sin x = 2u/(1+u^2)$, $\cos x = (1-u^2)/(1+u^2)$. Substitute, remembering to change limits of integration:

$$\int_0^{\pi/2}\frac{dx}{1+\sin x+\cos x} = \int_0^1 \frac{\dfrac{2\,du}{1+u^2}}{1 + \dfrac{2u}{1+u^2} + \dfrac{1-u^2}{1+u^2}}$$

$$= \int_0^1 \frac{2\,du}{1+u^2+2u+1-u^2} = \int_0^1\frac{du}{1+u}$$

$$= \ln|1+u|\Big|_0^1 = \ln 2$$

[See Section 14-5.]

PROBLEM 14-13 Find $\int x\sqrt{2x - x^2}\,dx$.

Solution: Complete the square under the radical and use an inverse trigonometric substitution:

$$\int x\sqrt{2x - x^2}\,dx = \int x\sqrt{1 - (x-1)^2}\,dx$$

With $x - 1 = \sin\theta$, $dx = \cos\theta\,d\theta$, you have

$$\begin{aligned}\int x\sqrt{1-(x-1)^2}\,dx &= \int (1+\sin\theta)\sqrt{1-\sin^2\theta}\cos\theta\,d\theta \\ &= \int (\cos^2\theta + \sin\theta\cos^2\theta)\,d\theta \\ &= \frac{1}{2}\int (1+\cos 2\theta)\,d\theta - \int \cos^2\theta(-\sin\theta)\,d\theta \\ &= \frac{\theta}{2} + \frac{\sin\theta\cos\theta}{2} - \frac{\cos^3\theta}{3} + C \\ &= \frac{1}{2}\sin^{-1}(x-1) + \frac{1}{2}(x-1)\sqrt{2x-x^2} - \frac{1}{3}(2x-x^2)^{3/2} + C\end{aligned}$$

[See Section 14-3.]

PROBLEM 14-14 Use a half-angle substitution to find $\int \frac{1}{(1+\tan x)}\,dx$.

Solution: Let $u = \tan(x/2)$, or $x = 2\tan^{-1}u$, with $dx = 2/(1+u^2)\,du$ and $\tan x = 2u/(1-u^2)$:

$$\begin{aligned}\int \frac{dx}{1+\tan x} &= \int \frac{2\,du}{(1+u^2)\left(1+\dfrac{2u}{1-u^2}\right)} \\ &= \int \frac{2(1-u^2)\,du}{(1+u^2)(1-u^2+2u)} \\ &= \int \left(\frac{u-1}{u^2-2u-1} - \frac{u-1}{u^2+1}\right)du \qquad \text{(by partial fractions)} \\ &= \frac{1}{2}\ln|u^2-2u-1| - \frac{1}{2}\ln|u^2+1| + \tan^{-1}u + C \\ &= \frac{1}{2}\ln\left|\tan^2\frac{x}{2} - 2\tan\frac{x}{2} - 1\right| - \frac{1}{2}\ln\left|\tan^2\frac{x}{2}+1\right| + \frac{x}{2} + C\end{aligned}$$

[See Section 14-5.]

PROBLEM 14-15 Find $\int \sqrt{x^2-4}\,dx$.

Solution: Use inverse trigonometric substitution: $x = 2\sec\theta$, $dx = 2\sec\theta\tan\theta\,d\theta$:

$$\begin{aligned}\int \sqrt{x^2-4}\,dx &= \int \sqrt{4(\sec^2\theta - 1)}\,2\sec\theta\tan\theta\,d\theta \\ &= 4\int \sec\theta\tan^2\theta\,d\theta \\ &= 4\int \sec\theta(\sec^2\theta - 1)\,d\theta \\ &= 4\left[\int \sec^3\theta\,d\theta - \int \sec\theta\,d\theta\right]\end{aligned}$$

To integrate $\sec^3\theta$, you integrate by parts. Let:

$$u = \sec\theta,\ du = \sec\theta\tan\theta\, d\theta \qquad dv = \sec^2\theta\, d\theta,\ v = \tan\theta$$

$$\int \sec^3\theta\, d\theta = \sec\theta\tan\theta - \int \sec\theta\tan^2\theta\, d\theta$$

You have

$$4\int \sec\theta\tan^2\theta\, d\theta = 4(\sec\theta\tan\theta) - 4\int \sec\theta\tan^2\theta\, d\theta - 4\int \sec\theta\, d\theta$$

$$8\int \sec\theta\tan^2\theta\, d\theta = 4\sec\theta\tan\theta - 4\ln|\sec\theta + \tan\theta| + C$$

$$4\int \sec\theta\tan^2\theta\, d\theta = 2\sec\theta\tan\theta - 2\ln|\sec\theta + \tan\theta| + C$$

and

$$\int \sqrt{x^2 - 4}\, dx = 2\frac{x\sqrt{x^2 - 4}}{4} - 2\ln\left|\frac{x + \sqrt{x^2 - 4}}{2}\right| + C$$

$$= \frac{1}{2}x\sqrt{x^2 - 4} - 2\ln\left|\frac{x + \sqrt{x^2 - 4}}{2}\right| + C$$

[See Sections 14-2 and 14-3.]

PROBLEM 14-16 Find $\int x^{1/3}/(1 + x)\, dx$.

Solution: Substitute $x = u^3$, with $dx = 3u^2\, du$, and get both a rational function (with the degree of the numerator equal to the degree of the denominator) and a factorable denominator. This means you divide first, then use partial fractions:

$$\int \frac{x^{1/3}}{1 + x}\, dx = \int \frac{u}{1 + u^3} 3u^2\, du = 3\int \frac{u^3}{1 + u^3}\, du$$

$$= 3\int\left(1 - \frac{1}{1 + u^3}\right) du$$

$$= 3\left[\int du - \int \frac{du}{(1 + u)(1 - u + u^2)}\right]$$

$$= 3\int du - 3\left[\int \frac{\frac{1}{3}}{1 + u}\, du + \int \frac{-\frac{1}{3}u + \frac{2}{3}}{1 - u + u^2}\, du\right]$$

$$= 3\int du - \int \frac{du}{1 + u} + \int \frac{u - 2}{1 - u + u^2}\, du$$

Do the first two integrals by formula. You must manipulate the third one to get the sum of a natural log and an inverse tangent. The derivative of the denominator is $2u - 1$; thus you want to find C_1 and C_2 so that

$$u - 2 = C_1(2u - 1) + C_2$$

$$= 2C_1 u + (C_2 - C_1)$$

Thus:

$$C_1 = \frac{1}{2} \quad \text{and} \quad C_2 = -2 + C_1 = -2 + \frac{1}{2} = -\frac{3}{2}$$

$$\int \frac{u - 2}{1 - u + u^2}\, du = \frac{1}{2}\int \frac{2u - 1}{1 - u + u^2}\, du - \frac{3}{2}\int \frac{du}{1 - u + u^2}$$

Finally, then, putting it all together:

$$\int \frac{x^{1/3}}{1+x}\,dx = 3u - \ln|1+u| + \frac{1}{2}\ln|1-u+u^2| - \frac{3}{2}\int \frac{du}{\left(u-\frac{1}{2}\right)^2 + \frac{3}{4}}$$

$$= 3u - \ln|1+u| + \frac{1}{2}\ln|1-u+u^2| - \sqrt{3}\tan^{-1}\left(\frac{2u-1}{\sqrt{3}}\right) + C$$

Returning to x's, with $u = x^{1/3}$:

$$\int \frac{x^{1/3}}{1+x}\,dx = 3x^{1/3} - \ln|1+x^{1/3}| + \frac{1}{2}\ln|1-x^{1/3}+x^{2/3}| - \sqrt{3}\tan^{-1}\left(\frac{2x^{1/3}-1}{\sqrt{3}}\right) + C$$

[See Chapter 13 and Section 14-4.]

PROBLEM 14-17 Find $\int (x+1)/\sqrt{x^2+6x+5}\,dx$.

Solution: Rewrite $x + 1 = \frac{1}{2}(2x+6) - 2$. $\left(\text{Notice that } \frac{d}{dx}(x^2+6x+5) = 2x+6.\right)$ Then:

$$\int \frac{x+1}{\sqrt{x^2+6x+5}}\,dx = \frac{1}{2}\int (x^2+6x+5)^{-1/2}(2x+6)\,dx - \int \frac{2}{\sqrt{x^2+6x+5}}\,dx$$

$$= \frac{1}{2}(2)(x^2+6x+5)^{1/2} - 2\int \frac{dx}{\sqrt{(x+3)^2-4}}$$

Let $x + 3 = 2\sec\theta$, with $dx = 2\sec\theta\tan\theta\,d\theta$:

$$\int \frac{dx}{\sqrt{(x+3)^2-4}} = \int \frac{2\sec\theta\tan\theta\,d\theta}{\sqrt{4(\sec^2\theta-1)}} = \int \sec\theta\,d\theta$$

$$= \ln|\sec\theta + \tan\theta| + C$$

$$= \ln\left|\frac{x+3+\sqrt{(x+3)^2-4}}{2}\right| + C$$

And finally:

$$\int \frac{x+1}{\sqrt{x^2+6x+5}}\,dx = (x^2+6x+5)^{1/2} - 2\ln\left|\frac{x+3+\sqrt{x^2+6x+5}}{2}\right| + C$$

[See Chapter 13 and Section 14-3.]

PROBLEM 14-18 Find $\int 1/[\cos x(\sin x - 1)^2]\,dx$.

Solution: Substitute $\sin x = u(x = \sin^{-1}u,\ dx = (1/\sqrt{1-u^2})\,du,\ \cos x = \sqrt{1-u^2})$:

$$\int \frac{dx}{\cos x(\sin x - 1)^2} = \int \frac{du}{\sqrt{1-u^2}\sqrt{1-u^2}\,(u-1)^2}$$

$$= \int \frac{du}{(1-u^2)(1-u)^2} = \int \frac{du}{(1+u)(1-u)^3}$$

$$= \int \left(\frac{1/2}{(1-u)^3} + \frac{1/4}{(1-u)^2} + \frac{1/8}{(1-u)} + \frac{1/8}{(1+u)}\right) du$$

$$= \frac{1}{4}\frac{1}{(1-u)^2} + \frac{1}{4}\frac{1}{(1-u)} - \frac{1}{8}\ln|1-u| + \frac{1}{8}\ln|1+u| + C$$

$$= \frac{1}{4}\frac{1}{(1-\sin x)^2} + \frac{1}{4}\left(\frac{1}{1-\sin x}\right) - \frac{1}{8}\ln|1-\sin x|$$

$$+ \frac{1}{8}\ln|1+\sin x| + C$$

[See Chapter 13 and Section 14-4.]

PROBLEM 14-19 Use integration by parts to derive the formula

$$\int x^n \sin x\, dx = -x^n \cos x + nx^{n-1}\sin x - n(n-1)\int x^{n-2}\sin x\, dx$$

Solution: You integrate by parts, choosing $u = x^n$, $dv = \sin x\, dx$:

$$\int x^n \sin x\, dx = -x^n \cos x + n\int x^{n-1}\cos x\, dx$$

Again, with $u = x^{n-1}$, $dv = \cos x\, dx$:

$$\int x^n \sin x\, dx = -x^n \cos x + nx^{n-1}\sin x - n(n-1)\int x^{n-2}\sin x\, dx$$

[See Section 14-1.]

Supplementary Exercises

14-20 $\int x/(x+1)\, dx$

14-21 $\int \sin^3 x \cos^2 x\, dx$

14-22 $\int e^{4x+5}/(e^{2x}+1)\, dx$

14-23 $\int \tan^4 x\, dx$

14-24 $\int x^2/\sqrt{9-4x^2}\, dx$

14-25 $\int 1/\sqrt{e^{2x}-1}\, dx$

14-26 $\int_0^6 \sqrt{36-x^2}\, dx$

14-27 $\int \sqrt{49-x^2}/x\, dx$

14-28 $\int \sqrt{4+25x^2}\, dx$

14-29 $\int 1/(x^2\sqrt{1-x^2})\, dx$

14-30 $\int_0^1 1/(1+x^2)^2\, dx$

14-31 $\int 1/(e^{2x}+4e^x+9)\, dx$

14-32 $\int 1/(x+3\sqrt{x}-4)\, dx$

14-33 $\int (x+6)/[(2x-1)(x^2+3)]\, dx$

14-34 $\int (6x+2)/(x^4-1)\, dx$

14-35 $\int (x^4+x+1)/(x^6+2x^4+x^2)\, dx$

14-36 $\int (3x+9)/(2x^3-x^2-5x-2)\, dx$

14-37 $\int \dfrac{x^4-2x^2-3x-3}{x^5+x^4+x^3}\, dx$

14-38 $\int_0^{\pi/4} \tan x/(1+\cos x)\, dx$

14-39 $\int_0^2 1/(e^x+1)\, dx$

14-40 $\int_{-1}^1 (x^3+5x+6)/(x^2-5x+6)\, dx$

14-41 $\int 4/(x^3-4x^2+4x)\, dx$

14-42 $\int \dfrac{x^3+x^2-6x-2}{x^3-3x^2+x-3}\, dx$

14-43 $\int \dfrac{x^4+9x^2+15}{(x-1)(x^2+4)^2}\, dx$

14-44 $\int_0^1 \sin^{-1} x\, dx$

14-45 $\int_0^2 \ln(4+x^2)\, dx$

14-46 $\int_2^4 (x^2 + 1)/(x^2 - 1)\, dx$

14-47 $\int_0^4 e^{\sqrt{x}}\, dx$

14-48 $\int x/\sqrt{4 + 4x - x^2}\, dx$

14-49 $\int x/(x^2 + 6x + 13)\, dx$

14-50 $\int_4^8 \sqrt{x - 4}/(x + 2)\, dx$

14-51 $\int x^2/\sqrt{4 - x^2}\, dx$

14-52 $\int 1/\sqrt{x^2 - 6x}\, dx$

14-53 $\int 1/(3 \sin x + 4 \cos x)\, dx$

14-54 $\int 1/(2 + \cos\theta)\, d\theta$

14-55 $\int x^2 e^x\, dx$

14-56 $\int \arctan x/(1 + x^2)\, dx$

14-57 $\int \arctan x\, dx$

14-58 $\int \sqrt{1 + \sqrt{x}}\, dx$

14-59 $\int x^2 \ln x\, dx$

14-60 $\int 1/(x\sqrt{\ln x})\, dx$

14-61 $\int \dfrac{4x^7 - 5x^4 + 1}{x^8 - 2x^5 + 2x - 5}\, dx$

14-62 $\int e^{2x} \sin 3x\, dx$

14-63 $\int x \arcsin x\, dx$

14-64 $\int \arctan x/(x^2)\, dx$

14-65 $\int \tan x\, dx$

14-66 $\int_0^{\pi/2} \sin x/(1 + \sqrt{\cos x})\, dx$

14-67 $\int \cos\sqrt{x}\, dx$

14-68 $\int x^9 \sqrt[3]{1 - x^5}\, dx$

14-69 $\int_0^1 2^x 2^{2^x}\, dx$

14-70 $\int x^3 e^{x^2}\, dx$

14-71 $\int (\ln x)/\sqrt{x}\, dx$

14-72 $\int e^x/(1 + e^{2x})\, dx$

14-73 $\int (1 + \sqrt{x})/(1 - \sqrt[3]{x})\, dx$

14-74 $\int x \sin x\, dx$

14-75 $\int_0^1 \sqrt{\arcsin x/(1 - x^2)}\, dx$

14-76 $\int x \arctan x\, dx$

Solutions to Supplementary Exercises

(14-20) (Improper rational function): $x - \ln|x + 1| + C$

(14-21) (Odd power of sin x): $\dfrac{-\cos^3 x}{3} + \dfrac{\cos^5 x}{5} + C$

(14-22) (Rational function after substitution $u = e^{2x}$): $(e^5/2)(e^{2x} - \ln|1 + e^{2x}|) + C$

(14-23) (Even power of sec x): $[(\tan^3 x)/3] - \tan x + x + C$

(14-24) (Inverse trigonometric substitution: $2x = 3\sin\theta$): $(9/16)[\sin^{-1}(2x/3) - (2x\sqrt{9-4x^2})/9] + C$

(14-25) (Inverse trigonometric substitution): $\sec^{-1}(e^x) + C$

(14-26) (Inverse trigonometric substitution: $x = 6\sin\theta$): $(36/2)[\sin^{-1}(x/6) + (x\sqrt{36-x^2})/36]\Big|_0^6 = 9\pi$

(14-27) (Inverse trigonometric substitution: $x = 7\sin\theta$): $7\ln|(7-\sqrt{49-x^2})/x| + \sqrt{49-x^2} + C$

(14-28) (Inverse trigonometric substitution: $5x = 2\tan\theta$):

$$\frac{2}{5}\left(\frac{5x\sqrt{4+25x^2}}{4} + \ln\left|5x + \sqrt{4+25x^2}\right|\right) + C$$

(14-29) (Inverse trigonometric substitution: $x = \sin\theta$): $-(\sqrt{1-x^2}/x) + C$

(14-30) (Inverse trigonometric substitution: $x = \tan\theta$; even power of $\cos\theta$):

$$\tfrac{1}{2}[\tan^{-1}x + x/(1+x^2)]\Big|_0^1 = \tfrac{1}{2}[(\pi/4) + \tfrac{1}{2}]$$

(14-31) ($u = e^x$, then partial fractions):

$$(1/9)x - (1/18)\ln(e^{2x} + 4e^x + 9) - (2/(9\sqrt{5}))\arctan((e^x+2)/\sqrt{5}) + C$$

(14-32) ($u = \sqrt{x}$, then partial fractions):

$$\int 2u/(u^2+3u-4)\,du = (8/5)\ln|\sqrt{x}+4| + (2/5)\ln|\sqrt{x}-1| + C$$

(14-33) (Partial fractions): $\ln|2x-1| - \frac{1}{2}\ln|x^2+3| + C$

(14-34) (Partial fractions):

$$\int 2/(x-1)\,dx + \int 1/(x+1)\,dx - \int (3x+1)/(x^2+1)\,dx$$
$$= 2\ln|x-1| + \ln|x+1| - (3/2)\ln|x^2+1| - \tan^{-1}x + C$$

(14-35) (Partial fractions; inverse trigonometric substitution):

$$\int (1/x)\,dx + \int (1/x^2)\,dx - \int x/(x^2+1)\,dx - \int (x+2)/[(x^2+1)^2]\,dx$$
$$= \ln|x| - (1/x) - (1/2)\ln|x^2+1| + 1/[2(x^2+1)] - \tan^{-1}x - [x/(1+x^2)] + C$$

(14-36) (Partial fractions): $2\ln|x+1| + \ln|x-2| - 3\ln|2x+1| + C$

(14-37) (Partial fractions; completing square):

$$\int\left(\frac{-3}{x^3} + \frac{1}{x} - \frac{1}{x^2+x+1}\right)dx = \frac{3}{2x^2} + \ln|x| - \frac{2}{\sqrt{3}}\tan^{-1}\left(\frac{2x+1}{\sqrt{3}}\right) + C$$

(14-38) ($\tan x = (\sin x)/\cos x$; $u = \cos x$; partial fractions):

$$-\int 1/(u^2+u)\,du = \ln|(u+1)/u|;\ \ln|(1+\cos x)/\cos x|\Big|_0^{\pi/4} = \ln(1+\sqrt{2}) - \ln 2$$

(14-39) ($u = e^x$, $dx = 1/u\,du$; partial fractions): $[x - \ln(e^x+1)]\Big|_0^2 = 2 - \ln(e^2+1) + \ln 2$

(14-40) (Improper rational function; partial fractions):

$$[\tfrac{1}{2}x^2 + 5x + 48\ln|x-3| - 24\ln|x-2|]\Big|_{-1}^1 = 10 - 48\ln 2 + 24\ln 3$$

(14-41) (Partial fractions): $\ln|x/(x-2)| - [2/(x-2)] + C$

(14-42) (Improper rational function; partial fractions):

$$x + (8/5)\ln|x-3| + (6/5)\ln|x^2+1| + (1/5)\tan^{-1}x + C$$

(14-43) (Partial fractions; inverse trigonometric substitution):

$$\ln|x-1| - \frac{1}{2(x^2+4)} + \frac{x}{8(x^2+4)} + \frac{1}{16}\tan^{-1}\left(\frac{x}{2}\right) + C$$

(14-44) (Integration by parts: $u = \sin^{-1}x$, $dv = dx$): $[x\sin^{-1}x + \sqrt{1-x^2}]\Big|_0^1 = (\pi/2) - 1$

(14-45) (Integration by parts: $u = \ln(4+x^2)$, $dv = dx$):

$$x\ln(4+x^2) - 2x + 4\tan^{-1}(x/2)\Big|_0^2 = 2\ln 8 - 4 + \pi$$

(14-46) (Improper rational function): $x + \ln|(x-1)/(x+1)|\Big|_2^4 = 2 + \ln(9/5)$

(14-47) (Substitution $t = \sqrt{x}$; integration by parts: $u = t$, $dv = e^t\,dt$): $2(te^t - e^t)\Big|_0^2 = 2(e^2+1)$

(14-48) (Complete the square: $4 + 4x - x^2 = 8 - (x-2)^2$; inverse trigonometric substitution: $x - 2 = 2\sqrt{2}\sin\theta$)

$$2\sin^{-1}\left(\frac{x-2}{2\sqrt{2}}\right) - \sqrt{4+4x-x^2} + C$$

(14-49) $\left(\text{Rational function: } \frac{x}{x^2+6x+13} = \left(\frac{1}{2}\right)\frac{2x+6}{x^2+6x+13} - \frac{3}{(x+3)^2+4}\right)$:

$$(1/2)\ln|x^2+6x+13| - (3/2)\tan^{-1}[(x+3)/2] + C$$

(14-50) (Substitute $u = \sqrt{x-4}$; improper rational function):

$$[2u - 2\sqrt{6}\tan^{-1}(u/\sqrt{6})]\Big|_0^2 = 4 - 2\sqrt{6}\tan^{-1}(2/\sqrt{6})$$

(14-51) (Inverse trigonometric substitution $x = 2\sin\theta$): $2[\sin^{-1}(x/2) - (x\sqrt{4-x^2})/4] + C$

(14-52) (Complete square: $x^2 - 6x = (x-3)^2 - 9$; inverse trigonometric substitution: $x - 3 = 3\sec\theta$):

$$\ln|x - 3 + \sqrt{x^2-6x}| + C$$

(14-53) (Rational function of $\sin x$, $\cos x$; partial fractions):

$$(1/5)\ln|2\tan(x/2)+1| - (1/5)\ln|\tan(x/2)-2| + C$$

(14-54) (Rational function of $\cos\theta$): $(2/\sqrt{3})\tan^{-1}[(1/\sqrt{3})\tan(x/2)] + C$

(14-55) (Integrate by parts): $x^2e^x - 2xe^x + 2e^x + C$

(14-56) ($u = \arctan x$): $\frac{1}{2}(\arctan x)^2 + C$

(14-57) (Integrate by parts): $x\arctan x - \frac{1}{2}\ln(1+x^2) + C$

(14-58) ($u = 1 + \sqrt{x}$): $(4/5)(1+\sqrt{x})^{5/2} - (4/3)(1+\sqrt{x})^{3/2} + C$

(14-59) (Integrate by parts): $(x^3/3)\ln x - (x^3/9) + C$

(14-60) ($u = \ln x$): $2\sqrt{\ln x} + C$

(14-61) ($u = x^8 - 2x^5 + 2x - 5$): $\frac{1}{2}\ln|x^8 - 2x^5 + 2x - 5| + C$

(14-62) (Integrate by parts twice): $(1/13)e^{2x}(2\sin 3x - 3\cos 3x) + C$

(14-63) (Integrate by parts, then use inverse trigonometric substitution):

$$\tfrac{1}{2}x^2\text{arc sin } x + \tfrac{1}{4}x\sqrt{1 - x^2} - \tfrac{1}{4}\text{ arc sin } x + C$$

(14-64) (Integrate by parts, then partial fractions): $\ln|x| - \frac{1}{2}\ln(1 + x^2) - [(\text{arc tan } x)/x] + C$

(14-65) ($u = \cos x$): $-\ln|\cos x| + C$

(14-66) ($u = 1 + \sqrt{\cos x}$): $2(\ln|1 + \sqrt{\cos x}| - \sqrt{\cos x})\Big|_0^{\pi/2} = 2 - 2\ln 2$

(14-67) ($u = \sqrt{x}$, then integrate by parts): $2\cos\sqrt{x} + 2\sqrt{x}\sin\sqrt{x} + C$

(14-68) ($u = 1 - x^5$): $(3/35)(1 - x^5)^{7/3} - (3/20)(1 - x^5)^{4/3} + C$

(14-69) ($u = 2^x$): $2^{2x}/[(\ln 2)^2]\Big|_0^1 = 2/[(\ln 2)^2]$

(14-70) (Integrate by parts: $u = x^2$, $dv = xe^{x^2}$): $\frac{1}{2}x^2e^{x^2} - \frac{1}{2}e^{x^2} + C$

(14-71) (Integrate by parts): $2\sqrt{x}\ln x - 4\sqrt{x} + C$

(14-72) ($u = e^x$): arc tan $e^x + C$

(14-73) ($u = x^{1/6}$, then partial fractions):

$$-6[(1/7)x^{7/6} + (1/5)x^{5/6} + \tfrac{1}{4}x^{2/3} + \tfrac{1}{3}x^{1/2} + \tfrac{1}{2}x^{1/3} + x^{1/6} + \ln|x^{1/6} - 1|] + C$$

(14-74) (Integrate by parts): $\sin x - x\cos x + C$

(14-75) ($u = \text{arc sin } x$): $(2/3)(\text{arc sin } x)^{3/2}\Big|_0^1 = (2/3)(\pi/2)^{3/2}$

(14-76) (Integrate by parts): $\frac{1}{2}x^2\text{ arc tan } x - \frac{1}{2}x + \frac{1}{2}\text{ arc tan } x + C$

15 NUMERICAL INTEGRATION

THIS CHAPTER IS ABOUT

- ☑ When to Use Numerical Integration
- ☑ Dividing the Interval
- ☑ The Trapezoidal Rule
- ☑ Simpson's Rule
- ☑ Error Estimates

15-1. When to Use Numerical Integration

To evaluate an integral using the techniques of Chapters 13 and 14, you must use the Fundamental Theorem of Calculus; that is, you must be able to antidifferentiate the function in order to integrate it. **Numerical integration** is a technique that you can use to approximate the value of the integral of a function that you can't antidifferentiate.

15-2. Dividing the Interval

In order to numerically integrate $\int_a^b f(x)\,dx$ by either of the methods of this chapter, you must first divide the interval $[a, b]$ into n intervals of equal length. Denote the endpoints of these intervals $x_0, x_1, \ldots, x_n$. If Δx is the width of these intervals, then $\Delta x = (b - a)/n$. Thus:

$$\begin{aligned}
x_0 &= a \\
x_1 &= x_0 + \Delta x = a + \Delta x \\
x_2 &= x_1 + \Delta x = a + 2\,\Delta x \\
&\vdots \\
x_i &= x_{i-1} + \Delta x = a + i\,\Delta x \\
&\vdots \\
x_n &= x_{n-1} + \Delta x = a + n\,\Delta x = b
\end{aligned}$$

EXAMPLE 15-1: Find the points that divide the interval $[2, 5]$ into $n = 6$ intervals of equal length.

Solution: First find Δx:

$$\Delta x = \frac{b - a}{n} = \frac{5 - 2}{6} = \frac{1}{2}$$

Now find the endpoints of the intervals:

$$\begin{aligned}
x_0 &= a = 2 \\
x_1 &= a + \Delta x = 2 + \frac{1}{2} = \frac{5}{2} \\
x_2 &= x_1 + \Delta x = \frac{5}{2} + \frac{1}{2} = 3
\end{aligned}$$

$$x_3 = x_2 + \Delta x = 3 + \frac{1}{2} = \frac{7}{2}$$

$$x_4 = x_3 + \Delta x = \frac{7}{2} + \frac{1}{2} = 4$$

$$x_5 = x_4 + \Delta x = 4 + \frac{1}{2} = \frac{9}{2}$$

$$x_6 = x_5 + \Delta x = \frac{9}{2} + \frac{1}{2} = 5$$

15-3. The Trapezoidal Rule

The **trapezoidal rule** is a method for integrating numerically. It is so named because the area described by the definite integral is approximated by a sum of areas of trapezoids. You approximate $\int_a^b f(x)\,dx$ by dividing $[a, b]$ into n intervals of equal length and then building trapezoids above each interval $[x_i, x_{i+1}]$ (see Figure 15-1). In general, you will obtain a better approximation to the actual area by using a larger n (i.e., more trapezoids, thinner trapezoids).

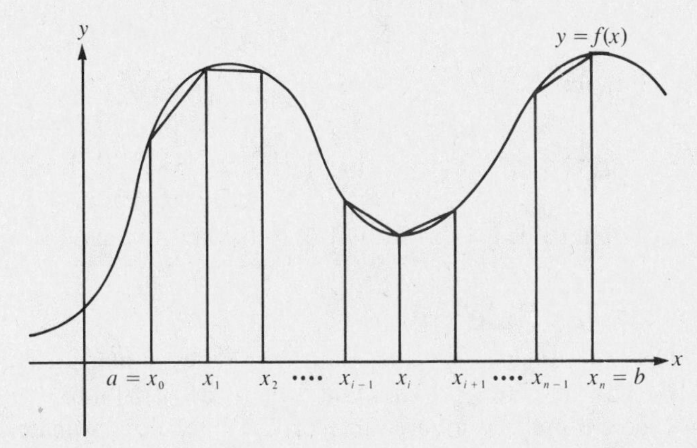

Figure 15-1

Fortunately, you do not have to draw trapezoids, or even compute areas of trapezoids, to use the trapezoidal rule. The sum of the areas of the trapezoids reduces to a simple formula:

TRAPEZOIDAL RULE

$$\int_a^b f(x)\,dx \approx \frac{b-a}{2n}\left[f(x_0) + 2f(x_1) + 2f(x_2) + \cdots + 2f(x_{n-1}) + f(x_n)\right] \qquad \textbf{(15-1)}$$

For the sake of simplicity, we have chosen two integrals that you can do more easily (and quickly) by the Fundamental Theorem of Calculus to serve as examples of the use of the trapezoidal rule. Thus, you can compare the actual value of the integral and the trapezoidal rule approximation.

EXAMPLE 15-2: Estimate $\int_1^3 1/x\,dx$ using the trapezoidal rule with $n = 4$ intervals.

Solution: You must first find the endpoints of the subintervals:

$$\Delta x = \frac{b-a}{n} = \frac{3-1}{4} = \frac{1}{2}$$

$$x_0 = a = 1$$

$$x_1 = x_0 + \Delta x = \frac{3}{2}$$

$$x_2 = x_1 + \Delta x = 2$$

$$x_3 = x_2 + \Delta x = \frac{5}{2}$$

$$x_4 = x_3 + \Delta x = 3$$

Now substitute these values into the trapezoidal rule formula, with $f(x) = 1/x$:

$$\int_1^3 \frac{1}{x}\,dx \approx \frac{3-1}{2(4)}\left[f(1) + 2f\left(\frac{3}{2}\right) + 2f(2) + 2f\left(\frac{5}{2}\right) + f(3)\right]$$

$$= \frac{1}{4}\left[1 + 2\left(\frac{2}{3}\right) + 2\left(\frac{1}{2}\right) + 2\left(\frac{2}{5}\right) + \frac{1}{3}\right] = 1.116\overline{66}$$

Note that the actual value of $\int_1^3 1/x\,dx$ is $\ln 3 = 1.0986\ldots$; the trapezoidal rule approximation is not bad. With more intervals (larger n), you could find a closer approximation.

EXAMPLE 15-3: Estimate $\int_0^1 x^2\,dx$ using the trapezoidal rule with $n = 5$ intervals.

Solution: First find the endpoints of the subintervals:

$$\Delta x = \frac{1}{5}$$

$$x_0 = 0 \qquad x_1 = \frac{1}{5} \qquad x_2 = \frac{2}{5} \qquad x_3 = \frac{3}{5} \qquad x_4 = \frac{4}{5} \qquad x_5 = 1$$

Now use the formula:

$$\int_0^1 x^2\,dx \approx \frac{1-0}{2(5)}\left[0 + 2\left(\frac{1}{25}\right) + 2\left(\frac{4}{25}\right) + 2\left(\frac{9}{25}\right) + 2\left(\frac{16}{25}\right) + 1\right] = 0.34$$

Note that the actual value is $\int_0^1 x^2\,dx = 1/3$, so the approximation is close.

15-4. Simpson's Rule

Simpson's rule replaces sums of areas of trapezoids with sums of areas under parabolas. To approximate $\int_a^b f(x)\,dx$, you first divide $[a, b]$ into n intervals of equal length (here n must be an *even* integer) and then approximate the curve above the intervals using second degree polynomials (parabolas). This method usually gives greater accuracy than the trapezoidal rule. Again, you needn't compute the parabolas to use Simpson's rule. The sum of the areas under the parabolas reduces to a simple formula:

SIMPSON'S RULE

$$\int_a^b f(x)\,dx \approx \frac{b-a}{3n}[f(x_0) + 4f(x_1) + 2f(x_2) + 4f(x_3) + \cdots + 2f(x_{n-2}) + 4f(x_{n-1}) + f(x_n)] \qquad \textbf{(15-2)}$$

Remember, n must be an even number for Simpson's rule.

EXAMPLE 15-4: Estimate $\displaystyle\int_1^3 \frac{1}{x}\,dx$ using Simpson's rule with $n = 4$ intervals.

Solution: First find Δx and the endpoints of the intervals:

$$\Delta x = \frac{b-a}{n} = \frac{3-1}{4} = \frac{1}{2}$$

$$x_0 = 1 \qquad x_1 = \frac{3}{2} \qquad x_2 = 2 \qquad x_3 = \frac{5}{2} \qquad x_4 = 3$$

Now substitute into the formula:

$$\int_1^3 \frac{1}{x}\,dx \approx \frac{3-1}{3(4)}\left[1 + 4\left(\frac{1}{3/2}\right) + 2\left(\frac{1}{2}\right) + 4\left(\frac{1}{5/2}\right) + \frac{1}{3}\right] = 1.1$$

EXAMPLE 15-5: Estimate $\int_0^{\pi} 1/(1 + \sin x)\,dx$ using Simpson's rule with $n = 6$ intervals.

Solution: First divide the interval $[0, \pi]$ into six intervals:

$$x_0 = 0 \qquad x_1 = \frac{\pi}{6} \qquad x_2 = \frac{\pi}{3} \qquad x_3 = \frac{\pi}{2} \qquad x_4 = \frac{2\pi}{3} \qquad x_5 = \frac{5\pi}{6} \qquad x_6 = \pi$$

Substitute into the formula:

$$\int_0^{\pi} \frac{1}{1 + \sin x}\,dx \approx \frac{\pi - 0}{3(6)}\left[f(0) + 4f\left(\frac{\pi}{6}\right) + 2f\left(\frac{\pi}{3}\right) + 4f\left(\frac{\pi}{2}\right) + 2f\left(\frac{2\pi}{3}\right) + 4f\left(\frac{5\pi}{6}\right) + f(\pi)\right]$$

$$= \frac{\pi}{18}\left[1 + 4\left(\frac{1}{1 + (1/2)}\right) + 2\left(\frac{1}{1 + (\sqrt{3}/2)}\right) + 4\left(\frac{1}{1+1}\right) + 2\left(\frac{1}{1 + (\sqrt{3}/2)}\right) + 4\left(\frac{1}{1 + (1/2)}\right) + 1\right]$$

$$= \frac{\pi}{18}\left[1 + 4\left(\frac{2}{3}\right) + \frac{4}{2 + \sqrt{3}} + 2 + \frac{4}{2 + \sqrt{3}} + 4\left(\frac{2}{3}\right) + 1\right]$$

$$= \frac{\pi}{18}\left[1 + \frac{8}{3} + (8 - 4\sqrt{3}) + 2 + (8 - 4\sqrt{3}) + \frac{8}{3} + 1\right]$$

$$= \frac{2\pi}{27}(19 - 6\sqrt{3}) = 2.003\,101\,6\ldots$$

15-5. Error Estimates

In addition to having an estimate for an integral, you should have some idea of how close the estimate is to the integral. An **error estimate** is a bound on the discrepancy between the estimate and the value of the integral.

A. Error estimate for the trapezoidal rule

Let T_n be the difference between $\int_a^b f(x)\,dx$ and the trapezoidal rule estimate for $\int_a^b f(x)\,dx$ with n intervals. Then T_n is the estimation error. If M_2 is the maximum value of $|f''(x)|$ on $[a, b]$, then

TRAPEZOIDAL RULE ERROR ESTIMATE

$$|T_n| \leqslant \frac{M_2(b - a)^3}{12n^2} \tag{15-3}$$

This estimate is valid only when $f''(x)$ is defined on all of $[a, b]$.

EXAMPLE 15-6: Find a bound on the error in estimating $\int_1^3 1/x\,dx$ using the trapezoidal rule with $n = 4$ intervals.

Solution: First find $f''(x)$:

$$f(x) = \frac{1}{x} = x^{-1}$$

$$f'(x) = -x^{-2}$$

$$f''(x) = 2x^{-3}$$

To find the maximum value of a function, you can use the techniques of Chapter 9. The maximum value of $|2x^{-3}|$ for $1 \leqslant x \leqslant 3$ is $M_2 = 2$, which occurs when $x = 1$. Thus you find:

$$|T_4| \leqslant \frac{M_2(b - a)^3}{12n^2} = \frac{2(3 - 1)^3}{12(4)^2} = \frac{1}{12}$$

This tells you that your trapezoidal rule estimate for $\int_1^3 1/x\,dx$ in Example 15-2 is within 1/12 of the actual value. Because in this example you were actually able to evaluate the integral, you can verify this error estimate.

EXAMPLE 15-7: Find a bound on the error in estimating $\int_0^1 x^2\,dx$ using the trapezoidal rule with $n = 5$ intervals.

Solution: First find M_2:

$$f(x) = x^2$$

$$f''(x) = 2$$

So the maximum value of $|f''(x)|$ on $[0, 1]$ is $M_2 = 2$. Thus,

$$|T_5| \leqslant \frac{2(1-0)^3}{12(5)^2} = \frac{1}{150}$$

EXAMPLE 15-8: How many intervals would you need in order to estimate $\int_1^4 \ln(1 + \sqrt{x})\,dx$ to within 0.01 using the trapezoidal rule?

Solution: First find M_2:

$$f(x) = \ln(1 + \sqrt{x})$$

$$f'(x) = \tfrac{1}{2}x^{-1/2}(1 + x^{1/2})^{-1}$$

$$f''(x) = -\tfrac{1}{4}x^{-3/2}(1 + x^{1/2})^{-1} + \tfrac{1}{2}x^{-1/2}(-1)(1 + x^{1/2})^{-2}(\tfrac{1}{2}x^{-1/2})$$

$$= \frac{-1}{4x^{3/2}(1 + x^{1/2})} - \frac{1}{4x(1 + x^{1/2})^2}$$

$$|f''(x)| = \frac{1}{4x^{3/2}(1 + x^{1/2})} + \frac{1}{4x(1 + x^{1/2})^2}$$

Evidently, as x increases, $|f''(x)|$ decreases. So the maximum value of $|f''(x)|$ for $1 \leqslant x \leqslant 4$ occurs when $x = 1$:

$$M_2 = \frac{1}{4(1)(1 + 1^{1/2})} + \frac{1}{4(1)(1 + 1^{1/2})^2} = \frac{3}{16}$$

So the error estimate is

$$|T_n| \leqslant \frac{M_2(b-a)^3}{12n^2} = \frac{(3/16)(4-1)^3}{12n^2} = \frac{81}{192n^2}$$

Choose n large enough so that

$$\frac{81}{192n^2} < 0.01$$

because this will force $|T_n| < 0.01$, the desired accuracy. If $81/(192n^2) < 0.01$, then

$$n^2 > \frac{81}{192(0.01)} = 42.1875$$

So, choose $n^2 > 42.1875$ (that is, $n \geqslant 7$) to obtain the desired accuracy.

B. Error estimate for Simpson's rule

Let S_n be the difference between $\int_a^b f(x)\,dx$ and the Simpson's rule estimate for $\int_a^b f(x)\,dx$ with n intervals. If M_4 is the maximum value of $|f^{(4)}(x)|$ on $[a, b]$, then

SIMPSON'S RULE ERROR ESTIMATE

$$|S_n| \leqslant \frac{M_4(b-a)^5}{180n^4} \qquad \textbf{(15-4)}$$

This estimate is valid only when $f^{(4)}(x)$ is defined on all of $[a, b]$.

EXAMPLE 15-9: Find a bound on the error in estimating $\int_{-1}^{2} x^6\,dx$ using Simpson's rule with eight intervals.

Solution: Find M_4:

$$f^{(4)}(x) = 360x^2$$

So the maximum value of $|f^{(4)}(x)|$ on $[-1, 2]$ is $M_4 = |f^{(4)}(2)| = 1440$. Thus,

$$|S_8| \leqslant \frac{1440[2 - (-1)]^5}{180(8)^4} = \frac{243}{512}$$

EXAMPLE 15-10: How many intervals would you need to estimate

$$\int_{-2}^{1} \left[x \arctan x - \tfrac{1}{2}\ln(1 + x^2) \right] dx$$

to within 0.0001 using Simpson's rule?

Solution: First find $f^{(4)}(x)$:

$$f(x) = x \arctan x - \frac{1}{2}\ln(1 + x^2)$$

$$f'(x) = x\frac{1}{1 + x^2} + \arctan x - \frac{1}{2}\left(\frac{2x}{1 + x^2}\right) = \arctan x$$

$$f''(x) = \frac{1}{1 + x^2} = (1 + x^2)^{-1}$$

$$f^{(3)}(x) = -2x(1 + x^2)^{-2}$$

$$f^{(4)}(x) = -2x(-2)(1 + x^2)^{-3}(2x) - 2(1 + x^2)^{-2} = \frac{6x^2 - 2}{(1 + x^2)^3}$$

To find the maximum value of $|f^{(4)}(x)|$ on $[-2, 1]$, examine $|f^{(4)}(x)|$ at the endpoints of the interval $[-2, 1]$ and at the points where the derivative of $f^{(4)}(x)$ is 0, i.e., where $f^{(5)}(x) = 0$. Find where $f^{(5)}(x) = 0$:

$$f^{(5)}(x) = (6x^2 - 2)(-3)(1 + x^2)^{-4}(2x) + 12x(1 + x^2)^{-3} = \frac{24x(1 - x^2)}{(1 + x^2)^4}$$

Consequently, $f^{(5)}(x) = 0$ when

$$24x(1 - x^2) = 0$$

$$x = 0 \qquad x = 1 \qquad x = -1$$

Now evaluate $f^{(4)}(x)$ at these points and at the endpoints of $[-2, 1]$:

$$|f^{(4)}(-1)| = |f^{(4)}(1)| = \frac{1}{2} \qquad |f^{(4)}(0)| = 2 \qquad |f^{(4)}(-2)| = \frac{22}{125}$$

So $M_4 = 2$ and the error estimate is

$$|S_n| \leqslant \frac{M_4(b - a)^5}{180n^4} = \frac{2[1 - (-2)]^5}{180n^4} = \frac{27}{10n^4}$$

In order that the error be less than 0.0001, choose n large enough so that

$$\frac{27}{10n^4} < 0.0001$$

$$n^4 > \frac{27}{10(0.0001)} = 27\,000$$

$$n > 12.81$$

Because n must be even, you need at least $n = 14$ intervals.

SUMMARY

1. Both the trapezoidal rule and Simpson's rule require that you divide the interval into n subintervals of equal length (although with Simpson's rule, n must be even).
2. The trapezoidal rule is used to estimate the value of an integral by approximating the area under a curve with sums of areas of trapezoids.
3. Simpson's rule is used to estimate the value of an integral by approximating the area under a curve with sums of areas under parabolas.
4. There are formulas that give you a bound on the error you've made in an approximation by the trapezoidal rule or Simpson's rule.

SOLVED PROBLEMS

PROBLEM 15-1 Use the trapezoidal rule with $n = 4$ intervals to estimate $\int_0^2 4^x\,dx$.

Solution: First divide the interval $[0, 2]$:

$$\Delta x = \frac{b-a}{n} = \frac{2-0}{4} = \frac{1}{2} \qquad x_2 = x_1 + \frac{1}{2} = 1$$

$$x_0 = a = 0 \qquad x_3 = x_2 + \frac{1}{2} = \frac{3}{2}$$

$$x_1 = x_0 + \Delta x = 0 + \frac{1}{2} = \frac{1}{2} \qquad x_4 = x_3 + \frac{1}{2} = 2$$

Now substitute into the trapezoidal rule ($f(x) = 4^x$):

$$\int_0^2 4^x\,dx \approx \frac{b-a}{2n}[f(x_0) + 2f(x_1) + 2f(x_2) + 2f(x_3) + f(x_4)]$$

$$= \frac{2-0}{2(4)}\left[f(0) + 2f\left(\frac{1}{2}\right) + 2f(1) + 2f\left(\frac{3}{2}\right) + f(2)\right]$$

$$= \frac{1}{4}[4^0 + 2(4^{1/2}) + 2(4^1) + 2(4^{3/2}) + 4^2]$$

$$= \frac{1}{4}[1 + 2(2) + 2(4) + 2(8) + 16] = \frac{45}{4}$$

[See Section 15-3.]

PROBLEM 15-2 Use the trapezoidal rule with $n = 5$ intervals to estimate $\displaystyle\int_{1/2}^1 \frac{1}{x^2}\,dx$.

Solution: Divide the interval $[\frac{1}{2}, 1]$:

$$\Delta x = \frac{1 - \frac{1}{2}}{5} = 0.1$$

$$x_0 = 0.5 \quad x_1 = 0.6 \quad x_2 = 0.7 \quad x_3 = 0.8 \quad x_4 = 0.9 \quad x_5 = 1$$

Substitute into the formula:

$$\int_{1/2}^1 \frac{1}{x^2}\,dx \approx \frac{1-\frac{1}{2}}{2(5)}\left[\frac{1}{(0.5)^2} + 2\left(\frac{1}{(0.6)^2}\right) + 2\left(\frac{1}{(0.7)^2}\right) + 2\left(\frac{1}{(0.8)^2}\right) + 2\left(\frac{1}{(0.9)^2}\right) + \frac{1}{1^2}\right]$$

$$= \frac{1}{20}\left[\frac{1}{0.25} + \frac{2}{0.36} + \frac{2}{0.49} + \frac{2}{0.64} + \frac{2}{0.81} + 1\right] = 1.011\,566\ldots$$

[See Section 15-3.]

PROBLEM 15-3 Use the trapezoidal rule with $n = 10$ intervals to estimate $\int_0^2 e^{-x^2}\,dx$.

Solution: Divide the interval [0, 2]: $\Delta x = (b - a)/n = 0.2$, so

$$x_0 = 0 \quad x_1 = 0.2 \quad x_2 = 0.4 \quad x_3 = 0.6 \quad \cdots \quad x_8 = 1.6 \quad x_9 = 1.8 \quad x_{10} = 2$$

Substitute into the formula:

$$\int_0^2 e^{-x^2}\,dx \approx \frac{2-0}{2(10)}[e^{-0^2} + 2e^{-(0.2)^2} + 2e^{-(0.4)^2} + \cdots + 2e^{-(1.8)^2} + e^{-2^2}]$$
$$= 0.881\,8388\ldots$$

[See Section 15-3.]

PROBLEM 15-4 Use Simpson's rule with $n = 4$ intervals to estimate $\int_0^2 (x^3 - 1)\,dx$.

Solution: Divide the interval:

$$\Delta x = \frac{2-0}{4} = \frac{1}{2}$$

$$x_0 = 0 \quad x_1 = \frac{1}{2} \quad x_2 = 1 \quad x_3 = \frac{3}{2} \quad x_4 = 2$$

Substitute into Simpson's rule:

$$\int_0^2 (x^3 - 1)\,dx \approx \frac{b-a}{3n}[f(x_0) + 4f(x_1) + 2f(x_2) + 4f(x_3) + f(x_4)]$$
$$= \frac{2-0}{3(4)}\left[f(0) + 4f\left(\frac{1}{2}\right) + 2f(1) + 4f\left(\frac{3}{2}\right) + f(2)\right]$$
$$= \frac{1}{6}\left[-1 + 4\left(\frac{-7}{8}\right) + 2(0) + 4\left(\frac{19}{8}\right) + 7\right] = 2$$

[See Section 15-4.]

PROBLEM 15-5 Use Simpson's rule with $n = 6$ intervals to estimate $\int_1^3 \sqrt{\ln x}\,dx$.

Solution: Divide the interval [1, 3]: $\Delta x = 1/3$, so

$$x_0 = 1 \quad x_1 = \frac{4}{3} \quad x_2 = \frac{5}{3} \quad x_3 = 2 \quad x_4 = \frac{7}{3} \quad x_5 = \frac{8}{3} \quad x_6 = 3$$

Substitute into the formula:

$$\int_1^3 \sqrt{\ln x}\,dx \approx \frac{3-1}{3(6)}\left[\sqrt{\ln 1} + 4\sqrt{\ln\frac{4}{3}} + 2\sqrt{\ln\frac{5}{3}} + 4\sqrt{\ln 2}\right.$$
$$\left. + 2\sqrt{\ln\frac{7}{3}} + 4\sqrt{\ln\frac{8}{3}} + \sqrt{\ln 3}\right]$$
$$= 1.528\,41\ldots$$

[See Section 15-4.]

PROBLEM 15-6 Use Simpson's rule with $n = 10$ intervals to estimate $\int_0^1 e^{x^2}\,dx$.

Solution: Divide the interval [0, 1]: $\Delta x = 0.1$, so

$$x_0 = 0 \quad x_1 = 0.1 \quad x_2 = 0.2 \quad x_3 = 0.3 \quad \cdots \quad x_9 = 0.9 \quad x_{10} = 1$$

Substitute into the formula:

$$\int_0^1 e^{x^2}\,dx \approx \frac{1-0}{3(10)}[e^{0^2} + 4e^{(0.1)^2} + 2e^{(0.2)^2} + 4e^{(0.3)^2} + \cdots + 4e^{(0.9)^2} + e^{1^2}]$$
$$= 1.462\,68\ldots$$

[See Section 15-4.]

PROBLEM 15-7 Find a bound on the error in estimating $\int_{1/2}^{1} 1/x^2\, dx$ using the trapezoidal rule with $n = 5$ intervals.

Solution: First find $f''(x)$:

$$f(x) = \frac{1}{x^2} = x^{-2}$$

$$f'(x) = -2x^{-3}$$

$$f''(x) = 6x^{-4} = \frac{6}{x^4}$$

Because $|f''(x)|$ decreases as x increases from $\frac{1}{2}$ to 1, the maximum value of $|f''(x)|$ occurs at $x = \frac{1}{2}$:

$$M_2 = \left|f''\left(\frac{1}{2}\right)\right| = \frac{6}{(\frac{1}{2})^4} = 96$$

Thus a bound on the error is

$$|T_5| \leqslant \frac{M_2(b-a)^3}{12n^2} = \frac{96(1-\frac{1}{2})^3}{12(5)^2} = \frac{1}{25}$$

[See Section 15-5.]

PROBLEM 15-8 Find a bound on the error in estimating $\int_0^2 4^x\, dx$ using the trapezoidal rule with $n = 4$ intervals.

Solution: First find $f''(x)$:

$$f(x) = 4^x$$

$$f'(x) = 4^x \ln 4$$

$$f''(x) = 4^x \ln^2 4$$

The maximum value of $|f''(x)|$ on $[0, 2]$ occurs at $x = 2$:

$$M_2 = |f''(2)| = 4^2 \ln^2 4 = 16 \ln^2 4$$

Thus a bound on the error is

$$|T_4| \leqslant \frac{16(\ln^2 4)(2-0)^3}{12(4)^2} = \frac{2}{3}\ln^2 4$$

[See Section 15-5.]

PROBLEM 15-9 Find a bound on the error in estimating $\int_0^3 e^{2x}\, dx$ using Simpson's rule with $n = 12$ intervals.

Solution: Find $f^{(4)}(x)$:

$$f(x) = e^{2x}$$

$$f'(x) = 2e^{2x}$$

$$f''(x) = 4e^{2x}$$

$$f^{(3)}(x) = 8e^{2x}$$

$$f^{(4)}(x) = 16e^{2x}$$

The maximum value of $|f^{(4)}(x)|$ on $[0, 3]$ is attained at $x = 3$:

$$M_4 = |f^{(4)}(3)| = 16e^6$$

A bound on the error is

$$|S_{12}| \leqslant \frac{M_4(b-a)^5}{180n^4} = \frac{16e^6(3-0)^5}{180(12)^4} = \frac{e^6}{960} \approx 0.420\,238$$

[See Section 15-5.]

PROBLEM 15-10 Find a bound on the error in estimating $\int_0^2 (x^3 - 1)\, dx$ using Simpson's rule with $n = 4$ intervals.

Solution: Find $f^{(4)}(x)$:

$$f(x) = x^3 - 1$$
$$f'(x) = 3x^2$$
$$f''(x) = 6x$$
$$f^{(3)}(x) = 6$$
$$f^{(4)}(x) = 0$$

The maximum value of $|f^{(4)}(x)|$ is $M_4 = 0$. So a bound on the error is

$$|S_4| \leqslant \frac{M_4(b-a)^5}{180(4)^4} = \frac{0(2)^5}{180(4)^4} = 0$$

The Simpson's rule estimate must be the actual value of the integral because the error is zero. In general, Simpson's rule (with any even number of intervals) will give the exact value of any integral of a cubic polynomial. [See Section 15-5.]

PROBLEM 15-11 How many intervals would you need for your trapezoidal rule estimate for $\int_4^7 \sqrt{x}\, dx$ to be accurate to within 0.001?

Solution: To find a bound on the error, calculate $f''(x)$:

$$f(x) = \sqrt{x} = x^{1/2}$$
$$f'(x) = \frac{1}{2}x^{-1/2}$$
$$f''(x) = -\frac{1}{4}x^{-3/2}$$

So the maximum value of $|f''(x)|$ on $[4, 7]$ occurs at $x = 4$:

$$M_2 = |f''(4)| = \frac{1}{4}(4)^{-3/2} = \frac{1}{32}$$

A bound on the error is

$$|T_n| \leqslant \frac{M_2(b-a)^3}{12n^2} = \frac{(1/32)(7-4)^3}{12n^2} = \frac{9}{128n^2}$$

In order that $|T_n|$ be less than 0.001, choose n large enough so that

$$\frac{9}{128n^2} < 0.001$$
$$n^2 > \frac{9}{128(0.001)} \approx 70.3$$
$$n > \sqrt{70.3} \approx 8.39$$

So you need nine or more intervals. [See Section 15-5.]

PROBLEM 15-12 How many intervals would you need for your Simpson's rule estimate for $\int_1^2 \ln x\, dx$ to be accurate to within 0.0001?

Solution: Find $f^{(4)}(x)$:

$$f(x) = \ln x$$
$$f'(x) = \frac{1}{x} = x^{-1}$$
$$f''(x) = -x^{-2}$$
$$f^{(3)}(x) = 2x^{-3}$$
$$f^{(4)}(x) = -6x^{-4}$$

The maximum value of $|f^{(4)}(x)|$ on $[1,2]$ occurs at $x = 1$:

$$M_4 = |f^{(4)}(1)| = 6$$

The error estimate is

$$|S_n| \leqslant \frac{M_4(b-a)^5}{180n^4} = \frac{6(2-1)^5}{180n^4} = \frac{1}{30n^4}$$

In order that your error be less than 0.0001, choose n large enough so that

$$\frac{1}{30n^4} < 0.0001$$

$$n^4 > \frac{1}{30(0.0001)} = 333.\overline{333}$$

$$n > 4.27$$

So six intervals will suffice. [See Section 15-5.]

PROBLEM 15-13 Use the trapezoidal rule to find an estimate for $\int_0^2 1/\sqrt{1+x}\,dx$ that is accurate to within 0.01.

Solution: First determine how many intervals to use. Find $f''(x)$:

$$f(x) = \frac{1}{\sqrt{1+x}} = (1+x)^{-1/2}$$

$$f'(x) = -\frac{1}{2}(1+x)^{-3/2}$$

$$f''(x) = \frac{3}{4}(1+x)^{-5/2}$$

The maximum value of $|f''(x)|$ on $[0,2]$ occurs at $x = 0$:

$$M_2 = |f''(0)| = \frac{3}{4}$$

So the error estimate is

$$|T_n| \leqslant \frac{M_2(b-a)^3}{12n^2} = \frac{\frac{3}{4}(2-0)^3}{12n^2} = \frac{1}{2n^2}$$

Choose n large enough so that

$$\frac{1}{2n^2} < 0.01$$

$$n^2 > \frac{1}{2(0.01)} = 50$$

$$n > 7.07$$

So use $n = 8$ intervals.

Divide the interval $[0,2]$ into eight intervals:

$$\Delta x = \frac{2-0}{8} = \frac{1}{4}$$

$$x_0 = 0 \qquad x_1 = \frac{1}{4} \qquad x_2 = \frac{1}{2} \qquad \cdots \qquad x_7 = \frac{7}{4} \qquad x_8 = 2$$

Plug into the trapezoidal rule:

$$\int_0^2 \frac{dx}{\sqrt{1+x}} \approx \frac{2-0}{2(8)}\left[f(0) + 2f\left(\frac{1}{4}\right) + 2f\left(\frac{1}{2}\right) + \cdots + 2f\left(\frac{7}{4}\right) + f(2)\right]$$

$$= \frac{1}{8}\left[1 + 2\frac{1}{\sqrt{5/4}} + 2\frac{1}{\sqrt{3/2}} + \cdots + 2\frac{1}{\sqrt{11/4}} + \frac{1}{\sqrt{3}}\right]$$

$$= 1.4662\ldots$$

[See Section 15-5.]

PROBLEM 15-14 Use Simpson's rule to find an estimate for $\int_0^{\pi/4} \ln(\sec x + \tan x)\,dx$ that is accurate to within 0.001.

Solution: First determine how many intervals to use:

$$f(x) = \ln(\sec x + \tan x)$$

$$f'(x) = \frac{\sec x \tan x + \sec^2 x}{\sec x + \tan x} = \sec x$$

$$f''(x) = \sec x \tan x$$

$$f^{(3)}(x) = \sec x \tan^2 x + \sec^3 x$$

$$\begin{aligned} f^{(4)}(x) &= \sec x \tan x \tan^2 x + \sec x\, 2 \tan x \sec^2 x + 3 \sec^2 x \sec x \tan x \\ &= \sec x \tan^3 x + 5 \sec^3 x \tan x \\ &= \frac{\sin x(\sin^2 x + 5)}{\cos^4 x} \end{aligned}$$

As x increases from 0 to $\pi/4$, the numerator of $|f^{(4)}(x)|$ increases and the denominator decreases. Thus the maximum value of $|f^{(4)}(x)|$ occurs at $x = \pi/4$:

$$M_4 = |f^{(4)}(\pi/4)| = \sec(\pi/4)\tan^3(\pi/4) + 5\sec^3(\pi/4)\tan(\pi/4) = 11\sqrt{2}$$

The error estimate is

$$|S_n| \leqslant \frac{M_4(b-a)^5}{180n^4} = \frac{11\sqrt{2}[(\pi/4) - 0]^5}{180n^4}$$

Choose n large enough so that

$$\frac{11\sqrt{2}(\pi/4)^5}{180n^4} < 0.001$$

$$n^4 > \frac{11\sqrt{2}(\pi/4)^5}{180(0.001)} \approx 25.828$$

$$n > 2.25$$

So use $n = 4$ intervals (recall that n must be even):

$$x_0 = 0 \qquad x_1 = \frac{\pi}{16} \qquad x_2 = \frac{\pi}{8} \qquad x_3 = \frac{3\pi}{16} \qquad x_4 = \frac{\pi}{4}$$

Plug into Simpson's rule:

$$\int_0^{\pi/4} \ln(\sec x + \tan x)\,dx \approx \frac{(\pi/4) - 0}{3(4)}\left[f(0) + 4f\left(\frac{\pi}{16}\right) + 2f\left(\frac{\pi}{8}\right) + 4f\left(\frac{3\pi}{16}\right) + f\left(\frac{\pi}{4}\right)\right]$$

$$= 0.326\,194\ldots$$

[See Section 15-5.]

Supplementary Exercises

In Problems 15-15 through 15-20 use the trapezoidal rule with n intervals to estimate the given integral.

15-15 $\int_1^3 \sqrt{\ln x\,dx}$ $\quad n = 6$

15-16 $\int_0^{\pi} \frac{dx}{1 + \sin x}$ $\quad n = 6$

15-17 $\int_1^3 e^{\sqrt{x}}\,dx$ $\quad n = 5$

15-18 $\int_1^3 \ln x\,dx$ $\quad n = 6$

15-19 $\int_{-1}^{0} \sqrt{2 + x^3}\, dx \qquad n = 8$

15-20 $\int_{0}^{\pi/2} \sqrt{1 + \cos x}\, dx \qquad n = 4$

In Problems 15-21 through 15-26 use Simpson's rule with n intervals to estimate the given integral.

15-21 $\int_{0}^{2} e^{-x^2}\, dx \qquad n = 4$

15-22 $\int_{0}^{1} \ln(1 + \sqrt{x})\, dx \qquad n = 6$

15-23 $\int_{0}^{1} x^x\, dx \qquad n = 4$

15-24 $\int_{1}^{3} e^{\sqrt{x}}\, dx \qquad n = 8$

15-25 $\int_{0}^{4} (x^3 - 3x^2 + 1)\, dx \qquad n = 6$

15-26 $\int_{-1}^{0} \sqrt{2 + x^3}\, dx \qquad n = 8$

In Problems 15-27 through 15-29 find a bound on the error in estimating the given integral using the trapezoidal rule with n intervals.

15-27 $\int_{-1}^{3} x^5\, dx \qquad n = 10$

15-28 $\int_{e}^{2e} \ln x\, dx \qquad n = 5$

15-29 $\int_{-2}^{0} [x \arctan x - \frac{1}{2}\ln(1 + x^2)]\, dx \qquad n = 8$

In Problems 15-30 through 15-32 find a bound on the error in estimating the given integral using Simpson's rule with n intervals.

15-30 $\int_{1}^{3} 1/x^2\, dx \qquad n = 8$

15-31 $\int_{0}^{2} 1/\sqrt{1 + x}\, dx \qquad n = 4$

15-32 $\int_{0}^{1} e^{x^2}\, dx \qquad n = 6$

In Problems 15-33 through 15-37 use the trapezoidal rule to estimate the given integral with the given accuracy.

15-33 $\int_{1}^{4} 3x^{4/3}\, dx \qquad \text{error} < 0.1$

15-34 $\int_{0}^{1} x^x\, dx \qquad \text{error} < 0.01$

15-35 $\int_{-1/2}^{1/2} x \arctan x\, dx \qquad \text{error} < 0.01$

15-36 $\int_{0}^{1} e^{x^2}\, dx \qquad \text{error} < 0.1$

15-37 $\int_{0}^{\pi/3} \ln(\sec x + \tan x)\, dx \qquad \text{error} < 0.01$

In Problems 15-38 through 15-42 use Simpson's rule to estimate the given integral with the given accuracy.

15-38 $\int_1^4 3x^{4/3}\,dx$ error < 0.01

15-39 $\int_1^3 \ln x\,dx$ error < 0.001

15-40 $\int_1^3 \sqrt{x}\,dx$ error < 0.0001

15-41 $\int_0^{\pi/4} \ln\cos x\,dx$ error < 0.001

15-42 $\int_0^{1/2} (x \arcsin x + \sqrt{1 - x^2})\,dx$ error $< 0.000\,01$

Solutions to Supplementary Exercises

(15-15) 1.506 188

(15-16) 2.044 72

(15-17) 8.279 0884

(15-18) 1.289 696

(15-19) 1.316 084

(15-20) 1.993 57

(15-21) 0.881 812

(15-22) 0.494 5259

(15-23) 0.788 8625

(15-24) 8.2754

(15-25) 4

(15-26) 1.318 001

(15-27) $M_2 = 540$ $|T_{10}| \leqslant 144/5$

(15-28) $M_2 = 1/e^2$ $|T_5| \leqslant e/300$

(15-29) $M_2 = 1$ $|T_8| \leqslant 1/96$

(15-30) $M_4 = 120$ $|S_8| \leqslant 1/192$

(15-31) $M_4 = 105/16$ $|S_4| \leqslant 7/1536$

(15-32) $M_4 = 76e$ $|S_6| \leqslant 19e/58\,320$

(15-33) 31.4182 $M_2 = 4/3$ $n = 6$

(15-34) 0.798 0916 $M_2 = 2$ $n = 5$

(15-35) 0.085 3263 $M_2 = 2$ $n = 5$

(15-36) 1.490 6789 $M_2 = 6e$ $n = 4$

(15-37) 0.613 166 $M_2 = 2\sqrt{3}$ $n = 6$

(15-38) 31.3705 $M_4 = 40/27$ $n = 4$

(15-39) 1.295 721 $M_4 = 6$ $n = 6$

(15-40) 2.797 428 $M_4 = 15/16$ $n = 8$

(15-41) $-0.086\,4441$ $M_4 = 16$ $n = 4$

(15-42) 0.5211 $M_4 = 16/(3\sqrt{3})$ $n = 4$

EXAM 6 (CHAPTERS 14 AND 15)

1. Antidifferentiate

(a) $\int x^2 \cos x \, dx$ (b) $\int \sec x \tan^3 x \, dx$

(c) $\int \frac{x^2}{(x^2 - 1)^{5/2}} \, dx$ (d) $\int \frac{2x^5 - 8x^4 + 26x^3 + 21x - 26}{x^4 - 4x^3 + 13x^2} \, dx$

2. Use Simpson's rule with $n = 4$ intervals to estimate $\int_{-1}^{1} 16^{x^2} \, dx$.

3. How many intervals would you need for a trapezoidal rule estimate that is accurate to within 0.01 for $\int_0^1 (x^2 + 1)^{5/2} \, dx$?

SOLUTIONS TO EXAM 6

1. **(a)** Integrate by parts. Let

$$u = x^2 \qquad dv = \cos x \, dx$$

$$du = 2x \, dx \qquad v = \sin x$$

$$\int x^2 \cos x \, dx = \int u \, dv = uv - \int v \, du = x^2 \sin x - \int \sin x \, (2x \, dx)$$

$$= x^2 \sin x - 2 \int x \sin x \, dx$$

Then, integrate $x \sin x$ by parts. Let

$$u = x \qquad dv = \sin x \, dx$$

$$du = dx \qquad v = -\cos x$$

$$\int x \sin x \, dx = \int u dv = uv - \int v du = -x \cos x + \int \cos x \, dx$$

$$= -x \cos x - \sin x + C$$

and so you have

$$\int x^2 \cos x \, dx = x^2 \sin x - 2(-x \cos x - \sin x) + C$$

$$= x^2 \sin x + 2x \cos x + 2 \sin x + C$$

(b) Because the integrand contains an odd power of $\tan x$, you can use the identity $\tan^2 x = \sec^2 x - 1$ to put the integrand in a manageable form:

$$\int \sec x \tan^3 x \, dx = \int \tan^2 x \sec x \tan x \, dx = \int (\sec^2 x - 1) \sec x \tan x \, dx$$

Now make the substitution $u = \sec x$, with $du = \sec x \tan x \, dx$:

$$\int \sec x \tan^3 x \, dx = \int (u^2 - 1) \, du = \frac{1}{3}u^3 - u + C = \frac{1}{3}\sec^3 x - \sec x + C$$

(c) Use the inverse trig substitution $\theta = \operatorname{arc\,sec} x$:

$$\sec \theta = x$$

$$dx = \sec \theta \tan \theta \, d\theta$$

$$\sqrt{x^2 - 1} = \tan \theta$$

$$\int \frac{x^2}{(x^2 - 1)^{5/2}} \, dx = \int \frac{\sec^2 \theta}{\tan^5 \theta} \sec \theta \tan \theta \, d\theta = \int \frac{\cos \theta}{\sin^4 \theta} \, d\theta$$

Make another substitution $u = \sin \theta$ with $du = \cos \theta \, d\theta$:

$$\int \frac{\cos \theta}{\sin^4 \theta} \, d\theta = \int \frac{1}{u^4} \, du = \frac{-1}{3} u^{-3} + C = \frac{-1}{3} \sin^{-3} \theta + C$$

$$= \frac{-1}{3}\left(\frac{\sqrt{x^2 - 1}}{x}\right)^{-3} = \frac{-x^3}{3(x^2 - 1)^{3/2}} + C$$

(d) The integrand is a rational function. Because the degree of the numerator exceeds the degree of the denominator, divide the denominator into the numerator to find:

$$\frac{2x^5 - 8x^4 + 26x^3 + 21x - 26}{x^4 - 4x^3 + 13x^2} = 2x + \frac{21x - 26}{x^4 - 4x^3 + 13x^2}$$

To find the partial fraction decomposition of $(21x - 26)/(x^4 - 4x^3 + 13x^2)$, first factor the denominator into irreducibles:

$$x^4 - 4x^3 + 13x^2 = x^2(x^2 - 4x + 13)$$

The form of the decomposition is

$$\frac{21x - 26}{x^2(x^2 - 4x + 13)} = \frac{A}{x} + \frac{B}{x^2} + \frac{Cx + D}{x^2 - 4x + 13}$$

Multiply both sides of the equation by $x^2(x^2 - 4x + 13)$ to clear all denominators:

$$21x - 26 = Ax(x^2 - 4x + 13) + B(x^2 - 4x + 13) + (Cx + D)x^2$$

$$= x^3(A + C) + x^2(-4A + B + D) + x(13A - 4B) + 13B$$

Equate the coefficients:

$$0 = A + C$$

$$0 = -4A + B + D$$

$$21 = 13A - 4B$$

$$-26 = 13B$$

and solve: $B = -2$, $A = 1$, $C = -1$, $D = 6$. Thus you have

$$\frac{21x - 26}{x^2(x^2 - 4x + 13)} = \frac{1}{x} - \frac{2}{x^2} + \frac{-x + 6}{x^2 - 4x + 13}$$

and so

$$\int \frac{2x^5 - 8x^4 + 26x^3 + 21x - 26}{x^4 - 4x^3 + 13x^2} \, dx = \int \left[2x + \frac{21x - 26}{x^2(x^2 - 4x + 13)}\right] dx$$

$$= \int \left(2x + \frac{1}{x} - \frac{2}{x^2} + \frac{-x + 6}{x^2 - 4x + 13}\right) dx$$

$$= x^2 + \ln x + \frac{2}{x} - \int \frac{x - 6}{x^2 - 4x + 13} \, dx$$

Now split the integrand into two parts: one yields a natural log term and the other, an arctangent:

$$\int \frac{x-6}{x^2-4x+13}\,dx = \frac{1}{2}\int \frac{x-4}{x^2-4x+13}\,dx - 4\int \frac{1}{x^2-4x+13}\,dx$$

$$= \frac{1}{2}\ln|x^2-4x+13| - 4\int \frac{1}{(x-2)^2+9}\,dx$$

$$= \frac{1}{2}\ln|x^2-4x+13| - \frac{4}{3}\arctan\left(\frac{x-2}{3}\right) + C$$

So the final answer is

$$x^2 + \ln|x| + \frac{2}{x} - \frac{1}{2}\ln|x^2-4x+13| + \frac{4}{3}\arctan\left(\frac{x-2}{3}\right) + C$$

2. Divide the interval $[-1, 1]$ into four intervals:

$$\Delta x = \frac{1-(-1)}{4} = \frac{1}{2}$$

$$x_0 = -1,\quad x_1 = -\frac{1}{2},\quad x_2 = 0,\quad x_3 = \frac{1}{2},\quad x_4 = 1$$

Use the formula for Simpson's rule:

$$\int_{-1}^{1} 16^{x^2}\,dx \approx \frac{b-a}{3n}[f(x_0) + 4f(x_1) + 2f(x_2) + 4f(x_3) + f(x_4)]$$

$$= \frac{2}{3\cdot 4}[16^{(-1)^2} + 4(16)^{(-1/2)^2} + 2(16)^{0^2} + 4(16)^{(1/2)^2} + 16^{1^2}$$

$$= \frac{1}{6}[16 + 4(16)^{1/4} + 2 + 4(16)^{1/4} + 16]$$

$$= \frac{1}{6}[16 + 4(2) + 2 + 4(2) + 16] = \frac{25}{3}$$

3. The formula for the bound on the error is

$$T_n \leqslant \frac{M_2(1-0)^3}{12n^2}$$

where M_2 is the maximum value of $|f''(x)|$ on $[-1, 1]$. You must find $f''(x)$:

$$f(x) = (x^2+1)^{5/2}$$

$$f'(x) = \frac{5}{2}(x^2+1)^{3/2}(2x) = 5x(x^2+1)^{3/2}$$

$$f''(x) = 5(x^2+1)^{3/2} + 5x\left(\frac{3}{2}\right)(x^2+1)^{1/2}(2x) = 5(x^2+1)^{3/2} + 15x^2(x^2+1)^{1/2}$$

$$= 5(x^2+1)^{1/2}(x^2+1+3x^2) = 5(x^2+1)^{1/2}(4x^2+1)$$

The maximum value of $|f''(x)|$ on $[0, 1]$ occurs at $x = 1$ (because $f''(x)$ is increasing for positive x):

$$M_2 = |f''(1)| = 25\sqrt{2}$$

Thus, the error is less than $25\sqrt{2}/(12n^2)$. To make this less than 0.01, you need:

$$\frac{25\sqrt{2}}{12n^2} < 0.01 \qquad \frac{2500\sqrt{2}}{12} < n^2 \qquad n > \frac{50\sqrt[4]{2}}{2\sqrt{3}} \approx 17.16$$

Therefore, use at least 18 intervals.

16 APPLICATIONS OF THE INTEGRAL

THIS CHAPTER IS ABOUT

☑ Area Between Curves
☑ Position, Velocity, and Acceleration
☑ Rates of Change

16-1. Area Between Curves

If $f(x)$ is greater than $g(x)$ for all x between a and b, then the area under the graph of $f(x)$ minus the area under the graph of $g(x)$ is the area between the curves (see Figure 16-1). In other words, the area between the curves $y = f(x)$ and $y = g(x)$ between $x = a$ and $x = b$ is

AREA BETWEEN CURVES

$$\int_a^b f(x)\,dx - \int_a^b g(x)\,dx = \int_a^b (f(x) - g(x))\,dx \tag{16-1}$$

as long as $f(x)$ is larger than $g(x)$ between $x = a$ and $x = b$.

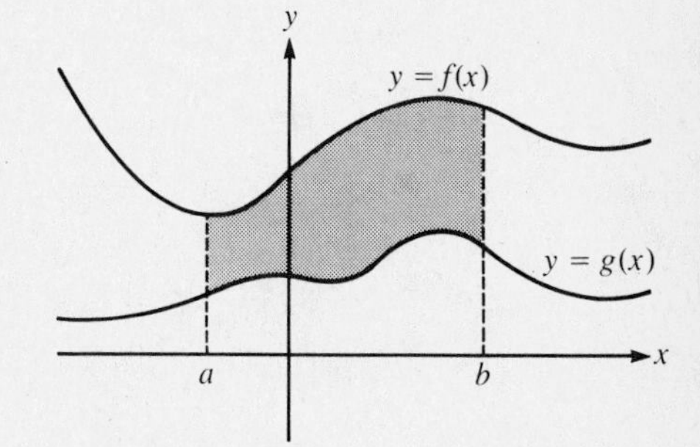

Figure 16-1

EXAMPLE 16-1: Find the area between the graphs of $f(x) = x^2 + 2$ and $g(x) = 1 - x$ between $x = 0$ and $x = 1$.

Solution: As you can see in Figure 16-2, $f(x)$ is greater than $g(x)$ between $x = 0$ and $x = 1$. Therefore the area you want is

$$\int_0^1 [(x^2 + 2) - (1 - x)]\,dx = \int_0^1 (x^2 + x + 1)\,dx = \left(\frac{x^3}{3} + \frac{x^2}{2} + x\right)\Bigg|_0^1$$

$$= \left(\frac{1}{3} + \frac{1}{2} + 1\right) - 0 = \frac{11}{6}$$

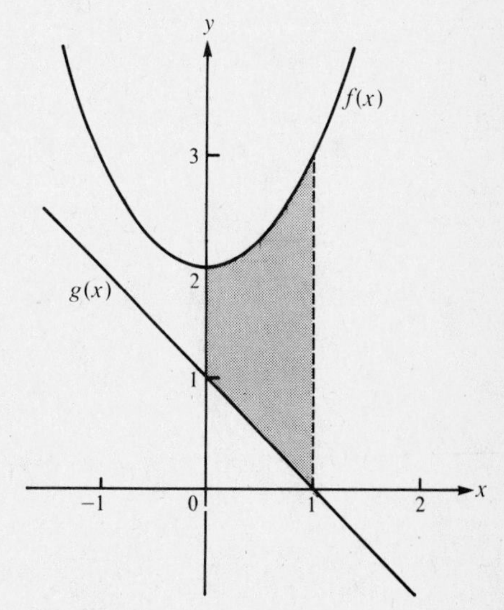

Figure 16-2
$f(x) = x^2 + 2$ and $g(x) = 1 - x$.

EXAMPLE 16-2: Find the area between the curves $y = x - 1$ and $y = 2x^3 - 1$ between $x = 1$ and $x = 2$.

Solution: You must first determine which function is larger. Between $x = 1$ and $x = 2$, $2x^3 - 1$ is larger than $x - 1$ (see Figure 16-3). The area is then

$$\int_1^2 [(2x^3 - 1) - (x - 1)]\,dx = \int_1^2 (2x^3 - x)\,dx = \left(\frac{1}{2}x^4 - \frac{1}{2}x^2\right)\Bigg|_1^2$$

$$= (8 - 2) - \left(\frac{1}{2} - \frac{1}{2}\right) = 6$$

EXAMPLE 16-3: Find the area of the region bounded by the graphs of $f(x) = (x - 1)^2$ and $g(x) = -x + 3$.

Solution: The graphs of these functions are shown in Figure 16-4. The region of interest extends from $x = -1$ to $x = 2$ and $g(x) = -x + 3$ is greater than

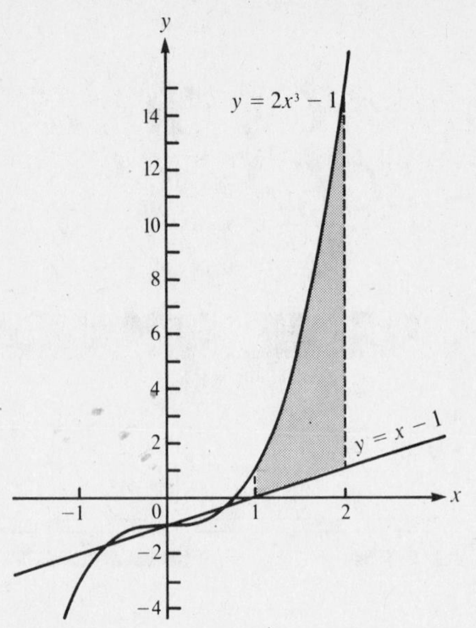

Figure 16-3
$y = x - 1$ and $y = 2x^3 - 1$.

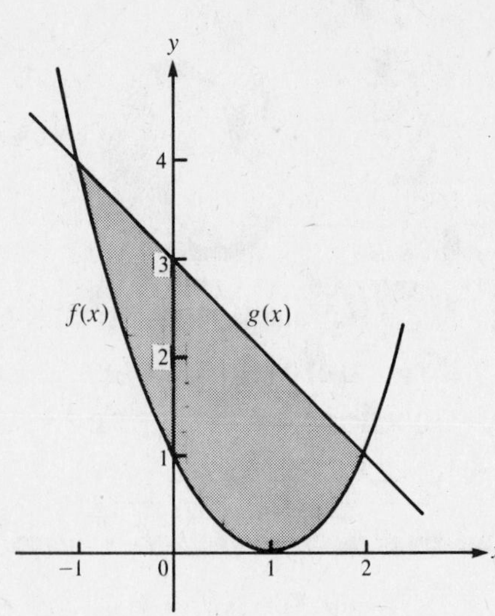

Figure 16-4
$f(x) = (x - 1)^2$ and $g(x) = -x + 3$.

$f(x) = (x - 1)^2$ in this region. So the area of the region is

$$\begin{aligned}\int_{-1}^{2} [(-x + 3) - (x - 1)^2]\, dx &= \int_{-1}^{2} (-x^2 + x + 2)\, dx \\ &= \left(\frac{-x^3}{3} + \frac{x^2}{2} + 2x\right)\bigg|_{-1}^{2} \\ &= \left(\frac{-8}{3} + 2 + 4\right) - \left(\frac{1}{3} + \frac{1}{2} - 2\right) \\ &= \frac{9}{2}\end{aligned}$$

You can do all of this without graphing f and g. To find the limits of integration, determine where the curves intersect. For what x is $f(x) = g(x)$?

$$\begin{aligned}(x - 1)^2 &= -x + 3 \\ x^2 - 2x + 1 &= -x + 3 \\ x^2 - x - 2 &= 0 \\ (x + 1)(x - 2) &= 0 \\ x = -1 \quad &\text{or} \quad x = 2\end{aligned}$$

To determine which function is larger in $[-1, 2]$, choose any point c in $[-1, 2]$ and find $f(c)$ and $g(c)$. Whichever function is larger at c will be larger on the entire interval $[-1, 2]$, as long as f and g are continuous on $[-1, 2]$. (Otherwise, the graphs would have to cross somewhere in $(-1, 2)$, and you know they only intersect at $x = -1$ and $x = 2$.)

If you choose $c = 0$, $g(0) = 3$ and $f(0) = 1$, so $g(x)$ is the larger function on $[-1, 2]$. If you err in determining which function is larger, your answer will be negative and the absolute value will yield the correct solution.

EXAMPLE 16-4: Find the area of the region bounded by the curves $y = x^3 - x^2 - 2x + 1$ and $y = x^2 - 2x + 1$.

Solution: Rather than graph these, simply find where they intersect and determine which is larger in the region of interest. To find where they intersect, equate the two functions and solve for x:

$$x^3 - x^2 - 2x + 1 = x^2 - 2x + 1$$
$$x^3 - 2x^2 = 0$$
$$x^2(x - 2) = 0$$
$$x = 0 \quad \text{or} \quad x = 2$$

So the region of interest is $[0, 2]$. Evaluate both at a point c in $[0, 2]$, say $c = 1$:

$$1^3 - 1^2 - 2(1) + 1 = -1$$
$$1^2 - 2(1) + 1 = 0$$

So $x^2 - 2x + 1$ is greater on $[0, 2]$. The area is therefore

$$\int_0^2 [(x^2 - 2x + 1) - (x^3 - x^2 - 2x + 1)]\, dx = \int_0^2 (2x^2 - x^3)\, dx$$
$$= \left(\frac{2}{3}x^3 - \frac{x^4}{4}\right)\Bigg|_0^2$$
$$= \left(\frac{16}{3} - 4\right) - (0 - 0) = \frac{4}{3}$$

EXAMPLE 16-5: Find the area of the region bounded by the graphs of $f(x) = x^3 - 3x + 2$ and $g(x) = x + 2$.

Solution: The graphs of $f(x)$ and $g(x)$ are shown in Figure 16-5. The region of interest consists of two portions, one for which $f(x) > g(x)$ and another for which $g(x) > f(x)$. You can solve this problem without graphing! Find where the graphs intersect:

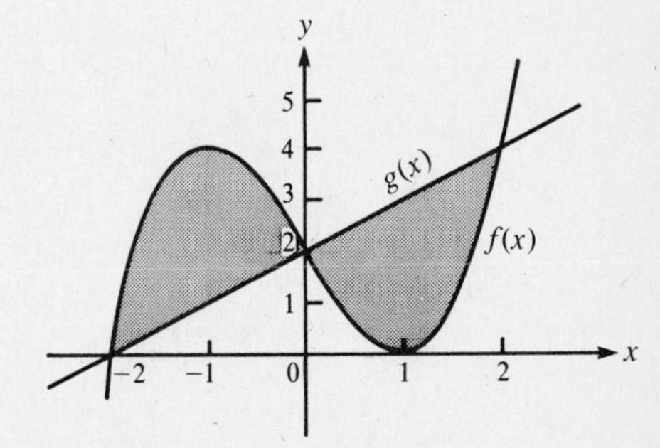

Figure 16-5
$f(x) = x^3 - 3x + 2$ and $g(x) = x + 2$.

$$x^3 - 3x + 2 = x + 2$$
$$x^3 - 4x = 0$$
$$x(x - 2)(x + 2) = 0$$

The graphs intersect at $x = -2$, $x = 0$, and $x = 2$. These values yield two intervals: $[-2, 0]$ and $[0, 2]$. On each interval determine which function is greater. On $[-2, 0]$ select $c = -1$:

$$f(-1) = 4 \qquad g(-1) = 1$$

So $f(x) \geqslant g(x)$ on $[-2, 0]$. On $[0, 2]$ select $c = 1$:

$$f(1) = 0 \qquad g(1) = 3$$

So $g(x) \geqslant f(x)$ on $[0, 2]$. The area is then the sum of the areas of the two portions:

$$\int_{-2}^0 (f(x) - g(x))\, dx + \int_0^2 (g(x) - f(x))\, dx$$
$$= \int_{-2}^0 [(x^3 - 3x + 2) - (x + 2)]\, dx + \int_0^2 [(x + 2) - (x^3 - 3x + 2)]\, dx$$
$$= \int_{-2}^0 (x^3 - 4x)\, dx + \int_0^2 (4x - x^3)\, dx$$
$$= \left(\frac{1}{4}x^4 - 2x^2\right)\Bigg|_{-2}^0 + \left(2x^2 - \frac{1}{4}x^4\right)\Bigg|_0^2$$
$$= 4 + 4 = 8$$

EXAMPLE 16-6: Find the area of the region between $x = -1$ and $x = 3$ bounded by the graph of $f(x) = x^2 - 4$ and the x axis.

Solution: You mustn't simply integrate $x^2 - 4$ between $x = -1$ and $x = 3$ because $x^2 - 4$ isn't always positive in that region. Find where $f(x)$ and the x axis intersect between $x = -1$ and $x = 3$:

$$x^2 - 4 = 0$$

$$x = \pm 2$$

So $x = 2$ is the only point of intersection in $[-1, 3]$. Between $x = -1$ and $x = 2$, the graph of $f(x)$ is below the x axis. Between $x = 2$ and $x = 3$, the graph of $f(x)$ is above the x axis. The area of the region is

$$\int_{-1}^{2} (-f(x))\, dx + \int_{2}^{3} f(x)\, dx = \int_{-1}^{2} (4 - x^2)\, dx + \int_{2}^{3} (x^2 - 4)\, dx$$

$$= \left(4x - \frac{x^3}{3}\right)\Bigg|_{-1}^{2} + \left(\frac{x^3}{3} - 4x\right)\Bigg|_{2}^{3} = \frac{34}{3}$$

16-2. Position, Velocity, and Acceleration

In Chapter 5 you saw that the derivative of the position function is the velocity function. It follows that you can recover the position function from the velocity function by antidifferentiation.

EXAMPLE 16-7: Suppose that a train passes the station at time $t = 0$ and that at time t hours the train has velocity $30 + 2t$ km/h. How far from the station is the train after $t = 2$ hours? After t hours? How far does the train travel between time $t = 1$ and $t = 2$?

Solution: If $f(t)$ is the distance from the station that the train has traveled at time t, then you know from Chapter 5 that $f'(t) = v(t)$, the velocity at time t. You are given that $f'(t) = 30 + 2t$ and you must find $f(t)$. You can antidifferentiate:

$$\int (30 + 2t)\, dt = 30t + t^2 + C$$

So $f(t) = 30t + t^2 + C$ for some constant C, but you need additional information, a so-called *initial condition*, to find C. You also know that $f(0) = 0$ because the train is at the station at time $t = 0$. This allows you to solve for C:

$$0 = f(0) = 30(0) + 0^2 + C = C$$

$$0 = C$$

So $f(t) = 30t + t^2$ is the distance the train has traveled by time t. After $t = 2$ hours, the train has traveled $f(2) = 64$ km. Between times $t = 1$ and $t = 2$, the train travels

$$f(2) - f(1) = 64 - 31 = 33 \text{ km}$$

Another interpretation is that $f(b) - f(a) = f(t)|_a^b$, and $f(t)$ is an antiderivative of $v(t)$. The Fundamental Theorem of Calculus says that the difference in positions at times $t = a$ and $t = b$ is $\int_a^b v(t)\, dt$.

EXAMPLE 16-8: A ball is thrown upward. The velocity of the ball at time t is $64 - 32t$ ft/s. How much higher is the ball after $t = 2$ seconds than after $t = 1$ second?

Solution: You can antidifferentiate the velocity function to find a position (or in this case altitude) function. Then find $f(2) - f(1)$:

$$f(2) - f(1) = \int_{1}^{2} v(t)\, dt = \int_{1}^{2} (64 - 32t)\, dt = (64t - 16t^2)\Bigg|_{1}^{2}$$

$$= 16 \text{ feet}$$

Because acceleration is the rate of change of velocity (i.e., the derivative of velocity), you can recover the velocity function from the acceleration function by antidifferentiation.

EXAMPLE 16-9: A ball is thrown upward. It experiences a constant acceleration of -32 ft/s^2. If the velocity of the ball is 50 ft/s at time $t = 0$, what is the velocity of the ball at time $t = 1$? If you further know that the altitude of the ball is 40 ft at time $t = 1$, find the altitude function.

Solution: The acceleration function is constant, $a(t) = -32$. Because $v'(t) = a(t)$, you know that $v(t)$ is an antiderivative of $a(t)$:

$$\int a(t)\,dt = \int(-32)\,dt = -32t + C$$

So $v(t) = 32t + C$ for some constant C. You can find C because you know an initial condition on v: $v(0) = 50$. So,

$$50 = v(0) = -32(0) + C = C$$
$$50 = C$$
$$v(t) = -32t + 50$$

So the velocity of the ball at $t = 1$ is $v(1) = 18$ ft/s. Find the position function by antidifferentiating velocity:

$$\int v(t)\,dt = \int(-32t + 50)\,dt = -16t^2 + 50t + C$$

So $f(t) = -16t^2 + 50t + C$ for some constant C. You have an initial condition on f that allows you to solve for this constant C: $f(1) = 40$. So,

$$40 = f(1) = -16(1)^2 + 50(1) + C = 34 + C$$
$$6 = C$$
$$f(t) = -16t^2 + 50t + 6$$

16-3. Rates of Change

If you are given the rate of change of a function, you can antidifferentiate to recover the function. You will need an initial condition to solve for the constant of antidifferentiation.

EXAMPLE 16-10: Air is being pumped into a balloon at the rate of $(4/3)t^{-2/3}$ in.^{3}/s at time t. If the balloon is empty at time $t = 0$, find $f(t)$, the amount of air in the balloon at time t.

Solution: The rate at which air is being pumped into the balloon is the rate of change of the amount of air in the balloon:

$$f'(t) = \frac{4}{3}t^{-2/3}$$

You find f by antidifferentiating f':

$$\int f'(t)\,dt = \int \frac{4}{3}t^{-2/3}\,dt = 4t^{1/3} + C$$

So $f(t) = 4t^{1/3} + C$ for some constant C. Your initial condition is that the balloon is empty at time $t = 0$, that is, $f(0) = 0$. So,

$$0 = f(0) = 4(0)^{1/3} + C = C$$
$$0 = C$$
$$f(t) = 4t^{1/3}$$

If you are given the rate of change of a function f, but no initial condition on f, you can still find the difference between two values of f, $f(b) - f(a)$:

$$f(b) - f(a) = \int_a^b f'(t)\, dt$$

EXAMPLE 16-11: The population of a city is growing at a rate of $25\,000t^{-1/2}$ citizens/year at time t. How much of an increase in population does the city experience between $t = 1$ year and $t = 4$ years?

Solution: Let $P(t)$ be the population at time t. You are given that $P'(t) = 25\,000t^{-1/2}$ and asked to find $P(4) - P(1)$:

$$P(4) - P(1) = \int_1^4 P'(t)\, dt = \int_1^4 25\,000t^{-1/2}\, dt$$

$$= 50\,000t^{1/2}\Big|_1^4 = 50\,000$$

SUMMARY

1. The area between two curves is an integral of the larger function minus the smaller function.
2. You can find the area between two continuous curves without graphing them. Find their points of intersection and determine which curve is above the other between successive points of intersection.
3. You can find the position function by antidifferentiating the velocity function. You'll need an initial condition to solve for the constant of integration.
4. You can find the velocity function by antidifferentiating the acceleration function and then using an initial condition.
5. To find a change in position between times a and b, integrate the velocity between a and b.
6. You can find a function by first antidifferentiating its rate of change function and then using the initial condition.

SOLVED PROBLEMS

PROBLEM 16-1 Find the area of the region between $x = 1$ and $x = 4$ bounded by the graphs of $f(x) = x^3 + 11$ and $g(x) = x^3 + 5$.

Solution: Because $f(x) \geqslant g(x)$, the area is

$$\int_1^4 (f(x) - g(x))\, dx = \int_1^4 [(x^3 + 11) - (x^3 + 5)]\, dx = \int_1^4 6\, dx$$

$$= 6x\Big|_1^4 = 24 - 6 = 18$$

[See Section 16-1.]

PROBLEM 16-2 Find the area of the region between $x = -1$ and $x = 1$ bounded by the graph of $f(x) = 3x^2 + 4$ and the x axis.

Solution: Because $f(x)$ is always positive, simply integrate f between $x = -1$ and $x = 1$:

$$\int_{-1}^1 f(x)\, dx = \int_{-1}^1 (3x^2 + 4)\, dx = (x^3 + 4x)\Big|_{-1}^1 = 5 - (-5) = 10$$

[See Section 16-1.]

PROBLEM 16-3 Find the area of the region bounded by the graphs of $f(x) = 3x^2 + 2x - 4$ and $g(x) = 2x^2 + 2x - 3$.

Solution: You can find the area without graphing f and g. Find where f and g intersect:

$$3x^2 + 2x - 4 = 2x^2 + 2x - 3$$

$$x^2 - 1 = 0$$

$$x = \pm 1$$

On the interval $[-1, 1]$, which function is larger? Choose a point c in $[-1, 1]$, say $c = 0$:

$$f(0) = -4 \qquad g(0) = -3$$

So $g(x) \geqslant f(x)$ on $[-1, 1]$. The area of the region of interest is

$$\int_{-1}^{1} (g(x) - f(x))\,dx = \int_{-1}^{1} [(2x^2 + 2x - 3) - (3x^2 + 2x - 4)]\,dx$$

$$= \int_{-1}^{1} (1 - x^2)\,dx = \left(x - \frac{x^3}{3}\right)\Bigg|_{-1}^{1} = \frac{4}{3}$$

[See Section 16-1.]

PROBLEM 16-4 Find the area of the region bounded by the curves $y = x^3 + x^2 - 2x + 1$ and $y = x^3 + 3x - 3$.

Solution: First find where the curves intersect:

$$x^3 + x^2 - 2x + 1 = x^3 + 3x - 3$$

$$x^2 - 5x + 4 = 0$$

$$(x - 1)(x - 4) = 0$$

$$x = 1 \quad \text{or} \quad x = 4$$

Now determine which function is greater on $[1, 4]$. Evaluate both at $c = 2$:

$$2^3 + 2^2 - 2(2) + 1 = 9$$

$$2^3 + 3(2) - 3 = 11$$

The second curve is greater on this interval. The area is

$$\int_{1}^{4} [(x^3 + 3x - 3) - (x^3 + x^2 - 2x + 1)]\,dx = \int_{1}^{4} (5x - x^2 - 4)\,dx$$

$$= \left(\frac{5}{2}x^2 - \frac{x^3}{3} - 4x\right)\Bigg|_{1}^{4} = \frac{9}{2}$$

[See Section 16-1.]

PROBLEM 16-5 Find the area of the region bounded by the graphs of $f(x) = x^4 + 2x^2 - 3x + 2$ and $g(x) = x^4 - x^3 + 2x^2 + 6x + 2$.

Solution: The curves meet where

$$x^4 + 2x^2 - 3x + 2 = x^4 - x^3 + 2x^2 + 6x + 2$$

$$x^3 - 9x = 0$$

$$x(x + 3)(x - 3) = 0$$

So you have two intervals: $[-3, 0]$ and $[0, 3]$. Evaluate f and g at $c = -1$:

$$f(-1) = 8 \qquad g(-1) = 0$$

So $f(x) \geqslant g(x)$ on $[-3, 0]$. Evaluate f and g at $c = 1$:

$$f(1) = 2 \qquad g(1) = 10$$

So $f(x) \leqslant g(x)$ on $[0, 3]$. The area of the region of interest is

$$\int_{-3}^{0} (f(x) - g(x))\,dx + \int_{0}^{3} (g(x) - f(x))\,dx$$

$$= \int_{-3}^{0} [(x^4 + 2x^2 - 3x + 2) - (x^4 - x^3 + 2x^2 + 6x + 2)]\,dx$$

$$+ \int_{0}^{3} [(x^4 - x^3 + 2x^2 + 6x + 2) - (x^4 + 2x^2 - 3x + 2)]\,dx$$

$$= \int_{-3}^{0} (x^3 - 9x)\,dx + \int_{0}^{3} (9x - x^3)\,dx = \left(\frac{1}{4}x^4 - \frac{9}{2}x^2\right)\Bigg|_{-3}^{0} + \left(\frac{9}{2}x^2 - \frac{1}{4}x^4\right)\Bigg|_{0}^{3}$$

$$= \frac{81}{2}$$

[See Section 16-1.]

PROBLEM 16-6 Find the area of the region bounded by the graphs of $y = x^3$ and $y = 3x^2$.

Solution: The curves intersect when

$$x^3 = 3x^2$$

$$x^2(x - 3) = 0$$

$$x = 0 \quad \text{or} \quad x = 3$$

The interval of integration is $[0, 3]$. Evaluate both functions at $c = 1$:

$$1^3 < 3(1)^2$$

So the area is

$$\int_{0}^{3} (3x^2 - x^3)\,dx = \left(x^3 - \frac{1}{4}x^4\right)\Bigg|_{0}^{3} = 27 - \frac{81}{4} = \frac{27}{4}$$

[See Section 16-1.]

PROBLEM 16-7 Find the area of the region bounded by the curves $y = x^2$, $y = 1$, and $x = 2$.

Solution: The curves $y = x^2$ and $y = 1$ meet at $x = 1$ and $x = -1$. You also have a bound of $x = 2$, so you have two intervals: $[-1, 1]$ and $[1, 2]$. On $[-1, 1]$, $x^2 \leqslant 1$; on $[1, 2]$, $x^2 \geqslant 1$. So the area is

$$\int_{-1}^{1} (1 - x^2)\,dx + \int_{1}^{2} (x^2 - 1)\,dx = \left(x - \frac{x^3}{3}\right)\Bigg|_{-1}^{1} + \left(\frac{x^3}{3} - x\right)\Bigg|_{1}^{2} = \frac{8}{3}$$

[See Section 16-1.]

PROBLEM 16-8 Find the area of the region bounded by the graphs of $f(x) = x^4 + x^2 - 3$ and $g(x) = 2x^2 - 3$.

Solution: Find where the curves intersect:

$$x^4 + x^2 - 3 = 2x^2 - 3$$

$$x^4 - x^2 = 0$$

$$x^2(x^2 - 1) = 0$$

$$x = 0, -1, 1$$

You have two intervals: $[-1, 0]$ and $[0, 1]$. Evaluate f and g at a point in each interval:

$$f\left(-\frac{1}{2}\right) = \frac{-43}{16} < \frac{-5}{2} = g\left(-\frac{1}{2}\right)$$

$$f\left(\frac{1}{2}\right) = \frac{-43}{16} < \frac{-5}{2} = g\left(\frac{1}{2}\right)$$

So $g(x) \geqslant f(x)$ on both intervals. Thus you can find the area using one integral:

$$\int_{-1}^{0} (g(x) - f(x))\,dx + \int_{0}^{1} (g(x) - f(x))\,dx = \int_{-1}^{1} (g(x) - f(x))\,dx$$

$$= \int_{-1}^{1} [(2x^2 - 3) - (x^4 + x^2 - 3)]\,dx$$

$$= \int_{-1}^{1} (x^2 - x^4)\,dx$$

$$= \left(\frac{x^3}{3} - \frac{x^5}{5}\right)\Bigg|_{-1}^{1} = \frac{4}{15}$$

[See Section 16-1.]

PROBLEM 16-9 Find the area of the region bounded by the graphs of $f(x) = x^4 - x^3 + x^2 + 1$ and $g(x) = 2x^4 - x^3 + 2x^2 + 1$ and the vertical line $x = 2$.

Solution: Find where the graphs of f and g intersect:

$$x^4 - x^3 + x^2 + 1 = 2x^4 - x^3 + 2x^2 + 1$$

$$-x^4 - x^2 = 0$$

$$-x^2(x^2 + 1) = 0$$

$$x = 0$$

So they intersect only at $x = 0$. Thus the interval of interest is $[0, 2]$. Evaluate at $c = 1$:

$$f(1) = 2 < 4 = g(1)$$

so $g(x) \geqslant f(x)$ on $[0, 2]$. The area is

$$\int_{0}^{2} [(2x^4 - x^3 + 2x^2 + 1) - (x^4 - x^3 + x^2 + 1)]\,dx = \int_{0}^{2} (x^4 + x^2)\,dx$$

$$= \left(\frac{x^5}{5} + \frac{x^3}{3}\right)\Bigg|_{0}^{2} = \frac{136}{15}$$

[See Section 16-1.]

PROBLEM 16-10 Find the area of the region bounded by the curves $x = (y - 1)^2$ and $x = 1$.

Solution: The curve $x = (y - 1)^2$ doesn't yield y as a function of x, as you can see in Figure 16-6. If you solve for x, you find

$$\pm\sqrt{x} = (y - 1)$$

$$y = 1 \pm \sqrt{x}$$

Between $x = 0$ and $x = 1$, the region of interest is bounded by $y = 1 + \sqrt{x}$ on the top and $y = 1 - \sqrt{x}$ on the bottom. The area is

$$\int_{0}^{1} [(1 + \sqrt{x}) - (1 - \sqrt{x})]\,dx$$

$$= \int_{0}^{1} 2\sqrt{x}\,dx = \frac{4}{3}x^{3/2}\Bigg|_{0}^{1} = \frac{4}{3}$$

[See Section 16-1.]

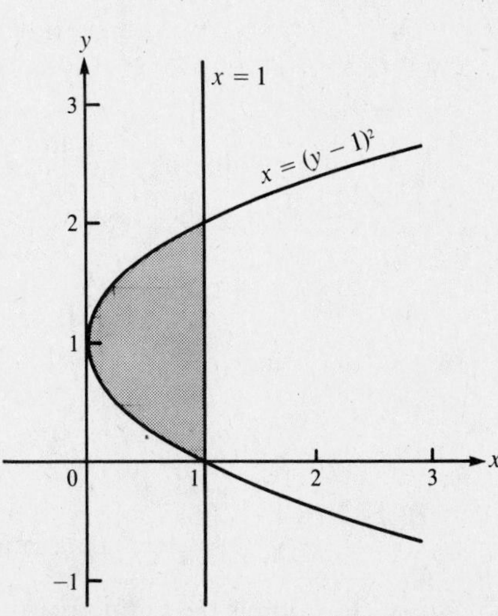

Figure 16-6
$x = (y - 1)^2$ and $x = 1$.

PROBLEM 16-11 Find the area of the region bounded by the curves $x + 1 = y^2$ and $y = 1 - x$.

Solution: The graphs of these curves are shown in Figure 16-7. You must integrate over two intervals: $[-1, 0]$ and $[0, 3]$. On $[-1, 0]$ the region is bounded above by $y = \sqrt{x+1}$ and below by $y = -\sqrt{x+1}$. On $[0, 3]$ the region is bounded above by $y = 1 - x$ and below by $y = -\sqrt{x+1}$. The area is

$$\int_{-1}^{0} [(\sqrt{x+1}) - (-\sqrt{x+1})]\, dx + \int_{0}^{3} [(1 - x) - (-\sqrt{x+1})]\, dx$$

$$= \int_{-1}^{0} 2\sqrt{x+1}\, dx + \int_{0}^{3} (1 - x + \sqrt{x+1})\, dx$$

$$= \frac{4}{3}(x+1)^{3/2}\Big|_{-1}^{0} + \left(x - \frac{1}{2}x^2 + \frac{2}{3}(x+1)^{3/2}\right)\Big|_{0}^{3}$$

$$= \frac{9}{2}$$

[See Section 16-1.]

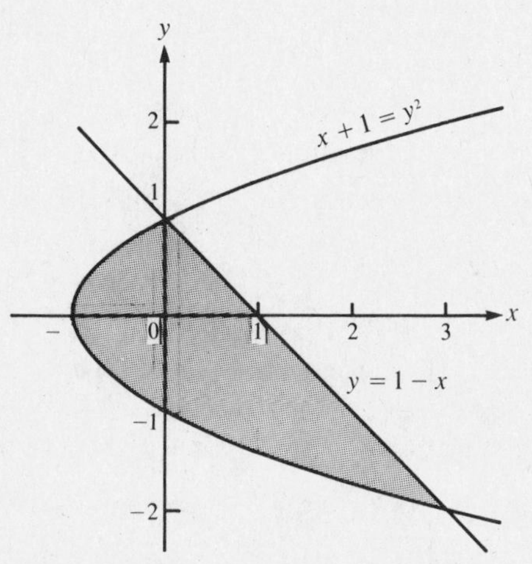

Figure 16-7
$x + 1 = y^2$ and $y = 1 - x$.

PROBLEM 16-12 A point moves along the x axis. At time t its velocity is $3t^2 - 16t + 5$. If the point is at the origin at time $t = 1$, find the position at time t.

Solution: If $P(t)$ is the position at time t, then $P'(t)$ is the velocity function: $P'(t) = 3t^2 - 16t + 5$. To find P, antidifferentiate P':

$$\int P'(t)\, dt = \int (3t^2 - 16t + 5)\, dt = t^3 - 8t^2 + 5t + C$$

So $P(t) = t^3 - 8t^2 + 5t + C$ for some constant C. You have an initial condition, $P(1) = 0$. This allows you to solve for C:

$$0 = P(1) = 1^3 - 8(1)^2 + 5(1) + C = -2 + C$$

$$2 = C$$

$$P(t) = t^3 - 8t^2 + 5t + 2$$

[See Section 16-2.]

PROBLEM 16-13 The velocity of a ball t seconds after being thrown is $80 - 32t$ ft/s. What is the position of the ball at time $t = 2$ if its position is 10 at time $t = 0$?

Solution: Position is the antiderivative of velocity:

$$\int (80 - 32t)\, dt = 80t - 16t^2 + C$$

So $P(t) = 80t - 16t^2 + C$ for some constant C. Solve for C:

$$10 = P(0) = 80(0) - 16(0)^2 + C = C$$

$$P(t) = 80t - 16t^2 + 10$$

$$P(2) = 80(2) - 16(2)^2 + 10 = 106 \text{ feet}$$

[See Section 16-2.]

PROBLEM 16-14 A stone is falling. Its velocity at time t is $-32t$. How far does the stone fall between $t = 1$ and $t = 3$?

Solution: If $f(t)$ is the altitude at time t, then $f'(t) = -32t$. You must find $f(3) - f(1)$:

$$f(3) - f(1) = \int_1^3 (-32t)\,dt = (-16t^2)\Big|_1^3 = -128 \text{ feet}$$

So at time $t = 3$, the stone is 128 feet lower (hence the minus sign) than at time $t = 1$.

[See Section 16-2.]

PROBLEM 16-15 A car leaves an intersection. Its acceleration at time t is $22\sqrt{t}$ ft/s^2. If the car has velocity 0 at time $t = 0$, how far does it travel between times $t = 0$ and $t = 4$?

Solution: You can find the velocity function by first antidifferentiating the acceleration function and then solving for C using the initial condition:

$$\int 22\sqrt{t}\,dt = \frac{44}{3}t^{3/2} + C$$

$$v(t) = \frac{44}{3}t^{3/2} + C$$

$$0 = v(0) = \frac{44}{3}(0)^{3/2} + C = C$$

$$v(t) = \frac{44}{3}t^{3/2}$$

The distance traveled between $t = 0$ and $t = 4$ is

$$\int_0^4 v(t)\,dt = \int_0^4 \frac{44}{3}t^{3/2}\,dt = \frac{88}{15}t^{5/2}\Big|_0^4 = \frac{2816}{15} \text{ feet}$$

[See Section 16-2.]

PROBLEM 16-16 A brushfire is spreading at a rate of $80 - 40t$ acres/h. The fire started at time $t = 0$. How much acreage has burned by time t?

Solution: If $f(t)$ is the acreage that has burned by time t, then $f'(t)$ is the rate of spread of the fire: $f'(t) = 80 - 40t$. You can recover f by antidifferentiating f':

$$\int f'(t)\,dt = \int (80 - 40t)\,dt = 80t - 20t^2 + C$$

So $f(t) = 80t - 20t^2 + C$ for some constant C. Because the fire started at $t = 0$, you have $f(0) = 0$. Use this to solve for C:

$$0 = f(0) = 80(0) - 20(0)^2 + C = C$$

$$f(t) = 80t - 20t^2 \text{ acres}$$

[See Section 16-3.]

PROBLEM 16-17 Water is pouring out of a faucet so that t minutes after the faucet is opened, the water flows at a rate of $6t + 5$ gallons/min. How much water leaves the faucet during the first three minutes?

Solution: If $f(t)$ is the amount of water that has left the faucet by time t, then you must find $f(3) - f(0)$. You are given that $f'(t) = 6t + 5$, so

$$f(3) - f(0) = \int_0^3 f'(t)\,dt = \int_0^3 (6t + 5)\,dt = (3t^2 + 5t)\Big|_0^3 = 42 \text{ gallons}$$

[See Section 16-3.]

PROBLEM 16-18 The area occupied by a bacterial culture on a nutrient solution is growing at a rate of $0.15\, e^{(1/2)t}$ cm^2/day at time t days. How much greater in area is the culture after $t = 4$ days than after $t = 2$ days?

Solution: If $f(t)$ is the area of the culture at time t, you must find $f(4) - f(2)$:

$$f(4) - f(2) = \int_2^4 f'(t)\, dt = \int_2^4 0.15\, e^{(1/2)t}\, dt$$

$$= 0.3\, e^{(1/2)t}\Big|_2^4 = 0.3(e^2 - e) \text{ cm}^2$$

[See Section 16-3.]

Supplementary Exercises

In Problems 16-19 through 16-45 find the area of the region bounded by the graphs of the given curves:

16-19 $f(x) = x^2$, the x axis, $x = 1$, $x = 3$

16-20 $f(x) = 4 - x^2$, the x axis

16-21 $f(x) = -8 + 2x^2$, the x axis

16-22 $f(x) = x^4 + 4$, $g(x) = x^2$, $x = -1$, $x = 1$

16-23 $y = 1 - x^3$, $y = 1 + x^3$, $x = 0$, $x = 2$

16-24 $y = x^2$, $y = x + 2$

16-25 $y = x^3$, $y = x^2$

16-26 $f(x) = x^3$, $g(x) = x$

16-27 $f(x) = \sqrt{x}$, $g(x) = x^2$

16-28 $f(x) = x^2 - 4x + 4$, the x axis, the y axis

16-29 $f(x) = x^3 + 2x - 2$, $g(x) = x^3 + x^2 + 2x - 3$

16-30 $f(x) = e^x$, $g(x) = e$, the y axis

16-31 $f(x) = \sqrt{x}$, $g(x) = \sqrt[3]{x}$

16-32 $y = (x + 1)^3$, $y = x + 1$

16-33 $f(x) = x^3 - 4x$, the x axis

16-34 $f(x) = x^3 + 3\sqrt{x} - 8x + 2$, $g(x) = 3\sqrt{x} + x + 2$

16-35 $y = x^4 + 2x^3 - 3x^2 + x - 1$, $y = x^4 + x^3 + 3x^2 - 7x - 1$

16-36 $y = \ln x$, $y = 1$, $x = 1$

16-37 $y = x^4$, $y = x^2$

16-38 $y = x^3 + x$, $y = 0$, $x = -1$, $x = 1$

16-39 $y = 2x + 3, y = -x + 6, x = 3$

16-40 $y = x(x^2 - 1)(x - 2) + 1, y = 1$

16-41 $y = 3 - x^2, y = 1 - x, x = -1, x = 1$

16-42 $y = x^2, y = 2 - x$

16-43 $x = y^2, x = 1$

16-44 $x = (y - 2)^2, y = 4 - x$

16-45 $x + 1 = (y - 1)^2, x = 0$

16-46 A stone is thrown upward. Its velocity is $25 - 9.8t$ meters/s, t seconds after it is thrown. If its altitude is 2 meters at time $t = 0$, what is its altitude at time t?

16-47 A train travels at $80t/\sqrt{t^2 + 1}$ mi/h, t hours after leaving the depot. How far does it travel in the first 2 hours after leaving the depot?

16-48 The velocity of an airplane increases steadily by 3 mi/min every minute. The plane starts with velocity $= 0$. How far does the plane travel during the first 5 minutes?

16-49 A ball is thrown upward with velocity 100 ft/s. It experiences constant acceleration due to gravity of -32 ft/s^2. Find the equation of motion for the ball (the altitude above initial position at time t).

16-50 Water flows down a river at the rate of $9 + t^{3/2}$ million ft^3/day, t days after a rain. How much water will flow past a given point during the first 4 days after a rain?

16-51 A helium balloon is inflated so that after t hours of being filled, the helium is entering at the rate $3t(t^2 + 9)^{1/2}$ yds^3/h. How much helium enters the balloon during the first 4 hours?

16-52 A radioactive substance disintegrates at the rate $\frac{1}{2} e^{-t/30}$ grams/year at time t. How much disintegrates between times $t = 0$ and $t = 60$ years?

Solutions to Supplementary Exercises

(16-19) 26/3

(16-20) 32/3

(16-21) 64/3

(16-22) 116/15

(16-23) 8

(16-24) 9/2

(16-25) 1/12

(16-26) $\frac{1}{2}$

(16-27) 1/3

(16-28) 8/3

(16-29) 4/3

(16-30) 1

(16-31) 1/12

(16-32) $\frac{1}{2}$

(16-33) 8

(16-34) 81/2

(16-35) 8

(16-36) $e - 2$

(16-37) 4/15

(16-38) 3/2

(16-39) 6

(16-40) 49/30

(16-41) 10/3

(16-42) 9/2

(16-43) 4/3

(16-44) 9/2

(16-45) 4/3

(16-46) $25t - 4.9t^2 + 2$ meters

(16-47) $80(\sqrt{5} - 1)$ miles

(16-48) 75/2 miles

(16-49) $-16t^2 + 100t$ ft

(16-50) 48.8 million ft^3

(16-51) 98 yds^3

(16-52) $15(1 - e^{-2})$ grams

17 IMPROPER INTEGRALS

THIS CHAPTER IS ABOUT

☑ Infinite Intervals of Integration
☑ Vertical Asymptotes of the Integrand

You can extend the concept of the integral to include two more possibilities called **improper integrals**: $\int_a^b f(x)\,dx$, where $a = -\infty$ or $b = \infty$, and $\int_a^b f(x)\,dx$, where f has a vertical asymptote in the interval $[a, b]$.

17-1. Infinite Intervals of Integration

An integral over an infinite interval is defined as a limit of integrals over finite intervals. The integral $\int_a^\infty f(x)\,dx$ is defined to be

$$\lim_{N\to\infty} \int_a^N f(x)\,dx \tag{17-1}$$

The integral $\int_{-\infty}^b f(x)\,dx$ is defined to be

$$\lim_{N\to\infty} \int_{-N}^b f(x)\,dx \quad \text{or} \quad \lim_{M\to-\infty} \int_M^b f(x)\,dx \tag{17-2}$$

If the limit in an improper integral doesn't exist (or is infinite) you say that the integral **diverges**. If the limit is a finite number L, the integral is said to **converge** to L.

EXAMPLE 17-1: Evaluate $\displaystyle\int_1^\infty 1/x^2\,dx$.

Solution: You must evaluate this integral by calculating a limit. According to the definition

$$\int_1^\infty \frac{1}{x^2}\,dx = \lim_{N\to\infty} \int_1^N \frac{1}{x^2}\,dx$$

Now find $\int_1^N 1/x^2\,dx$ and then take the limit as N approaches infinity:

$$\lim_{N\to\infty} \int_1^N \frac{1}{x^2}\,dx = \lim_{N\to\infty} \left(-\frac{1}{x}\bigg|_1^N\right) = \lim_{N\to\infty} \left(-\frac{1}{N} + 1\right) = 1$$

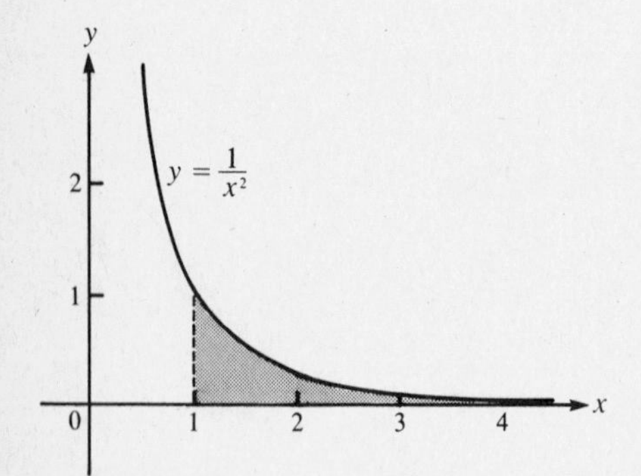

Figure 17-1

So this integral converges (to 1). You can interpret this as meaning that the area of the region bounded by the graph of $y = 1/x^2$ and the x axis to the right of $x = 1$ (and extending infinitely to the right) is 1 (see Figure 17-1).

EXAMPLE 17-2: Evaluate $\displaystyle\int_1^\infty 1/x\,dx$.

Solution: Find the integral by taking a limit of proper integrals:

$$\int_1^\infty \frac{1}{x}\,dx = \lim_{N\to\infty} \int_1^N \frac{1}{x}\,dx = \lim_{N\to\infty} \left(\ln|x|\bigg|_1^N\right) = \lim_{N\to\infty} (\ln|N| - \ln 1)$$

$$= \lim_{N\to\infty} (\ln|N|) = \infty$$

So this integral diverges.

EXAMPLE 17-3: Evaluate $\int_{-\infty}^{1} e^x\,dx$.

Solution: You must interpret this integral as a limit:

$$\int_{-\infty}^{1} e^x\,dx = \lim_{N\to\infty} \int_{-N}^{1} e^x\,dx = \lim_{N\to\infty} \left(e^x \Big|_{-N}^{1} \right) = \lim_{N\to\infty} (e - e^{-N}) = e$$

This integral converges (to e).

If the interval of integration is $(-\infty, \infty)$, then you must set up two limits:

$$\begin{aligned} \int_{-\infty}^{\infty} f(x)\,dx &= \int_{-\infty}^{0} f(x)\,dx + \int_{0}^{\infty} f(x)\,dx \\ &= \lim_{N\to\infty} \int_{-N}^{0} f(x)\,dx + \lim_{M\to\infty} \int_{0}^{M} f(x)\,dx \end{aligned} \tag{17-3}$$

EXAMPLE 17-4: Evaluate $\int_{-\infty}^{\infty} 1/(x^2 + 1)\,dx$.

Solution: Evaluate this integral using a pair of limits:

$$\begin{aligned} \int_{-\infty}^{\infty} \frac{dx}{x^2+1} &= \int_{-\infty}^{0} \frac{dx}{x^2+1} + \int_{0}^{\infty} \frac{dx}{x^2+1} \\ &= \lim_{N\to\infty} \int_{-N}^{0} \frac{dx}{x^2+1} + \lim_{M\to\infty} \int_{0}^{M} \frac{dx}{x^2+1} \\ &= \lim_{N\to\infty} \left(\arctan x \Big|_{-N}^{0} \right) + \lim_{M\to\infty} \left(\arctan x \Big|_{0}^{M} \right) \\ &= \lim_{N\to\infty} (\arctan 0 - \arctan(-N)) + \lim_{M\to\infty} (\arctan M - \arctan 0) \\ &= \lim_{N\to\infty} (-\arctan(-N)) + \lim_{M\to\infty} (\arctan M) \end{aligned}$$

To find $\lim_{M\to\infty} (\arctan M)$, consider the angles between $-\pi/2$ and $\pi/2$ that have very large tangent values. As x approaches $\pi/2$ from below, $\tan x$ approaches infinity. Thus $\lim_{M\to\infty} (\arctan M) = \pi/2$, and similarly you find $\lim_{N\to\infty} (-\arctan(-N)) = \pi/2$. You have

$$\int_{-\infty}^{\infty} \frac{dx}{x^2+1} = \lim_{N\to\infty} (-\arctan(-N)) + \lim_{M\to\infty} (\arctan M) = \frac{\pi}{2} + \frac{\pi}{2} = \pi$$

EXAMPLE 17-5: Evaluate $\int_{-\infty}^{\infty} x/(x^2 + 1)\,dx$.

Solution:

$$\begin{aligned} \int_{-\infty}^{\infty} \frac{x\,dx}{x^2+1} &= \int_{-\infty}^{0} \frac{x\,dx}{x^2+1} + \int_{0}^{\infty} \frac{x\,dx}{x^2+1} \\ &= \lim_{N\to\infty} \int_{-N}^{0} \frac{x\,dx}{x^2+1} + \lim_{M\to\infty} \int_{0}^{M} \frac{x\,dx}{x^2+1} \\ &= \lim_{N\to\infty} \left[\frac{1}{2}\ln(x^2+1) \Big|_{-N}^{0} \right] + \lim_{M\to\infty} \left[\frac{1}{2}\ln(x^2+1) \Big|_{0}^{M} \right] \\ &= \lim_{N\to\infty} \left[\frac{1}{2}\ln 1 - \frac{1}{2}\ln(N^2+1) \right] + \lim_{M\to\infty} \left[\frac{1}{2}\ln(M^2+1) - \frac{1}{2}\ln 1 \right] \\ &= \lim_{N\to\infty} \left[-\frac{1}{2}\ln(N^2+1) \right] + \lim_{M\to\infty} \left[\frac{1}{2}\ln(M^2+1) \right] = -\infty + \infty \end{aligned}$$

This integral is divergent. If any of the limits is infinite, the integral diverges.

Caution: You *shouldn't* evaluate $\int_{-\infty}^{\infty} f(x)\,dx$ using a single limit:

$$\lim_{N\to\infty}\int_{-N}^{N} f(x)\,dx.$$

The integral may diverge, and yet the quantities that make the two integrals infinite may cancel out, giving you the wrong answer!

17-2. Vertical Asymptotes of the Integrand

Another type of improper integral is $\int_a^b f(x)\,dx$, where $f(x)$ has a vertical asymptote at $x = c\ (a \leqslant c \leqslant b)$. If $f(x)$ has a vertical asymptote at $x = b$, then $\int_a^b f(x)\,dx$ is defined to be

$$\lim_{\varepsilon\to 0^+}\int_a^{b-\varepsilon} f(x)\,dx \qquad \textbf{(17-4)}$$

a limit of proper integrals. If $f(x)$ has a vertical asymptote at $x = a$, then $\int_a^b f(x)\,dx$ is defined to be

$$\lim_{\varepsilon\to 0^+}\int_{a+\varepsilon}^{b} f(x)\,dx \qquad \textbf{(17-5)}$$

If $f(x)$ has a vertical asymptote at $x = c$, where $a < c < b$, then $\int_a^b f(x)\,dx$ is defined to be

$$\int_a^c f(x)\,dx + \int_c^b f(x)\,dx = \lim_{\varepsilon\to 0^+}\int_a^{c-\varepsilon} f(x)\,dx + \lim_{\delta\to 0^+}\int_{c+\delta}^{b} f(x)\,dx \qquad \textbf{(17-6)}$$

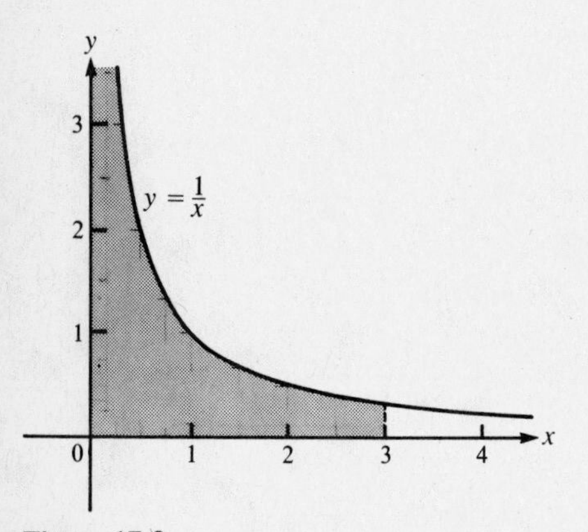

Figure 17-2

EXAMPLE 17-6: Evaluate $\displaystyle\int_0^3 1/x\,dx$.

Solution: The integrand, $f(x) = 1/x$, has a vertical asymptote at $x = 0$, the lower limit of integration:

$$\int_0^3 \frac{1}{x}\,dx = \lim_{\varepsilon\to 0^+}\int_\varepsilon^3 \frac{1}{x}\,dx = \lim_{\varepsilon\to 0^+}\left(\ln|x|\Big|_\varepsilon^3\right) = \lim_{\varepsilon\to 0^+}(\ln 3 - \ln\varepsilon) = \infty$$

This integral diverges. You can interpret this as saying that the region between $x = 0$ and $x = 3$ bounded by the graph of $y = 1/x$ and the x axis is infinite in area (see Figure 17-2).

EXAMPLE 17-7: Evaluate $\displaystyle\int_0^4 1/\sqrt{4-x}\,dx$.

Solution: The integrand has a vertical asymptote at $x = 4$, the upper limit of integration:

$$\int_0^4 \frac{dx}{\sqrt{4-x}} = \lim_{\varepsilon\to 0^+}\int_0^{4-\varepsilon}\frac{dx}{\sqrt{4-x}} = \lim_{\varepsilon\to 0^+}\left[-2(4-x)^{1/2}\Big|_0^{4-\varepsilon}\right]$$

$$= \lim_{\varepsilon\to 0^+}(-2\varepsilon^{1/2} + 4) = 4$$

This integral converges (to 4). The area of the region between the graphs of $y = 1/\sqrt{4-x}$ and the x axis and between $x = 0$ and $x = 4$ is 4.

EXAMPLE 17-8: Evaluate $\displaystyle\int_{-1}^1 1/x^2\,dx$.

Solution: The integrand has a vertical asymptote at $x = 0$, which is between the limits of integration:

$$\int_{-1}^1 \frac{dx}{x^2} = \int_{-1}^0 \frac{dx}{x^2} + \int_0^1 \frac{dx}{x^2} = \lim_{\varepsilon\to 0^+}\int_{-1}^{-\varepsilon}\frac{dx}{x^2} + \lim_{\delta\to 0^+}\int_\delta^1 \frac{dx}{x^2}$$

$$= \lim_{\varepsilon\to 0^+}\left(\frac{-1}{x}\Big|_{-1}^{-\varepsilon}\right) + \lim_{\delta\to 0^+}\left(\frac{-1}{x}\Big|_\delta^1\right) = \lim_{\varepsilon\to 0^+}\left(\frac{1}{\varepsilon} - 1\right) + \lim_{\delta\to 0^+}\left(-1 + \frac{1}{\delta}\right)$$

$$= \infty + \infty$$

This integral diverges.

If you had failed to notice that this is an improper integral, you might have evaluated this integral as follows:

$$\left(\frac{-1}{x}\right)\Bigg|_{-1}^{1} = -1 - (+1) = -2$$

This is certainly incorrect because the integrand is always positive. The area of the region bounded by the graph of $y = 1/x^2$ and the x axis and between $x = -1$ and $x = 1$ is infinite.

If the integrand has more than one vertical asymptote in the interval of integration, you'll have several integrals to evaluate.

EXAMPLE 17-9: Evaluate $\int_{-1}^{1} 1/\sqrt{1 - x^2}\, dx$.

Solution: Because the integrand has vertical asymptotes at $x = -1$ and $x = 1$, you must set up two limits. One way of doing this is

$$\int_{-1}^{1} \frac{dx}{\sqrt{1 - x^2}} = \int_{-1}^{0} \frac{dx}{\sqrt{1 - x^2}} + \int_{0}^{1} \frac{dx}{\sqrt{1 - x^2}}$$

because each of these two integrals has just one vertical asymptote in the interval of integration. Thus,

$$\int_{-1}^{1} \frac{dx}{\sqrt{1 - x^2}} = \lim_{\varepsilon \to 0^+} \int_{-1+\varepsilon}^{0} \frac{dx}{\sqrt{1 - x^2}} + \lim_{\delta \to 0^+} \int_{0}^{1-\delta} \frac{dx}{\sqrt{1 - x^2}}$$

$$= \lim_{\varepsilon \to 0^+} \left(\arcsin x \Big|_{-1+\varepsilon}^{0}\right) + \lim_{\delta \to 0^+} \left(\arcsin x \Big|_{0}^{1-\delta}\right)$$

$$= \lim_{\varepsilon \to 0^+} [0 - \arcsin(-1 + \varepsilon)] + \lim_{\delta \to 0^+} [\arcsin(1 - \delta) - 0]$$

$$= -\arcsin(-1) + \arcsin 1 = -\left(\frac{-\pi}{2}\right) + \frac{\pi}{2} = \pi$$

EXAMPLE 17-10: Evaluate $\int_{0}^{2} 1/(x \ln x)\, dx$.

Solution: The integrand, $f(x) = 1/(x \ln x)$, has vertical asymptotes at $x = 0$ and $x = 1$ ($\ln 1 = 0$). If you break up the integral,

$$\int_{0}^{2} \frac{dx}{x \ln x} = \int_{0}^{1/2} \frac{dx}{x \ln x} + \int_{1/2}^{2} \frac{dx}{x \ln x}$$

then these two integrals each have just one vertical asymptote in the interval of integration:

$$\int_{0}^{2} \frac{dx}{x \ln x} = \int_{0}^{1/2} \frac{dx}{x \ln x} + \int_{1/2}^{2} \frac{dx}{x \ln x}$$

$$= \lim_{\varepsilon \to 0^+} \int_{\varepsilon}^{1/2} \frac{dx}{x \ln x} + \lim_{\delta \to 0^+} \int_{1/2}^{1-\delta} \frac{dx}{x \ln x} + \lim_{\eta \to 0^+} \int_{1+\eta}^{2} \frac{dx}{x \ln x}$$

$$= \lim_{\varepsilon \to 0^+} \ln|\ln x| \Big|_{\varepsilon}^{1/2} + \lim_{\delta \to 0^+} \ln|\ln x| \Big|_{1/2}^{1-\delta} + \lim_{\eta \to 0^+} \ln|\ln x| \Big|_{1+\eta}^{2}$$

$$= \lim_{\varepsilon \to 0^+} \left(\ln\left|\ln \frac{1}{2}\right| - \ln|\ln \varepsilon|\right) + \lim_{\delta \to 0^+} \left[\ln|\ln(1 - \delta)| - \ln\left|\ln \frac{1}{2}\right|\right]$$

$$+ \lim_{\eta \to 0^+} [\ln|\ln 2| - \ln|\ln(1 + \eta)|]$$

$$= -\infty - \infty + \infty$$

So this integral diverges.

If you have an integral that is improper in both senses (i.e., infinite interval of integration and vertical asymptotes in that interval), then you should split the integral into integrals that are improper in just one sense.

EXAMPLE 17-11: Evaluate $\int_1^\infty 1/\sqrt{x-1}\,dx$.

Solution: If you split the integral,

$$\int_1^\infty \frac{dx}{\sqrt{x-1}} = \int_1^2 \frac{dx}{\sqrt{x-1}} + \int_2^\infty \frac{dx}{\sqrt{x-1}}$$

then each of the two resulting integrals is improper in just one sense. The first integral has a vertical asymptote at $x = 1$; the second integral has an infinite interval of integration. Evaluate these integrals by finding the appropriate limits:

$$\begin{aligned}\int_1^2 \frac{dx}{\sqrt{x-1}} + \int_2^\infty \frac{dx}{\sqrt{x-1}} &= \lim_{\varepsilon\to 0^+} \int_{1+\varepsilon}^2 \frac{dx}{\sqrt{x-1}} + \lim_{N\to\infty} \int_2^N \frac{dx}{\sqrt{x-1}} \\ &= \lim_{\varepsilon\to 0^+} \left(2\sqrt{x-1}\,\Big|_{1+\varepsilon}^{2}\right) + \lim_{N\to\infty} \left(2\sqrt{x-1}\,\Big|_{2}^{N}\right) \\ &= \lim_{\varepsilon\to 0^+} (2 - 2\sqrt{\varepsilon}) + \lim_{N\to\infty} (2\sqrt{N-1} - 2) \\ &= 2 + \infty\end{aligned}$$

This integral diverges.

SUMMARY

1. An integral over an infinite interval or an integral over an interval containing a vertical asymptote of the integrand is called an improper integral. You evaluate improper integrals by finding limits of proper integrals.
2. An integral over the interval $(-\infty, \infty)$ must be split into two integrals, and hence two limits must be calculated.
3. If you find any limits that diverge (or are infinite) when you are evaluating an improper integral, the integral diverges. Diverging limits don't cancel each other out!
4. Whenever you integrate, check for vertical asymptotes of the integrand! You may have an improper integral.
5. The rule for improper integrals with more than one vertical asymptote or with vertical asymptotes and an infinite interval of integration is as follows: Divide the interval of integration into subintervals so that the integral over each subinterval is improper for only one reason (i.e., one vertical asymptote or one infinite limit of integration).

SOLVED PROBLEMS

PROBLEM 17-1 Evaluate $\int_2^\infty 1/x^3\,dx$.

Solution: Because the interval of integration is infinite, this is an improper integral.

$$\int_2^\infty \frac{1}{x^3}\,dx = \lim_{N\to\infty}\int_2^N \frac{1}{x^3}\,dx = \lim_{N\to\infty}\left(-\frac{1}{2}x^{-2}\Big|_2^N\right)$$

$$= \lim_{N\to\infty}\left(\frac{-1}{2N^2} + \frac{1}{8}\right) = \frac{1}{8}$$

This integral converges to 1/8. [See Section 17-1.]

PROBLEM 17-2 Evaluate $\int_4^\infty 1/\sqrt{x}\,dx$.

Solution: You find that

$$\int_4^\infty \frac{1}{\sqrt{x}}\,dx = \lim_{N\to\infty}\int_4^N \frac{1}{\sqrt{x}}\,dx = \lim_{N\to\infty}\left(2x^{1/2}\Big|_4^N\right) = \lim_{N\to\infty}(2\sqrt{N} - 4) = \infty$$

This integral diverges. [See Section 17-1.]

PROBLEM 17-3 Evaluate $\int_0^\infty e^{-x}\,dx$.

Solution: You find that

$$\int_0^\infty e^{-x}\,dx = \lim_{N\to\infty}\int_0^N e^{-x}\,dx = \lim_{N\to\infty}\left(-e^{-x}\Big|_0^N\right) = \lim_{N\to\infty}(-e^{-N} + 1) = 1$$

[See Section 17-1.]

PROBLEM 17-4 Evaluate $\int_{-\infty}^{-1} 1/x^2\,dx$.

Solution: You find that

$$\int_{-\infty}^{-1} \frac{1}{x^2}\,dx = \lim_{N\to\infty}\int_{-N}^{-1} \frac{1}{x^2}\,dx = \lim_{N\to\infty}\left(\frac{-1}{x}\Big|_N^{-1}\right) = \lim_{N\to\infty}\left(1 + \frac{1}{N}\right) = 1$$

[See Section 17-1.]

PROBLEM 17-5 Evaluate $\int_{-\infty}^{-2} (\ln|x|)/x\,dx$.

Solution: You know that

$$\int_{-\infty}^{-2} \frac{\ln|x|}{x}\,dx = \lim_{N\to\infty}\int_{-N}^{-2} \frac{\ln|x|}{x}\,dx$$

To antidifferentiate $(\ln|x|)/x$, make the substitution $u = \ln|x|$:

$$= \lim_{N\to\infty}\int_{\ln|N|}^{\ln 2} u\,du = \lim_{N\to\infty}\left(\frac{u^2}{2}\Big|_{\ln|N|}^{\ln 2}\right)$$

$$= \lim_{N\to\infty}\left(\frac{\ln^2 2}{2} - \frac{\ln^2|N|}{2}\right) = -\infty$$

This integral diverges. [See Section 17-1.]

PROBLEM 17-6 Evaluate $\int_{-\infty}^\infty xe^{-x^2}\,dx$.

Solution: Split the integral in two:

$$\int_{-\infty}^\infty xe^{-x^2}\,dx = \int_{-\infty}^0 xe^{-x^2}\,dx + \int_0^\infty xe^{-x^2}\,dx$$

$$= \lim_{N\to\infty}\int_{-N}^0 xe^{-x^2}\,dx + \lim_{M\to\infty}\int_0^M xe^{-x^2}\,dx$$

Make the substitution $u = -x^2$ with $du = -2x\,dx$:

$$\int_{-\infty}^{\infty} xe^{-x^2}\,dx = \lim_{N\to\infty}\int_{-N^2}^{0} e^u\left(-\frac{1}{2}\,du\right) + \lim_{M\to\infty}\int_{0}^{-M^2} e^u\left(-\frac{1}{2}\,du\right)$$

$$= \lim_{N\to\infty}\left(-\frac{1}{2}e^u\Big|_{-N^2}^{0}\right) + \lim_{M\to\infty}\left(-\frac{1}{2}e^u\Big|_{0}^{-M^2}\right)$$

$$= \lim_{N\to\infty}\left(-\frac{1}{2} + \frac{1}{2}e^{-N^2}\right) + \lim_{M\to\infty}\left(-\frac{1}{2}e^{-M^2} + \frac{1}{2}\right)$$

$$= -\frac{1}{2} + \frac{1}{2} = 0$$

[See Section 17-1.]

PROBLEM 17-7 Evaluate $\int_{-\infty}^{\infty} (2x + 1)/(x^2 + x + 1)\,dx$.

Solution: You find that

$$\int_{-\infty}^{\infty} \frac{2x+1}{x^2+x+1}\,dx = \int_{-\infty}^{0} \frac{2x+1}{x^2+x+1}\,dx + \int_{0}^{\infty} \frac{2x+1}{x^2+x+1}\,dx$$

$$= \lim_{N\to\infty}\int_{-N}^{0} \frac{2x+1}{x^2+x+1}\,dx + \lim_{M\to\infty}\int_{0}^{M} \frac{2x+1}{x^2+x+1}\,dx$$

$$= \lim_{N\to\infty}\left(\ln|x^2+x+1|\Big|_{-N}^{0}\right) + \lim_{M\to\infty}\left(\ln|x^2+x+1|\Big|_{0}^{M}\right)$$

$$= \lim_{N\to\infty}(-\ln|N^2 - N + 1|) + \lim_{M\to\infty}(\ln|M^2 + M + 1|)$$

$$= -\infty + \infty$$

The integral diverges.

[See Section 17-1.]

PROBLEM 17-8 Evaluate $\int_0^4 1/\sqrt{x}\,dx$.

Solution: The integrand, $f(x) = 1/\sqrt{x}$, has a vertical asymptote at $x = 0$. So,

$$\int_0^4 \frac{dx}{\sqrt{x}} = \lim_{\varepsilon\to 0^+}\int_{\varepsilon}^{4} \frac{dx}{\sqrt{x}} = \lim_{\varepsilon\to 0^+}\left(2\sqrt{x}\Big|_{\varepsilon}^{4}\right) = \lim_{\varepsilon\to 0^+}(4 - 2\sqrt{\varepsilon}) = 4$$

[See Section 17-2.]

PROBLEM 17-9 Evaluate $\int_0^1 \ln x\,dx$.

Solution: The integrand has a vertical asymptote at $x = 0$. So,

$$\int_0^1 \ln x\,dx = \lim_{\varepsilon\to 0^+}\int_{\varepsilon}^{1} \ln x\,dx$$

Integrate by parts. Let $u = \ln x$, $du = 1/x\,dx$, $dv = dx$, and $v = x$:

$$\int_0^1 \ln x\,dx = \lim_{\varepsilon\to 0^+}\left(x\ln x\Big|_{\varepsilon}^{1} - \int_{\varepsilon}^{1} dx\right) = \lim_{\varepsilon\to 0^+}\left[(x\ln x - x)\Big|_{\varepsilon}^{1}\right]$$

$$= \lim_{\varepsilon\to 0^+}[-1 - \varepsilon\ln\varepsilon + \varepsilon] = -1 - \lim_{\varepsilon\to 0^+}(\varepsilon\ln\varepsilon) = -1 - \lim_{\varepsilon\to 0^+}\left(\frac{\ln\varepsilon}{1/\varepsilon}\right)$$

Now apply l'Hôpital's rule (see Chapter 21):

$$= -1 - \lim_{\varepsilon\to 0^+}\frac{(1/\varepsilon)}{(-1/\varepsilon^2)} = -1 - \lim_{\varepsilon\to 0^+}(-\varepsilon) = -1$$

[See Section 17-2.]

PROBLEM 17-10 Evaluate $\int_{-4}^{-2} 1/(x + 2)\,dx$.

Solution: There is a vertical asymptote at $x = -2$. So,

$$\int_{-4}^{-2} \frac{dx}{x+2} = \lim_{\varepsilon\to 0^+} \int_{-4}^{-2-\varepsilon} \frac{dx}{x+2} = \lim_{\varepsilon\to 0^+} \left(\ln|x+2| \Big|_{-4}^{-2-\varepsilon} \right)$$

$$\doteq \lim_{\varepsilon\to 0^+} (\ln|\varepsilon| - \ln 2) = -\infty$$

The integral diverges. [See Section 17-2.]

PROBLEM 17-11 Evaluate $\displaystyle\int_{-8}^{1} 1/\sqrt[3]{x}\, dx$.

Solution: The integrand has a vertical asymptote at $x = 0$, which is between the limits of integration, $x = -8$ and $x = 1$. So,

$$\int_{-8}^{1} \frac{dx}{\sqrt[3]{x}} = \int_{-8}^{0} \frac{dx}{\sqrt[3]{x}} + \int_{0}^{1} \frac{dx}{\sqrt[3]{x}} = \lim_{\varepsilon\to 0^+} \int_{-8}^{-\varepsilon} \frac{dx}{\sqrt[3]{x}} + \lim_{\delta\to 0^+} \int_{\delta}^{1} \frac{dx}{\sqrt[3]{x}}$$

$$= \lim_{\varepsilon\to 0^+} \left(\frac{3}{2} x^{2/3} \Big|_{-8}^{-\varepsilon} \right) + \lim_{\delta\to 0^+} \left(\frac{3}{2} x^{2/3} \Big|_{\delta}^{1} \right)$$

$$= \lim_{\varepsilon\to 0^+} \left[\frac{3}{2}(-\varepsilon)^{2/3} - 6 \right] + \lim_{\delta\to 0^+} \left[\frac{3}{2} - \frac{3}{2}(\delta)^{2/3} \right] = -6 + \frac{3}{2} = -\frac{9}{2}$$

[See Section 17-2.]

PROBLEM 17-12 Evaluate $\displaystyle\int_{0}^{\pi} \tan x\, dx$.

Solution: The integrand, $f(x) = \tan x$, has a vertical asymptote at $x = \pi/2$. So,

$$\int_{0}^{\pi} \tan x\, dx = \int_{0}^{\pi/2} \frac{\sin x}{\cos x}\, dx + \int_{\pi/2}^{\pi} \frac{\sin x}{\cos x}\, dx$$

$$= \lim_{\varepsilon\to 0^+} \int_{0}^{\pi/2-\varepsilon} \frac{\sin x}{\cos x}\, dx + \lim_{\delta\to 0^+} \int_{\pi/2+\delta}^{\pi} \frac{\sin x}{\cos x}\, dx$$

$$= \lim_{\varepsilon\to 0^+} \left(-\ln|\cos x| \Big|_{0}^{\pi/2-\varepsilon} \right) + \lim_{\delta\to 0^+} \left(-\ln|\cos x| \Big|_{\pi/2+\delta}^{\pi} \right)$$

$$= \lim_{\varepsilon\to 0^+} \left(-\ln\left|\cos\left(\frac{\pi}{2} - \varepsilon\right)\right| \right) + \lim_{\delta\to 0^+} \left(\ln\left|\cos\left(\frac{\pi}{2} + \delta\right)\right| \right) = \infty - \infty$$

The integral diverges. [See Section 17-2.]

PROBLEM 17-13 Evaluate $\displaystyle\int_{0}^{2} 1/(x(x-2))\, dx$.

Solution: There are two vertical asymptotes, $x = 0$ and $x = 2$. So,

$$\int_{0}^{2} \frac{dx}{x(x-2)} = \int_{0}^{1} \frac{dx}{x(x-2)} + \int_{1}^{2} \frac{dx}{x(x-2)}$$

$$= \lim_{\varepsilon\to 0^+} \int_{\varepsilon}^{1} \frac{dx}{x(x-2)} + \lim_{\delta\to 0^+} \int_{1}^{2-\delta} \frac{dx}{x(x-2)}$$

$$= \lim_{\varepsilon\to 0^+} \int_{\varepsilon}^{1} \frac{1}{2}\left(\frac{1}{x-2} - \frac{1}{x} \right) dx + \lim_{\delta\to 0^+} \int_{1}^{2-\delta} \frac{1}{2}\left(\frac{1}{x-2} - \frac{1}{x} \right) dx$$

$$= \lim_{\varepsilon\to 0^+} \left[\left(\frac{1}{2}\ln|x-2| - \frac{1}{2}\ln|x| \right) \Big|_{\varepsilon}^{1} \right] + \lim_{\delta\to 0^+} \left[\left(\frac{1}{2}\ln|x-2| - \frac{1}{2}\ln|x| \right) \Big|_{1}^{2-\delta} \right]$$

$$= \lim_{\varepsilon\to 0^+} \left(-\frac{1}{2}\ln|\varepsilon - 2| + \frac{1}{2}\ln|\varepsilon| \right) + \lim_{\delta\to 0^+} \left(\frac{1}{2}\ln|\delta| - \frac{1}{2}\ln|2-\delta| \right) = -\infty - \infty$$

The integral diverges. [See Section 17-2.]

PROBLEM 17-14 Evaluate $\int_0^{\pi} \csc^2 x\, dx$.

Solution: The integrand has vertical asymptotes at $x = 0$ and $x = \pi$. Split the integral into two integrals, each with just one vertical asymptote in the interval of integration:

$$\begin{aligned}\int_0^{\pi} \csc^2 x\, dx &= \int_0^{\pi/2} \csc^2 x\, dx + \int_{\pi/2}^{\pi} \csc^2 x\, dx \\ &= \lim_{\varepsilon \to 0^+} \int_{\varepsilon}^{\pi/2} \csc^2 x\, dx + \lim_{\delta \to 0^+} \int_{\pi/2}^{\pi-\delta} \csc^2 x\, dx \\ &= \lim_{\varepsilon \to 0^+} \left(-\cot x \Big|_{\varepsilon}^{\pi/2} \right) + \lim_{\delta \to 0^+} \left(-\cot x \Big|_{\pi/2}^{\pi-\delta} \right) \\ &= \lim_{\varepsilon \to 0^+} (\cot \varepsilon) + \lim_{\delta \to 0^+} (-\cot(\pi - \delta)) = \infty + \infty\end{aligned}$$

The integral diverges. [See Section 17-2.]

PROBLEM 17-15 Evaluate $\int_0^{\infty} \frac{e^{-\sqrt{x}}}{\sqrt{x}}\, dx$.

Solution: This integral is improper for two reasons: the integrand has a vertical asymptote at $x = 0$ and the interval of integration is infinite. So,

$$\begin{aligned}\int_0^{\infty} \frac{e^{-\sqrt{x}}}{\sqrt{x}}\, dx &= \int_0^{1} \frac{e^{-\sqrt{x}}}{\sqrt{x}}\, dx + \int_1^{\infty} \frac{e^{-\sqrt{x}}}{\sqrt{x}}\, dx \\ &= \lim_{\varepsilon \to 0^+} \int_{\varepsilon}^{1} \frac{e^{-\sqrt{x}}}{\sqrt{x}}\, dx + \lim_{N \to \infty} \int_1^{N} \frac{e^{-\sqrt{x}}}{\sqrt{x}}\, dx \\ &= \lim_{\varepsilon \to 0^+} \left(-2e^{-\sqrt{x}} \Big|_{\varepsilon}^{1} \right) + \lim_{N \to \infty} \left(-2e^{-\sqrt{x}} \Big|_{1}^{N} \right) \\ &= \lim_{\varepsilon \to 0^+} (-2e^{-1} + 2e^{-\sqrt{\varepsilon}}) + \lim_{N \to \infty} (-2e^{-\sqrt{N}} + 2e^{-1}) \\ &= -2e^{-1} + 2 + 2e^{-1} = 2\end{aligned}$$

[See Section 17-2.]

PROBLEM 17-16 Evaluate $\int_{-\infty}^{0} 1/x^5\, dx$.

Solution: You find that

$$\begin{aligned}\int_{-\infty}^{0} \frac{dx}{x^5} &= \int_{-\infty}^{-1} \frac{dx}{x^5} + \int_{-1}^{0} \frac{dx}{x^5} = \lim_{N \to \infty} \int_{-N}^{-1} \frac{dx}{x^5} + \lim_{\varepsilon \to 0^+} \int_{-1}^{-\varepsilon} \frac{dx}{x^5} \\ &= \lim_{N \to \infty} \left(-\frac{1}{4} x^{-4} \Big|_{-N}^{-1} \right) + \lim_{\varepsilon \to 0^+} \left(-\frac{1}{4} x^{-4} \Big|_{-1}^{-\varepsilon} \right) \\ &= \lim_{N \to \infty} \left(-\frac{1}{4} + \frac{1}{4N^4} \right) + \lim_{\varepsilon \to 0^+} \left(\frac{-1}{4\varepsilon^4} + \frac{1}{4} \right) = -\frac{1}{4} - \infty\end{aligned}$$

The integral diverges. [See Section 17-2.]

Supplementary Exercises

In Problems 17-17 through 17-50 evaluate the improper integral:

17-17 $\int_0^{\infty} 1/[(x + 1)^2]\, dx$

17-18 $\int_0^{\infty} 1/(x + 1)\, dx$

17-19 $\int_1^{\infty} e^{-x}\, dx$

17-20 $\int_1^{\infty} x^{-3/2}\, dx$

17-21 $\int_{-\infty}^{1} e^{2x}\,dx$

17-22 $\int_{-\infty}^{0} \sin x\,dx$

17-23 $\int_{-\infty}^{0} \sqrt[3]{x}/(\sqrt[3]{x^4}+1)\,dx$

17-24 $\int_{-\infty}^{0} e^x \sin x\,dx$

17-25 $\int_{-\infty}^{\infty} 1/(x^2+4)\,dx$

17-26 $\int_{-\infty}^{\infty} e^x/(1+e^{2x})\,dx$

17-27 $\int_{-\infty}^{\infty} x/[(x^2+1)^2]\,dx$

17-28 $\int_{-\infty}^{\infty} x/(x^2+4)\,dx$

17-29 $\int_{-\infty}^{\infty} xe^{-x^2+1}\,dx$

17-30 $\int_{0}^{\infty} xe^{-x}\,dx$

17-31 $\int_{-1}^{1} 1/\sqrt[3]{x+1}\,dx$

17-32 $\int_{-1}^{0} 1/x\,dx$

17-33 $\int_{0}^{1} (\ln x)/x\,dx$

17-34 $\int_{-3}^{0} 1/(x+3)^2\,dx$

17-35 $\int_{-1}^{0} 1/\sqrt{x+1}\,dx$

17-36 $\int_{1}^{e} 1/(x\sqrt{\ln x})\,dx$

17-37 $\int_{-1}^{1} 1/x\,dx$

17-38 $\int_{0}^{2\pi} (\cos x)/(1+\sin x)\,dx$

17-39 $\int_{-1}^{2} 1/(x^3+x)\,dx$

17-40 $\int_{-1}^{8} 1/(\sqrt[3]{x^2})\,dx$

17-41 $\int_{0}^{e} (\ln x)/(\sqrt{x})\,dx$

17-42 $\int_{-1}^{1} 1/(x^3-x)\,dx$

17-43 $\int_{0}^{4} 1/(\sqrt{x}-1)\,dx$

17-44 $\int_{0}^{\pi} \sec^2 x\,dx$

17-45 $\int_{-2}^{2} x/(\sqrt{4-x^2})\,dx$

17-46 $\int_{-1}^{1} \coth x\,dx$

17-47 $\int_{-\infty}^{\infty} e^{-\sqrt[3]{x^2}}/\sqrt[3]{x}\,dx$

17-48 $\int_{0}^{\infty} (\ln^2 x)/(x)\,dx$

17-49 $\int_{-\infty}^{\infty} x^{-4/3}\,dx$

17-50 $\int_{0}^{\infty} 1/(e^x-1)\,dx$

Solutions to Supplementary Exercises

(17-17) 1

(17-18) diverges

(17-19) e^{-1}

(17-20) 2

(17-21) $\frac{1}{2}e^2$

(17-22) diverges

(17-23) diverges

(17-24) $-\frac{1}{2}$

(17-25) $\pi/2$

(17-26) $\pi/2$

(17-27) 0

(17-28) diverges

(17-29) 0

(17-30) 1

(17-31) $(3/2)\sqrt[3]{4}$

(17-32) diverges

(17-33) diverges

(17-34) diverges

(17-35) 2

(17-36) 2

(17-37) diverges

(17-38) diverges

(17-39) diverges

(17-40) 9

(17-41) $-2\sqrt{e}$

(17-42) diverges

(17-43) diverges

(17-44) diverges

(17-45) 0

(17-46) diverges

(17-47) 0

(17-48) diverges

(17-49) diverges

(17-50) diverges

18 VOLUME

THIS CHAPTER IS ABOUT

☑ **Solids of Revolution**
☑ **Rotation about Other Axes**
☑ **Solids with Given Cross-Sectional Area**

18-1. Solids of Revolution

A. Disk method

1. Rotation about the x axis: If the region bounded by the graph of $y = f(x)$ and the x axis between $x = a$ and $x = b$ were rotated about the x axis, what would be the volume V of the resulting three-dimensional region (called a **solid of revolution**)?

To approximate this volume, divide the interval $[a, b]$ on the x axis into n subintervals of equal length, Δx:

$$\Delta x = \frac{b - a}{n}$$

$$x_0 = a, \ldots, x_i = a + i\,\Delta x, \ldots, x_n = b$$

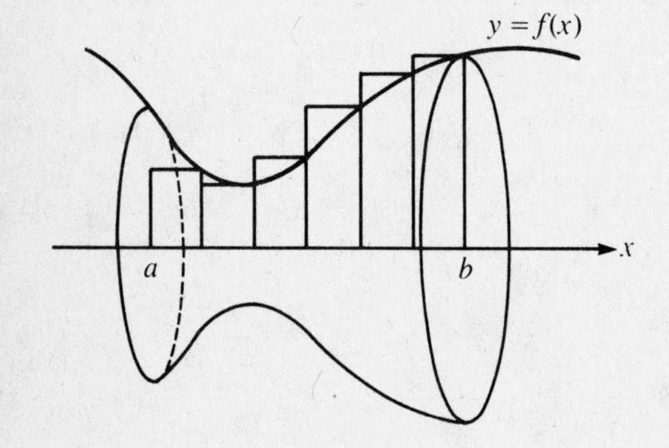

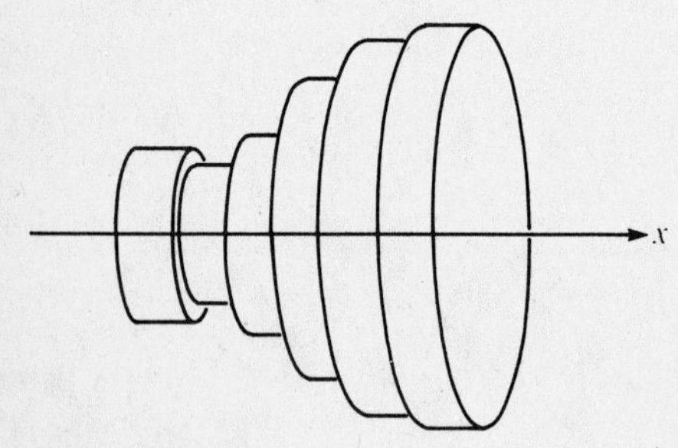

Figure 18-1
Disk method

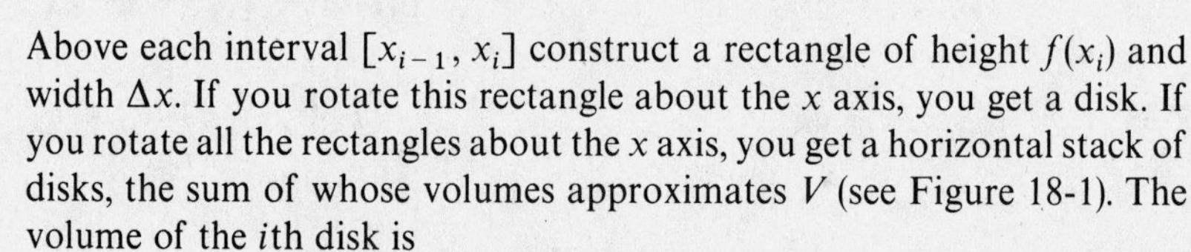

Above each interval $[x_{i-1}, x_i]$ construct a rectangle of height $f(x_i)$ and width Δx. If you rotate this rectangle about the x axis, you get a disk. If you rotate all the rectangles about the x axis, you get a horizontal stack of disks, the sum of whose volumes approximates V (see Figure 18-1). The volume of the ith disk is

$$\pi f^2(x_i)\,\Delta x$$

So the sum of the volumes of the disks is

$$\sum_{i=1}^{n} \pi f^2(x_i)\,\Delta x$$

If the function defining the curve is continuous, then the sum of the volumes of the disks will approach V as n increases:

$$V = \lim_{n \to \infty} \sum_{i=1}^{n} \pi f^2(x_i)\,\Delta x$$

This is a limit of Riemann sums (see Chapter 12), which you should recognize as an integral:

$$V = \pi \int_a^b f^2(x)\,dx \qquad \textbf{(18-1)}$$

EXAMPLE 18-1: The region bounded by the graph of $f(x) = x^2$ and the x axis between $x = 1$ and $x = 2$ is rotated about the x axis. What is the volume of the resulting solid of revolution?

Solution: Use the disk method. The volume is

$$\pi \int_1^2 f^2(x)\,dx = \pi \int_1^2 (x^2)^2\,dx$$
$$= \pi \int_1^2 x^4\,dx = \pi \frac{x^5}{5}\bigg|_1^2 = \frac{31\pi}{5}$$

Suppose that the region bounded by the graphs of $y = f(x)$ and $y = g(x)$ between $x = a$ and $x = b$ is rotated about the x axis. What is the volume V of the resulting solid? If $|f(x)| \geqslant |g(x)|$ on $[a, b]$, then V is the difference between the two volumes:

$$\pi \int_a^b f^2(x)\,dx - \pi \int_a^b g^2(x)\,dx = \pi \int_a^b [f^2(x) - g^2(x)]\,dx \quad \textbf{(18-2)}$$

EXAMPLE 18-2: The region bounded by the graphs of $f(x) = x$ and $g(x) = \frac{1}{2}x^2$ is rotated about the x axis. What is the volume of the resulting solid?

Solution: Since the two graphs intersect at (0, 0) and at (2, 2), the region is bounded on the left by $x = 0$ and on the right by $x = 2$. On the interval $[0, 2]$, $|f(x)| \geqslant |g(x)|$, so the volume is

$$\pi \int_0^2 [f^2(x) - g^2(x)]\,dx = \pi \int_0^2 \left[x^2 - \left(\frac{1}{2}x^2\right)^2\right] dx$$
$$= \pi \int_0^2 \left(x^2 - \frac{1}{4}x^4\right) dx$$
$$= \pi\left(\frac{x^3}{3} - \frac{1}{20}x^5\right)\bigg|_0^2 = \frac{16\pi}{15}$$

2. Rotation about the y axis: You can also use the disk method to find the volume of the solid obtained by rotating a region about the y axis. Your integrand will involve y's and dy instead of x's and dx.

   Consider the region between the horizontal lines $y = a$ and $y = b$ bounded on the left and right by curves C_1 and C_2 ($a < b$, and C_1 is farther from the y axis than C_2). To find the volume V of the solid obtained by rotating this region about the y axis, imagine subdividing the interval $[a, b]$ on the y axis. You construct horizontal rectangles (short and wide) that give horizontal disks when rotated about the y axis. You find

$$V = \pi \int_a^b [x_1^2(y) - x_2^2(y)]\,dy \quad \textbf{(18-3)}$$

   where $x_1(y)$ and $x_2(y)$ are the curves C_1 and C_2 (respectively), solved for x as a function of y.

EXAMPLE 18-3: The region bounded by the graphs of $y = \sqrt{x}$, $y = 2$, and the y axis is rotated about the y axis. What is the volume of the resulting solid?

Solution: This region (see Figure 18-2) is bounded above by $y = 2$ and below by $y = 0$. It is bounded on the left by the y axis and on the right by $y = \sqrt{x}$. Solve these curves for x as a function of y: $y = \sqrt{x}$ becomes $x = y^2$, and the y axis becomes $x = 0$. Because $x = y^2$ is further away from the y axis, the volume is

$$\pi \int_0^2 [(y^2)^2 - 0^2]\,dy = \pi \int_0^2 y^4\,dy = \pi \frac{y^5}{5}\bigg|_0^2 = \frac{32\pi}{5}$$

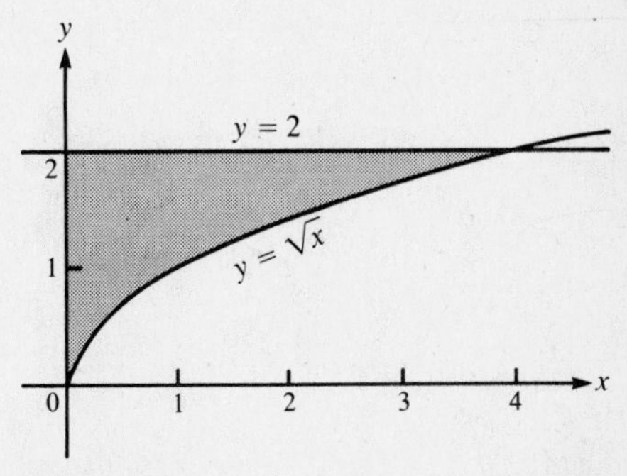

Figure 18-2
Example 18-3

EXAMPLE 18-4: The region bounded by the graphs of $y = (x - 1)^2 + 1$ and $y = 2$ is rotated about the y axis. Find the volume of the resulting solid.

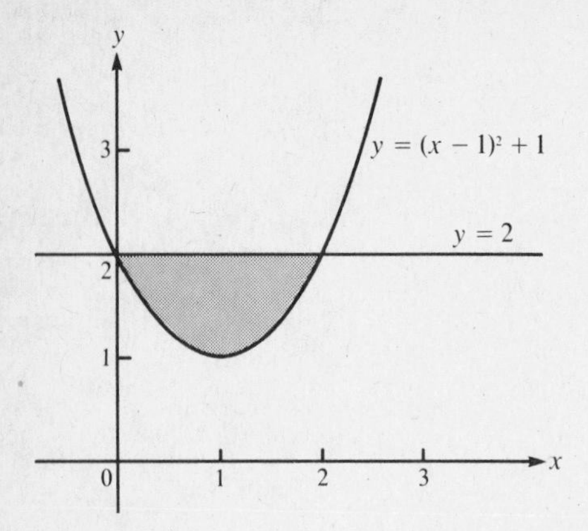

Figure 18-3
Example 18-4

Solution: This region is bounded above by $y = 2$ and below by $y = 1$. It is bounded on the left and right by $y = (x - 1)^2 + 1$ (see Figure 18-3). If you solve for x in terms of y,

$$y - 1 = (x - 1)^2$$
$$\pm\sqrt{y - 1} = x - 1$$
$$x = 1 \pm \sqrt{y - 1}$$

you find that the region is bounded on the left by $x = 1 - \sqrt{y - 1}$ and on the right by $x = 1 + \sqrt{y - 1}$. The volume is

$$\pi \int_1^2 [(1 + \sqrt{y - 1})^2 - (1 - \sqrt{y - 1})^2]\,dy = \pi \int_1^2 4\sqrt{y - 1}\,dy$$
$$= \frac{8\pi}{3}(y - 1)^{3/2}\Big|_1^2 = \frac{8\pi}{3}$$

B. Shell method

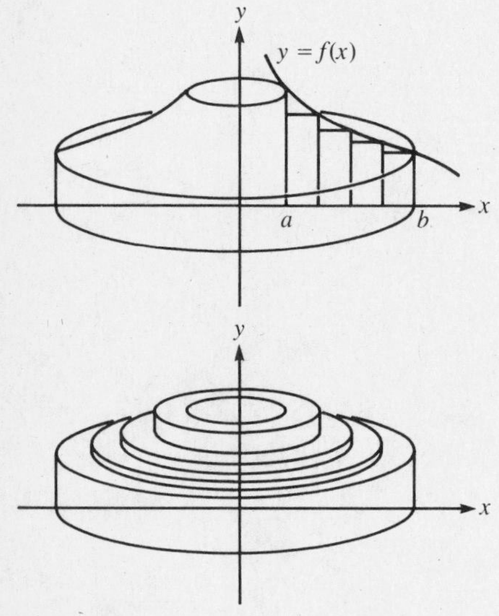

Figure 18-4
Shell method

1. Rotation about the y axis: The region bounded above by the graph of $y = f(x)$ and below by the x axis between the vertical lines $x = a$ and $x = b$ is rotated about the y axis. What is the volume V of the resulting solid of revolution? You may be able to solve this using the disk method, but the shell method is probably more appropriate.

Divide the interval $[a, b]$ of the x axis into n subintervals of equal length, Δx:

$$\Delta x = \frac{b - a}{n}$$

$$x_0 = a, \ldots, x_i = a + i\,\Delta x, \ldots, x_n = b$$

Above each interval $[x_{i-1}, x_i]$ construct a rectangle of height $f(x_i)$ and width Δx. If you rotate this rectangle about the y axis, you get a cylindrical shell. If you rotate all of the rectangles about the y axis, you get a spool of shells, the sum of whose volumes approximates V (see Figure 18-4). If $0 < a < b$ then the volume of the ith shell is

$$\pi f(x_i)(x_i^2 - x_{i-1}^2)$$

So the sum of the volumes of the shells is

$$\sum_{i=1}^{n} \pi f(x_i)(x_i^2 - x_{i-1}^2)$$

If the function defining the curve is continuous, then the sum of the volumes of the shells will approach V as n increases:

$$V = \lim_{n\to\infty} \sum_{i=1}^{n} \pi f(x_i)(x_i^2 - x_{i-1}^2)$$

This is a limit of Riemann sums, which (after some algebraic manipulation) you should recognize as an integral:

$$V = \lim_{n\to\infty} \sum_{i=1}^{n} \pi f(x_i)(x_i + x_{i-1})(x_i - x_{i-1})$$
$$= \lim_{n\to\infty} \sum_{i=1}^{n} \pi f(x_i)(x_i + x_{i-1})\,\Delta x$$
$$= \pi \int_a^b f(x)(x + x)\,dx = 2\pi \int_a^b xf(x)\,dx \qquad \textbf{(18-4)}$$

- If $a < b < 0$, then the volume is $\left|2\pi \int_a^b xf(x)\,dx\right| = -2\pi \int_a^b xf(x)\,dx.$

EXAMPLE 18-5: The region bounded by the graph of $f(x) = x^2$ and the x axis between $x = 1$ and $x = 2$ is rotated about the y axis. What is the volume of the resulting solid?

Solution: Use the shell method. The volume is

$$2\pi \int_1^2 xf(x)\,dx = 2\pi \int_1^2 xx^2\,dx = \frac{1}{2}\pi x^4 \Big|_1^2 = \frac{15\pi}{2}$$

Suppose that the region bounded by the graphs of $f(x)$ and $g(x)$ between $x = a$ and $x = b$ is rotated about the y axis. What is the volume V of the resulting solid? If $f(x) > g(x)$ on $[a, b]$ and $a > 0$, then V is the difference between the two volumes:

$$2\pi \int_a^b xf(x)\,dx - 2\pi \int_a^b xg(x)\,dx = 2\pi \int_a^b x[f(x) - g(x)]\,dx \quad \textbf{(18-5)}$$

EXAMPLE 18-6: The region bounded by the graphs of $f(x) = x$ and $g(x) = \frac{1}{2}x^2$ is rotated about the y axis. What is the volume of the resulting solid?

Solution: The region is bounded on the left by $x = 0$ and on the right by $x = 2$. On the interval $[0, 2]$, $f(x) > g(x)$, so the volume is

$$2\pi \int_0^2 x[f(x) - g(x)]\,dx = 2\pi \int_0^2 x\left(x - \frac{1}{2}x^2\right)dx = 2\pi \int_0^2 \left(x^2 - \frac{1}{2}x^3\right)dx$$

$$= 2\pi\left(\frac{x^3}{3} - \frac{x^4}{8}\right)\Big|_0^2 = \frac{4\pi}{3}$$

2. Rotation about the x axis: You can use the shell method to find volumes of solids obtained by rotating about the x axis, too. Your integrand will involve y's and dy instead of x's and dx.

Consider the region bounded on the right by the curve C_1 and on the left by the curve C_2, between the horizontal lines $y = a$ and $y = b$, where $b > a > 0$. To find the volume V of the solid obtained by rotating this region about the x axis, imagine subdividing the interval $[a, b]$ on the y axis. You construct horizontal rectangles (short and wide) that give shells when rotated about the x axis. You find

$$V = 2\pi \int_a^b y[x_1(y) - x_2(y)]\,dy \quad \textbf{(18-6)}$$

where $x_1(y)$ and $x_2(y)$ are the curves C_1 and C_2 (respectively) solved for x as a function of y.

EXAMPLE 18-7: The region bounded by the graphs of $2y = y^2 - x + 1$ and $x = 1$ is rotated about the x axis. Find the volume of the resulting solid of revolution using the shell method.

Solution: This region is shown in Figure 18-5. It is bounded above by $y = 2$ and below by $y = 0$. It is bounded on the right by $x = 1$ and on the left by $2y = y^2 - x + 1$. Solve these curves for x as a function of y: $2y = y^2 - x + 1$ becomes $x = (y - 1)^2$; $x = 1$ is already solved for x. The volume is

$$2\pi \int_0^2 y[1 - (y - 1)^2]\,dy = 2\pi \int_0^2 (2y^2 - y^3)\,dy$$

$$= 2\pi\left(\frac{2}{3}y^3 - \frac{1}{4}y^4\right)\Big|_0^2 = \frac{8\pi}{3}$$

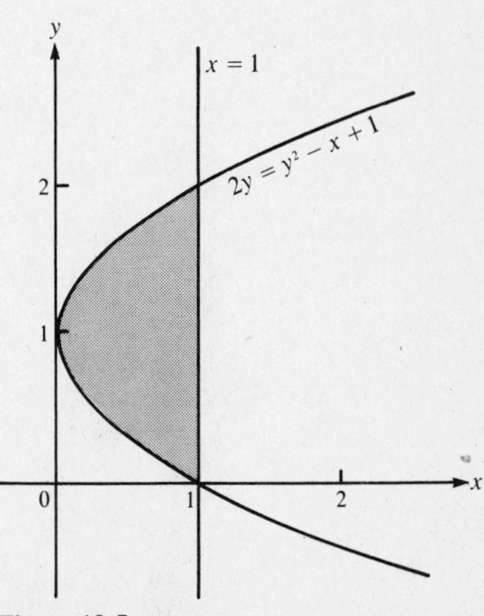

Figure 18-5
Example 18-7

C. Which method to use

To compute the volume of a solid of revolution, you simply use a formula. To determine which formula to use, you should think about rectangles. Both the shell and the disk method can be used for rotations about either coordinate axis. To decide which method to use:

1. Axis of subdivision: Is the region more easily described as
   - bounded above and below by curves and on the left and right by vertical lines, or
   - bounded on the left and right by curves and above and below by horizontal lines?

   In the first case, you should subdivide an interval of the x axis (your integrand will involve x's and dx). In the second case, you should subdivide an interval of the y axis (your integrand will involve y's and dy).
2. Shells or disks: If you subdivide the x axis, your rectangles will be tall and thin, as in Figure 18-6. If you subdivide the y axis, your rectangles will be short and wide, as in Figure 18-7.

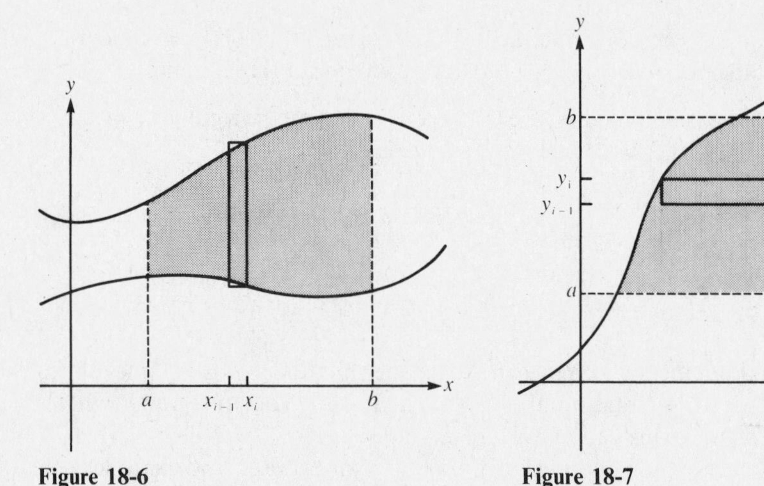

Figure 18-6

Figure 18-7

Picture your rectangles being rotated about the axis of rotation in the problem. If you are rotating about the axis you subdivided, you have disks. If you are rotating about the other axis, you have shells. Now you know which method to use. There are only two formulas to remember: one for the disk method (18-2) and one for the shell method (18-5). Subdividing the y axis just means that the integrand involves y's and dy.

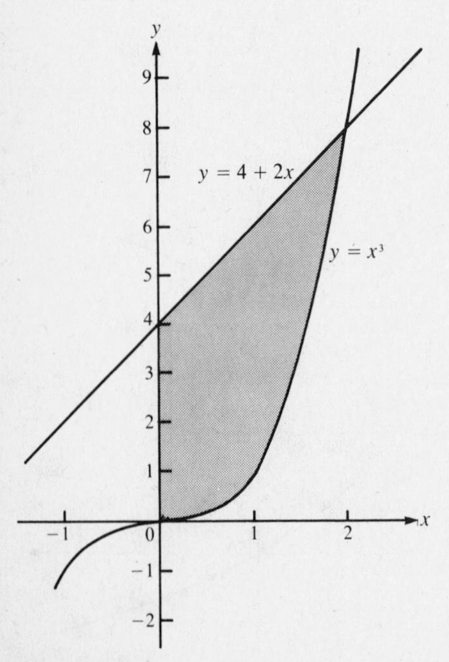

Figure 18-8
Example 18-8

EXAMPLE 18-8: The region bounded by the graphs of $y = x^3$, $y = 4 + 2x$, and the y axis is rotated about the y axis. Find the volume of the resulting solid.

Solution: This region is most easily described as bounded above and below by the curves $y = 4 + 2x$ and $y = x^3$, between the vertical lines $x = 0$ and $x = 2$ (see Figure 18-8). (If you try to describe it as bounded on the left and right by curves, the curve on the left is actually two different curves, the y axis and the line $y = 4 + 2x$. This would require two integrals, so avoid it.) Thus you subdivide the interval $[0, 2]$ on the x axis and obtain rectangles that are tall and thin. Rotating these about the y axis makes shells. The volume is

$$2\pi \int_0^2 x[(4 + 2x) - x^3]\, dx = 2\pi \int_0^2 (4x + 2x^2 - x^4)\, dx$$

$$= 2\pi\left(2x^2 + \frac{2}{3}x^3 - \frac{x^5}{5}\right)\Bigg|_0^2 = \frac{208\pi}{15}$$

EXAMPLE 18-9: The region bounded by the graphs of $y = \sqrt{x}$ and $y = \frac{1}{2}x$ below the line $y = 1$ is rotated about the y axis. Find the volume of the resulting solid of revolution.

Solution: You can most easily describe this region as bounded on the left by the curve $y = \sqrt{x}$ and on the right by $y = \frac{1}{2}x$, between the horizontal lines $y = 0$ and $y = 1$ (see Figure 18-9). (If you describe it as bounded above and below by curves and on the right and left by vertical lines, the curve that bounds it above would be $y = \sqrt{x}$ on $[0, 1]$ and $y = 1$ on $[1, 2]$. You would need two integrals, so try it the other way.) So you imagine subdividing the interval $[0, 1]$ on the y axis to obtain rectangles that are short and wide. These give disks when rotated about the y axis. So you use the disk method, integrating over the interval $[0, 1]$ on the y axis. You must solve the curves for x in terms of y: $x = y^2$ and $x = 2y$. The volume is

$$\pi \int_0^1 [(2y)^2 - (y^2)^2]\, dy = \pi\left(\frac{4}{3}y^3 - \frac{1}{5}y^5\right)\Bigg|_0^1 = \frac{17\pi}{15}$$

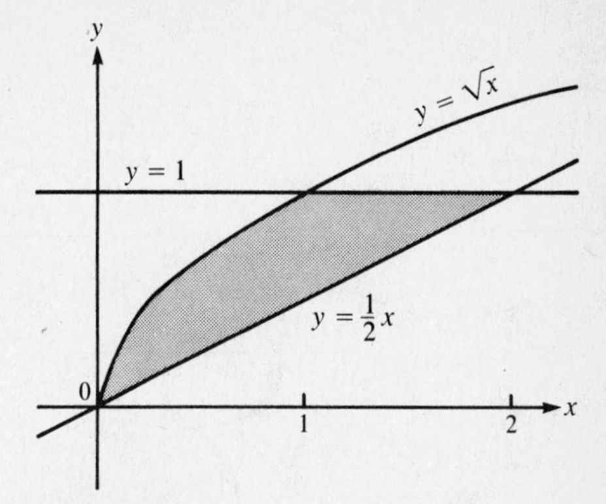

Figure 18-9
Example 18-9

18-2. Rotation about Other Axes

If a region in the plane is rotated about a horizontal line other than the x axis, you can find the volume of the resulting solid by shifting (translating) the region and the line up or down until the axis of revolution becomes the x axis.

EXAMPLE 18-10: The region bounded by the graphs of $y = x^2$ and $y = x$ is rotated about the line $y = -2$. Find the volume V of the resulting solid.

Solution: If you translate the region upward by two units and rotate about the x axis, you obtain a solid whose volume is V. To translate the region, translate the curves that bound the region:

$$y = x^2 \quad \text{becomes} \quad y = x^2 + 2$$

$$y = x \quad \text{becomes} \quad y = x + 2$$

These are shown in Figure 18-10. Find the volume using the disk method:

$$V = \pi \int_0^1 [(x + 2)^2 - (x^2 + 2)^2]\, dx = \pi \int_0^1 (4x - x^4 - 3x^2)\, dx$$

$$= \pi\left(2x^2 - \frac{x^5}{5} - x^3\right)\Bigg|_0^1 = \frac{4\pi}{5}$$

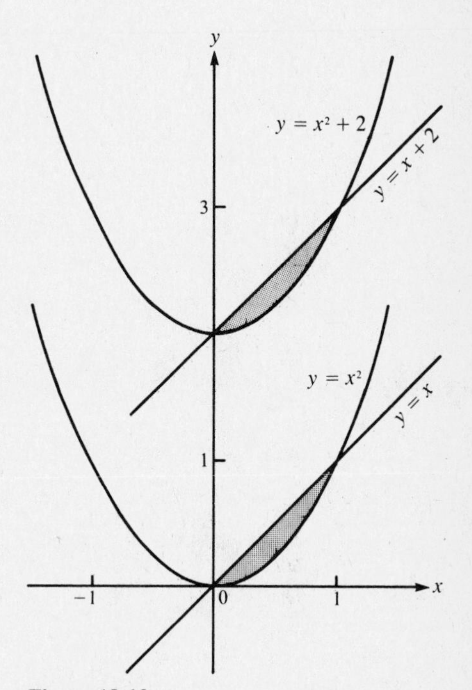

Figure 18-10
Example 18-10

If a region is to be rotated about a vertical line other than the y axis, translate the region and the line to the left or right until the axis of revolution becomes the y axis.

EXAMPLE 18-11: The region bounded by the graphs of $y = x^2$ and $y = x$ is rotated about the line $x = 2$. Find the volume of the resulting solid.

Solution: Translate the region (and hence the curves) to the left by two units:

$$y = x^2 \quad \text{becomes} \quad y = (x + 2)^2$$

$$y = x \quad \text{becomes} \quad y = x + 2$$

Use the shell method to find V:

$$-2\pi \int_{-2}^{-1} x[(x + 2) - (x + 2)^2]\, dx = 2\pi \int_{-2}^{-1} (x^3 + 3x^2 + 2x)\, dx$$

$$= 2\pi\left(\frac{1}{4}x^4 + x^3 + x^2\right)\Bigg|_{-2}^{-1} = \frac{\pi}{2}$$

The volume V is $\pi/2$. The minus sign arose in the integral because you used the shell method on the negative half of the axis.

Remember, to translate a curve upward by an amount a, rewrite the equation defining the curve with y replaced by $y - a$. To translate a curve to the right by an amount a, rewrite the equation defining the curve with x replaced by $x - a$.

18-3. Solids with Given Cross-Sectional Area

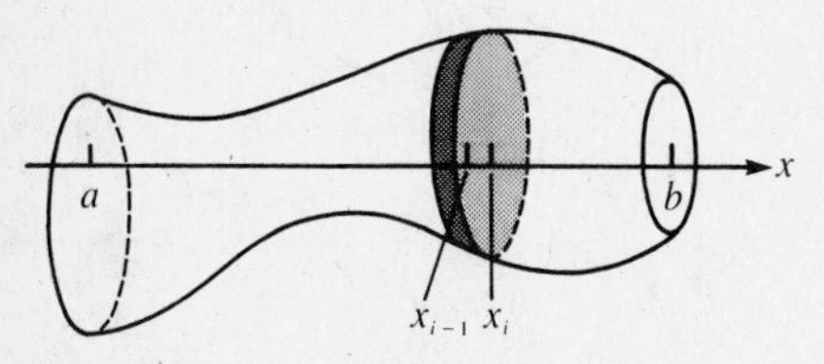

Figure 18-11

Another class of solids (in addition to solids of revolution) has volume that you can find by integration. These are the solids with the property that the cross-sectional area, in a plane perpendicular to the x axis at $x = x_0$, can be expressed as a function of x_0. Let S be such a solid, defined on the interval $[a, b]$. Divide $[a, b]$ into n subintervals of equal length, Δx, and slice the solid into "slabs," with the parallel planes at x_i for $i = 1, \ldots, n$ (see Figure 18-11). If the cross-sectional area in the plane at x_i is $A(x_i)$, then the volume of each slab could be approximated by

$$V_i = A(x_i)\,\Delta x$$

If the function $A(x)$ is continuous, then as n increases, the sum of the V_i terms will approach V, the volume of the solid:

$$V = \lim_{n \to \infty} \sum_{i=1}^{n} A(x_i)\,\Delta x$$

This is a limit of Riemann sums, which you should recognize as an integral:

$$V = \int_a^b A(x)\,dx \tag{18-7}$$

EXAMPLE 18-12: A pyramid is 45 ft high. A horizontal cross section x ft from the bottom is a square whose sides measure $90 - 2x$ ft. Find the volume of the pyramid.

Solution: Imagine an x axis running from $x = 0$ to $x = 45$. The cross-sectional area at x is the area of a square whose sides are $90 - 2x$:

$$A(x) = (90 - 2x)^2$$

The volume of the pyramid is

$$\int_0^{45} (90 - 2x)^2\,dx = \int_0^{45} (8100 - 360x + 4x^2)\,dx$$

$$= \left(8100x - 180x^2 + \frac{4}{3}x^3\right)\Bigg|_0^{45}$$

$$= 121\,500 \text{ ft}^3$$

EXAMPLE 18-13: The base of a solid S is a disk of radius 2. A line runs through the center of the disk. Any cross section of the solid by a plane perpendicular to the line is an equilateral triangle. Find the volume V of S.

Solution: Make the line the x axis, so that the disk is $x^2 + y^2 \leq 4$. A plane, perpendicular to the x axis at x, cuts the disk in a chord of length $2\sqrt{4 - x^2}$ (see Figure 18-12). This chord is the base of the cross section of the solid, the equilateral triangle. Thus $A(x)$ is the area of the equilateral triangle with side $2\sqrt{4 - x^2}$:

$$A(x) = \frac{1}{2}(\text{base})(\text{height}) = \frac{1}{2}(2\sqrt{4 - x^2})(\sqrt{3}\sqrt{4 - x^2})$$

$$= \sqrt{3}(4 - x^2)$$

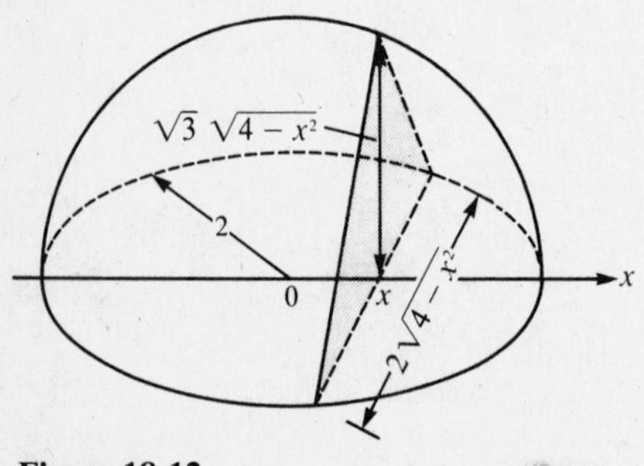

Figure 18-12
Example 18-13

The volume is

$$V = \int_{-2}^{2} A(x)\,dx = \int_{-2}^{2} \sqrt{3}(4 - x^2)\,dx$$

$$= \sqrt{3}\left(4x - \frac{x^3}{3}\right)\Bigg|_{-2}^{2} = \frac{32\sqrt{3}}{3}$$

SUMMARY

1. Use either the disk method or the shell method to find the volume of a solid of revolution.
2. You can use either method for rotations about either coordinate axis.
3. You choose which method to use by examining the region being rotated and determining whether it is easier to describe with vertical rectangles (integration with respect to x) or with horizontal rectangles (integration with respect to y). If you integrate along the axis of rotation—disk method; along the other axis—shell method.
4. You can rotate along axes other than the coordinate axes. Simply translate your region.
5. You can find the volume of a solid if you can find the cross-sectional areas.

SOLVED PROBLEMS

PROBLEM 18-1 The region bounded by the graph of $f(x) = e^x$ and the x axis between $x = 1$ and $x = 3$ is rotated about the x axis. Find the volume of the resulting solid of revolution.

Solution: Use the disk method. Integrate over the interval [1, 3] of the x axis:

$$\pi \int_1^3 f^2(x)\,dx = \pi \int_1^3 (e^x)^2\,dx = \pi \int_1^3 e^{2x}\,dx = \frac{\pi}{2} e^{2x}\Big|_1^3 = \frac{\pi}{2}(e^6 - e^2)$$

[See Section 18-1A.]

PROBLEM 18-2 The region bounded by the graph of $f(x) = 1 - (x - 1)^2$ and the x axis is rotated about the x axis. Find the volume of the resulting solid.

Solution: This region is bounded on the left by $x = 0$ and on the right by $x = 2$ because that is where the graph of f and the x axis intersect:

$$0 = 1 - (x - 1)^2$$
$$1 = (x - 1)^2$$
$$\pm 1 = (x - 1)$$
$$x = 1 \pm 1 = 0 \quad \text{or} \quad 2$$

By the disk method, the volume is

$$\pi \int_0^2 f^2(x)\,dx = \pi \int_0^2 [1 - (x - 1)^2]^2\,dx = \pi \int_0^2 (2x - x^2)^2\,dx$$

$$= \pi \int_0^2 (4x^2 - 4x^3 + x^4)\,dx = \pi\left(\frac{4}{3}x^3 - x^4 + \frac{x^5}{5}\right)\Big|_0^2 = \frac{16\pi}{15}$$

[See Section 18-1A.]

PROBLEM 18-3 The region bounded by the graphs of $y = (x - 1)^2$ and $y = 1 + x$ is rotated about the x axis. Find the volume of the resulting solid.

Solution: Find where the two curves intersect:

$$(x - 1)^2 = 1 + x$$
$$x^2 - 2x + 1 = 1 + x$$
$$x^2 - 3x = 0$$
$$x = 0 \quad \text{or} \quad x = 3$$

On the interval [0, 3],

$$|(x-1)^2| \leqslant |1+x|$$

By the disk method, the volume is

$$\pi\int_0^3 \{(1+x)^2 - [(x-1)^2]^2\}\,dx = \pi\left[\frac{(1+x)^3}{3} - \frac{(x-1)^5}{5}\right]\Bigg|_0^3 = \frac{72\pi}{5}$$ [See Section 18-1A.]

PROBLEM 18-4 The region bounded by the graph of $f(x) = x^2 + 1$ and the x axis between $x = 0$ and $x = 2$ is rotated about the y axis. Find the volume of the resulting solid.

Solution: Use the shell method. Integrate over the interval [0, 2] of the x axis:

$$2\pi\int_0^2 xf(x)\,dx = 2\pi\int_0^2 x(x^2+1)\,dx = 2\pi\int_0^2 (x^3+x)\,dx$$

$$= 2\pi\left(\frac{1}{4}x^4 + \frac{1}{2}x^2\right)\Bigg|_0^2 = 12\pi$$ [See Section 18-1B.]

PROBLEM 18-5 The region bounded by the graphs of $y = 2 - (x-1)^2$ and $y = 1$ is rotated about the y axis. Find the volume of the resulting solid.

Solution: Where do the curves intersect?

$$2 - (x-1)^2 = 1$$

$$(x-1)^2 = 1$$

$$x - 1 = \pm 1$$

$$x = 0 \quad \text{or} \quad x = 2$$

On the interval [0, 2], $2 - (x-1)^2 > 1$. By the shell method, the volume is

$$2\pi\int_0^2 x[2 - (x-1)^2 - 1]\,dx = 2\pi\int_0^2 (2x^2 - x^3)\,dx$$

$$= 2\pi\left(\frac{2}{3}x^3 - \frac{1}{4}x^4\right)\Bigg|_0^2 = \frac{8\pi}{3}$$ [See Section 18-1B.]

PROBLEM 18-6 The region bounded by the graphs of $y = e^x$, $y = 1$, and $x = 1$ is rotated about the x axis. Compute the volume of the resulting solid twice: once using the disk method and once using the shell method.

Solution: First graph the region, as in Figure 18-13. To use the disk method, integrate along the interval [0, 1] of the x axis:

$$\pi\int_0^1 [(e^x)^2 - 1^2]\,dx = \pi\left(\frac{1}{2}e^{2x} - x\right)\Bigg|_0^1$$

$$= \frac{\pi}{2}(e^2 - 3)$$

To use the shell method, integrate along the interval $[1, e]$ of the y axis. Solve $y = e^x$ for x: $x = \ln y$. On the interval $[1, e]$, $1 \geqslant \ln y$, so the volume is

$$2\pi\int_1^e y(1 - \ln y)\,dy$$

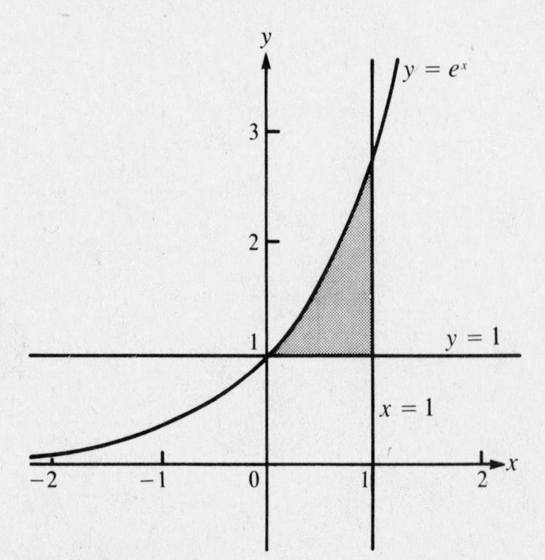

Figure 18-13
Problem 18-6

Integrate by parts:

$$2\pi \int_1^e y(1 - \ln y)\,dy = 2\pi\left[\frac{1}{2}y^2 - \left(\frac{1}{2}y^2 \ln y - \frac{1}{4}y^2\right)\right]\Bigg|_1^e$$

$$= 2\pi\left(\frac{3}{4}y^2 - \frac{1}{2}y^2 \ln y\right)\Bigg|_1^e = \frac{\pi}{2}(e^2 - 3) \quad \text{[See Section 18-1C.]}$$

PROBLEM 18-7 The region bounded by the graphs of $y = \sqrt{x}$, $y = x - 2$, and the x axis is rotated about the x axis. Find the volume of the resulting solid.

Solution: Graph the region, as in Figure 18-14. This region is best described as bounded on the left by the graph of $y = \sqrt{x}$ and on the right by the graph of $y = x - 2$, between the horizontal lines $y = 0$ and $y = 2$. The interval of integration is [0, 2] on the y axis. Subdivide this interval to get rectangles that are short and wide, which, when rotated about the x axis, become shells. Because you'll integrate along an interval of the y axis, solve the curves for x: $y = \sqrt{x}$ becomes $x = y^2$; $y = x - 2$ becomes $x = y + 2$. By the shell method, the volume is

$$2\pi \int_0^2 y(y + 2 - y^2)\,dy = 2\pi\left(\frac{1}{3}y^3 + y^2 - \frac{1}{4}y^4\right)\Bigg|_0^2 = \frac{16\pi}{3} \quad \text{[See Section 18-1C.]}$$

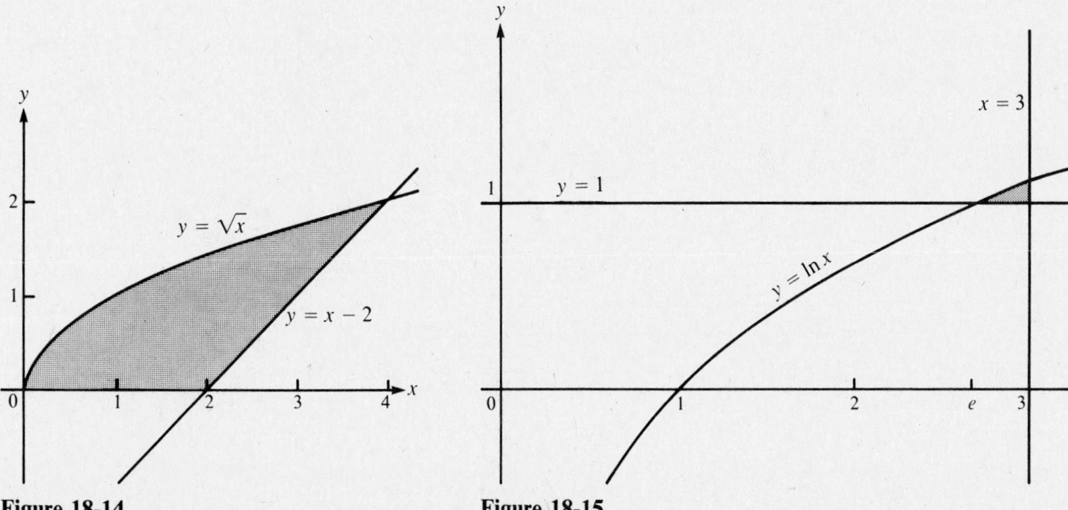

Figure 18-14
Problem 18-7

Figure 18-15
Problem 18-8

PROBLEM 18-8 The region bounded by the graphs of $y = \ln x$, $y = 1$, and $x = 3$ is rotated about the y axis. Find the volume of the resulting solid.

Solution: This region is shown in Figure 18-15. You can describe it as lying below $y = \ln x$ and above $y = 1$, between $x = e$ and $x = 3$. The volume, by the shell method, is

$$2\pi \int_e^3 x(\ln x - 1)\,dx = 2\pi\left(\frac{1}{2}x^2 \ln x - \frac{3}{4}x^2\right)\Bigg|_e^3 = 9\pi \ln 3 - \frac{27\pi}{2} + \frac{\pi}{2}e^2$$

You can also find this volume using the disk method. The region is bounded on the left by $y = \ln x$ and on the right by $x = 3$. It lies between the horizontal lines $y = 1$ and $y = \ln 3$. Subdivide the interval [1, ln 3] on the y axis. This gives rectangles that are short and wide and become disks when rotated about the y axis. The curve $y = \ln x$ becomes $x = e^y$ when solved for x. The curve $x = 3$ is further from the y axis the $x = e^y$, so the volume is

$$\pi \int_1^{\ln 3} [3^2 - (e^y)^2]\,dy = \pi\left(9y - \frac{1}{2}e^{2y}\right)\Bigg|_1^{\ln 3} = 9\pi \ln 3 - \frac{27\pi}{2} + \frac{\pi}{2}e^2$$

[See Section 18-1C.]

PROBLEM 18-9 The region bounded by the graphs of $y = x$ and $y = x^2$ is rotated about the y axis. Find the volume of the resulting solid in two ways.

Solution: You can subdivide the interval [0, 1] on the x axis and use the shell method:

$$2\pi \int_0^1 x(x - x^2)\,dx = 2\pi\left(\frac{x^3}{3} - \frac{x^4}{4}\right)\bigg|_0^1 = \frac{\pi}{6}$$

You can subdivide the interval [0, 1] on the y axis and use the disk method:

$$\pi \int_0^1 [(\sqrt{y})^2 - y^2]\,dy = \pi \int_0^1 (y - y^2)\,dy = \pi\left(\frac{1}{2}y^2 - \frac{1}{3}y^3\right)\bigg|_0^1 = \frac{\pi}{6}$$

[See Section 18-1C.]

PROBLEM 18-10 The region bounded by the graphs of $y = 3x$, $y = 2x$, and $y = 3$ is rotated about the y axis. Find the volume of the resulting solid.

Solution: Graph this region, as in Figure 18-16. The region is most easily described as bounded on the left and right by the curves $y = 3x$ and $y = 2x$, between the horizontal lines $y = 0$ and $y = 3$. You can compute the volume using an integral along the y axis from $y = 0$ to $y = 3$. Your rectangles run horizontally (short and wide). When rotated about the y axis, they yield disks. Since the integral is in the variable y, the curves bounding the region on the left and right must be solved for x: $x = \frac{1}{3}y$ and $x = \frac{1}{2}y$. The volume is

$$\pi \int_0^3 \left[\left(\frac{1}{2}y\right)^2 - \left(\frac{1}{3}y\right)^2\right] dy = \frac{5\pi}{36}\int_0^3 y^2\,dy = \frac{5\pi}{36}\left(\frac{y^3}{3}\right)\bigg|_0^3 = \frac{5\pi}{4}$$

[See Section 18-1C.]

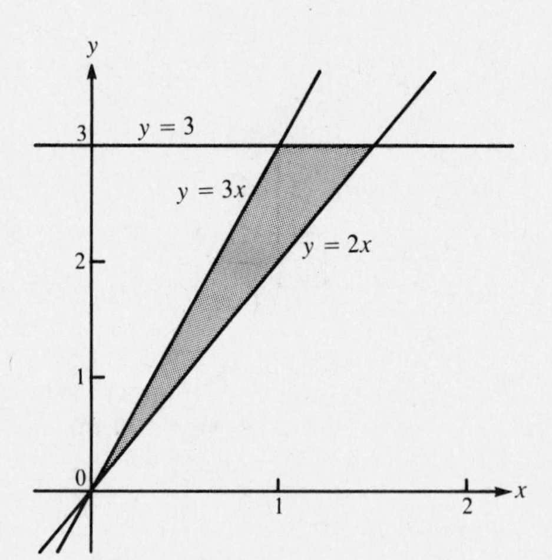

Figure 18-16
Problem 18-10

PROBLEM 18-11 The region bounded by the graph of $y = \sin x$ and the x axis between $x = 0$ and $x = \pi$ is rotated about the x axis. Find the volume of the resulting solid.

Solution: Use the disk method, integrating over the interval $[0, \pi]$ on the x axis:

$$\pi \int_0^\pi \sin^2 x\,dx = \pi \int_0^\pi \left(\frac{1 - \cos 2x}{2}\right) dx = \pi\left(\frac{1}{2}x - \frac{1}{4}\sin 2x\right)\bigg|_0^\pi = \frac{\pi^2}{2}$$

[See Section 18-1A.]

PROBLEM 18-12 The region bounded by the graphs of $y = e^x$, $y = 1$, and $x = -1$ is rotated about the x axis. Find the volume of the resulting solid in two ways.

Solution: Graph the region, as in Figure 18-17. The region is between the curves $y = 1$ and $y = e^x$ and between the vertical lines $x = -1$ and $x = 0$. Integrate along the interval $[-1, 0]$ of the x axis using the disk method:

$$\pi \int_{-1}^0 [1^2 - (e^x)^2]\,dx = \pi\left(x - \frac{1}{2}e^{2x}\right)\bigg|_{-1}^0 = \frac{\pi}{2}(1 + e^{-2})$$

You can also view the region as bounded by the curves $x = -1$ and $x = \ln y$, between the horizontal lines $y = e^{-1}$ and $y = 1$. Integrate along the interval $[e^{-1}, 1]$ on the y axis using the shell method. On this interval $\ln y \geqslant -1$, so the volume is

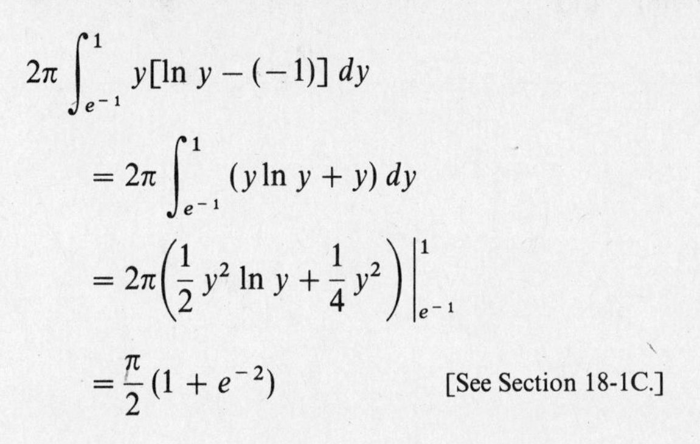

$$2\pi \int_{e^{-1}}^{1} y[\ln y - (-1)]\, dy$$

$$= 2\pi \int_{e^{-1}}^{1} (y \ln y + y)\, dy$$

$$= 2\pi\left(\frac{1}{2}y^2 \ln y + \frac{1}{4}y^2\right)\Bigg|_{e^{-1}}^{1}$$

$$= \frac{\pi}{2}(1 + e^{-2})$$

[See Section 18-1C.]

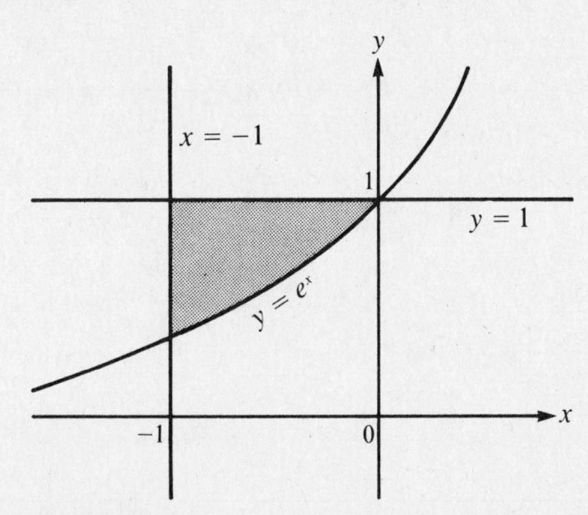

Figure 18-17
Problem 18-12

PROBLEM 18-13 The graph of the circle $x^2 + (y - 2)^2 = 1$ is rotated about the x axis. Find the volume of the resulting solid.

Solution: This circle extends from $x = -1$ to $x = 1$ and is bounded above and below by the graph of $x^2 + (y - 2)^2 = 1$. That is, the region is bounded above by the graph of $y = 2 + \sqrt{1 - x^2}$ and below by the graph $y = 2 - \sqrt{1 - x^2}$ (solve $x^2 + (y - 2)^2 = 1$ for y to find the two branches). By the disk method the volume is

$$\pi \int_{-1}^{1} [(2 + \sqrt{1 - x^2})^2 - (2 - \sqrt{1 - x^2})^2]\, dx = \pi \int_{-1}^{1} 8\sqrt{1 - x^2}\, dx$$

Use the inverse trigonometric substitution $\theta = \arcsin x$:

$$x = \sin\theta \qquad dx = \cos\theta\, d\theta \qquad \sqrt{1 - x^2} = \cos\theta$$

The volume is

$$8\pi \int_{-\pi/2}^{\pi/2} \cos^2\theta\, d\theta = 8\pi \int_{-\pi/2}^{\pi/2} \frac{1}{2}(1 + \cos 2\theta)\, d\theta = 4\pi\left(\theta + \frac{1}{2}\sin 2\theta\right)\Bigg|_{-\pi/2}^{\pi/2}$$

$$= 4\pi\left[\left(\frac{\pi}{2} + \frac{1}{2}\sin\pi\right) - \left(\frac{-\pi}{2} + \frac{1}{2}\sin(-\pi)\right)\right] = 4\pi^2$$

[See Section 18-1A.]

PROBLEM 18-14 Find the volume of the sphere of radius r.

Solution: Consider the region bounded by the graph of $y = \sqrt{r^2 - x^2}$ (the semicircle of radius r centered at the origin, above the x axis) and the x axis. If you rotate this region about the x axis, you obtain a sphere of radius r. By the disk method the volume is

$$\pi \int_{-r}^{r} (\sqrt{r^2 - x^2})^2\, dx = \pi\left(r^2 x - \frac{1}{3}x^3\right)\Bigg|_{-r}^{r} = \frac{4}{3}\pi r^3$$

[See Section 18-1A.]

PROBLEM 18-15 The region bounded by the graphs of $y = 1/x$, $x = 1$, and the x axis is rotated about the x axis. Find the volume of the resulting solid.

Solution: This region is bounded above by the graph of $y = 1/x$, below by the x axis, and on the left by $x = 1$. It is unbounded on the right. You can find the volume using the disk method with an improper integral:

$$\pi \int_{1}^{\infty} \left(\frac{1}{x}\right)^2 dx = \pi \lim_{N\to\infty} \int_{1}^{N} x^{-2}\, dx = \pi \lim_{N\to\infty} \left(\frac{-1}{x}\right)\Bigg|_{1}^{N}$$

$$= \pi \lim_{N\to\infty} \left(1 - \frac{1}{N}\right) = \pi$$

[See Section 18-1A.]

PROBLEM 18-16 The region bounded by the graphs of $y = x^2$ and $y = 1$ is rotated about the line $y = 2$. Find the volume of the resulting solid.

Solution: Translate the region downward two units and rotate about the x axis: $y = x^2$ becomes $y = x^2 - 2$; $y = 1$ becomes $y = -1$. The region bounded by the graphs of $y = x^2 - 2$ and $y = -1$ extends from $x = -1$ to $x = 1$. In this interval $|x^2 - 2| \geqslant |-1|$. Use the disk method to find the volume:

$$\pi \int_{-1}^{1} [(x^2 - 2)^2 - (-1)^2]\, dx = \pi \int_{-1}^{1} (x^4 - 4x^2 + 3)\, dx$$

$$= \pi\left(\frac{x^5}{5} - \frac{4}{3}x^3 + 3x\right)\Bigg|_{-1}^{1} = \frac{56\pi}{15}$$

[See Section 18-2.]

PROBLEM 18-17 The region bounded by the graph of $y = \sin x$ and the x axis between $x = 0$ and $x = \pi$ is rotated about the line $x = -2$. Find the volume of the resulting solid.

Solution: Translate the region two units to the right and rotate about the y axis. This new region is bounded above by the graph of $y = \sin(x - 2)$ and below by the x axis, between $x = 2$ and $x = 2 + \pi$. By the shell method, the volume is

$$2\pi \int_{2}^{2+\pi} x \sin(x - 2)\, dx$$

Integrate by parts: Let $u = x$, $du = dx$, $dv = \sin(x - 2)\, dx$, and $v = -\cos(x - 2)$:

$$2\pi \int_{2}^{2+\pi} x \sin(x - 2)\, dx = 2\pi\left[-x\cos(x - 2)\Bigg|_{2}^{2+\pi} + \int_{2}^{2+\pi} \cos(x - 2)\, dx\right]$$

$$= 2\pi[\sin(x - 2) - x\cos(x - 2)]\Bigg|_{2}^{2+\pi}$$

$$= 2\pi\{[\sin \pi - (2 + \pi)\cos \pi] - (\sin 0 - 2\cos 0)\}$$

$$= 2\pi(4 + \pi) = 8\pi + 2\pi^2$$

[See Section 18-2.]

PROBLEM 18-18 The base of a certain solid is a disk of radius 1. A line runs through the center of this disk. Any cross section of the solid by a plane perpendicular to this line is a square. Find the volume of this solid.

Solution: Let the line through the center be an x axis. Then the cross-sectional area at x is the area of a square with side extending from $\sqrt{1 - x^2}$ to $-\sqrt{1 - x^2}$ (see Figure 18-18). That is,

$$A(x) = (2\sqrt{1 - x^2})^2 = 4 - 4x^2$$

The volume of the region is

$$\int_{-1}^{1} A(x)\, dx = \int_{-1}^{1} (4 - 4x^2)\, dx$$

$$= \left(4x - \frac{4}{3}x^3\right)\Bigg|_{-1}^{1} = \frac{16}{3}$$

[See Section 18-3.]

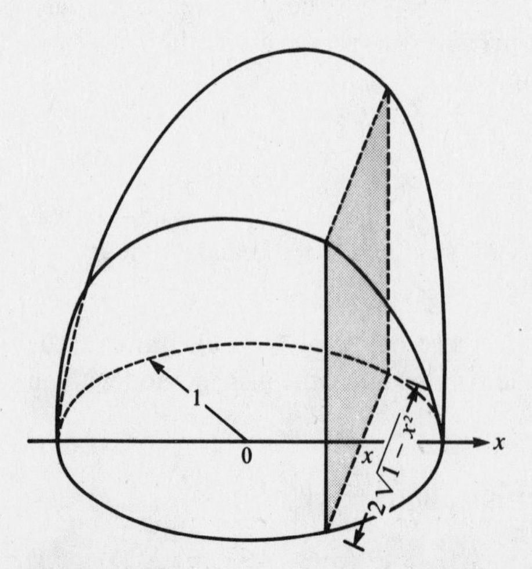

Figure 18-18
Problem 18-18

PROBLEM 18-19 Find the volume of the right circular cone of height h and radius r.

Solution: Let the axis of the cone be an x axis with $x = 0$ at the apex of the cone. The cross section at x is a circle of radius rx/h, as you can see in Figure 18-19. Thus,

$$A(x) = \pi\left(\frac{rx}{h}\right)^2 = \frac{\pi r^2}{h^2} x^2$$

and so the volume of the cone is

$$\int_0^h A(x)\,dx = \int_0^h \frac{\pi r^2}{h^2} x^2\,dx$$
$$= \frac{\pi r^2}{h^2}\left(\frac{x^3}{3}\right)\Bigg|_0^h = \frac{1}{3}\pi r^2 h$$

[See Section 18-3.]

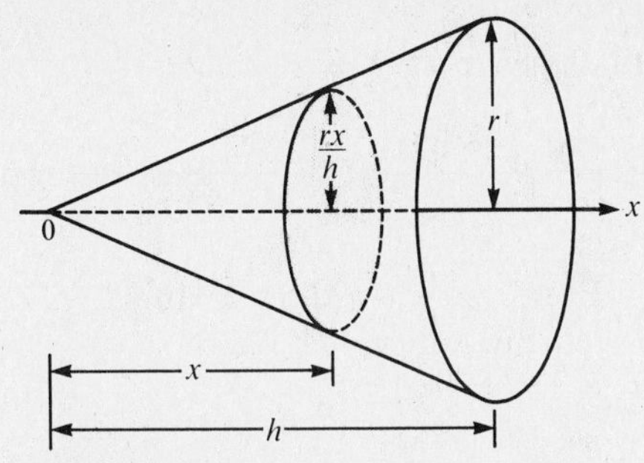

Figure 18-19
Problem 18-19

Supplementary Exercises

In Problems 18-20 through 18-27 the region bounded by the given curves is rotated about the x axis. Use the disk method to find the volume of the resulting solid:

18-20 $y = \sqrt{x}$, $x = 4$, x axis

18-21 $y = 1 + \cos x$, $x = 0$, $x = \pi/2$, x axis

18-22 $y = 1/\sqrt{1 + x^2}$, $x = -1$, $x = 1$, x axis

18-23 $y = x - x^2$, x axis

18-24 $y = x^4$, $y = x$

18-25 $y = \sqrt[3]{x}$, $y = x$

18-26 $y = 4 - x^2$, $y = 6 - 3x$

18-27 $y = x^3$, $x = -2$, x axis

In Problems 18-28 through 18-35, the region bounded by the given curve is rotated about the y axis. Use the shell method to find the volume of the resulting solid:

18-28 $y = \sqrt{x}$, $x = 4$, x axis

18-29 $y = 1/(1 + x^2)$, $x = 0$, $x = 1$, x axis

18-30 $y = \sin x$, $x = 0$, $x = \pi$, x axis

18-31 $y = x - x^2$, x axis

18-32 $y = x^4$, $y = x$

18-33 $y = e^{-x^2}$, $x = 0$, $x = 1$

18-34 $y = 4 - x^2$, $y = 6 - 3x$

18-35 $y = 2 - \frac{1}{2}x$, $x = 1$, $y = \frac{1}{2}$

In Problems 18-36 through 18-39 the region bounded by the given curves is rotated about the y axis. Use the disk method to find the volume of the resulting solid:

18-36 $y = x^4$, $y = x$

18-37 $y = x^2$, $x = 2$, x axis

18-38 $y = x^3$, $y = 1$, y axis

18-39 $y = x - 1$, $y = x - 3$, $y = -1$, $y = 1$

In Problems 18-40 through 18-43 the region bounded by the given curves is rotated about the x axis. Use the shell method to find the volume of the resulting solid:

18-40 $y = x^4, y = x$

18-41 $y = x^2, x = 2$, x axis

18-42 $y = x^3, y = 1$, y axis

18-43 $y = x - 1, y = x - 3, y = -1, y = 0$

In Problems 18-44 through 18-48 the region bounded by the given curves is rotated about the x axis. Find the volume of the resulting solid:

18-44 $y = 2x, y = x^2, x = 0, x = 1$

18-45 $y = x^2, y = 2x - 1, y = 4$

18-46 $y = \sqrt{1 - x^2}, y = 1 - x$

18-47 $y = e^x, y = e^{2x}, y = e^2$

18-48 $y = e^x, y = e^{2x}, x = 1$

In Problems 18-49 through 18-53 the region bounded by the given curves is rotated about the y axis. Find the volume of the resulting solid:

18-49 $y = 2x, y = x^2, x = 0, x = 1$

18-50 $y = x^2, y = 2x - 1, y = 4$

18-51 $y = \sqrt{1 - x^2}, y = 1 - x$

18-52 $y = e^x, y = e^{2x}, y = e^2$

18-53 $y = e^x, y = e^{2x}, x = 1$

18-54 The region bounded by the graphs of $y = x^2, x = 1$, and the x axis is rotated about the line $y = 1$. Find the volume of the resulting solid.

18-55 The region bounded by the graphs of $y = \ln x, x = e$, and the x axis is rotated about the line $y = -2$. Find the volume of the resulting solid.

18-56 The region bounded by the graphs of $y = x^2, x = 1$, and the x axis is rotated about the line $x = -1$. Find the volume of the resulting solid.

18-57 The region bounded by the graphs of $y = e^x, x = 1$, and $y = 1$ is rotated about the line $x = 1$. Find the volume of the resulting solid.

18-58 A monument stands 50 m high. A horizontal cross section x m from the top is an equilateral triangle with sides $x/5$ m. What is the volume of the monument?

18-59 The base of a certain solid is an elliptical region, $x^2 + 4y^2 \leqslant 4$. Any cross section of the solid by a plane perpendicular to the x axis is a half-disk. Find the volume of the solid.

18-60 The base of a certain solid is a disk of radius 1. A line runs through the center of the disk. Any cross section of the solid by a plane perpendicular to the line is an equilateral triangle. Find the volume of the solid.

Solutions to Supplementary Exercises

(18-20) $\pi \int_0^4 (\sqrt{x})^2\, dx = 8\pi$

(18-21) $\pi \int_0^{\pi/2} (1 + \cos x)^2\, dx = (3/4)\pi^2 + 2\pi$

(18-22) $\pi \int_{-1}^1 [1/(1 + x^2)]\, dx = \pi^2/2$

(18-23) $\pi \int_0^1 (x - x^2)^2\, dx = \pi/30$

(18-24) $\pi \int_0^1 (x^2 - x^8)\, dx = 2\pi/9$

(18-25) $\pi \int_0^1 (x^{2/3} - x^2)\, dx = 4\pi/15$

(18-26) $\pi \int_1^2 [(4 - x^2)^2 - (6 - 3x)^2]\,dx = 8\pi/15$

(18-27) $\pi \int_{-2}^0 (x^3)^2\,dx = 128\pi/7$

(18-28) $2\pi \int_0^4 x(\sqrt{x})\,dx = 128\pi/5$

(18-29) $2\pi \int_0^1 x[1/(1 + x^2)]\,dx = \pi \ln 2$

(18-30) $2\pi \int_0^\pi x \sin x\,dx = 2\pi^2$

(18-31) $2\pi \int_0^1 x(x - x^2)\,dx = \pi/6$

(18-32) $2\pi \int_0^1 x(x - x^4)\,dx = \pi/3$

(18-33) $2\pi \int_0^1 xe^{-x^2}\,dx = \pi(1 - e^{-1})$

(18-34) $2\pi \int_1^2 x[(4 - x^2) - (6 - 3x)]\,dx = \pi/2$

(18-35) $2\pi \int_1^3 x(2 - \frac{1}{2}x - \frac{1}{2})\,dx = 10\pi/3$

(18-36) $\pi \int_0^1 (y^{1/2} - y^2)\,dy = \pi/3$

(18-37) $\pi \int_0^4 (4 - y)\,dy = 8\pi$

(18-38) $\pi \int_0^1 (y^{1/3})^2\,dy = 3\pi/5$

(18-39) $\pi \int_{-1}^1 [(y + 3)^2 - (y + 1)^2]\,dy = 16\pi$

(18-40) $2\pi \int_0^1 y(y^{1/4} - y)\,dy = 2\pi/9$

(18-41) $2\pi \int_0^4 y(2 - \sqrt{y})\,dy = 32\pi/5$

(18-42) $2\pi \int_0^1 y(y^{1/3})\,dy = 6\pi/7$

(18-43) $-2\pi \int_{-1}^0 y[(y + 3) - (y + 1)]\,dy = 2\pi$

(18-44) $\pi \int_0^1 [(2x)^2 - (x^2)^2]\,dx = 17\pi/15$

(18-45) $2\pi \int_1^4 y\left(\frac{y + 1}{2} - \sqrt{y}\right)dy = \frac{37\pi}{10}$

(18-46) $\pi \int_0^1 [(1 - x^2) - (1 - x)^2]\,dx = \pi/3$

(18-47) $2\pi \int_1^{e^2} y(\ln y - \frac{1}{2}\ln y)\,dy = \frac{1}{4}\pi(1 + 3e^4)$

(18-48) $\pi \int_0^1 [(e^{2x})^2 - (e^x)^2]\,dx = \frac{1}{4}\pi(1 + e^4 - 2e^2)$

(18-49) $2\pi \int_0^1 x(2x - x^2)\,dx = 5\pi/6$

(18-50) $\pi \int_1^4 \left[\left(\frac{y + 1}{2}\right)^2 - y\right]dy = \frac{9\pi}{4}$

(18-51) $2\pi \int_0^1 x[\sqrt{1 - x^2} - (1 - x)]\,dx = \pi/3$

(18-52) $\pi \int_1^{e^2} [(\ln y)^2 - (\frac{1}{2}\ln y)^2]\,dy = (3\pi/2)(e^2 - 1)$

(18-53) $2\pi \int_0^1 x(e^{2x} - e^x)\,dx = (\pi/2)(e^2 - 3)$

(18-54) $\pi \int_0^1 [(-1)^2 - (x^2 - 1)^2]\,dx = 7\pi/15$

(18-55) $\pi \int_1^e [(2 + \ln x)^2 - 2^2]\,dx = \pi(2 + e)$

(18-56) $2\pi \int_1^2 x(x - 1)^2\,dx = 7\pi/6$

(18-57) $-2\pi \int_{-1}^0 x(e^{x+1} - 1)\,dx = \pi(2e - 5)$

(18-58) $\int_0^{50} (\sqrt{3}x^2/100)\,dx = (1250\sqrt{3}/3)\,m^3$

(18-59) $\int_{-2}^2 [\pi(4 - x^2)/8]\,dx = 4\pi/3$

(18-60) $\int_{-1}^1 (\sqrt{3}/4)\,4(1 - x^2)\,dx = 4\sqrt{3}/3$

19 OTHER APPLICATIONS OF THE INTEGRAL

THIS CHAPTER IS ABOUT

- ☑ Arc Length
- ☑ Surface Area of a Solid of Revolution
- ☑ Centroids
- ☑ Hydrostatic Pressure
- ☑ Work

19-1. Arc Length

The **length of the arc** of the curve $y = f(x)$ between $x = a$ and $x = b$ is the distance traveled by a particle that moves along the graph of $y = f(x)$ between $x = a$ and $x = b$. If $f'(x)$ is continuous on the interval $[a, b]$, then the length s of the arc of the curve $y = f(x)$ between $x = a$ and $x = b$ is

ARC LENGTH

$$s = \int_a^b \sqrt{1 + (f'(x))^2}\, dx \tag{19-1}$$

EXAMPLE 19-1: Find the length of the arc of the graph of $f(x) = 4x^{3/2}$ between $x = 0$ and $x = \frac{2}{3}$.

Solution: First find $f'(x)$:

$$f'(x) = 6x^{1/2}$$

Now use the formula. The length of the arc is

$$s = \int_0^{2/3} \sqrt{1 + (f'(x))^2}\, dx = \int_0^{2/3} \sqrt{1 + (6x^{1/2})^2}\, dx = \int_0^{2/3} \sqrt{1 + 36x}\, dx$$

$$= \frac{1}{36}\left(\frac{2}{3}\right)(1 + 36x)^{3/2}\Big|_0^{2/3} = \frac{1}{54}(125 - 1) = \frac{62}{27}$$

EXAMPLE 19-2: Find the length of $y = (x^3/6) + [1/(2x)]$ between $x = 1$ and $x = 3$.

Solution: First find dy/dx:

$$\frac{dy}{dx} = \frac{1}{2}x^2 - \frac{1}{2x^2}$$

Then find the length:

$$s = \int_1^3 \sqrt{1 + \left(\frac{dy}{dx}\right)^2}\, dx = \int_1^3 \sqrt{1 + \left(\frac{1}{2}x^2 - \frac{1}{2x^2}\right)^2}\, dx$$

$$= \int_1^3 \sqrt{1 + \frac{1}{4}x^4 - \frac{1}{2} + \frac{1}{4x^4}}\, dx = \int_1^3 \sqrt{\frac{1}{4}x^4 + \frac{1}{2} + \frac{1}{4x^4}}\, dx$$

$$= \int_1^3 \sqrt{\left(\frac{1}{2}x^2 + \frac{1}{2x^2}\right)^2}\, dx = \int_1^3 \left(\frac{1}{2}x^2 + \frac{1}{2x^2}\right) dx = \left(\frac{x^3}{6} - \frac{1}{2x}\right)\Big|_1^3 = \frac{14}{3}$$

EXAMPLE 19-3: Find the length of the curve $\cos x = e^y$ for x between $\pi/6$ and $\pi/3$.

Solution: Solving $\cos x = e^y$ for y, you find $y = \ln \cos x$. Find dy/dx:

$$\frac{dy}{dx} = \frac{1}{\cos x}(-\sin x) = -\tan x$$

The length of the arc is

$$s = \int_{\pi/6}^{\pi/3} \sqrt{1 + (-\tan x)^2}\, dx = \int_{\pi/6}^{\pi/3} \sqrt{1 + \tan^2 x}\, dx$$

$$= \int_{\pi/6}^{\pi/3} \sqrt{\sec^2 x}\, dx = \int_{\pi/6}^{\pi/3} \sec x\, dx = \ln|\sec x + \tan x|\Big|_{\pi/6}^{\pi/3}$$

$$= \ln|2 + \sqrt{3}| - \ln\left|\frac{2\sqrt{3}}{3} + \frac{\sqrt{3}}{3}\right| = \ln\left(\frac{2 + \sqrt{3}}{\sqrt{3}}\right)$$

$$= \ln\left(1 + \frac{2}{3}\sqrt{3}\right)$$

19-2. Surface Area of a Solid of Revolution

If the graph of $y = f(x)$ between $x = a$ and $x = b$ is rotated about the x axis, and if $f'(x)$ is continuous on $[a, b]$ and $f(x) \geqslant 0$ on $[a, b]$, then the **lateral surface area** (excluding the ends) of the resulting solid is

SURFACE AREA $$S = 2\pi \int_a^b f(x) \sqrt{1 + (f'(x))^2}\, dx \tag{19-2}$$

EXAMPLE 19-4: Find the lateral surface area of the solid obtained by rotating the graph of $y = 2\sqrt{x}$ between $x = 3$ and $x = 8$ about the x axis.

Solution: Simply apply the formula. First find y':

$$y' = x^{-1/2}$$

So, the surface area is

$$S = 2\pi \int_3^8 y\sqrt{1 + (y')^2}\, dx = 2\pi \int_3^8 2\sqrt{x}\sqrt{1 + (x^{-1/2})^2}\, dx$$

$$= 4\pi \int_3^8 \sqrt{x}\sqrt{1 + \frac{1}{x}}\, dx = 4\pi \int_3^8 \sqrt{x + 1}\, dx = 4\pi \frac{2}{3}(x + 1)^{3/2}\Big|_3^8$$

$$= \frac{8\pi}{3}(27 - 8) = \frac{152\pi}{3}$$

EXAMPLE 19-5: The region bounded by the graph of $y = x^3$, the x axis, and $x = \frac{1}{2}$ is rotated about the x axis. Find the lateral surface area of the resulting solid.

Solution: The surface area is

$$S = 2\pi \int_0^{1/2} y\sqrt{1 + (y')^2}\, dx = 2\pi \int_0^{1/2} x^3\sqrt{1 + (3x^2)^2}\, dx$$

$$= 2\pi \int_0^{1/2} x^3\sqrt{1 + 9x^4}\, dx = 2\pi \frac{1}{36}\left(\frac{2}{3}\right)(1 + 9x^4)^{3/2}\Big|_0^{1/2}$$

$$= \frac{\pi}{27}\left[\left(\frac{25}{16}\right)^{3/2} - 1\right] = \frac{61\pi}{1728}$$

Note: The procedure for parametric equations is similar. If $x = f(t)$ and $y = g(t)$ for $a \leqslant t \leqslant b$, then the formulas for arc length and surface area are as follows:

$$s = \int_a^b \sqrt{(f'(t))^2 + (g'(t))^2}\, dt \tag{19-3}$$

$$S = 2\pi \int_a^b g(t)\sqrt{(f'(t))^2 + (g'(t))^2}\, dt \tag{19-4}$$

19-3. Centroids

A. Center of mass of a horizontal rod

Consider a rod extending along the x axis from $x = a$ to $x = b$. Suppose that the density of the rod varies along its length. If the density of the rod at x is $\rho(x)$, then the **center of mass** $\bar{x}$ of the rod is

CENTER OF MASS

$$\bar{x} = \frac{M_y}{m} = \frac{\int_a^b x\rho(x)\, dx}{\int_a^b \rho(x)\, dx} \tag{19-5}$$

where the numerator is the **first moment** M_y of the rod about the y axis and the denominator is the mass m of the rod. (It may help you to think of the first moment about the y axis as a measure of the tendency of the rod to rotate about that axis.)

EXAMPLE 19-6: A 1-m metal rod has density $\rho(x) = 1/(1 + x^2)$ kg/m x meters from one end. Find the center of mass of the rod.

Solution: The first moment is

$$M_y = \int_0^1 x\rho(x)\, dx = \int_0^1 x\frac{1}{1 + x^2}\, dx$$

$$= \frac{1}{2}\ln(1 + x^2)\Big|_0^1 = \frac{1}{2}\ln 2 \text{ kg m}$$

The mass of the rod is

$$m = \int_0^1 \rho(x)\, dx = \int_0^1 \frac{1}{1 + x^2}\, dx$$

$$= \arctan x\Big|_0^1 = \frac{\pi}{4} \text{ kg}$$

The center of mass of the rod is

$$\bar{x} = \frac{M_y}{m} = \frac{\frac{1}{2}\ln 2}{\pi/4} = \frac{2}{\pi}\ln 2 \text{ m}$$

that is, 0.44 m from the specified end.

B. Centroid of a region in the plane

Suppose that a region in the plane is covered by a uniform thickness of material of constant density—a **lamina**. The center of mass of this object is the point in the plane from which you could suspend the region in perfect balance. This point is called the **centroid** of the plane region. If the region can be described as bounded above by the graph of $y = f(x)$ and below by the graph of $y = g(x)$, between $x = a$ and $x = b$, then the centroid $(\bar{x}, \bar{y})$ of the lamina is

CENTROID OF PLANE REGION

$$\bar{x} = \frac{\text{first moment about the } y \text{ axis}}{\text{area of the region}} = \frac{M_y}{A} = \frac{\int_a^b x[f(x) - g(x)]\,dx}{\int_a^b [f(x) - g(x)]\,dx} \tag{19-6}$$

$$\bar{y} = \frac{\text{first moment about the } x \text{ axis}}{\text{area of the region}} = \frac{M_x}{A} = \frac{\frac{1}{2}\int_a^b [f^2(x) - g^2(x)]\,dx}{\int_a^b [f(x) - g(x)]\,dx}$$

EXAMPLE 19-7: Find the centroid of the region bounded by the graph of $y = x^2$, the x axis, and the line $x = 2$.

Solution: This region is bounded above by the graph of $f(x) = x^2$, below by the x axis ($g(x) = 0$), on the left by $x = 0$, and on the right by $x = 2$. You must evaluate three integrals:

$$M_y = \int_0^2 xf(x)\,dx = \int_0^2 x(x^2)\,dx = \frac{1}{4}x^4\Big|_0^2 = 4$$

$$M_x = \frac{1}{2}\int_0^2 f^2(x)\,dx = \frac{1}{2}\int_0^2 (x^2)^2\,dx = \frac{1}{10}x^5\Big|_0^2 = \frac{16}{5}$$

$$A = \int_0^2 f(x)\,dx = \int_0^2 x^2\,dx = \frac{x^3}{3}\Big|_0^2 = \frac{8}{3}$$

$$\bar{x} = \frac{M_y}{A} = \frac{4}{8/3} = \frac{3}{2}$$

$$\bar{y} = \frac{M_x}{A} = \frac{16/5}{8/3} = \frac{6}{5}$$

The centroid is (3/2, 6/5).

EXAMPLE 19-8: Find the centroid of the region bounded by the graphs of $y = \sqrt{x}$ and $y = \frac{1}{2}x$.

Solution: This region is bounded above by the graph of $y = \sqrt{x}$ and below by $y = \frac{1}{2}x$, between $x = 0$ and $x = 4$. You must find M_x, M_y, and A.

$$M_y = \int_0^4 x\left(\sqrt{x} - \frac{1}{2}x\right)dx = \int_0^4 \left(x^{3/2} - \frac{1}{2}x^2\right)dx = \left(\frac{2}{5}x^{5/2} - \frac{1}{6}x^3\right)\Big|_0^4$$

$$= \frac{2}{5}(4)^{5/2} - \frac{1}{6}(4)^3 = \frac{32}{15}$$

$$M_x = \frac{1}{2}\int_0^4 \left[(\sqrt{x})^2 - \left(\frac{1}{2}x\right)^2\right]dx = \frac{1}{2}\int_0^4 \left(x - \frac{1}{4}x^2\right)dx$$

$$= \frac{1}{2}\left(\frac{1}{2}x^2 - \frac{x^3}{12}\right)\Big|_0^4 = \frac{4}{3}$$

$$A = \int_0^4 \left(\sqrt{x} - \frac{1}{2}x\right)dx = \left(\frac{2}{3}x^{3/2} - \frac{1}{4}x^2\right)\Big|_0^4 = \frac{4}{3}$$

$$\bar{x} = \frac{M_y}{A} = \frac{32/15}{4/3} = \frac{8}{5}$$

$$\bar{y} = \frac{M_x}{A} = \frac{4/3}{4/3} = 1$$

The centroid is (8/5, 1).

19-4. Hydrostatic Pressure

If a plate is submerged vertically in a liquid, then the total force on one side of the plate is the product of the pressure at the centroid of the plate with the area of the plate. The pressure at a given depth is the product of the depth with the density of the liquid. Thus, the force on one side of the vertical plate is

FORCE = (depth of centroid) × (area of plate) × (density of liquid) **(19-7)**

EXAMPLE 19-9: A trough is filled with water (density 62.5 lb/ft³). The ends of the trough are equilateral triangles, each side 3 ft. Find the total force on one end of the trough.

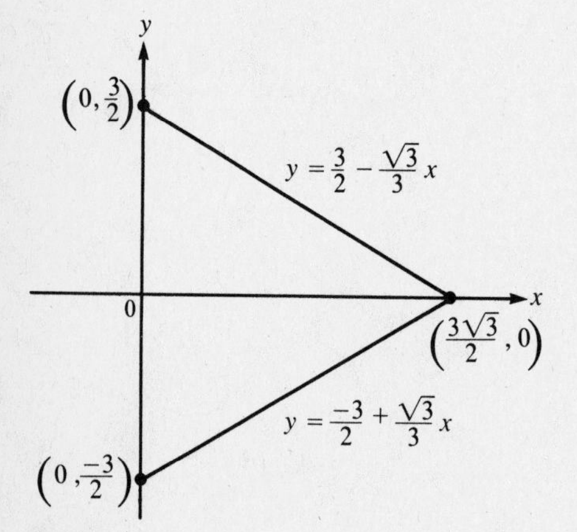

Figure 19-1
Example 19-9

Solution: Set up a coordinate system to find the centroid of an equilateral triangle. Try to choose one that will make your integrations as easy as possible, perhaps as in Figure 19-1. You must find M_y and A to find the depth of the centroid:

$$M_y = \int_0^{3\sqrt{3}/2} x\left[\left(\frac{3}{2} - \frac{\sqrt{3}}{3}x\right) - \left(\frac{-3}{2} + \frac{\sqrt{3}}{3}x\right)\right] dx$$

$$= \int_0^{3\sqrt{3}/2} \left(3x - \frac{2\sqrt{3}}{3}x^2\right) dx$$

$$= \left(\frac{3}{2}x^2 - \frac{2\sqrt{3}}{9}x^3\right)\Bigg|_0^{3\sqrt{3}/2} = \frac{27}{8}$$

$$A = \int_0^{3\sqrt{3}/2} \left[\left(\frac{3}{2} - \frac{\sqrt{3}}{3}x\right) - \left(\frac{-3}{2} + \frac{\sqrt{3}}{3}x\right)\right] dx$$

$$= \int_0^{3\sqrt{3}/2} \left(3 - \frac{2\sqrt{3}}{3}x\right) dx$$

$$= \left(3x - \frac{\sqrt{3}}{3}x^2\right)\Bigg|_0^{3\sqrt{3}/2}$$

$$= \frac{9\sqrt{3}}{4}$$

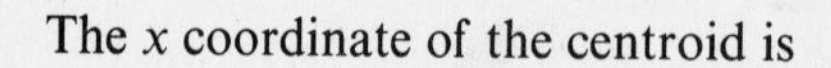

The x coordinate of the centroid is

$$\bar{x} = \frac{M_y}{A} = \frac{27/8}{9\sqrt{3}/4} = \frac{\sqrt{3}}{2}$$

So the depth of the centroid is $\sqrt{3}/2$ ft, and the area of an end of the trough is $A = 9\sqrt{3}/4$ ft². The force on an end of the trough is

$$F = (\text{depth of centroid}) \times (\text{area}) \times (\text{density})$$

$$= \left(\frac{\sqrt{3}}{2}\,\text{ft}\right)\left(\frac{9\sqrt{3}}{4}\,\text{ft}^2\right)\left(62.5\,\frac{\text{lb}}{\text{ft}^3}\right)$$

$$= 210.9375\ \text{lb}$$

EXAMPLE 19-10: A vertical dam across a river has parabolic shape. It is 40 ft across at water level and 10 ft deep at the deepest point. The density of water is 62.5 lb/ft³. Find the force exerted by the water on the dam.

Solution: Choose a convenient coordinate system, as in Figure 19-2. The parabola is $y = Cx^2$ for some C. Because the point (20, 10) lies on the parabola, you can find C:

$$10 = C(20)^2$$

$$C = \frac{1}{40}$$

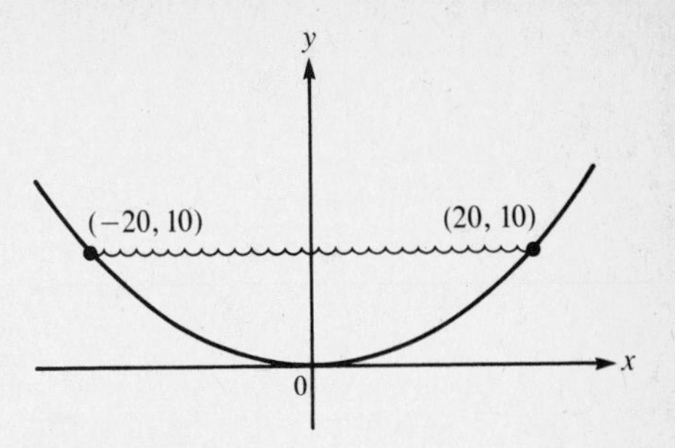

Figure 19-2
Example 19-10

So the parabola is $y = x^2/40$. To find the y coordinate of the centroid, you must find M_x and A:

$$M_x = \frac{1}{2}\int_{-20}^{20}\left[10^2 - \left(\frac{x^2}{40}\right)^2\right]dx = \frac{1}{2}\int_{-20}^{20}\left(100 - \frac{x^4}{1600}\right)dx$$

$$= \frac{1}{2}\left(100x - \frac{x^5}{8000}\right)\Bigg|_{-20}^{20} = 1600$$

$$A = \int_{-20}^{20}\left(10 - \frac{x^2}{40}\right)dx = \left(10x - \frac{x^3}{120}\right)\Bigg|_{-20}^{20} = \frac{800}{3}$$

$$\bar{y} = \frac{M_y}{A} = \frac{1600}{800/3} = 6$$

Because the y coordinate of the centroid is 6, the depth of the centroid is 4 ft. You've already found the area of the dam, $A = 800/3$. The force on the dam is

$$F = (4)\left(\frac{800}{3}\right)(62.5) = \frac{200\,000}{3}\text{ lb}$$

19-5. Work

Suppose that a variable force is applied along a straight line (make it the x axis). The **work** W done between $x = a$ and $x = b$ is defined to be

WORK $$W = \int_a^b F(x)\,dx \qquad \textbf{(19-8)}$$

where $F(x)$ is the force applied at x.

A. Springs

The force required to compress (or extend) a spring to a distance x from its natural length is proportional to x and is given by **Hooke's law**:

HOOKE'S LAW $$F(x) = Cx \qquad \textbf{(19-9)}$$

The constant C is called the **spring constant**, which varies from one spring to another.

EXAMPLE 19-11: A spring has a natural length of 14 in. It takes a force of 20 lb to compress the spring to a length of 10 in. How much work is done in compressing the spring from its natural length to a length of 10 in.?

Solution: First find the spring constant. To compress the spring from 14 in. to 10 in. requires 20 lb, so

$$20 = C(14 - 10)$$

$$C = 5$$

The spring constant is 5. The work done in compressing it 4 in. from its natural length to 10 in. is

$$\int_0^4 F(x)\,dx = \int_0^4 5x\,dx = \frac{5}{2}x^2\Bigg|_0^4 = 40\text{ in.-lb}$$

EXAMPLE 19-12: A spring has a natural length of 1 ft. When a 10 lb weight is suspended from the spring, it extends to a length of 1.5 ft. How much work is done in extending the spring from a length of 1.2 ft to a length of 1.6 ft?

Solution: Find the spring constant. A force of 10 lb extends the spring 0.5 ft, so

$$10 = C(0.5)$$
$$C = 20$$

The work done in extending it from 1.2 ft ($x = 0.2$) to 1.6 ft ($x = 0.6$) is

$$\int_{0.2}^{0.6} 20x\,dx = 10x^2\Big|_{0.2}^{0.6} = 3.2 \text{ ft-lb}$$

B. Other Systems

EXAMPLE 19-13: A cylindrical tank of radius 2 ft and height 4 ft is full of water. Water weighs 62.5 lb/ft^3. Find the work done in pumping the water out of the tank to a height 5 ft above the tank.

Solution: Think of the water in the tank as sliced horizontally into disks of width Δx. The weight of each disk is the volume of the disk times the density of water:

$$\pi 2^2\,\Delta x\,(62.5) = 250\pi\,\Delta x \text{ lb}$$

The disk x ft below the surface of the tank will be moved $5 + x$ ft. The work done in moving this disk of water is

$$W = 250\pi\,\Delta x\,(5 + x) \text{ lb}$$

Sum the work done in moving all the disks to obtain a Riemann sum for the integral:

$$W = 250\pi \int_0^4 (5 + x)\,dx = 250\pi\left(5x + \frac{1}{2}x^2\right)\Big|_0^4 = 7000\pi \text{ ft-lb}$$

EXAMPLE 19-14: A winch atop a 120 ft tall building pulls a cable attached to a 400 lb package. The cable weighs 2 lb/ft. How much work is done in lifting the package to a height of 60 ft?

Solution: When the package is x ft above the ground, the winch is pulling the package (400 lb) and $(120 - x)$ ft of cable that weighs $(120 - x)2$ lb. So the force on the winch is

$$F(x) = 400 + (120 - x)2 = 640 - 2x \text{ lb}$$

The work done is

$$\int_0^{60} F(x)\,dx = \int_0^{60} (640 - 2x)\,dx = (640x - x^2)\Big|_0^{60} = 34\,800 \text{ ft-lb}$$

SUMMARY

1. The graph of $f(x)$ over $[a, b]$ has length

$$s = \int_a^b \sqrt{1 + (f'(x))^2}\,dx$$

2. If you revolve the graph of $y = f(x)$ over the interval $[a, b]$ around the x axis, the surface area is

$$S = 2\pi \int_a^b f(x)\sqrt{1 + (f'(x))^2}\,dx$$

3. The centroid of the region in the plane bounded by the graphs of $y = f(x)$ and $y = 0$, between $x = a$ and $x = b$, is $(\bar{x}, \bar{y})$ where

$$\bar{x} = \frac{M_y}{A} = \frac{\int_a^b xf(x)\,dx}{\int_a^b f(x)\,dx}$$

$$\bar{y} = \frac{M_x}{A} = \frac{\frac{1}{2}\int_a^b f^2(x)\,dx}{\int_a^b f(x)\,dx}$$

4. The force on one side of a vertical plate submerged in a fluid is the area of the plate times the pressure at the centroid of the plate.
5. The work done by a variable force $F(x)$, applied along a line from $x = a$ to $x = b$, is

$$W = \int_a^b F(x)\,dx$$

SOLVED PROBLEMS

PROBLEM 19-1 Find the length of the arc of $f(x) = \frac{2}{3}(1 + x^2)^{3/2}$ between $x = 0$ and $x = 3$.

Solution: First differentiate $f(x)$: $f'(x) = (1 + x^2)^{1/2}(2x)$. Now use Equation 19-1:

$$s = \int_0^3 \sqrt{1 + (f'(x)^2)}\,dx = \int_0^3 \sqrt{1 + [(1 + x^2)^{1/2}(2x)]^2}\,dx$$

$$= \int_0^3 \sqrt{1 + (1 + x^2)4x^2}\,dx = \int_0^3 \sqrt{1 + 4x^2 + 4x^4}\,dx$$

$$= \int_0^3 \sqrt{(1 + 2x^2)^2}\,dx = \int_0^3 (1 + 2x^2)\,dx = \left(x + \frac{2}{3}x^3\right)\Big|_0^3 = 21$$

[See Section 19-1.]

PROBLEM 19-2 Find the length of the curve $y = (4/5)x^{5/4}$ between $x = 0$ and $x = 9$.

Solution: The arc length is

$$s = \int_0^9 \sqrt{1 + (x^{1/4})^2}\,dx = \int_0^9 \sqrt{1 + \sqrt{x}}\,dx$$

Let $u = 1 + \sqrt{x}$; then $x = (u - 1)^2$, and $dx = 2(u - 1)\,du$.

Make the substitution and solve for s:

$$s = \int_1^4 (\sqrt{u})2(u - 1)\,du = 2\int_1^4 (u^{3/2} - u^{1/2})\,du = 2\left(\frac{2}{5}u^{5/2} - \frac{2}{3}u^{3/2}\right)\Big|_1^4 = \frac{232}{15}$$

[See Section 19-1.]

PROBLEM 19-3 Find the length of the arc of $1 = e^y + x^2$ between $x = 0$ and $x = \frac{1}{2}$.

Solution: Solve for y in terms of x:

$$y = \ln(1 - x^2)$$

Differentiate:

$$y' = \frac{-2x}{1 - x^2}$$

$$1 + (y')^2 = 1 + \left(\frac{-2x}{1 - x^2}\right)^2 = 1 + \frac{4x^2}{(1 - x^2)^2} = \frac{x^4 - 2x^2 + 1 + 4x^2}{(1 - x^2)^2}$$

$$= \frac{x^4 + 2x^2 + 1}{(1 - x^2)^2} = \frac{(x^2 + 1)^2}{(1 - x^2)^2}$$

The arc length is

$$s = \int_0^{1/2} \sqrt{\frac{(x^2 + 1)^2}{(1 - x^2)^2}}\, dx = \int_0^{1/2} \left|\frac{x^2 + 1}{1 - x^2}\right| dx = \int_0^{1/2} \frac{x^2 + 1}{1 - x^2}\, dx = \int_0^{1/2} \frac{x^2 - 1 + 2}{1 - x^2}\, dx$$

$$= \int_0^{1/2} \frac{-(1 - x^2) + 2}{1 - x^2}\, dx = \int_0^{1/2} \left(-1 + \frac{2}{1 - x^2}\right) dx$$

$$= \left(-x + \ln\left|\frac{x + 1}{x - 1}\right|\right)\Bigg|_0^{1/2} = \ln 3 - \frac{1}{2}$$

[See Section 19-1.]

PROBLEM 19-4 Find the length of the curve $y = \frac{1}{2}x^2$ between $x = -\sqrt{3}$ and $x = 0$.

Solution: Plugging the given values into the formula gives you:

$$s = \int_{-\sqrt{3}}^{0} \sqrt{1 + (x)^2}\, dx = \left(\frac{1}{2}x\sqrt{1 + x^2} + \frac{1}{2}\ln|x + \sqrt{x^2 + 1}|\right)\Bigg|_{-\sqrt{3}}^{0}$$

$$= \sqrt{3} - \frac{1}{2}\ln(2 - \sqrt{3})$$

[See Section 19-1.]

PROBLEM 19-5 The curve $y = \sqrt{1 - x^2}$ between $x = -\frac{1}{2}$ and $x = \frac{1}{2}$ is rotated about the x axis. Find the surface area of the resulting solid.

Solution: First find y' and $1 + (y')^2$:

$$y' = \frac{-x}{\sqrt{1 - x^2}}$$

$$1 + (y')^2 = 1 + \frac{x^2}{1 - x^2} = \frac{1}{1 - x^2}$$

The surface area is

$$S = 2\pi \int_{-1/2}^{1/2} y\sqrt{1 + (y')^2}\, dx = 2\pi \int_{-1/2}^{1/2} \sqrt{1 - x^2}\sqrt{\frac{1}{1 - x^2}}\, dx$$

$$= 2\pi \int_{-1/2}^{1/2} dx = 2\pi$$

[See Section 19-2.]

PROBLEM 19-6 The graph of $y = (x/3)\sqrt{x + 3}$ between $x = 1$ and $x = 4$ is rotated about the x axis. Find the resulting surface area.

Solution: You find that

$$y' = \frac{1}{3}\sqrt{x + 3} + \frac{x}{3}\left(\frac{1}{2\sqrt{x + 3}}\right) = \frac{1}{3}\left(\sqrt{x + 3} + \frac{x}{2\sqrt{x + 3}}\right)$$

$$= \frac{1}{3}\left(\frac{2x + 6 + x}{2\sqrt{x + 3}}\right) = \frac{x + 2}{2\sqrt{x + 3}}$$

$$1 + (y')^2 = 1 + \left(\frac{x + 2}{2\sqrt{x + 3}}\right)^2 = 1 + \frac{x^2 + 4x + 4}{4(x + 3)} = \frac{x^2 + 8x + 16}{4(x + 3)} = \frac{(x + 4)^2}{4(x + 3)}$$

The surface area is

$$S = 2\pi \int_1^4 y\sqrt{1 + (y')^2}\,dx = 2\pi \int_1^4 \frac{x}{3}\sqrt{x+3}\sqrt{\frac{(x+4)^2}{4(x+3)}}\,dx$$

$$= 2\pi \int_1^4 \frac{x(x+4)}{6}\,dx = \frac{\pi}{3}\left(\frac{x^3}{3} + 2x^2\right)\Bigg|_1^4 = 17\pi$$

[See Section 19-2.]

PROBLEM 19-7 The graph of $y = (x^3/3) + [1/(4x)]$ between $x = 1$ and $x = 2$ is rotated about the x axis. Find the lateral surface area of the resulting solid.

Solution: Differentiating, you find

$$y' = x^2 - \frac{1}{4x^2} = \frac{4x^4 - 1}{4x^2}$$

$$1 + (y')^2 = 1 + \left(\frac{4x^4 - 1}{4x^2}\right)^2 = \frac{16x^4 + 16x^8 - 8x^4 + 1}{16x^4} = \frac{16x^8 + 8x^4 + 1}{16x^4} = \left(\frac{4x^4 + 1}{4x^2}\right)^2$$

The surface area is

$$S = 2\pi \int_1^2 y\sqrt{1 + (y')^2}\,dx = 2\pi \int_1^2 \left(\frac{x^3}{3} + \frac{1}{4x}\right)\sqrt{\left(\frac{4x^4 + 1}{4x^2}\right)^2}\,dx$$

$$= 2\pi \int_1^2 \left(\frac{x^3}{3} + \frac{1}{4x}\right)\left(\frac{4x^4 + 1}{4x^2}\right)dx = 2\pi \int_1^2 \left(\frac{x^5}{3} + \frac{x}{3} + \frac{1}{16}x^{-3}\right)dx$$

$$= 2\pi\left(\frac{x^6}{18} + \frac{x^2}{6} - \frac{1}{32}x^{-2}\right)\Bigg|_1^2 = \frac{515\pi}{64}$$

[See Section 19-2.]

PROBLEM 19-8 The graph of $y = e^x$ between $x = 0$ and $x = 1$ is rotated about the x axis. Find the surface area of the resulting solid.

Solution: The surface area is

$$S = 2\pi \int_0^1 y\sqrt{1 + (y')^2}\,dx = 2\pi \int_0^1 e^x\sqrt{1 + (e^x)^2}\,dx$$

Make the substitution $u = e^x$, with $du = e^x\,dx$:

$$S = 2\pi \int_1^e \sqrt{1 + u^2}\,du = \pi\left(u\sqrt{1 + u^2} + \ln|u + \sqrt{1 + u^2}|\right)\Bigg|_1^e$$

$$= \pi\left[e\sqrt{1 + e^2} - \sqrt{2} + \ln\left(\frac{e + \sqrt{1 + e^2}}{1 + \sqrt{2}}\right)\right]$$

[See Section 19-2.]

PROBLEM 19-9 A 3-ft rod has density $\rho(x) = e^{x/3}$ lb/ft x ft from one end. Find the center of mass of the rod.

Solution: The first moment is

$$\int_0^3 x\rho(x)\,dx = \int_0^3 xe^{x/3}\,dx = 3xe^{x/3}\Bigg|_0^3 - 3\int_0^3 e^{x/3}\,dx = (3xe^{x/3} - 9e^{x/3})\Bigg|_0^3 = 9$$

The mass of the rod is

$$\int_0^3 \rho(x)\,dx = \int_0^3 e^{x/3}\,dx = 3e^{x/3}\Bigg|_0^3 = 3(e - 1)$$

The center of mass of the rod is

$$\bar{x} = \frac{\text{first moment}}{\text{mass}} = \frac{9}{3(e-1)} = \frac{3}{e-1} \text{ ft from the specified end}$$

[See Section 19-3.]

PROBLEM 19-10 A 4-meter rod has density $\rho(x) = \sqrt{x}$ kg/m x meters from one end. Find the center of mass of the rod.

Solution: The first moment is

$$\int_0^4 x\rho(x)\,dx = \int_0^4 x\sqrt{x}\,dx = \frac{2}{5}x^{5/2}\Big|_0^4 = \frac{64}{5}$$

The mass of the rod is

$$\int_0^4 \rho(x)\,dx = \int_0^4 \sqrt{x}\,dx = \frac{2}{3}x^{3/2}\Big|_0^4 = \frac{16}{3}$$

The center of mass is

$$\bar{x} = \frac{64/5}{16/3} = \frac{12}{5} \text{ m from the specified end}$$

[See Section 19-3.]

PROBLEM 19-11 Find the centroid of the plane region bounded by the graph of $f(x) = e^x$ and the x axis between $x = 0$ and $x = 1$.

Solution: You must evaluate three integrals:

$$M_y = \int_0^1 xe^x\,dx = (xe^x - e^x)\Big|_0^1 = 1$$

$$A = \int_0^1 e^x\,dx = e^x\Big|_0^1 = e - 1$$

$$M_x = \frac{1}{2}\int_0^1 (e^x)^2\,dx = \frac{1}{4}e^{2x}\Big|_0^1 = \frac{1}{4}(e^2 - 1)$$

$$\bar{x} = \frac{M_y}{A} = \frac{1}{e-1}$$

$$\bar{y} = \frac{M_x}{A} = \frac{\frac{1}{4}(e^2 - 1)}{e - 1} = \frac{e+1}{4}$$

The centroid is $[1/(e - 1), (e + 1)/4]$.

[See Section 19-3.]

PROBLEM 19-12 Find the centroid of the region bounded by the graphs of $y = x^2 + 1$ and $y = x$ between $x = -1$ and $x = 1$.

Solution: You must evaluate three integrals:

$$M_y = \int_{-1}^1 x[(x^2 + 1) - x]\,dx = \int_{-1}^1 (x^3 + x - x^2)\,dx = \left(\frac{1}{4}x^4 + \frac{1}{2}x^2 - \frac{1}{3}x^3\right)\Big|_{-1}^1 = -\frac{2}{3}$$

$$M_x = \frac{1}{2}\int_{-1}^1 [(x^2 + 1)^2 - x^2]\,dx = \frac{1}{2}\int_{-1}^1 (x^4 + x^2 + 1)\,dx = \frac{1}{2}\left(\frac{x^5}{5} + \frac{x^3}{3} + x\right)\Big|_{-1}^1 = \frac{23}{15}$$

$$A = \int_0^1 [(x^2 + 1) - x]\,dx = \left(\frac{x^3}{3} + x - \frac{x^2}{2}\right)\Big|_{-1}^1 = \frac{8}{3}$$

$$\bar{x} = \frac{M_y}{A} = \frac{-2/3}{8/3} = -\frac{1}{4}$$

$$\bar{y} = \frac{M_x}{A} = \frac{23/15}{8/3} = \frac{23}{40}$$

The centroid is $(-1/4, 23/40)$.

[See Section 19-3.]

PROBLEM 19-13 Find the centroid of the region bounded by the graphs of $y = (x - 1)^2$ and $y = 1 + x$.

Solution: Find where the curves intersect:

$$(x-1)^2 = 1 + x$$
$$x^2 - 2x + 1 = 1 + x$$
$$x^2 - 3x = 0$$
$$x = 0, \qquad x = 3$$

Between $x = 0$ and $x = 3$, the region is bounded above by the graph of $y = 1 + x$ and below by $y = (x-1)^2$. You must evaluate three integrals:

$$M_y = \int_0^3 x[(1+x) - (x-1)^2]\,dx = \int_0^3 (3x^2 - x^3)\,dx = \left(x^3 - \frac{1}{4}x^4\right)\Big|_0^3 = \frac{27}{4}$$

$$M_x = \frac{1}{2}\int_0^3 [(1+x)^2 - (x-1)^4]\,dx = \frac{1}{2}\left[\frac{(1+x)^3}{3} - \frac{(x-1)^5}{5}\right]\Big|_0^3 = \frac{36}{5}$$

$$A = \int_0^3 [(1+x) - (x-1)^2]\,dx = \int_0^3 (3x - x^2)\,dx = \left(\frac{3}{2}x^2 - \frac{1}{3}x^3\right)\Big|_0^3 = \frac{9}{2}$$

$$\bar{x} = \frac{27/4}{9/2} = \frac{3}{2}$$

$$\bar{y} = \frac{36/5}{9/2} = \frac{8}{5}$$

The centroid is (3/2, 8/5). [See Section 19-3.]

PROBLEM 19-14 A circular patch of radius 3 in. is applied to a vertical wall of a vat. The top of the patch is 1 ft below fluid level. The fluid in the vat has density 48 lb/ft^3. Find the force on the patch.

Solution: The centroid of the patch is at its center, which is $1\frac{1}{4}$ ft below fluid level. The patch has area $\pi/16$ ft^2. The force on the patch is

$$F = (\text{depth of centroid}) \times (\text{area of patch}) \times (\text{density of fluid})$$
$$= \left(\frac{5}{4}\,\text{ft}\right)\left(\frac{\pi}{16}\,\text{ft}^2\right)(48\ \text{lb/ft}^3) = \frac{15\pi}{4}\ \text{lb}$$

[See Section 19-4.]

PROBLEM 19-15 The ends of a cylindrical barrel of oil are disks of diameter 4 ft. The oil has a density of 51 lb/ft^3. Determine the force on one end when the barrel is on its side and is half full.

Solution: Find the centroid of the submerged portion of an end. Set up a coordinate system, perhaps as in Figure 19-3. To find the $\bar{y}$ coordinate of the centroid, evaluate two integrals:

$$M_x = \frac{1}{2}\int_{-2}^{2} -(-\sqrt{4-x^2})^2\,dx$$
$$= -\frac{1}{2}\int_{-2}^{2} (4 - x^2)\,dx$$
$$= -\frac{1}{2}\left(4x - \frac{x^3}{3}\right)\Big|_{-2}^{2} = -\frac{16}{3}$$

$$A = \text{area of half-disk} = \int_{-2}^{2} \sqrt{4-x^2}\,dx = 2\pi$$

The centroid is located at $\bar{y} = \dfrac{-16/3}{2\pi} = \dfrac{-8}{3\pi}$,

so the depth of the centroid is $8/(3\pi)$ ft. The force on an end is

$$F = \left(\frac{8}{3\pi}\,\text{ft}\right)(2\pi\ \text{ft}^2)(51\ \text{lb/ft}^3) = 272\ \text{lb}$$

[See Section 19-4.]

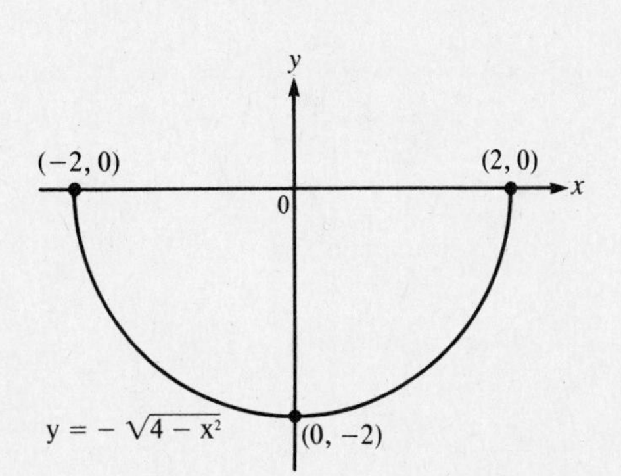

Figure 19-3
Problem 19-15

PROBLEM 19-16 A force of 120 lb compresses a spring 40 in. from its natural length. How much work is done in this compression?

Solution: Find the spring constant:

$$F(x) = Cx$$
$$120 = C(40)$$
$$C = 3$$

The work done is

$$\int_0^{40} F(x)\,dx = \int_0^{40} 3x\,dx = \frac{3}{2}x^2\Big|_0^{40} = 2400 \text{ in.-lb}$$

[See Section 19-5.]

PROBLEM 19-17 A force of 16 lb is required to extend a spring 2 ft beyond its natural length. How much work is done in extending the spring another foot (until it is 3 ft longer than its natural length)?

Solution: Find the spring constant:

$$F(x) = Cx$$
$$16 = C(2)$$
$$C = 8$$

The work done is

$$\int_2^{3} F(x)\,dx = \int_2^{3} 8x\,dx = 4x^2\Big|_2^{3} = 20 \text{ ft-lb}$$

[See Section 19-5.]

PROBLEM 19-18 A conical tank (apex down) has radius 2 ft and height 6 ft. The water in the tank is 4 ft deep. How much work is done in pumping the water out of the top of the tank?

Solution: Think of the water in the tank as sliced horizontally into disks of equal thickness Δy. Find the approximate volume of the disk that is a distance y from the bottom of the tank (see Figure 19-4). To do this, first find the radius r of a side of this disk by similar triangles:

$$\frac{2}{6} = \frac{r}{y}$$
$$r = \frac{y}{3}$$

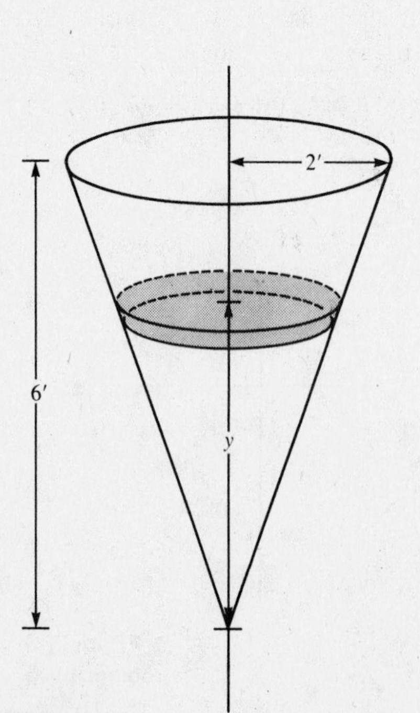

Figure 19-4
Problem 19-18

So the volume of the disk is approximately

$$\pi\left(\frac{y}{3}\right)^2 \Delta y = \frac{\pi y^2}{9}\Delta y$$

The weight of this disk is approximately

$$\text{(volume of disk)} \times \text{(density of water)}$$
$$= \frac{\pi y^2}{9}\Delta y\,(62.5)$$

The work in pumping this disk of water to the top of the tank is the weight of the water times the distance the water is lifted:

$$\frac{\pi y^2}{9}\Delta y\,(62.5)(6 - y)$$

Sum the work done in lifting all of the disks to obtain a Riemann sum for the integral:

$$\int_0^4 \frac{\pi y^2}{9}(62.5)(6 - y)\,dy = \frac{62.5}{9}\pi \int_0^4 (6y^2 - y^3)\,dy$$

$$= \frac{62.5}{9}\pi\left(2y^3 - \frac{1}{4}y^4\right)\Big|_0^4$$

$$= \frac{62.5}{9}\pi(64) = \frac{4000}{9}\pi \text{ ft-lb}$$

[See Section 19-5.]

PROBLEM 19-19 A rancher carries a 100-lb bag of feed up a vertical 20-ft ladder at a constant rate. The bag leaks at a constant rate, so when the bag reaches the top of the ladder, it weighs only 60 lb. How much work did the rancher do against gravity?

Solution: When the bag is x ft up the ladder, it weighs $100 - 2x$ lb. The work done is

$$\int_0^{20} F(x)\,dx = \int_0^{20} (100 - 2x)\,dx = (100x - x^2)\Big|_0^{20} = 1600 \text{ ft-lb}$$

[See Section 19-5.]

PROBLEM 19-20 At a distance x mi from the surface of the earth, the force of gravity on an object weighing p lb is approximately

$$p\left(\frac{4000}{4000 + x}\right)^2 \text{ lb}$$

Find the work done in raising a 20 lb object to a distance 2400 mi above the surface of the earth.

Solution: Let x be the distance to the surface of the earth. The force at x is

$$F(x) = 20\left(\frac{4000}{4000 + x}\right)^2$$

The work done in moving the object from the surface of the earth ($x = 0$) to $x = 2400$ is

$$\int_0^{2400} 20\left(\frac{4000}{4000 + x}\right)^2 dx = \frac{-20(4000)^2}{4000 + x}\Big|_0^{2400} = 30\,000 \text{ mi-lb}$$

[See Section 19-5.]

Supplementary Exercises

In Problems 19-21 through 19-28 find the length of the given curve on the given interval:

19-21 $y = 3 - \frac{1}{2}x$ $[0, 6]$

19-22 $y = \frac{1}{3}(x^2 + 2)^{3/2}$ $[0, 3]$

19-23 $y = (x^4/4) + [1/(8x^2)]$ $[1, 2]$

19-24 $y = (4 - x^{2/3})^{3/2}$ $[1, 8]$

19-25 $y = \cosh x$ $[0, 1]$

19-26 $y = (x^3/3) + [1/(4x)]$ $[1, 3]$

19-27 $y = 2(x - 1)^{3/2}$ $[1, 17/9]$

19-28 $y = \ln x$ $[1, \sqrt{3}]$

In Problems 19-29 through 19-34 find the lateral surface area of the surface obtained by rotating the graph of the given function on the given interval about the x axis:

19-29 $y = 3 - \frac{1}{2}x$ $[0, 6]$

19-30 $y = 4\sqrt{x - 1}$ $[1, 6]$

19-31 $y = (1 - x)^3$ $[\frac{1}{2}, 1]$

19-32 $y = \sin x$ $[0, \pi/2]$

19-33 $y = \cosh x$ $[0, 1]$

19-34 $y = \frac{1}{2}\ln x - \frac{1}{4}x^2$ $[1, e]$

In Problems 19-35 through 19-40 find the center of mass of the rod with given density and length:

19-35 $\rho(x) = 3x$ length 2

19-36 $\rho(x) = 1/(1 + x)$ length 3

19-37 $\rho(x) = \cos x$ length $\pi/2$

19-38 $\rho(x) = \arctan x$ length 1

19-39 $\rho(x) = 1/\sqrt{1 - x^2}$ length 1

19-40 $\rho(x) = \ln(x + 1)$ length $e - 1$

In Problems 19-41 through 19-50 find the centroid of the region bounded by the graphs of the given curves:

19-41 $y = \sqrt[3]{x}$, x axis, $x = 1$, $x = 8$

19-42 $y = x^2 + 2$, x axis, $x = -1$, $x = 1$

19-43 $y = e^x$, x axis, $x = -1$, $x = 0$

19-44 $y = x^2$, $y = 4$

19-45 $y = 1 + x^3$, x axis, $x = 1$

19-46 $y = 2x$, $y = x^2$

19-47 $y = x^2$, $y = x^3$

19-48 $y = \sqrt{1 - x^2}$, $y = 1 - x$

19-49 $y = e^x$, $x = 1$, $y = 1$

19-50 $y = 1/\sqrt{1 - x^2}$, x axis, $x = 0$, $x = \frac{1}{2}$

19-51 A vertical floodgate is 40 ft wide and holds back a depth of 12 ft of water. Assume that water weighs 62.5 lb/ft^3. Find the total force on the gate.

19-52 One vertical end of a trough is an isosceles triangle 2 ft × 3 ft × 3 ft. Find the force on one end of the trough when it is full of water.

19-53 The ends of a cylindrical barrel of oil are disks of diameter 4 ft. Oil has a density of 51 lb/ft^3. When the ends are vertical and the depth of oil in the barrel is 1 ft, what is the force on one end?

19-54 The vertical ends of a trough are trapezoids 4 ft across at the top, 2 ft across at the bottom, and 3 ft deep. Find the force on an end when the trough is full of water.

19-55 A dam across a river is a half-ellipse 50 ft across and 16 ft deep. Find the force on the dam.

19-56 A spring has a natural length of 16 in. A force of 10 lb will stretch it 2 in. How much work is done in stretching it to a length of 22 in.?

19-57 An 18-lb force will compress a spring 3 in. from its natural length. How much work is done in extending the spring 4 in. from its natural length?

19-58 Find the work done in pumping all of the water from a full conical tank (point down) of radius 10 ft and height 10 ft to a point 5 ft above the top of the tank.

19-59 Find the work done in pumping all of the water from a full cylindrical tank (flat ends are horizontal) of radius 10 ft and height 10 ft to a point 4 ft above the top of the tank.

19-60 Find the work done in pumping all of the water from a full hemispherical tank of radius 10 ft to a point 4 ft above the top of the tank.

19-61 A crane is lifting a 3000-lb car. The cable being used weighs 8 lb/ft. When the cable is attached to the car on the ground, 30 ft of cable hang from the crane. How much work is done in lifting the car 20 ft off the ground?

19-62 Find the work done in moving a 1-lb object from the surface of the earth to a point far away from the earth (infinity?). [See Problem 19-20.]

Solutions to Supplementary Exercises

(19-21) $3\sqrt{5}$

(19-22) 12

(19-23) 123/32

(19-24) 9

(19-25) $(\frac{1}{2})[e - (1/e)]$

(19-26) 53/6

(19-27) 52/27

(19-28) $2 - \sqrt{2} + \ln(1 + \sqrt{2}) - \frac{1}{2}\ln 3$

(19-29) $9\sqrt{5}\pi$

(19-30) $304\pi/3$

(19-31) $61\pi/1728$

(19-32) $\pi[\sqrt{2} + \ln(1 + \sqrt{2})]$

(19-33) $\frac{1}{4}\pi(e^2 - e^{-2} + 4)$

(19-34) $(\pi/16)(e^4 - 9)$

(19-35) 4/3

(19-36) $(3 - \ln 4)/\ln 4$

(19-37) $\frac{1}{2}\pi - 1$

(19-38) $(\pi - 2)/(\pi - 2\ln 2)$

(19-39) $2/\pi$

(19-40) $\frac{1}{4}(e^2 - 3)$

(19-41) (508/105, 62/75)

(19-42) (0, 83/70)

(19-43) $\left(\frac{2 - e}{e - 1}, \frac{e + 1}{4e}\right)$

(19-44) (0, 12/5)

(19-45) (1/5, 4/7)

(19-46) (1, 8/5)

(19-47) (3/5, 12/35)

(19-48) $\left(\frac{2}{3\pi - 6}, \frac{2}{3\pi - 6}\right)$

(19-49) $\left(\frac{1}{2(e - 2)}, \frac{e^2 - 3}{4(e - 2)}\right)$

(19-50) $\left(\frac{6 - 3\sqrt{3}}{\pi}, \frac{3\ln 3}{\pi}\right)$

(19-51) 180 000 lb

(19-52) 500/3 lb

(19-53) $62.5[3\sqrt{3} - (4\pi/3)]$ lb

(19-54) 750 lb

(19-55) 6250 lb

(19-56) 90 in.-lb

(19-57) 48 in.-lb

(19-58) $156\,250\pi$ ft-lb

(19-59) $562\,500\pi$ ft-lb

(19-60) $322\,916.\bar{6}\pi$ ft-lb

(19-61) 63 200 ft-lb

(19-62) 4000 ft-lb

20 POLAR COORDINATES

THIS CHAPTER IS ABOUT

- ☑ Polar Coordinates and Equations
- ☑ Graphs of Polar Equations
- ☑ Points of Intersection of Graphs
- ☑ Slope of a Tangent Line
- ☑ The Area Problem
- ☑ Arc Length

20-1. Polar Coordinates and Equations

A **polar coordinate system** consists of a fixed point O, called the **pole**, and a ray, emanating from O, called the **polar axis**. To locate a point P in the plane, you draw the segment OP. Let r be the length of OP, and let θ be the angle from the polar axis to OP, measured in a counterclockwise direction (see Figure 20-1). Then P may be described by any one of the following ordered pairs: (r, θ), $(r, \theta \pm 2k\pi)$, or $(-r, \theta \pm (2k - 1)\pi)$, where k is any integer.

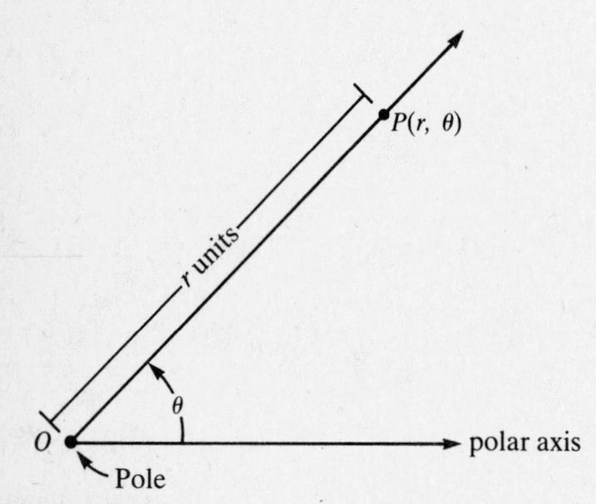

Figure 20-1
Polar coordinate system.

If you want to graph a given ordered pair (r, θ), you first draw θ, with its initial side along the polar axis and its vertex at the pole. Then, if $r > 0$, mark off r units along the terminal side of θ, starting from the pole. If r is negative, you first extend the terminal side of θ back through the pole (this is equivalent to drawing $\theta + \pi$) and then mark off r units (see Figure 20-2).

If you superimpose a polar system on a rectangular system, with the pole coinciding with the origin and the polar axis along the positive x axis, you can find the equations that relate (r, θ) of the polar system to (x, y) of the rectangular system (see Figure 20-3):

$$\begin{Bmatrix} x = r\cos\theta \\ y = r\sin\theta \end{Bmatrix} \quad \text{and} \quad \begin{Bmatrix} r^2 = x^2 + y^2 \\ \tan\theta = \dfrac{y}{x} \end{Bmatrix} \tag{20-1}$$

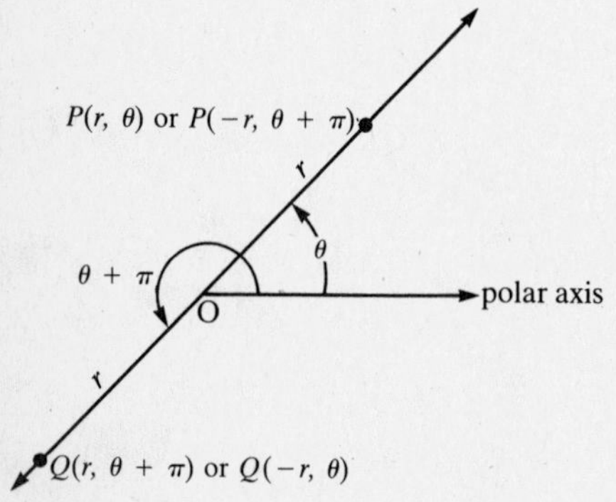

Figure 20-2

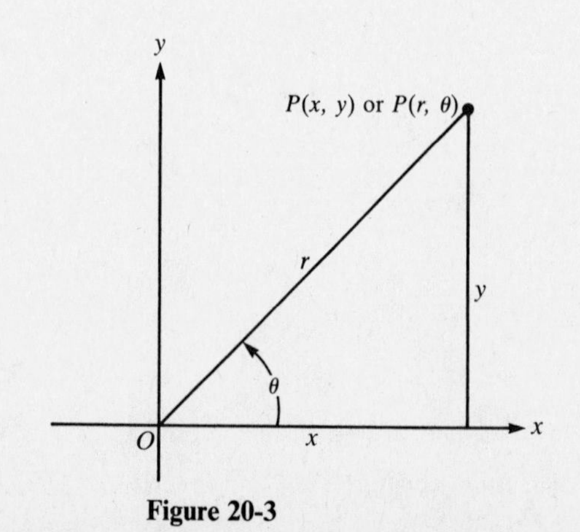

Figure 20-3

Note: To convert (x, y) to (r, θ), you must be careful in your choice of θ. Be sure that θ is in the proper quadrant!

EXAMPLE 20-1: A point P has polar coordinates $(3, \pi/6)$. Find its rectangular coordinates, and graph the point.

Solution: You have $r = 3$ and $\theta = \pi/6$. Substitute into Equation 20-1 for x and y:

$$x = r\cos\theta = 3\cos\frac{\pi}{6} = \frac{3\sqrt{3}}{2}$$

$$y = r\sin\theta = 3\sin\frac{\pi}{6} = \frac{3}{2}$$

P has rectangular coordinates $(3\sqrt{3}/2, 3/2)$. To graph P, draw $\theta = \pi/6$, and mark off three units along its terminal side (see Figure 20-4).

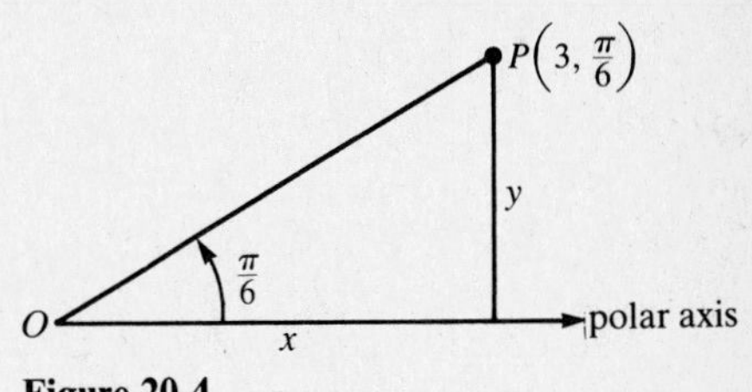

Figure 20-4

EXAMPLE 20-2: A point P has rectangular coordinates $(3, -4)$. Find three sets of polar coordinates for P.

Solution: Substitute $x = 3$ and $y = -4$ into Equation 20-1 for r and θ:

$$r = \sqrt{x^2 + y^2} = \sqrt{9 + 16} = 5$$

$$\tan\theta = \frac{y}{x} = -\frac{4}{3}$$

Your calculator tells you that $\theta \approx -0.93$ radians, which is in the fourth quadrant. Since $(3, -4)$ is in the fourth quadrant, you have one choice for θ; call it θ_1. Adding (or subtracting) any multiple of 2π will give you other choices; for example,

$$\theta_2 \approx -0.93 + 2(3.14) = 5.35 \text{ radians}$$

A third choice is found by taking $-r$, with $\theta + \pi$ (or $\theta \pm$ any odd multiple of π): $\theta_3 \approx -0.93 + 3.14 = 2.21$ radians. Three sets of polar coordinates for $P(3, -4)$ are approximately $(5, -0.93)$, $(5, 5.35)$, and $(-5, 2.21)$. The relationship is shown in Figure 20-5.

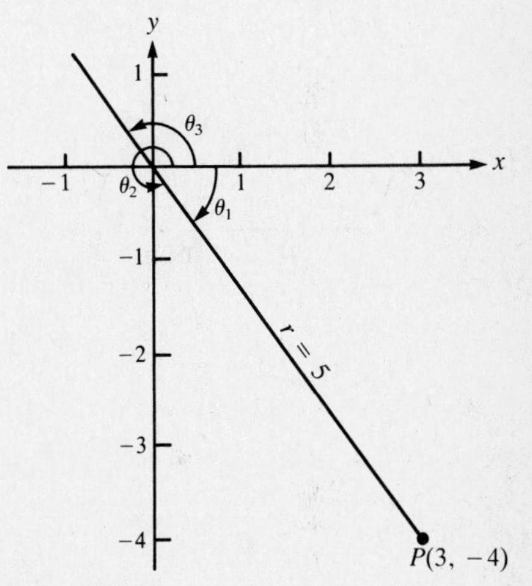

Figure 20-5
P has polar coordinates $(5, \theta_1)$, $(5, \theta_2)$, or $(-5, \theta_3)$.

You may transform an equation in x and y into an equation in r and θ (or vice versa) by using Equation 2-1.

EXAMPLE 20-3: A curve has equation $2x^2 + 3y^2 = 4$ in rectangular coordinates. Find its equation in polar coordinates.

Solution: Substitute:

$$x = r\cos\theta \qquad y = r\sin\theta$$

$$2r^2\cos^2\theta + 3r^2\sin^2\theta = 4$$

$$r^2 = \frac{4}{2\cos^2\theta + 3\sin^2\theta}$$

EXAMPLE 20-4: A curve has equation $r = 3\sin\theta$ in polar coordinates. Find its equation in rectangular coordinates.

Solution: To do this the easiest way, multiply by r:

$$r^2 = 3r\sin\theta$$

Now because $x^2 + y^2 = r^2$ and $y = r\sin\theta$, you have the desired equation:

$$x^2 + y^2 = 3y$$

20-2. Graphs of Polar Equations

The graph of an equation $r = f(\theta)$ is the set of ordered pairs (r, θ), found by assigning values to θ and finding corresponding values of r.

EXAMPLE 20-5: Graph $r = 2 \sin \theta$.

Solution: Take θ at intervals of $\pi/6$, starting with $\theta = 0$, and make a table of corresponding values of r:

θ	0	$\frac{\pi}{6}$	$\frac{\pi}{3}$	$\frac{\pi}{2}$	$\frac{2\pi}{3}$	$\frac{5\pi}{6}$	π	$\frac{7\pi}{6}$	$\frac{4\pi}{3}$	$\frac{3\pi}{2}$	$\frac{5\pi}{3}$	$\frac{11\pi}{6}$	2π
$r = 2\sin\theta$	0	1	$\sqrt{3}$	2	$\sqrt{3}$	1	0	-1	$-\sqrt{3}$	-2	$-\sqrt{3}$	-1	0

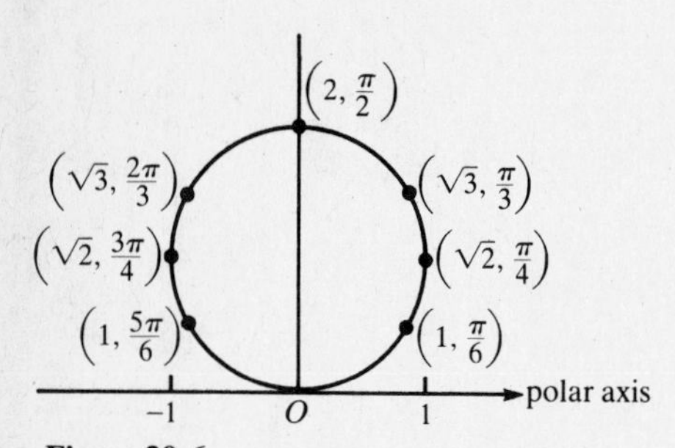

Figure 20-6
$r = 2 \sin \theta$.

Plot these points carefully. You'll notice that, since r is negative for $\pi < \theta < 2\pi$, you get the same points as you got for $0 < \theta < \pi$; the curve traces itself twice (see Figure 20-6).

EXAMPLE 20-6: Graph $r = 1 - 2 \cos \theta$.

Solution: Make a table of values as in Example 20-5:

θ	0	$\frac{\pi}{6}$	$\frac{\pi}{3}$	$\frac{\pi}{2}$	$\frac{2\pi}{3}$	$\frac{5\pi}{6}$	π	$\frac{7\pi}{6}$	$\frac{4\pi}{3}$	$\frac{3\pi}{2}$	$\frac{5\pi}{3}$	$\frac{11\pi}{6}$	2π
$r = 1 - 2\cos\theta$	-1	$1-\sqrt{3}$	0	1	2	$1+\sqrt{3}$	3	$1+\sqrt{3}$	2	1	0	$1-\sqrt{3}$	-1

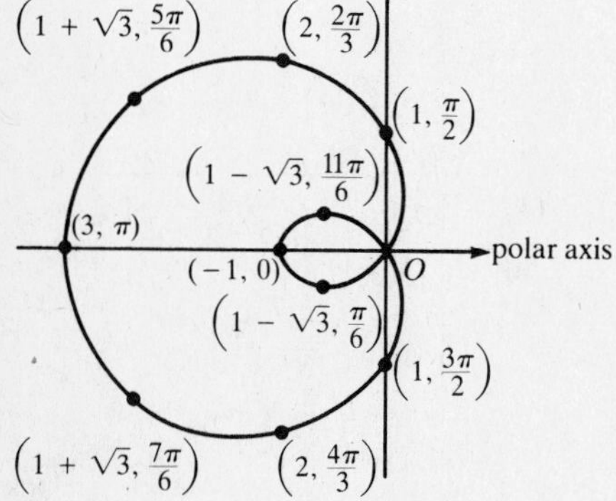

Figure 20-7
$r = 1 - 2 \cos \theta$.

Notice that when $\theta = 0$, $r = -1$: The first point is on the negative x axis. The second point is in the third quadrant: The terminal side of $\theta = \pi/6$ is extended back through the pole, and $r = 1 - \sqrt{3}$ is plotted. At $\theta = \pi/3$, $r = 0$ and the curve passes through the pole. When the points are all plotted and a smooth curve is drawn, you discover the graph has an inside loop (see Figure 20-7). This curve is known as a *limaçon.* Any equation of the form

$$r = a + b \cos \theta \quad \text{or} \quad r = a + b \sin \theta \qquad \textbf{(20-2)}$$

with $|b| \neq |a|$, will yield such a curve. If $|a| = |b|$, then the curve is a *cardioid.*

EXAMPLE 20-7: Graph $r^2 = \cos 2\theta$.

Solution: You see that $\cos 2\theta$ must be nonnegative ($r = \pm\sqrt{\cos 2\theta}$). Thus, you consider only those values of θ for which 2θ is in the first or fourth quadrant:

$$-\frac{\pi}{2} \leqslant 2\theta \leqslant \frac{\pi}{2} \quad \text{or} \quad \frac{3\pi}{2} \leqslant 2\theta \leqslant \frac{5\pi}{2}$$

$$-\frac{\pi}{4} \leqslant \theta \leqslant \frac{\pi}{4} \quad \text{or} \quad \frac{3\pi}{4} \leqslant \theta \leqslant \frac{5\pi}{4}$$

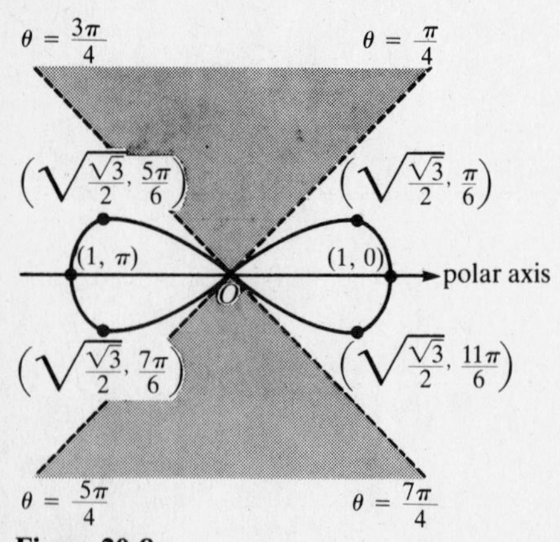

Figure 20-8
$r^2 = \cos 2\theta$.

There will be no points of the graph in the portions of the plane for which $\pi/4 < \theta < 3\pi/4$ or $5\pi/4 < \theta < 7\pi/4$ (the shaded area in Figure 20-8). Take θ in intervals of $\pi/8$, starting with $\theta = -\pi/4$ and considering only allowable values, and make a table of values:

θ	$\frac{-\pi}{4}$	$\frac{-\pi}{8}$	0	$\frac{\pi}{8}$	$\frac{\pi}{4}$	$\frac{3\pi}{4}$	$\frac{7\pi}{8}$	π	$\frac{9\pi}{8}$	$\frac{5\pi}{4}$
$r = \pm\sqrt{\cos 2\theta}$	0	$\pm\frac{1}{\sqrt[4]{2}}$	± 1	$\pm\frac{1}{\sqrt[4]{2}}$	0	0	$\pm\frac{1}{\sqrt[4]{2}}$	± 1	$\pm\frac{1}{\sqrt[4]{2}}$	0

Since r takes both positive and negative values for each θ, the curve is completely traced for $-\pi/4 < \theta < \pi/4$ and repeats itself when θ takes on values from $3\pi/4$ to $5\pi/4$.

EXAMPLE 20-8: Graph $r = \sin 2\theta$.

Solution: Make a table of values for $0 \leqslant \theta \leqslant 2\pi$, choosing intervals of length $\pi/4$:

θ	0	$\frac{\pi}{4}$	$\frac{\pi}{2}$	$\frac{3\pi}{4}$	π	$\frac{5\pi}{4}$	$\frac{3\pi}{2}$	$\frac{7\pi}{4}$	2π	$\frac{\pi}{8}$	$\frac{3\pi}{8}$
$r = \sin 2\theta$	0	1	0	-1	0	1	0	-1	0	$\frac{\sqrt{2}}{2}$	$\frac{\sqrt{2}}{2}$

The graph starts at the pole, $r = 0$; r then increases to one as θ increases to $\pi/4$, decreases to zero as θ increases to $\pi/2$. You've traced a closed loop. For $\pi/2 \leqslant \theta \leqslant \pi$, the values of r give a second loop, this time in the fourth quadrant, since r is negative. Continuing, you get a third loop, in the third quadrant since r is positive, and a fourth loop, in the second quadrant since r is negative for $3\pi/2 < \theta < 2\pi$. This graph is called a four-leaf rose (see Figure 20-9): $r = \sin n\theta$ (or $r = \cos n\theta$) gives a rose curve that has n leaves if n is odd, $2n$ leaves if n is even.

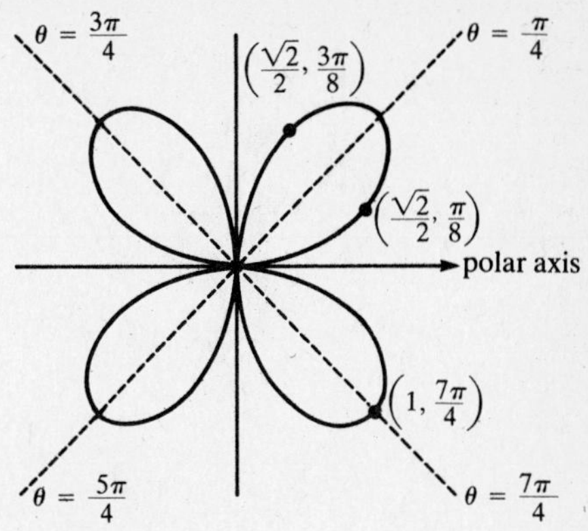

Figure 20-9
$r = \sin 2\theta$.

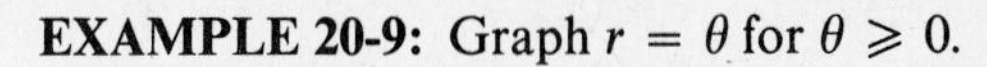

EXAMPLE 20-9: Graph $r = \theta$ for $\theta \geqslant 0$.

Solution: The graph differs from the preceding examples; as θ increases, r will continue to increase; the curve will spiral out from the origin (see Figure 20-10).

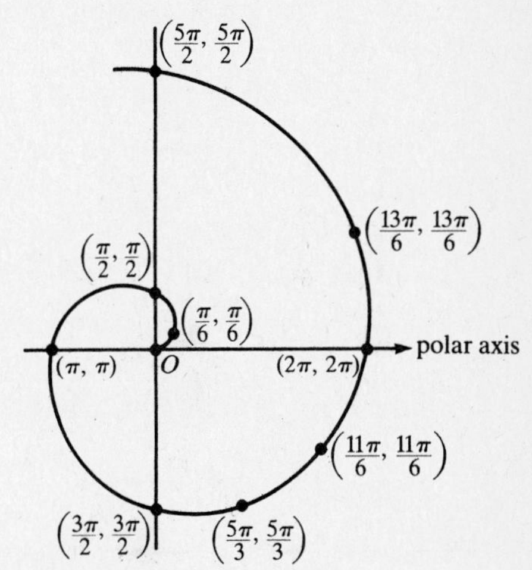

Figure 20-10
$r = \theta, \theta \geqslant 0$.

20-3. Points of Intersection of Graphs

The problem of finding points of intersection of two curves whose equations are in polar form is more complicated than the comparable problem in a rectangular coordinate system. The trouble is caused by the fact that a point in the plane has more than one set of polar coordinates.

To find all points of intersection of the graphs of $r = f(\theta)$ and $r = g(\theta)$, use the following steps:

1. Set $f(\theta) = g(\theta)$ and solve for θ. This may give all points of intersection: It will certainly yield those points (r, θ) that are the same on the two graphs.
2. If both curves pass through the pole, then that will surely be a point of intersection, even though $f(\theta)$ and $g(\theta)$ may equal zero for different values of θ.
3. Graph both equations and see if you've missed any points.

EXAMPLE 20-10: Find all points of intersection of $f(\theta) = 2 \sin \theta$ and $g(\theta) = 2 \cos \theta$.

Solution:

1. Solve for θ:

$$2 \sin \theta = 2 \cos \theta$$

$$\theta = \frac{\pi}{4}, \frac{5\pi}{4}$$

$$r = \sqrt{2}, -\sqrt{2}$$

   Notice that $(-\sqrt{2}, 5\pi/4)$ coincides with $(\sqrt{2}, \pi/4)$.
2. $2 \sin \theta = 0$ (at $\theta = 0, \pi$); $2 \cos \theta = 0$ (at $\theta = \pi/2, 3\pi/2$). Both graphs pass through the pole. It is the second point of intersection.
3. The graphs show that we have found all points of intersection (see Figure 20-11).

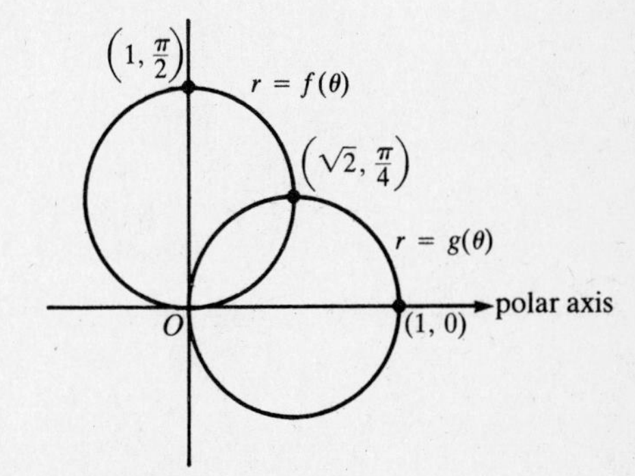

Figure 20-11
$f(\theta) = 2 \sin \theta, g(\theta) = 2 \cos \theta$.

EXAMPLE 20-11: Find all points of intersection of the graphs of $r = 1$ and $r^2 = 2 \sin 2\theta$ (called a *lemniscate*).

1. If $r = 1$, then $r^2 = 1$:

$$2 \sin 2\theta = 1 \qquad \sin 2\theta = \frac{1}{2}$$

$$2\theta = \frac{\pi}{6}, \frac{5\pi}{6}, \frac{13\pi}{6}, \frac{17\pi}{6}$$

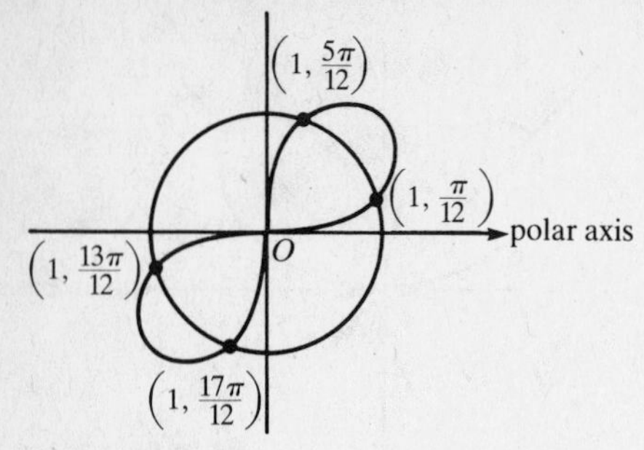

Figure 20-12
$r = 1, r^2 = 2 \sin 2\theta$.

Notice that if θ takes on values from zero to 2π, 2θ must take on values from zero to 4π:

$$\theta = \frac{\pi}{12}, \frac{5\pi}{12}, \frac{13\pi}{12}, \frac{17\pi}{12}$$

2. The pole isn't a point of intersection since on the graph of $r = 1$ you have $r \neq 0$ for all θ.
3. The graphs show that you have found all points of intersection (see Figure 20-12).

EXAMPLE 20-12: Find the points of intersection of $r = \frac{1}{2} \tan \theta$ and $r = \sin \theta$.

1. Solve for θ:

$$\frac{1}{2} \tan \theta = \sin \theta$$

$$\frac{\sin \theta}{\cos \theta} - 2 \sin \theta = 0$$

$$\sin \theta \left(\frac{1}{\cos \theta} - 2 \right) = 0$$

$$\sin \theta = 0 \qquad \cos \theta = \frac{1}{2}$$

$$\theta = 0, \pi \qquad \theta = \frac{\pi}{3}, \frac{5\pi}{3}$$

$$r = 0 \qquad r = \frac{\sqrt{3}}{2}, \frac{-\sqrt{3}}{2}$$

Notice that (0, 0) is the same point as $(0, \pi)$. You have found three points: (0, 0), $(\sqrt{3}/2, \pi/3)$, and $(-\sqrt{3}/2, 5\pi/3)$.

2. The pole is a point of intersection.
3. The graphs show you that you have found all points of intersection: the pole, $(\sqrt{3}/2, \pi/3)$, and $(-\sqrt{3}/2, 5\pi/3)$, as shown in Figure 20-13.

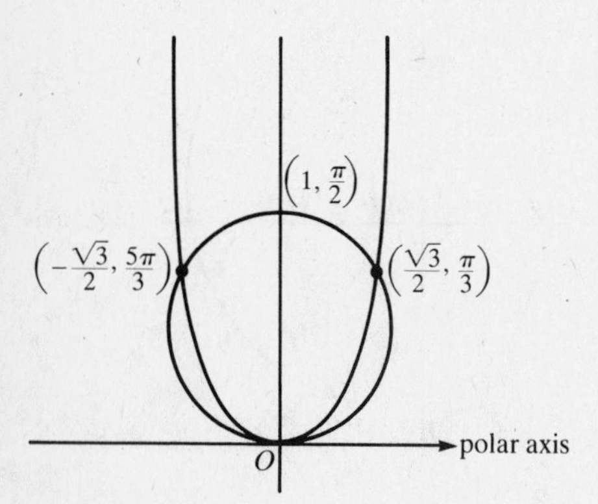

Figure 20-13
$r = \sin \theta, r = \frac{1}{2} \tan \theta$.

20-4. Slope of a Tangent Line

When the equation of a graph is given in rectangular coordinates, you know that the tangent line has slope $m = dy/dx$. There are two ways of finding the slope of the tangent line to a graph of an equation in polar form.

A. Use $m = dy/dx$ to derive a formula in r and θ.

You know that $y = r \sin \theta$ and $x = r \cos \theta$. Differentiate each of these, implicitly, with respect to θ:

$$\frac{dy}{d\theta} = r \cos \theta + \sin \theta \frac{dr}{d\theta}$$

$$\frac{dx}{d\theta} = -r \sin \theta + \cos \theta \frac{dr}{d\theta}$$

Noting that $\frac{dy}{dx} = \frac{dy/d\theta}{dx/d\theta}$ (one version of the chain rule), you have an expression for m in terms of r and θ:

$$m = \frac{r \cos \theta + \sin \theta \dfrac{dr}{d\theta}}{-r \sin \theta + \cos \theta \dfrac{dr}{d\theta}} \tag{20-3}$$

To find the slope of the tangent line to $r = f(\theta)$ at a given point, you find $dr/d\theta$, evaluate it at the given (r, θ), and substitute into the formula.

EXAMPLE 20-13: Find the slope of the tangent line to $r = 1 - 2\cos\theta$ (see Figure 20-7) at $(2, 2\pi/3)$.

Solution:

$$\left.\frac{dr}{d\theta}\right|_{(2,\,2\pi/3)} = 2\sin\theta\Big|_{\theta=2\pi/3} = 2\left(\frac{\sqrt{3}}{2}\right) = \sqrt{3}$$

Now substitute $r = 2$, $\theta = 2\pi/3$, and $dr/d\theta = \sqrt{3}$ into the formula:

$$m = \frac{2\cos\dfrac{2\pi}{3} + \left(\sin\dfrac{2\pi}{3}\right)\sqrt{3}}{-2\sin\dfrac{2\pi}{3} + \left(\cos\dfrac{2\pi}{3}\right)\sqrt{3}} = \frac{-1 + \dfrac{3}{2}}{-\sqrt{3} - \dfrac{\sqrt{3}}{2}} = \frac{1}{-3\sqrt{3}} = \frac{-\sqrt{3}}{9}$$

B. A second formula

Let (r, θ) be a point of a graph, with α the inclination of the tangent line at the point and β the angle between r and the tangent line (see Figure 20-14). You see that

$$\alpha = \beta + \theta \qquad \tan\alpha = \tan(\beta + \theta) = \frac{\tan\beta + \tan\theta}{1 - \tan\beta\tan\theta}$$

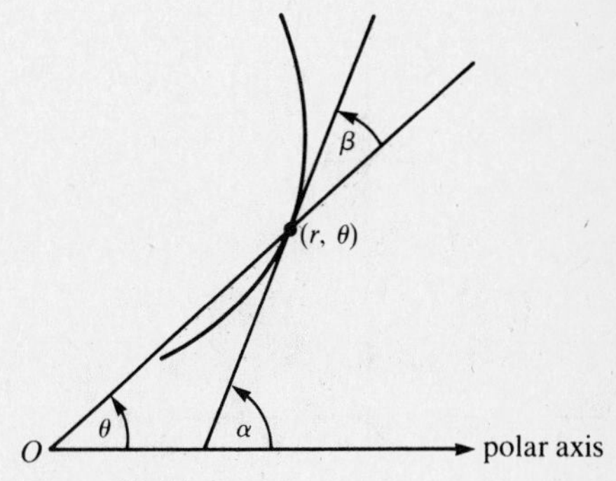

Figure 20-14

It follows that the slope of the tangent line at (r, θ) is

$$m = \tan(\beta + \theta) = \frac{\tan\beta + \tan\theta}{1 - \tan\beta\tan\theta} \tag{20-4}$$

It is easily shown that $\tan\beta = r/(dr/d\theta)$. Thus, at a given point of a given graph, you can easily evaluate $\tan\beta$ and $\tan\theta$, and hence you can find $\tan\alpha$.

EXAMPLE 20-14: Find the slope of the tangent line to $r = \sin 2\theta$ at $(1, \pi/4)$.

Solution: You have

$$\left.\frac{dr}{d\theta}\right|_{\theta=\frac{\pi}{4}} = 2\cos 2\theta\Big|_{\frac{\pi}{4}} = 0$$

This tells you that $\beta = \pi/2$ (you have $\tan\beta = r/(dr/d\theta)$). You know that $\alpha = \beta + \theta = (\pi/2) + (\pi/4) = 3\pi/4$. Hence the slope of the tangent line at $(1, \pi/4)$ is

$$\tan\alpha = \tan\frac{3\pi}{4} = -1$$

EXAMPLE 20-15: Find the slope of the tangent line to $r = 2 - 3\sin\theta$ at $(1/2, \pi/6)$.

Solution:

$$\tan\beta = \left.\frac{r}{\dfrac{dr}{d\theta}}\right|_{(\frac{1}{2},\frac{\pi}{6})}$$

$$= \frac{1/2}{-3\cos\dfrac{\pi}{6}} = \frac{-1}{3\sqrt{3}} = \frac{-\sqrt{3}}{9}$$

and

$$\tan\theta = \tan\frac{\pi}{6} = \frac{\sqrt{3}}{3}$$

You have

$$m = \tan(\beta + \theta) = \frac{\tan\beta + \tan\theta}{1 - \tan\beta\tan\theta} = \frac{\frac{-\sqrt{3}}{9} + \frac{\sqrt{3}}{3}}{1 + \frac{\sqrt{3}}{9}\left(\frac{\sqrt{3}}{3}\right)} = \frac{2\sqrt{3}}{10} = \frac{\sqrt{3}}{5}$$

20-5. The Area Problem

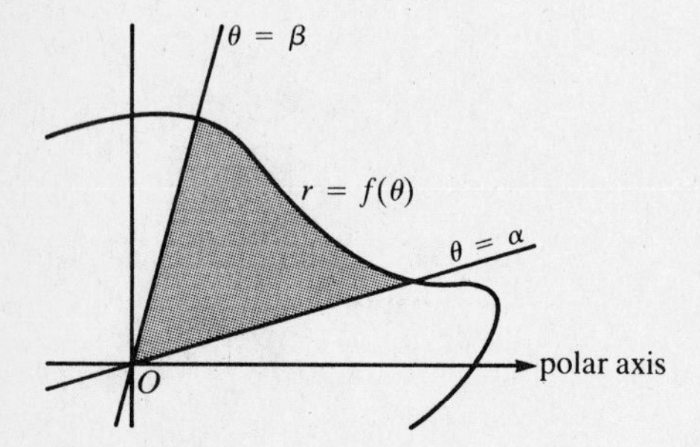

Figure 20-15

You can use an integral to find an area bounded by curves whose equations are given in polar form. The formula can be found from an approximation using sums of areas of circular sectors. As shown in Figure 20-15, the **area** bounded by the graph of $r = f(\theta)$ and by $\theta = \alpha$ and $\theta = \beta$ $(\alpha < \beta)$ is

AREA

$$A = \frac{1}{2}\int_{\alpha}^{\beta} [f(\theta)]^2\, d\theta \qquad \textbf{(20-5)}$$

EXAMPLE 20-16: Find the area of the region enclosed by $r = 1 + \sin\theta$.

Solution:
Draw the graph (see Figure 20-16) and note the symmetry of the region about the line $\theta = \pi/2$, that is, the area for which $-\pi/2 \leqslant \theta \leqslant \pi/2$ is equal to the area for which $\pi/2 \leqslant \theta \leqslant 3\pi/2$. You can write the area as

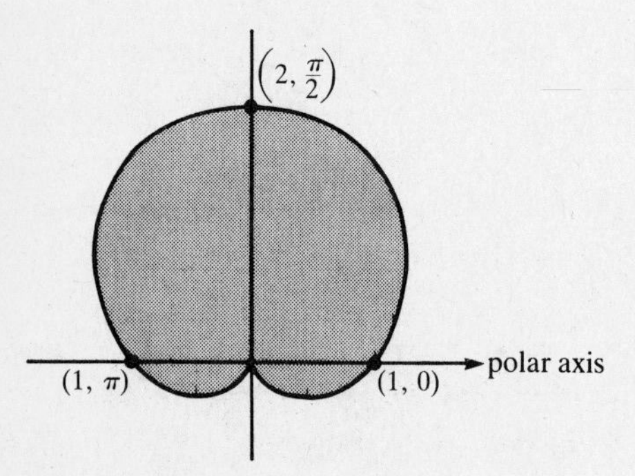

Figure 20-16
$r = 1 + \sin\theta$.

$$2\left[\frac{1}{2}\int_{-\pi/2}^{\pi/2} f^2(\theta)\, d\theta\right] = \int_{-\pi/2}^{\pi/2} (1 + \sin\theta)^2\, d\theta$$

To evaluate the integral, expand it and use whatever techniques you need from Chapters 13 and 14:

$$\begin{aligned}
&\int_{-\pi/2}^{\pi/2} (1 + 2\sin\theta + \sin^2\theta)\, d\theta \\
&= \int_{-\pi/2}^{\pi/2} (1 + 2\sin\theta)\, d\theta + \frac{1}{2}\int_{-\pi/2}^{\pi/2} (1 - \cos 2\theta)\, d\theta \\
&= (\theta - 2\cos\theta)\Big|_{-\pi/2}^{\pi/2} + \frac{1}{2}\left[\left(\theta - \frac{1}{2}\sin 2\theta\right)\Big|_{-\pi/2}^{\pi/2}\right] \\
&= (\pi - 0) + \frac{1}{2}(\pi - 0) = \frac{3\pi}{2}
\end{aligned}$$

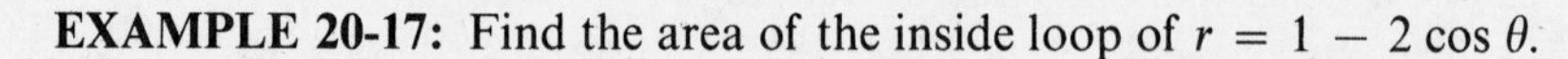

EXAMPLE 20-17: Find the area of the inside loop of $r = 1 - 2\cos\theta$.

Solution: Draw the graph (see Figure 20-7) and notice that the loop is described for $-\pi/3 \leqslant \theta \leqslant \pi/3$. Thus,

$$\begin{aligned}
A &= \frac{1}{2}\int_{-\pi/3}^{\pi/3} (1 - 2\cos\theta)^2\, d\theta = \frac{1}{2}\int_{-\pi/3}^{\pi/3} (1 - 4\cos\theta + 4\cos^2\theta)\, d\theta \\
&= \frac{1}{2}\int_{-\pi/3}^{\pi/3} (1 - 4\cos\theta)\, d\theta + \int_{-\pi/3}^{\pi/3} (1 + \cos 2\theta)\, d\theta \\
&= \frac{1}{2}(\theta - 4\sin\theta)\Big|_{-\pi/3}^{\pi/3} + \left(\theta + \frac{1}{2}\sin 2\theta\right)\Big|_{-\pi/3}^{\pi/3} \\
&= \frac{1}{2}\left(\frac{\pi}{3} - 4\frac{\sqrt{3}}{2}\right) - \frac{1}{2}\left(\frac{-\pi}{3} + 4\frac{\sqrt{3}}{2}\right) + \left[\frac{\pi}{3} + \left(\frac{1}{2}\right)\frac{\sqrt{3}}{2}\right] \\
&\quad - \left[\frac{-\pi}{3} + \frac{1}{2}\left(\frac{-\sqrt{3}}{2}\right)\right] \\
&= \pi - \frac{3\sqrt{3}}{2}
\end{aligned}$$

To find an area bounded by two polar curves, $f(\theta)$ and $g(\theta)$, you first find the points of intersection of the graphs (if any) and then write your desired area as a difference of two areas.

EXAMPLE 20-18: Find the area bounded by $r = 1$, $r = 1 + \cos\theta$, and $\theta = 0$ in the first quadrant.

Solution: Draw the graphs; the only point of intersection in the described region is $(1, \pi/2)$, as shown in Figure 20-17. The required area is the difference of the areas bounded by $r = 1 + \cos\theta$ and by $r = 1$, respectively, for $0 \leqslant \theta \leqslant \pi/2$:

$$\begin{aligned}
A &= \frac{1}{2}\int_0^{\pi/2} (1 + \cos\theta)^2\,d\theta - \frac{1}{2}\int_0^{\pi/2} 1^2\,d\theta \\
&= \frac{1}{2}\int_0^{\pi/2} (1 + 2\cos\theta + \cos^2\theta)\,d\theta - \frac{1}{2}\left(\frac{\pi}{2} - 0\right) \\
&= \frac{1}{2}\int_0^{\pi/2} (1 + 2\cos\theta)\,d\theta + \frac{1}{4}\int_0^{\pi/2} (1 + \cos 2\theta)\,d\theta - \frac{\pi}{4} \\
&= \frac{1}{2}\left[\left(\theta + 2\sin\theta\right)\Big|_0^{\pi/2}\right] + \frac{1}{4}\left[\left(\theta + \frac{1}{2}\sin 2\theta\right)\Big|_0^{\pi/2}\right] - \frac{\pi}{4} \\
&= \frac{1}{2}\left[\left(\frac{\pi}{2} + 2\right) - 0\right] + \frac{1}{4}\left(\frac{\pi}{2} + 0 - 0\right) - \frac{\pi}{4} = \frac{\pi}{8} + 1
\end{aligned}$$

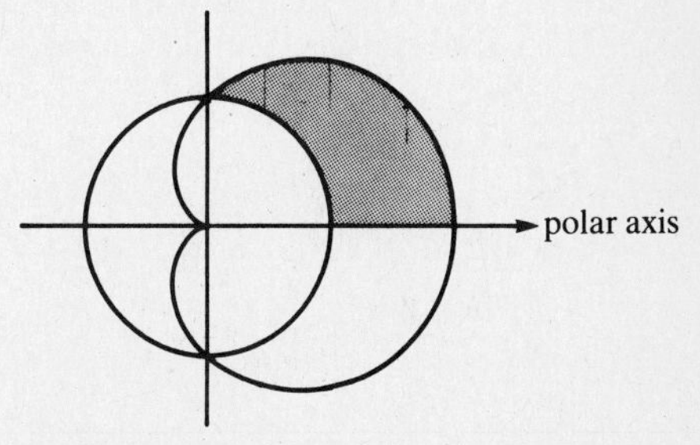

Figure 20-17
$r = 1$, $r = 1 + \cos\theta$.

EXAMPLE 20-19: Find the area enclosed by the graphs of $r = 2\sin 2\theta$ and $r = 2\cos 2\theta$.

Solution: Draw the graphs. Notice that the desired area consists of eight "petals" (see the shaded portion of Figure 20-18a) and that each of these petals consists of two pieces symmetric about the radial line, $\theta = \theta_0$, where θ_0 is the angle of a point of intersection. Find the first point of intersection and use that symmetry to find the area you want:

$$\begin{aligned}
2\sin 2\theta &= 2\cos 2\theta \\
\tan 2\theta &= 1 \\
2\theta &= \frac{\pi}{4} \\
\theta &= \frac{\pi}{8}
\end{aligned}$$

You could find the other points, but you don't need them. Thus,

$$\begin{aligned}
A &= \frac{8(2)}{2}\int_0^{\pi/8} (2\sin 2\theta)^2\,d\theta = 32\int_0^{\pi/8} \sin^2 2\theta\,d\theta \\
&= 16\int_0^{\pi/8} (1 - \cos 4\theta)\,d\theta \\
&= 16\left(\theta - \frac{\sin 4\theta}{4}\right)\Big|_0^{\pi/8} \\
&= 16\left[\left(\frac{\pi}{8} - \frac{1}{4}\right) - 0\right] \\
&= 2\pi - 4
\end{aligned}$$

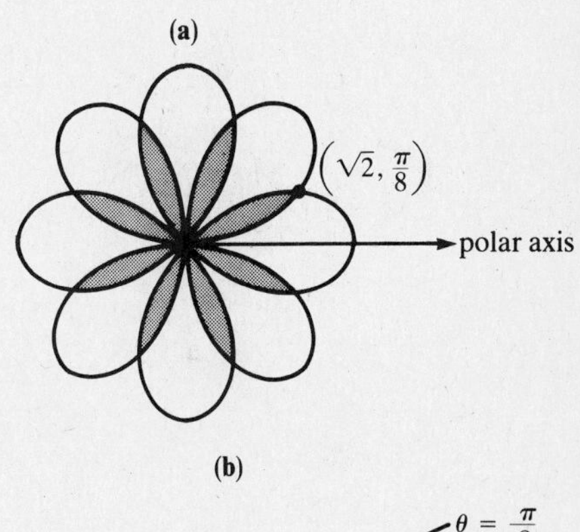

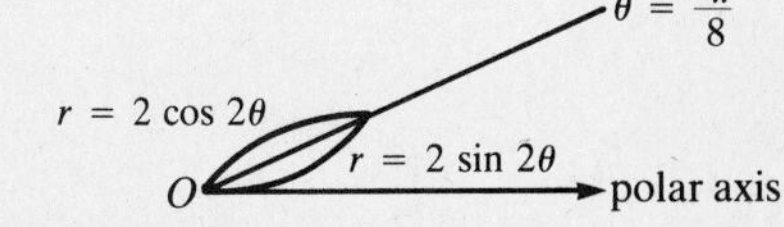

Figure 20-18
$r = 2\sin 2\theta$, $r = 2\cos 2\theta$.

Before you leave this problem, look at Figure 20-18b, an expanded sketch of one "petal" of the area. Notice that (if you don't take advantage of the symmetry) you would have to express the area as the sum of two parts: The bottom half is bounded by $r = 2\sin 2\theta$ for $0 \leqslant \theta \leqslant \pi/8$, and the top half by $r = 2\cos 2\theta$ for

$\pi/8 \leqslant \theta \leqslant \pi/4$. Thus the area of this "petal" is

$$\frac{1}{2}\int_0^{\pi/8} 4\sin^2 2\theta\, d\theta + \frac{1}{2}\int_{\pi/8}^{\pi/4} 4\cos^2 2\theta\, d\theta$$

20-6. Arc Length

You can find a formula for arc length in polar coordinates directly from the rectangular coordinate formula.

In Chapter 19 you were given the formula for the length of the arc of the graph of $y = f(x)$ over $[a, b]$:

$$s = \int_a^b \sqrt{1 + \left(\frac{dy}{dx}\right)^2}\, dx$$

In Section 20-4 you saw that

$$\frac{dy}{dx} = \frac{dy/d\theta}{dx/d\theta} = \frac{r\cos\theta + \sin\theta\dfrac{dr}{d\theta}}{-r\sin\theta + \cos\theta\dfrac{dr}{d\theta}}$$

where $r = r(\theta)$, was the polar form of the graph, and the x interval $[a, b]$ corresponds to a θ interval $[\alpha, \beta]$. If you substitute for dy/dx and for dx in the arc length formula and do the algebraic simplification of squaring dy/dx and adding one, you find the polar equivalent of the formula:

ARC LENGTH

$$s = \int_\alpha^\beta \sqrt{r^2 + \left(\frac{dr}{d\theta}\right)^2}\, d\theta \qquad \textbf{(20-6)}$$

EXAMPLE 20-20: Find the length of $r = 1 + \sin\theta$.

Solution: This graph, which you get by choosing values of θ over an interval of length 2π, is symmetric about the line $\theta = \pi/2$ (see Figure 20-16). Thus, an integration over the interval $[-\pi/2, \pi/2]$ will give you half the desired arc length, and

$$\begin{aligned} s &= 2\int_{-\pi/2}^{\pi/2} \sqrt{(1 + \sin\theta)^2 + \cos^2\theta}\, d\theta \\ &= 2\int_{-\pi/2}^{\pi/2} \sqrt{2 + 2\sin\theta}\, d\theta = 2\sqrt{2}\int_{-\pi/2}^{\pi/2} \sqrt{1 + \sin\theta}\, d\theta \end{aligned}$$

You have a choice of two methods of finding the antiderivative. You may use the half-angle substitution of Section 14-5 ($u = \tan(\theta/2)$, or $\theta = 2\tan^{-1}u$), or you can save yourself some labor by what amounts to a rotation of the plane.

Substitute $\phi = \theta - (\pi/2)$, with $d\theta = d\phi$, and the θ interval $[-\pi/2, \pi/2]$ becomes the ϕ interval $[-\pi, 0]$. Then $\sin\theta = \sin[\phi + (\pi/2)] = \cos\phi$, and using the trigonometric identity $\cos(\alpha/2) = \sqrt{(1 + \cos\alpha)/2}$, you get

$$\sqrt{1 + \sin\theta} = \sqrt{1 + \sin\left(\phi + \frac{\pi}{2}\right)} = \sqrt{1 + \cos\phi} = \sqrt{2}\cos\frac{\phi}{2}$$

Thus,

$$\begin{aligned} 2\sqrt{2}\int_{-\pi/2}^{\pi/2} \sqrt{1 + \sin\theta}\, d\theta &= 2\sqrt{2}\int_{-\pi}^{0} \sqrt{2}\cos\frac{\phi}{2}\, d\phi \\ &= 8\sin\frac{\phi}{2}\bigg|_{-\pi}^{0} = 8[0 - (-1)] = 8 \end{aligned}$$

EXAMPLE 20-21: Find the length of the spiral $r = \theta$ for $0 \leqslant \theta \leqslant \pi/2$.

Solution: Substituting $r = \theta$ and $dr/d\theta = 1$ into the formula, you have

$$s = \int_0^{\pi/2} \sqrt{\theta^2 + 1}\, d\theta$$

Use the inverse trigonometric substitution $\theta = \tan t$, $d\theta = \sec^2 t$, with the corresponding change of the limits of integration ($0 \leqslant \theta \leqslant \pi/2$ leads to $0 \leqslant t \leqslant \tan^{-1}(\pi/2)$). You get

$$\int_0^{\pi/2} \sqrt{\theta^2 + 1}\, d\theta = \int_0^{\tan^{-1}(\pi/2)} \sqrt{1 + \tan^2 t}\, \sec^2 t\, dt$$

$$= \int_0^{\tan^{-1}(\pi/2)} \sec^3 t\, dt$$

Do you remember how to do this? Look back at Section 14-2; you need one integration by parts, with $u = \sec t$ and $dv = \sec^2 t\, dt$, followed by a use of the identity $\tan^2 t = \sec^2 t - 1$. You get

$$\int_0^{\tan^{-1}(\pi/2)} \sec^3 t\, dt = \frac{1}{2}[\sec t \tan t + \ln|\sec t + \tan t|]\Bigg|_0^{\tan^{-1}(\pi/2)}$$

$$= \frac{1}{2}\left[\frac{(\sqrt{\pi^2 + 4})(\pi)}{4} + \ln\left|\frac{\sqrt{\pi^2 + 4} + \pi}{2}\right|\right]$$

SUMMARY

1. A polar coordinate system locates points in the plane by an ordered pair (r, θ).
2. You use the equation $x = r\cos\theta$, $y = r\sin\theta$:
   - to find the (x, y) coordinates of a point (r, θ)
   - to find the equation in polar form of an equation in rectangular form
3. You use the equation $r^2 = x^2 + y^2$, $\tan\theta = y/x$:
   - to find (r, θ) coordinates of a point (x, y)
   - to find the rectangular form of an equation given in polar form
4. To graph $r = f(\theta)$, you assign values of θ and find corresponding values of r and then plot the points (r, θ): Positive values of θ correspond to counterclockwise rotation from the polar axis, negative values to clockwise rotation; positive values of r are measured as distances from the pole along the terminal side of θ, negative values as distances along the terminal side of $(\theta + \pi)$.
5. You may need more than one technique to find all points of intersection of two curves whose equations are given in polar form.
6. You solve problems of slope of a tangent line, area, and arc length when the equation of a graph is in polar form by the use of formulas that are derived from their analogs in rectangular form.

SOLVED PROBLEMS

PROBLEM 20-1 Find rectangular coordinates for each of the points (r, θ): **(a)** $(2, \pi/6)$; **(b)** $(1, 7\pi/4)$; **(c)** $(-2, \pi/2)$.

Solution: Use $x = r\cos\theta$ and $y = r\sin\theta$:

(a) $x = 2\cos(\pi/6) = \sqrt{3}$

$y = 2\sin(\pi/6) = 1$

$(\sqrt{3}, 1)$

(b) $x = 1\cos(7\pi/4) = \sqrt{2}/2$

$y = 1\sin(7\pi/4) = -\sqrt{2}/2$

$(\sqrt{2}/2, -\sqrt{2}/2)$

(c) $x = -2\cos(\pi/2) = 0$

$y = -2\sin(\pi/2) = -2$

$(0, -2)$

[See Section 20-1.]

PROBLEM 20-2 Find polar coordinates (r, θ), with $r > 0$ and $0 \leqslant \theta < 2\pi$, for each of the points (x, y): **(a)** $(-\sqrt{3}, 1)$; **(b)** $(-1, -1)$; **(c)** $(2, 2\sqrt{3})$.

Solution: Use $r = \sqrt{x^2 + y^2}$, $\tan\theta = y/x$, being sure you choose θ in the correct quadrant:

(a) $r = \sqrt{3 + 1} = 2$

$\tan\theta = -1/\sqrt{3}$, in the second quadrant

$\theta = 5\pi/6$

$(2, 5\pi/6)$

(b) $r = \sqrt{1 + 1} = \sqrt{2}$

$\tan\theta = 1$, in the third quadrant

$\theta = 5\pi/4$

$(\sqrt{2}, 5\pi/4)$

(c) $r = \sqrt{4 + 12} = 4$

$\tan\theta = \sqrt{3}$, in the first quadrant

$\theta = \pi/3$

$(4, \pi/3)$

[See Section 20-1.]

Find the equation in rectangular coordinates of each of the following:

PROBLEM 20-3 $r = a\cos\theta + b\sin\theta$.

Solution: You know that $x = r\cos\theta$ and $y = r\sin\theta$. Use this as a hint to yourself that this problem will be much easier if you first multiply both sides of the equation by r. You have

$$r^2 = (a)r\cos\theta + (b)r\sin\theta$$

and so $x^2 + y^2 = ax + by$, or

$$\left(x - \frac{a}{2}\right)^2 + \left(y - \frac{b}{2}\right)^2 = \frac{a^2 + b^2}{4}$$

Notice that you will get the same result if you don't put in a factor of r first; you're using $r^2 = x^2 + y^2$, $\tan\theta = y/x$, so that $\cos\theta = x/\sqrt{x^2 + y^2}$ and $\sin\theta = y/\sqrt{x^2 + y^2}$. Thus,

$$\sqrt{x^2 + y^2} = (a)\frac{x}{\sqrt{x^2 + y^2}} + (b)\frac{y}{\sqrt{x^2 + y^2}}$$

and multiplying by $\sqrt{x^2 + y^2}$ brings you to the same result as before: $x^2 + y^2 = ax + by$.

[See Section 20-1.]

PROBLEM 20-4 $r^2 = 4\cos 2\theta$.

Solution: The complicating term here is the $\cos 2\theta$ term. Use an identity to get rid of it: $\cos 2\theta = \cos^2\theta - \sin^2\theta$. You have

$$r^2 = 4\cos^2\theta - 4\sin^2\theta$$

Now multiply by a factor of r^2 (in the same spirit as in Problem 20-3):

$$r^4 = 4r^2\cos^2\theta - 4r^2\sin^2\theta$$

and you get

$$(x^2 + y^2)^2 = 4x^2 - 4y^2$$

[See Section 20-1.]

PROBLEM 20-5 $r = 1/(2 + 3\sin\theta)$.

Solution: Multiply by $(2 + 3 \sin \theta)$ and then substitute:

$$2r + 3r \sin \theta = 1$$

$$2\sqrt{x^2 + y^2} + 3y = 1$$

$$4x^2 - 5y^2 + 6y - 1 = 0$$

[See Section 20-1.]

Find an equation in polar coordinates for each of the following:

PROBLEM 20-6 $x^2 + y^2 = 2y$.

Solution: Using $r^2 = x^2 + y^2$, $x = r \cos \theta$, and $y = r \sin \theta$ you get

$$r^2 = 2r \sin \theta \quad \text{or} \quad r = 2 \sin \theta$$

[See Section 20-1.]

PROBLEM 20-7 $y^2 = 2x$.

Solution: Substituting as in Problem 20-5: $r^2 \sin^2\theta = 2r \cos \theta$, $r \sin^2\theta = 2 \cos \theta$, or

$$r = \frac{2 \cos \theta}{\sin^2\theta}$$

You can get an alternative form of this equation using two identities:

$$\frac{\cos \theta}{\sin \theta} = \cot \theta \quad \text{and} \quad \frac{1}{\sin \theta} = \csc \theta$$

You have

$$r = 2\frac{\cos \theta}{\sin \theta}\left(\frac{1}{\sin \theta}\right) = 2 \cot \theta \csc \theta$$

[See Section 20-1.]

PROBLEM 20-8 **(a)** $x = 2$; **(b)** $y = 3$.

Solution: **(a)** You have $r \cos \theta = 2$, or $r = 2 \sec \theta$.

(b) Similarly, $r \sin \theta = 3$, or $r = 3 \csc \theta$.

[See Section 20-1.]

Graph each of the following:

PROBLEM 20-9 $r + 2 = \sin \theta$.

Solution: Notice two things before you start to graph: First, since $r = \sin \theta - 2$ and $\sin \theta \leqslant 1$ for all θ, you will have only negative values of r. Remember that to plot a point for which r is negative, you draw the terminal side of θ and then extend it back through the pole; you are measuring r along the terminal side of $\theta + \pi$. The points in the table of values all lie in the second and third quadrants.

Secondly, the values of $\sin \theta$ are the same in the first and second quadrants, $\sin[(\pi/2) - \theta] = \sin[(\pi/2) + \theta]$, and in the third and fourth quadrants, $\sin[(3\pi/2) - \theta] = \sin[(3\pi/2) + \theta]$. Thus your graph is symmetric about a vertical line through the pole; make a table of values for $-\pi/2 \leqslant \theta \leqslant \pi/2$ and obtain the rest of the graph by symmetry (see Figure 20-19):

θ	$-\frac{\pi}{2}$	$-\frac{\pi}{3}$	$-\frac{\pi}{6}$	0	$\frac{\pi}{6}$	$\frac{\pi}{3}$	$\frac{\pi}{2}$
$\sin \theta$	-1	$-\frac{\sqrt{3}}{2}$	$-\frac{1}{2}$	0	$\frac{1}{2}$	$\frac{\sqrt{3}}{2}$	1
$r = \sin \theta - 2$	-3	$-2 - \frac{\sqrt{3}}{2}$	-2.5	-2	-1.5	$-2 + \frac{\sqrt{3}}{2}$	-1

[See Section 20-2.]

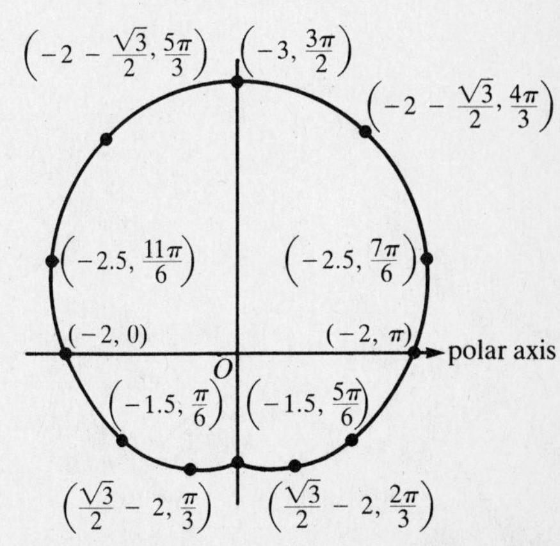

Figure 20-19
$r + 2 = \sin \theta$.

PROBLEM 20-10 $r = 2\cos 3\theta$.

Solution: This is a three-leaf rose. You find the radial lines about which the leaves are centered by finding where $\cos 3\theta = 1$, remembering that if θ varies from 0 to 2π, you will have 3θ between 0 and 6π. Thus, $\cos 3\theta = 1$ if $3\theta = 0, 2\pi, 4\pi, 6\pi$, or $\theta = 0, 2\pi/3, 4\pi/3, 2\pi$; the leaves center on the lines $\theta = 0$, $\theta = 2\pi/3$, and $\theta = 4\pi/3$. To find how "fat" the leaves are, you find where $\cos 3\theta = 0$; that is, at $3\theta = \pi/2, 3\pi/2, 5\pi/2, 7\pi/2, 9\pi/2, 11\pi/2$, or at $\theta = \pi/6, \pi/2, 5\pi/6, 7\pi/6, 3\pi/2, 11\pi/6$. The leaves come into the pole tangent to these lines (see Figure 20-20).

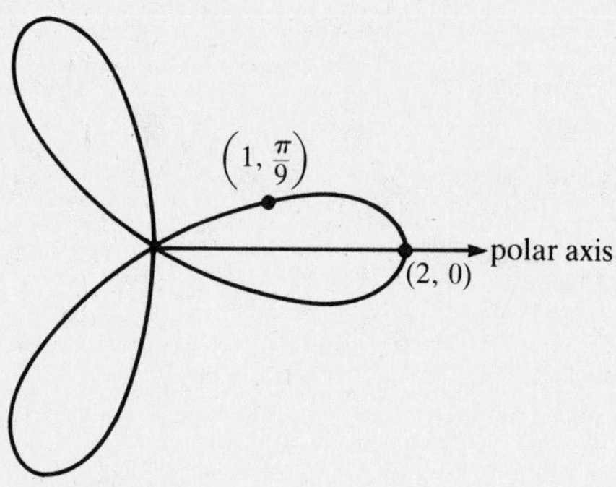

Figure 20-20
$r = 2\cos 3\theta$.

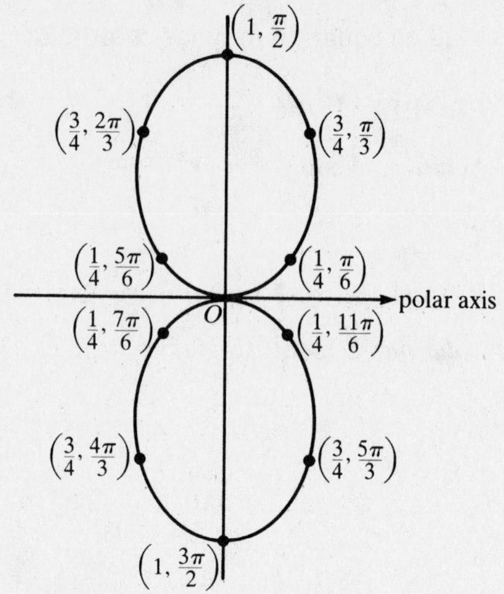

Figure 20-21
$r = \sin^2\theta$.

PROBLEM 20-11 $r = \sin^2\theta$.

Solution: Make a table of values for this one, including only values of θ in the first quadrant; r will take on the same set of values in each quadrant (see Figure 20-21):

θ	0	$\frac{\pi}{6}$	$\frac{\pi}{4}$	$\frac{\pi}{3}$	$\frac{\pi}{2}$
$\sin\theta$	0	$\frac{1}{2}$	$\frac{1}{\sqrt{2}}$	$\frac{\sqrt{3}}{2}$	1
$r = \sin^2\theta$	0	$\frac{1}{4}$	$\frac{1}{2}$	$\frac{3}{4}$	1

[See Section 20-2.]

PROBLEM 20-12 Find all points of intersection of the graphs of $r = f_1(\theta) = \sin\theta$ and $r = f_2(\theta) = 1 - \sin\theta$.

Solution: Set $f_1(\theta)$ equal to $f_2(\theta)$ and solve:

$$\sin\theta = 1 - \sin\theta$$

$$2\sin\theta = 1$$

$$\sin\theta = \frac{1}{2}$$

$$\theta = \frac{\pi}{6}, \frac{5\pi}{6}$$

Note that both graphs go through the pole. Finally, the graph shows that you have found all the points of intersection: the pole, $(1/2, \pi/6)$, and $(1/2, 5\pi/6)$ (see Figure 20-22).

[See Section 20-3.]

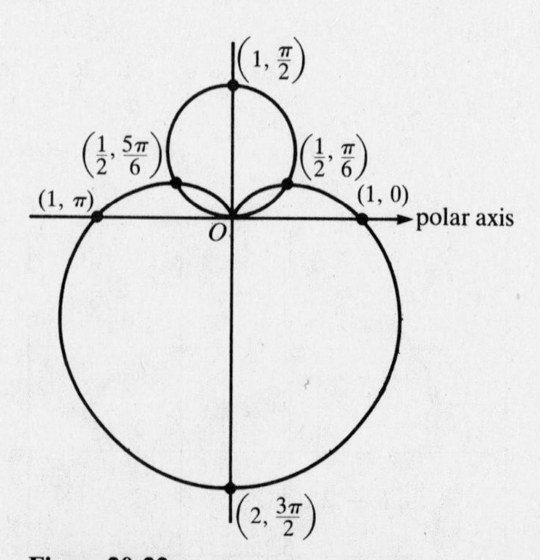

Figure 20-22
$r = \sin\theta$, $r = 1 - \sin\theta$.

PROBLEM 20-13 Find all points of intersection of $r = f_1(\theta) = \cos\theta$ and $r = f_2(\theta) = \cos 2\theta$.

Solution: Set $f_1(\theta) = f_2(\theta)$ and use the double angle identity $\cos 2\theta = 2\cos^2\theta - 1$:

$$\cos\theta = \cos 2\theta = 2\cos^2\theta - 1$$

$$2\cos^2\theta - \cos\theta - 1 = 0$$

Factoring, you get

$$(2\cos\theta + 1)(\cos\theta - 1) = 0$$

$$\cos\theta = -\frac{1}{2} \quad \text{or} \quad \theta = \frac{2\pi}{3}, \frac{4\pi}{3}$$

$$\cos\theta = 1 \quad \text{or} \quad \theta = 0$$

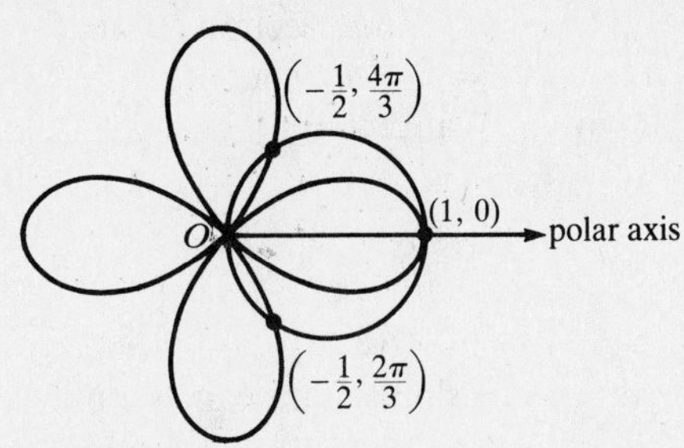

Figure 20-23
$r = \cos 2\theta$, $r = \cos\theta$.

Notice that the points $(-1/2, 2\pi/3)$ and $(-1/2, 4\pi/3)$ are in the first and fourth quadrants and that $(1, 0)$ is on the polar axis. This is consistent with what you know about the graphs: $r = \cos\theta$ is a circle, centered on the polar axis, passing through the pole; $r = \cos 2\theta$ is a four-leaf rose, with leaves centered on the polar axis and on the radial lines $\theta = \pi/2, \pi, 3\pi/2$. The only points of intersection are $(1, 0)$, $(-1/2, 2\pi/3)$, $(-1/2, 4\pi/3)$, and the pole (see Figure 20-23).

[See Section 20-3.]

PROBLEM 20-14 Find the slope of the tangent line to the graph of $r = 1 + 2\cos\theta$ at $(2, \pi/3)$.

Solution: Use the formula:

$$m = \tan\alpha = \frac{dy}{dx} = \frac{r\cos\theta + \sin\theta\,\dfrac{dr}{d\theta}}{-r\sin\theta + \cos\theta\,\dfrac{dr}{d\theta}}$$

with $r = 2$, $\theta = \pi/3$. You have

$$m = \frac{2\cos\frac{\pi}{3} + \sin\frac{\pi}{3}\left(-2\sin\frac{\pi}{3}\right)}{-2\sin\frac{\pi}{3} + \cos\frac{\pi}{3}\left(-2\sin\frac{\pi}{3}\right)} = \frac{1 - \frac{3}{2}}{-\sqrt{3} - \frac{\sqrt{3}}{2}} = \frac{1 - \frac{3}{2}}{\frac{-3\sqrt{3}}{2}} = \frac{\sqrt{3}}{9}$$

[See Section 20-4.]

PROBLEM 20-15 Find the equation (in rectangular coordinates) of the tangent line to the graph of $r = 3\cos 2\theta$ at $(3/2, \pi/6)$.

Solution: You need the point in rectangular coordinates and the slope. You have

$$x = r\cos\theta = \frac{3}{2}\cos\frac{\pi}{6} = \frac{3\sqrt{3}}{4}$$

$$y = r\sin\theta = \frac{3}{2}\sin\frac{\pi}{6} = \frac{3}{4}$$

$$m = \frac{\frac{3}{2}\cos\frac{\pi}{6} + \sin\frac{\pi}{6}\left(-6\sin\frac{\pi}{3}\right)}{-\frac{3}{2}\sin\frac{\pi}{6} + \cos\frac{\pi}{6}\left(-6\sin\frac{\pi}{3}\right)}$$

$$= \frac{\frac{3\sqrt{3}}{4} - \frac{6\sqrt{3}}{4}}{-\frac{3}{4} - \frac{9}{2}} = \frac{\frac{-3\sqrt{3}}{4}}{\frac{-21}{4}} = \frac{\sqrt{3}}{7}$$

The tangent line has equation $y - (3/4) = (\sqrt{3}/7)[x - (3\sqrt{3}/4)]$ or $y = (\sqrt{3}/7)x + (3/7)$.

PROBLEM 20-16 Use the angle β to find the slope of the tangent line to the spiral $r = \theta$ at $(\pi/4, \pi/4)$.

Solution: Recall (see Section 20-4B) that α (the inclination of the tangent line) is related to θ by the equation $\alpha = \beta + \theta$, where $\tan \beta = r/(dr/d\theta)$. Use this pair of equations with $r = \theta$, $dr/d\theta = 1$ and evaluate at $(\pi/4, \pi/4)$. You get

$$\tan \beta = \frac{\frac{\pi}{4}}{1}$$

and thus the slope of the tangent line is

$$\tan \alpha = \tan(\theta + \beta) = \frac{\tan \theta + \tan \beta}{1 - \tan \theta \tan \beta}$$

$$= \frac{\tan \frac{\pi}{4} + \frac{\pi}{4}}{1 - \left(\tan \frac{\pi}{4}\right)\frac{\pi}{4}} = \frac{1 + \frac{\pi}{4}}{1 - \frac{\pi}{4}} = \frac{4 + \pi}{4 - \pi}$$

[See Section 20-4.]

PROBLEM 20-17 Use the angle β to find the slope of the tangent line to $r = \sin \theta \tan \theta$ at $(\sqrt{2}/2, \pi/4)$. Write the equation of the tangent line in rectangular coordinates.

Solution: If $r = \sin \theta \tan \theta$, then $dr/d\theta = \sin \theta \sec^2\theta + \cos \theta \tan \theta$, and

$$\tan \beta = \frac{r}{\frac{dr}{d\theta}} = \frac{\sin \theta \tan \theta}{\sin \theta \sec^2\theta + \cos \theta \tan \theta} = \frac{\tan \theta}{\sec^2\theta + 1}$$

At $\theta = \pi/4$ you have $\tan \beta = \frac{1}{3}$. Thus,

$$\tan \alpha = \frac{1 + \frac{1}{3}}{1 - \frac{1}{3}} = 2$$

Find the rectangular coordinates of the point $(\sqrt{2}/2, \pi/4)$:

$$x = r \cos \theta = \frac{\sqrt{2}}{2} \cos \frac{\pi}{4} = \frac{1}{2}$$

$$y = r \sin \theta = \frac{1}{2}$$

The equation of the tangent line at $r = \sqrt{2}/2$, $\theta = \pi/4$ is

$$y - \frac{1}{2} = 2\left(x - \frac{1}{2}\right)$$

$$y = 2x - \frac{1}{2}$$

[See Section 20-4.]

PROBLEM 20-18 Find the area enclosed by the graph of $r = 2(1 + \sin \theta)$.

Solution: Use the formula:

$$A = \frac{1}{2}\int_0^{2\pi} f^2(\theta)\, d\theta = \frac{1}{2}\int_0^{2\pi} [2(1 + \sin \theta)]^2\, d\theta$$

$$= 2\int_0^{2\pi} (1 + 2 \sin \theta + \sin^2\theta)\, d\theta$$

$$= 2(\theta - 2 \cos \theta)\Big|_0^{2\pi} + \int_0^{2\pi} (1 - \cos 2\theta)\, d\theta$$

$$= 4\pi + \left(\theta - \frac{1}{2} \sin 2\theta\right)\Big|_0^{2\pi} = 6\pi$$

[See Section 20-5.]

PROBLEM 20-19 Find the area inside both of the curves $r = f(\theta) = 1$ and $r = g(\theta) = 2\sin\theta$.

Solution: The graph of $r = 1$ is a circle, center at the pole, radius 1, while $r = 2\sin\theta$ is a circle, center on the radial line $\theta = \pi/2$, radius 1 (see Figure 20-24). The points of intersection are in the first two quadrants:

$$2\sin\theta = 1$$

$$\sin\theta = \frac{1}{2}$$

$$\theta = \frac{\pi}{6}, \frac{5\pi}{6}$$

The area you seek is divided into three parts: the areas determined by $r = 2\sin\theta$ for $0 \leqslant \theta \leqslant \pi/6$ and for $5\pi/6 \leqslant \theta \leqslant \pi$, and the area determined by $r = 1$ for $\pi/6 \leqslant \theta \leqslant 5\pi/6$. By symmetry you have

$$A = 2\left[\frac{1}{2}\int_0^{\pi/6} (2\sin\theta)^2\, d\theta + \frac{1}{2}\int_{\pi/6}^{\pi/2} 1^2\, d\theta\right]$$

$$= 4\int_0^{\pi/6} \sin^2\theta\, d\theta + \left(\frac{\pi}{2} - \frac{\pi}{6}\right)$$

$$= \frac{4}{2}\int_0^{\pi/6} (1 - \cos 2\theta)\, d\theta + \frac{\pi}{3}$$

$$= 2\left(\theta - \frac{1}{2}\sin 2\theta\right)\Bigg|_0^{\pi/6} + \frac{\pi}{3}$$

$$= 2\left(\frac{\pi}{6} - \frac{\sqrt{3}}{4}\right) + \frac{\pi}{3} = \frac{2\pi}{3} - \frac{\sqrt{3}}{2}$$

[See Section 20-5.]

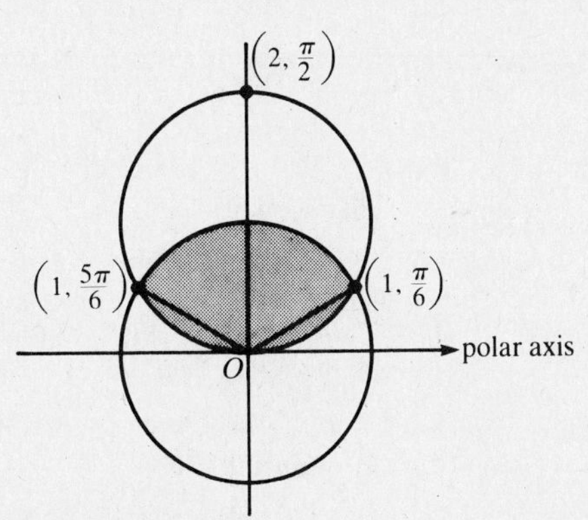

Figure 20-24
$r = 2\sin\theta$, $r = 1$.

PROBLEM 20-20 Find the area inside $r = f(\theta) = 1 + \sin\theta$, outside $r = g(\theta) = 1 - \cos\theta$, and above the line $\theta = 0$.

Solution: These graphs intersect at the pole, and at points where $1 + \sin\theta = 1 - \cos\theta$, i.e., where $\tan\theta = -1$, or

$$\theta = \frac{3\pi}{4}, \frac{7\pi}{4}$$

Draw the graphs carefully (see Figure 20-25) and notice that the area in question may be described as a difference of two areas for $0 \leqslant \theta \leqslant 3\pi/4$. Substituting into the area formulas, you have

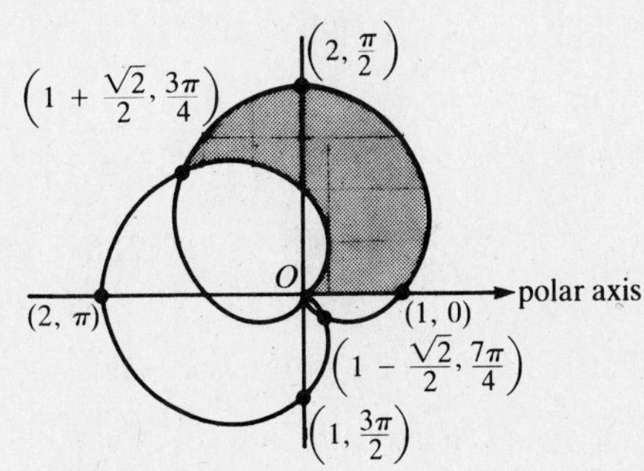

Figure 20-25
$r = 1 + \sin\theta$, $r = 1 - \cos\theta$.

$$A = \frac{1}{2}\int_0^{3\pi/4} (1 + \sin\theta)^2\, d\theta - \frac{1}{2}\int_0^{3\pi/4} (1 - \cos\theta)^2\, d\theta$$

Writing as a simple integral, expanding, and collecting terms, you have

$$A = \frac{1}{2}\int_0^{3\pi/4} (2\sin\theta + 2\cos\theta + \sin^2\theta - \cos^2\theta)\, d\theta$$

$$= \int_0^{3\pi/4} \left(\sin\theta + \cos\theta - \frac{1}{2}\cos 2\theta\right) d\theta$$

$$= \left(-\cos\theta + \sin\theta - \frac{1}{4}\sin 2\theta\right)\Bigg|_0^{3\pi/4}$$

$$= \left[\frac{\sqrt{2}}{2} + \frac{\sqrt{2}}{2} - \frac{1}{4}(-1)\right] - [-1] = \sqrt{2} + \frac{5}{4}$$

[See Section 20-5.]

PROBLEM 20-21 Find the area between the two loops of the graph of $r = 1 + 2\cos\theta$.

Solution: The outside loop is described by $-2\pi/3 \leqslant \theta \leqslant 2\pi/3$. The inner loop is described by $2\pi/3 \leqslant \theta \leqslant 4\pi/3$. The graph is shown in Figure 20-26. The area is

$$\frac{1}{2}\int_{-2\pi/3}^{2\pi/3} (1 + 2\cos\theta)^2\, d\theta - \frac{1}{2}\int_{2\pi/3}^{4\pi/3} (1 + 2\cos\theta)^2\, d\theta$$

$$= \frac{1}{2}\int_{-2\pi/3}^{2\pi/3} (1 + 4\cos\theta + 4\cos^2\theta)\, d\theta - \frac{1}{2}\int_{2\pi/3}^{4\pi/3} (1 + 4\cos\theta + 4\cos^2\theta)\, d\theta$$

$$= \frac{1}{2}\int_{-2\pi/3}^{2\pi/3} (1 + 4\cos\theta + 2 + 2\cos 2\theta)\, d\theta - \frac{1}{2}\int_{2\pi/3}^{4\pi/3} (1 + 4\cos\theta + 2 + 2\cos 2\theta)\, d\theta$$

$$= \frac{1}{2}(3\theta + 4\sin\theta + \sin 2\theta)\Big|_{-2\pi/3}^{2\pi/3} - \frac{1}{2}(3\theta + 4\sin\theta + \sin 2\theta)\Big|_{2\pi/3}^{4\pi/3}$$

$$= \frac{1}{2}(4\pi + 4\sqrt{3} - \sqrt{3}) - \frac{1}{2}(2\pi - 4\sqrt{3} + \sqrt{3}) = \pi + 3\sqrt{3}$$

[See Section 20-5.]

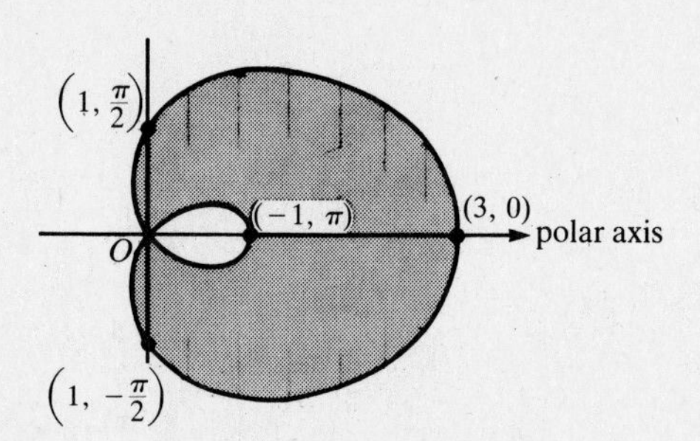

Figure 20-26
$r = 1 + 2\cos\theta$.

PROBLEM 20-22 Find the length of the graph of $r = 2(1 + \cos\theta)$.

Solution: You have $dr/d\theta = -2\sin\theta$, so

$$r^2 + \left(\frac{dr}{d\theta}\right)^2 = 4(1 + \cos\theta)^2 + 4\sin^2\theta = 8(1 + \cos\theta)$$

Using the symmetry of the graph about the polar axis, you have

$$s = 2\int_0^{\pi} \sqrt{8(1 + \cos\theta)}\, d\theta = 2\int_0^{\pi} \sqrt{16\cos^2(\theta/2)}\, d\theta$$

$$= 8\int_0^{\pi} \cos(\theta/2)\, d\theta = 16\sin(\theta/2)\Big|_0^{\pi} = 16$$

[See Section 20-6.]

PROBLEM 20-23 Find the length of the spiral $r = 2\theta$ for $0 \leqslant \theta \leqslant 2\pi$.

Solution: Substituting into the formula, you get

$$\sqrt{r^2 + \left(\frac{dr}{d\theta}\right)^2} = \sqrt{4\theta^2 + 4} = 2\sqrt{\theta^2 + 1}$$

Thus,

$$s = 2\int_0^{2\pi} \sqrt{\theta^2 + 1}\, d\theta$$

An inverse trigonometric substitution $\theta = \tan u$, $d\theta = \sec^2 u\, du$, gives you

$$s = 2\int_0^{\arctan 2\pi} \sec^3 u\, du = (\sec u \tan u + \ln|\sec u + \tan u|)\Big|_0^{\arctan 2\pi}$$

$$= 2\pi\sqrt{1 + 4\pi^2} + \ln|\sqrt{1 + 4\pi^2} + 2\pi|$$

[See Section 20-6.]

Supplementary Exercises

20-24 Find rectangular coordinates for each of the following points (given in polar coordinates (r, θ)): **(a)** $(3, \pi/2)$; **(b)** $(1, \pi)$; **(c)** $(-3, \pi/4)$; **(d)** $(\frac{1}{2}, 1)$.

20-25 Find two sets of polar coordinates for each of the following points (given in rectangular coordinates): **(a)** $(-1, 0)$; **(b)** $(3, -3)$; **(c)** $(-1, -\sqrt{3})$; **(d)** $(4, 2)$.

In Problems 20-26 through 20-30 change the equation to rectangular form:

20-26 $r = 2 \sec \theta$

20-27 $r = 3 \sin \theta$

20-28 $r^2 = \sin \theta$

20-29 $r = \sin 3\theta$ (***Hint:*** Use identities first; $\sin 3\theta = \sin(\theta + 2\theta)$).

20-30 $r = 1/(1 - 2 \cos \theta)$

In Problems 20-31 through 20-35 change the equation to polar form:

20-31 $x^2 + y^2 = 2x - y$

20-32 $x^2 = x - 2y$

20-33 $x^2 - y^2 = 4$

20-34 $x^2 + y^2 = y/x$

20-35 $x - y = 1$

In Problems 20-36 through 20-43 draw the graph of the given polar equation:

20-36 $r = 2 + 3 \sin \theta$

20-37 $r = 3$

20-38 $\theta = \pi/4$

20-39 $r = 4(1 - \cos \theta)$

20-40 $r = 3 + 2 \cos \theta$

20-41 $r^2 = \sin \theta$

20-42 $r = -\cos \theta$

20-43 $r = 2 \cos 2\theta$

In Problems 20-44 through 20-46 find the points of intersection of the given curves:

20-44 $r = 1$ and $r = 2 \cos 2\theta$

20-45 $r = 1 - \cos \theta$ and $r = 1 + \sin \theta$

20-46 $r = 2 \cos \theta$ and $r = 2(1 - \cos \theta)$

In Problems 20-47 through 20-50 find the equation of the tangent line to the given curve at the given point:

20-47 $r = 1 + 2 \sin \theta$ at $(2, \pi/6)$

20-48 $r = 2 + 2 \cos \theta$ at $(3, \pi/3)$

20-49 $r = 2\theta$ at $(\pi, \pi/2)$

20-50 $r = e^{\theta}$ at $(e^{\pi/2}, \pi/2)$

In Problems 20-51 through 20-54 find the slope of the tangent line at the given point using $\alpha = \theta + \beta$:

20-51 $r^2 = \cos 2\theta$ at $(1/\sqrt{2}, \pi/6)$

20-52 $r = 9 - 3 \csc \theta$ at $(3, \pi/6)$

20-53 $r = 5 \sin 3\theta$ at $(5/\sqrt{2}, \pi/12)$

20-54 $r = 2 + 3 \sec \theta$ at $(8, \pi/3)$

In Problems 20-55 through 20-61 find the area:

20-55 The area enclosed by $r = \cos 2\theta$

20-56 The area inside $r = 2(1 + \sin\theta)$, outside $r = 3$

20-57 The area inside both $r = 3 + 3\cos\theta$ and $r = -3\cos\theta$

20-58 The area inside $r = 3 + 3\cos\theta$, outside $r = 3\cos\theta$

20-59 The area inside $r = 1 + \cos\theta$, outside $r = 1$

20-60 The area inside $r = 2$, outside $r^2 = 4\cos 2\theta$

20-61 The area inside $r^2 = \sin 2\theta$

In Problems 20-62 through 20-66 find the arc length:

20-62 The length of $r = 5\sin\theta$

20-63 The length of $r = \theta^2$ for $0 \leqslant \theta \leqslant 2\pi$

20-64 The length of $r = \sin^3(\theta/3)$ for $0 \leqslant \theta \leqslant \pi$

20-65 The length of $r = e^{a\theta}$ for $0 \leqslant \theta \leqslant 2\pi$

20-66 The length of $r = 2^{\theta}$ for $0 \leqslant \theta \leqslant 2\pi$

Solutions to Supplementary Exercises

(20-24) **(a)** $(0, 3)$, **(b)** $(-1, 0)$, **(c)** $(-3\sqrt{2}/2, -3\sqrt{2}/2)$, **(d)** $(\frac{1}{2}\cos 1, \frac{1}{2}\sin 1)$

(20-25) **(a)** $(1, \pi), (-1, 0)$; **(b)** $(3\sqrt{2}, -\pi/4), (-3\sqrt{2}, 3\pi/4)$; **(c)** $(2, 4\pi/3), (-2, \pi/3)$; **(d)** $(2\sqrt{5}, \arctan\frac{1}{2})$, $(2\sqrt{5}, \arctan(\frac{1}{2}) + 2\pi)$

(20-26) $x = 2$

(20-27) $x^2 + y^2 = 3y$

(20-28) $(x^2 + y^2)^{3/2} = y$, or $(x^2 + y^2)^3 = y^2$

(20-29) $(x^2 + y^2)^2 = 3x^2y - y^3$

(20-30) $\sqrt{x^2 + y^2} - 2x = 1$, or $x^2 + y^2 = (2x + 1)^2$

(20-31) $r^2 = 2r\cos\theta - r\sin\theta$, or $r = 2\cos\theta - \sin\theta$

(20-32) $r\cos^2\theta = \cos\theta - 2\sin\theta$

(20-33) $r^2(\cos^2\theta - \sin^2\theta) = 4$, or $r^2 = 4\sec 2\theta$

(20-34) $r^2 = \tan\theta$

(20-35) $r = 1/(\cos\theta - \sin\theta)$

(20-36) See Figure 20-27.

(20-37) See Figure 20-28.

(20-38) See Figure 20-29.

(20-39) See Figure 20-30.

(20-40) See Figure 20-31.

(20-41) See Figure 20-32.

(20-42) See Figure 20-33.

(20-43) See Figure 20-34.

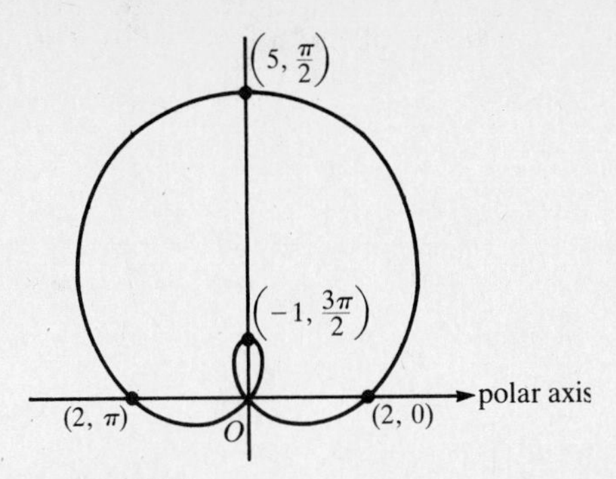

Figure 20-27
$r = 2 + 3 \sin \theta$.

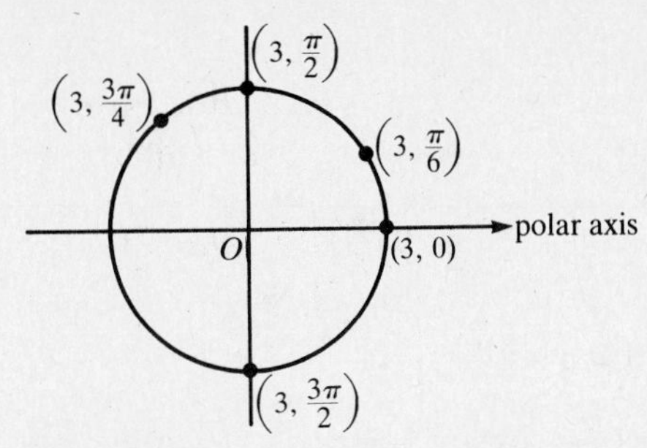

Figure 20-28
$r = 3$.

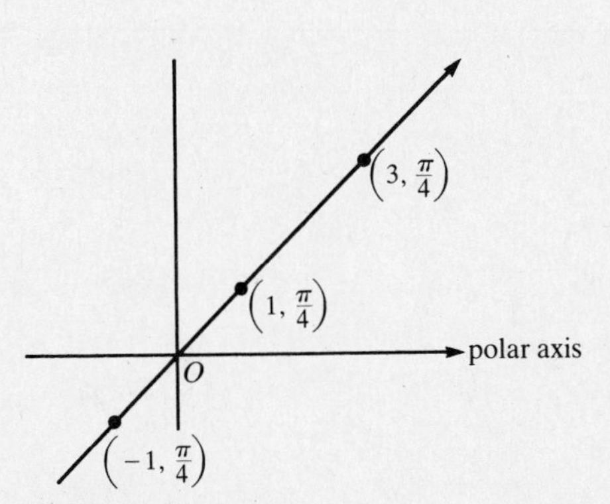

Figure 20-29
$\theta = \pi/4$.

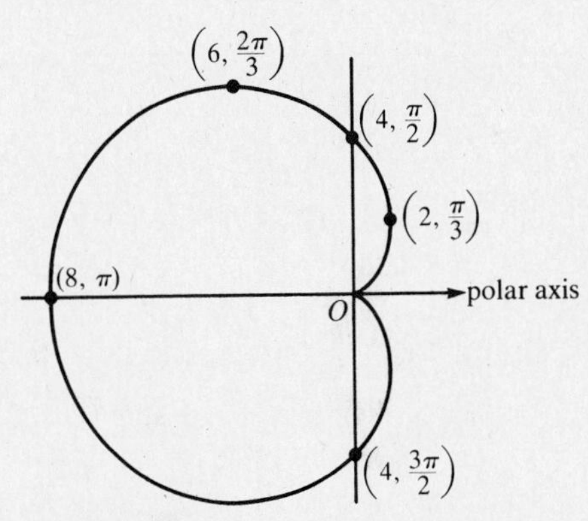

Figure 20-30
$r = 4(1 - \cos \theta)$.

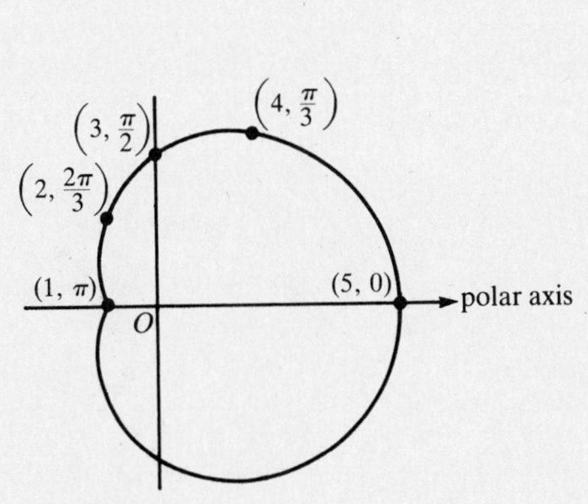

Figure 20-31
$r = 3 + 2 \cos \theta$.

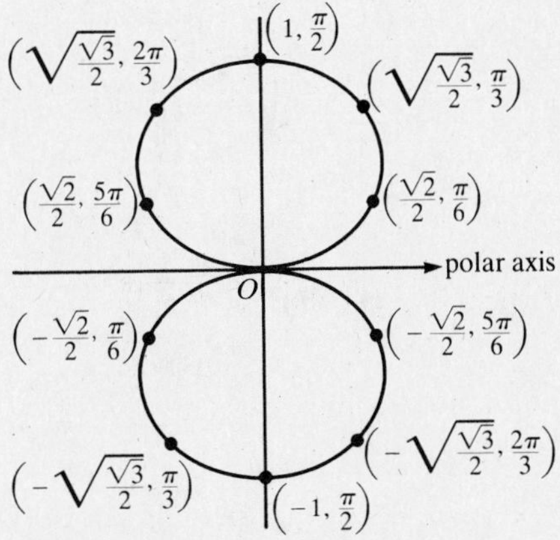

Figure 20-32
$r^2 = \sin \theta$.

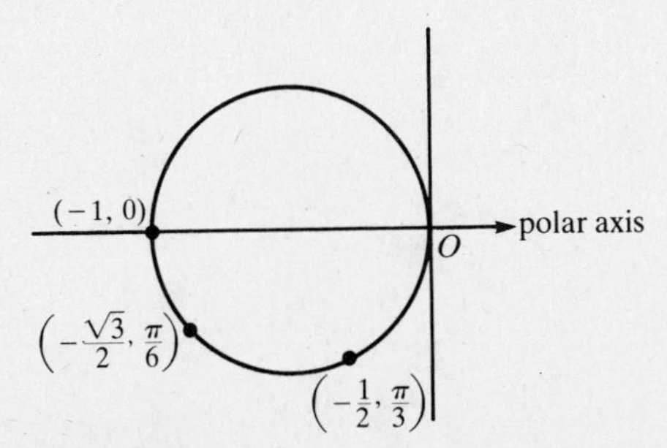

Figure 20-33
$r = -\cos \theta$.

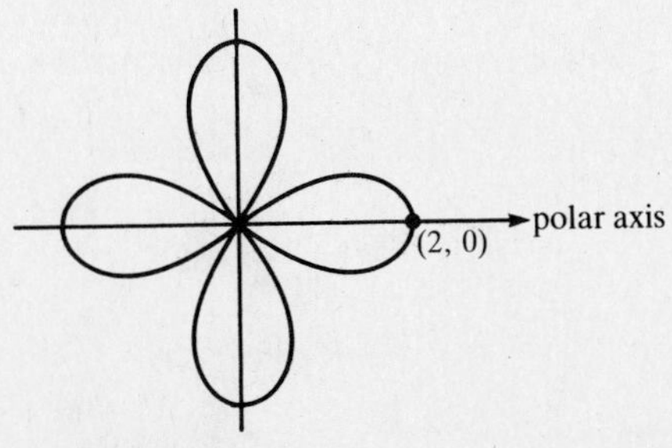

Figure 20-34
$r = 2 \cos 2\theta$.

(20-44) A four-leaf rose and a circle (center at origin) intersecting at the pole and at eight points $(1, \theta)$, where $\theta = \pi/6, \pi/3, 2\pi/3, 5\pi/6, 7\pi/6, 4\pi/3, 5\pi/3, 11\pi/6$.

(20-45) Two cardioids (one symmetric about $\theta = \pi$, one about $\theta = \pi/2$) intersecting at the pole, $((2 + \sqrt{2})/2, 3\pi/4)$, and $((2 - \sqrt{2})/2, 7\pi/4)$.

(20-46) A circle (center on the polar axis and radius 1), and a cardioid (symmetric about $\theta = \pi$) intersecting at the pole, $(1, \pi/3)$, and $(1, -\pi/3)$.

(20-47) $y - 1 = 3\sqrt{3}(x - \sqrt{3})$

(20-48) $y = 3\sqrt{3}/2$

(20-49) $y = (-2/\pi)x + \pi$

(20-50) $y = -x + e^{\pi/2}$

(20-51) $m = 0$

(20-52) $m = 3\sqrt{3}/5$

(20-53) $m = 5\sqrt{3} - 8$

(20-54) $m = -13\sqrt{3}/3$

(20-55) A four-leaf rose: $A = 4\int_{-\pi/4}^{\pi/4} \frac{1}{2}\cos^2 2\theta \, d\theta = \pi/2$

(20-56) Area between $r = 3$ and $r = 2(1 + \sin\theta)$ for $\pi/6 \leqslant \theta \leqslant 5\pi/6$; symmetric about $\theta = \pi/2$: $A = (2)\frac{1}{2}\int_{\pi/6}^{\pi/2} \{[2(1 + \sin\theta)]^2 - (3^2)\}\, d\theta = (9\sqrt{3}/2) - \pi$

(20-57) Area inside cardioid (symmetric about $\theta = 0$) and inside circle (center on $\theta = \pi$, passing through the pole): $A = \int_{\pi/2}^{2\pi/3} (9\cos^2\theta) d\theta + \frac{1}{2}\int_{2\pi/3}^{\pi} (3 + 3\cos\theta)^2\, d\theta = (21\pi/4) - 9\sqrt{3}$

(20-58) Area inside cardioid (symmetric about $\theta = 0$) and outside circle (center at $\theta = 0$, passing through the pole): $A = \frac{1}{2}\int_0^{2\pi} (3 + 3\cos\theta)^2\, d\theta - \frac{1}{2}\int_{-\pi/2}^{\pi/2} (-3\cos\theta)^2\, d\theta = 45\pi/4$

(20-59) Area inside cardioid (symmetric about $\theta = 0$) and outside circle (center at pole): $A = 2\int_0^{\pi/2} [\frac{1}{2}(1 + \cos\theta)^2 - \frac{1}{2}(1)^2] d\theta = 2 + (\pi/4)$

(20-60) Area of two quarter circles: $r = 2$, $\pi/4 \leqslant \theta \leqslant 3\pi/4$, and $5\pi/4 \leqslant \theta \leqslant 7\pi/4$; and area of region inside circle, outside lemniscate, for $-\pi/4 \leqslant \theta \leqslant \pi/4$ and $3\pi/4 \leqslant \theta \leqslant 5\pi/4$:

$$A = 2\left[\frac{\pi(2^2)}{4}\right] + 4\int_0^{\pi/4} [\tfrac{1}{2}(2)^2 - \tfrac{1}{2}(4\cos 2\theta)]\, d\theta = 4\pi - 4$$

(20-61) Area enclosed by lemniscate: $A = 2\int_0^{\pi/2} \frac{1}{2}\sin 2\theta\, d\theta = 1$

(20-62) $s = 2\int_0^{\pi/2} \sqrt{(5\sin\theta)^2 + (5\cos\theta)^2}\, d\theta = 5\pi$

(20-63) $s = \int_0^{2\pi} \sqrt{(\theta^2)^2 + (2\theta)^2}\, d\theta = (8/3)[(\pi^2 + 1)^{3/2} - 1]$

(20-64) $s = \int_0^{\pi} \sqrt{[\sin^3(\theta/3)]^2 + [3\sin^2(\theta/3)\cos(\theta/3)\frac{1}{3}]^2}\, d\theta = (4\pi - 3\sqrt{3})/8$

(20-65) $s = \int_0^{2\pi} \sqrt{e^{2a\theta} + a^2 e^{2a\theta}}\, d\theta = (\sqrt{1 + a^2}/a)(e^{2\pi a} - 1)$

(20-66) $s = \int_0^{2\pi} \sqrt{2^{2\theta} + 2^{2\theta}(\ln 2)^2}\, d\theta = (\sqrt{1 + (\ln 2)^2}/\ln 2)(2^{2\pi} - 1)$

EXAM 7 (CHAPTERS 16–20)

1. Find the area of the region bounded by the graphs of $f(x) = x^3 + 2\sqrt{x} - 5x + 1$ and $g(x) = 2\sqrt{x} + 4x + 1$.
2. A radioactive substance disintegrates at the rate $(1/3)\,e^{-t/10}$ grams/year at time t (in years). How much disintegrates between $t = 10$ and $t = 20$ years?
3. Integrate **(a)** $\displaystyle\int_0^{\pi} \sec^2 x\,dx$ **(b)** $\displaystyle\int_0^{\infty} \frac{e^x}{1 + e^{2x}}\,dx$
4. Let R be on the region bounded by the graphs of $y = e^x$, $y = 1$, and $x = 1$.
   (a) Set up and evaluate two integrals that express the resulting volume when R is rotated about the x axis.
   (b) Do the same for R rotated about the y axis.
5. Find the length of the arc of the graph of $y = 2(x + 2)^{3/2}$ between $x = -2$ and $x = 2/3$.
6. Find the work done in pumping all of the water (62.5 lb/ft^3) from a full conical tank of height 10 ft and diameter 6 ft (vertex down) to a point 2 ft above the top of the tank.
7. Graph the curve (in polar coordinates) $r = 1 + 3\sin\theta$ and find the equation of the tangent line (in rectangular coordinates) at $(-1/2, -\pi/6)$.

SOLUTIONS TO EXAM 7

1. Find where the graphs intersect:

$$x^3 + 2\sqrt{x} - 5x + 1 = 2\sqrt{x} + 4x + 1$$

$$x^3 - 9x = 0$$

$$x(x + 3)(x - 3) = 0$$

Although the graphs appear to intersect at $x = 0$, $x = -3$, and $x = 3$, they are only defined for $x \geqslant 0$. The interval of interest is $[0, 3]$. Determine which function is greater in $[0, 3]$ by evaluating both functions at an arbitrary point in the interval, say at $x = 1$: $f(1) = -1 < g(1) = 6$. The area is

$$\int_0^3 [g(x) - f(x)]\,dx = \int_0^3 [(2\sqrt{x} + 4x + 1) - (x^3 + 2\sqrt{x} - 5x + 1)]\,dx$$

$$= \int_0^3 (9x - x^3)\,dx = \left(\frac{9}{2}x^2 - \frac{x^4}{4}\right)\Bigg|_0^3$$

$$= \frac{81}{4}$$

2. If $W(t)$ is the weight of the substance at time t, then $W'(t)$ is the rate of change of the weight. That is:

$$W'(t) = \frac{-1}{3}e^{-t/10}$$

The amount that disintegrates between $t = 10$ and $t = 20$ is $W(10) - W(20)$. By the fundamental theorem of calculus,

$$W(10) - W(20) = \int_{20}^{10} W'(t)\,dt = \int_{20}^{10} \frac{-1}{3} e^{-t/10}\,dt = \frac{10}{3} e^{-t/10}\Big|_{20}^{10}$$

$$= \frac{10}{3}(e^{-1} - e^{-2}) \text{ grams}$$

3. **(a)** This is an improper integral because the integrand has a vertical asymptote at $x = \pi/2$, which lies in the interval of integration. You have

$$\int_0^{\pi} \sec^2 x\,dx = \int_0^{\pi/2} \sec^2 x\,dx + \int_{\pi/2}^{\pi} \sec^2 x\,dx$$

$$= \lim_{\varepsilon \to 0^+} \int_0^{\pi/2-\varepsilon} \sec^2 x\,dx + \lim_{\eta \to 0^+} \int_{\pi/2+\eta}^{\pi} \sec^2 x\,dx$$

$$= \lim_{\varepsilon \to 0^+} \left(\tan x \Big|_0^{\pi/2-\varepsilon} \right) + \lim_{\eta \to 0^+} \left(\tan x \Big|_{\pi/2+\eta}^{\pi} \right)$$

$$= \lim_{\varepsilon \to 0^+} \tan(\pi/2 - \varepsilon) + \lim_{\eta \to 0^+} \tan(\pi/2 + \eta)$$

$$= \infty + (-\infty)$$

so the integral diverges.

(b) Because the interval of integration $[0, \infty)$ is infinite, this is an improper integral:

$$\int_0^{\infty} \frac{e^x}{1 + e^{2x}}\,dx = \lim_{N \to \infty} \int_0^{N} \frac{e^x}{1 + e^{2x}}\,dx$$

Make the substitution $u = e^x$ to find

$$\int_0^{\infty} \frac{e^x}{1 + e^{2x}}\,dx = \lim_{N \to \infty} \int_1^{e^N} \frac{1}{1 + u^2}\,du = \lim_{N \to \infty} \left(\arctan u \Big|_1^{e^N} \right)$$

$$= \lim_{N \to \infty} (\arctan e^N - \arctan 1) = \frac{\pi}{2} - \frac{\pi}{4} = \frac{\pi}{4}$$

4. The region bounded by the graphs of $y = e^x$, $y = 1$, and $x = 1$ is shown in the figure.

(a) If you divide the x axis, you have rectangles that are thin and tall, and therefore yield disks. The volume is

$$\pi \int_0^1 [(e^x)^2 - 1^2]\,dx = \pi \int_0^1 (e^{2x} - 1)\,dx = \pi\left(\frac{1}{2}e^{2x} - x\right)\Big|_0^1$$

$$= \pi\left(\frac{1}{2}e^2 - 1\right) - \pi\left(\frac{1}{2}\right) = \frac{\pi}{2}e^2 - \frac{3\pi}{2}$$

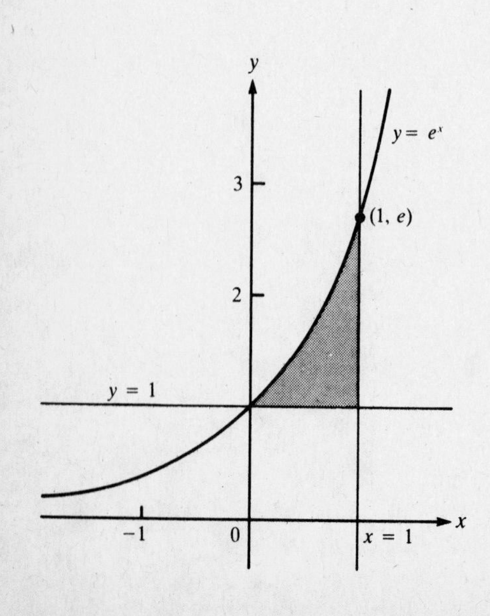

If you divide the y axis, you have shells. Solve $y = e^x$ for x: $x = \ln y$. The volume is

$$2\pi \int_1^e y(1 - \ln y)\,dy = 2\pi \int_1^e y\,dy - 2\pi \int_1^e y \ln y\,dy$$

$$= \pi y^2 \Big|_1^e - 2\pi \int_1^e y \ln y\,dy$$

$$= \pi e^2 - \pi - 2\pi \int_1^e y \ln y\,dy$$

Integrate by parts. Let

$$u = \ln y \qquad dv = y\,dy$$

$$du = \frac{1}{y}\,dy \qquad v = \frac{1}{2}y^2$$

$$\int_1^e y \ln y\,dy = \int u\,dv = uv - \int v\,du = \frac{1}{2}y^2 \ln y\Big|_1^e - \int_1^e \left(\frac{1}{2}y^2\right)\left(\frac{1}{y}\,dy\right)$$

$$= \frac{1}{2}y^2 \ln y\Big|_1^e - \frac{1}{2}\int_1^e y\,dy = \left(\frac{1}{2}y^2 \ln y - \frac{1}{4}y^2\right)\Big|_1^e$$

$$= \frac{1}{2}e^2 - \frac{1}{4}e^2 + \frac{1}{4} = \frac{1}{4}e^2 + \frac{1}{4}$$

Thus the volume is

$$\pi e^2 - \pi - 2\pi\left(\frac{1}{4}e^2 + \frac{1}{4}\right) = \frac{\pi}{2}e^2 - \frac{3\pi}{2}$$

(b) If you divide the x axis, you use the shell method:

$$2\pi \int_0^1 x(e^x - 1)\,dx = 2\pi\left[\int_0^1 xe^x\,dx - \int_0^1 x\,dx\right]$$

$$= 2\pi \int_0^1 xe^x\,dx - \pi x^2\Big|_0^1$$

$$= 2\pi \int_0^1 xe^x\,dx - \pi$$

Now integrate by parts. Let

$$u = x \qquad dv = e^x\,dx$$

$$du = dx \qquad v = e^x$$

$$\int_0^1 xe^x\,dx = xe^x\Big|_0^1 - \int_0^1 e^x\,dx = (xe^x - e^x)\Big|_0^1 = (e - e) - (0 - 1) = 1$$

So the volume is

$$2\pi(1) - \pi = \pi$$

If you divide the y axis, you use the disk method:

$$\pi \int_1^e (1^2 - \ln^2 x)\,dx = \pi x\Big|_1^e - \pi \int_1^e \ln^2 x\,dx = \pi(e - 1) - \pi \int_1^e \ln^2 x\,dx$$

Integrate by parts. Let

$$u = \ln^2 x \qquad dv = dx$$

$$du = \frac{2 \ln x}{x}\,dx \qquad v = x$$

$$\int_1^e \ln^2 x\,dx = x \ln^2 x\Big|_1^e - \int_1^e x \cdot \frac{2 \ln x}{x}\,dx = x \ln^2 x\Big|_1^e - 2\int_1^e \ln x\,dx$$

Let

$$u = \ln x \qquad dv = dx$$

$$du = \frac{1}{x}\,dx \qquad v = x$$

$$\int_1^e \ln^2 x\,dx = x \ln^2 x\Big|_1^e - 2\left[x \ln x\Big|_1^e - \int_1^e x\left(\frac{1}{x}\,dx\right)\right]$$

$$= (x \ln^2 x - 2x \ln x + 2x)\Big|_1^e = (e - 2e + 2e) - (2) = e - 2$$

The volume is

$$\pi(e - 1) - \pi(e - 2) = \pi$$

5. Find $1 + (y')^2$:

$$y' = 2\left(\frac{3}{2}\right)(x + 2)^{1/2} = 3(x + 2)^{1/2}$$

$$1 + (y')^2 = 1 + 9(x + 2) = 9x + 19$$

The arc length is

$$\int_{-2}^{2/3} \sqrt{1 + (y')^2}\, dx = \int_{-2}^{2/3} \sqrt{9x + 19}\, dx$$

Let $u = 9x + 19$ with $du = 9\, dx$:

$$\int_{-2}^{2/3} \sqrt{9x + 19}\, dx = \int_{1}^{25} \sqrt{u}\left(\frac{1}{9}\, du\right) = \frac{1}{9}\cdot\frac{2}{3}u^{3/2}\Big|_{1}^{25} = \frac{2}{27}(125 - 1) = \frac{248}{27}$$

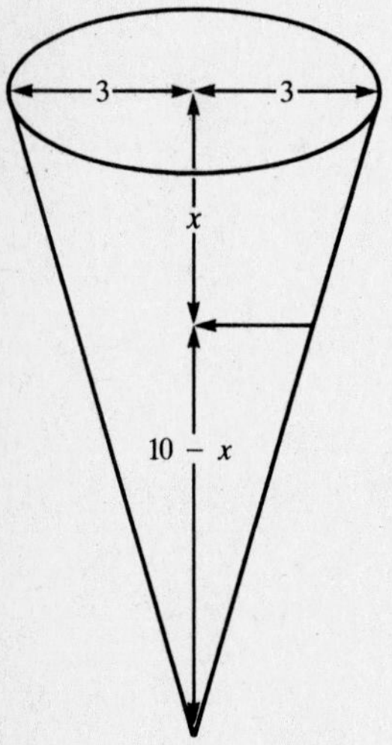

6. Set up an x axis, perhaps as in the figure. The radius of the cross section at x is, by similar triangles,

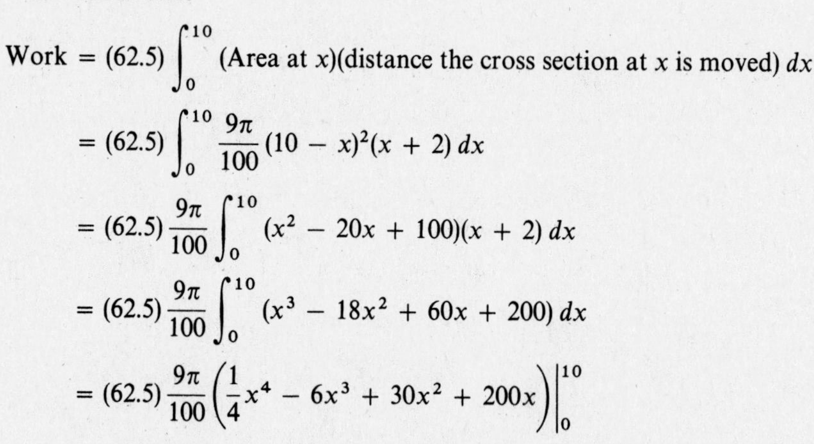

$$\frac{10}{3} = \frac{10 - x}{r}$$

$$r = \frac{3}{10}(10 - x)$$

The area of the cross section is

$$\pi r^2 = \pi\left[\frac{3}{10}(10 - x)\right]^2 = \frac{9\pi}{100}(10 - x)^2$$

The work done is

$$\text{Work} = (62.5)\int_0^{10} (\text{Area at } x)(\text{distance the cross section at } x \text{ is moved})\, dx$$

$$= (62.5)\int_0^{10} \frac{9\pi}{100}(10 - x)^2(x + 2)\, dx$$

$$= (62.5)\frac{9\pi}{100}\int_0^{10} (x^2 - 20x + 100)(x + 2)\, dx$$

$$= (62.5)\frac{9\pi}{100}\int_0^{10} (x^3 - 18x^2 + 60x + 200)\, dx$$

$$= (62.5)\frac{9\pi}{100}\left(\frac{1}{4}x^4 - 6x^3 + 30x^2 + 200x\right)\Big|_0^{10}$$

$$= (62.5)\frac{9\pi}{100}(2500 - 6000 + 3000 + 2000) = 8437.5\pi \text{ ft-lb}$$

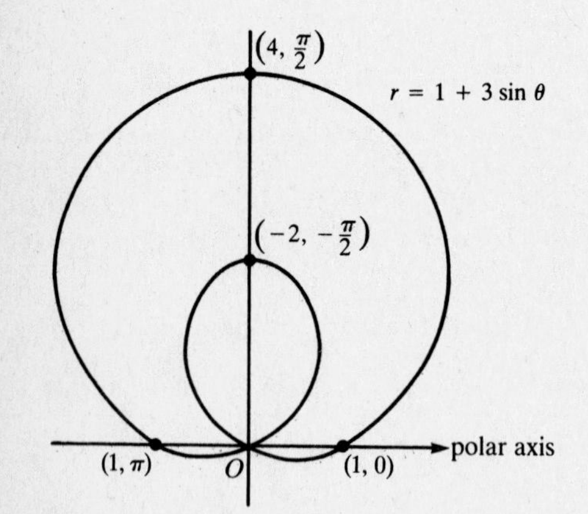

7. The graph of $r = 1 + 3\sin\theta$ is shown in the figure. The tangent line at (r, θ) has slope

$$m = \frac{r\cos\theta + \sin\theta\dfrac{dr}{d\theta}}{-r\sin\theta + \cos\theta\dfrac{dr}{d\theta}}$$

$$= \frac{r\cos\theta + \sin\theta(3\cos\theta)}{-r\sin\theta + \cos\theta(3\cos\theta)}$$

So the slope at $(-1/2, -\pi/6)$ is

$$m = \frac{-\frac{1}{2}\cos\left(\frac{-\pi}{6}\right) + 3\sin\left(\frac{-\pi}{6}\right)\cos\left(\frac{-\pi}{6}\right)}{\frac{1}{2}\sin\left(\frac{-\pi}{6}\right) + 3\cos^2\left(\frac{-\pi}{6}\right)}$$

$$= \frac{-\frac{1}{2}\left(\frac{\sqrt{3}}{2}\right) + 3\left(-\frac{1}{2}\right)\left(\frac{\sqrt{3}}{2}\right)}{\frac{1}{2}\left(-\frac{1}{2}\right) + 3\left(\frac{\sqrt{3}}{2}\right)^2}$$

$$= \frac{-\sqrt{3} - 3\sqrt{3}}{-1 + 9} = \frac{-4\sqrt{3}}{8} = \frac{-\sqrt{3}}{2}$$

The tangent line passes through $(-1/2, -\pi/6)$ which has rectangular coordinates:

$$x = r\cos\theta = -\frac{1}{2}\left(\frac{\sqrt{3}}{2}\right) = \frac{-\sqrt{3}}{4}$$

$$y = r\sin\theta = -\frac{1}{2}\left(\frac{-1}{2}\right) = \frac{1}{4}$$

The tangent line has slope $-\sqrt{3}/2$ and passes through $(-\sqrt{3}/4, 1/4)$:

$$y - \frac{1}{4} = \frac{-\sqrt{3}}{2}\left(x + \frac{\sqrt{3}}{4}\right)$$

$$y = \frac{-\sqrt{3}}{2}x - \frac{1}{8}$$

21 L'HÔPITAL'S RULE

THIS CHAPTER IS ABOUT

☑ The Indeterminate Form 0/0
☑ The Indeterminate Form ∞/∞
☑ The Indeterminate Forms ∞ − ∞ and 0 · ∞
☑ The Indeterminate Forms 0^0, ∞^0, and 1^∞

21-1. The Indeterminate Form 0/0

A limit of the form $\lim_{x\to a} f(x)/g(x)$ where $\lim_{x\to a} f(x) = 0$ and $\lim_{x\to a} g(x) = 0$ has the **indeterminate form 0/0** at $x = a$. It's called indeterminate because such a limit might be any real number, ∞, $-\infty$, or it might not exist. In Chapter 3 you saw a few techniques for finding such limits. **L'Hôpital's rule** is yet another powerful technique. If $\lim_{x\to a} f(x) = 0$ and $\lim_{x\to a} g(x) = 0$, then

L'HÔPITAL'S RULE

$$\lim_{x\to a}\frac{f(x)}{g(x)} = \lim_{x\to a}\frac{f'(x)}{g'(x)} \tag{21-1}$$

provided the limit on the right exists or is ∞ or $-\infty$. This is valid for one-sided limits and two-sided limits, when a is a real number, or when $a = \infty$ or $a = -\infty$.

EXAMPLE 21-1: Find $\lim_{x\to 0} (x^2 - 3x)/(2x^2 + x)$.

Solution: Both the numerator and the denominator approach zero as x approaches zero, so this is the indeterminate form 0/0. By l'Hôpital's rule,

$$\lim_{x\to 0}\frac{x^2 - 3x}{2x^2 + x} = \lim_{x\to 0}\frac{\dfrac{d}{dx}(x^2 - 3x)}{\dfrac{d}{dx}(2x^2 + x)} = \lim_{x\to 0}\frac{2x - 3}{4x + 1} = \frac{0 - 3}{0 + 1} = -3$$

The first equality is valid only because the second limit exists. If the second limit had not existed, you would have tried another technique.

EXAMPLE 21-2: Find $\lim_{x\to 1} (x^3 + x - 2)/(x^2 - 1)$.

Solution: Both the numerator and the denominator approach zero, so you can use l'Hôpital's rule:

$$\lim_{x\to 1}\frac{x^3 + x - 2}{x^2 - 1} = \lim_{x\to 1}\frac{\dfrac{d}{dx}(x^3 + x - 2)}{\dfrac{d}{dx}(x^2 - 1)}$$

$$= \lim_{x\to 1}\frac{3x^2 + 1}{2x} = \frac{3 + 1}{2} = 2$$

EXAMPLE 21-3: Find $\lim_{x\to\pi/2} \cos x/[(x - \frac{1}{2}\pi)^3]$.

Solution: This is the indeterminate form 0/0, so you can use l'Hôpital's rule:

$$\lim_{x\to\pi/2}\frac{\cos x}{(x-\frac{1}{2}\pi)^3}=\lim_{x\to\pi/2}\frac{\frac{d}{dx}\cos x}{\frac{d}{dx}(x-\frac{1}{2}\pi)^3}=\lim_{x\to\pi/2}\frac{-\sin x}{3(x-\frac{1}{2}\pi)^2}$$

In this new limit the numerator approaches -1 and the denominator approaches 0. This form, $-1/0$, isn't indeterminate. Indeed, it approaches $-\infty$. Therefore, you have

$$\lim_{x\to\pi/2}\frac{\cos x}{(x-\frac{1}{2}\pi)^3}=-\infty$$

EXAMPLE 21-4: Find $\lim_{x\to 0}(e^x-1)/(x^2+x)$.

Solution: If you've verified that this is the indeterminate form 0/0, you can use l'Hôpital's rule:

$$\lim_{x\to 0}\frac{e^x-1}{x^2+x}=\lim_{x\to 0}\frac{\frac{d}{dx}(e^x-1)}{\frac{d}{dx}(x^2+x)}=\lim_{x\to 0}\frac{e^x}{2x+1}=\frac{e^0}{1}=1$$

EXAMPLE 21-5: Find $\lim_{x\to-\infty}e^x/e^{2x}$.

Solution: Because $\lim_{x\to-\infty}e^x=0$ and $\lim_{x\to-\infty}e^{2x}=0$, you can try l'Hôpital's rule:

$$\lim_{x\to-\infty}\frac{\frac{d}{dx}e^x}{\frac{d}{dx}e^{2x}}=\lim_{x\to-\infty}\frac{e^x}{2e^{2x}}$$

It is no easier to find this limit than it is to find the original limit. Don't forget that l'Hôpital's rule isn't the only method you have for finding limits. In Chapter 3 you saw other techniques, most of which were just algebraic manipulations:

$$\frac{e^x}{e^{2x}}=e^{-x}$$

$$\lim_{x\to-\infty}\frac{e^x}{e^{2x}}=\lim_{x\to-\infty}e^{-x}=\infty$$

You may have to use l'Hôpital's rule more than once to find a limit.

EXAMPLE 21-6: Find $\lim_{x\to 0}(\sin x-x)/x^2$.

Solution: First verify that this is the indeterminate form 0/0. Now apply l'Hôpital's rule:

$$\lim_{x\to 0}\frac{\sin x-x}{x^2}=\lim_{x\to 0}\frac{\frac{d}{dx}(\sin x-x)}{\frac{d}{dx}x^2}=\lim_{x\to 0}\frac{\cos x-1}{2x}$$

Again you have the indeterminate form 0/0, so apply l'Hôpital's rule again:

$$\lim_{x\to 0}\frac{\cos x-1}{2x}=\lim_{x\to 0}\frac{\frac{d}{dx}(\cos x-1)}{\frac{d}{dx}2x}=\lim_{x\to 0}\frac{-\sin x}{2}=\frac{0}{2}=0$$

21-2. The Indeterminate Form ∞/∞

A limit of the form $\lim_{x\to a} f(x)/g(x)$ where $\lim_{x\to\infty} f(x) = \pm\infty$ and $\lim_{x\to\infty} g(x) = \pm\infty$ has the **indeterminate form** ∞/∞ at $x = a$. L'Hôpital's rule also applies in this situation, provided the limit on the right exists or is ∞ or is $-\infty$. This is valid for one-sided limits and two-sided limits, when a is a real number or when $a = \infty$ or $a = -\infty$.

EXAMPLE 21-7: Find $\lim_{x\to\infty} (x + 1)/(3x - 4)$.

Solution: Because $\lim_{x\to\infty} (x + 1) = \infty$ and $\lim_{x\to\infty} (3x - 4) = \infty$, this is the indeterminate form ∞/∞. Apply l'Hôpital's rule:

$$\lim_{x\to\infty} \frac{x+1}{3x-4} = \lim_{x\to\infty} \frac{\frac{d}{dx}(x+1)}{\frac{d}{dx}(3x-4)} = \lim_{x\to\infty} \frac{1}{3} = \frac{1}{3}$$

EXAMPLE 21-8: Find $\lim_{x\to 0^+} \ln x/(1/\sqrt{x})$.

Solution: This is the indeterminate form ∞/∞, so you can use l'Hôpital's rule:

$$\lim_{x\to 0^+} \frac{\ln x}{\frac{1}{\sqrt{x}}} = \lim_{x\to 0^+} \frac{\frac{d}{dx}\ln x}{\frac{d}{dx}\frac{1}{\sqrt{x}}} = \lim \frac{\frac{1}{x}}{-\frac{1}{2}x^{-3/2}}$$

Now simplify,

$$\frac{\frac{1}{x}}{-\frac{1}{2}x^{-3/2}} = -2x^{1/2}$$

and find the limit:

$$\lim_{x\to 0^+} \frac{\frac{1}{x}}{-\frac{1}{2}x^{-3/2}} = \lim_{x\to 0^+} (-2x^{1/2}) = -2\sqrt{0} = 0$$

EXAMPLE 21-9: Find $\lim_{x\to\infty} x^2/e^x$.

Solution: Be sure to check that l'Hôpital's rule applies (indeterminate form 0/0 or ∞/∞), and then

$$\lim_{x\to\infty} \frac{x^2}{e^x} = \lim_{x\to\infty} \frac{\frac{d}{dx}x^2}{\frac{d}{dx}e^x} = \lim_{x\to\infty} \frac{2x}{e^x}$$

This again is the indeterminate form ∞/∞, so apply l'Hôpital's rule again:

$$\lim_{x\to\infty} \frac{2x}{e^x} = \lim_{x\to\infty} \frac{\frac{d}{dx}2x}{\frac{d}{dx}e^x} = \lim_{x\to\infty} \frac{2}{e^x} = 0$$

EXAMPLE 21-10: Find $\lim_{x\to\infty} (x - 1)/(2x + \cos x)$.

Solution: Although this is the indeterminate form ∞/∞, l'Hôpital's rule doesn't apply. Indeed,

$$\lim_{x\to\infty} \frac{\frac{d}{dx}(x-1)}{\frac{d}{dx}(2x+\cos x)} = \lim_{x\to\infty} \frac{1}{2-\sin x}$$

which doesn't exist. This does not mean that $\lim_{x\to\infty} (x - 1)/(2x + \cos x)$ doesn't exist. Just use another technique to find the limit: Divide numerator and denominator by x:

$$\lim_{x\to\infty} \frac{x-1}{2x+\cos x} = \lim_{x\to\infty} \frac{(x-1)\frac{1}{x}}{(2x+\cos x)\frac{1}{x}} = \lim_{x\to\infty} \frac{1-\frac{1}{x}}{2+\frac{\cos x}{x}}$$

As x approaches infinity, $1/x$ and $(\cos x)/x$ both approach zero, so you have

$$\lim_{x\to\infty} \frac{x-1}{2x+\cos x} = \frac{1-0}{2+0} = \frac{1}{2}$$

21-3. The Indeterminate Forms $\infty - \infty$ and $0 \cdot \infty$

A limit of the form $\lim_{x\to a} (f(x) - g(x))$ where $\lim_{x\to a} f(x) = \infty$ and $\lim_{x\to a} g(x) = \infty$ has the **indeterminate form** $\infty - \infty$. To find such a limit, try to combine $f(x)$ and $g(x)$. Often you will be able to write $f(x) - g(x)$ in a form that allows you to use l'Hôpital's rule (0/0 or ∞/∞).

EXAMPLE 21-11: Find $\lim_{x\to -1} [2/(1 - x^2) - 1/(1 + x)]$.

Solution: This is the indeterminate form $\infty - \infty$. Combine the fractions:

$$\lim_{x\to -1}\left(\frac{2}{1-x^2} - \frac{1}{1+x}\right) = \lim_{x\to -1}\left(\frac{2}{1-x^2} - \frac{1-x}{1-x^2}\right) = \lim_{x\to -1} \frac{2-(1-x)}{1-x^2}$$

$$= \lim_{x\to -1} \frac{1+x}{1-x^2}$$

You may either factor $1 + x$ out of the numerator and denominator and cancel or you may use l'Hôpital's rule because this is the indeterminate form 0/0:

$$\lim_{x\to -1} \frac{1+x}{1-x^2} = \lim_{x\to -1} \frac{1}{-2x} = \frac{1}{2}$$

EXAMPLE 21-12: Find $\lim_{x\to 0}[(1/x) - (1/\sin x)]$.

Solution: This is the indeterminate form $\infty - \infty$, so try to rewrite it in a form that allows you to use l'Hôpital's rule:

$$\lim_{x\to 0}\left(\frac{1}{x} - \frac{1}{\sin x}\right) = \lim_{x\to 0} \frac{\sin x - x}{x \sin x}$$

This is the indeterminate form 0/0, so

$$\lim_{x\to 0}\left(\frac{\sin x - x}{x\sin x}\right) = \lim_{x\to 0} \frac{\frac{d}{dx}(\sin x - x)}{\frac{d}{dx}(x \sin x)}$$

$$= \lim_{x\to 0} \frac{\cos x - 1}{\sin x + x\cos x}$$

This again is the indeterminate form 0/0, so apply l'Hôpital's rule again:

$$\lim_{x\to 0} \frac{\cos x - 1}{\sin x + x\cos x} = \lim_{x\to 0} \frac{\frac{d}{dx}(\cos x - 1)}{\frac{d}{dx}(\sin x + x\cos x)} = \lim_{x\to 0} \frac{-\sin x}{2\cos x - x\sin x}$$

$$= -\frac{0}{2} = 0$$

A limit of the form $\lim_{x\to a} f(x)\,g(x)$ where $\lim_{x\to a} f(x) = 0$ and $\lim_{x\to a} g(x) = \pm\infty$ has the **indeterminate form $0 \cdot \infty$**. You can usually find this limit by rewriting it in one of two ways:

$$\lim_{x\to a} \frac{g(x)}{\dfrac{1}{f(x)}} \qquad \text{or} \qquad \lim_{x\to a} \frac{f(x)}{\dfrac{1}{g(x)}}$$

These are the indeterminate forms ∞/∞ and $0/0$, respectively, so you can use l'Hôpital's rule.

EXAMPLE 21-13: Find $\lim_{x\to 0^+} x \ln x$.

Solution: Because $\lim_{x\to 0^+} x = 0$ and $\lim_{x\to 0^+} \ln x = -\infty$, this is the indeterminate form $0 \cdot \infty$. Rewriting,

$$\lim_{x\to 0^+} x \ln x = \lim_{x\to 0^+} \frac{\ln x}{1/x}$$

This is the indeterminate form ∞/∞, so you can use l'Hôpital's rule:

$$\lim_{x\to 0^+} \frac{\ln x}{\dfrac{1}{x}} = \lim_{x\to 0^+} \frac{\dfrac{d}{dx}\ln x}{\dfrac{d}{dx}\dfrac{1}{x}} = \lim_{x\to 0^+} \frac{\dfrac{1}{x}}{\dfrac{-1}{x^2}} = \lim_{x\to 0^+} -x = 0$$

EXAMPLE 21-14: Find $\lim_{x\to \pi/2} (\pi - 2x)\sec x$.

Solution: This is the indeterminate form $0 \cdot \infty$. You can rewrite it as

$$\lim_{x\to \pi/2} (\pi - 2x)\sec x = \lim_{x\to \pi/2} \frac{\pi - 2x}{\cos x}$$

Now that it is in the indeterminate form $0/0$, you can use l'Hôpital's rule:

$$\lim_{x\to \pi/2} \frac{\pi - 2x}{\cos x} = \lim_{x\to \pi/2} \frac{-2}{-\sin x} = \frac{2}{\sin(\pi/2)} = 2$$

21-4. The Indeterminate Forms 0^0, ∞^0, and 1^∞

A. The form 0^0

A limit of the form $\lim_{x\to a} f(x)^{g(x)}$ where $\lim_{x\to a} f(x) = 0$ and $\lim_{x\to a} g(x) = 0$ has the **indeterminate form 0^0**. The trick to finding such a limit is to let $y = f(x)^{g(x)}$ and find $\lim_{x\to a} \ln y$. This will be the indeterminate form $0 \cdot \infty$, so you can use the methods of Section 21-3 to find this limit. Finally, you can find the limit of y by applying the exponential function to the limit of $\ln y$.

EXAMPLE 21-15: Find $\lim_{x\to 0^+} x^x$.

Solution: This is the indeterminate form 0^0. Let $y = x^x$. Find $\lim_{x\to 0^+} \ln y$:

$$\lim_{x\to 0^+} \ln y = \lim_{x\to 0^+} \ln x^x = \lim_{x\to 0^+} x \ln x$$

This is the indeterminate form $0 \cdot \infty$. Indeed, you saw this limit in Example 21-13:

$$\lim_{x\to 0^+} \ln y = \lim_{x\to 0^+} \frac{\ln x}{\dfrac{1}{x}} = \lim_{x\to 0^+} \frac{\dfrac{1}{x}}{\dfrac{-1}{x^2}} = \lim_{x\to 0^+} -x = 0$$

So you have found that $\ln y$ approaches zero. For $\ln y$ to approach zero, y must approach one. To see this, recognize that

$$y = e^{\ln y}$$

so that if $\ln y$ approaches zero, y approaches $e^0 = 1$:

$$\lim_{x\to 0^+} y = \lim_{x\to 0^+} x^x = 1$$

B. The form ∞^0

A limit of the form $\lim_{x\to a} f(x)^{g(x)}$ where $\lim_{x\to a} f(x) = \infty$ and $\lim_{x\to a} g(x) = 0$ has the **indeterminate form ∞^0**. You can find this limit by using the same technique as in Section A; that is, you find $\lim_{x\to a} \ln f(x)^{g(x)}$ and raise e to that power.

EXAMPLE 21-16: Find $\lim_{x\to\infty} (e^{2x} + 3)^{1/x}$.

Solution: This is the indeterminate form ∞^0. Let $y = (e^{2x} + 3)^{1/x}$. Then you have

$$\lim_{x\to\infty} \ln y = \lim_{x\to\infty} \ln(e^{2x} + 3)^{1/x} = \lim_{x\to\infty} \frac{1}{x}\ln(e^{2x} + 3)$$

$$= \lim_{x\to\infty} \frac{\ln(e^{2x} + 3)}{x}$$

This is the indeterminate form ∞/∞, so use l'Hôpital's rule:

$$\lim_{x\to\infty} \frac{\ln(e^{2x} + 3)}{x} = \lim_{x\to\infty} \frac{\dfrac{2e^{2x}}{e^{2x} + 3}}{1} = \lim_{x\to\infty} \frac{2}{1 + \dfrac{3}{e^{2x}}} = \frac{2}{1 + 0} = 2$$

Because $\ln y$ approaches two, y approaches e^2, so

$$\lim_{x\to\infty} (e^{2x} + 3)^{1/x} = e^2$$

C. The form 1^∞

A limit of the form $\lim_{x\to a} f(x)^{g(x)}$ where $\lim_{x\to a} f(x) = 1$ and $\lim_{x\to a} g(x) = \infty$ has the **indeterminate form 1^∞**. To solve limits in this form, use the same technique as in Sections A and B.

EXAMPLE 21-17: Find $\lim_{x\to\infty} [1 + (1/x)]^{x^2}$.

Solution: This is the indeterminate form 1^∞, so let $y = [1 + (1/x)]^{x^2}$ and find $\lim_{x\to\infty} \ln y$:

$$\lim_{x\to\infty} \ln y = \lim_{x\to\infty} \ln\left(1 + \frac{1}{x}\right)^{x^2} = \lim_{x\to\infty} x^2 \ln\left(1 + \frac{1}{x}\right)$$

$$= \lim_{x\to\infty} \frac{\ln\left(1 + \dfrac{1}{x}\right)}{\dfrac{1}{x^2}} = \lim_{x\to\infty} \frac{\dfrac{d}{dx}\ln\left(1 + \dfrac{1}{x}\right)}{\dfrac{d}{dx}\dfrac{1}{x^2}}$$

$$= \lim_{x\to\infty} \frac{\dfrac{1}{\left(1 + \dfrac{1}{x}\right)}\left(\dfrac{-1}{x^2}\right)}{\dfrac{-2}{x^3}} = \lim_{x\to\infty} \frac{x}{2\left(1 + \dfrac{1}{x}\right)} = \infty$$

Because $\ln y$ approaches infinity, y approaches infinity ($e^\infty = \infty$, if you will).

SUMMARY

1. You can use l'Hôpital's rule on the indeterminate forms 0/0 and ∞/∞.
2. There are occasions when, although applicable, l'Hôpital's rule doesn't lead to a solution.
3. You should manipulate the indeterminate forms $0 \cdot \infty$ and $\infty - \infty$ into either the indeterminate form 0/0 or ∞/∞; then use l'Hôpital's rule.
4. If $\lim_{x\to a} f(x)$ is the indeterminate form $0^0, 1^\infty$, or ∞^0, you should find $\lim_{x\to a} \ln f(x)$ (which you can write in the form 0/0 or ∞/∞) using l'Hôpital's rule. If this limit is a finite number b, then $\lim_{x\to a} f(x) = e^b$. If $b = \infty$, then $\lim_{x\to a} f(x) = \infty$. If $b = -\infty$, then $\lim_{x\to a} f(x) = 0$.

SOLVED PROBLEMS

PROBLEM 21-1 Find $\lim_{x\to 3} (x - 3)/(e^x - e^3)$.

Solution: This is the indeterminate form 0/0, so apply l'Hôpital's rule:

$$\lim_{x\to 3} \frac{x-3}{e^x - e^3} = \lim_{x\to 3} \frac{\dfrac{d}{dx}(x-3)}{\dfrac{d}{dx}(e^x - e^3)} = \lim_{x\to 3} \frac{1}{e^x} = \frac{1}{e^3} = e^{-3}$$

[See Section 21-1.]

PROBLEM 21-2 Find $\lim_{x\to 1} \ln x/(x - 1)$.

Solution: This is the indeterminate form 0/0, so apply l'Hôpital's rule:

$$\lim_{x\to 1} \frac{\ln x}{x-1} = \lim_{x\to 1} \frac{\dfrac{d}{dx}\ln x}{\dfrac{d}{dx}(x-1)} = \lim_{x\to 1} \frac{\frac{1}{x}}{1} = 1$$

[See Section 21-1.]

PROBLEM 21-3 Find $\lim_{x\to 0} (\tan x - x)/\sin x$.

Solution: This is the indeterminate form 0/0 because $\tan 0 = 0$ and $\sin 0 = 0$. Apply l'Hôpital's rule:

$$\lim_{x\to 0} \frac{\tan x - x}{\sin x} = \lim_{x\to 0} \frac{\dfrac{d}{dx}(\tan x - x)}{\dfrac{d}{dx}\sin x} = \lim_{x\to 0} \frac{\sec^2 x - 1}{\cos x} = \frac{1-1}{1} = 0$$

[See Section 21-1.]

PROBLEM 21-4 Find $\lim_{x\to\infty} (e^{1/x} - 1)/\sin(2/x)$.

Solution: This is the indeterminate form 0/0, so,

$$\lim_{x\to\infty} \frac{e^{1/x} - 1}{\sin\left(\frac{2}{x}\right)} = \lim_{x\to\infty} \frac{\left(\frac{-1}{x^2}\right) e^{1/x}}{\left(\frac{-2}{x^2}\right)\cos\left(\frac{2}{x}\right)}$$

$$= \lim_{x\to\infty} \frac{e^{1/x}}{2\cos\left(\frac{2}{x}\right)} = \frac{1}{2}$$

[See Section 21-1.]

PROBLEM 21-5 Find $\lim_{x\to 0} [\ln(1 + x^2)]/(\cos x - 1)$.

Solution: Before you use l'Hôpital's rule, check that this is the indeterminate form 0/0 or ∞/∞.

$$\frac{\ln(1 + 0^2)}{\cos 0 - 1} = \frac{0}{0}$$

$$\lim_{x\to 0} \frac{\ln(1 + x^2)}{\cos x - 1} = \lim_{x\to 0} \frac{\dfrac{d}{dx}\ln(1 + x^2)}{\dfrac{d}{dx}(\cos x - 1)} = \lim_{x\to 0} \frac{\dfrac{2x}{1 + x^2}}{-\sin x}$$

This is the indeterminate form 0/0, so apply l'Hôpital's rule again:

$$\lim_{x\to 0} \frac{\left(\dfrac{2x}{1 + x^2}\right)}{-\sin x} = \lim_{x\to 0} \frac{\dfrac{d}{dx}\left(\dfrac{2x}{1 + x^2}\right)}{\dfrac{d}{dx}(-\sin x)} = \lim_{x\to 0} \frac{\dfrac{2 - 2x^2}{(1 + x^2)^2}}{-\cos x} = -2 \qquad \text{[See Section 21-1.]}$$

PROBLEM 21-6 Find $\lim_{x\to 0} (xe^x - x)/\sin^2 x$.

Solution: Verify that this is the indeterminate form 0/0. Now use l'Hôpital's rule:

$$\lim_{x\to 0} \frac{xe^x - x}{\sin^2 x} = \lim_{x\to 0} \frac{\dfrac{d}{dx}(xe^x - x)}{\dfrac{d}{dx}(\sin^2 x)} = \lim_{x\to 0} \frac{e^x + xe^x - 1}{2 \sin x \cos x}$$

Again you have the form 0/0:

$$\lim_{x\to 0} \frac{e^x + xe^x - 1}{2 \sin x \cos x} = \lim_{x\to 0} \frac{\dfrac{d}{dx}(e^x + xe^x - 1)}{\dfrac{d}{dx}(2 \sin x \cos x)}$$

$$= \lim_{x\to 0} \frac{2e^x + xe^x}{2 \cos^2 x - 2 \sin^2 x} = \frac{2 + 0}{2 - 0} = 1 \qquad \text{[See Section 21-1.]}$$

PROBLEM 21-7 Find $\lim_{x\to \infty} \ln x/(x^2)$.

Solution: This is the indeterminate form ∞/∞, so apply l'Hôpital's rule:

$$\lim_{x\to \infty} \frac{\ln x}{x^2} = \lim_{x\to \infty} \frac{\dfrac{d}{dx}\ln x}{\dfrac{d}{dx}x^2} = \lim_{x\to \infty} \frac{\dfrac{1}{x}}{2x} = \lim_{x\to \infty} \frac{1}{2x^2} = 0 \qquad \text{[See Section 21-2.]}$$

PROBLEM 21-8 Find $\lim_{x\to 0^+} e^{1/x}/\ln x$.

Solution: As x approaches zero from the right, $1/x$ approaches infinity. Thus $e^{1/x}$ approaches infinity. Because this is the indeterminate form ∞/∞, you can use l'Hôpital's rule:

$$\lim_{x\to 0^+} \frac{e^{1/x}}{\ln x} = \lim_{x\to 0^+} \frac{\dfrac{d}{dx}e^{1/x}}{\dfrac{d}{dx}\ln x}$$

$$= \lim_{x\to 0^+} \frac{\left(\dfrac{-1}{x^2}\right)e^{1/x}}{\dfrac{1}{x}} = \lim_{x\to 0^+} \frac{-e^{1/x}}{x} = -\infty \qquad \text{[See Section 21-2.]}$$

PROBLEM 21-9 Find $\lim_{x\to\infty} 3^x/(x2^x)$.

Solution: To find this limit, rewrite it as $\lim_{x\to\infty} (3/2)^x/x$. This is in the indeterminate form ∞/∞, so

$$\lim_{x\to\infty} \frac{\left(\frac{3}{2}\right)^x}{x} = \lim_{x\to\infty} \frac{\frac{d}{dx}\left(\frac{3}{2}\right)^x}{\frac{d}{dx}x} = \lim_{x\to\infty} \frac{\left(\frac{3}{2}\right)^x \ln\left(\frac{3}{2}\right)}{1} = \infty$$

[See Section 21-2.]

PROBLEM 21-10 Find $\lim_{x\to 0} (\sec x - x)/(\csc x + x)$.

Solution: This isn't an indeterminate form. It is the form $1/\infty$, so

$$\lim_{x\to 0} \frac{\sec x - x}{\csc x + x} = 0$$

Remember not to use l'Hôpital's rule until you've checked that it is the indeterminate form 0/0 or ∞/∞.

[See Section 21-2.]

PROBLEM 21-11 Find $\lim_{x\to 0^+} (\csc x)/\ln x$.

Solution: This is the indeterminate form ∞/∞, so use l'Hôpital's rule:

$$\lim_{x\to 0^+} \frac{\csc x}{\ln x} = \lim_{x\to 0^+} \frac{-\csc x \cot x}{1/x} = \lim_{x\to 0^+} \frac{-x\cos x}{\sin^2 x}$$

This is the indeterminate form 0/0, so apply l'Hôpital's rule again:

$$\lim_{x\to 0^+} \frac{-x\cos x}{\sin^2 x} = \lim_{x\to 0^+} \frac{x\sin x - \cos x}{2\sin x\cos x}$$

The numerator approaches -1; the denominator approaches 0 from above, so

$$\lim_{x\to 0^+} \frac{\csc x}{\ln x} = -\infty$$

[See Section 21-2.]

PROBLEM 21-12 Find $\displaystyle\lim_{x\to 0^+} \left(\frac{1}{x} - \frac{1}{1-\cos x}\right)$.

Solution: This is the indeterminate form $\infty - \infty$, so rewrite it:

$$\lim_{x\to 0^+} \left(\frac{1}{x} - \frac{1}{1-\cos x}\right) = \lim_{x\to 0^+} \frac{1-\cos x - x}{x(1-\cos x)}$$

This is now in the indeterminate form 0/0:

$$\lim_{x\to 0^+} \frac{1-\cos x - x}{x(1-\cos x)} = \lim_{x\to 0^+} \frac{\frac{d}{dx}(1-\cos x - x)}{\frac{d}{dx}[x(1-\cos x)]}$$

$$= \lim_{x\to 0^+} \frac{\sin x - 1}{1 - \cos x + x\sin x} = -\infty$$

[See Section 21-3.]

PROBLEM 21-13 Find $\lim_{x\to\pi/2} (\tan x - \sec x)$.

Solution: Rewrite to get

$$\lim_{x\to\pi/2} (\tan x - \sec x) = \lim_{x\to\pi/2} \left(\frac{\sin x}{\cos x} - \frac{1}{\cos x}\right) = \lim_{x\to\pi/2} \left(\frac{\sin x - 1}{\cos x}\right)$$

This is the indeterminate form 0/0, so apply l'Hôpital's rule:

$$\lim_{x\to\pi/2} \left(\frac{\sin x - 1}{\cos x}\right) = \lim_{x\to\pi/2} \frac{\cos x}{-\sin x} = \frac{0}{-1} = 0$$

[See Section 21-3.]

PROBLEM 21-14 Find $\lim_{x\to 1^+}\left(\frac{1}{\ln x}-\frac{1}{x^2-1}\right)$.

Solution: Rewrite to find

$$\lim_{x\to 1^+}\left(\frac{1}{\ln x}-\frac{1}{x^2-1}\right)=\lim_{x\to 1^+}\frac{x^2-1-\ln x}{(x^2-1)\ln x}$$

This is the indeterminate form 0/0, so

$$\lim_{x\to 1^+}\frac{x^2-1-\ln x}{(x^2-1)\ln x}=\lim_{x\to 1^+}\frac{2x-\frac{1}{x}}{2x\ln x+(x^2-1)\frac{1}{x}}=\infty$$

[See Section 21-3.]

PROBLEM 21-15 Find $\lim_{x\to 0^+} xe^{1/x}$.

Solution: This is the indeterminate form $0\cdot\infty$. Rewrite this so that you can use l'Hôpital's rule:

$$\lim_{x\to 0^+} xe^{1/x}=\lim_{x\to 0^+}\frac{e^{1/x}}{\frac{1}{x}}=\lim_{x\to 0^+}\frac{\left(\frac{-1}{x^2}\right)e^{1/x}}{\frac{-1}{x^2}}=\lim_{x\to 0^+} e^{1/x}=\infty$$

[See Section 21-3.]

PROBLEM 21-16 Find $\lim_{x\to\infty}[1+(1/x)]^{3x}$.

Solution: This is the indeterminate form 1^∞, so let $y=[1+(1/x)]^{3x}$ and find $\lim_{x\to\infty}\ln y$:

$$\lim_{x\to\infty}\ln y=\lim_{x\to\infty}\ln\left(1+\frac{1}{x}\right)^{3x}=\lim_{x\to\infty}3x\ln\left(1+\frac{1}{x}\right)$$

$$=\lim_{x\to\infty}\frac{3\ln\left(1+\frac{1}{x}\right)}{\frac{1}{x}}=\lim_{x\to\infty}\frac{\frac{d}{dx}3\ln\left(1+\frac{1}{x}\right)}{\frac{d}{dx}\frac{1}{x}}$$

$$=\lim_{x\to\infty}\frac{\left(\frac{-1}{x^2}\right)\frac{3}{\left(1+\frac{1}{x}\right)}}{\frac{-1}{x^2}}=\lim_{x\to\infty}\frac{3}{1+\frac{1}{x}}=3$$

Because $\ln y$ approaches 3, y must approach e^3. [See Section 21-4.]

PROBLEM 21-17 Find $\lim_{x\to\infty}(\ln x)^{1/x}$.

Solution: This is the indeterminate form ∞^0, so let $y=(\ln x)^{1/x}$ and find $\lim_{x\to\infty}\ln y$.

$$\lim_{x\to\infty}\ln y=\lim_{x\to\infty}\ln(\ln x)^{1/x}=\lim_{x\to\infty}\frac{\ln(\ln x)}{x}$$

$$=\lim_{x\to\infty}\frac{\frac{d}{dx}[\ln(\ln x)]}{\frac{d}{dx}x}=\lim_{x\to\infty}\frac{\frac{1}{\ln x}\left(\frac{1}{x}\right)}{1}=0$$

Because $\ln y$ approaches zero, y must approach $e^0=1$. [See Section 21-4.]

PROBLEM 21-18 Find $\lim_{x\to\infty}(1/x)^{e^{-x}}$.

Solution: This is the indeterminate form 0^0, so let $y = (1/x)^{e^{-x}}$ and find $\lim_{x\to\infty} \ln y$:

$$\lim_{x\to\infty} \ln y = \lim_{x\to\infty} \ln\left(\frac{1}{x}\right)^{e^{-x}} = \lim_{x\to\infty} e^{-x} \ln\left(\frac{1}{x}\right) = \lim_{x\to\infty} \frac{\ln\left(\frac{1}{x}\right)}{e^x}$$

$$= \lim_{x\to\infty} \frac{\frac{d}{dx}(-\ln x)}{\frac{d}{dx}e^x} = \lim_{x\to\infty} \frac{\frac{-1}{x}}{e^x} = 0$$

Because $\ln y$ approaches zero, y approaches $e^0 = 1$. [See Section 21-4.]

PROBLEM 21-19 Find $\lim_{x\to 0^+} (x + 1)^{\csc x}$.

Solution: This is the indeterminate form 1^∞, so let $y = (x + 1)^{\csc x}$ and find $\lim_{x\to 0^+} \ln y$:

$$\lim_{x\to 0^+} \ln y = \lim_{x\to 0^+} \ln(x + 1)^{\csc x} = \lim_{x\to 0^+} \csc x \ln(x + 1)$$

$$= \lim_{x\to 0^+} \frac{\ln(x + 1)}{\sin x} = \lim_{x\to 0^+} \frac{\frac{1}{x + 1}}{\cos x} = 1$$

Because $\ln y$ approaches 1, y approaches $e^1 = e$. [See Section 21-4.]

PROBLEM 21-20 Find $\lim_{x\to\infty} (e^x + e^{2x})^{1/x}$.

Solution: Let $y = (e^x + e^{2x})^{1/x}$:

$$\lim_{x\to\infty} \ln y = \lim_{x\to\infty} \ln(e^x + e^{2x})^{1/x} = \lim_{x\to\infty} \frac{\ln(e^x + e^{2x})}{x}$$

$$= \lim_{x\to\infty} \frac{\frac{e^x + 2e^{2x}}{e^x + e^{2x}}}{1} = \lim_{x\to\infty} \frac{e^{-x} + 2}{e^{-x} + 1} = 2$$

Because $\ln y$ approaches 2, y approaches e^2. [See Section 21-4.]

PROBLEM 21-21 Find $\lim_{x\to 0^+} x^{\sin x}$.

Solution: This is the indeterminate form 0^0. Let $y = x^{\sin x}$:

$$\lim_{x\to 0^+} \ln y = \lim_{x\to 0^+} \ln x^{\sin x} = \lim_{x\to 0^+} \sin x \ln x = \lim_{x\to 0^+} \frac{\ln x}{\csc x}$$

$$= \lim_{x\to 0^+} \frac{1/x}{-\csc x \cot x} = \lim_{x\to 0^+} \frac{-\sin^2 x}{x \cos x}$$

$$= \lim_{x\to 0^+} \frac{-2 \sin x \cos x}{\cos x - x \sin x} = 0$$

Because $\ln y$ approaches zero, y approaches $e^0 = 1$. [See Section 21-4.]

Supplementary Exercises

In Problems 21-22 through 21-76 find the limit:

21-22 $\lim_{x\to 0} (\sin x)/\sin 2x$

21-23 $\lim_{x\to 1} [(x - 1)^2]/\ln x$

21-24 $\lim_{x\to 0} x/\arcsin x$

21-25 $\lim_{x\to 0} (\cos x - e^{-2x})/(3x^2 + 5x)$

21-26 $\lim_{x\to 0^+} \sin x/(1 - \cos x)$

21-27 $\lim_{x\to 0} (\cos x - 1)/x^2$

21-28 $\lim_{x \to \infty} (\ln x)/\sqrt{x}$

21-29 $\lim_{x \to 1} \ln x/(\sqrt{x} - 1)$

21-30 $\lim_{x \to \infty} (2x + \sqrt{x})/\sqrt{3x^2 - 1}$

21-31 $\lim_{x \to 0} (\sin x - x)/x^3$

21-32 $\lim_{x \to \infty} x^3/e^{2x}$

21-33 $\lim_{x \to 0^+} \sqrt{x}/(\sqrt{x + 1} - 1)$

21-34 $\lim_{x \to \pi/2} \cos^2 x/(\sin x - 1)$

21-35 $\lim_{x \to \infty} 2^x/(2^x - x^2)$

21-36 $\lim_{x \to 0} (2^x - e^x)/(3^x - 1)$

21-37 $\lim_{x \to \pi/2^-} (\sec x)/\ln[(\pi/2) - x]$

21-38 $\lim_{x \to \infty} (\ln^2 x)/x$

21-39 $\lim_{x \to 0^+} (1 - \cos\sqrt{x})/\ln(1 + x)$

21-40 $\lim_{x \to 0} (1 - \cos 3x)/(1 - \cos 2x)$

21-41 $\lim_{x \to \infty} (e^x - 5x^2)/(x^3 + 2e^x)$

21-42 $\lim_{x \to 0} \ln^2(1 - x)/(\cos x - 1)$

21-43 $\lim_{x \to \infty} 2^x/(2^x - x^2)$

21-44 $\lim_{x \to 0} (x \sin x)/x^2$

21-45 $\lim_{x \to 0} x \sin x/(\cos x - 1)$

21-46 $\lim_{x \to 0} [\ln(1 + x) - x]/x^2$

21-47 $\lim_{x \to 0} (e^x - x - 1)/(\cos x - 1)$

21-48 $\lim_{x \to 0} (\ln \cos x)/\sin^2 x$

21-49 $\lim_{x \to 0^+} e^{1/x}/\csc x$

21-50 $\displaystyle \lim_{x \to 0} \left(\frac{\sin x}{x^3} - \frac{1}{x^2}\right)$

21-51 $\lim_{x \to 0} (\cot x - \csc x)$

21-52 $\displaystyle \lim_{x \to \infty} \left(\frac{\ln x}{\sqrt{x}} - \frac{1}{x}\right)$

21-53 $\lim_{x \to 0^+} (\csc x - \csc 2x)$

21-54 $\displaystyle \lim_{x \to 0^+} \left(\csc x - \frac{1}{\ln(1 + x)}\right)$

21-55 $\displaystyle \lim_{x \to 0^+} \left(\frac{1}{x} - \frac{1}{\ln(1 + x)}\right)$

21-56 $\lim_{x \to 0} x \csc x$

21-57 $\lim_{x \to \infty} \sin(1/x)\ln x$

21-58 $\displaystyle \lim_{x \to \infty} e^x\left(\text{arc tan } x - \frac{\pi}{2}\right)$

21-59 $\lim_{x \to 0^+} \csc x \ln(1 - \sin x)$

21-60 $\lim_{x \to 0^+} x \ln(\csc x)$

21-61 $\lim_{x \to \infty} e^{2x} \ln(1 + e^{-x})$

21-62 $\lim_{x \to \infty} [1 + (1/x)]^x$

21-63 $\lim_{x \to 0^+} (\csc x)^x$

21-64 $\lim_{x \to \infty} (x + e^x)^{1/x}$

21-65 $\lim_{x \to 0^+} x^{e^{-1/x}}$

21-66 $\lim_{x \to \infty} (1 + e^{-x})^{e^{2x}}$

21-67 $\lim_{x \to \pi/2^-} (\tan x)^{\cos x}$

21-68 $\lim_{x \to 0^+} (1 - \sin x)^{\csc x}$

21-69 $\lim_{x \to \pi/2^+} [x - (\pi/2)]^{\cos x}$

21-70 $\lim_{x \to 1^+} (\ln x)^{\ln x}$

21-71 $\lim_{x \to 0^+} (e^{-x} + 5x)^{1/x}$

21-72 $\lim_{x \to \infty} [1 - (1/x^2)]^{x^2 + 1}$

21-73 $\lim_{x \to 0^+} (\cos x + \sin x)^{\cot x}$

21-74 $\lim_{x \to \infty} x^{\sin(1/x)}$

21-75 $\lim_{x \to \infty} [(\ln x)/x]^{1/x}$

21-76 $\lim_{x \to 0^+} (1 + x)^{-\ln x}$

Solutions to Supplementary Exercises

(21-22) $\frac{1}{2}$

(21-23) 0

(21-24) 1

(21-25) 2/5

(21-26) ∞

(21-27) $-\frac{1}{2}$

(21-28) 0

(21-29) 2

(21-30) $2/\sqrt{3}$

(21-31) $-1/6$

(21-32) 0

(21-33) ∞

(21-34) -2

(21-35) 1

(21-36) $(\ln 2 - 1)/\ln 3$

(21-37) $-\infty$

(21-38) 0

(21-39) $\frac{1}{2}$

(21-40) $9/4$

(21-41) $\frac{1}{2}$

(21-42) -2

(21-43) 1

(21-44) 1

(21-45) -2

(21-46) $-\frac{1}{2}$

(21-47) -1

(21-48) $-\frac{1}{2}$

(21-49) ∞

(21-50) $-1/6$

(21-51) 0

(21-52) 0

(21-53) $\frac{1}{2}$

(21-54) $-\frac{1}{2}$

(21-55) $-\frac{1}{2}$

(21-56) 1

(21-57) 0

(21-58) $-\infty$

(21-59) -1

(21-60) 0

(21-61) ∞

(21-62) e

(21-63) 1

(21-64) e

(21-65) 1

(21-66) ∞

(21-67) 1

(21-68) e^{-1}

(21-69) 1

(21-70) 1

(21-71) e^4

(21-72) e^{-1}

(21-73) e

(21-74) 1

(21-75) 1

(21-76) 1

22 SEQUENCES AND SERIES

THIS CHAPTER IS ABOUT

- ☑ **Sequences**
- ☑ **Series**
- ☑ **Geometric Series**
- ☑ **Series with Nonnegative Terms**
- ☑ **Alternating Series**

22-1. Sequences

A **sequence** of real numbers is an infinite list of numbers. The numbers on the list are called the *terms* of the sequence. For example, $\{\frac{1}{2}, \frac{1}{4}, \frac{1}{8}, \ldots, \frac{1}{2}^n, \ldots\}$ is an infinite sequence whose nth term is $\frac{1}{2}^n$. If the nth term of a sequence is a_n, then the sequence is often denoted $\{a_n\}$.

EXAMPLE 22-1: Write the first few terms of the sequences $\{1/n\}$, $\{n^2/(n^2 + 1)\}$, and $\{\cos \pi n\}$.

Solution:

$$\left\{\frac{1}{n}\right\} = \left\{\frac{1}{1}, \frac{1}{2}, \frac{1}{3}, \frac{1}{4}, \ldots\right\}$$

$$\left\{\frac{n^2}{n^2 + 1}\right\} = \left\{\frac{1^2}{1^2 + 1}, \frac{2^2}{2^2 + 1}, \frac{3^2}{3^2 + 1}, \frac{4^2}{4^2 + 1}, \ldots\right\}$$

$$= \left\{\frac{1}{2}, \frac{4}{5}, \frac{9}{10}, \frac{16}{17}, \ldots\right\}$$

$$\{\cos \pi n\} = \{\cos \pi, \cos 2\pi, \cos 3\pi, \cos 4\pi, \ldots\}$$

$$= \{-1, 1, -1, 1, \ldots\}$$

You say that the sequence $\{a_n\}$ **converges to L** (a finite number) if the terms of the sequence get (and stay) arbitrarily close to L as n gets large. We write $\lim_{n\to\infty} a_n = L$.

The formal definition is as follows: The sequence $\{a_n\}$ converges to L if, for every $\varepsilon > 0$, there exists an integer N such that whenever $n > N$, you have $|a_n - L| < \varepsilon$.

A. Rules for convergence

Suppose that $\{a_n\}$ and $\{b_n\}$ are convergent sequences. Then

1. $\{ca_n\}$ converges and $\lim_{n\to\infty} ca_n = c \lim_{n\to\infty} a_n$ for any constant c
2. $\{a_n + b_n\}$ converges and $\lim_{n\to\infty} (a_n + b_n\} = \lim_{n\to\infty} a_n + \lim_{n\to\infty} b_n$
3. $\{a_nb_n\}$ converges and $\lim_{n\to\infty} a_nb_n = (\lim_{n\to\infty} a_n)(\lim_{n\to\infty} b_n)$
4. $\left\{\frac{a_n}{b_n}\right\}$ converges and $\lim\limits_{n\to\infty} \frac{a_n}{b_n} = \frac{\lim_{n\to\infty} a_n}{\lim_{n\to\infty} b_n}$, provided $\lim_{n\to\infty} b_n = 0$

B. Limit of a sequence

If the nth term in the sequence $\{a_n\}$ is $f(n)$ for a function f, which is defined not just on the positive integers but on the positive real numbers, and if $\lim_{x\to\infty} f(x) = L$ (L a finite number), then

$$\lim_{n\to\infty} a_n = \lim_{x\to\infty} f(x) = L \qquad \textbf{(22-1)}$$

This often enables you to find the limit of a sequence using the techniques from Chapters 3 and 21.

EXAMPLE 22-2: Find the limit of the sequence $\{3/n\}$.

Solution: You can find this limit by finding the limit of the function $f(x) = 3/x$:

$$\lim_{n\to\infty} \frac{3}{n} = \lim_{x\to\infty} \frac{3}{x} = 0$$

EXAMPLE 22-3: Find the limit of the sequence $\{1/2, 4/5, 9/10, 16/17, 25/26, \ldots\}$.

Solution: This is the sequence $\{n^2/(n^2 + 1)\}$, so evaluate the limit

$$\lim_{x\to\infty} \frac{x^2}{x^2 + 1}$$

You can find this limit using l'Hôpital's rule. You should also recognize the limit of a rational function as the quotient of the leading coefficients when the numerator and denominator have the same degree. Either way, you have

$$\lim_{n\to\infty} \frac{n^2}{n^2 + 1} = \lim_{x\to\infty} \frac{x^2}{x^2 + 1} = \frac{1}{1} = 1$$

EXAMPLE 22-4: Find the limit of the sequence $\{(\ln n)/n\}$.

Solution: Use l'Hôpital's rule because this is the indeterminate form ∞/∞:

$$\lim_{n\to\infty} \frac{\ln n}{n} = \lim_{x\to\infty} \frac{\ln x}{x} = \lim_{x\to\infty} \frac{\frac{d}{dx}\ln x}{\frac{d}{dx}x} = \lim_{x\to\infty} \frac{\frac{1}{x}}{1} = 0$$

EXAMPLE 22-5: Find the limit of the sequence $\{e^n/n\}$.

Solution: You can easily find the limit:

$$\lim_{x\to\infty} \frac{e^x}{x} = \lim_{x\to\infty} \frac{\frac{d}{dx}e^x}{\frac{d}{dx}x} = \lim_{x\to\infty} \frac{e^x}{1} = \infty$$

Because the function $f(x) = e^x/x$ approaches infinity, the sequence $\{e^n/n\}$ grows without bound, that is, the sequence $\{e^n/n\}$ **diverges**.

EXAMPLE 22-6: Find the limit of the sequence $\{\sin \pi n\}$.

Solution: If $f(x) = \sin \pi x$, then

$$\lim_{x\to\infty} f(x) = \lim_{x\to\infty} \sin \pi x$$

which doesn't exist. Indeed, as x gets larger and larger through the real numbers, $f(x) = \sin \pi x$ cycles through all numbers between -1 and 1—it never settles down. However, this doesn't mean that the sequence $\{\sin \pi n\}$ diverges. You are only interested in what happens to $\sin \pi n$ as n gets larger through integer values. Just write a few terms in the sequence:

$$\{\sin \pi n\} = \{0, 0, 0, \ldots\}$$

and you'll see that the sequence converges to zero.

EXAMPLE 22-7: Find the limit of the sequence $\{(-1)^n/n\}$.

Solution: It may be difficult to find a function $f(x)$ such that $f(n) = (-1)^n/n$ at the integers. (What would the value of such a function be at $\frac{1}{2}$?) Just consider the terms $(-1)^n/n$. The denominator gets large as n gets large, and regardless of whether the numerator is 1 or -1, these terms get closer and closer to zero. The sequence $\{(-1)^n/n\}$ approaches zero.

22-2. Series

An infinite **series** is an infinite *sum* of numbers, such as

$$a_1 + a_2 + a_3 + \cdots$$

or, in summation notation,

$$\sum_{k=1}^{\infty} a_k$$

To make sense of an infinite sum, consider a finite sum:

$$S_n = \sum_{k=1}^{n} a_k = a_1 + a_2 + a_3 + \cdots + a_n$$

So S_n is the sum of the first n terms—the ***n*th partial sum** of the series. The infinite series $\sum_{k=1}^{\infty} a_k$ is said to **converge** to L (written $\sum_{k=1}^{\infty} a_k = L$) if the sequence of partial sums $\{S_n\}$ converges to L; if the sequence $\{S_n\}$ diverges, then the series $\sum_{k=1}^{\infty} a_k$ is **divergent**.

To see why this makes sense, imagine yourself adding an infinite list of numbers. You find the sum of the first two terms, then the sum of the first three terms, then the sum of the first four terms, and so on. That is, you find S_2, S_3, S_4, etc. If this "running total" approaches L, then the infinite sum approaches L.

It may seem to you that an infinite sum of numbers must be infinite, but you know of examples to the contrary. Imagine cutting a pie in half, and then cutting one of the halves in half, and then cutting one of the quarters in half, and so on, ad infinitum. What size pieces of pie do you have? You have one half of a pie, one quarter of a pie, one eighth of a pie, and so on. That is, $1 = \frac{1}{2} + \frac{1}{4} + \frac{1}{8} + \cdots +$.

Another easy example of a convergent infinite series is referred to as a *telescoping series*:

EXAMPLE 22-8: Does the series $\displaystyle\sum_{k=1}^{\infty}\left(\frac{1}{k} - \frac{1}{k+1}\right)$ converge?

Solution: Find the sequence of partial sums, S_n:

$$S_1 = \sum_{k=1}^{1}\left(\frac{1}{k} - \frac{1}{k+1}\right) = \left(\frac{1}{1} - \frac{1}{2}\right) = \frac{1}{2}$$

$$S_2 = \sum_{k=1}^{2}\left(\frac{1}{k} - \frac{1}{k+1}\right) = \left(\frac{1}{1} - \frac{1}{2}\right) + \left(\frac{1}{2} - \frac{1}{3}\right) = 1 - \frac{1}{3} = \frac{2}{3}$$

$$\begin{aligned} S_n &= \sum_{k=1}^{n}\left(\frac{1}{k} - \frac{1}{k+1}\right) \\ &= \left(1 - \frac{1}{2}\right) + \left(\frac{1}{2} - \frac{1}{3}\right) + \left(\frac{1}{3} - \frac{1}{4}\right) + \cdots + \left(\frac{1}{n} - \frac{1}{n+1}\right) \\ &= 1 + \left(\frac{-1}{2} + \frac{1}{2}\right) + \left(\frac{-1}{3} + \frac{1}{3}\right) + \cdots + \left(\frac{-1}{n} + \frac{1}{n}\right) - \frac{1}{n+1} \\ &= 1 - \frac{1}{n+1} \end{aligned}$$

To find the number to which the series converges, find the limit of the sequence $\{S_n\} = \{1 - [1/(n + 1)]\}$:

$$\lim_{n \to \infty} \left(1 - \frac{1}{n + 1}\right) = 1$$

So the series $\sum_{k=1}^{\infty} \left(\frac{1}{k} - \frac{1}{k + 1}\right)$ converges to one.

Remember the difference between a sequence and a series. A sequence is an infinite list of numbers a_n that you examine to find what happens when n gets large. A series is an infinite *sum* of numbers.

22-3. Geometric Series

A. Finite geometric series

A **finite geometric series** is a finite sum of the form $\sum_{k=0}^{n} r^k$ for some number r. A formula that will save you labor in summing such a series is

$$\sum_{k=0}^{n} r^k = \frac{1 - r^{n+1}}{1 - r} \quad \text{for } r \neq 1 \tag{22-2}$$

EXAMPLE 22-9: Sum the finite series $1 + (1/2) + (1/4) + (1/8) + \cdots + (1/512)$.

Solution: Write this in summation notation and use the formula:

$$1 + \frac{1}{2} + \frac{1}{4} + \cdots + \frac{1}{512} = \sum_{k=0}^{9} \left(\frac{1}{2}\right)^k = \frac{1 - (\frac{1}{2})^{10}}{1 - \frac{1}{2}}$$

$$= \frac{1 - 1/1024}{1/2} = \frac{1023}{512}$$

EXAMPLE 22-10: Sum the finite series $1 + e^2 + e^4 + \cdots + e^{20}$.

Solution: To recognize that this is geometric, notice that the ratio between successive terms is e^2. Now write this in summation notation as a geometric series with $r = e^2$:

$$1 + e^2 + e^4 + \cdots + e^{20} = \sum_{k=0}^{10} (e^2)^k$$

$$= \frac{1 - (e^2)^{11}}{1 - e^2} = \frac{1 - e^{22}}{1 - e^2}$$

EXAMPLE 22-11: Sum the finite series

$$(4/3) - (8/9) + (16/27) - (32/81) + (64/243).$$

Solution: Notice that the ratio between successive terms is $-2/3$. This will be a geometric series with ratio $-2/3$. Factor out the first term and write in summation notation:

$$\frac{4}{3} - \frac{8}{9} + \frac{16}{27} - \frac{32}{81} + \frac{64}{243} = \frac{4}{3}\left(1 - \frac{2}{3} + \frac{4}{9} - \frac{8}{27} + \frac{16}{81}\right)$$

$$= \frac{4}{3} \sum_{k=0}^{4} \left(\frac{-2}{3}\right)^k = \left(\frac{4}{3}\right) \frac{1 - (-2/3)^5}{1 - (-2/3)}$$

$$= \left(\frac{4}{3}\right) \frac{1 + (32/243)}{5/3} = \frac{4[1 + (32/243)]}{5}$$

$$= \frac{220}{243}$$

B. Infinite geometric series

An **infinite geometric series** is a series of the form $\sum_{k=0}^{\infty} r^k$ for some number r. To determine whether or not a geometric series converges, consider the sequence of partial sums, $\{S_n\}$:

$$S_n = \sum_{k=0}^{n} r^k = \frac{1 - r^{n+1}}{1 - r}$$

The series $\sum_{k=0}^{\infty} r^k$ will converge whenever the sequence $\{S_n\}$ converges. If $|r| < 1$, then

$$\lim_{n \to \infty} S_n = \lim_{n \to \infty} \frac{1 - r^{n+1}}{1 - r} = \frac{1 - 0}{1 - r} = \frac{1}{1 - r}$$

If $|r| \geqslant 1$, then $\lim_{n \to \infty} (1 - r^{n+1})/(1 - r)$ diverges because $\{r^{n+1}\}$ doesn't converge. Putting this together you have

$$\sum_{k=0}^{\infty} r^k = \begin{cases} 1/(1 - r) & \text{if } |r| < 1 \\ \text{diverges} & \text{if } |r| \geqslant 1 \end{cases} \tag{22-3}$$

EXAMPLE 22-12: Does $\sum_{k=0}^{\infty} \left(\frac{-1}{3}\right)^k$ converge?

Solution: From Equation 22-3,

$$\sum_{k=0}^{\infty} \left(\frac{-1}{3}\right)^k = \frac{1}{1 - \left(-\frac{1}{3}\right)} = \frac{3}{4}$$

so the series converges to 3/4.

EXAMPLE 22-13: Does (5/2) − (15/8) + (45/32) − (135/128) + ⋯ converge?

Solution: Recognize this as a constant times a geometric series:

$$\frac{5}{2} - \frac{15}{8} + \frac{45}{32} - \cdots = \frac{5}{2}\left(1 - \frac{3}{4} + \frac{9}{16} - \frac{27}{64} + \cdots\right) = \frac{5}{2} \sum_{k=0}^{\infty} \left(-\frac{3}{4}\right)^k$$

$$= \left(\frac{5}{2}\right) \frac{1}{1 - \left(-\frac{3}{4}\right)} = \left(\frac{5}{2}\right) \frac{1}{\frac{7}{4}} = \frac{10}{7}$$

This series converges to 10/7.

EXAMPLE 22-14: Express the repeating decimal $0.162\,162\,\overline{162}$ as a quotient of integers.

Solution: Recognize the repeating decimal as a series:

$$0.162\,162\,\overline{162} = 0.162 + 0.000\,162 + 0.000\,000\,162 + \cdots$$

$$= \frac{162}{10^3} + \frac{162}{10^6} + \frac{162}{10^9} + \cdots$$

$$= \frac{162}{10^3}\left(1 + \frac{1}{10^3} + \frac{1}{10^6} + \cdots\right) = \frac{162}{10^3} \sum_{k=0}^{\infty} \left(\frac{1}{10^3}\right)^k$$

$$= \frac{162}{10^3}\left(\frac{1}{1 - \frac{1}{10^3}}\right) = \frac{162}{1000}\left(\frac{1}{\frac{999}{1000}}\right) = \frac{162}{999} = \frac{6}{37}$$

22-4. Series with Nonnegative Terms

You'll often be asked to determine whether a given series converges or diverges. (It is a much more difficult problem—not addressed in this book—to determine the number to which it converges.) This section covers several tests for determining whether or not a series with nonnegative terms converges.

A. Integral test

If f is a continuous, decreasing, and positive function for $x > a$, then

1. $\sum_{k=a}^{\infty} f(k)$ converges if $\int_a^{\infty} f(x)\,dx$ converges.

2. $\sum_{k=a}^{\infty} f(k)$ diverges if $\int_a^{\infty} f(x)\,dx$ diverges.

EXAMPLE 22-15: Does $\sum_{k=1}^{\infty} 1/k^2$ converge?

Solution: Because the function $f(x) = 1/x^2$ is continuous, decreasing, and positive for $x > 1$, you can try the integral test:

$$\int_1^{\infty} \frac{1}{x^2}\,dx = \lim_{N\to\infty} \int_1^N \frac{1}{x^2}\,dx = \lim_{N\to\infty} \left(\frac{-1}{x}\right)\bigg|_1^N = \lim_{N\to\infty} \left(\frac{-1}{N} + 1\right) = 1$$

Because the integral converges, the corresponding series converges.

> ***Caution:*** This doesn't imply that $\sum_{k=1}^{\infty} 1/k^2 = 1$; it doesn't! This only implies that $\sum_{k=1}^{\infty} 1/k^2$ converges.

EXAMPLE 22-16: Does $\sum_{k=1}^{\infty} 1/k$ converge?

Solution: You can employ the integral test:

$$\int_1^{\infty} \frac{1}{x}\,dx = \lim_{N\to\infty} \int_1^N \frac{1}{x}\,dx = \lim_{N\to\infty} (\ln|x|)\bigg|_1^N = \lim_{N\to\infty} (\ln|N| - \ln|1|)$$
$$= \lim_{N\to\infty} \ln N = \infty$$

Because the integral diverges, the series $\sum_{k=1}^{\infty} 1/k$ diverges.

This is an important example, called the *harmonic series*. Notice that although the terms of the series approach zero ($1/k$ approaches zero), the series diverges. Apparently it isn't enough that the terms approach zero; they must approach zero sufficiently quickly if the series is to converge. Of course if the terms don't approach zero, then certainly the sum can't be finite, that is, the series diverges.

EXAMPLE 22-17: Does $\sum_{k=3}^{\infty} 1/(\ln k)^k$ converge?

Solution: Although $f(x) = 1/(\ln x)^x$ is a continuous, decreasing, positive function for $x > 3$, the integral test will not help you unless you can antidifferentiate $f(x)$, which seems unlikely in this case. Some of the later tests will be useful in determining whether or not this series converges (see Example 22-26).

EXAMPLE 22-18: Does $\sum_{k=3}^{\infty} 1/(k\sqrt{\ln k})$ converge?

Solution: Apply the integral test:

$$\int_3^{\infty} \frac{dx}{x\sqrt{\ln x}} = \lim_{N\to\infty} \int_3^N \frac{dx}{x\sqrt{\ln x}}$$

Make the substitution $u = \ln x$, with $du = 1/x\,dx$:

$$\lim_{N\to\infty} \int_3^N \frac{dx}{x\sqrt{\ln x}} = \lim_{N\to\infty} \int_{\ln 3}^{\ln N} \frac{du}{\sqrt{u}}$$
$$= \lim_{N\to\infty} 2u^{1/2}\bigg|_{\ln 3}^{\ln N}$$
$$= \lim_{N\to\infty} (2\sqrt{\ln N} - 2\sqrt{\ln 3}) = \infty$$

Because the improper integral diverges, you see that the series $\sum_{k=3}^{\infty} 1/(k\sqrt{\ln k})$ diverges.

B. Comparison test

If $\sum_{k=1}^{\infty} b_k$ is a convergent series and if $b_k \geqslant a_k \geqslant 0$ for all k (or at least for all sufficiently large k), then $\sum_{k=1}^{\infty} a_k$ converges. If $\sum_{k=1}^{\infty} b_k$ is a divergent series and if $0 \leqslant b_k \leqslant a_k$ for all k (or at least for all sufficiently large k), then $\sum_{k=1}^{\infty} a_k$ diverges.

In other words, a nonnegative series that is term by term smaller than a known convergent series is convergent. A series that is term by term larger than a known divergent series is divergent.

EXAMPLE 22-19: Does $\sum_{k=3}^{\infty} 1/\ln k$ converge?

Solution: You know from Example 22-16 that $\sum_{k=1}^{\infty} 1/k$ diverges. But $1/k < 1/\ln k$ because $\ln k < k$. By comparison test, $\sum_{k=3}^{\infty} 1/\ln k$ diverges because it is term by term larger than the divergent series $\sum_{k=1}^{\infty} 1/k$.

EXAMPLE 22-20: Does the series $\sum_{k=1}^{\infty} k/3^k$ converge?

Solution: The terms of the series, $k/3^k$, are smaller than $1/2^k$: $k/3^k < 1/2^k$ because $k2^k < 3^k$ for $k \geqslant 1$. But you know that $\sum_{k=1}^{\infty} (\frac{1}{2})^k$ is a convergent series because it is geometric with $|r| < 1$. Thus $\sum_{k=1}^{\infty} k/3^k$ converges.

> ***Note:*** A common first attempt on this problem is the observation $k/3^k > 1/3^k$ and that $\sum_{k=1}^{\infty} (\frac{1}{3})^k$ converges (geometric with $r = \frac{1}{3}$). However, knowing that a series is term by term larger than a known convergent series is of no help. To show convergence by the comparison test, you must show that it is term by term *less* than a convergent series!

You'll most commonly use geometric series of the form $\sum_{k=0}^{\infty} r^k$, which converge for $|r| < 1$ and diverge for $|r| \geqslant 1$, and series of the form $\sum_{k=1}^{\infty} 1/k^s$, which converge for $s > 1$ and diverge for $s \leqslant 1$ (by the integral test) for comparison.

EXAMPLE 22-21: Does the series $\sum_{k=1}^{\infty} (\sqrt{k} - 1)/(k^2 + 1)$ converge?

Solution: Observe that $(\sqrt{k} - 1)/(k^2 + 1) < 1/k^{3/2}$ and that $\sum_{k=1}^{\infty} 1/k^{3/2}$ converges ($\sum_{k=1}^{\infty} 1/k^s$, with $s = 3/2$), and hence you conclude that $\sum_{k=1}^{\infty} (\sqrt{k} - 1)/(k^2 + 1)$ converges.

C. Limit comparison test

The **limit comparison test** will sometimes help you avoid the messy inequalities of the comparison test: If $\sum_{k=1}^{\infty} a_k$ and $\sum_{k=1}^{\infty} b_k$ are series with $a_k \geqslant 0$ and $b_k \geqslant 0$ and if $\lim_{k\to\infty} (a_k/b_k) = c$, where c is a finite positive number ($c > 0$), then either both series converge or both series diverge.

EXAMPLE 22-22: Does the series $\sum_{k=1}^{\infty} (k + 1)/(k^2 - 2)$ converge?

Solution: The idea is to find a series with known behavior that "looks like" this series. You may find this series by examining the highest-degree terms in the numerator and denominator, as these will dominate the lower degree terms. In this case the highest-degree term in the numerator is k; in the denominator, k^2. So this series "looks like" $\sum_{k=1}^{8} k/k^2 = \sum_{k=1}^{\infty} 1/k$, the harmonic series. Now that you have made your guess, use the limit comparison test, with $a_k = (k + 1)/(k^2 - 2)$ and $b_k = 1/k$:

$$\lim_{k\to\infty} \frac{a_k}{b_k} = \lim_{k\to\infty} \frac{\dfrac{k+1}{k^2-2}}{\dfrac{1}{k}} = \lim_{k\to\infty} \frac{k^2 + k}{k^2 - 2} = 1$$

Because the limit is a finite number that is greater than zero, either both series ($\sum_{k=1}^{\infty} a_k$ and $\sum_{k=1}^{\infty} b_k$) converge or both diverge. But $\sum_{k=1}^{\infty} b_k = \sum_{k=1}^{\infty} 1/k$, which diverges, so $\sum_{k=1}^{\infty} a_k = \sum_{k=1}^{\infty} (k+1)/(k^2-2)$ must diverge.

EXAMPLE 22-23: Does the series $\sum_{k=1}^{\infty} (\sqrt{k} + \ln k)/(k^2+1)$ converge?

Solution: A good guess is that this series "looks like" $\sum_{k=1}^{\infty} \sqrt{k}/k^2 = \sum_{k=1}^{\infty} 1/k^{3/2}$. Try the limit comparison test:

$$\lim_{k\to\infty} \frac{\dfrac{\sqrt{k}+\ln k}{k^2+1}}{\dfrac{1}{k^{3/2}}} = \lim_{k\to\infty} \frac{\sqrt{k}+\ln k}{k^{1/2}+k^{-3/2}} = \lim_{k\to\infty} \frac{(\sqrt{k}+\ln k)\left(\dfrac{1}{\sqrt{k}}\right)}{(k^{1/2}+k^{-3/2})\left(\dfrac{1}{\sqrt{k}}\right)}$$

$$= \lim_{k\to\infty} \frac{1+\dfrac{\ln k}{\sqrt{k}}}{1+\dfrac{1}{k^2}}$$

By l'Hôpital's rule:

$$\lim_{k\to\infty} \frac{\ln k}{\sqrt{k}} = \lim_{x\to\infty} \frac{\ln x}{\sqrt{x}} = \lim_{x\to\infty} \frac{\dfrac{1}{x}}{\dfrac{1}{2\sqrt{x}}} = \lim_{x\to\infty} \frac{2\sqrt{x}}{x} = 0$$

$$\lim_{k\to\infty} \frac{1+\dfrac{\ln k}{\sqrt{k}}}{1+\dfrac{1}{k^2}} = \frac{1+0}{1+0} = 1$$

Because this limit is nonzero and finite and $\sum_{k=1}^{\infty} 1/k^{3/2}$ converges, $\sum_{k=1}^{\infty} (\sqrt{k}+\ln k)/(k^2+1)$ must converge.

D. Ratio test

If $\sum_{k=1}^{\infty} a_k$ is a series with $a_k \geqslant 0$ for all k, then you perform the **ratio test** by finding

$$\lim_{k\to\infty} \frac{a_{k+1}}{a_k} = c$$

If $c > 1$ or if $c = +\infty$, then the series $\sum_{k=1}^{\infty} a_k$ diverges. If $c < 1$, then the series $\sum_{k=1}^{\infty} a_k$ converges. If $c = 1$, then the ratio test is inconclusive—you should try another test.

EXAMPLE 22-24: Does the series $\sum_{k=1}^{\infty} 1/k!$ converge?

Solution: Find the limit of the ratio between the $(k+1)$st term and the kth term:

$$\lim_{k\to\infty} \frac{a_{k+1}}{a_k} = \lim_{k\to\infty} \frac{\dfrac{1}{(k+1)!}}{\dfrac{1}{k!}} = \lim_{k\to\infty} \frac{k!}{(k+1)!} = \lim_{k\to\infty} \frac{k(k-1)\cdots 3(2)}{(k+1)k(k-1)\cdots 3(2)}$$

$$= \lim_{k\to\infty} \frac{1}{k+1} = 0$$

Because the limit is less than one, the series $\sum_{k=1}^{\infty} 1/k!$ converges.

EXAMPLE 22-25: Does the series $\sum_{k=2}^{\infty} k^k/(k-1)!$ converge?

Solution: Try the ratio test:

$$\lim_{k\to\infty}\left[\frac{\dfrac{(k+1)^{k+1}}{(k+1-1)!}}{\dfrac{k^k}{(k-1)!}}\right] = \lim_{k\to\infty}\left[\frac{(k+1)^{k+1}}{k!}\left(\frac{(k-1)!}{k^k}\right)\right]$$

$$= \lim_{k\to\infty}\frac{(k+1)^{k+1}}{k(k)^k} = \lim_{k\to\infty}\left(\frac{k+1}{k}\right)^{k+1}$$

$$= \lim_{k\to\infty}\left(1+\frac{1}{k}\right)^{k+1} = e$$

Because $e > 1$, the series $\sum_{k=2}^{\infty} k^k/(k-1)!$ diverges.

E. Root test

If $\sum_{k=1}^{\infty} a_k$ is a series with $a_k \geqslant 0$ for all k, then you perform the **root test** by finding

$$\lim_{k\to\infty} (a_k)^{1/k} = c$$

If $c > 1$ (including $c = +\infty$), then the series $\sum_{k=1}^{\infty} a_k$ diverges. If $c < 1$, then the series $\sum_{k=1}^{\infty} a_k$ converges. If $c = 1$, then the root test is inconclusive—you should try another test.

EXAMPLE 22-26: Does the series $\sum_{k=3}^{\infty} 1/(\ln k)^k$ converge?

Solution: Find the limit of the kth root of the kth term:

$$\lim_{k\to\infty} (a_k)^{1/k} = \lim_{k\to\infty}\left[\frac{1}{(\ln k)^k}\right]^{1/k} = \lim_{k\to\infty}\frac{1}{\ln k} = 0$$

Because the limit is less than one, the series $\sum_{k=3}^{\infty} 1/(\ln k)^k$ converges.

EXAMPLE 22-27: Does the series $\sum_{k=1}^{\infty} 2^k/k^3$ converge?

Solution: Try the root test:

$$\lim_{k\to\infty} (a_k)^{1/k} = \lim_{k\to\infty}\left(\frac{2^k}{k^3}\right)^{1/k} = \lim_{k\to\infty}\frac{2}{k^{3/k}} = 2\lim_{k\to\infty} k^{-3/k}$$

This is the indeterminate form ∞^0, so let $y = k^{-3/k}$ and find $\lim_{k\to\infty} \ln y$:

$$\lim_{k\to\infty} \ln y = \lim_{k\to\infty} \ln k^{-3/k} = \lim_{k\to\infty}\left(\frac{-3}{k}\right)\ln k = \lim_{k\to\infty}\frac{-3\ln k}{k} = \lim_{x\to\infty}\frac{-3\ln x}{x}$$

$$= \lim_{x\to\infty}\frac{\dfrac{d}{dx}(-3\ln x)}{\dfrac{d}{dx}x} = \lim_{x\to\infty}\frac{\dfrac{-3}{x}}{1} = 0$$

Because $\ln y$ approaches zero, y approaches $e^0 = 1$. Thus,

$$\lim_{k\to\infty} (a_k)^{1/k} = 2\lim_{k\to\infty} k^{-3/k} = 2\lim_{k\to\infty} y = 2(1) = 2$$

Because $2 > 1$, the series $\sum_{k=1}^{\infty} 2^k/k^3$ diverges.

F. Which test to use

For a given series, you may find several tests that will tell you whether or not the series converges:

- If the terms involve a factorial sign, the ratio test may be appropriate.
- If there is a power of k, the root test may be appropriate.

- If the terms are integrable, try the integral test.
- If the terms are a quotient of polynomials or roots of polynomials, try the limit comparison test.

There is no set of rules that tells you which test to use. There is no substitute for experience in making that decision.

22-5. Alternating Series

A series $\sum_{k=1}^{\infty} a_k$ is called **absolutely convergent** in case the series $\sum_{k=1}^{\infty} |a_k|$ converges. If a series converges absolutely, then it converges; that is, if $\sum_{k=1}^{\infty} |a_k|$ converges, then $\sum_{k=1}^{\infty} a_k$ converges. This also means that if $\sum_{k=1}^{\infty} a_k$ diverges, then $\sum_{k=1}^{\infty} |a_k|$ diverges.

EXAMPLE 22-28: Does the series $\sum_{k=1}^{\infty} (\sin k)/k^2$ converge?

Solution: Examine the series $\sum_{k=1}^{\infty} |(\sin k)/k^2| = \sum_{k=1}^{\infty} |\sin k|/k^2$. This series has positive terms and $|\sin k|/k^2 < 1/k^2$. So by the comparison test, $\sum_{k=1}^{\infty} |(\sin k)/k^2|$ converges because $\sum_{k=1}^{\infty} 1/k^2$ converges. Thus $\sum_{k=1}^{\infty} (\sin k)/k^2$ is absolutely convergent and hence convergent.

A series in which successive terms have opposite signs is called an **alternating series**. For example,

$$\sum_{k=1}^{\infty} \frac{(-1)^{k+1}}{k} = 1 - \frac{1}{2} + \frac{1}{3} - \frac{1}{4} + \frac{1}{5} - \cdots$$

is an alternating series. Once you recognize that a series is alternating, it is usually very easy to determine whether or not the series converges: If $\sum_{k=1}^{\infty} a_k$ is an alternating series and $|a_1| \geqslant |a_2| \geqslant |a_3| \geqslant \ldots$, then $\sum_{k=1}^{\infty} a_k$ converges if $\lim_{k\to\infty} |a_k| = 0$, and diverges if $\lim_{k\to\infty} |a_k| \neq 0$.

In other words, an alternating series with decreasing terms will converge if the terms approach zero. This differs greatly from the general (nonalternating) case. Remember from Example 22-16 that $\sum_{k=1}^{\infty} 1/k$, the harmonic series, diverges even though $\lim_{k\to\infty} 1/k = 0$. So, you have just one test for convergence for an alternating series whose terms decrease in absolute value.

With an alternating series, you'll be asked not only whether or not $\sum_{k=1}^{\infty} a_k$ converges, but also whether or not $\sum_{k=1}^{\infty} |a_k|$ converges. There are three possibilities:

1. If $\sum_{k=1}^{\infty} |a_k|$ converges, then you know that $\sum_{k=1}^{\infty} a_k$ converges ($\sum_{k=1}^{\infty} a_k$ is absolutely convergent).
2. If $\sum_{k=1}^{\infty} |a_k|$ diverges and $\sum_{k=1}^{\infty} a_k$ converges, then $\sum_{k=1}^{\infty} a_k$ is **conditionally convergent**.
3. If $\sum_{k=1}^{\infty} a_k$ diverges, then $\sum_{k=1}^{\infty} |a_k|$ also diverges ($\sum_{k=1}^{\infty} a_k$ is divergent).

EXAMPLE 22-29: Determine whether $\sum_{k=1}^{\infty} (-1)^k/k$ converges absolutely, converges conditionally, or diverges.

Solution: This is an alternating series, so you must determine whether $\sum_{k=1}^{\infty} (-1)^k/k$ converges and whether $\sum_{k=1}^{\infty} |(-1)^k/k|$ converges. The only test you need for the alternating series $\sum_{k=1}^{\infty} (-1)^k/k$ is to examine the limit:

$$\lim_{k\to\infty} |a_k| = \lim_{k\to\infty} \left|\frac{(-1)^k}{k}\right| = \lim_{k\to\infty} \frac{1}{k} = 0$$

Because the terms approach zero, the series $\sum_{k=1}^{\infty} (-1)^k/k$ converges.

Now examine the series $\sum_{k=1}^{\infty} |(-1)^k/k| = \sum_{k=1}^{\infty} 1/k$. You have already seen that this series diverges (by the integral test), and so you have the situation where $\sum_{k=1}^{\infty} |a_k|$ diverges and $\sum_{k=1}^{\infty} a_k$ converges. That is, $\sum_{k=1}^{\infty} a_k = \sum_{k=1}^{\infty} (-1)^k/k$ is conditionally convergent.

EXAMPLE 22-30: Determine whether $\sum_{k=1}^{\infty} (-1)^k/k^2$ converges absolutely, converges conditionally, or diverges.

Solution: You must determine whether $\sum_{k=1}^{\infty} (-1)^k/k^2$ converges and whether $\sum_{k=1}^{\infty} |(-1)^k/k^2|$ converges. Examine the latter series:

$$\sum_{k=1}^{\infty} \left|\frac{(-1)^k}{k^2}\right| = \sum_{k=1}^{\infty} \frac{1}{k^2}$$

You have seen in Example 22-15 that this series converges. Because $\sum_{k=1}^{\infty} |a_k|$ converges, you know that $\sum_{k=1}^{\infty} a_k$ converges. That is, $\sum_{k=1}^{\infty} a_k = \sum_{k=1}^{\infty} (-1)^k/k^2$ converges absolutely.

EXAMPLE 22-31: Determine whether $\sum_{k=1}^{\infty} (-1)^k k^2/2^k$ converges absolutely, converges conditionally, or diverges.

Solution: Examine the series $\sum_{k=1}^{\infty} |(-1)^k k^2/2^k| = \sum_{k=1}^{\infty} k^2/2^k$. This is a series with positive terms, and you can use the integral test, the comparison test, the ratio test, or the root test to find whether or not it converges. Try the ratio test:

$$\lim_{k\to\infty} \frac{b_{k+1}}{b_k} = \lim_{k\to\infty} \frac{\dfrac{(k+1)^2}{2^{k+1}}}{\dfrac{k^2}{2^k}} = \lim_{k\to\infty} \frac{(k+1)^2}{2^{k+1}}\left(\frac{2^k}{k^2}\right) = \lim_{k\to\infty} \frac{(k+1)^2}{2k^2} = \frac{1}{2}$$

Because this limit is less than one, the series $\sum_{k=1}^{\infty} k^2/2^k$ converges. Because $\sum_{k=1}^{\infty} |(-1)^k k^2/2^k|$ converges, $\sum_{k=1}^{\infty} (-1)^k k^2/2^k$ is absolutely convergent.

EXAMPLE 22-32: Determine whether $\sum_{k=3}^{\infty} (-1)^k \sqrt{k}/\ln k$ converges absolutely, converges conditionally, or diverges.

Solution: This is an alternating series. You must determine whether $\sum_{k=3}^{\infty} (-1)^k \sqrt{k}/\ln k$ converges and whether $\sum_{k=3}^{\infty} |(-1)^k \sqrt{k}/\ln k|$ converges. To determine whether $\sum_{k=3}^{\infty} (-1)^k \sqrt{k}/\ln k$ converges, you need only examine the limit:

$$\lim_{k\to\infty} \left|\frac{(-1)^k\sqrt{k}}{\ln k}\right| = \lim_{k\to\infty} \frac{\sqrt{k}}{\ln k} = \lim_{x\to\infty} \frac{\sqrt{x}}{\ln x} = \lim_{x\to\infty} \frac{\dfrac{d}{dx}\sqrt{x}}{\dfrac{d}{dx}\ln x}$$

$$= \lim_{x\to\infty} \frac{\dfrac{1}{2\sqrt{x}}}{\dfrac{1}{x}} = \lim_{x\to\infty} \frac{x}{2\sqrt{x}} = \infty$$

Because this limit is not zero, the series $\sum_{k=3}^{\infty} (-1)^k \sqrt{k}/\ln k$ diverges. This also means that $\sum_{k=3}^{\infty} |(-1)^k \sqrt{k}/\ln k|$ diverges.

SUMMARY

1. To find the limit of a sequence $\{a_n\}$, find a function f for which $f(n) = a_n$ and find $\lim_{x\to\infty} f(x)$. If this limit exists, then $\lim_{n\to\infty} a_n = \lim_{x\to\infty} f(x)$.
2. A series converges if the sequence of partial sums converges.
3. Use the integral test, comparison test, limit comparison test, ratio test, or root test to determine whether or not a series with positive terms converges.
4. An absolutely convergent series converges.
5. An alternating series is either absolutely convergent, conditionally convergent, or divergent.

SOLVED PROBLEMS

PROBLEM 22-1 Find the limit of the sequence $\{(2 - n^2)/(n^2 + 1)\}$.

Solution: You can find this limit by finding the limit of the function $f(x) = (2 - x^2)/(x^2 + 1)$:

$$\lim_{n\to\infty} \frac{2 - n^2}{n^2 + 1} = \lim_{x\to\infty} \frac{2 - x^2}{x^2 + 1} = -1$$

[See Section 21-1.]

PROBLEM 22-2 Find the limit of the sequence $\{[1 + (2/n)]^{3n}\}$.

Solution: Find the limit: $\lim_{n\to\infty} [1 + (2/n)]^{3n} = \lim_{x\to\infty} [1 + (2/x)]^{3x}$, which is the indeterminate form 1^∞, so let $y = [1 + (2/x)]^{3x}$ and find $\lim_{x\to\infty} \ln y$:

$$\lim_{x\to\infty} \ln y = \lim_{x\to\infty} \ln\left(1 + \frac{2}{x}\right)^{3x} = \lim_{x\to\infty} 3x \ln\left(1 + \frac{2}{x}\right) = \lim_{x\to\infty} \frac{3\ln\left(1 + \frac{2}{x}\right)}{\frac{1}{x}}$$

$$= \lim_{x\to\infty} \frac{\frac{d}{dx} 3\ln\left(1 + \frac{2}{x}\right)}{\frac{d}{dx}\frac{1}{x}} = \lim_{x\to\infty} \frac{\frac{3}{1 + (2/x)}\left(-\frac{2}{x^2}\right)}{\left(-\frac{1}{x^2}\right)} = \lim_{x\to\infty} \frac{6}{1 + \frac{2}{x}} = 6$$

Because $\ln y$ approaches 6, y approaches e^6. Thus the limit of the sequence $\{[1 + (2/n)]^{3n}\}$ is e^6.

[See Section 22-1.]

PROBLEM 22-3 Find the limit of the sequence $\{\cos k\pi\}$.

Solution: You won't find this limit by finding $\lim_{x\to\infty} \cos x\pi$ because this limit doesn't exist. However, examine a few terms in the sequence:

$$\{\cos k\pi\} = \{\cos \pi, \cos 2\pi, \cos 3\pi, \cos 4\pi, \ldots\}$$
$$= \{-1, 1, -1, 1, -1, 1, \ldots\}$$

From this it's apparent that the sequence diverges. The terms don't get closer and closer to any number; they just bounce back and forth between -1 and 1.

[See Section 22-1.]

PROBLEM 22-4 Find the limit of the sequence $\{k^2/k!\}$.

Solution: It's difficult to find a function f defined for $x \geqslant 1$ for which $f(x) = x^2/x!$ because the factorial is only defined for integers. Simply examine the terms in the sequence:

$$\frac{k^2}{k!} = \frac{k}{(k-1)!} = \frac{k}{(k-1)(k-2)\cdots 3(2)}$$

As k grows, the denominator grows much more rapidly than the numerator! Thus the sequence has limit zero.

[See Section 22-1.]

PROBLEM 22-5 Does the series $\sum_{k=2}^{\infty} \left(\frac{1}{\ln k} - \frac{1}{\ln(k+1)}\right)$ converge?

Solution: Find the nth partial sum:

$$S_n = \sum_{k=2}^{n} \left(\frac{1}{\ln k} - \frac{1}{\ln(k+1)}\right) = \left(\frac{1}{\ln 2} - \frac{1}{\ln 3}\right) + \left(\frac{1}{\ln 3} - \frac{1}{\ln 4}\right) + \cdots + \left(\frac{1}{\ln n} - \frac{1}{\ln(n+1)}\right)$$

$$= \frac{1}{\ln 2} + \left(\frac{-1}{\ln 3} + \frac{1}{\ln 3}\right) + \left(\frac{-1}{\ln 4} + \frac{1}{\ln 4}\right) + \cdots + \left(\frac{-1}{\ln n} + \frac{1}{\ln n}\right) - \frac{1}{\ln(n+1)}$$

$$= \frac{1}{\ln 2} - \frac{1}{\ln(n+1)}$$

Now find the limit of the sequence of partial sums, $\{S_n\} = \left\{\frac{1}{\ln 2} - \frac{1}{\ln(n+1)}\right\}$:

$$\lim_{n\to\infty}\left(\frac{1}{\ln 2} - \frac{1}{\ln(n+1)}\right) = \frac{1}{\ln 2} - 0$$

Thus the series converges to $1/\ln 2$. [See Section 22-2.]

PROBLEM 22-6 To what number does the series $\sum_{k=1}^{\infty} 1/(k^2 + 3k + 2)$ converge?

Solution: Use partial fractions to recognize this as a telescoping series:

$$\frac{1}{k^2+3k+2} = \frac{1}{k+1} - \frac{1}{k+2}$$

$$S_n = \sum_{k=1}^{n}\left(\frac{1}{k+1} - \frac{1}{k+2}\right) = \left(\frac{1}{2} - \frac{1}{3}\right) + \left(\frac{1}{3} - \frac{1}{4}\right) + \cdots + \left(\frac{1}{n+1} - \frac{1}{n+2}\right)$$

$$= \frac{1}{2} - \frac{1}{n+2}$$

The sequence $\{S_n\}$ approaches $\frac{1}{2}$, so the series $\sum_{k=1}^{\infty} 1/(k^2 + 3k + 2)$ converges to $\frac{1}{2}$.
[See Section 22-2.]

PROBLEM 22-7 Sum the finite series $1 - (2/3) + (4/9) - (8/27) + (16/81) - (32/243)$.

Solution: Write it in summation notation and use the formula for finite geometric series, $\sum_{k=1}^{n} r^k = (1 - r^{n+1})/(1 - r)$:

$$1 - \frac{2}{3} + \frac{4}{9} - \cdots - \frac{32}{243} = \sum_{k=0}^{5}\left(\frac{-2}{3}\right)^k = \frac{1 - (-2/3)^6}{1 - (-2/3)} = \frac{1 - (2/3)^6}{5/3}$$

$$= \frac{3^6 - 2^6}{3^6}\left(\frac{3}{5}\right) = \frac{133}{243}$$

[See Section 22-3.]

PROBLEM 22-8 Sum the finite series

$$(2/\pi) + (\pi/2) + (\pi^3/8) + (\pi^5/32) + (\pi^7/128) + (\pi^9/512).$$

Solution: To recognize that this is a geometric series, notice that the ratio between successive terms is $\pi^2/4$. This will be a geometric series with ratio $r = \pi^2/4$. Factor out the first term and write it in summation notation:

$$\frac{2}{\pi} + \frac{\pi}{2} + \cdots + \frac{\pi^9}{512} = \frac{2}{\pi}\left(1 + \frac{\pi^2}{4} + \cdots + \frac{\pi^{10}}{1024}\right)$$

$$= \frac{2}{\pi}\sum_{k=0}^{5}\left(\frac{\pi^2}{4}\right)^k = \left(\frac{2}{\pi}\right)\frac{1 - (\pi^2/4)^6}{1 - (\pi^2/4)}$$

$$= \frac{\pi^{12} - 4096}{512\pi(\pi^2 - 4)}$$

[See Section 22-3.]

PROBLEM 22-9 To what number does the series $\sum_{k=0}^{\infty} (-2/7)^k$ converge?

Solution: This is an infinite geometric series with $|r| < 1$, so $\sum_{k=0}^{\infty} r^k = 1/(1 - r)$:

$$\sum_{k=0}^{\infty}\left(\frac{-2}{7}\right)^k = \frac{1}{1 - (-2/7)} = \frac{7}{9}$$

[See Section 22-3.]

PROBLEM 22-10 To what number does the series $(7/2) - (7/6) + (7/18) - (7/54) + \cdots$ converge?

Solution: The ratio between successive terms is $\frac{-1}{3}$, so this is a geometric series with $r = \frac{-1}{3}$:

$$\frac{7}{2} - \frac{7}{6} + \frac{7}{18} - \cdots = \frac{7}{2}\left(1 - \frac{1}{3} + \frac{1}{9} - \frac{1}{27} + \cdots\right) = \frac{7}{2}\sum_{k=0}^{\infty}\left(\frac{-1}{3}\right)^k$$

$$= \left(\frac{7}{2}\right)\frac{1}{1-(-\frac{1}{3})} = \left(\frac{7}{2}\right)\frac{1}{4/3} = \frac{21}{8}$$

[See Section 22-3.]

PROBLEM 22-11 Express the repeating decimal $0.515\,1\overline{51}$ as a quotient of integers.

Solution: Recognize the repeating decimal as a geometric series:

$$0.515\,1\overline{51} = 0.51 + 0.0051 + 0.000\,051 + \cdots$$

$$= \frac{51}{10^2} + \frac{51}{10^4} + \frac{51}{10^6} + \cdots = \frac{51}{100}\left(1 + \frac{1}{10^2} + \frac{1}{10^4} + \cdots\right)$$

$$= \frac{51}{100}\sum_{k=0}^{\infty}\left(\frac{1}{10^2}\right)^k = \left(\frac{51}{100}\right)\frac{1}{1-(1/100)}$$

$$= \frac{51}{99} = \frac{17}{33}$$

[See Section 22-3.]

PROBLEM 22-12 Does the series $\sum_{k=3}^{\infty} 1/(k \ln^2 k)$ converge?

Solution: Try the integral test: Does the improper integral $\int_3^{\infty} 1/(x \ln^2 x)\, dx$ converge?

$$\int_3^{\infty} \frac{dx}{x \ln^2 x} = \lim_{N\to\infty} \int_3^N \frac{dx}{x \ln^2 x}$$

Make the substitution $u = \ln x$, with $du = 1/x\, dx$:

$$\lim_{N\to\infty} \int_3^N \frac{dx}{x \ln^2 x} = \lim_{N\to\infty} \int_{\ln 3}^{\ln N} \frac{1}{u^2}\, du = \lim_{N\to\infty} \left(\frac{-1}{u}\right)\Bigg|_{\ln 3}^{\ln N}$$

$$= \lim_{N\to\infty} \left(\frac{-1}{\ln N} + \frac{1}{\ln 3}\right) = \frac{1}{\ln 3}$$

The integral converges, so by the integral test, the series $\sum_{k=3}^{\infty} 1/(k \ln^2 k)$ converges.

[See Section 22-4.]

PROBLEM 22-13 Does the series $\sum_{k=3}^{\infty} 1/(\sqrt{k} \ln k)$ converge?

Solution: Try the comparison test. If, for example $1/(\sqrt{k} \ln k) > 1/k$, then you can show that the series diverges by comparison with the harmonic series. Because $\sqrt{k} > \ln k$ for all k, you have: $k = \sqrt{k}\sqrt{k} > \sqrt{k} \ln k$, and so $1/k < 1/(\sqrt{k} \ln k)$. Thus $\sum_{k=3}^{\infty} 1/(\sqrt{k} \ln k)$ is term by term larger than the divergent series $\sum_{k=3}^{\infty} 1/k$, so $\sum_{k=3}^{\infty} 1/(\sqrt{k} \ln k)$ diverges.

[See Section 22-4.]

PROBLEM 22-14 Does the series $\sum_{k=1}^{\infty} \frac{(k^2-2)}{(k^3+k+2)}$ converge?

Solution: Try the limit comparison test. The terms in the series "look like" $k^2/k^3 = 1/k$, so compare with the harmonic series:

$$\lim_{k\to\infty} \frac{\dfrac{k^2-2}{k^3+k+2}}{\dfrac{1}{k}} = \lim_{k\to\infty} \frac{k^3-2k}{k^3+k+2} = 1$$

Because the limit is a finite nonzero number, the two series $\sum_{k=1}^{\infty} (k^2-2)/(k^3+k+2)$ and $\sum_{k=1}^{\infty} 1/k$ either both converge or both diverge. You know that $\sum_{k=1}^{\infty} 1/k$ diverges, so $\sum_{k=1}^{\infty} (k^2-2)/(k^3+k+2)$ must also diverge.

[See Section 22-4.]

PROBLEM 22-15 Does $\sum_{k=1}^{\infty} 1/\sqrt{k!}$ converge?

Solution: Try the ratio test:

$$\lim_{k\to\infty} \frac{a_{k+1}}{a_k} = \lim_{k\to\infty} \frac{\dfrac{1}{\sqrt{(k+1)!}}}{\dfrac{1}{\sqrt{k!}}} = \lim_{k\to\infty} \sqrt{\frac{k!}{(k+1)!}} = \lim_{k\to\infty} \sqrt{\frac{1}{k+1}} = 0$$

Because this limit is less than one, the series $\sum_{k=1}^{\infty} 1/\sqrt{k!}$ converges. [See Section 22-4.]

PROBLEM 22-16 Does $\sum_{k=3}^{\infty} (\ln k)^{2k}/k^k$ converge?

Solution: Try the root test:

$$\lim_{k\to\infty} (a_k)^{1/k} = \lim_{k\to\infty} \left[\frac{(\ln k)^{2k}}{k^k}\right]^{1/k} = \lim_{k\to\infty} \frac{(\ln k)^2}{k}$$

You are trying to find the limit of a sequence. Find the limit of the function $f(x) = (\ln x)^2/x$:

$$\lim_{x\to\infty} f(x) = \lim_{x\to\infty} \frac{(\ln x)^2}{x} = \lim_{x\to\infty} \frac{\dfrac{d}{dx}(\ln x)^2}{\dfrac{d}{dx}x} = \lim_{x\to\infty} \frac{2(\ln x)\dfrac{1}{x}}{1}$$

$$= \lim_{x\to\infty} \frac{\dfrac{d}{dx}2\ln x}{\dfrac{d}{dx}x} = \lim_{x\to\infty} \frac{\dfrac{2}{x}}{1} = 0$$

Because this limit is less than one, the series $\sum_{k=3}^{\infty} (\ln k)^{2k}/k^k$ converges. [See Section 22-4.]

PROBLEM 22-17 Does the series $\sum_{k=1}^{\infty} \dfrac{k + (1/k)}{\sqrt{k^5 + \ln k}}$ converge?

Solution: Try the limit comparison test. The terms in the series "look like" $k/\sqrt{k^5} = 1/k^{3/2}$, so compare with the series $\sum_{k=1}^{\infty} 1/k^{3/2}$:

$$\lim_{k\to\infty} \left[\frac{\dfrac{k + \dfrac{1}{k}}{\sqrt{k^5 + \ln k}}}{\dfrac{1}{k^{3/2}}}\right] = \lim_{k\to\infty} \frac{k^{5/2} + k^{1/2}}{\sqrt{k^5 + \ln k}}$$

$$= \lim_{k\to\infty} \frac{(k^{5/2} + k^{1/2})\left(\dfrac{1}{k^{5/2}}\right)}{\sqrt{k^5 + \ln k}\left(\dfrac{1}{k^{5/2}}\right)} = \lim_{k\to\infty} \frac{1 + \left(\dfrac{1}{k^2}\right)}{\sqrt{1 + \left(\dfrac{\ln k}{k^5}\right)}} = 1$$

Because this limit is a nonzero finite number, the two series either both converge or both diverge. But you know that $\sum_{k=1}^{\infty} (1/k^{3/2})$ converges ($\sum_{k=1}^{\infty} 1/k^s$, with $s > 1$), so $\sum_{k=1}^{\infty} (k + (1/k))/\sqrt{k^5 + \ln k}$ converges. [See Section 22-4.]

PROBLEM 22-18 Does the series $\sum_{k=1}^{\infty} [1 - (1/k)]^{k^2}$ converge?

Solution: Try the root test:

$$\lim_{k\to\infty} (a_k)^{1/k} = \lim_{k\to\infty} \left[\left(1 - \frac{1}{k}\right)^{k^2}\right]^{1/k} = \lim_{k\to\infty} \left(1 - \frac{1}{k}\right)^k$$

This is the indeterminate form 1^∞, so let $y = [1 - (1/k)]^k$ and find $\lim_{k\to\infty} \ln y$:

$$\lim_{k\to\infty} \ln y = \lim_{k\to\infty} \ln\left(1 - \frac{1}{k}\right)^k = \lim_{x\to\infty} \ln\left(1 - \frac{1}{x}\right)^x = \lim_{x\to\infty} x \ln\left(1 - \frac{1}{x}\right)$$

$$= \lim_{x\to\infty} \frac{\ln\left(1 - \frac{1}{x}\right)}{\frac{1}{x}} = \lim_{x\to\infty} \frac{\frac{d}{dx}\ln\left(1 - \frac{1}{x}\right)}{\frac{d}{dx}\frac{1}{x}}$$

$$= \lim_{x\to\infty} \frac{\frac{1}{x^2}\frac{1}{\left(1 - \frac{1}{x}\right)}}{\frac{-1}{x^2}} = \lim_{x\to\infty} \frac{-1}{1 - \frac{1}{x}} = -1$$

Because $\ln y$ approaches -1, y approaches e^{-1}. Because $\lim_{k\to\infty}(a_k)^{1/k} = e^{-1}$, which is less than one, the series converges by the root test. [See Section 22-4.]

PROBLEM 22-19 Does the series $\sum_{k=0}^{\infty} k/e^{k^2}$ converge?

Solution: Try the integral test:

$$\int_0^\infty \frac{x}{e^{x^2}}\,dx = \lim_{N\to\infty}\int_0^N xe^{-x^2}\,dx$$

Make the substitution $u = -x^2$, with $du = -2x\,dx$:

$$\lim_{N\to\infty}\int_0^N xe^{-x^2}\,dx = \lim_{N\to\infty}\int_0^{-N^2} e^u\left(-\frac{1}{2}\,du\right) = \lim_{N\to\infty}\left(-\frac{1}{2}e^u\right)\Bigg|_0^{-N^2}$$

$$= \lim_{N\to\infty}\left(-\frac{1}{2}e^{-N^2} + \frac{1}{2}\right) = \frac{1}{2}$$

Because this improper integral converges, you can conclude that the series $\sum_{k=0}^{\infty} k/e^{k^2}$ converges. [See Section 22-4.]

PROBLEM 22-20 Does the series $\sum_{k=1}^{\infty} k^{2k}/(2k)!$ converge?

Solution: The factorial sign leads you to try the ratio test:

$$\lim_{k\to\infty}\frac{a_{k+1}}{a_k} = \lim_{k\to\infty}\left[\frac{\dfrac{(k+1)^{2(k+1)}}{[2(k+1)]!}}{\dfrac{k^{2k}}{(2k)!}}\right]$$

$$= \lim_{k\to\infty}\frac{(k+1)^{2(k+1)}}{[2(k+1)]!}\left(\frac{(2k)!}{k^{2k}}\right)$$

$$= \lim_{k\to\infty}\frac{(k+1)^{2k+2}(2k)(2k-1)\cdots 3(2)}{k^{2k}(2k+2)(2k+1)(2k)\cdots 3(2)}$$

$$= \lim_{k\to\infty}\frac{(k+1)^{2k+2}}{k^{2k}(2k+2)(2k+1)} = \lim_{k\to\infty}\frac{(k+1)^{2k+1}}{k^{2k}2(2k+1)}$$

$$= \lim_{k\to\infty}\frac{k+1}{2(2k+1)}\left(1 + \frac{1}{k}\right)^{2k}$$

But $\lim_{k\to\infty}[1 + (1/k)]^k = e$, so $\lim_{k\to\infty}[1 + (1/k)]^{2k} = e^2$ and you have

$$\lim_{k\to\infty}\frac{a_{k+1}}{a_k} = \lim_{k\to\infty}\frac{k+1}{2(2k+1)}e^2 = \frac{1}{4}e^2 > 1$$

Because the limit is greater than one, the series $\sum_{k=1}^{\infty} k^{2k}/(2k)!$ diverges. [See Section 22-4.]

PROBLEM 22-21 Does the series $\sum_{k=3}^{\infty} (\ln k)/k^2$ converge?

Solution: Try the comparison test:

$$\ln k < \sqrt{k} \qquad \frac{\ln k}{k^2} < \frac{\sqrt{k}}{k^2} = \frac{1}{k^{3/2}}$$

Because the series is term by term less than the convergent series $\sum_{k=3}^{\infty} 1/k^{3/2}$, you conclude that $\sum_{k=3}^{\infty} (\ln k)/k^2$ converges. [See Section 22-4.]

PROBLEM 22-22 Does the series $\sum_{k=3}^{\infty} (-1)^k/(k \ln k)$ converge absolutely, converge conditionally, or diverge?

Solution: First examine the series $\sum_{k=3}^{\infty} |(-1)^k/(k \ln k)| = \sum_{k=3}^{\infty} 1/(k \ln k)$. Try the integral test:

$$\int_3^{\infty} \frac{dx}{x \ln x} = \lim_{N\to\infty} \int_3^N \frac{dx}{x \ln x} = \lim_{N\to\infty} \ln \ln x \Big|_3^N$$

$$= \lim_{N\to\infty} (\ln \ln N - \ln \ln 3) = \infty$$

Because the integral diverges, the series $\sum_{k=3}^{\infty} 1/(k \ln k)$ diverges. Now examine the series $\sum_{k=3}^{\infty} (-1)^k/k \ln k$. This is an alternating series, so to test for convergence consider the limit:

$$\lim_{k\to\infty} |a_k| = \lim_{k\to\infty} \left|\frac{(-1)^k}{k \ln k}\right| = \lim_{k\to\infty} \frac{1}{k \ln k} = 0$$

Because the terms approach zero, the alternating series converges. You already saw that the series isn't absolutely convergent ($\sum_{k=3}^{\infty} |(-1)^k/(k \ln k)|$ diverges), so the series is conditionally convergent. [See Section 22-5.]

PROBLEM 22-23 Does the series $\sum_{k=1}^{\infty} (-1)^k[1 - (1/k)]^k$ converge absolutely, converge conditionally or diverge?

Solution: To determine whether this alternating series converges, consider the limit:

$$\lim_{k\to\infty} \left|(-1)^k\left(1 - \frac{1}{k}\right)^k\right| = \lim_{k\to\infty}\left(1 - \frac{1}{k}\right)^k = \lim_{x\to\infty}\left(1 - \frac{1}{x}\right)^x$$

This is the indeterminate form 1^∞, so let $y = [1 - (1/x)]^x$ and find $\lim_{x\to\infty} \ln y$:

$$\lim_{x\to\infty} \ln y = \lim_{x\to\infty} \ln\left(1 - \frac{1}{x}\right)^x = \lim_{x\to\infty} \frac{\ln\left(1 - \frac{1}{x}\right)}{\frac{1}{x}}$$

$$= \lim_{x\to\infty} \frac{\frac{d}{dx}\ln\left(1 - \frac{1}{x}\right)}{\frac{d}{dx}\frac{1}{x}} = \lim_{x\to\infty} \frac{\frac{1}{(1 - 1/x)}\left(\frac{1}{x^2}\right)}{\frac{-1}{x^2}}$$

$$= \lim_{x\to\infty} \frac{-1}{1 - \frac{1}{x}} = -1$$

Because $\ln y$ approaches -1, y approaches e^{-1}. Because the terms in the series $\sum_{k=1}^{\infty} (-1)^k[1 - (1/k)]^k$ don't approach zero, the series diverges. [See Section 22-5.]

PROBLEM 22-24 Does the series $\sum_{k=0}^{\infty}$ (cos $k\pi$) (arc tan k)/$(1 + k^2)$ converge absolutely, converge conditionally, or diverge?

Solution: Notice that cos $k\pi$ is $+1$ if k is even, -1 if k is odd. Thus this is an alternating series. To examine the series $\sum_{k=0}^{\infty}$ |(cos $k\pi$)(arc tan k)/$(1 + k^2)$| = $\sum_{k=0}^{\infty}$ (arc tan k)/$(1 + k^2)$, you can use the integral test:

$$\int_0^\infty \frac{\arctan x}{1+x^2}\,dx = \lim_{N\to\infty}\int_0^N \frac{\arctan x}{1+x^2}\,dx = \lim_{N\to\infty}\frac{1}{2}(\arctan x)^2\Big|_1^N$$

$$= \lim_{N\to\infty}\left[\frac{1}{2}(\arctan N)^2 - \frac{1}{2}(\arctan 1)^2\right] = \frac{1}{2}\left(\frac{\pi}{2}\right)^2 - \frac{1}{2}\left(\frac{\pi}{4}\right)^2$$

Because the integral converges, the series $\sum_{k=0}^{\infty} |(\cos k\pi)(\arctan k)/(1+k^2)|$ converges, and hence $\sum_{k=0}^{\infty} (\cos k\pi)(\arctan k)/(1+k^2)$ converges absolutely.

Supplementary Exercises

In Problems 22-25 through 22-40 find the limit of the sequence if it exists:

22-25 $\{(n^2-3)/(2+2n^2)\}$

22-26 $\{(\ln n)/n\}$

22-27 $\{\sqrt{n-3}/\sqrt{4n+2}\}$

22-28 $\{[7-(1/\ln n)]^2\}$

22-29 $\{\sin[(n/2)\pi]\}$

22-30 $\{(\sin n)/\ln(n+1)\}$

22-31 $\{\sqrt{n}-\sqrt{n-1}\}$

22-32 $\{2^n/3^{n-2}\}$

22-33 $\{n\ln[1-(3/n)]\}$

22-34 $\{n^n/n!\}$

22-35 $\{(\sin n)/\sqrt{n}\}$

22-36 $\{2^n/n!\}$

22-37 $\{\ln\ln n\}$

22-38 $\{n^{1/n}\}$

22-39 $\{n\sin n\pi\}$

22-40 $\{n^{\sin(\pi/2)n}\}$

In Problems 22-41 through 22-45 find the sum:

22-41 $1+(1/3)+(1/9)+(1/27)+(1/81)+(1/243)$

22-42 $1-3+9-27+81-243$

22-43 $6-3+(3/2)-(3/4)+(3/8)-(3/16)+(3/32)-(3/64)+(3/128)-(3/256)$

22-44 $\pi^3-\pi^4+\pi^5-\pi^6+\cdots+\pi^{31}$

22-45 $(\pi/e^2)-(1/e\pi)+(1/\pi^3)-(e/\pi^5)+(e^2/\pi^7)-\cdots+(e^5/\pi^{13})$

In Problems 22-46 through 22-52 sum the given geometric series:

22-46 $\sum_{k=0}^{\infty}(\frac{1}{2})^k$

22-47 $\sum_{k=0}^{\infty}(-3/5)^k$

22-48 $1-(4/9)+(16/81)-(64/729)+\cdots$

22-49 $3+(4/3)+(16/27)+(64/243)+\cdots$

22-50 $(7/9)-(1/3)+(1/7)-(3/49)+(9/343)-\cdots$

22-51 $\pi + e + (e^2/\pi) + (e^3/\pi^2) + \cdots$

22-52 $2 - \sqrt{3} + (3/2) - (3\sqrt{3}/4) + (9/8) - (9\sqrt{3}/16) + \cdots$

In Problems 22-53 through 22-56 express the repeating decimal as a quotient of integers:

22-53 $0.727\,2\overline{72}$

22-54 $0.177\,177\,\overline{177}$

22-55 $0.191\,191\,\overline{191}$

22-56 $0.810\,981\,\overline{098\,109}$

In Problems 22-57 through 22-90 determine whether the series converges or diverges.

22-57 $\sum_{k=3}^{\infty} 1/(k \ln^3 k)$

22-58 $\sum_{k=3}^{\infty} (\ln k)/3^k$

22-59 $\sum_{k=2}^{\infty} (k^3 + 1)/(k^5 - 2)$

22-60 $\sum_{k=1}^{\infty} 2^k/k!$

22-61 $\sum_{k=3}^{\infty} [(\ln k)/k]^k$

22-62 $\sum_{k=1}^{\infty} \sqrt{k}/k!$

22-63 $\sum_{k=0}^{\infty} \arctan k$

22-64 $\sum_{k=1}^{\infty} \sin(1/k)$

22-65 $\sum_{k=1}^{\infty} e^{1/k}/k^2$

22-66 $\sum_{k=1}^{\infty} [1 - (1/k)]^k$

22-67 $\sum_{k=0}^{\infty} k/2^k$

22-68 $\sum_{k=1}^{\infty} k^k/(k!)^2$

22-69 $\sum_{k=0}^{\infty} 1/(k^2 + 1)$

22-70 $\sum_{k=1}^{\infty} k^k/k!$

22-71 $\sum_{k=1}^{\infty} k^k/(2k)!$

22-72 $\sum_{k=3}^{\infty} 1/(\sqrt[3]{k} \ln k)$

22-73 $\sum_{k=3}^{\infty} (k + \ln k)/(k^2 + \ln k)$

22-74 $\sum_{k=3}^{\infty} k^3/(\ln k)^k$

22-75 $\sum_{k=2}^{\infty} \sqrt{k^2 + 1}/\sqrt{k^5 + k - 2}$

22-76 $\sum_{k=3}^{\infty} (\ln k)/k$

22-77 $\sum_{k=1}^{\infty} (k - 1)/\sqrt{k!}$

22-78 $\sum_{k=2}^{\infty} \sqrt[3]{1/(k^4 - 1)}$

22-79 $\sum_{k=2}^{\infty} \sqrt[3]{1/(k^2 - 1)}$

22-80 $\sum_{k=3}^{\infty} 1/(k\sqrt[3]{\ln k})$

22-81 $\sum_{k=1}^{\infty} (2^k)^k/(k^2)^k$

22-82 $\sum_{k=0}^{\infty} k^3/e^{k^4}$

22-83 $\sum_{k=0}^{\infty} 3^{k^2}/2^{2^k}$

22-84 $\sum_{k=3}^{\infty} \sqrt{k + 3}/(k + \ln k)$

22-85 $\sum_{k=1}^{\infty} 1/\sinh^2 k$

22-86 $\sum_{k=1}^{\infty} k^2/(2k)!$

22-87 $\sum_{k=1}^{\infty} (\sqrt{k} - \sqrt{k - 1})^k$

22-88 $\sum_{k=2}^{\infty} k/[(1 + k^2)\ln(1 + k^2)]$

22-89 $\sum_{k=10}^{\infty} (\sqrt{k} + 3)^3/(\sqrt{k} - 3)^6$

22-90 $\sum_{k=1}^{\infty} \sqrt{k!}/k^k$

In Problems 22-91 through 22-105 determine whether the series converges absolutely, converges conditionally, or diverges:

22-91 $\sum_{k=1}^{\infty} (-1)^k/\sqrt{k}$

22-92 $\sum_{k=3}^{\infty} (-1)^k/(k \ln^2 k)$

22-93 $\sum_{k=1}^{\infty} (-2)^k/k^2$

22-94 $\sum_{k=1}^{\infty} (-2)^k/k!$

22-95 $\sum_{k=0}^{\infty} (-1)^k k^3/e^{k^4}$

22-96 $\sum_{k=0}^{\infty} (-1)^k k^3/(k^4 + 2)$

22-97 $\sum_{k=3}^{\infty} (-1)^k/(\sqrt{k} \ln k)$

22-98 $\sum_{k=3}^{\infty} (\ln k)/(-3)^k$

22-99 $\sum_{k=1}^{\infty} (2k)!/(-k)^k$

22-100 $\sum_{k=1}^{\infty} (-1)^k k^2/k!$

22-101 $\sum_{k=1}^{\infty} (-1)^k[1 - (1/k)]^{k^2}$

22-102 $\sum_{k=3}^{\infty} (-1)^k/(k\sqrt{\ln k})$

22-103 $\sum_{k=1}^{\infty} \sin[(-1)^k/k]$

22-104 $\sum_{k=1}^{\infty} (-k)^k/k!$

22-105 $\sum_{k=0}^{\infty} (-1)^k \sqrt{k}/(k + 1)$

Solutions to Supplementary Exercises

(22-25) $\frac{1}{2}$

(22-26) 0

(22-27) $\frac{1}{2}$

(22-28) 49

(22-29) diverges

(22-30) 0

(22-31) 0

(22-32) 0

(22-33) -3

(22-34) diverges

(22-35) 0

(22-36) 0

(22-37) diverges

(22-38) 1

(22-39) 0

(22-40) diverges

(22-41) 364/243

(22-42) -182

(22-43) 1023/256

(22-44) $\pi^3(1 + \pi^{29})/(1 + \pi)$

(22-45) $(\pi^{16} - e^8)/[e^2\pi^{13}(e + \pi^2)]$

(22-46) 2

(22-47) 5/8

(22-48) 9/13

(22-49) 27/5

(22-50) 49/90

(22-51) $\pi^2/(\pi - e)$

(22-52) $4(2 - \sqrt{3})$

(22-53) 8/11

(22-54) 59/333

(22-55) 191/999

(22-56) 901/1111

(22-57) converges (by integral test)

(22-58) converges (by comparison with $\sum 1/2^k$)

(22-59) converges (by limit comparison with $\sum 1/k^2$)

(22-60) converges (by ratio test)

(22-61) converges (by root test)

(22-62) converges (by ratio test)

(22-63) diverges (by integral test)

(22-64) diverges (by limit comparison with $\sum 1/k$)

(22-65) converges (by integral or comparison test)

(22-66) diverges (by root test)

(22-67) converges (by integral test, comparison test, or root test)

(22-68) converges (by ratio test)

(22-69) converges (by integral test or comparison test)

(22-70) diverges (by ratio test)

(22-71) converges (by ratio test)

(22-72) diverges (by comparison with $\sum 1/k$)

(22-73) diverges (by limit comparison with $\sum 1/k$)

(22-74) converges (by root test)

(22-75) converges (by limit comparison with $\sum 1/k^{3/2}$)

(22-76) diverges (by integral test or comparison with $\sum 1/k$)

(22-77) converges (by ratio test)

(22-78) converges (by limit comparison with $\sum 1/k^{4/3}$)

(22-79) diverges (by limit comparison with $\sum 1/k^{2/3}$)

(22-80) diverges (by integral test)

(22-81) diverges (by root test)

(22-82) converges (by integral test)

(22-83) converges (by ratio test)

(22-84) diverges (by limit comparison with $\sum 1/\sqrt{k}$)

(22-85) converges (by integral test)

(22-86) converges (by ratio test)

(22-87) converges (by root test)

(22-88) diverges (by integral test)

(22-89) converges (by limit comparison with $\sum 1/k^{3/2}$)

(22-90) converges (by ratio test)

(22-91) converges conditionally

(22-92) converges absolutely

(22-93) diverges

(22-94) converges absolutely

(22-95) converges absolutely

(22-96) converges conditionally

(22-97) converges conditionally

(22-98) converges absolutely

(22-99) diverges

(22-100) converges absolutely

(22-101) converges absolutely

(22-102) converges conditionally

(22-103) converges conditionally

(22-104) diverges

(22-105) converges conditionally

23 TAYLOR POLYNOMIALS AND POWER SERIES

THIS CHAPTER IS ABOUT

☑ Power Series
☑ Taylor and Maclaurin Polynomials
☑ Power Series Expansions of Functions
☑ Estimation with Power Series

23-1. Power Series

A *power series in x* is a series of the form

MACLAURIN SERIES

$$\sum_{k=0}^{\infty} a_k x^k = a_0 + a_1 x + a_2 x^2 + a_3 x^3 + \cdots \tag{23-1}$$

also called a **Maclaurin series** or a *power series expansion about the origin.* A *power series in $x - a$* is a series of the form

TAYLOR SERIES

$$\sum_{k=0}^{\infty} a_k (x-a)^k = a_0 + a_1(x-a) + a_2(x-a)^2 + a_3(x-a)^3 + \cdots \tag{23-2}$$

also called a **Taylor series** or a *power series expansion about $x = a$.* Notice that a Maclaurin series is just a Taylor series with $a = 0$.

One way to think of a power series is as a polynomial of infinite degree. In particular, you should think of a power series as a function of x.

A. Radius of convergence

Your first concern with a power series will be to determine the values of x for which the series converges.

EXAMPLE 23-1: For what values of x does $\sum_{k=0}^{\infty} x^k$ converge?

Solution: This is just the geometric series, which converges for $|x| < 1$ (as you saw in Section 22-3).

For each power series $\sum_{k=0}^{\infty} a_k(x-a)^k$ there is a number $R \geqslant 0$ (R can be infinity) such that:

1. If $|x - a| < R$, then the series converges.
2. If $|x - a| > R$, then the series diverges.

This number R is the **radius of convergence** of the power series.

To find the radius of convergence of the power series $\sum_{k=0}^{\infty} a_k(x-a)^k$, find either limit:

$$\lim_{k\to\infty} \left|\frac{a_{k+1}}{a_k}\right| = L \quad \text{or} \quad \lim_{k\to\infty} |a_k|^{1/k} = L$$

You then find R as follows:

1. If $L = \infty$, then $R = 0$; the series converges only for $x = a$.
2. If $L = 0$, then $R = \infty$; the series converges for all x.
3. Otherwise, $R = 1/L$.

These methods are derived from the ratio and root tests of Chapter 22, and they are called the **ratio** and **root tests** for the radius of convergence. When the expression "root test" or "ratio test" is used, you will understand from the context whether it's a test for convergence of a series or a method for finding the radius of convergence of a power series.

EXAMPLE 23-2: Find the radius of convergence of the power series $\sum_{k=0}^{\infty} (2k/3^k)x^k$.

Solution: You can find either limit, whichever is easier. Here the ratio test seems easier:

$$\lim_{k\to\infty}\left|\frac{a_{k+1}}{a_k}\right| = \lim_{k\to\infty}\left|\frac{\dfrac{2(k+1)}{3^{k+1}}}{\dfrac{2k}{3^k}}\right| = \lim_{k\to\infty}\left[\frac{2(k+1)}{3^{k+1}}\left(\frac{3^k}{2k}\right)\right]$$

$$= \lim_{k\to\infty}\frac{2(k+1)}{3(2k)} = \frac{1}{3}$$

Because $L = \frac{1}{3}$, the radius of convergence is $R = 1/L = 3$.

B. Interval of convergence

Once you know the radius of convergence of a power series $\sum_{k=0}^{\infty} a_k(x-a)^k$, you must still determine whether or not the series converges on the radius of convergence, i.e., when $x = a + R$ and $x = a - R$. To do this, simply use the techniques of Chapter 22. The set of values of x for which the power series converges is the **interval of convergence**.

EXAMPLE 23-3: Find the interval of convergence of $\sum_{k=1}^{\infty} x^k/(k2^k)$.

Solution: First find the radius of convergence:

$$\lim_{k\to\infty}\left|\frac{a_{k+1}}{a_k}\right| = \lim_{k\to\infty}\left|\frac{\dfrac{1}{(k+1)2^{k+1}}}{\dfrac{1}{k2^k}}\right|$$

$$= \lim_{k\to\infty}\frac{k2^k}{(k+1)2^{k+1}} = \lim_{k\to\infty}\frac{k}{(k+1)2} = \frac{1}{2}$$

Thus the radius of convergence is $R = 1/L = 1/\frac{1}{2} = 2$. So this series converges for $|x| < 2$ and diverges for $|x| > 2$. You must still determine whether or not it converges for $|x| = 2$. For $x = 2$, the series is $\sum_{k=1}^{\infty} 2^k/(k2^k) = \sum_{k=1}^{\infty} 1/k$, which diverges. For $x = -2$, the series is $\sum_{k=1}^{\infty} (-2)^k/(k2^k) = \sum_{k=1}^{\infty} (-1)^k/k$. This is an alternating series whose terms approach zero, so this series converges. You now know for which values of x this series converges: The interval of convergence of this power series is $[-2, 2)$.

EXAMPLE 23-4: Find the interval of convergence of $\sum_{k=2}^{\infty} x^k/(\ln k)^k$.

Solution: To find the radius of convergence, try the root test:

$$\lim_{k\to\infty}|a_k|^{1/k} = \lim_{k\to\infty}\left|\frac{1}{(\ln k)^k}\right|^{1/k} = \lim_{k\to\infty}\frac{1}{\ln k} = 0$$

Because this limit is zero, the radius of convergence is infinity. Thus the interval of convergence for this series is $(-\infty, \infty)$.

EXAMPLE 23-5: Find the interval of convergence of $\sum_{k=1}^{\infty} (x-3)^k/k^2$.

Solution: First find the radius of convergence. Try the ratio test:

$$\lim_{k\to\infty}\left|\frac{a_{k+1}}{a_k}\right| = \lim_{k\to\infty}\left|\frac{\dfrac{1}{(k+1)^2}}{\dfrac{1}{k^2}}\right| = \lim_{k\to\infty}\frac{k^2}{(k+1)^2} = 1$$

Thus the radius of convergence is $R = 1/L = 1/1 = 1$. So this series converges for $|x-3| < 1$, i.e., for $2 < x < 4$, and diverges for $|x-3| > 1$. What happens when $|x-3| = 1$, i.e., when $x = 2$ and $x = 4$? At $x = 4$ the series is $\sum_{k=1}^{\infty} (4-3)^k/k^2 = \sum_{k=1}^{\infty} 1/k^2$, which converges. At $x = 2$ the series becomes $\sum_{k=1}^{\infty} (2-3)^k/k^2 = \sum_{k=1}^{\infty} (-1)^k/k^2$, which converges (absolutely). Thus the interval of convergence is [2, 4].

23-2. Taylor and Maclaurin Polynomials

The **Taylor polynomial** of degree n about the point $x = a$ for the function f is

TAYLOR POLYNOMIAL

$$P_n(x) = f(a) + f'(a)(x-a) + \frac{f''(a)}{2!}(x-a)^2 + \cdots + \frac{f^{(n)}(a)}{n!}(x-a)^n \qquad \textbf{(23-3)}$$

It can be shown that $P_n(x)$ is the polynomial of degree n which best approximates $f(x)$ in the vicinity of $x = a$. If $a = 0$, the polynomial is the **degree n Maclaurin polynomial** for f:

MACLAURIN POLYNOMIAL

$$P_n(x) = f(0) + f'(0)x + \frac{f''(0)}{2!}x^2 + \cdots + \frac{f^{(n)}(0)}{n!}x^n \qquad \textbf{(23-4)}$$

EXAMPLE 23-6: Find the Taylor polynomial of degree three for $f(x) = \sqrt{x}$ about $a = 4$.

Solution: Simply use the formula. You'll need to know $f(4)$, $f'(4)$, $f''(4)$, and $f^{(3)}(4)$:

$$f(x) = \sqrt{x} \qquad f(4) = \sqrt{4} = 2$$

$$f'(x) = \frac{1}{2}x^{-1/2} \qquad f'(4) = \frac{1}{2}(4)^{-1/2} = \frac{1}{4}$$

$$f''(x) = -\frac{1}{4}x^{-3/2} \qquad f''(4) = -\frac{1}{4}(4)^{-3/2} = \frac{-1}{32}$$

$$f^{(3)}(x) = \frac{3}{8}x^{-5/2} \qquad f^{(3)}(4) = \frac{3}{8}(4)^{-5/2} = \frac{3}{256}$$

Now substitute these values into Equation 23-3 for $P_3(x)$:

$$\begin{aligned}P_3(x) &= f(4) + f'(4)(x-4) + \frac{f''(4)}{2!}(x-4)^2 + \frac{f^{(3)}(4)}{3!}(x-4)^3\\ &= 2 + \frac{1}{4}(x-4) + \frac{-1/32}{2}(x-4)^2 + \frac{3/256}{6}(x-4)^3\\ &= 2 + \frac{1}{4}(x-4) - \frac{1}{64}(x-4)^2 + \frac{1}{512}(x-4)^3\end{aligned}$$

EXAMPLE 23-7: Find the fourth-degree Maclaurin polynomial for $f(x) = e^{-2x}$

Solution: You'll need to find derivatives of f, evaluated at $a = 0$:

$$\begin{aligned} f(x) &= e^{-2x} & f(0) &= 1 \\ f'(x) &= -2e^{-2x} & f'(0) &= -2 \\ f''(x) &= 4e^{-2x} & f''(0) &= 4 \\ f^{(3)}(x) &= -8e^{-2x} & f^{(3)}(0) &= -8 \\ f^{(4)}(x) &= 16e^{-2x} & f^{(4)}(0) &= 16 \end{aligned}$$

Now use Equation 23-4:

$$\begin{aligned} P_4(x) &= f(0) + f'(0)x + \frac{f''(0)}{2!}x^2 + \frac{f^{(3)}(0)}{3!}x^3 + \frac{f^{(4)}(0)}{4!}x^4 \\ &= 1 + (-2)x + \frac{4}{2!}x^2 + \frac{-8}{3!}x^3 + \frac{16}{4!}x^4 \\ &= 1 - 2x + 2x^2 - \frac{4}{3}x^3 + \frac{2}{3}x^4 \end{aligned}$$

A. Remainder term and error estimates

A Taylor polynomial approximates f. For this to be useful, you'll need an estimate for the error you've made in this approximation.

Let f be a function with $f, f', \ldots, f^{(n)}$ continuous. If $P_n(x)$ is the degree-n Taylor polynomial for $f(x)$ about $x = a$, then for each x it can be shown that there is a number c between a and x such that

$$f(x) = P_n(x) + R_{n+1}(x) \tag{23-5}$$

where

$$R_{n+1}(x) = \frac{f^{(n+1)}(c)}{(n+1)!}(x-a)^{n+1} \tag{23-6}$$

and R_{n+1} is the **remainder term**. The error in estimating $f(b)$ by Taylor polynomial $P_n(b)$ is bounded by

$$\frac{M_{n+1}}{(n+1)!}|b-a|^{n+1} \tag{23-7}$$

where M_{n+1} is the maximum value of $|f^{(n+1)}(x)|$ on the interval $[a, b]$ (or $[b, a]$ if $b < a$).

EXAMPLE 23-8: Use the third-degree Taylor polynomial for $f(x) = \sqrt{x}$ about $a = 4$ to estimate $\sqrt{4.2}$. Find a bound on the error in this approximation.

Solution: Recall from Example 23-6 that this degree-three Taylor polynomial is

$$P_3(x) = 2 + \frac{1}{4}(x-4) - \frac{1}{64}(x-4)^2 + \frac{1}{512}(x-4)^3$$

Thus you can estimate $\sqrt{4.2}$ by $P_3(4.2)$ because $\sqrt{4.2} = f(4.2) \approx P_3(4.2)$:

$$\begin{aligned} P_3(4.2) &= 2 + \frac{1}{4}(4.2-4) - \frac{1}{64}(4.2-4)^2 + \frac{1}{512}(4.2-4)^3 \\ &= 2 + \frac{0.2}{4} - \frac{(0.2)^2}{64} + \frac{(0.2)^3}{512} \\ &= 2.049\,390\,625 \end{aligned}$$

The error (the difference between this value and the actual value of $\sqrt{4.2}$) is bounded by

$$\frac{M_{n+1}}{(n+1)!}|b-a|^{n+1} = \frac{M_4}{4!}|4.2-4|^4$$

You must find M_4, the maximum value of $|f^{(4)}(x)|$ on [4, 4.2], where

$$f^{(4)}(x) = \frac{-15}{16}x^{-7/2}$$

The maximum value of $|f^{(4)}(x)| = (15/16)x^{-7/2}$ on [4, 4.2] occurs at $x = 4$ (because $x^{-7/2}$ is a decreasing function):

$$M_4 = \frac{15}{16}(4)^{-7/2} = \frac{15}{2048}$$

Thus the error is bounded by

$$\frac{M_4}{4!}|4.2-4|^4 = \frac{15/2048}{24}(0.2)^4 \approx 5 \times 10^{-7}$$

So $\sqrt{4.2}$ is approximately 2.049 390 625, and this is accurate to within 5×10^{-7}.

EXAMPLE 23-9: Find an approximation for $e^{0.1}$ that is accurate to within 10^{-4}.

Solution: The idea is to find a Taylor polynomial for $f(x) = e^x$. You should choose a to be a number *close* to 0.1 at which $f(a), \ldots, f^{(n)}(a)$ can be found because the Taylor approximation is best in the vicinity of a. In this case $a = 0$ is the logical choice—find a Maclaurin polynomial.

Once you've found $P_n(x)$, you have $e^{0.1} = f(0.1) \approx P_n(0.1)$. You must first determine what degree Maclaurin polynomial to use. The level of accuracy dictates the degree n. First see if $n = 2$ will suffice. If $n = 2$, then the error is bounded by

$$\frac{M_3}{3!}|b-a|^3 = \frac{M_3}{6}(0.1-0)^3 = \frac{M_3}{6000}$$

Find M_3, the maximum value of $|f^{(3)}(x)|$ on $[a, b] = [0, 0.1]$:

$$f(x) = e^x \qquad f'(x) = e^x \qquad f''(x) = e^x \qquad f^{(3)}(x) = e^x$$

The maximum value of $|f^{(3)}(x)| = e^x$ on [0, 0.1] is $e^{0.1}$. If you knew the value of $e^{0.1}$, you wouldn't be doing this problem! However, you can at least say something about $e^{0.1}$, for example, $e^{0.1} < 2$. This means that $M_3 < 2$, and the error is bounded by

$$\frac{M_3}{6000} < \frac{2}{6000} = \frac{1}{3000}$$

This isn't sufficiently accurate ($1/3000 > 10^{-4}$). That is, a Maclaurin polynomial of degree two isn't sufficiently accurate. Try $n = 3$. The error is then bounded by

$$\frac{M_4}{4!}|b-a|^4 = \frac{M_4}{24}(0.1-0)^4 = \frac{M_4}{240\,000}$$

The maximum value of $|f^{(4)}(x)| = e^x$ on [0, 0.1] is $e^{0.1} < 2$. So the error is bounded by

$$\frac{M_4}{240\,000} < \frac{2}{240\,000} = \frac{1}{120\,000} < 10^{-4}$$

Thus $n = 3$ will supply ample accuracy. Now find $P_3(x)$:

$$P_3(x) = f(0) + f'(0)x + \frac{f''(0)}{2!}x^2 + \frac{f^{(3)}(0)}{3!}x^3$$

$$= 1 + x + \frac{1}{2}x^2 + \frac{1}{6}x^3$$

Finally you have the approximation:

$$e^{0.1} = f(0.1) \approx P_3(0.1) = 1 + 0.1 + \frac{1}{2}(0.1)^2 + \frac{1}{6}(0.1)^3$$

$$= 1.105\,1\overline{66}$$

23-3. Power Series Expansions of Functions

A. Taylor and Maclaurin expansions

A *power series expansion* of a function $f(x)$ about $x = a$ is a power series in $x - a$ that agrees with $f(x)$ inside the radius of convergence of the power series. It is a power series $\sum_{k=0}^{\infty} a_k(x-a)^k$ that satisfies

$$f(x) = \sum_{k=0}^{\infty} a_k(x-a)^k \quad \text{for} \quad |x - a| < R$$

For example, $\sum_{k=0}^{\infty} x^k = 1/(1-x)$ for $|x| < 1$. Thus $\sum_{k=0}^{\infty} x^k$ is a power series expansion for $f(x) = 1/(1-x)$ about $x = 0$. Don't be concerned that $f(x) = 1/(1-x)$ makes sense for $|x| > 1$ while the series $\sum_{k=0}^{\infty} x^k$ doesn't.

Given a function $f(x)$, how do you find a power series expansion $\sum_{k=0}^{\infty} a_k(x-a)^k$ for $f(x)$? You can find the **Taylor series expansion**—a Taylor polynomial of infinite degree:

TAYLOR SERIES EXPANSION

$$f(x) = f(a) + f'(a)(x-a) + \frac{f''(a)}{2!}(x-a)^2 + \cdots + \frac{f^{(n)}(a)}{n!}(x-a)^n + \cdots$$

$$= \sum_{k=0}^{\infty} \frac{f^{(k)}(a)}{k!}(x-a)^k \quad \text{for} \quad |x-a| < R \qquad \textbf{(23-8)}$$

EXAMPLE 23-10: Find a Maclaurin series expansion for $f(x) = e^x$.

Solution: A **Maclaurin series expansion** is a Taylor series expansion about $a = 0$:

$$\sum_{k=0}^{\infty} \frac{f^{(k)}(0)}{k!} x^k$$

You'll need a general formula for $f^{(k)}(0)$:

$$\begin{aligned} f(x) &= e^x & f(0) &= e^0 = 1 \\ f'(x) &= e^x & f'(0) &= e^0 = 1 \\ &\vdots & &\vdots \\ f^{(n)}(x) &= e^x & f^{(n)}(0) &= e^0 = 1 \end{aligned}$$

So the Maclaurin series for $f(x) = e^x$ is

$$\sum_{k=0}^{\infty} \frac{f^{(k)}(0)}{k!} x^k = \sum_{k=0}^{\infty} \frac{1}{k!} x^k$$

To find the values of x for which this power series converges, find the radius of convergence. Use the ratio test:

$$\lim_{k\to\infty} \left| \frac{a_{k+1}}{a_k} \right| = \lim_{k\to\infty} \left| \frac{\frac{1}{(k+1)!}}{\frac{1}{k!}} \right| = \lim_{k\to\infty} \frac{k!}{(k+1)!}$$

$$= \lim_{k\to\infty} \frac{1}{k+1} = 0$$

You conclude that the radius of convergence is infinity, so $f(x) = \sum_{k=0}^{\infty} (1/k!)\, x^k$ for all x.

B. Manipulating power series

If $f(x) = \sum_{k=0}^{\infty} a_k(x - a)^k$ for $|x - a| < R$, then

$$f'(x) = \sum_{k=0}^{\infty} ka_k(x - a)^{k-1} \quad \text{for} \quad |x - a| < R$$

and

$$\int f(x)\,dx = \sum_{k=0}^{\infty} \frac{a_k(x - a)^{k+1}}{k + 1} + C \quad \text{for} \quad |x - a| < R$$

In other words, you can integrate and differentiate power series term by term inside the radius of convergence.

EXAMPLE 23-11: Find a Maclaurin series for $f(x) = \ln(1 - x)$.

Solution: First recall the Maclaurin series for $1/(1 - x)$:

$$\frac{1}{1 - x} = \sum_{k=0}^{\infty} x^k \quad \text{for} \quad |x| < 1$$

Now integrate:

$$\int \frac{1}{1 - x}\,dx = \int \sum_{k=0}^{\infty} x^k\,dx$$

$$-\ln|1 - x| = \sum_{k=0}^{\infty} \frac{x^{k+1}}{k + 1} + C \quad \text{for} \quad |x| < 1$$

So for $|x| < 1$ you have

$$\ln(1 - x) = \sum_{k=0}^{\infty} \left(\frac{-1}{k + 1}\right) x^{k+1} + C = -x - \frac{x^2}{2} - \frac{x^3}{3} - \cdots + C$$

$$= \sum_{k=1}^{\infty} \left(\frac{-1}{k}\right) x^k + C$$

To find the value of C, substitute a value of x for which you can easily evaluate both sides, say $x = 0$:

$$0 = \ln(1 - 0) = \sum_{k=1}^{\infty} \left(\frac{-1}{k}\right) 0^k + C = C$$

You now have the Maclaurin series for $f(x) = \ln(1 - x)$:

$$\ln(1 - x) = \sum_{k=1}^{\infty} \left(\frac{-1}{k}\right) x^k \quad \text{for} \quad |x| < 1$$

You could also have found this series from the formula for Maclaurin polynomials

$$\sum_{k=0}^{\infty} \frac{f^{(k)}(0)}{k!} x^k,$$

but the preceding method seems easier than finding all those derivatives.

You may find the following series useful in finding expansions for other functions:

$$e^x = 1 + x + \frac{x^2}{2!} + \frac{x^3}{3!} + \cdots \quad \text{for all} \quad x \qquad \textbf{(23-9)}$$

$$\frac{1}{1 - x} = 1 + x + x^2 + x^3 + \cdots \quad \text{for all} \quad |x| < 1 \qquad \textbf{(23-10)}$$

$$\sin x = x - \frac{x^3}{3!} + \frac{x^5}{5!} - \frac{x^7}{7!} + \cdots \quad \text{for all} \quad x \qquad \textbf{(23-11)}$$

$$\cos x = 1 - \frac{x^2}{2!} + \frac{x^4}{4!} - \frac{x^6}{6!} + \cdots \quad \text{for all} \quad x \qquad \textbf{(23-12)}$$

$$\ln(1 + x) = x - \frac{x^2}{2} + \frac{x^3}{3} - \frac{x^4}{4} + \cdots \quad \text{for all} \quad |x| < 1 \qquad \textbf{(23-13)}$$

You can find these expansions for yourself using the formula

$$\sum_{k=0}^{\infty} \frac{f^{(k)}(a)}{k!}(x - a)^k$$

You can use these series to find power series expansions for other functions.

EXAMPLE 23-12: Find a Maclaurin series for $f(x) = (\sin x)/x$.

Solution: The series for $\sin x$ is

$$\sin x = x - \frac{x^3}{3!} + \frac{x^5}{5!} - \frac{x^7}{7!} + \cdots \quad \text{for all} \quad x$$

So the series for $(\sin x)/x$ is

$$\frac{\sin x}{x} = \frac{1}{x}\left(x - \frac{x^3}{3!} + \frac{x^5}{5!} - \frac{x^7}{7!} + \cdots\right) = 1 - \frac{x^2}{3!} + \frac{x^4}{5!} - \frac{x^6}{7!} + \cdots \quad \text{for all } x$$

Of course $(\sin x)/x$ is not defined at $x = 0$, but

$$\lim_{x \to 0} (\sin x)/x = 1 = [1 - (x^2/3!) + (x^4/5!) - \cdots]\big|_{x=0}$$

EXAMPLE 23-13: Find a Maclaurin series for $f(x) = \arctan x$.

Solution: You can either find a formula for $f^{(k)}(0)$ (which is difficult) or manipulate series that you already have (which is much easier):

$$\frac{1}{1 - x} = 1 + x + x^2 + x^3 + \cdots \quad \text{for} \quad |x| < 1$$

$$\frac{1}{1 + x^2} = \frac{1}{1 - (-x^2)} = 1 + (-x^2) + (-x^2)^2 + (-x^2)^3 + \cdots$$

$$= 1 - x^2 + x^4 - x^6 + \cdots \quad \text{for} \quad |-x^2| < 1$$

$$\text{(i.e.,} \quad \text{for} \quad |x| < 1)$$

$$\int \frac{1}{1 + x^2}\,dx = \int (1 - x^2 + x^4 - x^6 + \cdots)\,dx$$

$$\arctan x = x - \frac{x^3}{3} + \frac{x^5}{5} - \frac{x^7}{7} + \cdots + C \quad \text{for} \quad |x| < 1$$

To find C, plug in $x = 0$:

$$0 = \arctan 0 = 0 - \frac{0^3}{3} + \frac{0^5}{5} - \cdots + C = C$$

So you have

$$\arctan x = x - \frac{x^3}{3} + \frac{x^5}{5} - \frac{x^7}{7} + \cdots \quad \text{for} \quad |x| < 1$$

23-4. Estimation with Power Series

You can estimate $f(b)$ by truncating the power series for $f(x)$ and evaluating at $x = b$. This is nothing more than a Taylor polynomial approximation because a truncated Taylor series is a Taylor polynomial.

EXAMPLE 23-14: Estimate the sine of $1/10$ radian using the first two nonzero terms of the Maclaurin series for $\sin x$.

Solution: Use the Maclaurin series for $f(x) = \sin x$:

$$\sin x = x - \frac{x^3}{3!} + \frac{x^5}{5!} - \cdots$$

$$\sin\frac{1}{10} = \frac{1}{10} - \frac{(1/10)^3}{3!} + \frac{(1/10)^5}{5!} - \cdots$$

$$\approx \frac{1}{10} - \frac{(1/10)^3}{3!} = \frac{1}{10} - \frac{1}{6000} = \frac{599}{6000} = 0.099\,8\overline{33}$$

To find the accuracy in a power series approximation, you can use Equation 23-7. There is another handy formula for error estimates of alternating series: If $\sum_{k=0}^{\infty} b_k$ is an alternating series with $|b_k| > |b_{k+1}|$ for all k, then the error in the approximation

$$\sum_{k=0}^{\infty} b_k \approx \sum_{k=0}^{n} b_k$$

is bounded by $|b_{n+1}|$.

EXAMPLE 23-15: Find a bound on the error in the approximation for $\sin(1/10)$ from Example 23-14.

Solution: The series for $\sin(1/10)$ is an alternating series:

$$\sin\frac{1}{10} = \frac{1}{10} - \frac{(-1/10)^3}{3!} + \frac{(-1/10)^5}{5!} - \cdots$$

If you estimate $\sin(1/10)$ by truncating after two terms,

$$\sin\frac{1}{10} \approx \frac{1}{10} - \frac{(1/10)^3}{3!} = 0.099\,833$$

then the error is bounded by the absolute value of the next nonzero term:

$$\left|\frac{(-1/10)^5}{5!}\right| = \frac{1}{12\,000\,000} = 8.\overline{33} \times 10^{-8}$$

So the approximation $\sin(1/10) \approx 0.099\,8\overline{33}$ is accurate to within $8.\overline{33} \times 10^{-8}$.

EXAMPLE 23-16: Find an estimate for $\int_0^1 (1 - \cos x)/x\,dx$ that is accurate to within 10^{-5}.

Solution: Find a series for $f(x) = (1 - \cos x)/x$:

$$\cos x = 1 - \frac{x^2}{2!} + \frac{x^4}{4!} - \frac{x^6}{6!} + \cdots$$

$$1 - \cos x = \frac{x^2}{2!} - \frac{x^4}{4!} + \frac{x^6}{6!} - \cdots$$

$$\frac{1 - \cos x}{x} = \frac{x}{2!} - \frac{x^3}{4!} + \frac{x^5}{6!} - \cdots$$

Now integrate. Although $\int_0^1 (1 - \cos x)/x\,dx$ is improper (because $(1 - \cos x)/x$ is discontinuous at $x = 0$), the integrand doesn't have a vertical asymptote at $x = 0$, so you can integrate without a limit ($\lim_{\varepsilon\to 0^+} \int_\varepsilon^1$):

$$\int_0^1 \frac{1 - \cos x}{x}\,dx = \int_0^1 \left(\frac{x}{2!} - \frac{x^3}{4!} + \frac{x^5}{6!} - \cdots\right) dx$$

$$= \left(\frac{x^2}{2(2!)} - \frac{x^4}{4(4!)} + \frac{x^6}{6(6!)} - \cdots\right)\Bigg|_0^1$$

$$= \frac{1}{2(2!)} - \frac{1}{4(4!)} + \frac{1}{6(6!)} - \frac{1}{8(8!)} + \cdots$$

This is an alternating series with decreasing terms, so if you truncate after three terms,

$$\int_0^1 \frac{1-\cos x}{x}\,dx \approx \frac{1}{2(2!)} - \frac{1}{4(4!)} + \frac{1}{6(6!)}$$
$$= 0.239\,81\overline{48}$$

A bound on the error is supplied by the absolute value of the next (the fourth) nonzero term:

$$\left|\frac{-1}{8(8!)}\right| = \frac{1}{322\,560} < 10^{-5}$$

SUMMARY

1. To find the interval of convergence of $\sum_{k=0}^{\infty} a_k(x-a)^k$, find the radius of convergence R and then examine the power series at the two values of x for which $|x-a| = R$.
2. The Taylor polynomial of degree n about $x = a$ for $f(x)$ is the polynomial of degree n that best approximates $f(x)$ for x close to a.
3. When you estimate the value of a function using a Taylor polynomial, you can also find a bound on the error.
4. The Taylor expansion about $x = a$ for a function f is a power series in $x - a$ that has the same value as $f(x)$ for x inside the radius of convergence. You can find this power series from the formula
$$\sum_{k=0}^{\infty} \frac{f^{(k)}(a)}{k!}(x-a)^k$$
5. You can sometimes find a power series expansion for f by manipulating known power series.
6. The error you incur by truncating an alternating series whose terms decrease in absolute value is bounded by the absolute value of the first term following the truncation.

SOLVED PROBLEMS

PROBLEM 23-1 Find the interval of convergence of $\sum_{k=1}^{\infty} x^k/(\sqrt{k}\,4^k)$.

Solution: First find the radius of convergence. Try the ratio test:

$$\lim_{k\to\infty}\left|\frac{a_{k+1}}{a_k}\right| = \lim_{k\to\infty}\left|\frac{\dfrac{1}{\sqrt{k+1}\,4^{k+1}}}{\dfrac{1}{\sqrt{k}\,4^k}}\right|$$
$$= \lim_{k\to\infty}\frac{\sqrt{k}\,4^k}{\sqrt{k+1}\,4^{k+1}} = \lim_{k\to\infty}\frac{\sqrt{k}}{\sqrt{k+1}\,4} = \frac{1}{4} = L$$

The radius of convergence is $R = 1/L = 4$. This series converges for $|x| < 4$ and diverges for $|x| > 4$. You must determine whether or not the series converges for $|x| = 4$. When $x = 4$, the series is $\sum_{k=1}^{\infty} 4^k/(\sqrt{k}\, 4^k) = \sum_{k=1}^{\infty} 1/\sqrt{k}$, which diverges. When $x = -4$, the series is $\sum_{k=1}^{\infty} (-4)^k/(\sqrt{k}\, 4^k) = \sum_{k=1}^{\infty} (-1)^k/\sqrt{k}$. This is an alternating series whose terms decrease in absolute value and approach zero, so this series converges. Putting it all together, the interval of convergence of $\sum_{k=1}^{\infty} x^k/(\sqrt{k}\, 4^k)$ is $[-4, 4)$. [See Section 23-1.]

PROBLEM 23-2 Find the interval of convergence of $\sum_{k=1}^{\infty} (k^k/2^k)\, x^k$.

Solution: To find the radius of convergence, try the root test:

$$\lim_{k\to\infty} |a_k|^{1/k} = \lim_{k\to\infty} \left|\frac{k^k}{2^k}\right|^{1/k} = \lim_{k\to\infty} \frac{k}{2} = \infty$$

Because the limit is infinity, the radius of convergence is $R = 0$. Thus the only value of x for which the series converges is $x = 0$. [See Section 23-1.]

PROBLEM 23-3 Find the interval of convergence of $\sum_{k=0}^{\infty} \sqrt{k}(x - 5)^k$.

Solution: Find the radius of convergence using the ratio test:

$$\lim_{k\to\infty} \left|\frac{a_{k+1}}{a_k}\right| = \lim_{k\to\infty} \frac{\sqrt{k+1}}{\sqrt{k}} = 1$$

Thus the radius of convergence is $R = 1/L = 1$. This series converges for $|x - 5| < 1$ and diverges for $|x - 5| > 1$. What happens when $|x - 5| = 1$? When $x = 6$, the series is $\sum_{k=0}^{\infty} \sqrt{k}(6 - 5)^k = \sum_{k=0}^{\infty} \sqrt{k}$, which diverges. When $x = 4$, the series is $\sum_{k=0}^{\infty} \sqrt{k}(4 - 5)^k = \sum_{k=0}^{\infty} (-1)^k \sqrt{k}$, which diverges. The interval of convergence for $\sum_{k=0}^{\infty} \sqrt{k}(x - 5)^k$ is $(4, 6)$. [See Section 23-1.]

PROBLEM 23-4 Find the interval of convergence of $\sum_{k=0}^{\infty} (2x + 1)^k/3^k$.

Solution: You must first recognize this as a series of the form $\sum_{k=0}^{\infty} a_k(x - a)^k$ so you'll know what a_k and a are:

$$\sum_{k=0}^{\infty} \frac{(2x+1)^k}{3^k} = \sum_{k=0}^{\infty} \frac{2^k(x + \frac{1}{2})^k}{3^k}$$

Now use the root test to find R:

$$\lim_{k\to\infty} |a_k|^{1/k} = \lim_{k\to\infty} \left|\frac{2^k}{3^k}\right|^{1/k} = \lim_{k\to\infty} \frac{2}{3} = \frac{2}{3}$$

So the radius of convergence is 3/2 and the series converges for $|x - (-\frac{1}{2})| < 3/2$. What happens on the radius of convergence (i.e., at $x = -2$ and $x = 1$)? At $x = -2$ the series is $\sum_{k=0}^{\infty} 2^k(-2 + \frac{1}{2})^k/3^k = \sum_{k=0}^{\infty} (-1)^k$, which diverges. At $x = 1$ the series is $\sum_{k=0}^{\infty} 2^k(1 + \frac{1}{2})^k/3^k = \sum_{k=0}^{\infty} 1$, which diverges. The interval of convergence is $(-2, 1)$. [See Section 23-1.]

PROBLEM 23-5 Find the interval of convergence of $\sum_{k=3}^{\infty} (-1)^k(x + 1)^k/(k\sqrt{\ln k})$.

Solution: Find R using the ratio test:

$$\lim_{k\to\infty} \left[\frac{\dfrac{1}{(k+1)\sqrt{\ln(k+1)}}}{\dfrac{1}{k\sqrt{\ln k}}}\right] = \lim_{k\to\infty} \frac{k\sqrt{\ln k}}{(k+1)\sqrt{\ln(k+1)}} = 1$$

The radius of convergence is one. The series converges for $|x + 1| < 1$. At $x = 0$ the series is $\sum_{k=3}^{\infty} (-1)^k/(k\sqrt{\ln k})$, which converges because it is alternating and the terms approach zero. At $x = -2$ the series is $\sum_{k=3}^{\infty} 1/(k\sqrt{\ln k})$. This series diverges by the integral test, so the interval of convergence is $(-2, 0]$. [See Section 23-1.]

PROBLEM 23-6 Find the degree-two Maclaurin polynomial for $f(x) = \sqrt{1 - x}$.

Solution: The Maclaurin polynomial is a Taylor polynomial with $a = 0$, so you'll need to find $f(0)$, $f'(0)$, and $f''(0)$:

$$f(x) = \sqrt{1-x} \qquad f(0) = 1$$
$$f'(x) = -\tfrac{1}{2}(1-x)^{-1/2} \qquad f'(0) = -\tfrac{1}{2}$$
$$f''(x) = -\tfrac{1}{4}(1-x)^{-3/2} \qquad f''(0) = -\tfrac{1}{4}$$

Now plug into the formula:

$$P_2(x) = f(0) + f'(0)x + \frac{f''(0)}{2!}x^2$$
$$= 1 + \left(-\frac{1}{2}\right)x + \frac{-\frac{1}{4}}{2}x^2$$
$$= 1 - \frac{1}{2}x - \frac{1}{8}x^2$$

[See Section 23-2.]

PROBLEM 23-7 Find the degree-three Taylor polynomial about $a = \pi/3$ for $f(x) = \cos x$.

Solution: You'll need to find $f(\pi/3)$, $f'(\pi/3)$, $f''(\pi/3)$, and $f^{(3)}(\pi/3)$:

$$f(x) = \cos x \qquad f\left(\frac{\pi}{3}\right) = \cos\frac{\pi}{3} = \frac{1}{2}$$
$$f'(x) = -\sin x \qquad f'\left(\frac{\pi}{3}\right) = -\sin\frac{\pi}{3} = \frac{-\sqrt{3}}{2}$$
$$f''(x) = -\cos x \qquad f''\left(\frac{\pi}{3}\right) = -\frac{1}{2}$$
$$f^{(3)}(x) = \sin x \qquad f^{(3)}\left(\frac{\pi}{3}\right) = \frac{\sqrt{3}}{2}$$

$$P_3(x) = f\left(\frac{\pi}{3}\right) + f'\left(\frac{\pi}{3}\right)\left(x - \frac{\pi}{3}\right) + \frac{f''\left(\frac{\pi}{3}\right)}{2!}\left(x - \frac{\pi}{3}\right)^2 + \frac{f^{(3)}\left(\frac{\pi}{3}\right)}{3!}\left(x - \frac{\pi}{3}\right)^3$$
$$= \frac{1}{2} + \frac{-\sqrt{3}}{2}\left(x - \frac{\pi}{3}\right) + \frac{-\frac{1}{2}}{2!}\left(x - \frac{\pi}{3}\right)^2 + \frac{\sqrt{3}/2}{3!}\left(x - \frac{\pi}{3}\right)^3$$
$$= \frac{1}{2} - \frac{\sqrt{3}}{2}\left(x - \frac{\pi}{3}\right) - \frac{1}{4}\left(x - \frac{\pi}{3}\right)^2 + \frac{\sqrt{3}}{12}\left(x - \frac{\pi}{3}\right)^3$$

[See Section 23-2.]

PROBLEM 23-8 Find the degree-three Maclaurin polynomial for $f(x) = \arcsin x$.

Solution: Find $f(0)$, $f'(0)$, $f''(0)$, and $f^{(3)}(0)$:

$$f(x) = \arcsin x \qquad f(0) = 0$$
$$f'(x) = (1-x^2)^{-1/2} \qquad f'(0) = 1$$
$$f''(x) = x(1-x^2)^{-3/2} \qquad f''(0) = 0$$
$$f^{(3)}(x) = (1-x^2)^{-3/2} + 3x^2(1-x^2)^{-5/2} \qquad f^{(3)}(0) = 1$$

$$P_3(x) = f(0) + f'(0)x + \frac{f''(0)}{2!}x^2 + \frac{f^{(3)}(0)}{3!}x^3$$
$$= 0 + 1x + \frac{0}{2!}x^2 + \frac{1}{3!}x^3 = x + \frac{1}{6}x^3$$

[See Section 23-2.]

PROBLEM 23-9 Find the degree-four Taylor polynomial about $a = \pi/4$ for $f(x) = -\ln \cos x$.

Solution: Find $f(\pi/4), \ldots, f^{(4)}(\pi/4)$:

$$f(x) = -\ln\cos x \qquad f(\pi/4) = \tfrac{1}{2}\ln 2$$

$$f'(x) = \tan x \qquad f'(\pi/4) = 1$$

$$f''(x) = \sec^2 x \qquad f''(\pi/4) = 2$$

$$f^{(3)}(x) = 2\sec^2 x \tan x \qquad f^{(3)}(\pi/4) = 4$$

$$f^{(4)}(x) = 4\sec^2 x \tan^2 x + 2\sec^4 x \qquad f^{(4)}(\pi/4) = 16$$

$$P_4(x) = \frac{1}{2}\ln 2 + 1\left(x - \frac{\pi}{4}\right) + \frac{2}{2!}\left(x - \frac{\pi}{4}\right)^2 + \frac{4}{3!}\left(x - \frac{\pi}{4}\right)^3 + \frac{16}{4!}\left(x - \frac{\pi}{4}\right)^4$$

$$= \frac{1}{2}\ln 2 + \left(x - \frac{\pi}{4}\right) + \left(x - \frac{\pi}{4}\right)^2 + \frac{2}{3}\left(x - \frac{\pi}{4}\right)^3 + \frac{2}{3}\left(x - \frac{\pi}{4}\right)^4$$

[See Section 23-2.]

PROBLEM 23-10 Use the degree-two Maclaurin polynomial for $f(x) = \sqrt{1 - x}$ to estimate $\sqrt{0.9}$ and find a bound on the accuracy of this estimate.

Solution: Recall from Problem 23-6:

$$P_2(x) = 1 - \frac{1}{2}x - \frac{1}{8}x^2$$

You have

$$\sqrt{0.9} = \sqrt{1 - 0.1} = f(0.1) \approx P_2(0.1) = 1 - \frac{1}{2}(0.1) - \frac{1}{8}(0.1)^2 = 0.948\,75$$

The error in this estimate is bounded by

$$\frac{M_3}{3!}|b - a|^3 = \frac{M_3}{6}|0.1 - 0|^3 = \frac{M_3}{6000}$$

M_3 is the maximum value of $|f^{(3)}(x)|$ on $[0, 0.1]$:

$$|f^{(3)}(x)| = \frac{3}{8}(1 - x)^{-5/2}$$

The maximum value of $|f^{(3)}(x)|$ on $[0, 0.1]$ is $M_3 = (3/8)(1 - 0.1)^{-5/2} = 3/[8(0.9)^{5/2}] < 0.5$
So the error estimate is

$$\frac{M_3}{6000} < \frac{0.5}{6000} = \frac{1}{12\,000} = 8.\overline{33} \times 10^{-5}$$

The estimate $\sqrt{0.9} \approx 0.948\,75$ is accurate to within $8.\overline{33} \times 10^{-5}$. [See Section 23-2.]

PROBLEM 23-11 Use the third-degree Taylor polynomial about $a = \pi/3$ for $f(x) = \cos x$ to estimate $\cos[(\pi/3) + 0.2]$ and find a bound on the accuracy of this estimate.

Solution: In Problem 23-7 you found $P_3(x)$:

$$P_3(x) = \frac{1}{2} - \frac{\sqrt{3}}{2}\left(x - \frac{\pi}{3}\right) - \frac{1}{4}\left(x - \frac{\pi}{3}\right)^2 + \frac{\sqrt{3}}{12}\left(x - \frac{\pi}{3}\right)^3$$

$$\cos\left(\frac{\pi}{3} + 0.2\right) = f\left(\frac{\pi}{3} + 0.2\right) \approx P_3\left(\frac{\pi}{3} + 0.2\right)$$

$$= \frac{1}{2} - \frac{\sqrt{3}}{2}(0.2) - \frac{1}{4}(0.2)^2 + \frac{\sqrt{3}}{12}(0.2)^3 = 0.317\,9496\ldots$$

The error in this estimate is bounded by

$$\frac{M_4}{4!}|b - a|^4 = \frac{M_4}{24}\left|\frac{\pi}{3} + 0.2 - \frac{\pi}{3}\right|^4 = \frac{M_4}{15\,000}$$

M_4 is the maximum value of $|f^{(4)}(x)| = |\cos x|$ on $[\pi/3, (\pi/3) + 0.2]$. This maximum value occurs at $x = \pi/3$,

$$M_4 = \left|f^{(4)}\left(\frac{\pi}{3}\right)\right| = \frac{1}{2}$$

so the error is bounded by

$$\frac{\frac{1}{2}}{15\,000} = 3.\overline{33} \times 10^{-5}$$

[See Section 23-2.]

PROBLEM 23-12 Find an estimate for $\tan[(\pi/4) - 0.1]$ that is accurate to within 10^{-2}.

Solution: The idea is to find a Taylor polynomial for $f(x) = \tan x$. Use $a = \frac{1}{4}\pi$ because $\frac{1}{4}\pi$ is close to $\frac{1}{4}\pi - 0.1$, and you know how to evaluate the trigonometric functions at $\frac{1}{4}\pi$. The degree of accuracy will determine the degree of the Taylor polynomial. See if $n = 2$ will give sufficient accuracy. The error will be bounded by

$$\frac{M_3}{3!}\left|\frac{\pi}{4} - 0.1 - \frac{\pi}{4}\right|^3 = \frac{M_3}{6000}$$

M_3 is the maximum value of $|f^{(3)}(x)|$ on $[(\pi/4) - 0.1, \pi/4]$.

$$f(x) = \tan x$$

$$f'(x) = \sec^2 x$$

$$f''(x) = 2\sec^2 x \tan x$$

$$f^{(3)}(x) = 4\sec^2 x \tan^2 x + 2\sec^4 x$$

The maximum value of $|f^{(3)}(x)|$ on $[(\pi/4) - 0.1, \pi/4]$ occurs at $x = \pi/4$:

$$M_4 = 4\sec^2\frac{\pi}{4}\tan^2\frac{\pi}{4} + 2\sec^4\frac{\pi}{4} = 16$$

When $n = 2$, the error is bounded by

$$\frac{M_3}{6000} = \frac{16}{6000} = \frac{8}{3000} = 2.\overline{66} \times 10^{-3}$$

So $n = 2$ is good enough! Find $P_2(x)$:

$$f\left(\frac{\pi}{4}\right) = \tan\frac{\pi}{4} = 1$$

$$f'\left(\frac{\pi}{4}\right) = \sec^2\frac{\pi}{4} = 2$$

$$f''\left(\frac{\pi}{4}\right) = 2\sec^2\frac{\pi}{4}\tan\frac{\pi}{4} = 4$$

$$P_2(x) = f\left(\frac{\pi}{4}\right) + f'\left(\frac{\pi}{4}\right)\left(x - \frac{\pi}{4}\right) + \frac{f''(\pi/4)}{2!}\left(x - \frac{\pi}{4}\right)^2$$

$$= 1 + 2\left(x - \frac{\pi}{4}\right) + 2\left(x - \frac{\pi}{4}\right)^2$$

$$\tan\left(\frac{\pi}{4} - 0.1\right) = f\left(\frac{\pi}{4} - 0.1\right) \approx P_2\left(\frac{\pi}{4} - 0.1\right)$$

$$= 1 + 2\left(\frac{\pi}{4} - 0.1 - \frac{\pi}{4}\right) + 2\left(\frac{\pi}{4} - 0.2 - \frac{\pi}{4}\right)^2$$

$$= 1 - 2(0.1) + 2(0.1)^2 = 0.82$$

[See Section 23-2.]

PROBLEM 23-13 Find a Taylor series for $f(x) = \ln x$ about $a = 1$.

Solution: The series will be $\sum_{k=0}^{\infty} (f^{(k)}(1)/k!)(x-1)^k$, so you'll need a formula for $f^{(k)}(1)$:

$$\begin{aligned}
f(x) &= \ln x & f(1) &= \ln 1 = 0 \\
f'(x) &= \frac{1}{x} = x^{-1} & f'(1) &= 1 \\
f''(x) &= -x^{-2} & f''(1) &= -1 \\
f^{(3)}(x) &= 2x^{-3} & f^{(3)}(1) &= 2 \\
f^{(4)}(x) &= -6x^{-4} & f^{(4)}(1) &= -6 \\
&\vdots & &\vdots \\
f^{(k)}(x) &= (-1)^{k+1}(k-1)!\,x^{-k} & f^{(k)}(1) &= (-1)^{k+1}(k-1)!
\end{aligned}$$

So the series is

$$\begin{aligned}
\ln x &= 0 + (x-1) + \frac{-1}{2!}(x-1)^2 + \frac{2}{3!}(x-1)^3 + \cdots + \frac{(-1)^{k+1}(k-1)!}{k!}(x-1)^k + \cdots \\
&= (x-1) - \frac{1}{2}(x-1)^2 + \frac{1}{3}(x-1)^3 + \cdots + \frac{(-1)^k}{k}(x-1)^k + \cdots \\
&= \sum_{k=1}^{\infty} \frac{(-1)^{k+1}}{k}(x-1)^k
\end{aligned}$$

[See Section 23-3.]

PROBLEM 23-14 Find a Taylor series about $x = \ln 2$ for $\cosh x$.

Solution: The series will be $\sum_{k=0}^{\infty} \frac{f^{(k)}(\ln 2)}{k!}(x-\ln 2)^k$, so you'll need a formula for $f^{(k)}(\ln 2)$:

$$\begin{aligned}
f(x) &= \cosh x & f(\ln 2) &= \cosh(\ln 2) = \frac{e^{\ln 2} + e^{-\ln 2}}{2} = \frac{5}{4} \\
f'(x) &= \sinh x & f'(\ln 2) &= \sinh(\ln 2) = \frac{e^{\ln 2} - e^{-\ln 2}}{2} = \frac{3}{4} \\
f''(x) &= \cosh x & f''(\ln 2) &= \frac{5}{4} \\
&\vdots & &\vdots
\end{aligned}$$

The series is

$$\begin{aligned}
\cosh x &= f(\ln 2) + f'(\ln 2)(x - \ln 2) + \frac{f''(\ln 2)}{2!}(x-\ln 2)^2 + \cdots \\
&= \frac{5}{4} + \frac{3}{4}(x - \ln 2) + \frac{5}{4(2!)}(x-\ln 2)^2 + \frac{3}{4(3!)}(x - \ln 2)^3 + \cdots
\end{aligned}$$

[See Section 23-3.]

PROBLEM 23-15 Find the Maclaurin series for $f(x) = xe^{x^2}$.

Solution: Rather than find a formula for $f^{(k)}(0)$ (which seems quite difficult), you can manipulate the series for e^x:

$$\begin{aligned}
e^x &= 1 + x + \frac{x^2}{2!} + \frac{x^3}{3!} + \frac{x^4}{4!} + \cdots \\
e^{x^2} &= 1 + x^2 + \frac{(x^2)^2}{2!} + \frac{(x^2)^3}{3!} + \frac{(x^2)^4}{4!} + \cdots \\
&= 1 + x^2 + \frac{x^4}{2!} + \frac{x^6}{3!} + \frac{x^8}{4!} + \cdots \\
xe^{x^2} &= x + x^3 + \frac{x^5}{2!} + \frac{x^7}{3!} + \frac{x^9}{4!} + \cdots = \sum_{k=0}^{\infty} \frac{x^{2k+1}}{k!} \quad \text{for all } x
\end{aligned}$$

[See Section 23-3.]

PROBLEM 23-16 Find a Maclaurin series for cosh x.

Solution: Recall that $\cosh x = (e^x + e^{-x})/2$:

$$e^x = 1 + x + \frac{x^2}{2!} + \frac{x^3}{3!} + \frac{x^4}{4!} + \cdots \quad \text{for all} \quad x$$

$$e^{-x} = 1 + (-x) + \frac{(-x)^2}{2!} + \frac{(-x)^3}{3!} + \frac{(-x)^4}{4!} + \cdots$$

$$= 1 - x + \frac{x^2}{2!} - \frac{x^3}{3!} + \frac{x^4}{4!} - \cdots$$

$$e^x + e^{-x} = 2 + 2\left(\frac{x^2}{2!}\right) + 2\left(\frac{x^4}{4!}\right) + \cdots$$

$$\cosh x = \frac{e^x + e^{-x}}{2} = 1 + \frac{x^2}{2!} + \frac{x^4}{4!} + \frac{x^6}{6!} + \cdots \quad \text{for all} \quad x$$

[See Section 23-3.]

PROBLEM 23-17 Use the first four nonzero terms of the Maclaurin series for $\ln(1 + x)$ to estimate $\ln(1.1)$ and find a bound on the accuracy of this estimate.

Solution: The Maclaurin series for $\ln(1 + x)$ is

$$\ln(1 + x) = x - \frac{x^2}{2} + \frac{x^3}{3} - \frac{x^4}{4} + \cdots$$

$$\ln(1.1) = \ln(1 + 0.1) = 0.1 - \frac{(0.1)^2}{2} + \frac{(0.1)^3}{3} - \frac{(0.1)^4}{4} + \cdots$$

$$\approx 0.1 - \frac{(0.1)^2}{2} + \frac{(0.1)^3}{3} - \frac{(0.1)^4}{4} = 0.095\,308\,\overline{33}$$

The series for $\ln(1.1)$ is alternating, with the terms decreasing in absolute value, so the error in truncating after four terms is bounded by the absolute value of the fifth term:

$$\left|\frac{(0.1)^5}{5}\right| = 2 \times 10^{-6}$$

[See Section 23-4.]

PROBLEM 23-18 Find an estimate for e^{-1} that is accurate to within 10^{-2}.

Solution: Use the Maclaurin series for e^x:

$$e^x = 1 + x + \frac{x^2}{2!} + \frac{x^3}{3!} + \frac{x^4}{4!} + \cdots$$

$$e^{-1} = 1 - 1 + \frac{(-1)^2}{2!} + \frac{(-1)^3}{3!} + \frac{(-1)^4}{4!} + \frac{(-1)^5}{5!} + \cdots$$

$$= \frac{1}{2!} - \frac{1}{3!} + \frac{1}{4!} - \frac{1}{5!} + \cdots$$

You can estimate e^{-1} by truncating this series. Just be sure to truncate at a term where the error is less than 10^{-2}. That is, the absolute value of the next term must be less than 10^{-2}. What is the first term whose absolute value is less than 10^{-2}?

$$\frac{1}{5!} = \frac{1}{120} < 10^{-2}$$

So you obtain the necessary accuracy by truncating after the third term:

$$e^{-1} \approx \frac{1}{2!} - \frac{1}{3!} + \frac{1}{4!} = 0.375$$

[See Section 23-4.]

PROBLEM 23-19 Find an estimate for $\int_0^1 e^{-x^2}\,dx$ that is accurate to within 10^{-3}.

Solution: Find a Maclaurin series for e^{-x^2}:

$$e^x = 1 + x + \frac{x^2}{2!} + \frac{x^3}{3!} + \cdots$$

$$e^{-x^2} = 1 + (-x^2) + \frac{(-x^2)^2}{2!} + \frac{(-x^2)^3}{3!} + \frac{(-x^2)^4}{4!} + \cdots$$

$$= 1 - x^2 + \frac{x^4}{2!} - \frac{x^6}{3!} + \frac{x^8}{4!} - \cdots$$

$$\int_0^1 e^{-x^2}\,dx = \int_0^1 \left(1 - x^2 + \frac{x^4}{2!} - \frac{x^6}{3!} + \frac{x^8}{4!} - \cdots\right) dx$$

$$= \left(x - \frac{x^3}{3} + \frac{x^5}{5(2!)} - \frac{x^7}{7(3!)} + \frac{x^9}{9(4!)} - \cdots\right)\Bigg|_0^1$$

$$= 1 - \frac{1}{3} + \frac{1}{5(2!)} - \frac{1}{7(3!)} + \frac{1}{9(4!)} - \frac{1}{11(5!)} + \cdots$$

Because this is an alternating series whose terms decrease in absolute value, you must find the first term that is less than 10^{-3} in absolute value:

$$\frac{1}{9(4!)} \approx 4.6 \times 10^{-3} \qquad \frac{1}{11(5!)} \approx 7.6 \times 10^{-4} < 10^{-3}$$

So if you truncate after five terms,

$$\int_0^1 e^{-x^2}\,dx \approx 1 - \frac{1}{3} + \frac{1}{5(2!)} - \frac{1}{7(3!)} + \frac{1}{9(4!)} = 0.747\,486\ldots$$

you have an approximation with the desired accuracy. [See Section 23-4.]

Supplementary Exercises

In Problems 23-20 through 23-45 find the interval of convergence:

23-20 $\sum_{k=1}^{\infty} (2^k/k!)\, x^k$

23-21 $\sum_{k=0}^{\infty} x^k/(k^2 + 1)$

23-22 $\sum_{k=0}^{\infty} x^k/(k + 2)$

23-23 $\sum_{k=1}^{\infty} (kx)^k$

23-24 $\sum_{k=0}^{\infty} (x - 1)^k/4^k$

23-25 $\sum_{k=2}^{\infty} (x + 3)^k/(\sqrt{k}\ln k)$

23-26 $\sum_{k=0}^{\infty} (x + 1)^k/\sqrt{k^3 + 1}$

23-27 $\sum_{k=1}^{\infty} (\ln k/k)(4 - x)^k$

23-28 $\sum_{k=2}^{\infty} (x - 1)^k/[(k - 1)2^k]$

23-29 $\sum_{k=0}^{\infty} (-3x)^k$

23-30 $\sum_{k=0}^{\infty} (2x - 5)^k/(2k - 5)^k$

23-31 $\sum_{k=0}^{\infty} [(5 + 2k)/(3 + k)]^k(x - 2)^k$

23-32 $\sum_{k=2}^{\infty} x^k/[k^k(\ln k)^{k/2}]$

23-33 $\sum_{k=1}^{\infty} (x^k \sin k)/k^2$

23-34 $\sum_{k=0}^{\infty} [k!/(2k)!]\, x^k$

23-35 $\sum_{k=2}^{\infty} (x \ln k)^k$

23-36 $\sum_{k=1}^{\infty} (-x)^k/(k5^k)$

23-37 $\sum_{k=1}^{\infty} [(k^2 + \ln k)/(k + \ln k)](x + 1)^k$

23-38 $\sum_{k=2}^{\infty} x^k/(2^k \ln k)$

23-39 $\sum_{k=2}^{\infty} (x + 1)^k/\ln k$

23-40 $\sum_{k=1}^{\infty} \dfrac{[1 + (-1)^k](x + \frac{1}{2})^k}{k^2}$

23-41 $\sum_{k=0}^{\infty} k(x - 2)^k/(k^2 + 1)$

23-42 $\sum_{k=2}^{\infty} x^k/(k \ln^2 k)$

23-43 $\sum_{k=0}^{\infty} (3x + 2)^k$

23-44 $\sum_{k=1}^{\infty} (xk)^k/(2k)!$

23-45 $\sum_{k=1}^{\infty} (k!/k^{k/2})x^k$

In Problems 23-46 through 23-55 find the Taylor polynomial of degree n for $f(x)$ about $x = a$:

23-46 $f(x) = e^{3x}$ $\quad a = 0 \quad n = 3$

23-47 $f(x) = x^{3/2}$ $\quad a = 1 \quad n = 3$

23-48 $f(x) = x^3 + 5x^2 - 2$ $\quad a = 0 \quad n = 4$

23-49 $f(x) = \cos x^2$ $\quad a = 0 \quad n = 4$

23-50 $f(x) = \sin x$ $\quad a = \pi/4 \quad n = 5$

23-51 $f(x) = \cosh x$ $\quad a = 0 \quad n = 4$

23-52 $f(x) = \ln(x + 3)$ $\quad a = -2 \quad n = 4$

23-53 $f(x) = \sqrt[3]{x}$ $\quad a = 8 \quad n = 2$

23-54 $f(x) = e^{x^2}$ $\quad a = 0 \quad n = 2$

23-55 $f(x) = \cos^2 x$ $\quad a = 0 \quad n = 4$

In Problems 23-56 through 23-60 use a Taylor polynomial to estimate the number with given accuracy:

23-56 $\sin[(\pi/6) - 0.2]$ $\quad$ error $< 10^{-4}$

23-57 e $\quad$ error $< 10^{-3}$

23-58 $\sin[(\pi/4) - 0.1]$ $\quad$ error $< 10^{-5}$

23-59 $\sqrt[3]{9}$ $\quad$ error $< 10^{-3}$

23-60 $\cosh \frac{1}{2}$ $\quad$ error $< 10^{-3}$

In Problems 23-61 through 23-69 find the Taylor series for $f(x)$ about $x = a$:

23-61 $f(x) = \sinh x$ $\quad a = \ln 3$

23-62 $f(x) = e^{2x}$ $\quad a = 0$

23-63 $f(x) = \sinh x$ $\quad a = 0$

23-64 $f(x) = \cos x^2$ $\quad a = 0$

23-65 $f(x) = \ln(2 + x)$ $\quad a = -1$

23-66 $f(x) = (e^{x^2} - 1)/x^2$ $\quad a = 0$

23-67 $f(x) = 1/(1 - x)^2$ $\quad a = 0$

23-68 $f(x) = \sin x$ $\quad a = \pi/4$

23-69 $f(x) = e^x$ $\quad a = \ln 2$

In Problems 23-70 through 23-77 estimate the number with the given accuracy:

23-70 $\cos(0.2)$ $\quad$ error $< 10^{-4}$

23-71 $e^{-0.1}$ $\quad$ error $< 10^{-5}$

23-72 $\sin(0.3)$ $\quad$ error $< 10^{-7}$

23-73 $\int_0^1 (\sin x)/x \, dx$ $\quad$ error $< 10^{-4}$

23-74 $\int_0^{1/2} \sin x^2 \, dx$ $\quad$ error $< 10^{-6}$

23-75 $\int_0^{1/2} 1/(1 + x^5) \, dx$ $\quad$ error $< 10^{-4}$

23-76 $\int_0^1 e^{-x^3} \, dx$ $\quad$ error $< 10^{-3}$

23-77 $\int_{-1}^0 (e^{x^3} - 1)/x^3 \, dx$ $\quad$ error $< 10^{-3}$

Solutions to Supplementary Exercises

(23-20) $(-\infty, \infty)$

(23-21) $[-1, 1]$

(23-22) $[-1, 1)$

(23-23) 0

(23-24) $(-3, 5)$

(23-25) $[-4, 5)$

(23-26) $[-2, 0]$

(23-27) $(3, 5]$

(23-28) $[-1, 3)$

(23-29) $(-\frac{1}{3}, \frac{1}{3})$

(23-30) $(-\infty, \infty)$

(23-31) $(\frac{3}{2}, \frac{5}{2})$

(23-32) $(-\infty, \infty)$

(23-33) $[-1, 1]$

(23-34) $(-\infty, \infty)$

(23-35) 0

(23-36) $(-5, 5]$

(23-37) $(-2, 0)$

(23-38) $[-2, 2)$

(23-39) $[-2, 0)$

(23-40) $[-\frac{1}{2}, \frac{3}{2}]$

(23-41) $[1, 3)$

(23-42) $[-1, 1)$

(23-43) $(-1, -\frac{1}{3})$

(23-44) $(-\infty, \infty)$

(23-45) 0

(23-46) $P_3(x) = 1 + 3x + (9/2)x^2 + (9/2)x^3$

(23-47) $P_3(x) = 1 + (3/2)(x - 1) + (3/8)(x - 1)^2 - (1/16)(x - 1)^3$

(23-48) $P_4(x) = -2 + 5x^2 + x^3$

(23-49) $P_4(x) = 1 - \frac{1}{2}x^4$

(23-50) $P_5(x) = (\sqrt{2}/2) + (\sqrt{2}/2)(x - \pi/4) - (\sqrt{2}/4)(x - \pi/4)^2 - (\sqrt{2}/12)(x - \pi/4)^3 + (\sqrt{2}/48)(x - \pi/4)^4$

(23-51) $P_4(x) = 1 + \frac{1}{2}x^2 + (1/24)x^4$

(23-52) $P_4(x) = (x + 2) - \frac{1}{2}(x + 2)^2 + \frac{1}{3}(x + 2)^3 - \frac{1}{4}(x + 2)^4$

(23-53) $P_2(x) = 2 + (1/12)(x - 8) - (1/288)(x - 8)^2$

(23-54) $P_2(x) = 1 + x^2$

(23-55) $P_4(x) = 1 - x^2 + \frac{1}{3}x^4$

(23-56) 0.3180

(23-57) 2.718

(23-58) 0.632 98

(23-59) 2.080

(23-60) 1.1276

(23-61) $\sinh x = \frac{4}{3} + \frac{5}{3}(x - \ln 3) + \frac{4/3}{2!}(x - \ln 3)^2 + \frac{5/3}{3!}(x - \ln 3)^3 + \cdots$

(23-62) $e^{2x} = 1 + 2x + (2^2/2!)x^2 + (2^3/3!)x^3 + (2^4/4!)x^4 + \cdots$

(23-63) $\sinh x = x + \frac{x^3}{3!} + \frac{x^5}{5!} + \frac{x^7}{7!} + \cdots$

(23-64) $\cos x^2 = 1 - \frac{x^4}{2!} + \frac{x^8}{4!} - \frac{x^{12}}{6!} + \cdots$

(23-65) $\ln(2 + x) = (x + 1) - \frac{1}{2}(x + 1)^2 + \frac{1}{3}(x + 1)^3 - \frac{1}{4}(x + 1)^4 + \cdots$

(23-66) $(e^{x^2} - 1)/x^2 = 1 + x^2/2! + x^4/3! + x^6/4! + x^8/5! + \cdots$

(23-67) $1/(1 - x)^2 = 1 + 2x + 3x^2 + 4x^3 + 5x^4 + \cdots$

(23-68) $\sin x = \frac{\sqrt{2}}{2} + \frac{\sqrt{2}}{2}\left(x - \frac{\pi}{4}\right) - \frac{\sqrt{2}}{2(2!)}\left(x - \frac{\pi}{4}\right)^2 - \frac{\sqrt{2}}{2(3!)}\left(x - \frac{\pi}{4}\right)^3 + \frac{\sqrt{2}}{2(4!)}\left(x - \frac{\pi}{4}\right)^4 + \cdots$

(23-69) $e^x = 2 + 2(x - \ln 2) + (2/2!)(x - \ln 2)^2 + (2/3!)(x - \ln 2)^3 + (2/4!)(x - \ln 2)^4 + \cdots$

(23-70) 0.9801

(23-71) 0.904 84

(23-72) 0.295 520 25

(23-73) 0.946 11

(23-74) 0.041 4806

(23-75) 0.4974

(23-76) 0.808

(23-77) 0.895

EXAM 8 (CHAPTERS 21–23)

1. Compute the following limits:

(a) $\lim_{x\to 5} \frac{\sin x - \sin 5}{x - 5}$ **(c)** $\lim_{x\to 0^+} \frac{1 + \cos x}{\sin x}$

(b) $\lim_{x\to\infty} \frac{e^x}{x^2}$ **(d)** $\lim_{x\to\pi/2^-} (\tan x)^{\cos x}$

2. Sum the series $\frac{5}{2} - \frac{5}{6} + \frac{5}{18} - \frac{5}{54} + \cdots$.

3. Determine whether the following series converge or diverge:

(a) $\sum_{k=2}^{\infty} \frac{(k+3)(k-5)}{k^3 + k - 4}$ **(b)** $\sum_{k=0}^{\infty} \frac{2^{k/2}}{(k+1)!}$ **(c)** $\sum_{k=1}^{\infty} \frac{k}{k + \ln k}$

4. Determine whether the following series converge absolutely, converge conditionally, or diverge:

(a) $\sum_{k=0}^{\infty} \frac{(-1)^k}{\sqrt{k+1}}$ **(b)** $\sum_{k=1}^{\infty} \frac{(-1)^k k^2}{k!}$

5. Find the interval of convergence:

(a) $\sum_{k=1}^{\infty} \frac{x^k}{\sqrt{k!}}$ **(b)** $\sum_{k=1}^{\infty} \frac{2^k}{k^2}(x-1)^k$

6. Use the Maclaurin series for $f(x) = \frac{e^{-x} - 1}{x}$ to estimate $\int_0^1 \frac{e^{-x} - 1}{x}\,dx$ with error less than 10^{-2}.

SOLUTIONS TO EXAM 8

1. **(a)** This is the indeterminate form 0/0, so use l'Hôpital's rule:

$$\lim_{x\to 5} \frac{\sin x - \sin 5}{x - 5} = \lim_{x\to 5} \frac{\frac{d}{dx}(\sin x - \sin 5)}{\frac{d}{dx}(x - 5)} = \lim_{x\to 5} \frac{\cos x}{1} = \cos 5$$

(b) This is the indeterminate form ∞/∞, so use l'Hôpital's rule:

$$\lim_{x\to\infty} \frac{e^x}{x^2} = \lim_{x\to\infty} \frac{\frac{d}{dx}e^x}{\frac{d}{dx}x^2} = \lim_{x\to\infty} \frac{e^x}{2x} = \lim_{x\to\infty} \frac{\frac{d}{dx}e^x}{\frac{d}{dx}2x} = \lim_{x\to\infty} \frac{e^x}{2} = \infty$$

(c) The numerator approaches $\cos 0 + 1 = 2$ and the denominator approaches $\sin 0 = 0$ from the positive side. Thus, the form is 2/0, which isn't indeterminate:

$$\lim_{x\to 0^+} \frac{1 + \cos x}{\sin x} = \infty$$

(d) This is the indeterminate form ∞^0, so let $y = (\tan x)^{\cos x}$ and find $\lim_{x\to\pi/2^-} \ln y$:

$$\lim_{x\to\pi/2^-} \ln(\tan x)^{\cos x} = \lim_{x\to\pi/2^-} \cos x \ln(\tan x) = \lim_{x\to\pi/2^-} \frac{\ln(\tan x)}{\sec x}$$

$$= \lim_{x\to\pi/2^-} \frac{\frac{d}{dx}\ln(\tan x)}{\frac{d}{dx}\sec x} = \lim_{x\to\pi/2^-} \frac{\frac{\sec^2 x}{\tan x}}{\sec x \tan x}$$

$$= \lim_{x\to\pi/2^-} \frac{\cos x}{\sin^2 x} = 0$$

Because $\ln y$ approaches 0, y must approach $e^0 = 1$:

$$\lim_{x\to\pi/2^-} (\tan x)^{\cos x} = 1$$

2. The series is geometric with ratio $-1/3$:

$$\frac{5}{2} - \frac{5}{6} + \frac{5}{18} - \frac{5}{54} + \cdots = \frac{5}{2}\left(1 - \frac{1}{3} + \frac{1}{9} - \frac{1}{27} + \cdots\right)$$

$$= \frac{5}{2}\sum_{k=0}^{\infty}\left(\frac{-1}{3}\right)^k = \frac{5}{2}\cdot\frac{1}{1-(-1/3)}$$

$$= \frac{5}{2}\cdot\frac{1}{(4/3)} = \frac{15}{8}$$

3. **(a)** Use the limit comparison test. Compare with $\sum_{k=2}^{\infty} 1/k$:

$$\lim_{k\to\infty}\left[\frac{(k+3)(k-5)}{k^3+k-4}\bigg/\frac{1}{k}\right] = \lim_{k\to\infty}\frac{k(k+3)(k-5)}{k^3+k-4}$$

$$= \lim_{k\to\infty}\frac{k^3-2k^2-15k}{k^3+k-4} = 1$$

Because the limit is a positive number and $\sum_{k=2}^{\infty} 1/k$ (the harmonic series) diverges, you know that $\sum_{k=2}^{\infty}\frac{(k+3)(k-5)}{k^3+k-4}$ diverges.

(b) Try the ratio test:

$$\lim_{k\to\infty}\left[\frac{a_{k+1}}{a_k}\right] = \lim_{k\to\infty}\left[\frac{2^{(k+1)/2}}{(k+2)!}\bigg/\frac{2^{k/2}}{(k+1)!}\right] = \lim_{k\to\infty}\frac{2^{(k+1)/2}(k+1)!}{2^{k/2}(k+2)!} = \lim_{k\to\infty}\frac{2^{1/2}}{k+2} = 0$$

Because this limit is less than 1, the series converges.

(c) The terms don't approach zero,

$$\lim_{x\to\infty}\frac{k}{k+\ln k} = \lim_{x\to\infty}\frac{x}{x+\ln x} = \lim_{x\to\infty}\frac{\frac{d}{dx}x}{\frac{d}{dx}(x+\ln x)} = \lim_{x\to\infty}\frac{1}{1+\frac{1}{x}} = 1$$

so the series diverges.

4. **(a)** Determine whether

$$\sum_{k=0}^{\infty}\left|\frac{(-1)^k}{\sqrt{k+1}}\right| = \sum_{k=0}^{\infty}\frac{1}{\sqrt{k+1}}$$

converges. Try the integral test:

$$\int_0^{\infty}\frac{1}{\sqrt{x+1}}\,dx = \lim_{N\to\infty}\int_0^N\frac{1}{\sqrt{x+1}}\,dx = \lim_{N\to\infty}\left(2\sqrt{x+1}\,\Big|_0^N\right) = \lim_{N\to\infty}(2\sqrt{N+1}-2) = \infty$$

Because the integral diverges, the series $\sum_{k=0}^{\infty}\left|\frac{(-1)^k}{\sqrt{k+1}}\right|$ diverges. To determine whether the alternating series $\sum_{k=0}^{\infty}\frac{(-1)^k}{\sqrt{k+1}}$ converges, examine the limit:

$$\lim_{k\to\infty}\left|\frac{(-1)^k}{\sqrt{k+1}}\right| = \lim_{k\to\infty}\frac{1}{\sqrt{k+1}} = 0$$

Because the terms alternate and decrease in absolute value to zero, the series $\sum_{k=0}^{\infty}\frac{(-1)^k}{\sqrt{k+1}}$ converges. Thus $\sum_{k=0}^{\infty}\frac{(-1)^k}{\sqrt{k+1}}$ converges conditionally.

(b) Determine whether $\sum_{k=1}^{\infty}\left|\frac{(-1)^k k^2}{k!}\right|$ converges. Try the ratio test:

$$\lim_{k\to\infty}\left[\frac{a_{k+1}}{a_k}\right] = \lim_{k\to\infty}\left[\frac{(k+1)^2}{(k+1)!}\Big/\frac{k^2}{k!}\right] = \lim_{k\to\infty}\frac{(k+1)^2 k!}{k^2(k+1)!}$$

$$= \lim_{k\to\infty}\frac{(k+1)^2}{k^3} = 0$$

So the series converges absolutely.

5. **(a)** Try the ratio test for the radius of convergence.

$$\lim_{k\to\infty}\left|\frac{a_{k+1}}{a_k}\right| = \lim_{k\to\infty}\left[\frac{1}{\sqrt{(k+1)!}}\Big/\frac{1}{\sqrt{k!}}\right]$$

$$= \lim_{k\to\infty}\sqrt{\frac{k!}{(k+1)!}} = \lim_{k\to\infty}\sqrt{\frac{1}{k+1}} = 0$$

Thus the radius of convergence is $R = \infty$, so the interval of convergence is $(-\infty, \infty)$.

(b) Try the root test:

$$\lim_{k\to\infty}\left|\frac{2^k}{k^2}\right|^{1/k} = \lim_{k\to\infty}\frac{2}{k^{2/k}} = 2$$

Thus, the radius of convergence is $\mathbf{R} = 1/2$. You know that the series converges for $|x-1| < 1/2$ (i.e., for $1/2 < x < 3/2$). Next, determine whether the series converges for $x = 1/2$ and $x = 3/2$. At $x = 1/2$, the series is

$$\sum_{k=1}^{\infty}\frac{2^k(\frac{1}{2}-1)^k}{k^2} = \sum_{k=1}^{\infty}\frac{(-1)^k}{k^2}$$

which converges (alternating series whose terms decrease in absolute value to zero). At $x = 3/2$, the series is

$$\sum_{k=1}^{\infty}\frac{2^k(\frac{3}{2}-1)^k}{k^2} = \sum_{k=1}^{\infty}\frac{1}{k^2}$$

which converges (by the integral test). The interval of convergence is $[1/2, 3/2]$.

6. The Maclaurin series for e^x is:

$$e^x = 1 + x + \frac{x^2}{2!} + \frac{x^3}{3!} + \frac{x^4}{4!} + \cdots$$

so the series for e^{-x} is

$$e^{-x} = 1 + (-x) + \frac{(-x)^2}{2!} + \frac{(-x)^3}{3!} + \frac{(-x)^4}{4!} + \cdots$$

$$= 1 - x + \frac{x^2}{2!} - \frac{x^3}{3!} + \cdots$$

so the series for $e^{-x} - 1$ is

$$e^{-x} - 1 = -x + \frac{x^2}{2!} - \frac{x^3}{3!} + \frac{x^4}{4!} - \frac{x^5}{5!} + \cdots$$

and the Maclaurin series for $(e^{-x} - 1)/x$ is

$$\frac{e^{-x} - 1}{x} = -1 + \frac{x}{2!} - \frac{x^2}{3!} + \frac{x^3}{4!} - \frac{x^4}{4!} + \cdots$$

Integrate term by term:

$$\int_0^1 \frac{e^{-x} - 1}{x}\,dx = \int_0^1 \left(-1 + \frac{x}{2!} - \frac{x^2}{3!} + \frac{x^3}{4!} - \frac{x^4}{5!} + \cdots\right) dx$$

$$= \left(-x + \frac{x^2}{2 \cdot 2!} - \frac{x^3}{3 \cdot 3!} + \frac{x^4}{4 \cdot 4!} - \frac{x^5}{5 \cdot 5!} + \cdots\right)\Bigg|_0^1$$

$$= -1 + \frac{1}{2 \cdot 2!} - \frac{1}{3 \cdot 3!} + \frac{1}{4 \cdot 4!} - \frac{1}{5 \cdot 5!} + \cdots$$

Find the first term with magnitude less than 10^{-2}:

$$\frac{1}{5 \cdot 5!} = \frac{1}{600} < 10^{-2}$$

Thus, because this series alternates, the estimate

$$\int_0^1 \frac{e^{-x} - 1}{x}\,dx \approx -1 + \frac{1}{2 \cdot 2!} - \frac{1}{3 \cdot 3!} + \frac{1}{4 \cdot 4!} = \frac{-229}{288}$$

is accurate to within $1/600 < 10^{-2}$.

FINAL EXAM (Chapters 11–23)

1. Find the limits:

(a) $\lim_{x \to 3} \dfrac{x-3}{e^x - e^3}$ (b) $\lim_{x \to \infty} (\ln x)^{1/x}$

2. Do the following converge or diverge?

(a) $\sum_{k=2}^{\infty} \dfrac{e^{k+1}}{(\ln k)^k}$ (b) $\sum_{k=2}^{\infty} \dfrac{k^2+1}{\sqrt{k^7 - 2}}$

3. Find the centroid of the region bounded by the graphs of $f(x) = 2x$ and $g(x) = x^2$.

4. Integrate:

(a) $\displaystyle\int \frac{x+1}{x^2(x-1)}\, dx$

(b) $\displaystyle\int \arctan x \, dx$

(c) $\displaystyle\int_0^{\infty} \frac{\sqrt{\arctan x}}{1+x^2}\, dx$

(d) $\displaystyle\int \frac{x+3}{x^2+4x+13}\, dx$

(e) $\displaystyle\int_{-1}^{1} \frac{1}{x^2\sqrt{9-x^2}}\, dx$

5. Find the area enclosed by the graph of $r = 2\cos\theta$ between $\theta = 0$ and $\theta = \pi/4$.

6. Do the following converge absolutely, converge conditionally, or diverge?

(a) $\sum_{k=1}^{\infty} \dfrac{(-2)^k}{3^{k+1}}$ (b) $\sum_{k=1}^{\infty} \dfrac{(-k)^k}{\sqrt{k^3+1}}$

7. Find the interval of convergence:

(a) $\sum_{k=1}^{\infty} \dfrac{x^k (k!)^k}{k^{2k}}$ (b) $\sum_{k=1}^{\infty} \dfrac{(2x+1)^k}{\sqrt{2k+1}}$

8. Estimate $\sqrt[3]{9}$ with error less than 10^{-3}.

SOLUTIONS TO FINAL EXAM

1. **(a)** This is the indeterminate form 0/0, so use l'Hôpital's rule:

$$\lim_{x \to 3} \frac{x-3}{e^x - e^3} = \lim_{x \to 3} \frac{\dfrac{d}{dx}(x-3)}{\dfrac{d}{dx}(e^x - e^3)} = \lim_{x \to 3} \frac{1}{e^x} = e^{-3}$$

(b) This is the indeterminate form ∞^0, so let $y = (\ln x)^{1/x}$ and find $\lim_{x\to\infty} \ln y$:

$$\lim_{x\to\infty} \ln[(\ln x)^{1/x}] = \lim_{x\to\infty} \frac{\ln \ln x}{x} = \lim_{x\to\infty} \frac{\frac{d}{dx}\ln \ln x}{\frac{d}{dx}x}$$

$$= \lim_{x\to\infty} \frac{\frac{1}{x \ln x}}{1} = 0$$

Because $\ln y$ approaches 0, y approaches $e^0 = 1$:

$$\lim_{x\to\infty} (\ln x)^{1/x} = 1$$

2. **(a)** Try the root test:

$$\lim_{k\to\infty} \left[\frac{e^{k+1}}{(\ln k)^k}\right]^{1/k} = \lim_{k\to\infty} \frac{e^{1+(1/k)}}{\ln k} = 0 < 1$$

Thus, the series converges.

(b) Try the limit comparison test with $\sum_{k=2}^{\infty} \frac{1}{k^{3/2}}$, a known convergent series.

$$\lim_{k\to\infty} \left[\frac{k^2+1}{\sqrt{k^7-2}} \Big/ \frac{1}{k^{3/2}}\right] = \lim_{k\to\infty} \frac{k^{7/2}+k^{3/2}}{\sqrt{k^7-2}} = \lim_{k\to\infty} \frac{1+(1/k^2)}{\sqrt{1-(2/k^7)}} = 1$$

Thus the series $\sum_{k=2}^{\infty} \frac{k^2+1}{\sqrt{k^7-2}}$ converges.

3. The region is bounded on the left by $x = 0$ and on the right by $x = 2$, above by the graph of $f(x) = 2x$ and below by $g(x) = x^2$.

$$M_y = \int_0^2 x[f(x) - g(x)]\,dx = \int_0^2 x(2x - x^2)\,dx = \int_0^2 (2x^2 - x^3)\,dx$$

$$= \left(\frac{2}{3}x^3 - \frac{1}{4}x^4\right)\Big|_0^2 = \frac{4}{3}$$

$$M_x = \frac{1}{2}\int_0^2 [f^2(x) - g^2(x)]\,dx = \frac{1}{2}\int_0^2 [(2x)^2 - (x^2)^2]\,dx$$

$$= \frac{1}{2}\int_0^2 (4x^2 - x^4)\,dx = \frac{1}{2}\left(\frac{4}{3}x^3 - \frac{1}{5}x^5\right)\Big|_0^2 = \frac{32}{15}$$

$$m = \int_0^2 [f(x) - g(x)]\,dx = \int_0^2 (2x - x^2)\,dx = \left(x^2 - \frac{1}{3}x^3\right)\Big|_0^2 = \frac{4}{3}$$

$$\bar{x} = \frac{M_y}{m} = \frac{4/3}{4/3} = 1 \qquad \bar{y} = \frac{M_x}{m} = \frac{32/15}{4/3} = \frac{8}{5}$$

The centroid is (1, 8/5).

4. **(a)** Find the partial fraction decomposition:

$$\frac{x+1}{x^2(x-1)} = \frac{A}{x} + \frac{B}{x^2} + \frac{C}{x-1}$$

$$x + 1 = Ax(x-1) + B(x-1) + Cx^2 = x^2(A + C) + x(-A + B) - B$$

Equate coefficients

$$0 = A + C$$

$$1 = B - A$$

$$1 = -B$$

and solve: $B = -1, A = -2, C = 2$. Now integrate:

$$\int \frac{x+1}{x^2(x-1)}\,dx = \int \left(\frac{-2}{x} - \frac{1}{x^2} + \frac{2}{x-1}\right) dx$$

$$= -2\ln|x| + \frac{1}{x} + 2\ln|x-1| + C$$

(b) Integrate by parts. Let

$$u = \text{arc tan } x \qquad dv = dx$$

$$du = \frac{1}{1+x^2}\,dx \qquad v = x$$

$$\int \text{arc tan } x\,dx = x \text{ arc tan } x - \int (x)\frac{1}{1+x^2}\,dx$$

$$= x \text{ arc tan } x - \frac{1}{2}\ln(1+x^2) + C$$

$$\int_0^\infty \frac{\sqrt{\text{arc tan } x}}{1+x^2}\,dx = \lim_{N\to\infty}\int_0^N \frac{\sqrt{\text{arc tan } x}}{1+x^2}\,dx$$

(c) Let $u = \text{arc tan } x$ with $du = 1/(1+x^2)\,dx$:

$$\int_0^\infty \frac{\sqrt{\text{arc tan } x}}{1+x^2}\,dx = \lim_{N\to\infty}\int_0^{\text{arc tan } N} u^{1/2}\,du = \lim_{N\to\infty}\left(\frac{2}{3}u^{3/2}\Big|_0^{\text{arc tan } N}\right)$$

$$= \lim_{N\to\infty}\frac{2}{3}(\text{arc tan } N)^{3/2} = \frac{2}{3}\left(\frac{\pi}{2}\right)^{3/2}$$

(d)

$$\int \frac{x+3}{x^2+4x+13}\,dx = \frac{1}{2}\int \frac{2x+4}{x^2+4x+13}\,dx + \int \frac{dx}{x^2+4x+13}$$

$$= \frac{1}{2}\ln|x^2+4x+13| + \frac{1}{3}\text{arc tan}\left(\frac{x+2}{3}\right) + C$$

(e) This is an improper integral because the integrand has a vertical asymptote at $x = 0$.

$$\int_{-1}^1 \frac{dx}{x^2\sqrt{9-x^2}} = \int_{-1}^0 \frac{dx}{x^2\sqrt{9-x^2}} + \int_0^1 \frac{dx}{x^2\sqrt{9-x^2}}$$

$$= \lim_{\varepsilon\to 0^+}\int_{-1}^{-\varepsilon} \frac{dx}{x^2\sqrt{9-x^2}} + \lim_{\eta\to 0^+}\int_\eta^1 \frac{dx}{x^2\sqrt{9-x^2}}$$

Make the inverse trig substitution $\theta = \text{arc sin}(x/3)$:

$$x = 3\sin\theta \qquad dx = 3\cos\theta\,d\theta \qquad \sqrt{9-x^2} = 3\cos\theta$$

$$\int \frac{dx}{x^2\sqrt{9-x^2}} = \int \frac{3\cos\theta\,d\theta}{(3\sin\theta)^2 3\cos\theta} = \frac{1}{9}\int \csc^2\theta\,d\theta = \frac{-1}{9}\cot\theta + C$$

$$= \frac{-1}{9}\cot(\text{arc sin}(x/3)) + C = \frac{-1}{9}\cdot\frac{\sqrt{9-x^2}}{x} + C$$

Thus you have

$$\int_{-1}^1 \frac{dx}{x^2\sqrt{9-x^2}} = \lim_{\varepsilon\to 0^+}\left(\frac{-1}{9}\cdot\frac{\sqrt{9-x^2}}{x}\Big|_{-1}^{-\varepsilon}\right) + \lim_{\eta\to 0^+}\left(\frac{-1}{9}\cdot\frac{\sqrt{9-x^2}}{x}\Big|_\eta^1\right)$$

$$= \lim_{\varepsilon\to 0^+}\left(\frac{\sqrt{9-\varepsilon^2}}{9\varepsilon} - \frac{\sqrt{8}}{9}\right) + \lim_{\eta\to 0^+}\left(\frac{-1}{9}\sqrt{8} + \frac{\sqrt{9-\eta^2}}{9\eta}\right)$$

$$= \infty + \infty$$

The integral diverges.

5. The area is

$$\frac{1}{2}\int_0^{\pi/4} f^2(\theta)\,d\theta = \frac{1}{2}\int_0^{\pi/4} (2\cos\theta)^2\,d\theta = 2\int_0^{\pi/4} \cos^2\theta\,d\theta$$

$$= \int_0^{\pi/4} (1+\cos 2\theta)\,d\theta = \left(\theta + \frac{1}{2}\sin 2\theta\right)\Big|_0^{\pi/4}$$

$$= \frac{\pi}{4} + \frac{1}{2}$$

6. **(a)** Notice that $\sum_{k=1}^{\infty}\left|\frac{(-2)^k}{3^{k+1}}\right| = \sum_{k=1}^{\infty}\frac{2^k}{3^{k+1}}$ is a geometric series with ratio 2/3, which converges. Thus $\sum_{k=1}^{\infty}\frac{(-2)^k}{3^{k+1}}$ converges absolutely.

   (b) Because the terms don't approach zero, the series diverges.

7. **(a)** Try the root test:

$$\lim_{k\to\infty}|a_k|^{1/k} = \lim_{k\to\infty}\left[\frac{(k!)^k}{k^{2k}}\right]^{1/k} = \lim_{k\to\infty}\frac{k!}{k^2} = \infty$$

   Thus the radius of convergence is 0, and so the series converges only for $x = 0$.

   (b)

$$\sum_{k=1}^{\infty}\frac{(2x+1)^k}{\sqrt{2k+1}} = \sum_{k=1}^{\infty}\frac{2^k(x+\frac{1}{2})^k}{\sqrt{2k+1}}$$

   Try the ratio test:

$$\lim_{k\to\infty}\left[\frac{a_{k+1}}{a_k}\right] = \lim_{k\to\infty}\left[\frac{2^{k+1}}{\sqrt{2(k+1)+1}}\Bigg/\frac{2^k}{\sqrt{2k+1}}\right]$$

$$= \lim_{k\to\infty} 2\sqrt{\frac{2k+1}{2k+3}} = 2$$

   Thus, the radius of convergence is 1/2. Now determine whether the series converges for $x = -1$ and $x = 0$. At $x = -1$,

$$\sum_{k=1}^{\infty}\frac{(-2+1)^k}{\sqrt{2k+1}} = \sum_{k=1}^{\infty}\frac{(-1)^k}{\sqrt{2k+1}}$$

   This is an alternating series whose terms decrease in absolute value to zero. Thus, it converges. At $x = 0$ the series is $\sum_{k=1}^{\infty} 1/\sqrt{2k+1}$, which diverges (integral test). The interval of convergence is $[-1, 0)$.

8. Estimate $\sqrt[3]{9}$ using the Taylor polynomial for $f(x) = \sqrt[3]{x}$ at $a = 8$. To obtain the desired accuracy, use a polynomial of degree n where

$$\frac{M_{n+1}}{(n+1)!}|b-a|^{n+1} < 10^{-3}$$

   because this is a bound on the error. Recall, M_{n+1} is the maximum value of $|f^{(n+1)}(x)|$ for x between a and b (here, $a = 8$, $b = 9$).

$$f(x) = x^{1/3}$$

$$f'(x) = \frac{1}{3}x^{-2/3}$$

$$f''(x) = \frac{-2}{9}x^{-5/3}$$

$$f^{(3)}(x) = \frac{10}{27}x^{-8/3}$$

The maximum value of $|f^{(3)}(x)|$ on $[8, 9]$ is

$$M_3 = |f^{(3)}(8)| = \frac{10}{27}(8)^{-8/3} = \frac{5}{3456}$$

Thus, if you use a degree-two Taylor polynomial, the error is less than

$$\frac{M_3}{3!}|9 - 8|^3 = \frac{5}{20\,736} < 10^{-3}$$

The degree-two Taylor polynomial for $f(x) = x^{1/3}$ at $a = 8$ is

$$P_2(x) = f(8) + f'(8)(x - 8) + \frac{f''(8)}{2!}(x - 8)^2$$

$$= 8^{1/3} + \frac{1}{3}(8)^{-2/3}(x - 8) + \frac{(-2/9)(8)^{-5/3}}{2!}(x - 8)^2$$

$$= 2 + \frac{x - 8}{12} - \frac{(x - 8)^2}{288}$$

Finally, you have

$$9 = f(9) \approx P_2(9) = 2 + \frac{9 - 8}{12} - \frac{(9 - 8)^2}{288}$$

$$= 2 + \frac{1}{12} - \frac{1}{288} = \frac{599}{288}$$

INDEX

Page numbers in italic refer to problems.

3
4
5
6
H 7
I 8
J 9